W9-BXL-255

ENCYCLOPEDIA OF ETHICS IN SCIENCE AND TECHNOLOGY

ENCYCLOPEDIA OF ETHICS IN SCIENCE AND TECHNOLOGY

NIGEL BARBER, Ph.D.

☑® Facts On File, Inc.

ENCYCLOPEDIA OF ETHICS IN SCIENCE AND TECHNOLOGY

Facts On File, Inc.
132 West 31st Street
New York NY 10001

Library of Congress Cataloging-in-Publication Data

Barber, Nigel
Encyclopedia of ethics in science and technology / Nigel Barber.
p. cm.
Includes bibliographical references and index.
ISBN 0-8160-4314-0
1. Science—Moral and ethical aspects—Encyclopedias. 2. Technology—Moral and ethical aspects—Encyclopedias.
I. Title.
Q175.35.B37 2002
174'.95'03—dc21   2001040832

Facts On File books are available at special discounts when purchased in bulk quantities for businesses, associations, institutions or sales promotions. Please call our Special Sales Department in New York at 212/967-8800 or 800/322-8755.

You can find Facts On File on the World Wide Web at http://www.factsonfile.com

Text design by Erika K. Arroyo
Cover design by Cathy Rincon
Illustrations by Dale Williams

Printed in the United States of America

VB FOF 10 9 8 7 6 5 4 3 2

This book is printed on acid-free paper.

This book is dedicated to the memory of Freda Barber-Rowntree,
the environmentalist of the family.

CONTENTS

ACKNOWLEDGMENTS
ix

INTRODUCTION
xi

ENTRIES A–Z
1

APPENDIX
Ethics Organizations, Resources, and Websites
369

INDEX
371

ACKNOWLEDGMENTS

I am grateful for the help received from many people at various stages in the planning and writing of this volume. The idea for the book was originally proposed by Anthony Serafini who had to bow out of writing it for health reasons. I wish him well in his retirement. My literary agent Elizabeth Frost-Knappman encouraged me to submit a proposal and I am grateful to her and her partner at New England Publishing Associates, Edward W. Knappman, and their assistant Kristine A. Schiavi. My research efforts have been greatly facilitated by the work of the Interlibrary Loan Department at Portland Public Library. The interlibrary loan staff, Rita Gorham, Eileen Mac-Adam, and Anne Ball, were unfailingly cheerful and always delivered the goods. I am grateful for their professional dedication. I was also helped by Maja Keech of the Library of Congress Prints and Photographic Division and by Gayle T. Harris of Research Unlimited.

My editor at Facts On File, Frank K. Darmstadt, and his colleague, Tracy Bradbury, both provided me with numerous challenging queries that forced me to reexamine my facts and ideas at many points, thus improving the work. They were diligent about preventing me from taking a one-sided view of many of the complex and controversial topics that are raised in this volume, although any remaining biases are my responsibility alone. I have benefited from their personal enthusiasm for the project and interest in the ethical complexity of life in a scientifically and technologically advanced society. Trudy Callaghan has also provided editorial assistance and has particularly helped in eradicating any unintended gender bias in the work. She was fascinated by many of the entries and professed to learn a great deal. I am grateful for her support.

INTRODUCTION

John Stuart Mill observed that a person's freedom to swing their fist ended at a neighbor's nose. Although this is far from being the only approach to ethical decisions, it does grasp the dynamic quality of ethics. If my neighbor chooses to flinch, then, according to Mill, my right to swing my fist is correspondingly enlarged. Mill grasped the basic notion that ethical dilemmas are created by adversarial situations and was in favor of a judicious weighing of costs and benefits to each party so as to achieve a balance at which the greater good was served.

Given that ethical problems are produced by our interactions with other people and with the environment, it follows that changes in the nature of these interactions will produce new ethical dilemmas. We live in a golden age of ethics because technology has greatly expanded the scope of what we may do and science has given us a keener appreciation of the potentially deleterious consequences of our actions.

Today, it is not only the right to swing a fist but also the right to smoke that ends with a neighbor's nose. Given the knowledge that passive inhalation of tobacco smoke may cause cancer, it is no longer possible to permit smoking in public buildings. Science thus alters the definition of ethical conduct.

On the technological side, we are constantly being bombarded with ethical issues for which we have no ready solution because the problems themselves have no precedent. Two such frontiers are medical ethics and the Internet. Within medical ethics, some of the most startling issues to surface in recent years are connected with advances in reproductive technologies. One consequence has been the fractionation of the roles of parents. In one remarkable legal case, a couple, John and Luanne Buzzanco, arranged to have a baby using the sperm and egg of two other people. The egg was fertilized in vitro and implanted into a surrogate mother. After the couple had divorced, no one wanted to take responsibility for the baby. The courts had to decide whether the infant had the usual two biological parents, or three (including the surrogate), or five (including the social parents who had planned the pregnancy). In the end, the couple who had arranged the pregnancy were deemed to be the legal parents even though they had no biological connection to the infant whatever. Technological innovation in the biomedical field thus propels lawyers, and everyone else, into previously unexplored ethical territory.

Another great technological revolution that is changing our lives is the Internet. Sometimes described as the information super highway, it is potentially far more than a medium for sending information around the world. After all, this is what postal services have done for centuries and what telephones, radios, and televisions have done more efficiently in more recent times. What the Internet promises, or threatens, to do is to shrink all of the people on the planet into a single community interconnected in new and important ways. At its most mundane, our refrigerators might be connected to the Internet so that depleted food stocks will be automatically reordered. At its most exotic, people might wear microprocessors close to their brains, allowing them to communicate with the rest of the world without the necessity for keyboards or other input devices. If this ever happens, it will be potentially threatening to Western notions of individuality and freedom and will transform ethics.

Apart from transforming the way in which transactions are conducted at present, the Internet will facilitate entirely new kinds of transactions. One interesting example is the Hunger Site, at which people may donate free food to feed the starving populations of the world. The food is paid for by sponsors who are rewarded by having their logo presented before the eyes of the nominal donor.

Perhaps the most heated ethical problem generated by the emergence of the Internet relates to invasion of privacy in all of its many forms. Child molesters, for example, have used chat lines to arrange meetings with potential victims. Identity thieves have used the Internet to obtain information on their victims. Stalkers have purchased personal information from on-line personal report services that has been used to harass their victims. Hackers have devoted themselves to crossing the electronic firewalls set up by companies to protect the privacy of their intellectual property, sensitive business information, and client data. Perhaps the most insidious threat to on-line privacy and freedom is the development of profiling techniques by advertisement companies that potentially allow them to track a person's every move on the Internet and to use that information for marketing purposes.

Even if we ignored the burning issues in medical ethics and information technology, applied ethics has undergone remarkable change in the past few decades resulting in the emergence of several new branches. Technological innovation has changed what we do and how we live in so many ways that new ethical problems have been generated. To take a simple example, the use of steel girders as a support system for buildings has permitted the construction of high-rise buildings. Skyscrapers have proven to be a useful innovation in business, allowing thousands of employees of a corporation to work in close proximity. The same design has turned out to be a great deal more problematic in the case of residential housing. High-rise apartment buildings for low-income residents have produced a breeding ground for criminal activities. In some cities, including St. Louis, Missouri, and Newark, New Jersey, the social problems in public housing projects were so bad that they had to be razed to the ground. Perhaps the insight that the design of housing affects the quality of life of residents is not new. What is new is the sense of responsibility of architects for the human consequences of their buildings and knowledge of such consequences flows from advances in the social sciences.

The emergence of architectural ethics can be seen as part of a pervasive ethical concern with the consequences of human activities for the environment in which we live. Environmental ethicists have raised concerns about the consequences of human industrial activity in terms of environmental pollution and destruction of ecosystems. The rapid pace of global change, as exemplified in the unprecedented rate of destruction of rain forests along with their rich and unique flora and fauna, has prompted environmental ethicists to ruminate about our obligations to future generations and has resuscitated long-dormant ethical theories of value. If ecosystems can be thought of as having inherent value, as some prominent environmental ethicists maintain, then the problem arises as to how the value of the rain forests, for example, stacks up against the desperate need of the world's starving millions for new agricultural land.

In addition to destruction of habitat and annihilation of resident species, our obligation to future generations includes the need to preserve historical sites and artifacts. In the past, archaeology was a minority interest enjoyed, like other scientific pursuits, by a handful of gentleman amateurs who provided their own funding and generally did pretty much as they pleased. Archaeology today is a great deal more careful, responsible, systematic, scientific, and labor intensive. Archaeologists are much more aware that the information they unearth concerning previous civilizations is uniquely valuable and that their handling of artifacts is open to public scrutiny.

The rise of professionalism in science has been accompanied by an increased investment of labor and capital in many scientific endeavors. This process has reached an extreme in the case of sciences like physics, chemistry, and biochemistry, in which the sophisticated equipment necessary to do state-of-the-art research means that there must be a huge infusion of funding by government, private foundations, and commercial corporations. The capital-intensive nature of modern research has placed scientists under increasing external pressures that may increase the likelihood of unethical conduct. The publish-or-perish environment of modern universities can be a factor in the publication of trivial research, or even fraudulent research, and it can lead to the exploitation of junior researchers. Commercialism in science can have a baneful influence on scientific practice. For example, if the product of scientific research is potentially patentable, researchers may deliberately obscure, or even falsify, details of their procedure to prevent the patent being developed by a rival laboratory.

Commercialism in the field of genetic engineering has led to American applications for patents on genetically engineered life forms. Some such patents have been granted but there is considerable ethical controversy over whether life forms can be considered property. Once again, the creation of new frontiers in science and technology opens up ethical problems for which our previous history as a species has not prepared us.

SCOPE OF THE WORK

This encyclopedia brings together for the first time the central concepts, people, facts, phenomena, and controversies in the ethics of science and technology, as well as the seminal figures and formative ideas in each of these fields. Its center of gravity lies in the ethical issues affecting advanced industrialized countries which gives the book a Western focus, although the influence of other traditions on Western science is acknowledged. In addition, the classical ethical concepts of both East and West are included.

Ethical problems in the biomedical area are included only if they are related to advances in technology. Using this guideline, abortion, for example, is discussed only in the context of contraceptive technologies that now include abortifacient drugs such as RU-486.

The study of ethical issues in science and technology should properly include some treatment of the ideas of those who have criticized both the practice of science and the potentially harmful consequences of technological advances. For example, the Luddites objected to the displacement of people by machines and there is a great deal of suspicion today about the potential misuse of information technology by businesses and by governments. There is also much concern about toxins in the environment and about the safety of the food supply. Even though some of these concerns have a "lunatic fringe" flavor, as reflected in the ramblings of Unabomber Theodore Kaczynski, for example, it is also both reasonable and responsible to consider the disadvantages of scientific and technological advances as well as their benefits.

A more subtle kind of critique of scientific practice has been advanced by feminists and others who either question the objectivity of current scientific knowledge or make the more general case that objectivity is impossible in principle. Such critiques are not always beside the point or unconstructive. For example, feminist primatologists have provided an important corrective to male-biased field work that often focused exclusively on the behavior of males. The general guideline for including such critiques in the encyclopedia is that they must have been discussed in peer-reviewed professional journals.

Similarly, there is room to discuss phenomena found at the fringes of science but may have important ethical implications, such as voodoo death (as described by reputable scholars) and experimental research into extrasensory perception (as published in peer-reviewed journals) and controversy over UFO sightings.

TARGET READERS

The primary readers of this work will be users of high school, public, and college libraries, as well as the collections of foundations and other organizations that study science and technology as well as those that focus primarily on ethics. The book will be helpful to teachers at the high school and college levels, as well as to undergraduate and graduate students. It can be used as a secondary textbook for college courses in ethics, women's studies, the history of science, as well as graduate-level courses in professional ethics. Ethical issues inevitably tend toward controversy and this volume embraces many controversial topics while striving for balance and fairness.

ORGANIZATION

An encyclopedia written by a single author has the great advantage of flexibility in format so that larger topics can be presented in longer pieces and smaller ones in short pieces. If I want to understand what a Trojan horse virus does, for example, I may not want to read a chapter on Internet protocol.

In the main body of this encyclopedia, topics are arranged in alphabetical order. Entries range from a paragraph long to several book pages and include some 20 substantial integrative essays dealing with foundational issues. These essays include extensive cross-referencing and bibliography. Most entries that are longer than a paragraph are supported by selective bibliographies that focus on reliable sources that are in print and up to date. The Index consists of an alphabetical listing of the main terms occurring in the body of the work.

abortion   *See* CONTRACEPTION.

academic vigilantism

This term was used in reference to attacks on EDWARD O. WILSON's book *Sociobiology* (1975) by the left-wing pressure group SCIENCE FOR THE PEOPLE. The term implies that academics are not free to express their political views and is a forerunner of the term "political correctness."

See also SCIENCE FOR THE PEOPLE; WILSON, EDWARD O.

acid rain

Acid rain is created when air pollutants, particularly sulfur and nitrogen dioxides, combine with moisture in the air. When this moisture falls as rain, hail, snow, or fog, it is more acidic than normal precipitation, having a Ph value of 4 or lower compared to the normal Ph of 5. Its ethical significance is that it provides a clear example of the damaging effects of industry on the environment with long-term adverse consequences for vulnerable ecosystems and human health.

Acid rain can be very damaging to bodies of water, plant life, and soil. It has been blamed for the disappearance of fish from lakes in the Adirondacks and for the loss of forests in European mountains. Approximately a quarter of the forests in Sweden, Germany, and other parts of Western Europe have been damaged by acid rain. It can also damage buildings and infrastructure. Acidity from air pollution can be very damaging to all agricultural plants, an effect that has disrupted farming in the Los Angeles basin. Sulfur dioxide combines with available water from rain, or from the plants themselves, to produce sulfuric acid that has a corrosive effect on plant tissues.

Some local ecologies are much more vulnerable than others. Regions with plentiful limestone neutralize the acidity when acid rain reacts chemically with the limestone, thereby protecting groundwater, plants, and animals. The Adirondacks region of New York is not rich in limestone, however, which explains why it has become much more acidic than other regions of the state. Even with the reduction in acid emissions due to stricter environmental regulations, these fragile ecologies are not guaranteed to bounce back because the acidity remains in bodies of water making it impossible for fish stocks to recover. Acid rain also depletes soil of essential nutrients such as calcium and magnesium.

Although some variation in the acidity of rain may be natural, there is no doubt that large-scale ecological damage produced by acid rain first emerged in the 20th century due to increased combustion of oil and coal. The major culprits have been sulfur dioxide produced by the burning of coal and nitrogen dioxide from car exhausts.

Acidity is also increased by animal wastes that produce nitric acid in the soil.

Recent decades have seen substantial decreases in emissions of "acid rain" gases in the United States, Europe, and other regions following legislative changes. Thus, U.S. emissions of sulfur dioxide have been reduced by over 30% since passage of the CLEAN AIR ACT OF 1970. CATALYTIC CONVERTERS have also reduced the emission of nitrogen oxides in vehicle exhausts.

Sulfur dioxide emissions from industrial smokestacks have been reduced in several ways. For example, the electric utility industry, a major source of acid rain pollutants, has switched to a different form of coal that is lower in sulfur. This change was partly motivated by an innovative emissions trading system that allowed the electricity companies to lower costs by reducing emissions.

Limestone scrubbers can remove up to 90% of the sulfur dioxide from a smokestack but they consume about 5% of a power plant's electricity output. A limestone scrubber consists of a large vessel that receives exhaust gases from steam boilers. As the waste gases enter the vessel, they are met by a pressurized spray of fine limestone-based slurry. The limestone reacts with the sulfur dioxide producing calcium sulfite that is then converted to calcium sulfate, or gypsum. Unfortunately, the gypsum sludge cannot be reused and this toxic waste must be placed in a landfill after it is dried. One reason that limestone scrubbers require so much energy is that the limestone must be crushed into a fine powder. A single scrubber may crush as much as 36 tons of rock per hour.

The fact that electric utilities are now held legally responsible for the amount of pollution they send into the atmosphere is an important development in the control of acid rain. Yet it is a stop-gap solution. The more profound technological and ethical question is whether it would be possible to generate enough electricity by cleaner means without polluting the atmosphere at all. Many environmentalists emphasize development of cleaner energy technologies such as wind power, solar cells, and water power.

See also AIR POLLUTION; CATALYTIC CONVERTER; CLEAN AIR ACT OF 1970; NUCLEAR POWER.

advertising and science

Advertising can be defined as any communication (other than face-to-face interaction) that is designed primarily to sell products or services. Advertising media include print (newspapers, magazines, brochures, coupon books, billboards, etc.) and electronic or electromagnetic channels (radio, television, Internet). To the extent that advertising campaigns are based on scientific research, they are sometimes seen as an abuse of science in the sense that scientific knowledge is being used to serve the interests of commercial organizations at the expense of the public. Recent research in the field of advertising generally does not support the view that it is threatening to the interests of consumers.

Criticisms of advertising have raised numerous ethical concerns the most important of which are:

1. Advertising adds cost to products without providing any benefit to consumers. (This seems to be true although companies have little choice about using advertising if they want to defend their market share).
2. Advertising expenses create a barrier against small companies bringing new products to market and therefore deprives consumers of the benefits of many competitive brands from which to choose.
3. It creates needs for products, leading people to waste their money or purchase items that are either defective or harmful.
4. It debases the media that carry it. For example, a great work of cinematic art that lasts 90 minutes is interpolated with 30 minutes of television commercials.
5. Advertisements clutter up periodicals. They create an unnecessary form of interference on the Internet. Many advertisements are criticized because of their reliance on sexually explicit imagery that promotes an objectionable lifestyle. Reliance of television news programs on commercial sponsors may also bias the content of television journalism and the same is true of magazines and other media.
6. It is aesthetically objectionable creating what might be described as media litter. For example, billboards at the outskirts of American cities are held to create an unsightly, and unseemly, impression of crass materialism.
7. We are exposed to advertising against our will, making it an unethical form of coercion.
8. Advertising dupes people into making purchases through false and misleading claims.

Academic research on advertising suggests that most, or all, of these fears concerning advertising are unfounded. Advertising is not really capable of causing people to make mindless purchases of unneeded products.

CREATING NEEDS

The biggest single argument against the view that advertisers create needs is the finding that, for products in mature markets, increased advertising budgets within a category do not translate into increased sales. The goal of advertising from the point of view of product manufacturers is not a war waged against consumers but a battle among competing brands. One British study of the relationship between advertising expenditures and sales of 10 food products between 1985 and 1994 found that even though advertising costs rose by 17% during the period, sales declined by 10%. This demonstrates that increased advertising does not inevitably generate increased sales.

The surprising lack of effect of advertising on sales of a product category is also illustrated by the ban on television advertising of tobacco products that went into effect in the United States in 1971. At the time, huge amounts of money were being spent on cigarette advertising. If advertisements really create needs and stimulate sales, then the banning of television advertising should have produced a steep drop in tobacco sales. Yet cigarette sales began to fall only in 1981, 10 years after the big drop in advertising budgets. Ironically, this money went straight back into the coffers of the tobacco companies making a highly profitable industry even more so. The data from other countries that have stopped television advertising of tobacco have been broadly similar.

Given that advertising cannot boost category sales, its primary goal is to boost market share for a particular brand within a product category. Consumer behavior is characterized by a steady pattern of purchasing a favored brand most often while occasionally purchasing other brands. In mature markets, where the dominant brands have been established for some time, changes in buying behavior can be affected by sales promotions and advertising campaigns but such changes tend to be short-lived.

It is true that sales of a new brand can be boosted by making consumers aware that it exists. Most new brands fail, however, (around four out of five) and there is no evidence that a large advertising budget makes a product more likely to succeed, which punches another hole in the argument that people are easily manipulated by advertising. In summary, advertising does not produce dramatic changes in purchasing behavior. This might seem odd because of the obvious potential of slick commercials to alter our emotions.

USING SEX TO SELL

Marketing professionals seem to believe that sex sells. There are two good reasons why sexual provocation is a staple ingredient of television advertising. One is that it draws people's attention. The other is that it can evoke pleasurable feelings which get associated with the product, making it seem more desirable and thereby making us want to purchase it. Associating a brand with pleasant feelings in this way is similar to IVAN PAVLOV's experiment in which a dog learns to make a connection between the sounding of a buzzer and receiving food. Such association can be considered a form of Pavlovian conditioning. The use of Pavlovian conditioning techniques in advertising is attributable to the early influence of JOHN B. WAT-

SON, a leading academic exponent of behaviorism who was forced to leave his academic position after his love letters to a graduate student were published in the *Baltimore Sun*. Watson brought his prodigious talents and energies to the field of advertising.

Neuroscientific research on the chemical responses of male rats' brains to anticipated sexual intercourse has found that it is the same as that produced by the sexual activity itself. Generalizing to humans, advertising that transmits the expectation of pleasure could be transmitting pleasure itself. The perceived value of the product being marketed may be enhanced by associating it with a pleasurable experience. If people pay attention to a sexy advertisement, their liking for the product can increase.

Students of advertising effectiveness encounter an interesting paradox. Although advertising should work, most people believe it doesn't work on them. This issue is illuminated by a simple pioneering experiment dealing with the role played by attractive models in car advertisements in the 1960s. Psychologists George Smith and Rayme Engel of Rutgers University asked men and women to evaluate a sports car depicted in a magazine advertisement. Participants were randomly divided into two equivalent groups. One group saw only the car. The others saw the car with the picture of an attractive scantily clad model snipped from *Playboy* glued next to it. When the car was accompanied by the model, men and women saw it as faster, more dangerous, and more expensive. That is, it was seen as a better sports car. Interestingly, the participants denied in the strongest possible terms that their evaluation of the sports car had been affected by the presence of the model! Advertising can therefore work at an unthinking emotional level, at least in terms of changing our attitudes.

Not all advertising uses pleasant images of nubile young women and handsome men, of course. Other primary emotional reactions are stimulated by cute children and cuddly animals. The emotion is not always obviously pleasant, either. Advertisements aimed at young adults, particularly young men, often emphasize danger. Younger people tend to perceive dangerous situations as more of a thrill and less of a threat than older adults do. Moreover, evoking fear and then depicting a safe conclusion produces a pleasurable sense of relief. The product will then become paired with excitement as well as relief.

THE REAL OBJECTIVE OF ADVERTISING

Although simple laboratory experiments have demonstrated that commercials may change our feelings toward a product and can increase our appraisal of its quality, in the real world advertising promotions do not produce substantial changes in people's evaluation of a brand. This conclusion is intuitively puzzling but might have an elementary explanation. One important difference between a laboratory experiment and the real world is that the experiment is virtually assured the undivided attention of the participants. An advertisement, on the other hand, is likely to be ignored.

If people do not pay attention to a given commercial, then it is unlikely that it will affect their attitude toward the advertised brand, even after dozens of exposures. To their chagrin, advertisers have found that consumers generally pay attention only to commercials for brands that they already buy. Moreover, even though marketers had assumed that changing consumer attitudes was the key to stimulating sales of a brand, the actual chain of causation goes in the other direction. If people buy a product and have a good experience with it, then their liking for the brand increases, making them more likely to buy it in future.

Contrary to the Big Brother scenario of people being effortlessly manipulated by clever ad campaigns, so that they desire, and purchase, products that they do not need, there is surprisingly little room for the sort of thought control that opponents fear. This does not deny that television, as a medium, can have enormous influence on consumption by revealing how other people live and thereby boosting expectations and purchasing behavior. Advertising, as such, is used primarily to make consumers feel good about a product they already buy. In this way, their purchase frequency can be increased, or at least protected against attrition by other major brands currently being purchased by the same consumers.

Since consumers repeatedly purchase the same products, they come to know them very well. Their purchasing behavior is thus a great deal more rational than would be assumed by many critics of advertising. Customers continue to buy products because they are happy with the quality of the product, not because they have been seduced by cunning advertisers. By the same token, false or deceptive advertising is unlikely to be effective. Consumers may be deceived into thinking that brand Y is a great deal better than it actually is and may make a trial purchase. When the product fails to live up to its advertising, they will be disappointed and unlikely to make the same mistake again. Contrary to the myth of all-powerful advertising techniques and consumers who are both mindless and pliable—a myth often propagated by opponents of television advertising—customers evidently build up rational purchasing habits that are extraordinarily resistant to deliberate change by advertisers.

CONCLUSIONS

The foregoing discussion addresses some, but not all, of the ethical criticisms that have been laid against modern advertising. Thus, advertising is often described as aesthetically objectionable, although apologists might argue that television advertisements are usually made with more art than the regular viewing that they interrupt. One response to the esthetic objections is to say that people may choose to ignore, or completely avoid, advertising that they find objectionable on esthetic or other grounds. After all, they are not compelled to watch commercial TV channels. The issue of unsightly billboards is more difficult because they cannot be avoided in practice. Whatever aesthetic pain they may cause to the individual is arguably balanced by the contribution they make to the prosperity of local businesses.

Critics also object to advertising on the basis that its great cost imposes a barrier to introducing new brands to the market to the ultimate detriment of the customer. This objection is also mistaken on several grounds. Advertising comprises only a fraction of the huge cost of bringing a new brand to market and is thus not a real barrier to new products. Moreover, product categories that have a lot of advertising are also the ones that have a lot of competition. Producers are not actually inhibited from introducing new brands into highly competitive markets. In most such product categories, consumers have far more different brands than they could possibly want or need. In such cases, differences in real quality and differences in price may be small.

Advertising campaigns clearly benefit from scientific research. This has often been interpreted as an unethical abuse of science for the benefit of commercial organizations and to the detriment of the public. Research documenting the limited effectiveness of advertisement campaigns has tended to defuse these ethical concerns.

See also PORNOGRAPHY; TELEVISION VIOLENCE.

Further Reading

Jones, J. P., ed. *How Advertising Works*. Thousand Oaks, Calif.: Sage, 1998.
Smith, G. H., and R. Engel. "Influence of a Female Model on Perceived Characteristics of an Automobile." *Proceedings of the 76th Annual Convention of the American Psychological Association*, 3 (1968): 681–82.

aerosol sprays   *See* AIR POLLUTION, OZONE DEPLETERS.

affirmative action in science and technology careers

Science and technology careers are a comparatively recent historical development. Before the rise of industrialism in Europe in the 18th century, scientists were mainly amateurs who conducted their study and investigations without monetary compensation. Since the conduct of scientific investigation implied a high level of education, science was inevitably elitist. Rather than being a profession, it was the pastime of gentleman amateurs. Moreover, the pace of technological innovation was slow so that technological change provided new careers at a low rate.

Not only was science the province of the privileged elite but also it was almost exclusively confined to men. Affirmative action in science and technology attempts to improve the education, employment, working conditions, and opportunities of women and ethnic minorities in these fields.

The few women who have made important contributions to science before the 20th century were conscious of going against sex role expectations. Even wealthy women were less likely to receive a good education than men. If they did, their instruction was much more likely to focus on skills that would have improved their marriage prospects, such as music, dancing, embroidery, and etiquette, than on rigorous instruction in science and mathematics.

The lack of opportunity for women in the pre-professional era of science has carried over to modern times in that there are far fewer women scientists and engineers today than there are men. The underrepresentation of ethnic minorities in the United States and other multi-ethnic societies is even more extreme (see table). In 1993, persons with disabilities constituted 20% of the U.S. population but only 6% of scientists and engineers. There was no change in the degree of underrepresentation of any of these groups between 1993 and 1997, the most recent year for which data are available. Science and technology careers have become increasingly specialized and complex and almost always require advanced degrees as the price of admission. Ethnic minorities tend to get screened out of scientific and technological careers because they do not have the educational credentials to enter undergraduate and graduate fields of study. Yet there are interesting exceptions. For example, Chinese Americans are overrepresented in technical and scientific careers in America for the past half-century both because they are well prepared academically and because they are more likely to have aspirations in these areas. This might constitute a side effect engendered by the difficulty in acquiring requisite skills in English by this group, which has made it difficult for them to penetrate professions like law and medicine in which verbal skills in the English language are critical.

Whatever the fundamental causes of underrepresentation of particular demographic groups in science and technology careers, there is a sense of obligation among those who place students in graduate programs, and among those who hire scientists, engineers, and other technical specialists, to ensure that such groups are given an opportunity to participate on an equal basis in these fields. There are two primary motives. The first is political. The public funding received by an educational institution may be affected by the extent to which it is perceived as living up to achieving a diverse student body. The second is vocational. Increasing the representation of women and ethnic minorities in science and technology has the potential for introducing fresh perspectives in these fields.

Equality of opportunity in science, basic and applied, is important because it promotes the ethics of intellectual freedom and freedom of inquiry and dispels the notion that academic pursuits in these fields are the province of a privileged elite of white males. It can also be argued that diversity promotes scientific objectivity. For example, the scientific study of sex differences, one of the most controversial endeavors in modern science, is likely to produce biased and inaccurate conclusions if it is carried out only by men. Thus early students of moral development found that men typically reached a higher level of moral reasoning than women. This conclusion has been challenged by CAROL GILLIGAN, who argues that men and women differ in their moral values, with men preferring a more abstract moral code and women expressing a more social form of ethical reasoning. This debate has enriched the study of moral development. Similar arguments can be made for involving homosexuals in the scientific study of sexual orientation and racial minorities in the scientific study of ethnic group differences.

Freedom of opportunity in science can also be justified on the basis of a political ethic such as that defended by JOHN RAWLS, who favors not only egalitarian entry to careers but also efforts to correct economic inequality that he holds is largely responsible for unequal entry to careers in the first place. Needless to say, these views are highly controversial and recent events in California and other U.S. states, which have rolled back affirmative action in colleges, suggest that the tide of history may be turning in the other direction.

If the underrepresentation of women and ethnic minorities in science careers is to be redressed, vigorous affirmative action is required. Simply allowing women and minorities equality of opportunity in these technical careers is not going to change the status quo. Part of the problem is that people often aspire to careers that they see as suitable for them. The dearth of black physicists and women engineers who are known to young people and can serve as role models may limit the aspirations of some groups in these fields. There is currently a very determined effort to recruit women to the engineering professions, fields in which men outnumber women by more than 10 to one.

The most vigorous approach to affirmative action consists in using gender or ethnic group as a qualifying criterion for college and graduate school admissions. This means, for example, that African Americans might be admitted to graduate study in science having lower test scores than were necessary for admitting European Americans. At an extreme, a quota might be imposed according to which a certain proportion of all recruits would have to be African Americans.

Several ethical problems are raised by such vigorous affirmative action techniques. If a demographic group is used as a criterion in recruitment and hiring, it inevitably means that the limited number of careers in basic and applied sciences are occupied by people of lesser merit than would otherwise be the case. This holds back the advancement of science because less qualified people are likely to do less distinguished research. Moreover, undistinguished scientists do not make effective role models, so that the goal of recruitment of disadvantaged groups is not helped.

Not only is the profession dragged down but there are potentially undesirable consequences for all the people directly affected. Individuals who are favored by the policy may not only feel overwhelmed by their position but they might also experience a lack of professional respect. Moreover, members of the favored group who happen to hold their position due to merit may feel that their personal achievements have been devalued, that they are regarded as "token" people. The use of gender and ethnic criteria in recruitment has the unintended consequence of perpetuating the stereotype that some groups are intellectually inferior. Just as bad is the situation of qualified white men who are excluded from careers purely on the basis of skin color and gender. Such reverse discrimination strikes many ethicists as morally indefensible.

There are several convincing ethical problems with using stronger affirmative action measures to achieve the desirable goal of greater diversity in science. A weaker approach would be to use demographic criteria to break ties between equally qualified candidates for graduate school placement or jobs, for example. Such techniques are helpful in increasing the availability of role models who can influence career aspirations in scientific fields. It must be recognized, however, that the underrepresentation of some groups in graduate programs reflects problems at lower levels of the educational system. Moreover, it is a mistake to assume that unequal representation of different groups in science and technology careers, particularly the difference between men and women, is entirely due to inequality of opportunity. Such inequality also reflects different individual choices and preferences. There is a growing consensus among psychologists that men and women have average differences in interests that are not attributable to social learning. If this is correct, then equal participation of men and women in careers, like engineering, where these interests are salient may never be realized. This does not mean that the goal of greater diversity in these fields cannot be accomplished.

Affirmative action programs in the United States are a comparatively recent attempt to redress historical patterns of discrimination against women and minorities in education and occupations. The Equal Employment Opportunity Act of 1972 required federal contractors and subcontractors, state and local governments, and institutions such as universities to initiate plans designed to raise the proportion of female and minority employees to

LABOR FORCE PARTICIPATION OF WOMEN AND ETHNIC MINORITIES IN AMERICAN SCIENCE AND ENGINEERING (S/E) CAREERS IN 1993

DEMOGRAPHIC GROUP	PROPORTION OF U.S. POPULATION (%)	LABOR FORCE PROPORTION (%)	PROPORTION S/E CAREERS (%)	REPRESENTATION IN S/E CAREERS (%)
Women	51.2	45.5	22.4	49
Men	48.8	54.5	77.6	142
African Americans	11.9	10.9	3.5	32
Male	5.6	5.4	2.3	43
Female	6.3	5.5	1.2	22
Hispanic Americans	9.8	8.0	2.8	35
Male	5.0	4.8	2.1	44
Female	4.8	3.2	0.7	22

Women, African Americans, and Hispanic Americans are underrepresented in science and technology careers. Representation, in the fourth column, is calculated by dividing the third column by the second. Women for example make up 22.4% of science and technology careers but constitute 45.5% of the U.S. work force so that there are only 49% as many female scientists and engineers as predicted from their representation in the labor force.

(Source: Women, Minorities, and Persons with Disabilities in Science and Engineering, 1996)

match their proportion in the labor force. Over the years, the U.S. Supreme Court issued several rulings that weakened or narrowed the effect of the 1972 act. One of the most important of these was the case of *University of California v. Bakke* (1978), which struck down, as unconstitutional, affirmative action plans based on racial quotas. The Civil Rights Act of 1991 restored and strengthened legal protections for minorities in the 1972 act. Following the Republican congressional election victory of 1994, there have been calls for an end to affirmative action programs that favor racial minorities. In 1996, California's voters passed Proposition 209, making it illegal for the state to discriminate against individuals, or grant them preferential treatment, based on race, sex, color, ethnicity, or national origin. Affirmative action programs have been suspended in education in several states, including California, Texas, and Michigan.

Despite the outlawing of affirmative action, there has not been a reduction in commitment to the goal of diversity in education and changes in the law have not produced a sharp decline in minority admissions to colleges. Affirmative action is not the only means of achieving diversity in science and technology careers. Other possible methods include improving the quality of science education so that a more diverse group of people will be qualified for entry into science and technology careers. Another approach that has being adopted by some colleges is to recruit from the top 10% of high school students, which has the effect of including more minority students than would otherwise be the case.

See also ASSOCIATION OF WOMEN IN SCIENCE; GILLIGAN, CAROL; SCIENCE CAREERS; SCIENTIFIC RACISM.

Further Reading

De George, R. *Business Ethics.* Englewood Cliffs, N.J.: Prentice Hall, 1995.
National Science Foundation. *Women, Minorities and Persons with Disabilities in Science and Engineering.* (NSF 96-311). Arlington, Va.: National Science Foundation, 1996.
National Science Foundation. *Women, Minorities and Persons with Disabilities in Science and Engineering.* (NSF 00-327). Arlington, Va.: National Science Foundation, 2000.
Resnik, D. B. *The Ethics of Science.* New York: Routledge, 1998.
Sadler, A., ed. *Affirmative Action.* San Diego: Greenhaven Press, 1995.

Agent Orange

Agent orange is the code name for 2,4,5-T, a widely used herbicide, that was used by the U.S. military during the Vietnam War as a defoliant, namely, an agent for removing the leaves from plants. Evidently the drums in which it was shipped were painted with an orange stripe. American soldiers frequently came under attack by guerrilla units that operated with relative impunity by using the cover of jungle vegetation. Defoliants caused leaves to wither so that hostile units became visible and were forced to move on. It also ruined rice crops, causing food scarcity. The herbicide itself was not particularly harmful to humans but it contained small quantities of dioxin (tetrachlorodibenzo-p-dioxin, or TCDD) as a contaminant. This dioxin trace was an unwanted by-product of the manufacturing process.

The use of Agent Orange has engendered many ethical controversies ranging from the health consequences for soldiers and civilians to environmental pollution. It belongs in the category of chemical warfare that is often regarded as morally unacceptable. Moral outrage against the use of Agent Orange (and other chemicals such as napalm) was so strong that the United States stopped using it in 1971, four years before the end of American military involvement in Vietnam. Yet, it is important to point out that Agent Orange was not used with the intention of causing health problems to humans. Much ethical controversy also surrounds the subsequent behavior of the U.S. government, particularly its unwillingness to take responsibility for the possible health consequences for Americans and Vietnamese alike.

Dioxin is one of the most dangerous toxins known whose potent carcinogenic effects have been clearly established in animal studies conducted after the Vietnam War. In view of this fact, it might seem strange that there has been so much controversy about the health effects of exposure to Agent Orange among the two exposed groups, residents of Vietnam (both soldiers and civilians) and U.S. soldiers. This anomaly is partly resolved by the fact that exposures were quantitatively small, being measured in parts per trillion (ppt) rather than the more familiar parts per million (ppm) normally used to measure toxins.

The small quantities of dioxin found in human tissue mean that the tests for dioxin exposure are technically demanding and prohibitively expensive for a poor country like Vietnam. One result is that the kind of high quality research necessary to establish health consequences of Agent Orange generally could not be done for the largest exposed population, Vietnamese residents.

Careful research conducted on the most heavily exposed American soldiers, who were compared to a control group of nonexposed air force soldiers, has generally not supported initial concerns about many of the health consequences of Agent Orange exposure. Suggested adverse effects of dioxin exposure had included birth defects, cancers, immune suppression, liver dysfunction, and peripheral neuropathy. The only health problem that has been definitively linked to Agent Orange is skin cancer (basal cell carcinomas), which had a 50% higher incidence in the exposed group. Although Vietnam veterans had a 50% higher incidence of non-Hodgkins lymphoma, this could not be accounted for in terms of Agent Orange exposure.

Even though the research does not show that Agent Orange caused a variety of ailments that veterans had attributed to the herbicide, many years of litigation against Dow Chemical and Monsanto corporations, who manufactured it, have had an impact. A class action settlement provided $180 million to 20,000 veterans. Moreover, the Veterans Administration (currently the Department of Veterans' Affairs) agreed to compensate Agent Orange–exposed troops for a number of ailments in which the herbicide might be implicated. Compensatable illnesses include skin diseases like chloracne, respiratory cancers, soft tissue sarcoma, and prostate cancer. Compensated veterans may receive up to $1,989 per month. By 1989, 270,000 veterans had registered with the Agent Orange program and 6,000 had qualified for benefits.

Whatever the health problems produced among American veterans, the brunt of the problem has been borne by the Vietnamese because more of them have been exposed. The dioxin from Agent Orange has lingered in the environment and contaminated food chains in much the same way as DDT does. TCDD, the dioxin found in Agent Orange, accumulates in fatty tissues. In areas of Vietnam unsprayed with Agent Orange, the average TCDD concentration was 0.6 ppt. Concentrations in the sprayed areas ranged from 14.7 ppt to 103 ppt among southern Vietnamese, according to a Red Cross–sponsored study conducted during the 1970s and 1980s. Particularly troubling is the fact that such high levels of dioxin were found in young people as well as old since this indicates that the contamination is not due to the direct effects of spraying, which occurred before their births, but comes from lingering toxicity within the food chain. Dioxin persists in blood tissue and in breast milk, which means that children are exposed to it from an early age, even before birth.

Attempts to quantify the health effects of Agent Orange on the Vietnamese population are unlikely to rise beyond pure guesswork, but these effects are inevitably much larger than the effects on U.S. soldiers if only because so many more Vietnamese were exposed. Dioxin is known to be a potent carcinogen in studies of laboratory animals and is known to cause skin cancers in humans. While it is reasonable to assume that it might also contribute to other cancers and health problems, there is little scientific evidence at present to support this. Yet, health workers in Vietnam have been impressed by the emergence of unusual, unexplained illnesses among former soldiers. These diseases include rare forms of cancer, severe diarrhea, immune-deficiency diseases, and a chronic form of malaria that does not respond to drug treatment.

One of the most serious problems to which Agent Orange is linked is an increased incidence of birth defects. Studies of laboratory animals have shown that dioxin exposure increases the rate of deformities of offspring. According to a Red Cross survey, five times as many children with birth defects (more than 5%) were fathered by Vietnamese soldiers from areas heavily exposed to Agent Orange as were fathered by those who avoided exposure by remaining in North Vietnam. This evidence suggests that Agent Orange might have increased birth defects; however, if that were true, then the same finding should have been produced in the conceptually similar study of American veterans.

One possible explanation for the discrepancy in findings might be traced to different levels of exposure. If the Vietnamese soldiers were exposed to higher levels of dioxin, then it might be argued that the U.S. veterans were generally below a theoretical threshold at which chromosomal damage and birth defects are produced. Yet, the tissue concentrations of dioxin in American veterans of 10–150 ppt are almost exactly the same as those for Vietnamese troops. On balance, it seems probable that the high rate of birth defects among children of Vietnamese soldiers exposed to Agent Orange must have other causes, although it is not clear why birth defects should be less common for troops who stayed in North Vietnam. It is worth pointing out that the Veterans' Administration indirectly acknowledged that Agent Orange may cause birth defects by adding spina bifida (a disorder in which the spinal column is split in two) to its list of compensable diseases in 1996.

See also BHOPAL; BIOLOGICAL AND CHEMICAL WARFARE; DDT; SEVESO; TIMES BEACH.

Further Reading
Dreyfuss, R. "Apocalypse Still." *Mother Jones* 25, no. 1 (2000): 42–51.
Institute of Medicine. *Veterans and Agent Orange.* Washington, D.C.: National Academy Press, 1996.
Liss, J. L., and F. DeStefano. "Effects of Agent Orange Exposure." *JAMA* 273 (1995): 1494–5.

air pollution
Air pollution is the buildup in the atmosphere of extraneous substances that either threaten human health or have measurable effects on other organisms or the physical environment. Air is a fundamental need of human beings. Since we constantly breathe the air, there is no practical escape from the impurities it may contain. Individuals who are concerned about the quality of drinking water may choose to avoid pollutants by drinking bottled water but it is unlikely that many people will want to rely on bottled air even though traffic police in Tokyo breathe air from canisters to protect their lungs from air pollution.

Recent decades have seen an alarming increase in rates of respiratory diseases and other ailments that are caused or aggravated by pollutants in the air. Although air pollution may be caused by natural phenomena, such as volcanoes and forest fires, most of the problems are caused by human industrial activity. For this reason, air pollution is most intense in large urban centers, such as New Orleans and Tokyo, where it is visible as SMOG.

Once pollutants are released into the atmosphere, they can have extremely complex, but usually deleterious effects on the quality of the air that is breathed as well as affecting global climate as in the case of GLOBAL WARMING produced by excess carbon dioxide and other gases in the atmosphere. Some idea of the complexity of these interactions may be gleaned from the fact that ozone at ground level is a dangerous constituent of smog. The same substance (O_3) is a natural and necessary constituent in the higher atmosphere that is being depleted by pollutant gases, particularly tetra fluorocarbons that used to be present in refrigerant systems and aerosol cans.

Air pollution raises important and complex ethical questions. From the point of view of controlling air pollution, no method of purifying the air is particularly effective or practical. This means that air pollution must be addressed at the point of emission, whether this is through a factory smokestack, a car exhaust, a domestic heating system, a strip mine, or the spraying of pesticides on farms. Although air pollution is almost entirely a product of the INDUSTRIAL REVOLUTION, and could be abated by scaling back industry, most reasonable people accept that there must be some compromise. Atmospheric pollution is a price paid for industrial development. Yet, it is a heavy price, which is likely to be compounded over time. It is therefore essential to seek technological solutions to a technological problem. In this respect, some important improvements have been made in smokestack designs and in the use of catalytic converters to minimize pollution from cars.

SOURCES OF AIR POLLUTION

The major sources of air pollution through human activities involve the burning of fossil fuels (oil, coal, natural gas) as well as other fuels (wood, peat, etc.) and the incineration of solid wastes. Fuels are burned to provide electric power and domestic heating as well as to facilitate industrial processes. Another major source of air pollution is the fuel that is expended for transportation. These processes all release both gaseous and particulate (solid) pollutants. Industries such as mining and agriculture, which disturb topsoil, allow particulate matter to be carried into the air by wind. If this material contains toxins, it can have harmful effects on animal populations that are at a great distance from the source of the pollution, as illus-

trated by the case of Antarctic penguins that are affected by airborne DDT. Likely sources of this DDT include countries in Africa and South America where the pesticide is still used to control mosquitoes. Such potential for distant action provides an important ethical argument in favor of restricting pesticide use. Mining techniques that are particularly likely to release particulates into the air include STRIP MINING and MOUNTAINTOP REMOVAL.

The major air pollutants include carbon dioxide, carbon monoxide, nitrogen oxides, sulfur dioxide, particulates, hydrocarbons like unburned gasoline, and photochemical oxidants that are produced by the effects of light on chemicals contained in the air. Industrial pollution in cities is visible as smog, or a combination of smoke and fog. In the early days of the industrial revolution, smog was produced by condensation of water droplets on smoke particles in the air. The smog of today is largely of petrochemical origin. Unburned gasoline combined with other air pollutants, including sulfur dioxide, is exposed to solar radiation, which causes a complex chain of chemical reactions that result in the production of numerous irritant substances such as ozone, organic acids, aldehydes, peroxyacetyl nitrate, organic particles, ketones, and other oxidants. In addition to photochemical oxidants, car exhausts contain the toxic gas carbon monoxide and account for many particulates in the air.

CONSEQUENCES OF POLLUTING THE AIR

Air pollution is not only unsightly but also has grave health consequences, some of which are not yet well understood. In addition, air pollution can having damaging effects on animal and plant life and can have serious effects on the earth's atmosphere, such as GLOBAL WARMING and depletion of the ozone layer.

The city of Los Angeles has had serious smog problems since the 1940s. Although originally produced by heavy industry, the smog of today is largely created by car exhausts. Smog has seriously damaged the pine forests in neighboring mountains and has been detrimental to agricultural crops in the Los Angeles basin. These effects are produced by the action of chemicals in smog that react with plant tissues causing cellular damage.

Global effects of pollution are less obvious but potentially much more important. Industrial development has raised the level of carbon dioxide and other gases to produce a greenhouse effect. Light rays from the sun penetrate the greenhouse gases and the heat that is absorbed by the earth is prevented from escaping by the gases, just as heat is trapped in a real greenhouse. Although there has been much controversy about whether industrial activity has begun to warm up the earth via a greenhouse effect, the preponderance of the evidence now indicates that global warming has begun. The most direct evidence for this has been the increase in average global temperatures throughout the 20th century and the fact that this is correlated with increases in atmospheric carbon dioxide.

The other major concern in relation to the global consequences of air pollution has to do with depletion of the ozone layer. The ozone layer is vulnerable even to the small quantity of chlorofluorocarbons emitted from aerosol cans (where such propellants have not been banned) and leaking from the compressors of refrigerating systems. The current hole in the ozone layer has shifted away from the source of these pollutants in industrialized areas of Western Europe and North America and is located close to the North Pole. The ozone layer plays an important protective function because it absorbs ultraviolet radiation from sunlight. Ozone depletion means that sunlight is dangerous because of the potential for increased frequency of skin cancer among humans exposed to high levels of ultraviolet radiation.

Another widespread climatic problem associated with air pollution is ACID RAIN. The burning of fossil fuels releases sulfur dioxide and nitrogen oxide, and these combine with water vapor. When the combination falls as rain, it can be very damaging to bodies of water, plant life, and soil. For example, acid rain has been blamed for the disappearance of fish from lakes in the Adirondacks, and for damage to forests in European mountains. It can also damage buildings, which deteriorate from the effects of corrosive acids, and it is hazardous to health.

Troubling as the global consequences of air pollution are, the health effects are even more worrying, if only because they are more immediate. The connection between air pollution and mortality rates has been recognized for at least half a century. Thus an exceptionally severe coal-smoke fog in London in 1952 was associated with some 4,000 excess deaths. Recent research has attempted to fill in the gaps by identifying pollutants that have the most important health effects and accounting for the greater vulnerability of some individuals compared to others. A 1993 study comparing pollutants in American cities over a 14–16-year period, the Harvard Six Cities Study, found that 26% more people died prematurely in the city with the highest pollution levels compared to the one with the lowest air pollution. This study was important for two reasons. First, it controlled for individual risk factors, such as smoking, that had operated as a wild card in previous studies. Second, it identified fine particles (less than 2.5 micrometers in diameter) produced by burning of fuels as the most serious health risk. Such fine particulate matter can penetrate deeply into the lungs and stay there for long periods, thus increasing vulnerability to chronic obstructive pulmonary disease, asthma, and cardiovascular disease. The Harvard study also identified sulfates, another

product of fuel combustion, as contributing to the risk of early death.

Air pollution has alarming effects on the pulmonary health of children. Thus, countries with air pollution problems, of which there are many in Europe, have seen a huge increase in asthma rates among children. Asthma is now so common in England that many large public schools have "puffer" rooms in which children may treat themselves for asthma attacks using inhalers. Approximately one English child in seven suffers from the disease. The elderly are also extremely vulnerable to asthma attacks when exposed to polluted air. The velocity with which the incidence of asthma is increasing is staggering. In 1997, 10% of the U.S. population suffered from asthma, which was over 40% higher than the number just a decade earlier.

The primary symptom of asthma is difficulty in breathing caused by the narrowing of airways in the lungs due to the combined effects of muscular constriction, inflammation, and increased phlegm. Most asthmatics suffer from nasal symptoms, or allergic rhinitis. Asthma attacks are often triggered by an allergic reaction to pollen, dust mites, and animal dander. Current evidence suggests that air pollutants, such as ozone, sulfur dioxide, and nitrogen dioxide, are irritating to the lungs, which makes an asthma attack both more likely and more severe. For example, a 1993 California study showed, by means of lung biopsies, that exercising in air with high ozone levels can cause pulmonary inflammation even in healthy people. Children are at a greater risk than adults because they spend more time outdoors during summer when ozone levels are highest and because their respiration rate is higher and their lungs are developing.

Even though environmental factors are clearly implicated in the increased risk of asthma among children in developed countries, it is important to realize that there are many other environmental factors that could play a role in addition to air pollution. They include: exposure to indoor allergens, which are worsened by carpeting (which harbors mites and molds) and home insulation techniques making houses tightly sealed; electric heating that circulates dry air and dust; increased exposure to junk food that is high in chemical preservatives; reduced exposure to viral infections at an early age because of vaccines; and reduced incidence and duration of breast-feeding.

Even though some traffic pollutants, such as nitrogen dioxide, are known to trigger asthma attacks, the link between air pollution levels and asthma symptoms has not been consistently demonstrated. For example, the great London smog of 1952 did not exacerbate the symptoms of asthmatic children. Researchers in the Netherlands have suggested that airborne allergens, like pollen, rather than pollutants, may be the primary cause of asthma. In many cases there is a correlation between weather conditions that trap smog and weather conditions that contribute to high levels of pollen in the air, but this relationship is unreliable. This problem was illustrated by a study conducted in Barcelona, Spain, a city that experienced epidemics of asthma symptoms. In the beginning, it was found that asthma was correlated with the level of nitrogen dioxide in the air. It has now been established that the asthma epidemics are due to soybean allergen from ships being unloaded in the city's harbor.

A role for air pollution in causing asthma attacks is suggested by the fact that children living in the vicinity of a busy road were more likely to suffer from asthma. This had been attributed to particulate matter from car exhausts and many researchers thought that diesel exhaust was the most likely culprit. Recent evidence shows that cars throw up a lot of fine dust and that this dust is carried through the air into the homes of children living nearby. Moreover, it has been shown that fine particulate matter (less that 10 micrometers in diameter) contains potent allergens, such as mold and pollen particles, that are capable of triggering asthma attacks. If this interpretation is correct, then it appears that car exhaust and other industrial pollutants play a secondary role in asthma. They do not cause it but they irritate the lungs, which makes symptoms more severe.

Even though traffic exhaust may thus play an indirect role in producing asthma symptoms, it seems unlikely that the current asthma epidemic can be greatly alleviated through controls on vehicle emissions. Irritants such as nitrogen dioxide and ozone may aggravate asthma symptoms, or they may even trigger an attack, but the primary cause of asthma seems to be responsiveness to allergens. Switching to fuel-efficient HYBRID-POWERED CARS of the future that have very low emissions may improve the lives of asthmatics but probably will not reduce the number of people affected by the disease.

One disorder that is known to have been produced by vehicle exhausts is LEAD POISONING. Lead used to be a constituent of gasoline and it was emitted with car exhaust. Much of the lead settled close to roadways and children became contaminated by playing on surfaces or in soil close to roadways. Lead is a neurotoxin that accumulates in the nervous system causing learning disabilities. Researchers have estimated that lead poisoning has shaved several IQ points off the scores of inner-city children whose academic performance has also been adversely affected by correlates of poverty. These effects are due to toxic interference with brain development and are therefore irreversible. Reducing the lead content of gasoline has helped to lessen the problem, but it has not solved it because lead-contaminated areas have remained contaminated.

Apart from the atmospheric pollution resulting from routine daily activities in advanced industrial economies, another troubling type of contaminant has been nuclear fallout from aboveground nuclear tests (see DOWNWINDERS). Even more pervasive effects have been observed in the context of the nuclear power industry. The partial meltdown of the CHERNOBYL nuclear power plant in the Ukraine, for example, scattered radioactive dust all around the globe with significant long-term health implications.

REGULATING AIR POLLUTION
Once pollutants have been released into the air it is almost impossible to clean them up even though some interesting methods have been developed using biofiltration in which microorganisms growing on a porous surface convert pollutants into harmless products. For this reason, governmental regulation concentrates on reducing emissions. Many of these efforts have been successful in reducing levels of dangerous pollutants in the air, particularly over cities. The U.S. CLEAN AIR ACT OF 1970 set standards of air quality in relation to six pollutants: sulfur oxides, nitrogen oxides, carbon monoxide, hydrocarbons, photochemical oxidants, and particulates. States were given responsibility for implementing these standards. In regions where the standards were not met, new industrial development, or expansion, was to be halted. Amendments to the Clean Air Act in 1990 generally tightened standards for automobile and industrial emissions. These regulations have helped to reduce air pollution. For example, emissions of sulfur dioxide, a key component of acid rain, have been reduced by approximately one-third since passage of the act. Despite these improvements, most of the people living in the United States are exposed to dangerously high levels of ozone pollution on at least some days of the year.

Improvements have been achieved largely through the use of new technologies to reduce waste emissions. Thus, the use of CATALYTIC CONVERTERS has reduced the emission of nitrogen oxides, carbon monoxide, and hydrocarbons in vehicle exhausts. The catalytic converters oxidize unburned gases. Emissions have also been brought down by lowering engine combustion temperatures. In addition, efforts have been made to reduce smokestack emissions from industry. Some of these have been quite simple and relatively inexpensive. For example, the electric utility industry, a major source of acid rain pollutants, has switched to a different form of coal that is lower in sulfur. The utilities were motivated to make this change by an innovative government-initiated emissions trading system that allowed the electricity companies to lower costs by reducing emissions.

Apart from such rational changes in energy source, reduction of sulfur dioxide is both difficult and expensive. Limestone scrubbers are capable of removing 90% of the sulfur dioxide from a smokestack but they consume about 5% of a power plant's electricity output. They also generate huge amounts of calcium sulfite sludge, which creates another disposal problem. Ideally, the pollutants would be separated in a form that made them reusable. Thus the sulfur might be used to make sulfuric acid.

A variety of methods have been developed to remove particles from smoke. The simplest is the wet scrubber in which particles are washed out of the exhaust stream using jets of water. The electrostatic precipitator removes particles by giving them an electric charge and attracting them to charged plates. The cyclone separator works on a centrifuge principle with the heaviest particles being whirled to the outside for removal. The baghouse traps particles in filters in a manner that is similar to domestic vacuum cleaners. The baghouse method is most effective at removing particles from the exhaust stream but the electrostatic precipitator is used more often because it is cheaper.

THE ETHICS OF AIR POLLUTION

Air pollution is a TRAGEDY OF THE COMMONS because the selfish interests of the individual, or company, motivate them to behave in ways that cause damage to the vital shared resource of the air that everyone breathes. The immediate benefits of being a polluter often seem to outweigh the delayed cost of irremediable environmental destruction that are borne by large numbers of people. The tragedy of environmental pollution occurs because the benefits of polluting go mainly to the polluter whereas the costs are distributed among the vast numbers of people whose health and well-being are compromised by breathing polluted air.

The tragedy of the commons idea suggests that individuals and companies will not take pains to avoid polluting the atmosphere unless their behavior is regulated through government action. Moreover, viewed from an international perspective, it is unlikely that independent states can be relied upon to protect the global environment unless their behavior is regulated by global treaties that have real powers of enforcement.

The central ethical questions regarding air pollution are (a) how much air pollution is morally justifiable? (b) who has the right to pollute the atmosphere? and (c) what are the obligations of polluters to the victims of air pollution? The first question is couched in terms of amount rather than asking if *any* pollution is justified because it is assumed that all human industrial and domestic activities are in some way dependent on environmental pollution. The car a person drives to work is constantly polluting the air. Even if motorists converted to exhaustless fuel-cell vehicles, the mere fabrication of the car would have resulted in the burning of large

amounts of fuel and thus in the emission of carbon dioxide, carbon monoxide, and sulfur dioxide, thereby exacerbating acid rain, ozone pollution, and the greenhouse effect. The same is true of a person sitting in their heated home in winter. Whether their heating system is powered by electricity from the local utility or there is an oil-fired furnace in their basement, staying warm in winter probably pollutes the air because it calls for the burning of solid fuel. This connection may seem inevitable but it does not have to be. Clearly, there is a good environmental rationale for promoting the use of energy sources, including hydroelectricity, wind power, and solar power, that produce little or no air pollution. To do otherwise is to set out on a course of progressive environmental degradation and eventual ecological collapse.

Given that some level of environmental pollution follows from our everyday activities, the question is whether every individual should be thought of as personally responsible for the atmospheric damage produced by their individual activities. For example, is it immoral to drive a large fuel-inefficient vehicle when your transportation needs could be met by a more modest car, or even a bicycle? Is it ethically justifiable to own a large home that requires thousands of gallons of oil to heat each year when a more modest home would suffice? What about ownership of a gas-guzzling speedboat?

These examples of wasteful uses of fuel are also examples of status symbols. In democratic societies, wealthy people have the economic resources to expend large quantities of energy and frequently choose to do so. In that sense, it is clear that wealth gives people the opportunity to engage in activities that pollute the atmosphere. Moreover, the greatest sources of environmental pollution have traditionally been large powerful corporations such as electric utilities, steel producers, and chemical companies. This relationship no longer holds because information technology that is a primary source of wealth generation in modern economies is much cleaner than the industries of the past. As information technology grows in the West, dirtier industries like chemical manufacturing and steel production get shifted to developing countries.

Even though some heavily polluting industries may be exported by wealthy nations, this does not get rid of the problem. On the contrary, it means that the global atmosphere is under threat from more different places simultaneously and therefore more likely to suffer lasting damage. Not all polluting industries will leave the most developed countries. Some, like electric utilities, have to be located near their customers due to the difficulty of efficient transportation of energy over long distances. Even though government regulations have begun to restrict the amount of pollution that is permissible for electric utilities, there is clearly an implicit understanding that electricity is essential to the way we live and that utilities therefore have the right to pollute the atmosphere. Of course, this implicit moral right has been made into an explicit legal one in the United States through the purchase of options from the government to emit a ton of pollutants into the atmosphere. This is part of an ingenious attempt to use free-market principles to reduce air pollution. Yet it conveys the implication that some companies have the right to pollute the atmosphere and that this right can be conferred, or abridged, by government.

Instead of working to make the air cleaner, many corporations have a history of fighting government regulations that would have protected the atmosphere. American car manufacturers have been accused of delaying regulation of car design to reduce environmental pollution. Thus, unleaded gasoline was not introduced until 1986 even though the harmful effects of LEAD POISONING from gasoline additives were well known in the car industry for more than 40 years. Design improvements, such as unleaded gasoline, that could have improved the quality of the air were arguably resisted for years and decades evidently to protect profit margins. This means that the air is more polluted than it might otherwise be due to corporate self interest.

In general it would appear that major environmental polluters feel little sense of moral or legal liability for the adverse effects of pollution in terms of health problems and ecological damage. Once the pollutants have escaped from the smokestack, the owners of the plant are not expected to feel any responsibility for them with one exception. Local residents are sometimes given partial protection from air pollution through the construction of tall chimneys that release pollutants higher up and therefore allow them to be more widely dispersed. This helps new industrial plants to receive planning permission but it does nothing for the global atmosphere. Once pollutants have escaped into the air, it is virtually impossible to clean them up.

See also ACID RAIN; CLEAN AIR ACT; DOWNWINDERS; WATER POLLUTION.

Further Reading

Burney, P. "Air Pollution and Asthma: The Dog That Doesn't Always Bark." *Lancet* 353 (1999): 859.

Cotton, P. "'Best Data Yet' Say Air Pollution Kills Below Levels Currently Considered Safe." *JAMA* 269 (1993): 3087–88.

De Nevers, N. *Air Pollution Control Engineering.* New York: McGraw Hill, 1999.

Devinny, J. S., M. A. Deshusses, and T. S. Webster. *Biofiltration for Air Pollution Control.* Cherry Hill, N.J.: Lewis, 1998.

Doyle, J. T. *Taken for a Ride: Detroit's Big Three and the Politics of Air Pollution.* New York: Four Walls Eight Windows, 2000.

Heck, R. M., and R. J. Farrauto. *Catalytic Air Pollution Control: Commercial Technology.* New York: John Wiley, 1997.

Holgate, S. T., H. S. Koren, and R. L. Maynard, eds. *Air Pollution and Health.* New York: Academic, 1999.

Raloff, J. "Traffic May Worsen Hay Fever and Asthma." *Science News* 156 (November 1999): 325.

Wark, K., C. F. Warner, and W. T. Davis. *Air Pollution: Its Origin and Control.* Reading, Mass.: Addison Wesley Longman, 1997.

alar scare

Alar is the trade name of an apple growth regulator, daminozide, which was produced by Uniroyal Chemical Company. Daminozide itself is not considered to be dangerous but cooking the juice to produce concentrate releases UDMH, which has been shown to cause malignant blood vessel tumors in laboratory animals. Alar illustrates the complexity of regulating food based on scientific findings. In addition, the federal regulators were found to have a conflict of interest because of consultancy work for chemical companies. The resulting loss of consumer confidence in the safety of apples has had severe economic consequences for apple growers.

In 1985, the Environmental Protection Agency (EPA) first became aware of animal studies indicating that daminozide and UDMH may cause cancer and proposed banning Alar. Then the EPA's Science Advisory Panel found that the animal studies were scientifically flawed. The ban was suspended awaiting the results of new animal studies. When these studies were completed, in 1989, daminozide was found to be noncarcinogenic. However, UDMH was found to be highly carcinogenic. The EPA concluded that UDMH at the dietary levels found in 1986 could cause 45 cancers per million people exposed over their lifetimes. This risk was far lower than the 8,300 people per million estimated to develop cancer from alar consumption according to a 1987 National Academy of Sciences report. Since any cancer risk greater than one per million justifies regulatory action, the EPA again decided that alar should be banned. In the end, the cancer risk posed by alar was found to be minimal. The early risk estimates, based on faulty extrapolation from laboratory rodents, grossly exaggerated the dangers of exposure to small quantities of a potential carcinogen.

The alar scare reached its highest pitch with a CBS *60 Minutes* story aired on February 26, 1989. Viewers were told that apples were giving them cancer and that children were most vulnerable as high-volume consumers of apples and apple products, particularly juice. The program shattered public confidence in the safety of apples.

Following the adverse publicity about the carcinogenicity of their product, American apple growers announced a voluntary ban on alar use and supermarket chains announced that they had stopped buying apples from growers who used the growth regulator on their apples. Nevertheless, shoppers stopped buying apples and other fruits and some even threw out what they had already purchased. Many consumers took an increased interest in carefully washing their fruit in the hope that toxins would be removed from the skin. The losses in sales of apples due to the alar scare were exacerbated by a deterioration in the cosmetic properties of the fruit. Some varieties, particularly Red Delicious and Macintosh, had a loss of color quality, firmness, and texture.

The alar scare proved to be greatly overblown. In the end, people who ate plenty of apples and other fruit, alar and all, were less, not more, likely to develop cancer indicating that possible protective agents in the fruit outweighed the risk of chemical contaminants.

At the time of the alar scare, a troubling conflict of interest within the EPA's Science Advisory Panel emerged. It turned out that seven of the eight members of the panel were consultants in the chemical industry. One member of the panel subsequently accepted a job with Uniroyal Chemical Company, the producer of alar. Another member of the panel was paid to speak on behalf of a different Uniroyal product that had been reviewed by the EPA. Such evident conflicts of interest undermine public trust in the scientific advisory panel and are particularly troubling when its decisions overrule scientific evidence, however flawed this might be, in favor of the interests of a chemical company. None of the panel members was charged with violating government ethics laws.

See also AGENT ORANGE; BHOPAL CHEMICAL ACCIDENT; DDT.

Further Reading
"Bad Apples." *Consumer Reports* 54, no. 5 (1989): 288–92.
Henry, S. "An Apple a Day." *Technology Review* 91, no. 2 (1988): 11–13.
Saxton, L. "Post-alar Effects on Apples Surface." *Supermarket News,* August 26, 1991.
Spencer, P. "Risky Assumptions." *Consumers' Research Magazine* 82, no. 3 (1999) 43.

altruism, evolution and

Darwinian evolution assumes a struggle for existence that can be interpreted quite literally in terms of competition over food, shelter, and opportunities to reproduce. The expectation that nature should be red in tooth and claw has had important ethical ramifications in the context of SOCIAL DARWINISM, the view that a similar level of competition is to be expected among human beings in their economic and political behavior. Darwin himself believed, however, that human altruism is partly instinctive, based on the love of praise and fear of condemnation, and was prepared to concede that nonhuman animals may also

have a primitive moral sense based on analogous social instincts.

Modern evolutionists have made a compelling case that altruistic behavior of two distinct types could have been produced by natural selection. These are kin-directed altruism (or kin selection) and reciprocal altruism between nonrelatives. A third, GROUP SELECTION, is based on the idea that individuals sacrifice themselves for the good of the species, or social group. Thus, when lemmings drown during population explosions, they are seen to "commit suicide" to save the species from the adverse effects of overpopulation. In general, biologists do not believe that such self-sacrificial altruism can evolve among complex animals simply because individuals that behaved selfishly instead would have a reproductive advantage and selfish tendencies would thus be propagated into future generations.

KIN SELECTION
One of the most exciting theoretical advances in modern evolutionary theory has been the development of kin selection theory and its application to problems of altruism such as those posed by the social insects. Thus, female worker bees, which are sterile, spend their lives taking care of the hive and raising offspring for their mother, the queen. This phenomenon puzzled Darwin greatly because he could not understand how such altruism could exist in the midst of the harshly competitive struggle for existence that he saw as the foundation of evolution by natural selection.

Kin selection theory, developed in its modern mechanistic form by English biologist William Hamilton in 1964, offers a simple and compelling solution to this riddle. Hamilton showed that if individuals can increase their genetic representation in the next generation more effectively by helping relatives than by helping themselves, then evolution will favor the development of altruistic behavior directed at relatives. Hamilton's analysis involved gene-level selectionism in which reproduction is described in terms of passing on copies of one's genes. Since relatives share many of the same genes, an individual can propagate genes into future generations either by reproducing themselves or by helping blood relatives to produce offspring. An individual's direct reproductive success, or number of offspring, is referred to as their fitness. Reproductive success measured in terms of the total number of copies of their genotype transmitted into future generations, including those produced indirectly through the reproduction of relatives, is referred to as inclusive fitness.

Hamilton's gene selectionism puts the evolutionary account of kin-directed altruism on a secure mechanistic basis and thus helps to explain how altruism could have emerged in the harshly competitive arena of Darwinian evolution. Although kin-selected altruism might appear to be teleological, or based on conscious intentions (an impression that is encouraged by the use of metaphors such as "selfish" genes), its theoretical importance is that it explains the evolution of moral behavior in completely mechanistic terms.

This point can be appreciated by imagining that in a world of complete Darwinian competitiveness, a mutation arises that makes individuals behave altruistically to all other members of their species by giving up their food, territory, and mating opportunities. The owner of such a mutation would be taken advantage of by competitors, would fail to reproduce, or even survive for long and the altruism mutation would disappear in the first generation. Now imagine a mutation that produces altruistic behavior directed exclusively toward close relatives. Such altruism might reduce direct fitness but, if it makes enough of a contribution to inclusive fitness, it can be propagated into future generations. At an extreme case, an altruistic individual who fails to reproduce personally but who is responsible for the survival of four young nieces and nephews will likely have sent the altruistic gene into future generations. Of course it is simplistic to imagine that a single gene can be responsible for any complex pattern of behavior, or that the behavior is genetically determined, but however many genes are implicated, or however probabilistic the relationship between genes and behavior, the same logic can be applied and the same conclusion holds.

The case of altruistic behavior among nonreproductive worker bees (and related hymenopteran social insects such as ants and wasps) provides a rather startling confirmation of Hamilton's kin selectionist account of altruism, at least in the case of queens that mate only once in their lives. Hymenopterans have a haplo-diploid reproductive system, which means that males have only one set of chromosomes whereas females have two sets. Females are produced in the conventional manner when the eggs of the mother are fertilized by the sperm of a drone. Drones themselves are produced from unfertilized eggs and thus receive only a single set of chromosomes, those of their mothers.

These unusual reproductive mechanisms lead to some strange genetic relationships. Daughters receive half of their genes from the queen and are therefore 50% related to her. On average, the daughters share 25% of the genes received from the mother, just as two human siblings with a relatedness of 50% derive half of this similarity (25%) from the mother (and half from the father). Among hymenopterans, it is the relatedness through the father that is the wild card because all fertilized eggs receive the entire genotype of the father, which means that daughters are related 50% through the identical genetic information received from their father. Adding

both components, the degree of relatedness of sister worker bees is 75%. This means that workers get more of their genes into the next generation by helping their mother produce another worker, that is 75% related to them, than if they were to produce a daughter that would be only 50% related to them.

This analysis omits the many complex and fascinating details of how workers come to be sterile and how queens are produced. However, the logic of the analysis does not require these details to be filled in. It turns out that larvae that are fed a special food, a highly nutritious secretion from the pharyngeal gland of workers known as royal jelly, develop into sexually mature queens capable of reproducing while those that do not receive this food become sterile workers.

Superficially it might seem that the queen is exploiting the workers to subserve her own reproduction, but it is equally plausible to conclude that the queen is an egg-laying factory through which workers indirectly produce many thousand copies of themselves. Which description is used is largely a matter of semantics. It is clear that the unusual reproductive system of hymenopterans lends itself to a high degree of social organization with sterile worker castes because these phenomena have evolved independently among social insects approximately thirteen times, but not in any other insects except for termites.

The simplest case of kin-selected altruism is provided by parental care of the young, which is well developed in birds of both sexes, female mammals, and male fishes. Parents feed and protect their offspring because if they failed to do so the young would not survive to reproduce and their inclusive fitness would be adversely affected. Phrased in more mechanistic, less teleological terms, a gene promoting parental care in some animals is evolutionarily stable because the offspring who are helped are more likely to survive and reproduce and since they are likely to carry the gene, they propagate it into future generations.

Parental care is particularly well developed among human beings because of the long period of juvenile dependency. Human males are much more involved in the care of offspring than is true of most male mammals and primates, reflecting the difficulty of successfully raising human children to maturity. To say that human parental care is a product of the evolutionary mechanism of kin selection is not to say that adoptive parents cannot develop close parental relationships with their unrelated adoptive children.

Kin-selected altruism affects behavior outside the immediate family but is generally weaker. Political leaders are likely to channel wealth and privileges to their extended family, a practice known as nepotism that is observed in all sorts of societies regardless of complexity, and the bequeathal of property after death is usually to blood relatives whose reproductive success would increase the inclusive fitness of the deceased. Among the Yanomamo horticulturalists of South America, villages typically consist of a few hundred people, which, when they grow larger, become unstable and split into two daughter villages. When the new villages are formed, the average degree of relatedness of residents is higher than it was in the original village, indicating that in this subsistence economy people prefer to live next to close relatives.

There are many everyday social transactions in which altruism is not greatly influenced by relatedness. Among hunter-gatherer societies, like the !Kung of the Kalahari Desert in South Africa, hunting is cooperative and a successful hunter often has the honor of dividing up the kill. The meat is usually distributed with scrupulous fairness and the hunter does not favor his own family. This is clearly an example of altruism that goes beyond the helping of kin. Evolutionists interpret this kind of altruism as arising out of a sort of insurance system. Hunting is a very uncertain activity and on most days hunters do not bring down large prey animals. This could be serious if the hunter and his family went for several days without meat. Under the insurance system, a successful hunter who has more meat than he or his family can eat, and no reliable method of storing the excess, redistributes it to the community because he knows that when he returns from the hunt with nothing, the favor will be repaid. This system of mutual aid among nonrelatives is referred to as reciprocal altruism. In this example, the reciprocal altruism system would probably break down if there were blatant kin biases.

RECIPROCAL ALTRUISM

That human beings are capable of reciprocal altruism is hardly surprising because we can easily grasp the underlying contingencies in an abstract fashion. We know that if we want other people to share with us in our time of need we must be prepared to share with them in theirs. Reciprocal altruism among humans can be stable because it solves a practical problem in a simple and mutually beneficial way. To make the case that reciprocal altruism could have been produced by evolution through natural selection rather than emerging from human rationality, it would be necessary to demonstrate clear examples of reciprocity among unrelated individuals in other species.

Recent study of nonhuman primates, including rhesus monkeys, baboons, and CHIMPANZEES, has revealed many examples of long-term alliances between unrelated individuals for their mutual benefit. Reciprocal altruism is particularly pervasive in the lives of chimpanzees. Chimpanzee allies share food, groom each other, help each other to fight off attacks by enemies, and comfort each other in times of distress. Even so, it could be argued that chimpanzees and other primates are intelligent enough to understand reciprocal interactions. If so, the emergence of reciprocity might not require an evolutionary interpretation for them any more than for humans. Hence, the theoretical importance of establishing reciprocal altruism among animals that are probably less intelligent.

What looks like a clear example of reciprocal altruism has been described in the case of vampire bats. These mammals acquire blood, which is their only food, from cattle, horses, and other large animals. Bats are not always successful on their nightly blood-sucking raids. Since they have a high metabolic rate, vampire bats are vulnerable to starvation. Bats that return to the cave with empty stomachs often receive regurgitated blood from others. Although much of the giving is from mothers to young, much of it is between unrelated pairs of individuals. These recognize each other by means of special contact calls. When they are not sharing blood, they may express their friendship by grooming each other.

Altruistic behavior may thus emerge due to two distinct evolutionary mechanisms. This conclusion softens the impression that nature is necessarily violent and competitive and provides two different routes for the emergence of moral sentiments. On the one hand, there is the self-sacrificial altruism of parents caring for their young and the selfless emotion on which this is based. On the other, there are the emotions that have to do with living up to social obligations built up by a cycle of giving and receiving help. One might imagine that the individuals in a hunter-gatherer band would be very upset if a hunter did not live up to his obligation to share meat. Evolutionary anthropologist Robert Trivers has argued that virtually all of the emotions underlying moral behavior, such as hatred, love, moralistic aggression, sympathy, gratitude, suspicion, and guilt, can be understood as adaptations that regulate the expression of reciprocal altruism among humans.

See also DARWIN, CHARLES; GROUP SELECTION; SOCIAL DARWINISM.

Further Reading
Chagnon, N. A. "Terminological Kinship, Genealogical Relationship, and Village Fissioning among the Yanomamo Indians." In *Natural Selection and Social Behavior*, eds. R. D. Alexander and D. W. Tinkle, 490–508. New York: Chiron Press, 1981.
de Waal, F. *Chimpanzee Politics.* Baltimore: Johns Hopkins University Press, 1982.
Hamilton, W. D. "The Genetical Evolution of Social Behavior, Parts I, II." *Journal of Theoretical Biology* 7 (1964): 1–52.
Ruse, M. *Darwinism Defended.* London: Addison Wesley, 1982.
Trivers, R. L. "The Evolution of Reciprocal Altruism." *Quarterly Review of Biology* 46 (1971): 35–56.
Wilkinson, G. S. "Reciprocal Food Sharing in the Vampire Bat." *Nature* 308 (1984): 181–84.

American Association for the Advancement of Science

The American Association for the Advancement of Science (AAAS) promotes scientific research, science education, academic freedom, and ethical science in all fields. Founded in 1848, with headquarters in Washington, D.C., the AAAS has 130,000 members and is the umbrella organization for 290 scientific societies.

amniocentesis

In amniocentesis, a small quantity of the amniotic fluid that surrounds a fetus is removed using a fine needle. (The technique was first developed in 1950 by British physician Douglas Bevis to test fetuses for Rh-factor incompatibility). Fetal cells from the fluid are cultured in the lab and tested either to detect genetic disorders or to determine the sex of the unborn child. (Other techniques of testing fetal cells include chorionic villus sampling and fetal blood sampling). Conducted in the second trimester of pregnancy, amniocentesis results may lead to abortion of fetuses having serious disorders or constituting the sex not desired by parents. Disorders detected by amniocentesis include Down's syndrome and other chromosomal anomalies, neural tube defects, sex-linked conditions like hemophilia, and numerous single-gene biochemical abnormalities.

Amniocentesis provides parents with information that was unavailable prior to modern scientific medicine and forces them to make tough ethical decisions. These decisions balance the potential quality of the life of the fetus, the needs of parents, and the impact on communities. In general, amniocentesis that is used primarily to screen for genetic disorders is less controversial than when it is used primarily in sex selection because children with genetic disorders frequently have a very poor quality of life.

In some Asian countries, notably India, amniocentesis has been widely used since the 1970s for the purpose of sex selection. In such countries, males are more highly valued than females, who are seen as an economic drain on family resources. For this reason, amniocentesis was used to facilitate selective abortion of female fetuses. The proportion of male to female births was 114 in China, in 1990, compared to a world average of 105. Amniocentesis has now been abandoned as a technique of sex selection in favor of ultrasound imaging that can identify the sex of a fetus in the first trimester of pregnancy.

Feminists generally support a woman's right to choose an abortion but reject sex selection as discrimination against females and therefore as an unethical expression of freedom of choice. Some ethicists also reject sex selection on utilitarian grounds. In a society with too few women, many men will inevitably fail to marry and will thus suffer reduced happiness. Moreover, an extreme scarcity of women can mean that their lives are highly controlled by men and can provide an environment in which coercive controls over female sexuality, including prostitution, may be common. Alarmed about the possibility of a shortage of brides in the future, China, South Korea, and India have outlawed sex-selective abortions. However, legal bans have not had an impact on the proportion of female births in these countries.

See also CONTRACEPTION; REPRODUCTIVE TECHNOLOGIES.

Further Reading

Chadwick, R. F., ed. *Ethics, Reproduction, and Genetic Control.* London: Routledge, 1987.
MacFarquhar, E. "The War Against Women." *U.S. News & World Report,* March 28, 1994.
Mangla, B. "India: Missing Women." *The Lancet* 338 (1991): 685–86.
Tuljapurker, S., N. Li, and M. W. Feldman. "High Sex Ratios in China's Future." *Science* 267 (1995): 874–76.

animal research, ethics of

Ethical concerns about the use of animals as research subjects have arisen mainly in the past few decades and these have been stimulated, in part, by the efforts of animal rights activists. The whole concept of animal rights is a recent product of utilitarian philosophy. Peter Singer, a leading ideologue of the animal rights movement, takes the perspective that humans and other animals share a key moral quality, namely, the capacity to feel pain. According to this perspective, if it is wrong to inflict pain on human participants in research, then it is equally wrong to inflict pain in animal experiments. This is a minority view among scientists and ethicists, the majority of whom believe animal research has been a necessary step in medical research, which has produced treatments for afflictions that include measles, polio, diabetes, smallpox, heart disease, and serious burns. Most philosophers and researchers rate the suffering of humans as being more important than that of animals and argue that research which saves the life of a single human being is justified in sacrificing many nonhuman animals. Moreover, just as there is a different calculus for human and nonhuman pain and suffering, so many ethicists would sacrifice a gnat more readily than a mouse, a mouse more readily than a rat, a rat more readily than a dog, and a dog more readily than a chimpanzee. These distinctions may not be very clearly drawn or very objectively defined but the imputation of different levels of moral worth to members of different species is clearly of great importance to how people make ethical judgments about animal experiments. Thus, high school physiology experiments that are restricted to the use of worms and frogs are more easily defended than if they involve the use of rabbits, cats, or dogs (see chart).

Animals used in research are normally maintained under very strict regulations and experimental protocols must be approved by animal care and use committees. Animal research at institutions receiving support from the U.S. government is tightly regulated in accordance with guidelines drafted by the National Institutes of Health (NIH). Ethically questionable animal experiments have nevertheless received much attention due to the efforts of political pressure groups like People for the Ethical Treatment of Animals (PETA).

Federal regulations ensure that animals are housed in adequate facilities having good ventilation and climate control. Animal facilities are cleaned and disinfected regularly and kept free of infestations. Different species for which contact could be a source of stress or disease, may not be kept in the same room. Animal rights activists have publicized cases in which they felt that animal regulations were not being observed, such as the SILVER SPRING MONKEYS in which animals with wounds were not bandaged according to regulations. Apart from possible breaches of animal regulations, many animal rights activists feel that the mere fact of maintaining wild animals in captivity is an unjustifiable form of cruelty.

The goal of maintaining the health of research animals is sometimes placed at a premium over their social needs. For this reason, laboratory rats, for example, that belong to a highly social species are housed in isolation. Such stringent housing guidelines greatly increase the cost of maintaining animal facilities at colleges and universities. Researchers generally favor solitary housing because it reduces unwanted variability in their outcome measures and increases the likelihood of conducting successful experiments. Apart from social needs, there is a growing awareness that the sterile laboratory housing required to prevent disease and infestations gives animals a limited opportunity to engage in normal, species-typical behavior and thus arguably detracts from their quality of life.

Animal rights activists and others who take an interest in the ethical treatment of research animals are generally most concerned about the use of experimental protocols that are surgically invasive or otherwise stressful. Experimental protocols are stringently regulated. Institutions receiving federal funding in the United States are required to maintain Laboratory Animal Care Committees that perform a role analogous to that of the INSTITUTIONAL REVIEW BOARDS for research involving human participants. These committees usually consist of veterinarians and members of the local community as well as scientists. All animal research must be conducted strictly in accordance with a detailed experimental protocol that has been approved by the Laboratory Animal Care Committee.

The committee has the power, and the obligation, to prevent any research that is deemed to involve the unethical infliction of pain. Under the NIH guidelines, procedures that cause pain to animals, or inflict any other harm on them, must be justified in terms of their likely beneficial results to human beings whether in terms of contribution to scientific knowledge or in the development of practical applications, such as the preliminary testing of a drug. According to this perspective, research that inflicts pain, discomfort, or harm on animal subjects is not necessarily prohibited but it has to be balanced by the theoretical importance or likely practical benefits of the research.

Most animal care committees operate under the assumption that the pain and suffering of human beings counts for much more than the pain and discomfort of nonhumans. If they did not, then much research in the biomedical field would not be carried out to the likely detriment of human health. Although the use of electric shock in behavioral research on animal subjects was fairly common several decades ago, shock generators are very rarely used today. This reflects in part the difficulty in obtaining approval for such research and in part the theoretical turning away from aversive techniques in the behavioral laboratory and in psychology in general.

In addition to regulatory control by governments, scientists are also bound by the ethical guidelines of the professions in which they operate. In America, such guidelines are consistent with federal guidelines in most respects but they often add helpful details that guide the research activities of scientists in different disciplines. Thus the American Psychological Association issues its own "Guidelines for Ethical Conduct in the Care and Use of Animals." These guidelines require that all research on animals must be monitored by a person who is expert in the care and use of laboratory animals. Moreover, a veterinarian must be available for consultation while the research is being conducted and to provide whatever postcare might be required.

Similarly, the Society for Neuroscience guidelines for animal research require that experimental procedures should be designed so as to minimize an animal's pain and discomfort. For example, invasive surgical procedures should never be carried out without anesthesia unless there is an overwhelming scientific rationale for this. Moreover, if animals are to be immobilized using paralytic drugs, they must also be anesthetized to minimize the stress of not being able to move. Surgical procedures must be carried out under sterile conditions and wounds must be protected against the possibility of infection during the experimental procedure and afterward. If there is no intention that the animal should survive the procedure, then it must be kept under anesthetic throughout the experimental procedure and until it is killed. Physical restraint of animals must generally be avoided if alternative procedures are possible. If physical restraint is used, animals must be gradually habituated to the restraint device and must be given reasonable periods of rest that are defined in the experimental protocol.

Many such guidelines have been formulated with the intention of preventing, or at least discouraging, procedures that were seen as ethically questionable when they were carried out. For example, Joseph Brady's research in the 1950s, in which monkeys were immobilized by physical restraint while receiving random electric shocks for six hours at a stretch, would have difficulty being approved by an animal care committee today. Brady was studying the effects of stress on stomach ulcers. His discovery of "executive stress" received much attention but could not be replicated, apparently reflecting design flaws in the original research. Ethically questionable early animal research of this type makes a compelling case for the need of oversight in animal research. Regulatory oversight increases the cost of research in various ways, uses up the time of scientists, discourages creativity, and arguably dissuades some of the brightest people from entering careers involving animal research.

Even though animal research is highly regulated by laws and professional regulations, some research, much of it in the biomedical area, involves infliction of pain and distress on animals. For example, the toxicity of many drugs is evaluated using a Lethal Dose-50 (LD 50) measure. This is the dosage of the drug that kills 50% of animal subjects (often mice). Although cruel, such techniques provide an essential objective tool that can be used to compare the toxicity of different medications. Many human lives are thereby saved. Some utilitarians argue that the pain and suffering of animal subjects is more than balanced by the great potential benefits to human beings. Peter Singer, however, sees such calculations as immoral SPECIESISM.

Even when research passes ethical review at the institution or company that sponsors it, this is not a guarantee either that the research is beyond ethical question or that it will escape the attention of animal rights activists who are prepared to invest a great deal of effort in preventing it. All of these complexities are illustrated in the case of the Silver Spring monkeys, in which the research on movement of neuroscientist Edward Taub was cut short by the activism of PETA.

COMPARATIVE CRITERIA OF MORAL WORTH FOR DIFFERENT SPECIES
(based on Psychology and Neuroscience)

X = Present
O = Absent
? = Probably

SPECIES	FEELS PAIN	MORAL EMOTIONS: SYMPATHY, LOVE, GUILT	FOLLOWS MORAL RULES	ABSTRACT REASONING ABILITY	SELF-AWARENESS	LANGUAGE ABILITY
Human						
- adult	X	X	X	X	X	X
- infant	X	X?	O	O	O	O
Chimp	X	X	X?	X	X	X
Monkey	X	X?	O?	O?	O	O
Dog	X	X?	X?	O?	O	O
Rat	X	X?	X?	O?	O	O
Pigeon	X	X?	O?	O?	O	O
Fish	X	X?	O?	O?	O	O
Frog	X	X?	O?	O?	O	O
Fruit fly	X?	O	O	O	O	O
Earthworm	X?	O	O	O	O	O
Amoeba	O	O	O	O	O	O

Note: X? = probably present, O? = probably absent.

Apart from their moral objections to the use of animals in research on the grounds that animals should not be forced to endure pain, opponents of animal research often argue that the use of animals in biomedical research is inappropriate because other animals are not good models for studying human disease processes. With this in mind, they point to human clinical studies of typhoid, hepatitis, anesthesiology, hyperthyroidism, appendicitis, and immunology that have engendered progress in medicine. Yet, the existence of such important human studies does not challenge the importance of animal research in medicine. For example, most of the Nobel laureates in medicine actually conducted animal research.

The use of animal experiments to study disease processes is sometimes referred to as a causal analogy. That is, if some agent causes a particular disease in rats, the assumption is often made that it will produce the same effect in humans. At its simplest, such a causal analogy is highly fallible. Rats may simply differ from human beings in some important way that is relevant to development of the disease. For example, saccharin used to be classified as a carcinogen on the basis that it caused bladder cancer in laboratory rats. Further research suggested that the development of cancer is probably due to the effect of high concentrations of a protein in the urine of rats that results in a buildup of toxic crystals. The absence of high concentrations of this protein in humans means that saccharin is not a human carcinogen. Such findings do not mean that rodents should not be used to study human disease processes, although some animal rights activists make precisely this case. Rather, such research suggests that it is wise to verify results using different animal species, particularly those that are more closely related to humans, such as monkeys.

The use of primates in biomedical research raises some of the toughest ethical issues. In the minds of many ethical philosophers, as well as the general public, animals that are more closely related to human beings have higher moral value. The criteria on which the moral worth of animals is evaluated are not entirely clear, although there seems to be reasonably good consensus as to the outcome. Most people would agree that a dog has more moral worth than a frog, for example, and that a frog merits more ethical consideration than an amoeba. That said, it is clear that the most important determinant of moral worth is the capacity to feel pain. Animals that do not feel any pain, such as single-celled amoebae, are of minimal moral concern, at least as individuals (see table on preceding page).

The table ranks some common research animals in descending order of moral worth according to frequently invoked criteria, including the capacity to feel pain, experience moral emotions, follow moral rules, think abstractly, be self-aware, and use language. Research has shown that chimpanzees are self-aware and that they can solve problems using abstract reasoning. Observation of chimpanzees in captivity suggests that they are capable of appreciating and enforcing "rules" of expected conduct. For example, when two of the chimpanzees delayed the evening meal of the others by staying out late, Frans de Waal suggests that the community got together to punish the wrongdoers. Many psychologists are wary of accepting such interpretations, however. Although extensive training may bring chimpanzees to the point of using symbols to communicate, such performances may not constitute language. Moreover, chimpanzees in the wild are not known to use language. Language is of particular importance to determination of moral value because it is the medium through which moral concepts and contracts are communicated and mediated.

The scales presented in the table can be read as suggesting that the moral worth of human infants is equivalent to that of monkeys and rats and lower than that of dogs. Most people would disagree, of course. This suggests that our evaluation of the moral worth of human infants may be based on different criteria from those used for common research animals. In particular, it may be that even though human infants cannot speak and are not self-aware we evaluate them in terms of their potential capabilities concerning these criteria. According to Peter Singer, this would be speciesist. Singer's view is that much of animal research is predicated on an unethical lack of concern for nonhuman animals.

See also BATTERY HENS; CHIMPANZEES; SILVER SPRING MONKEYS; PEOPLE FOR THE ETHICAL TREATMENT OF ANIMALS; SPECIESISM.

Further Reading

Barnard, N., and S. Kaufman. "Animal Research Is Wasteful and Misleading." *Scientific American* 276 no. 2 (1997): 80–82.
Botting, J., and A. Morrison. "Animal Research Is Vital to Medicine." *Scientific American* 276 no. 2 (1997): 83–85.
LaFollette, H., and N. Shanks. *Brute Science.* New York: Routledge, 1996.
Miller, N. E. "The Value of Behavioral Research on Animals." *American Psychologist* 40 (1985): 423–40.
Resnik, D. B. *The Ethics of Science.* New York: Routledge, 1998.
Singer, P. *Animal Liberation.* New York: Random House, 1975.

applied ethics

Applied ethics deals with practical ethical problems and it constitutes the counterpart of theoretical ethics, which deals with abstract principles like CONSEQUENTIALISM and DEONTOLOGY. The explosive growth of interest in modern ethics is almost entirely driven by the host of practical ethical problems that we confront today. In part, this reflects the rapid pace of technological change that presents new kinds of ethical challenge, such as those posed by genetic engineering and the Internet. It is also arguable that people today have greater ethical sensitivity, which is illustrated by increasing concern over issues such as sexual harassment in the workplace, affirmative action, environmental pollution, protection of archaeological sites, world hunger, health insurance, and so forth. The increased modern interest in applied ethics may reflect higher levels of affluence and education today than at any other time in history. It may also reflect the ethnic and religious diversity of modern life. Other examples of issues in applied ethics include: abortion, ADVERTISING, artificial insemination, birth control, capital punishment, CASTRATION, chemical weapons, drug legalization, earthquake-proof buildings, ENGINEERING ETHICS, EUGENICS, EUTHANASIA, fetal research, GLOBAL WARMING, HAZARDOUS WASTE, herbicides, industrial accidents, insecticides, NUCLEAR POWER, ORGAN TRANSPLANTATION, PORNOGRAPHY, social welfare, sexual behavior, and WHISTLE-BLOWING in industry. Many of these practical ethical concerns are peculiar to modern life because they were produced by technological innovation. Others may come to be the focus of ethical discussion because traditional formulae (e.g., sex before marriage is wrong) are no longer used.

See also CONSEQUENTIALISM; DEONTOLOGY; MORAL PHILOSOPHY.

Further Reading
Chadwick, R., ed. *Encyclopedia of Applied Ethics.* New York: Academic, 1997.

archaeological ethics

Archaeology is one of the most recent sciences to acquire an interest in ethics. As recently as 1970, there was virtually no literature on ethics in the field. During the 1970s and 1980s, archeologists found themselves under attack for appropriating the artifacts of indigenous peoples around the globe. Such conduct was seen as Eurocentric, neocolonialist, and even sacrilegious. Some early archaeologists were little more than treasure hunters who not only took the artifacts that they discovered but also damaged the sites through careless excavations, thereby destroying much potentially valuable scientific data.

Stung by such criticisms, archaeologists have written extensively on ethical topics and the Society for American Archaeology (SAA) recently adopted a set of eight ethical principles to guide their professional conduct. The principles can be summarized as follows:

1. Stewardship: protecting sites.
2. Accountability to the public.
3. Commercialization of artifacts is discouraged because this damages the archaeological record.

4. Public education is promoted in respect to the archaeological record, archaeological methods, and site preservation.
5. Intellectual property in the form of access to original materials and documents must be shared among scholars.
6. Public reporting of investigations should be to as wide an audience as possible.
7. Records should be preserved and made as accessible to interested persons as possible.
8. To avoid damaging sites, digs should only be conducted by people with adequate training.

These principles are noteworthy in the emphasis that is placed on public relations as a vital ingredient in pursuing archaeological objectives. It is also noted that, although archaeologists are enjoined to discourage commercialism in relation to artifacts, they are not explicitly prohibited from selling them. The language of Principle No. 3 is: "Archeologists should therefore carefully weigh the benefits to scholarship of a project against the costs of potentially enhancing the commercial value of archaeological objects." This would appear to suggest that artifacts may be sold if there is no perceived threat to scholarship by so doing.

One interesting omission to the ethical principles of archaeologists is that, although scientific openness is strongly encouraged, there is no explicit consideration of scientific dishonesty of any kind. Fraud is not mentioned, for example, even though there seems to be a flourishing trade in fake artifacts, some of which could end up in museum collections. The possibility of scientific dishonesty in taking credit for important discoveries is another interesting omission. This problem is well illustrated in relation to the discovery of the ancient city of Troy at Hisarlik in modern Turkey. Nineteenth-century archaeologist Heinrich Schliemann claimed credit for this discovery even though a recent book makes a compelling case that the credit should go to the British archaeologist Frank Calvert, who passed on some critical information to Schliemann because he was personally unable to conduct the investigation. Schliemann publicized the discovery without giving any credit to Calvert. Presumably future editions of the ethical principles for archaeologists will provide some guidance on such matters of SCIENTIFIC DISHONESTY.

See also DARWIN/WALLACE AND CO-DISCOVERY OF EVOLUTION; ENVIRONMENTAL ETHICS; SCIENTIFIC DISHONESTY.

Further Reading
Allen, S. H. *Finding the Walls of Troy.* Los Angeles: University of California Press, 1999.
Society for American Archaeology. "Society for American Archaeology Principles of Archaeological Ethics." *American Antiquity* 61 (1996): 451–52.

Vitelli, K. D., ed. *Archaeological Ethics*. Walnut Creek, Calif.: Altamira Press, 1996.

Archimedes

(c. 298–212 B.C.)
Italian
Mathematician, inventor

Archimedes was an important mathematician and inventor of the Greco-Roman period. A native of Syracuse, in Sicily, he died during its capture by the Romans in the Second Punic War. Archimedes was unusual in being devoted both to the highly theoretical and the highly practical, which sets him apart from another great contemporary mathematician, EUCLID, who disdained practical applications of his geometry.

Archimedes was the beneficiary of a flourishing tradition in mathematics and drew on the work of Eudoxus, who had lived a century earlier and who developed a crude form of integral calculus known as the method of exhaustion. Thus, if you wish to measure the area of a circle, you might begin by fitting a hexagon inside it and using the area of the hexagon as an approximation. This crude calculation can be improved by increasing the number of sides, say to 12. In this way, you can devise ever closer approximations to the true area of the circle. Archimedes used this method to produce an approximation of pi, the ratio between the circumference of a circle and its diameter. Using two polygons with 96 sides drawn outside, and within a circle, he proved that pi is less than 3 1/7 and greater than 3 10/71. Archimedes used the former value, or 22/7 as an approximation in his calculations. Even though there was nothing original about any of the methods used by Archimedes to calculate pi, it is interesting that his estimate was good enough to be useful to engineers into the 20th century.

Archimedes did a lot of work on computing areas of flat surfaces and areas and volumes of curved surfaces, interests that are reflected in the majority of his surviving treatises, which include: *Measurement of a Circle*; *On the Sphere and the Cylinder*; *On Conoids and Spheroids*; *On Spirals*; *On Plane Equilibriums*; and *Quadrature of the Parabola*; as well as *On Floating Bodies*; *Stomachion* (a fragment); and *The Sand Reckoner*. Archimedes' works are characterized by originality, clear demonstrations, and careful computation. He was proudest of his demonstration that the volume of a sphere is two-thirds the volume of a circumscribed cylinder, and he requested that a diagram representing this accomplishment be inscribed on his tomb.

Even though Archimedes is often thought of as a mathematician, his practical inventions and scientific contributions were as striking as his mathematics. He studied the centers of gravity of plane figures and solids. His interest in the relationship between gravity and flotation culminated in the formulation of Archimedes, principle, which states that the apparent loss in weight of a floating body is equivalent to the weight of liquid it displaces. This important discovery is the subject of an anecdote. Archimedes is supposed to have had a flash of insight while lying in his bath, which prompted him to run naked through the streets crying *Eureka* (or "I have found it"). This story has no basis in fact but it continues to be told apparently because it corresponds to widely shared beliefs about the nature of scientific genius and scientific discovery. Archimedes' practical inventions include a burning mirror used in the defense of Syracuse. Whether or not he invented Archimedes' screw, a device used for elevating water and other fluids by rotating a spiral tube fixed to a crank, is debatable.

See also INVENTIONS, THEORIES OF; SCIENCE, HISTORY OF; SCIENCE-TECHNOLOGY DISTINCTION TECHNOLOGY, HISTORY OF.

Further Reading
Russell, B. *A History of Western Philosophy*. New York: Simon and Schuster, 1972.
Stein, S. *Archimedes: What Did He Do Besides Cry Eureka?* Washington, D.C.: Mathematical Association of America, 1999.

architecture and crime

One of the more provocative ideas of modern architectural theory is that the design of a building can make crime more likely to occur there. This principle is well illustrated in the 1972 dynamiting of the Pruitt-Igoe housing project in St. Louis, Missouri. Although structurally sound, the development was so overrun by organized crime that the housing authority decided to raze it and begin again. This spectacular waste of public funds was repeated in other American cities, including Newark, New Jersey. The housing authorities were tacitly admitting that the flawed architectural design of the buildings fostered crime. Can a constructed environment really foster or prevent crime?

Architect Oscar Newman, writing in 1972, was the first to propose that urban buildings could be designed so as to inhibit crime. His central concept was referred to as "defensible space." This refers to semi-public spaces around housing developments where residents often spend time and feel in charge. Residents, in occupying areas around their buildings, create an environment that generally inhibits crime because criminals like to function in private. Observers can easily tip police off that a crime is in progress. Moreover, they notice the presence of strangers and may even anticipate crimes before they happen.

Examples of defensible spaces include the stoops in front of apartment buildings, where residents often like to sit, and the playing fields, playgrounds, and picnic areas that often separate urban high rises. By using these spaces, residents provide informal surveillance services. Moreover, their mere presence makes it less likely that criminal elements will hang around.

Newman argued that some buildings and neighborhoods, particularly public housing projects, do not include defensible spaces. The absence of such resident-controlled areas creates an impersonal environment in which crime can flourish. The situation was so bad in one Newark, New Jersey, housing project that residents were frequently attacked in the hallways of their buildings. Residents were often charged by thugs for the privilege of using the elevator. The design of buildings in this project apparently made them vulnerable to invasion by criminals. The decision to raze the building implies recognition that the problem was inherent to the buildings and could not be solved by adequate policing.

It is, of course, misleading to say that the buildings "caused" crime. A more accurate interpretation is that they provided an impersonal environment that discouraged residents from occupying public areas, thus encouraging criminals to congregate and facilitating crime.

Newman's ideas about the relationship between the built environment and crime are intuitive generalizations and have generated considerable controversy, particularly among those who reject the environmental DETERMINISM inherent in the theory of defensible space. It is therefore important that Newman and the Institute for Community Design Analysis have been given the opportunity to put these ideas into effect. The city of Dayton, Ohio, hired him as a consultant after the Five Oaks area, formerly a peaceful community, had been overrun by violent criminals. Newman supervised the installation, in 1992, of 56 gates that eliminated through traffic and divided the area up into small neighborhoods ending in cul-de-sacs. At the very modest cost of $693,000, this intervention produced dramatic reductions in crime. In the first year, crime overall fell by approximately a quarter and violent crime was down by a half. Residents began to feel safer and spent more time outdoors, getting to know their neighbors better. Children were allowed to play on the streets. Property values increased as people took more care of their homes. A sense of community returned.

Whether the reduced crime rates can be attributed to an increase in defensible space, as conceived by Newman, is difficult to determine. A simpler explanation is that the gates made it difficult to navigate through the neighborhood by car. Through traffic was actually reduced by 60%. A reduction in the number of people passing through the community would inevitably reduce the crime rate even if there was no other intrinsic change in the community. Despite these qualifications, the concept of defensible space does help to account for the effectiveness of the fencing project in Dayton, particularly in respect to the enhanced sense of community.

Architects generally accept that the design of buildings can influence what people choose to do in them. This overarching principle can be extended to criminal behavior, a type of activity that is more often analyzed in terms of the unethical propensities of perpetrators. Evidence that the problem of crime in public housing projects is largely a function of the buildings themselves rather than the characteristics of the inhabitants comes from another natural experiment in Yonkers, New York, in which Newman served as a consultant.

The city of Yonkers had been found guilty of segregating its public housing. As part of a court-ordered remedy, it placed 200 new public housing units dispersed in seven sites in affluent white neighborhoods over the protests of local residents. What is remarkable about this experiment is that the much-dreaded increase in crime failed to occur. Secure in their enclosed communities, the low-income residents were much less vulnerable to crime. They planted flowers, erected picket fences, and purchased play equipment for children that could be left out in their fenced yards. In Newman's terms, they thus "defended" the spaces around their homes.

The ethical significance of the well-established relationship between architecture and crime is clear. Architects, and others who participate in the planning of public housing that fosters crime, incur some moral responsibility just as they would if the building was structurally unsound in the event of an earthquake. Even though the construction of public housing operates under severe budgetary restrictions that prevent the principle of defensible space from being adequately expressed in residential designs, a more enlightened approach to government funding decisions would allow the bonus of reduced expenses from crime to be added to the housing budget.

See also BRIDGE COLLAPSE; EARTHQUAKES AND CONSTRUCTION CODES; GREEN ARCHITECTURE.

Further Reading
Newman, O. *Creating Defensible Space.* Upland, Pa.: Diane, 1996.

Aristotle
(384–322 B.C.)
Greek
Philosopher

Aristotle is arguably one of the most influential philosophers who ever lived, and it is Plato who provides his only real competition. Plato, his teacher, proved far more appealing to Christian and Islamic theologians and his works were read and studied by clerics from the third century on. Long regarded as heretical, Aristotle was revived by the scholastic tradition of learning that came into full flower in the 13th century with the development of universities in medieval Europe. Aristotle's thoughts fit more in tune with modern scientific thinking than does Plato's mystical idealism, and his influence has grown as that of Plato's has declined.

Aristotle's most important contribution may be that he formulated rules of logical inference according to which valid conclusions are defined in science as well as philosophy. Despite his influence on subsequent generations, he was not a bold or original thinker. His writings were important as a vehicle through which Hellenic philosophy and learning were transmitted to future generations. Aristotle espoused a virtue approach to ethics according to which the point of ethical conduct is to develop good character. This has always been a minority viewpoint, but it has recently become more popular among philosophers.

ARISTOTLE'S LIFE

Aristotle was born in Stagira in northern Greece. His father, Nicomachus, was a physician who had close ties to the court of King Phillip II of Macedonia. His father's profession may have been partly responsible for Aristotle's subsequent interests in anatomy and natural science. At the age of 17, Aristotle joined Plato's Academy in Athens and stayed on as a teacher until Plato's death. Plato's academy resembled a modern research university in the sense that a group of brilliant individuals were encouraged to cultivate learning in a wide variety of different fields. They were united only by the desire to organize and extend human knowledge. Aristotle's scholarly tone and astonishing breadth of learning are evidently due to this experience.

Following Plato's death, Aristotle moved on to the court of the tyrant Hermias of Atameus and married the ruler's niece or sister. Four years later, at the age of 41, he obtained an important job in the court of Phillip II where he assumed responsibility for educating the royal heir, the future Alexander the Great. History preserves no record of the relationship between Aristotle and Alexander but the mere fact that Aristotle does not mention his famous student in his writings suggests that the headstrong pupil and the pedantic tutor may have had little in common. When this term of royal service was complete, Aristotle returned to Athens to found his own school, the Lyceum. The Lyceum differed from Plato's academy mainly in pursuing a more diverse range of interests. In particular, more attention was given to the systematic study of the natural world. Following the death of

Alexander the Great, there was a wave of anti-Macedonian sentiment that forced Aristotle to leave Athens for Chalcis, where he died, a year later, in 323.

THE ARISTOTELIAN TRADITION

Following centuries of rejection and neglect, Aristotle's works were revived by medieval European Scholasticism in the 12th century. Scholasticism was a method of learning that consisted of careful reading of texts and open discussion of a problem or question arising from the reading according to strict rules of logic. It elevated the work of authorities such as Euclid in geometry, GALEN and AVICENNA in medicine, Cicero in rhetoric, and Aristotle in logic.

There are many reasons why Aristotle's works were appealing to medieval scholars. He not only formulated the rules of logic on which scholastic discussion was based but he also practiced them. Moreover, he was organized and professorial in his hierarchical arrangement of topics within a book. Aristotle had an exceptionally high opinion of his own civilization and had an unusually complacent view of the importance of philosophy and philosophers in the world. He was patrician, elitist, and well connected, and therefore in an ideal position to write with authority. Aristotle also covered an astonishing range of topics from logic to metaphysics, ethics and politics to theology, and from rhetoric and theater to physics, astronomy, meteorology, and zoology. At least 22 books have been attributed to him although most seem to have been substantially revised by subsequent editors. So great was Aristotle's authority that scholars developing new ideas in virtually any field felt compelled to review what he had said about their subject and to demonstrate why Aristotle's pronouncements needed to be replaced. In a scholarly tradition that valued authority more than originality he was *the* authority.

ARISTOTLE'S LOGIC

Aristotle's most important contribution to logic was his invention of the syllogism. A syllogism is a form of reasoning that has three parts, a major premise, a minor premise, and a conclusion. For example:

> *All humans are mortal (major premise).*
> *Plato is human (minor premise).*
> *Therefore Plato is mortal (conclusion).*

Syllogistic reasoning can be quite powerful but it has received considerable criticism from contemporary logicians. For one thing, however straightforward it may seem, syllogistic reasoning can contain subtle semantic traps. For example, the minor premise, "Plato is human," could be replaced with another minor premise, "Fictional characters are human." This minor premise is not logically equivalent to the first, even though it might seem to be, and it certainly does not lead to the same logical conclusion.

In his book *Prior and Posterior Analytics*, Aristotle attempted to work out the conditions that needed to be met for a scientific statement to be considered valid. He concluded that although the appreciation of scientific truth relies on empirical information, or experiences, it is not itself a matter of experience but a matter of reason. In other words, when we observe some set of phenomena, their governing principles can be grasped by reason. Simply appreciating scientific principles, or generalizations, did not amount to scientific understanding in Aristotle's eyes. It was also necessary to understand why the scientific truths held. He therefore saw the scientific enterprise as a deductive building process in which self-evident truths based on observation were tied together.

ARISTOTLE'S INFLUENCE ON THE HISTORY OF SCIENCE

Philosopher and mathematician Bertrand Russell describes Aristotle's philosophy as a mixture of Plato and common sense and acidly points out that this is not a good mix, making for a certain amount of vagueness, obscurity, and inconsistency in Aristotle's thinking. One important point of divergence between Aristotle and Plato is Aristotle's rejection of the Platonic theory of ideas, or "forms." Plato held that every object in the perceptible world is a reflection, or shadow, of the ideal world. A particular cat that we might see, for example, is an imperfect copy of the ideal form of a cat. That copy is less real than the ideal form from which it is derived. According to Plato, the perceptual world is like a shadow, or dream, which provides us with a very imperfect impression of the underlying reality that consists of ideas.

This interpretation of reality has been quite popular among religious mystics through the ages but it could not be more antagonistic to scientific investigation. For scientists, reality is equivalent to data and scientists acquire data through their sensory systems.

Aristotle takes issue with Plato's theory of forms, relying on some of the arguments presented in Plato's *Parmenides*. One of the strongest logical rebuttals of Plato relates to the "third man" argument. If a man is a man purely because he resembles the ideal man, then there must be a more ideal man that both the ordinary man and the ideal one each resemble. Without getting into the finer points of Aristotelian metaphysics, it is clear that, contrary to Plato, he considered that individuals are both real and worthy of our attention. By rejecting this aspect of Platonic idealism, Aristotle opened the way for empirical science in which he had a personal interest and which was encouraged at the Lyceum. He is the founder of biol-

ogy, and Charles Darwin considered him the most important contributor in this field after over two millennia. Aristotle was, in fact, primarily interested in zoology, and he produced four books dealing with animals (*History of Animals; Generation of Animals; Parts of Animals; and Motion of Animals*).

ARISTOTLE'S ETHICS

Aristotle's ethical ideas are presented in *Nicomachean Ethics*. He sees happiness of the soul as the primary ethical objective. This objective can be achieved only through a good life composed of virtuous actions. The view that moral actions are those that turn us into happy people is a departure from the two main branches of moral theory, DEONTOLOGY, and CONSEQUENTIALISM, that derive moral actions from abstract principles and from consequences, respectively, and that is referred to as the virtue approach to ethics.

Aristotle does not provide any clear guidelines for defining which actions are moral and which are immoral. One interesting clue is provided by the doctrine of the golden mean. Virtuous action is never extreme. Generosity, for example, is a virtue that involves giving neither too little nor too much.

In Aristotle's system, the soul was divided into two parts, rational and irrational. The moral virtues corresponding to the irrational part consisted of good habits. The intellectual virtues corresponding to the rational part are derived from education and learning. Aristotle saw contemplation, rather than right action, as the highest good. Since contemplation was facilitated by a life of leisure, he advised readers to avoid careers requiring a lot of work.

It is clear from this that the highest virtue is possible only for the leisured aristocracy. Ancient Greece was a slave-owning society and Aristotle, like Plato before him, was very comfortable with the notion that the good life, moral as well as material, can be lived only by an educated elite supported by servants and captives. This view is very different from the Christian tradition of morality, according to which all individuals, regardless of their political and economic status, enjoy equality of opportunity in the moral realm.

It is probably unfair to judge the ethical and political views of Aristotle and his teacher in the light of the entire development of Western civilization but there is a lack of sympathy for the needs and sufferings of the underprivileged that is quite jarring to the modern reader. It is difficult to accept Aristotle's ethics without also accepting a very great degree of social inequality, such as slavery, a view that is scarcely possible for modern readers.

Unlike the Christian tradition that conceives humility as a virtue, Aristotle considered humility to be a vice.

Rather, he saw pride as a virtue, at least if it did not go to the extreme of vanity. His ideal of the magnanimous man who walked, talked, and acted like royalty could not be more distant from the ascetic ideal promoted by Christianity. This patrician ideal provided some support for German philosopher FRIEDRICH NIETZSCHE'S 19th-century argument that Christianity promotes a slave mentality. nietzsche's philosophy of pride and self-assertion was an important influence in the development of 20th-century German fascism.

See also AVICENNA; CONSEQUENTIALISM; DEONTOLOGY; MORAL PHILOSOPHY; SCIENTIFIC METHOD.

Further Reading
Barnes, J., ed. *The Cambridge Companion to Aristotle.* New York: Cambridge University Press, 1995.
Irwin, T. H. *Aristotle's First Principles.* New York: Oxford University Press, 1990.
Russell, B. *A History of Western Philosophy.* New York: Simon and Schuster, 1972.

artificial growth hormone   *See* BOVINE SOMATOTROPIN.

artificial insemination   *See* REPRODUCTIVE TECHNOLOGIES.

artificial tissue, organs   *See* ORGAN TRANSPLANTATION.

asbestos and cancer risk

Asbestos, a hydrated silicate, is produced by crushing rock, such as quartz, to liberate the fibers. The fibers can then be woven and spun into textiles or matted to produce durable sheets. Unlike organic fibers, which tend to be highly flammable, asbestos resists fire and has excellent properties as a heat insulator. It is also an electrical insulator and is chemically stable. Asbestos was generally dug from open-pit mines. Leading producers included Canada and South Africa. Long considered a miracle fiber, asbestos has emerged as arguably the most dangerous industrial carcinogen.

ASBESTOS EXPOSURE IN INDUSTRY

Considered as a carcinogen, asbestos has some extremely unusual properties that make it particularly hazardous. One is that even a single exposure to high levels of asbestos dust can have extremely deleterious consequences for health. Another is that the health consequences are typically delayed for at least 20 years after exposure, reflecting the fact that asbestos fibers remain trapped in the lungs during this period. Frequently fatal, and often producing a rapid decline following diagnosis, several asbestos-related ailments are currently considered incurable.

Even though asbestos is highly carcinogenic, this effect is produced only when it comes in contact with the lungs. Thus, even though over 3,000 products are estimated to contain asbestos, the health dangers may be minor if the product is not broken or cut so as to release airborne fragments. Persons at greatest risk of exposure included industrial workers, construction workers, and asbestos miners.

Common products that contain asbestos include: roofing sheets and shingles, pipe insulation, curtains, flooring felt, gaskets, clutch pads, brake linings, flower pots, floor tiles, cement pipes, wallboard, putty, caulk, and electrical insulation. Industrial workers who have used materials containing asbestos have been at much higher risk of dangerous levels of exposure than users of these products because fabrication and manufacturing processes are likely to break asbestos fibers and send them into the air allowing entry into the lungs.

The first exposure to dangerous levels of asbestos occurred in the case of miners and millers in the 1930s and 1940s. Others who were very heavily exposed were tradespeople (almost exclusively male) who used asbestos products in pipefitting, insulation, and shipbuilding. The first study to make a compelling case for the association between asbestos exposure and lung diseases involved a 1964 study of World War II shipbuilders in America. Exposure to airborne asbestos was predictive of pulmonary disorders emerging some 20 years later.

Even though only a small number of shipyard workers engaged in the dangerous task of installing insulation in ship hulls and pipes, it is estimated that their activities placed several million other shipyard workers at risk. A large number of construction workers have also been exposed to asbestos dust. Moreover, family members could have been affected indirectly through contact with contaminated clothing. Living close to asbestos mines and mills also elevates the risk of exposure to asbestos although this is less important in the developed world today because of a large reduction in asbestos use.

In 1971, asbestos became the first substance to be regulated by the Occupational Safety and Health Administration (OSHA) of the United States. It is estimated that between 1940 and 1979, over 27 million American workers were exposed to asbestos at their place of work. The OSHA mandated a 94% reduction in asbestos use over a seven-year period and outlawed its use in brake linings and in all building materials. This legislation spurred manufacturers to form the Alternative Materials Institute in 1987 in an effort to produce viable alternatives to asbestos. The OSHA also introduced regulations designed to reduce exposure levels of asbestos workers. In 1986, the Asbestos Hazard Emergency Response Act required the removal of asbestos from schools and other public buildings. Similar measures have been introduced in other developed countries. This means that health-related consequences of asbestos exposure are expected to peak around 2020 and then decline steeply.

HEALTH CONSEQUENCES OF ASBESTOS EXPOSURE

Asbestos fibers come in two forms, a straight type known as amphibole and a coiled type known as serpentine. The straight thin amphibole type is found in crocidolite (blue asbestos), tremolite, amosite (brown asbestos), and anthophyllite. Serpentine fibers are found in chrysotile (white asbestos). The structure of the fiber has important consequences because the straight fibers can penetrate to the deepest recesses of the lungs, accumulating in peripheral airways from which they are impossible to detach. The coiled chrysotile type tends to accumulate outside the lungs in the respiratory tract. Moreover, chrysotile is somewhat soluble and is naturally cleared from the respiratory system.

Fortunately, 90% of the asbestos used in the United States has been of the less dangerous chrysotile variety, also known as white asbestos. Nevertheless, straight amphibole fibers are an almost universal feature of the environment in the sense that they can be found in the lungs of persons with no workplace exposure to asbestos. Even such minor exposure to asbestos can cause disease in vulnerable individuals. People who are exposed to substantial levels of asbestos at work may accumulate as many as one million amphibole fibers per gram of dried lung tissue.

The accumulation of asbestos fibers in the lung causes or contributes to three serious conditions: asbestosis, mesothelioma, and lung cancer. Asbestosis is a nonmalignant condition marked by difficulty in breathing and chest pain. This is characterized by the presence of asbestos bodies (fibers coated by protein and ferritin). The formation of asbestos bodies is believed to cause fibrosis (or increased fibrous tissue) in the walls of respiratory bronchioles. Asbestosis manifests itself more than 20 years after exposure. It is much more likely in the case of workers exposed to high levels of asbestos. Asbestosis is treated primarily with oxygen therapy to improve breathing effectiveness.

Mesothelioma is a rare form of cancer with an incidence of approximately two persons per million population per year. It occurs most frequently in people occupationally exposed to high levels of asbestos. Yet, a third of those suffering from mesothelioma have no history of occupational exposure to asbestos. In malignant mesothelioma, a bulky tumor encases the lung. It is an aggressive disease for which there is usually no effective therapy. Survival following diagnosis is measured in months rather than years.

Asbestos exposure increases the probability of developing lung cancer by a factor of two or three. The risk is greatly increased in the case of workers who are also heavy smokers. Cigarette smoking interacts strongly with asbestos exposure. Heavy smokers exposed to asbestos are more than 80 times as likely to develop lung cancer as the average person. The increased rate of cancer among asbestos workers applies not just to lung cancer but also to cancers of the larynx, pharynx, gastrointestinal tract, kidneys, gallbladder, pancreas, and bile ducts.

Asbestos use has been banned in many countries but the long latency for development of pulmonary complications of asbestos exposure means that the health consequences will be felt for many more decades. It has been estimated that in the next 35 years a quarter of a million men from Western Europe will die of asbestos-related cancer. In the United States, the annual number of asbestos-related cancers, according to the Environmental Protection Agency, falls somewhere between 3,000 and 12,000 cases.

ISSUES OF CULPABILITY AND LIABILITY

The emergence of asbestos as a significant public health problem evidently took the asbestos industry by surprise. Once the health threat became clear to industry leaders, neither asbestos workers nor the general public received the kind of detailed information that would have helped to minimize exposure and reduce the health consequences.

Definitive evidence indicating that asbestos is a potent carcinogen emerged only in 1963 with the publication of data compiled by Thomas Mancuso of the University of Pittsburgh on the elevated risk of mesothelioma and lung cancer among workers at a brake lining factory. The first evidence, based on case histories, emerged approximately 30 years earlier. Asbestosis was described early in the 20th century and the British government had implemented asbestos regulation as early as 1931. The association between lung cancer and blue asbestos (crocidolite) was recognized by 1950. By about 1959, physicians knew about the increased risk of mesothelioma for asbestos workers.

Appreciated by many physicians in the 1930s, the health risks of asbestos were not widely known to the public, or to workers who used asbestos products each day. Even by 1963, when the evidence of health problems was incontrovertible, no safety measure were introduced by leading companies in the asbestos industry. The U.S.

government bears some responsibility because it did not implement safety regulations for another five years. Regulations were not strictly enforced. Thus, one shipyard worker, Paul Safchuk, who developed asbestosis following a 40-year stretch of employment at a Bethlehem Steel facility in Baltimore, Maryland, claims that he received no information about the dangers of asbestos. Workers ate in a room covered with thick asbestos dust. They sat on dust-covered decks to eat their lunch and even placed their sandwiches right on top of the dust. Workers were not advised that they should wear respirators even though the dust was so thick in the engine rooms that it resembled a blizzard.

A variety of explanations have been proposed to explain why safety regulations were implemented so slowly in the United States. These include: the inertia of the regulatory process itself; the status of asbestos as a miracle material that was making a fortune for the industries that used it; the importance of asbestos in military applications, such as warships, that were being sent to protracted international conflicts spanning the period between World War II and the Vietnam War. Whether any or all of these interpretations is correct is difficult to determine.

Manufacturers were often grossly negligent in communicating what they have about the dangers of asbestos in their products. The clearest case involved the production, between 1952 and 1956, of Kent cigarettes containing asbestos filters manufactured by R. Lorillard Company. The company initially claimed that the filters removed most of the tars and nicotine from the cigarette. They were so enthralled by the new "Micronite" filter that they even took out full-page adds in the *Journal of the American Medical Association* extolling its virtues. Then, in 1954, the company's electron microscope tests revealed an uncomfortable truth. Smokers were inhaling asbestos. As they smoked, tiny fragments of asbestos fibers were sucked into their lungs, along with the tar and nicotine that they had trapped. This might have seemed like an excellent time to discontinue the filters but the company allowed 18 months to pass before doing so, apparently because Kent was a top-selling brand. Moreover, even after the asbestos had been removed from the filters, Lorillard still failed to inform consumers that they had been exposed to asbestos. This kind of secrecy has been typical of the entire tobacco industry's suppression of evidence concerning the harmful effects of tobacco (see TOBACCO COMPANY LITIGATION).

Lorillard has been sued on behalf of smokers and workers whose health might have been compromised by contact with the asbestos filter. The cases of two of the workers have been settled out of court, the terms of the settlement remaining secret. Thousands of claims have been filed against the asbestos industry in the United States. A landmark 1987 California case ruled that insur-

ance companies, rather than asbestos manufacturers, should bear the financial liability in these cases.

As of 1999, over 2,000 South African residents, miners and people living in the vicinity of mines, were pursuing cases against Cape PLC, the owner of Cape Asbestos, a mining and milling operation that ceased production in 1979. South Africa used to be the world's biggest producer of the highly dangerous blue asbestos and the South African companies have paid only nominal damages of less than $100 per asbestosis sufferer under a statutory settlement. Hence the decision to sue in Britain where much higher compensation is possible. All asbestos mining in South Africa stopped in 1995 and the export of asbestos ended about three years later.

See also AIR POLLUTION; TOBACCO COMPANY LITIGATION.

Further Reading

Bonn, D. "Asbestos—The Legacy Lives On." *The Lancet* 353 (1999): 1336.

Castleman, B. J., and S. L. Berger, *Asbestos: Medical and Legal Aspects.* New York: Aspen, 1996.

Henderson, C. W. "Mesothelioma (epidemiology): Europe Said Facing Asbestos-Linked Cancer Epidemic." *Impotence & Male Health Weekly Plus,* February 1, 1999.

Lordi, G. M., and L. B. Reichman. "Pulmonary Complications of Asbestos Exposure." *American Family Physician* 48 (1993): 1471–478.

Roggli, V. L., S. D. Greenberg, and P. C. Pratt. *The Pathology of Asbestos-Associated Diseases.* New York: Little Brown, 1992.

asbestos mining

The world's biggest producers of asbestos halted mining operations in the 1990s on account of clear evidence that asbestos produces a number of serious health problems, including cancer. Evidence of the dangers of asbestos included the fact that miners, their families, and people who lived close to asbestos mines developed more health problems.

Inactive asbestos mines may still pose a health threat because of windborne asbestos fibers. Moreover, mines that are not in the business of asbestos production may release asbestos into the air. Hundreds of residents of Libby, Montana, a small town adjacent to a defunct vermiculite mine, have been diagnosed with asbestos-related illnesses. When the mine was active, rocks were crushed, releasing asbestos dust into the air. Soil samples taken in the town have found high levels of tremolite, a dangerous type of asbestos that has the straight (amphibole) fibers that cause lung diseases. Many of the ill Libby residents did not work in the W. R. Grace mine, which closed in 1990. Children may have been exposed to asbestos dust simply by playing around their homes.

See also ASBESTOS AND CANCER.

Further Reading
Vollers, M., and A. Barnett. "Libby's Deadly Grace." *Mother Jones* (May 2000).

Association of Women in Science

The Association of Women in Science was founded in 1971 to promote career opportunities and advancement of women in science. It has over 5,000 members, some of whom are men. The multinational organization provides information to women planning careers in science, monitors the status of women in science, and supports equal opportunity legislation.

See also AFFIRMATIVE ACTION IN SCIENCE AND TECHNOLOGY CAREERS; CURIE, MARIE.

Avicenna (Ibn Sina)

(980–1037)
Persian
Philosopher

Avicenna, to use the Westernized version of his Arabic name, Ibn Sina, was a Persian philosopher who made his living as a physician and scholar at several Islamic courts. He was not an original thinker but became extremely important as an interpreter of, and a vehicle for, ancient Greek and Roman scholarship, particularly the philosophy of ARISTOTLE and medical learning. Like other Muslim scholars of his period, he attempted to reconcile classical philosophy with the teachings of Islam. In addition to his summaries of Aristotle's works, Avicenna compiled a philosophical encyclopedia. This was little read in the Islamic world because of the opposition by theologians to secular philosophy, but it was influential in Western Europe, where it became available in a Latin translation. Another important Islamic interpreter of Aristotle was Averroës (or Ibn Rushd, 1126–98).

Avicenna was well known in the West because of his medical scholarship. His most important medical text was the *Canon of Medicine*. This was based on Greco-Roman texts, and it drew very heavily on Galen. It was astonishingly popular in Europe, occupying a central role in medical curricula from the 12th to the 17th centuries.

See also ARISTOTLE; SCIENCE, HISTORY OF.

Further Reading
Goodman, L. E. *Avicenna.* London: Routledge, 1992.
Russell, B. *A History of Western Philosophy.* New York: Simon and Schuster, 1972.

B

Bacon, Sir Francis
(1561–1626)
English
Scientist, Philosopher, Lawyer, Essayist

Bacon's greatest contribution was as a philosopher of science (see illustration). He was also an essayist, lawyer, and statesman. His studies at Trinity College, Cambridge, began at the age of 12, after which he entered Gray's Inn to study law. He was elected to Parliament for the first time in 1584. Bacon was a leading figure of his day and held high offices such as attorney general and Lord Chancellor as a result of the patronage of King James I, who knighted him in 1603. His life of privilege came to a sudden end in 1621, when Bacon was found guilty of accepting bribes. This allowed him to devote his energies to writing and scientific work.

Bacon's own scientific work was not very important but he occupies a unique position in the history of modern science because of both his articulation of the aims and methods of science and his visionary ideas about the potential importance of science for improving the lot of human beings. Bacon is often credited with coining the phrase "knowledge is power." Whether this is true or not, it certainly captures his point of view. He recognized that human beings could harness the forces of nature through scientific discovery and the creation of inventions. His

own lack of scientific intuition is revealed by the fact that he missed the significance of some of the important advances made by contemporaries. He was unconvinced by Kepler's work on planetary motion, and he seems to have been unaware of WILLIAM HARVEY's discovery of the circulation of the blood, which is quite astonishing considering that Harvey was his personal physician.

EPISTEMOLOGY AND SCIENTIFIC METHOD
Bacon had a clear conception of the scientific method that contains most of the important ingredients by which science is defined today even though his emphasis on data gathering (or induction) over theory (or deduction) seems strangely biased today. These ideas are presented in one of his most important books, *The Advancement of Learning*.

One of Bacon's key contributions to epistemology was the notion that matters of religion and faith were best kept separate from matters of reason and science. Bacon belonged to the Church of England, and, like most of his contemporaries, was deeply religious. A conservative and defender of the status quo he had no desire to challenge orthodox religious views.

According to the doctrine of double truth advocated by Bacon, a conflict between religion and science is not inevitable because each occupies a separate mental realm. Bacon disagreed with the habit of some scholars of his

Francis Bacon, a leading English philosopher of science from the Elizabethan period. (LIBRARY OF CONGRESS)

day who intermingled religious discourse and reason. Philosophy ought to be a separate realm that depended only on reason. Religion, on the other hand, was a matter of divine revelation. Religious truths could only be experienced and could not be arrived at through rationalization. Bacon even went so far as to suggest that the triumph of faith is most commendable when it pertains to some phenomenon that to the eye of a coldly logical philosopher seems ridiculous. It is difficult to know exactly what Bacon's own religious temperament was but the double truth doctrine provided a formula that allowed for the development of a purely secular science without challenging the central role of religious belief in contemporary life. By the same token, many modern scientists entertain religious beliefs that are not consistent with natural science.

Bacon was of a skeptical temperament and set out to identify bad mental habits that often lead people into making errors. He referred to these as "idols." Five types of mental error were identified: those inherent in human nature, such as expecting reality to be a lot tidier and more orderly than it actually is; individual prejudices; false inferences based on semantic problems; overreliance on rules and formulae; and the difficulty of shaking off

the shackles of received ideas. He was well aware that accepted truths are often not actually true. What is more, invalid accepted truths (or "phantoms") can cloud our minds, making it difficult for valid ideas to be grasped.

A strikingly modern aspect of Bacon's thought was his elevation of data gathering (induction) over reasoning (or deduction) as a method of obtaining truth. His conception of the scientific method consisted primarily of induction. For example, he believed that when an object was heated, its finer particles began to move erratically. His research on temperature consisted of compiling lists of the characteristics of hot objects and lists of the qualities of cold objects. He hoped that these lists would reveal a pattern of qualities that revealed distinctions between them so that generalizations, or laws, could be formulated that helped him to understand what caused temperature change. Beyond such simple induction, Bacon believed that scientific laws could themselves be combined into a simpler, and more general, principle.

This example illustrates a key weakness of Bacon's scientific philosophy. If you simply collect observations without a well-developed theory explaining why you are collecting them, you are unlikely to make any important discovery. Simple induction is not an aggressive knowledge-building tool. Bacon seriously underestimated the role of deductive thinking in science, reflecting his personal antagonism to classical philosophers such as Aristotle. In particular, he underestimated the importance of coming up with a good hypothesis. A hypothesis, or guess, guides scientific observations by motivating the scientist to collect data that can be used to support, or reject, it. The whole chain of reasoning by which a scientist comes up with a testable hypothesis and then develops a procedure that is capable of testing it is one of the most difficult and important aspects of the scientific method. It involves intuition and creativity and is the basis for achieving rapid progress in science.

Another of Bacon's blind spots was his dismissal of mathematics, seeing it as too removed from empiricism to be very useful to scientists. Modern science tests most of its hypotheses using complex mathematical tools and, without statistics, many sciences would fail to progress. Despite such limitations, his conception of science as a professional and human activity was astonishingly prescient of modern scientific research.

SCIENCES AS A HUMAN ACTIVITY

Bacon recognized that science is a social activity and understood the great importance of scientific communication in generating progress. In *The New Instrument*, he presented a model of the scientific enterprise that included not just empiricism and hypothesis testing but that endorsed public communication of results. Scientific research should be open to public criticism and discussion that would help to dispel fallacious thinking and dogmatic claims. Science was also seen as helpful in solving practical problems and improving the quality of life.

Bacon was an advocate of scientific associations that would provide a forum for scientists to get together and discuss their data, hypotheses, and theories. These aspirations were realized in 1662 with the formation of the ROYAL SOCIETY OF LONDON, which functioned very much as Bacon had envisaged. The Royal Society was the first scientific society in the world and its official journal *Philosophical Transactions* constituted the first scientific periodical in the world. It is still published today. The Royal Society can thus be considered a fitting epitaph to Bacon's scientific philosophy, which emphasized the value of communication in pursuing the practical goals of science. It was founded 40 years after his death, which was brought on by a final experiment he conducted on refrigeration. The experiment involved stuffing chickens with snow and Bacon developed a chill that proved fatal.

See also ROYAL SOCIETY; SCIENTIFIC METHOD.

Further Reading
Resnik, D. B. *The Ethics of Science.* New York: Routledge, 1998.
Russell, B. *A History of Western Philosophy.* New York: Simon and Schuster, 1972.

Bacon, Roger
(1214–1292)
English
Philosopher, Scientist

Roger Bacon was an exceptionally learned Franciscan monk who differed from other clergymen of his day by constantly getting into trouble on suspicion of heresy and black magic. The suspicion that he practiced the dark arts derived from contemporary interpretation of his scientific research, which involved optics and chemistry (then known as alchemy). The seriousness of Bacon's scientific work has been questioned but he was certainly a man of vision and foresight who looked ahead to a future that contained horseless carriages, explosives, flying machines, and telescopes.

Bacon is of interest to modern philosophers and historians of science because he preached the value of obtaining knowledge directly by empirical observation and experimentation. He constantly railed against the errors that are produced by accepting the conclusions of authorities, a position that was somewhat weakened by his own use of authorities to support his assertions.

Bacon's life is of interest to ethicists because of the important role that the lack of academic freedom played in his life. His stinging attacks on what he saw as the appalling ignorance of clerics of his day were so offensive, even to the relatively tolerant Franciscans, that they banned his books and banished him from Oxford to Paris to live under a type of house arrest. During the same period, the pope, pulling rank on Saint Bonaventura, the general of the Franciscan order, commanded him to commit his philosophical ideas to paper. The three books he produced as a result (*Opus Maius, Opus Minus,* and *Opus Tertium*) were evidently well received because he was allowed to return to Oxford. In these works, Bacon argued for a reform of education, advocating that examination of the natural world by means of precise measurement and experimentation was the most effective means by which to know its creator.

See also GALILEO GALILEI; SCIENCE, HISTORY OF; SCIENTIFIC METHOD.

Further Reading
Russell, B. *A History of Western Philosophy.* New York: Simon and Schuster, 1972.

Baltimore Affair, the
One of the most publicized cases of alleged scientific misconduct swirled around Nobel laureate David Baltimore. Baltimore was a coauthor of a paper published in the journal *Cell* on April 25, 1986. The paper claimed to have produced experimental evidence showing that insertion of a foreign gene into a mouse can stimulate the mouse genes to produce antibodies the same as those produced by the foreign gene. This finding was important because it suggested that gene insertion could be used to control the immune system, making it produce any antibody that was lacking. The experiments were funded by the National Institutes of Health (NIH) and carried out at the Whitehead Institute, which is affiliated with Tufts University and the Massachusetts Institute of Technology. The number of institutions involved played a role in the Byzantine quality of subsequent investigation into allegations of scientific misconduct. Suspicions about the research arose when other researchers, who were excited by the findings, could not replicate them. Suspicions were also aroused within the Whitehead Institute when Margot O'Toole, a graduate student working under the supervision of Thereza Imanishi-Kari, one of the *Cell* paper's authors, discovered what seemed like a smoking gun—17 pages of Imanishi-Kari's notes that were inconsistent with the published findings. O'Toole attempted without success to replicate some of the experiments described in the paper and she formed the impression that much of the work reported in the *Cell* paper either had not yielded the stated results or had never been done. O'Toole reported her suspicions to eth-

ical review boards at Tufts and MIT. Both of these bodies launched investigations. Both found that the research contained errors, but they did not see evidence of intentional scientific misconduct. O'Toole's whistle-blowing activities resulted in her one-year contract not being renewed. Branded as a troublemaker, she experienced difficulty in finding work.

The matter might have rested there were it not for the efforts of two scientists at the NIH, Walter Stewart and Ned Feder, who brought the matter to the attention of Congressman John Dingell, a Michigan Democrat. Dingell further publicized the affair in hearings on "Fraud in NIH Grant Programs" conducted in April 1988.

The NIH appointed an investigative panel in January 1989 that also cleared Imanishi-Kari of wrongdoing. The investigation was reopened by the new Office of Scientific Integrity (OSI) of the NIH. The OSI tentatively concluded that Imanishi-Kari was guilty of scientific fraud. The investigation was then passed on to its successor agency, the OFFICE OF RESEARCH INTEGRITY (ORI).

At the same time, spurred on by John Dingell, the Congressional House Oversight and Investigations Committee held two hearings on the case. For the first time, the congressional investigation examined Imanishi-Kari's notebooks with the benefit of expert forensic analysis by the Secret Service. Investigators concluded that dates in the notebooks had been altered. Moreover, records had been kept on different pieces of paper using different types of ink. Much of the research was not actually done at the time that it was purported to have been done. The condition of the notebooks suggested that Imanishi-Kari had assembled the data retrospectively after questions had been raised about the research. In 1994, the ORI found that Imanishi-Kari had not only falsified, or changed, experimental results, but that she had fabricated, or made up, results without conducting the relevant research. Tufts University asked her to take a leave of absence.

Despite being found guilty of academic misconduct, Imanishi-Kari maintained that she was innocent. She was vindicated on June 21, 1996, by a research integrity appeals panel of the Department of Health and Human Services consisting of two lawyers and one scientist. Tufts University reinstated her. The panel also criticized the ORI for irresponsible handling of the case.

There was little disagreement about the facts of the case. Differences between the NIH panel and the ORI revolved around interpretation of the evidence. The ORI took the sorry state of Imanishi-Kari's notebooks as evidence that she had probably committed intentional scientific fraud. The appeals panel argued that the evidence pointed just as logically to the simple fact of sloppy recordkeeping without fraudulent intent. They were particularly persuaded by the fact that some of the notebook data tended to cast doubt on the published experimental findings. Why would she fabricate data that called her published results into question? Moreover, the panel was impressed by the fact that the challenged data was of relatively little importance to the main conclusions of the paper. Why would a researcher go to the trouble of fabricating data that had little bearing on the main outcome of the research and leave the centrally important data unchanged?

It is ironic that after 10 years and five formal investigations matters ended up largely where they had stood after the first investigation. Many scientists would argue that the three governmental investigations were inappropriate and that it is better if scientists are left to police themselves. Many found the use of Secret Service investigators to analyze the jottings of Imanishi-Kari to be unduly intrusive if not downright threatening to academic liberty.

The tangled history of the Baltimore Affair shows that American society is still very far from having an effective procedure for dealing with scientific fraud. Moreover, the differing interpretations of the notebook evidence suggest that it would be very difficult, even in principle, to establish retrospectively that data falsification or fabrication had occurred. What is more, there are few ethical guidelines telling scientists what they should do in such cases. Should Baltimore have checked Imanishi-Kari's data given that the paper carried his name? Should O'Toole have blown the whistle without consulting with Baltimore or the other authors? Once she did, should she have been given more professional protection?

David Baltimore, whose name will forever be associated with this controversy, was not accused of fraud but he did feel compelled to resign as president of Rockefeller University. Baltimore defended Imanishi-Kari throughout the controversy and he felt that she had been the victim of a scientific witch hunt. Consistent with the practice of ethical researchers, Baltimore and his coauthors removed the errors in the original paper by publishing a correction in *Cell*.

See also SCIENTIFIC DISHONESTY.

Further Reading

Dresser, R. "Giving Scientists Their Due: The Imanishi-Kari Decision." *The Hastings Center Report* 27 (1997): 26–29.
Resnik, D. B. *The Ethics of Science*. New York: Routledge, 1998.

battery chickens

Battery chickens are laying hens that are confined, in groups of three to nine in small cages arranged in tiers in large sheds that often contain over 100,000 birds. The battery system is a type of factory farming in which scientific methods are used to produce the maximum possible number of eggs in the minimum time and at minimal cost. Animal rights activists argue that modern egg-production techniques are unethical because they cause pain and suffering to the chickens, damage their health, and deny them the opportunity to engage in the normal species-typical behaviors enjoyed by active, or free-range, hens.

One example of cruel treatment is the de-beaking process in which the beak of a young bird is removed with a hot blade. The blade passes through the sensitive tissue of the beak and may result in pain and sensitivity as the bird feeds and attempts to preen itself. It is also possible that the operation results in chronic pain analogous to phantom limb pain in human amputees. De-beaking is carried out to prevent the continuous pecking that is part of the natural behavioral repertoire of birds whose ancestors spent their day disturbing the soil in search of food. De-beaked birds are less active, eat less, and fling their food less, thereby promoting food efficiency of the egg-production operation.

Modern poultry farms employ some fairly extreme biological manipulation of hens. The Leghorn chickens used in most egg farms are selectively bred to produce around 250 eggs per year compared with the approximately two dozen produced by their wild ancestors, Burmese jungle fowl. This unnatural overproduction of eggs may cause inflammation of the oviduct together with a variety of painful reproductive diseases, including prolapse of the uterus, and damage to nonreproductive organs, such as enlargement of the liver.

Battery chickens are maintained in a fairly unsanitary environment because of the large number of birds defecating in a small space. Coupled with the general discomfort of living in a small cage without being able to exercise or engage in most normal behaviors, this is threatening to the health of the birds. For this reason, they are routinely dosed with antibiotics. The antibiotics not only control disease but also are used to manipulate the efficiency of feed conversion to eggs laid, improve shell quality, and stimulate egg formation. Such indiscriminate use of antibiotics in farming creates conditions under which resistant strains of disease-causing bacteria evolve.

The lack of exercise due to confinement in cramped cages, combined with the large drain of calcium for egg production, can produce severe problems for bones, including osteoporosis and brittleness. Calcium-depleted birds often suffer paralysis and may die of thirst and hunger just inches from water and food.

Free range chickens are active animals and spend much of their time engaged in species-typical activities such as scratching, pecking, and preening feathers. Being kept on a wire surface, in factory farming, eliminates much normal foraging activity. De-beaking prevents chickens from pecking because their beak stumps are too painful. For the same reason, they find it difficult to preen their feathers.

Animal rights activists often point to the painfully cramped life of the battery chicken as an extreme example of the inhumane conditions to which animals are subjected in factory farming. However, such complaints have not produced any practical changes in the lives of battery chickens. None of the routine practices of chicken farmers, which activists see as cruel, has been restricted by laws or government regulations.

In Britain, many consumers have expressed their ethical disapproval of modern egg-production methods by choosing to buy free range eggs at an increased cost of some 50%. Some factory farmers are suspected of making huge profits by selling their battery chicken–produced eggs in boxes labeled "free range," and in the United Kingdom, the Ministry of Agriculture, Fisheries, and Food is conducting ongoing investigations into these allegations.

See also ANIMAL RESEARCH, ETHICS OF; SINGER, PETER.

Further Reading
Davis, K. *Prisoned Chickens, Poisoned Eggs.* New York: The Book Publishing Company, 1997.
"Shoppers 'Duped' in Egg Con." *BBC News,* (January 21, 1998).

Belmont Report, the

Whereas the DECLARATION OF HELSINKI is the primary ethical document guiding biomedical research around the world, the Belmont Report is the primary ethical document for research at institutions that receive U.S. federal funding. It is named after the Belmont Conference Center at the Smithsonian Institution in Washington, D.C., where the initial discussions leading up to its formulation were conducted. Produced by the federally mandated National Commission for the Protection of Human Subjects of Biomedical and Behavioral Research (created by the National Research Act of 1974), the Belmont Report first appeared in 1971 and was revised in 1991. It has been used to draft the extremely detailed DHSS GUIDELINES FOR THE PROTECTION OF HUMAN SUBJECTS. Its primary sources are the NUREMBERG CODE and the Declaration of Helsinki.

The Belmont Report generally goes beyond the Nuremberg Code (and, arguably, beyond Helsinki) in terms of protecting the interests of research subjects. For example there is an explicit principle of beneficence in contrast to the general HIPPOCRATIC maxim of "do no harm" that influenced the Nuremberg Code. The researcher is obligated not only to avoid doing harm but also to make efforts to secure the well-being of subjects.

The Belmont Report addresses situations in which application of its ethical principles is difficult. For example, under some circumstances, providing fully informed consent may undermine the purposes of the research. In this eventuality, the research can be conducted only if there are no undisclosed risks that are more than would be expected outside the research (the MINIMAL RISK CRITERION) and if subjects are fully debriefed at the conclusion of the research. In the Helsinki document, the ethical review committee is given much more latitude to permit risky therapeutic research without informed consent.

In addition to the principle of beneficence, the Belmont Report emphasizes respect for persons and the principle of justice (or fairness) in apportioning the costs and benefits of research. Respect for persons involves a moral requirement both to acknowledge autonomy and to protect people with diminished autonomy, such as young children and those incapacitated by old age or illness. The principle of justice points out that it is unethical for some groups, such as the poor, ethnic minorities, or prisoners, to bear the brunt of research participation if they do not enjoy its benefits. The Belmont Report thus articulates a political agenda with the addition of an affirmative action requirement. This constitutes a clear departure from both the Nuremberg Code and the Helsinki Declaration, which confine themselves to articulating the relationship between researchers and subjects rather than to addressing issues aimed at solving larger societal problems.

See also DECLARATION OF HELSINKI; DHHS REGULATIONS FOR THE PROTECTION OF HUMAN SUBJECTS; INFORMED CONSENT; MINIMAL RISK CRITERION; NUREMBERG CODE.

Bentham, Jeremy
(1748–1832)
English
Philosopher, Political Reformer

Trained as a lawyer, Jeremy Bentham never practiced law but his interest in the foundations of jurisprudence led him to ethics and politics. Although self-effacing and diffident, he led the philosophical radicals, a group including James and John Stuart Mill, who set out to reform English government and institutions.

Bentham's book *An Introduction to the Principles of Morals and Legislation,* which was published in 1789, contained a clear formulation of the principles of UTILITARIANISM. Bentham considered that pleasure (or happiness) is the only good and that pain (or unhappiness) is bad. Each individual always pursues what he believes to be his own happiness. It is the function of laws to make sure that people do not pursue happiness in ways that are socially disruptive, such as by stealing. This objective is accomplished by arranging for unpleasant consequences of theft. Thus, it serves the happiness of the individual to avoid stealing the property of others and the pursuit of happiness by the individual is brought into harmony with the interests of the community.

In more general terms, the objective of the legislative system is to promote the greatest happiness of the greatest number of people and this phrase more than any other summarizes utilitarian philosophy. Note that utilitarianism is fundamentally egalitarian because it implies that the happiness of one person is as important as the happiness of any other. Bentham believed that happiness could be measured, at least in principle, and proposed that moral decisions could be made on a quantitative basis. If you wished to evaluate the morality of an action, you could measure the good, or happiness-promoting consequences and the bad, or painful ones. If happiness exceeded pain in the imaginary calculation, then the action was ethical. This hedonistic calculus idea never lived up to Bentham's hopes but a more imprecise method of weighing up potential good and had consequences of an action remains the focus of utilitarian ethics. Bentham was influenced by the ideas of Scottish philosopher David Hume (1711–76) in formulating his principles of utilitarianism and his philosophical heir was JOHN STUART MILL.

See also CONSEQUENTIALISM; MORAL PHILOSOPHY; MILL, JOHN STUART.

Further Reading
Harrison, R. *Bentham.* New York: Routledge, 1999.
Russell, B. A *History of Western Philosophy.* New York: Simon and Schuster, 1972.

Bhopal, India, chemical accident
The site of an explosion in a chemicals factory on December 3, 1984, Bhopal was the site of the worst industrial accident the world had ever seen. Since that time, the magnitude of the disaster has arguably been exceeded by the CHERNOBYL nuclear power plant explosion in 1986, which has engendered global consequences for health and the environment. It is estimated that at least 2,500 people died at the time of the Bhopal explosion and that several hundred others died subsequently from their injuries. (Estimates of the final death toll vary between 3,800 from Union Carbide and 6,600 from the Indian government). Many thousands were permanently disabled following exposure to a poisonous cloud of methyl isocyanate (MIC) gas. MIC is a highly toxic and unstable substance that was being used in the production of a pesticide known as SEVIN (carbaryl), which is itself a stable and relatively safe product. The Bhopal accident raises

many ethical questions concerning the safety of the chemicals industry, the obligation of manufacturers to protect the welfare of the surrounding population, the responsibility of multinational corporations for foreign subsidiaries, and the role of politics in helping, or hindering, relief for the victims.

The immediate cause of the Bhopal accident is quite clear. It resulted from an act of sabotage by a disgruntled worker in the employ of Union Carbide India Limited (UCIL). The worker, whose name is known but was never published, evidently wanted to put water in a tank of MIC to destroy the next batch of pesticide being produced. The worker removed the pressure gauge from MIC storage tank 610. He then placed a hose in the opening where the gauge had been and turned on the water. Although he knew that water was not supposed to come in contact with the MIC the saboteur almost certainly did not wish to cause the ensuing explosion because his own family would likely have been living nearby. A vigorous chemical reaction ensued that within a few hours blew the safety valve off the storage tank and released a cloud of toxic gas that spread over a 25-square-mile area downwind of the UCIL plant.

MIC is an extremely dangerous substance because it reacts violently with water in bodily tissues causing burning. People exposed to the MIC gas first experience burning in the lungs and respiratory failure was a major cause of death. Among survivors, symptoms of choking and difficulty in breathing reflected severe burning of lung tissue. Damage to the eyes also occurred rapidly. Over time, MIC exposure causes damage to internal organs. There is no antidote for MIC and when thousands of people began to show up in Indian hospitals, little could be done for them apart from providing oxygen to assist breathing and administering eye drops to relieve irritation.

The Bhopal incident was apparently caused by an act of sabotage that went far beyond what the perpetrator had intended. Closer scrutiny, following the tragedy, revealed an astonishing disregard for safety in just about everything that was done at the Bhopal pesticide factory beginning with the construction of the plant and the design of the chemical processes. These safety lapses not only allowed the sabotage to take place but also contributed to the scale of the ensuing destruction.

SAFETY PROBLEMS AT BHOPAL

When the Bhopal plant was built in the 1970s, UCIL applied for permits for a 10,000-ton facility even though there existed a market for only 2,000 tons of pesticide. Inflating the permits was a strategy for keeping other multinationals out of the market. UCIL intended to build a much smaller plant but this plan was overruled by the engineering department of Union Carbide USA, which operated from Charleston, West Virginia. The Americans

also prevailed in building huge storage tanks to house the lethal MIC despite the objections of UCIL that it would have been more economical, and safer, to store it in 55-gallon drums. The exploding tank was so big that it discharged some 40 tons of MIC into the atmosphere.

The size of the tank clearly determined the magnitude of the disaster but it is not the only reason that the act of a saboteur had such dire consequences for the inhabitants of Bhopal. Had the tanks been maintained properly, and had the safety procedures designed to manage them been functional, it is questionable whether there would ever have been an explosion. Among its other problems, the Bhopal plant had been hit by recession. Against a backdrop of steadily declining sales, UCIL had been forced to cut back their expenses. One area that suffered greatly was routine maintenance.

MIC has a low boiling point (102 degree Fahrenheit or 39 degrees Celsius). Safe storage requires a refrigeration system. Although the MIC tank had been refrigerated, the refrigeration system had been shut off long before the day of the explosion. Moreover, the temperature and pressure gauges were so unreliable that workers ignored early warning signs of trouble. A gas scrubber that would have inactivated escaping MIC had been closed down for maintenance. Its role may have been academic because it could not have managed the very high gas pressures generated by the explosion. The flare tower that would have burned off additional MIC gas had been turned off pending replacement of a corroded pipe. Post-accident investigators concluded that the flare tower would not have been able to handle the volume of gas released in the explosion. The final tier of containment consisted of a water curtain or system of high-pressure water spray that was supposed to neutralize any gas escaping from the flare tower. This had a serious design problem because it did not extend to the top of the flare tower from which the MIC gas was escaping and therefore might as well not have been installed.

Neglect of maintenance in the safety equipment, including unreliable temperature and pressure gauges, probably contributed to the scale of the disaster, but an even more obvious question to ask is how the disaffected worker could have had time to remove the pressure gauge from the tank and run water into it through the opening without being observed. Where were the supervisors? All had evidently been taking a break together contrary to company policy and actual instructions.

RESPONSIBILITY TO THE SURROUNDING COMMUNITY

It seems strange that such a hazardous operation as the manufacture of SEVIN should have been conducted in the heart of a densely populated city without any public awareness of the danger. Why was the pesticide plant not built in a sparsely populated region in which accidents would have had less catastrophic results? From the perspective of Union Carbide, it was more cost effective to produce the pesticide in India, where it was to be used as part of India's GREEN REVOLUTION, rather than to manufacture it in America and export the pesticide. Primary considerations were the high cost of transportation and the cheapness of Indian labor. The decision to locate the plant in Bhopal seems to have been political in the sense that the Indian government wanted to stimulate the depressed economy of the region and to this end leased land to Union Carbide for $40 an acre in Bhopal.

There was very little perception that the Bhopal plant was dangerous perhaps because it was designed much the same as pesticide factories in the United States. Not only was Bhopal a populous city of around a million people, but employees brought their extended families to live near them in the vicinity of the plant, thereby increasing the number of people affected by the explosion. Moreover, an army of homeless people lived in shanty towns adjacent to the UCIL plant. Some were literally built up against the walls of the factory.

When the explosion occurred, most of the neighborhood residents had no idea that their lives were in danger. In fact many ran toward the plant to find out what had happened instead of fleeing to safety. Deficient as they were, all attempts at safety clearly stopped at the factory gate. If the population had known that there was a danger of toxic gas being released and that they could have protected themselves by merely wearing a wet cloth around their faces, hundreds of lives could have been saved.

LIABILITY AFTER THE EVENT
It is easy to misinterpret the Bhopal accident as a case of what happens when a large and powerful corporation sets up its "dirty industry" in an unsuspecting third world country. Yet the truth is a great deal more complex and more interesting than this. Pesticide can be produced safely and is produced safely in America and other developed countries. If UCIL had managed its plant in the same way as Union Carbide's American plants were managed, the incident would probably never have happened, or at least would not have had such terrible consequences.

Failure of UCIL to live up to safety regulations set the scene in which the irresponsible actions of a disgruntled worker produced unprecedented loss of life in an industrial accident. Why did Union Carbide not insist that the Bhopal plant should live up to proper safety standards? The answer is that they did but their warnings went unheeded. A Union Carbide safety inspection was carried out two years before the explosion. Many problems in

safety procedures were identified and a list of recommended changes was drawn up but these were never implemented. The safety problems at Bhopal were well known to the Indian government and to the reading public because the Union Carbide report had been obtained by local journalist Raj Kumar Keswani who published, in the local newspaper, a series of articles on the dangers posed by the plant.

Even though UCIL was largely an Indian-owned company (with 25% of the stock owned by the Indian government and 49% owned by Indian citizens) with no remaining American executives or employees, Union Carbide decided to take full moral responsibility for the accident and immediately agreed that the victims would be compensated.

POLITICAL AND LEGAL RAMIFICATIONS
The cooperative attitude of Union Carbide in connection with the Bhopal tragedy may not be what many people have come to expect from powerful multinational corporations in their dealings with developing countries but it has to be remembered that the whole U.S. chemicals industry was still reeling from the adverse publicity surrounding the LOVE CANAL toxic dump. Union Carbide seems to have been strongly motivated to demonstrate good corporate citizenry in respect to the Bhopal tragedy. Warren B. Anderson, the chief executive officer and chairman of the board of Union Carbide, left for Bhopal with a budget of up to $5 million to pay for help at the site of the disaster and to find out what the company could do to help the victims. Instead of being allowed to help, Anderson was arrested by Indian authorities as soon as he got off the plane at Bhopal. He was kept under house arrest for a few days and all of his attempts to interview the plant managers and communicate with political leaders were frustrated.

It was an election year and Indian politicians were making political capital out of the sufferings of the victims, which could be conveniently blamed on an American corporation, instead of examining the details of what had actually happened and the actual chain of managerial failures that led up to the tragedy. Union Carbide set up a vocational training school to help Bhopal residents escape from poverty but the center was closed by the Indian government. Unambiguously rebuffed from helping the victims in public, Union Carbide attempted to help them in private. It funded Arizona State University to create and run a rehabilitation unit for the injured. After the rehabilitation center had become fully functional, the Indian government learned that it had been funded by Union Carbide money. They sent in bulldozers to level the building. This action appears to have been a political response to popular anger over the catastrophe. By preventing the company from discharging any moral obligations to Bhopal's victims, the government might also have hoped for a larger financial settlement.

Litigation on behalf of the victims of the explosion was tortuous and lasted for five years. These proceedings were clearly influenced by political calculations. Much of the litigation revolved around whether Union Carbide or its Indian subsidiary was liable for damages. India's High Court ruled that Union Carbide held jurisdiction over the UCIL plant, despite the fact that there were no American executives there.

The litigation began in America where Union Carbide was sued on behalf of the victims. The Indian government wanted the case tried in America so that victims would have the advantage of more stringent safety and environmental standards and so that they could benefit from a tradition of larger awards for damages in such cases. Moreover, the plaintiffs could obtain those advantages without any necessity for the Indian government to clean up its own act by applying similar standards at home. Union Carbide was successful in its argument that since India was the site of the accident, and since the evidence and witness were located there, the trial should be appropriately conducted in India.

As soon as the trial was returned to India, the government effectively took charge of it by invalidating existing lawsuits and assuming the sole right to speak on behalf of victims. Once the trial was conducted on home soil, the Indian government adopted a different strategy. Officials no longer had any desire to obtain a large financial settlement from Union Carbide possibly because such harsh treatment of a multinational corporation by the Indian government could paint a picture of India as a bad investment risk and thus dry up the inflow of desperately needed investment capital. Moreover, it was estimated that the litigation might drag on for as long as another 20 years and, with another election coming on, the legal proceedings were viewed as a liability. The Indian chief justice interrupted the trial to recommend a deal in which Union Carbide would pay $470 million in damages to the Indian government for distribution to victims. This was a small sum in comparison to the $3.3 billion amount of the lawsuit, or the $5 billion paid to plaintiffs in the case of the Exxon Valdez oil spill. Families received $3,000 in compensation for each death. This suggested to many people that when life is lost to an industrial mishap in a developing nation it can be cheap.

BHOPAL AND SELF-REGULATION BY THE CHEMICALS INDUSTRY

Industrial fiascoes on the scale of Bhopal usually stimulate responsible people to work on safety procedures that greatly diminish the likelihood of such incidents recur-

ring. The American trade organization, Chemical Manufacturer's Association (CMA) quickly realized the importance of Bhopal for the entire chemicals industry in the sense that the possibility of similar accidents occurring at other sites puts the industry under a cloud of suspicion and makes the implementation of highly restrictive safety regulations more likely. The most constructive way for the chemicals industry to respond to this challenge is to implement a vigorous policy of self-regulation in which chemicals producers not only diminish the threat of their operations to local populations but also communicate more effectively with communities concerning safety issues.

One of the most startling conclusions to be drawn from the Bhopal story is that despite journalistic criticisms of safety procedures in the plant, local residents had virtually no awareness of the potential threat that the chemicals posed for their lives and health. With this problem in mind, the CMA drafted in 1985 a set of proposals known as Community Awareness and Emergency Response (CAER). The CAER requirements were designed to promote greater transparency in dealings between chemical manufacturers and the local community. The public was to be kept informed of any potential dangers posed by the plant and the safety measures that had been adopted for their protection. Moreover, the company was to initiate a dialogue with fire fighters, police, and other emergency services to discuss what kind of accidents might potentially happen and what could be done to deal with them most effectively while minimizing deaths and injuries. CAER also committed companies to conducting a disaster drill at least once every year. One objective of the drill was to ensure that employees were fully conversant with all emergency procedures.

The original draft of CAER has been modified to make it both more general and more binding. Inspired by the work of the Canadian Chemical Producers Association, it now includes all aspects of the chemicals industry, not just manufacturing safety but also research, transportation, distribution, health and safety, and toxic waste disposal. Signing on to the CAER requirements is now a precondition for membership of the CMA. These changes are an impressive ethical response to the widespread perception that the chemicals industry is dirty and socially irresponsible. They also reflect a commitment to action over words. That is, the industry has decided to commit its collective resources to improving safety practices rather than to a public relations campaign designed to improve public perception of the industry. The rationale is that if practices are genuinely changed public perception will inevitably improve.

See also CHERNOBYL NUCLEAR DISASTER; LOVE CANAL; SEVESO CHEMICAL ACCIDENT.

Further Reading
Cassels, J. *The Uncertain Promise of Law: Lessons from Bhopal.* Toronto: University of Toronto Press, 1993.
Newton, L. H., and C. R. Dillingham. *Watersheds 2: Ten Cases in Environmental Ethics.* Belmont, Calif.: Wadsworth. 1997.

Binet, Alfred

(1857–1911)
French
Psychologist

Alfred Binet is remembered as the author of the first modern intelligence test. Binet was director of the psychological laboratory at the Sorbonne, Paris. The French Ministry of Public Instruction appointed a commission in 1904 to study the problem of children in public schools who did not have the cognitive ability to benefit from their education. As a member of the commission, Binet, along with psychiatrist Theodore Simon, undertook to develop an intelligence test that would predict which children could benefit from ordinary public education. The Binet test consisted of a series of 30 problems arranged in order of increasing difficulty. The problems tested reasoning ability, judgment, and comprehension. Binet and Simon standardized their test in terms of the average performance of children of various ages. Thus, if a five-year-old child solved problems typically mastered by children at the age of seven years, he or she was said to have a mental age of seven. Binet and Simon's test was further developed in the United States as the Stanford-Binet Test. Intelligence tests have proved to be useful for predicting academic performance but their use has presented ethical problems, including the potential for racial bias in immigration and education because of the incorrect belief that there are substantial fixed differences between ethnic groups in IQ scores due to biological differences.

See also INTELLIGENCE TESTS, ETHICS OF.

biodiversity and industrialization

Biodiversity refers to the number of different species that can survive on earth. Environmental ethicists are very concerned about the potential damage caused by human industrial activities to ECOSYSTEMS and their constituent species. Some ecologists, such as E. O. WILSON, claim that we are in the middle of a mass extinction event. Unlike the 10 other mass extinctions of species on the planet, this one is the result of human industrial activities. Tropical rain forests are of particular concern because they house an unusually diverse array of species and because the forests are being cut down at an unprecedented rate to make way for agriculture (which quickly turns their fragile ecosystems into deserts), provide lumber for construction, and serve as a fuel for mining and other industries.

Loss of biodiversity in the rain forest and elsewhere includes the loss of many thousands of species that have never been examined by science. These plants and animals probably contain substances that would have been of incalculable benefit to medicine. According to DEEP ECOLOGISTS the real loss consists of inherently valuable information stored in ecosystems rather than any such loss to medicine or science. Other major threats to ecosystems and biodiversity include GLOBAL WARMING that expands deserts, industrial pollution, and ACID RAIN.

See also ACID RAIN; CATASTROPHISM; ECOSYSTEM; DDT; DEEP ECOLOGY; GLOBAL WARMING.

Further Reading
Wilson, E. O. *The Diversity of Life.* New York: W. W. Norton, 1993.

biological and chemical warfare

In chemical warfare, chemicals are used to injure or kill plants and animals as well as human beings. Biological warfare uses living organisms to cause illness and death. Chemical and biological warfare are regarded with a particular horror despite the fact that they have rarely been used to any tactical advantage. Both have been effectively banned by international treaties and are used today primarily by terrorists and rogue states that do not recognize international treaties. It has recently been revealed that both the United States and the former Soviet Union had built up enormous stockpiles of chemical and biological agents and their delivery vehicles.

CHEMICAL AND BIOLOGICAL AGENTS UP TO WORLD WAR I

There are two good reasons for not deploying chemical and biological weapons. One is that they are difficult to use effectively against enemies. The other is that they are dangerous to the people who develop and use them.

The potential for a biological agent to get out of control is clearly demonstrated in the case of one of the earliest examples of biological warfare. In 1347, at the siege of Caffa (now Feodosiya in the Ukraine), Mongol attackers hurled the corpses of plague victims over the walls at their Genoese enemies. When the Genoese returned to Italy, they carried the plague with them and are given the credit for beginning the huge epidemic known as the Black Death that decimated the entire population of Europe. More recent attempts to spread disease among enemy populations have never had much success and this bears testimony to the extreme virulence of the bubonic plague itself rather than any accomplishment of its users.

During World War I, the Germans used germ warfare against cavalry horses and livestock. They infected Romanian horses and U.S. farm animals earmarked for shipment to the Allies in Europe with glanders, a bacterial disease. Germ warfare has never been used successfully against people in the 20th century.

Chemical warfare was first used to serious effect in World War I (1914–18). Although the tactical advantage of using chemical weapons can be questioned, there is no doubt that their use contributed to the horror of war for the frontline troops. The use of chemicals was initiated by the Germans, who, in January 1915, used chlorine, a heavy toxic gas, against the Russians in Poland. The chlorine was released from cylinders, which meant that changing winds could easily blow it back on the releasers. The Allies responded by producing their own chlorine and developing gas masks to protect troops. Chlorine was followed by phosgene, a more toxic kind of choking gas. This was delivered using artillery shells.

In 1917, the Germans introduced mustard gas, a blistering agent that produced severe skin burns. Mustard gas was used extensively by both sides in the final year of the war. It was greatly feared because inhaling it produced permanent, and sometimes fatal, damage to the lungs. It is estimated that some 800,000 soldiers were injured by phosgene or mustard gas during World War I, although most of these survived. The heaviest fatalities occurred in the Russian army with an estimated loss of 60,000 men to chemical attacks.

MODERN BIOLOGICAL AND CHEMICAL WARFARE

One of the most extensive applications of chemical warfare was the use of AGENT ORANGE, a defoliant, in the Vietnam War. American troops had been constantly exposed to guerilla attacks in the jungle and used this herbicide to destroy the vegetation that was providing cover for Viet Cong soldiers. Large swaths of jungle were sprayed from the air causing the foliage to wither. Deprived of concealment, the enemy soldiers were forced to move on.

Direct contact with Agent Orange caused the skin to blister but apparently posed no major health threats. Unfortunately, Agent Orange contained small quantities of dioxin as a contaminant. Dioxin is one of the most potent carcinogens known and it is believed to be responsible for cancer, birth defects, and a variety of health problems among Vietnamese residents who have spent decades living in a dioxin-polluted environment. Agent orange increased the rates of skin cancer among U.S. troops and may have produced other health problems as well. Yet, it is important to realize that the dioxin in Agent Orange was inadvertent and was not deliberately used to injure people.

Highly lethal nerve agents, or neurotoxins, were developed in the period just before and during World War II. These belonged to a group of compounds known as the organophosphorous nerve agents. The Germans developed the first of these, tabun, in the late 1930s and produced a large stockpile during 1942–45 in a plant at Dyhemfürth. After the Soviets seized this plant at the end of the war, they began producing nerve gases of their own. It is not known why Germany did not use this deadly nerve agent but some historians believe that its use was inhibited by fear of chemical counterattacks against civilians. Other organophosphorous nerve agents include sarin, which was produced in large quantities by the U.S. military, and soman, the agent favored by the Soviet military. Both sarin and soman were developed in the 1930s by I. G. Farben, the German chemical company, as an offshoot of their research on pesticides.

The devastating effects of sarin gas were illustrated in a 1995 attack on the Tokyo subway, which killed 12 people and injured over 5,000. The compound, which had been made and released by members of the Aum Shinrikyo religious cult, turned out to be impure. The pure form is a great deal more dangerous.

Sarin and soman can enter the body through the lungs or the skin and are fatal at very small doses. They act at the nerve-muscle junction causing paralysis, convulsions, and immediate death. These nerve agents are far more toxic than the chemicals used in World War I but they have rarely been used in warfare. A probable exception is the use of tabun (along with mustard gas) by Iraq in its war against Iran.

Possible chemical warfare agents encompass not just synthetic chemicals but also naturally occurring poisons. These include tetrodotoxin (from fish), saxitoxin (from shellfish), and botulin from the bacterium *Clostridium botulinum*. Botulin is exceptionally toxic and minute quantities cause death.

The development of strains of microorganisms that could be used as biological warfare agents has proceeded in many countries. By the end of World War II, at least six countries had established biological warfare laboratories. They included Canada, France, Japan, and Britain, as well as the United States and the USSR. Species that have been developed for use against people include bacteria, such as anthrax, tularemia, and plague, viruses such as Venezuelan equine encephalomyelitis, and fungi such as coccidioidomycosis. In addition, a large spectrum of agents has been developed for use against economically important animals and plants. These include fowl plague, foot-and-mouth disease, black-stem rust that attacks cereals, potato blight, and rice blast.

During the cold war period, America and its allies collaborated in research on psychedelic drugs that were envisaged as a humanitarian alternative to deadly weapons. If enemy soldiers could be rendered confused and disoriented, it would be unnecessary to kill them. Unethical experiments were conducted using American and British servicemen as guinea pigs. Many did not know that they had received a hallucinogenic drug and some developed severe emotional problems possibly as a consequence of their participation. In the end, psychedelics were considered to be too unpredictable and too unreliable for use in chemical warfare. They were replaced by BZ, a drug that has the advantage of being deliverable as an airborne spray that is dispersed from bombs. This produces nausea, blurred vision, speech difficulty, confusion, and stupor that may last for up to four days.

LIMITING THE USE OF BIOLOGICAL AND CHEMICAL WEAPONS

There are at least two important reasons that chemical and biological weapons arsenals are rarely used. One is that the weapons are capable of producing such horrifying consequences that no warring nation wants to initiate their use because in doing so they are exposing their own soldiers and citizens to retaliation in kind. There is little point in exposing your enemy to bubonic plague if you know that the enemy will reply in kind, possibly responding with an even deadlier virus, or with some recently produced, genetically engineered agent against which you are defenseless. The other reason is ethical. Most countries cannot contemplate exposing even their bitter enemies to such horrifying weapons for humanitarian reasons or because they cannot accept the stigma of being the first to use an appalling agent.

Chemical and biological weapons resemble nuclear warfare in that the only real value in developing them is to serve as a deterrent. NUCLEAR WEAPONS were used in World War II because they were possessed by the United States. Since that time, they have not be used because of the principle of MUTUAL ASSURED DESTRUCTION. This means that the initial user sets in motion a chain of retaliatory actions that seal their own doom. Since chemical and biological weapons are practically useless, and are dangerous to develop and maintain, it is in the profound self-interest of all countries to join treaties designed to get rid of them.

Under the leadership of President Richard Nixon, the United States decided in 1969 to unilaterally destroy its biological weapons stockpile. The facility at Fort Detrick, Maryland, where the weapons had been produced, was turned into a cancer research laboratory. Conducted in secrecy, America's biological weapons program was a large-scale operation that employed some 4,000 people at its peak. Pathogens were tested on 2,000 human volunteers and bacterial aerosols were sprayed in American cities exposing citizens to poten-

tially harmful effects without their knowledge or consent. In the course of their biological weapons research, the U.S. government used the data obtained from the horribly unethical experiments conducted on Chinese prisoners of war by the Japanese military. The true nature and extent of the U.S. program has been revealed in recently declassified documents. By the time of its cancellation, the biological weapons program had weaponized three lethal biological agents and toxins and four incapacitating agents. These weapons had unpredictable and uncontrollable aspects that apparently made them unacceptably risky. There was a real fear that biological weapons could unleash deadly plagues that might spread globally.

During the same period, the Soviet biological weapons program was much larger, employing over 60,000 people at over 100 facilities. The extent of their interest in germ warfare has only recently come to light. Major American cities were targeted in a contingency plan. The Soviets are believed to have deployed biological weapons in Afghanistan. They also played a role in the development of biological warfare programs in other countries, including Cuba and Libya. A mysterious outbreak of anthrax in Sverdlovsk, Russia, in 1979, was apparently due to the accidental release of a biological warfare agent from a nearby military laboratory but this has not been acknowledged by the Russian government.

In 1972, the United Nations introduced the Convention on the Prohibition of Biological and Toxin Weapons, which was eventually approved by over 100 countries, including the United States. The convention prohibited the production or stockpiling of chemical and biological weapons but it was considerably weakened by a provision that allowed for defensive research. The idea was that countries should still be able to protect themselves by developing antidotes for chemical poisons and vaccines to protect against germ warfare. Since the United States signed the treaty, in 1974, the Department of Defense has greatly scaled back its research on biological agents but its operations have stretched the definition of defensive research. Thus, new toxins may have been developed along with the development of antidotes. Research designed to produce vaccines through genetic engineering may have resulted in the production of more virulent strains. It is difficult, even in principle, to separate defensive research in this field from the development of hazardous agents.

The absurdity of stockpiling expensive weapons that are unlikely to be used and pose a threat to their owners applies to chemical agents as well as biological weapons. In 1987, the Soviet Union, under the leadership of Mikhail Gorbachev, announced that it was unilaterally suspending the production of chemical weapons. In the same year, however, the United States resumed production of nerve gases. These were contained in binary shells. The idea is that two less dangerous substances, contained in separate compartments, are mixed together at the moment of the shell's explosion to make a lethal nerve agent. This technical improvement meant that personnel producing the bombs were at reduced risk.

The United States and the Soviet Union paved the way for a chemical weapons treaty in 1989 by both agreeing to inspections of their chemical weapons plants. The Chemical Weapons Convention was signed in 1993 by representatives of 131 countries. This committed signatory nations to halt production of all chemical weapons, as well as their use, sale, or stockpiling. Moreover the convention called for the destruction of all existing chemical weapons by the year 2005.

The destruction of all chemical weapons by this deadline is quite unlikely. One problem is that some developing countries, such as Iraq and Libya, may persist in developing chemical weapons. These are comparatively cheap to produce relative to NUCLEAR WEAPONS and import controls relating to the raw materials are easy to circumvent. The scope of the problem was well illustrated by the failure of the United Nations to locate and destroy Iraq's chemical and biological weapons despite years of inspections.

Another problem is the very great expense of destroying weapons. It is estimated that the cost of neutralizing the huge chemical arsenal inherited by Russia from the USSR will be in the vicinity of $10 billion. With its economy in a shambles following the 1999 collapse of the ruble, it seems unlikely that Russia can devote extensive economic resources to an operation that produces no economic return. In the United States, money is less of an impediment, but the proposal to destroy chemical weapons has ironically produced great opposition from environmentalists. Incineration, the method of destruction preferred by the U.S. Army, is opposed by environmentalists. In view of the poor safety record of chemical incinerators, such as that at TIMES BEACH, Missouri, which spewed the carcinogen dioxin over the surrounding countryside necessitating the permanent evacuation of a whole town, the environmentalist perspective is understandable. Local communities everywhere have adopted a "not-in-my-backyard" stance that makes it extremely difficult to find sites for the incinerators needed to destroy chemical weapons.

See also AGENT ORANGE; MUTUAL ASSURED DESTRUCTION; NUCLEAR WEAPONS; PSYCHEDELIC DRUG RESEARCH; TIMES BEACH.

Further Reading
Alibek, K. W. *Biohazard: The Chilling True Story of the Largest Covert Biological Weapons Program in the World.* New York: Random House, 1999.
Harris, S. H. *Factories of Death: Japanese Biological Warfare, 1932–1945, and the American Cover-Up.* New York: Routledge, 1995.
Lederberg, J. S., ed. *Biological Weapons: Limiting the Threat.* Cambridge: MIT Press, 1999.
Regis, E. *The Biology of Doom: America's Secret Germ Warfare Program.* New York: Henry Holt, 1999.
Somani, S. *Chemical Warfare Agents.* New York: Academic, 1992.
Zilinskas, R. A., ed. *Biological Warfare: Modern Offense and Defense.* Boulder, Colo.: Lynne Rienner Publishers, 1999.

biological psychiatry

Modern biological psychiatry dawned in 1952 with the development of chlorpromazine, the first drug that was effective in treating the hallucinations and delusions of schizophrenia. Earlier biological treatments for mental disorders had been crude and ethically questionable. Thus, Austrian psychiatrist Julius Wagner von Juaregg successfully treated the symptoms of general paresis (a brain manifestation of syphilis) by infecting patients with malaria in 1917. When treated with chlorpromazine, schizophrenics who had been completely out of touch with reality quickly began to act like sane and rational people and were released from psychiatric hospitals in large numbers. Biological psychiatry is based on the premise that mental illness is due to brain disorders and that it is treatable using drugs, surgery, and other techniques like electroconvulsive shock therapy (ECT). Previously, mental illness was attributed to moral weakness and carried a heavy social stigma.

The drug revolution in psychiatry was greeted with delight as well as amazement but since that first wave of enthusiasm, a number of problems have emerged. Astonishingly successful though some drug treatments for mental disorders have been, they have not turned out to be quite the magic bullet that they seemed. One problem is that psychiatric medications often make people feel ill with symptoms like nausea, dry mouth, irregular heartbeat, and uncontrollable twitching. Taken for long periods even more troubling side effects may emerge. One of the worst is tardive dyskinesia, an irreversible movement disorder, characterized by involuntary tremors and by smacking of the lips, that emerges after years of medication for schizophrenia. The risk of tardive dyskinesia is reduced by gradually weaning patients off high doses of medication, and drugs have been developed that are less likely to produce these symptoms. The use of lithium salts to treat manic depression incurs a risk of actual toxicity that requires careful calibration of the dosage.

There are several other ethical problems with biological psychiatry. One is that people who were released from psychiatric hospitals after their condition was improved by medication did not receive adequate aftercare. Many who stopped taking their medication experienced a return of psychotic symptoms. Some critics have pointed out that the net effect of releasing people from psychiatric hospitals has been to swell the ranks of the mentally disturbed homeless, who are a depressing feature of modern life in cities. This raises the issue of whether they ought to be rehospitalized for their own good.

Another complaint about drug therapies is that they tend to treat symptoms without addressing underlying causes. Thus if someone suffers from anxiety, they may be prescribed addictive tranquilizing drugs like Valium, or Xanax, without addressing any of the lifestyle issues that might be causing the problem. Long-term use of medication is bad not only because it develops dependence and has the potential for troubling side effects but also because it is expensive and creates a huge burden on those who pay for health care. Psychological interventions may be much less expensive and produce better quality of life for the patient. For this reason, contemporary drug treatments are often combined with psychotherapy.

Even though drug therapies can be highly effective for conditions such as schizophrenia and severe clinical depression that were previously untreatable, the science underlying their use is often sketchy at best. Thus almost all of the major biological therapies were discovered by accident and, in many cases, subsequent investigation has not revealed how they work. In part, this reflects a patchy knowledge about brain function. This point is illustrated by the case of ECT, which was first used to treat schizophrenia on the mistaken notion that epileptics do not suffer from this disease. ECT is somewhat more effective for treating depression than the best drugs but the theoretical understanding of why it works can be compared to the logic of pounding the top of a television set to improve the picture.

Not only does biological psychiatry have trouble providing a detailed explanation of how therapies work, but, in some cases, there is even controversy about whether the disorder being treated is real. This is true of attention deficit disorder, which is characterized by an inability or unwillingness of children to remain stationary and follow directions. Such children are a disruptive presence in the classroom. Moreover, their problem can be helped by prescriptions of Ritalin and other central nervous system stimulants that produce a paradoxical calming effect. The problem is that attention deficit disorder has not been described in terms of a brain disorder and many experts doubt that it is one. It can be argued that many children may have behavioral problems in school because they have not been socialized to accept the kind of discipline that is required in this setting. Even if difficulty in following instructions is exacerbated by brain conditions like lead poisoning and fetal alcohol syndrome, it does not follow that the underlying cause is brain pathology. Another interesting, and troubling, feature of Ritalin prescription for attention deficit disorder is that while its usage is commonplace among American children, it is much less prescribed in other developed countries, presumably reflecting different beliefs about the nature of the problem.

Far from being the magic bullet that it seemed at the outset, biological psychiatry does not always work. Thus, for persons suffering from clinical depression, less than two-thirds are helped by drug treatment. Although the benefits of drug treatments are usually ascribed to the intended pharmacological effect of the drug, this is often a shaky inference because most medications recruit a placebo effect according to which patients get better in part because they expect to improve. Placebo effects in psychiatry can be substantial, occasionally rivaling the effect of the pharmacological agent. In conclusion, although biological psychiatry has helped to relieve the misery of millions of people suffering from mental illness, it has proved to be a mixed blessing. Many ethical problems have been raised. One of the most important of these is the questionable quality of the underlying science.

See also ELECTROCONVULSIVE THERAPY; DIAGNOSTIC AND STATISTICAL MANUAL OF MENTAL DISORDERS; KRAFFT-EBING, RICHARD VON; PLACEBO-CONTROLLED STUDY; PREFRONTAL LOBOTOMY.

Further Reading
Hedaya, R. J. *Understanding Biological Psychiatry.* New York: W. W. Norton, 1996.
Ross, C. A., and P. Alvin. *Pseudoscience in Biological Psychiatry: Blaming the Body.* New York: John Wiley, 1994.

Biosphere

The Biosphere 2 project was an unsuccessful attempt to simulate the earth's ecosystem (or Biosphere 1) by constructing a self-contained model Earth within a huge greenhouse in Oracle, Arizona. The original plan was for Biosphere 2 to comprise a self-sustaining ecosystem complete with human inhabitants. The experiment captured the popular imagination for several reasons, one of which is that it can be considered a prototype for building human colonies in space. Although originally scorned by scientists, the Biosphere experiment provided an important practical lesson in the complexity and fragility of ecosystems that has clear relevance for environmental ethics. Despite its highly publicized scientific failures, the Biosphere 2 structure has been maintained as a research and teaching facility and tourist attraction.

Construction of Biosphere 2 began in 1986, thanks to the financial support of Texas billionaire Edward P. Bass, who spent an estimated $200 million on the design and building of the 3-acre 91-foot-tall structure. Inside this tightly sealed greenhouse, thousands of species of plants and animals were installed in what was conceived of as a miniature version of the earth's ecosystem. In addition to a half-acre garden for growing food to feed human inhabitants, the Biosphere contained five wilderness areas: a savanna with grasses from three continents; a marsh, transported from the Florida everglades; a rain forest; a desert; and an "ocean" complete with a coral reef.

In September 1991, excess carbon dioxide was removed from the Biosphere atmosphere using chemical "scrubbers" and eight human occupants passed through the airlock for a two-year stint. Life in the Biosphere was arduous and the Biosphereans confronted one ecological nightmare after another. Shortly after the crew had been sealed in, their computers detected a worrying decline of oxygen levels in the air they were breathing. It was later discovered that the oxygen was being consumed by microbes in the specially enriched soil of the agricultural plot. The decline was to continue until the oxygen level had fallen from a normal 24% of the air to 14.5% in February 1993. At that point, conditions resembled those found on a very high mountain and the inhabitants found it difficult to work in the thin air. Their lives were also at risk. Contrary to the original intent of the project, which was to maintain a sealed environment, it was decided to pump in oxygen gas. Carbon dioxide levels were also too high and these were reduced using the scrubbers.

Another unanticipated problem with the atmosphere was the unusually high level of nitrous oxide, or laughing gas. In the normal course of events, nitrous oxide is broken up by ultraviolet rays from the sun. These rays were blocked by the glass of the greenhouse. Concentrations of laughing gas were high enough to interfere with vitamin B-12 synthesis, exposing Biosphereans to the risk of nervous system damage.

In other respects, life in the Biosphere was difficult. Residents spent almost all of their time working to support themselves and found that they had little time for doing the scientific work they had planned. In addition to growing all of their own food, slaughtering animals, and other domestic chores, they were responsible for complex computer systems and maintaining a complicated plumbing system.

The Biosphereans soon learned that they could not produce enough food to sustain themselves. The light level inside the structure was lower than expected and the crops suffered from a heavy infestation of mites. Vines, such as morning glory, spread aggressively and required constant weeding. Extreme weight loss occurred with one individual losing 42% of his body weight, or 110 pounds. The Biosphereans were obsessed with eating and had arguments when food was stolen.

Difficult as life may have been for the human occupants, at least they all survived the ordeal, which is more than can be said for other animal species in the experiment. Of 25 vertebrate species originally present, only six escaped extinction. The insects fared no better. All of the plant pollinators were lost, which meant that many plant species could not reproduce and would die out after a single generation. Insects that prospered included ants and cockroaches, which further reduced the quality of life of the human occupants.

Following the ecological nightmare experienced by the first crew, a second one began a one-year mission in 1994. Two months later, members of the first crew smashed windows of the Biosphere structure claiming that they were concerned about the safety of the new Biosphereans. This brought the experiment to an unscheduled end.

Although the Biosphere had obviously failed in its mission to be a self-sustaining ecological system, the structure was taken over by Columbia University in 1996 and is now being used as a research facility specializing in earth science. The former living quarters of the Biosphereans have been sealed off with glass walls and are used as a museum that commemorates the Biosphere 2 project. This museum contains an interactive exhibit dealing with global climate change. It draws some 200,000 visitors each year.

The Biosphere project failed in its original mission largely because it proved impossible to simulate some key aspects of earth's complex atmosphere. The simulated atmosphere was too simple, but this simplicity is nevertheless appealing to many research scientists because it allows for manipulation and control. Researchers can study the effects of differing levels of atmospheric carbon dioxide on plant growth or on the growth of coral reefs, and they can study the thermal tolerance of rain forest vegetation. Many of the research projects currently being conducted in the Biosphere facility are of great relevance for understanding the impact of industrial development on climate and biodiversity.

The scientific failure of the Biosphere experiment allows one simple but profound conclusion to be drawn. Even though the Biosphere lacked support from mainstream scientists and arguably made avoidable mistakes, the truth is that we do not yet know how to create an artificial environment that can sustain human life. Whether this will ever be accomplished is an open question. Yet attempts to create such systems can provide enormous insight into the functioning of the only biosphere that can currently support us, the natural life-supporting ecosystems of the earth.

See also GAIA HYPOTHESIS; GLOBAL WARMING; RAIN FOREST DEPLETION.

Further Reading
Cohen, J. E., and D. Tilman. "Biosphere and Biodiversity: The Lessons so Far." *Science* 274 (1996): 1150–51.
Erickson, K. "Class Under Glass." *E* (March 1999): 16.
Maxson, G. "Second Chance for Biosphere." *Popular Science* 250 (1997): 56–60.

Blondlot, Rene   *See* N-RAYS.

Bouchard, Thomas   *See* MINNESOTA TWIN STUDY.

bovine somatotropin (BST)

Bovine somatropin (BST) is an artificial hormone that is fed to cows to boost their milk production. Produced by transgenic techniques (i.e., genetic engineering), the drug is capable of boosting milk production by around 10 pounds per cow per day. BST is used by approximately 13,000 American farmers generating annual sales of some $200 million for the Monsanto Company that manufactures it.

BST is banned in Canada and its use has been placed under a moratorium by the European Union. European scientists argue that the milk of BST-treated cows contains insulin-like proteins that are associated with prostate and breast cancer. The American Food and Drug Administration (FDA) concluded, in November 1993, that the drug was safe. In doing so, the FDA was not swayed by a high-dose study showing that BST is taken up by the bloodstream in rats leading to the development of thyroid cysts. This study affected the Canadian Health Ministry's decision to ban the drug on the grounds that it can cause harm to animals. Recent research has shown that cows in herds where BST is used are as healthy as those in which it is not used.

Restriction of BST use in Europe is one of many examples of a greater regulatory caution about the use of genetically engineered products in foods. This caution does not extend to food products in general and probably represents the greater political salience of environmental issues in Europe. Recent public relations attempts by Monsanto to promote the safety of transgenic foods have been a dismal failure for the company, which has presented itself as a target for protests by environmental activists.

In the United States, the state of Vermont briefly required grocers to label dairy products with three blue dots if they came from BST-treated cows. This law was motivated by concern for the safety of the milk and the desire to provide an advantage to Vermont's small dairy farmers, most of whom do not use BST. On August 29, 1996, the one-year-old ban was struck down by a U.S. District Court injunction. The court cited studies showing that milk from BST-treated cows is indistinguishable from the milk of untreated cows.

See also GENETIC ENGINEERING; GENETICALLY MODIFIED FOOD.

Further Reading
Ramey, J. "Vermont's BST Labeling Law Lifted." *Supermarket News*, September 9, 1996.
Scott, A. "Monsanto, Brussels Clash over BST." *Chemical Week*, March 31, 1999.

Boyle, Robert
(1627–1691)
Irish
Chemist and Philosopher

Boyle was the seventh son of the first Earl of Cork and received a privileged upbringing and education. He cultivated an interest in the new science as represented by the work of GALILEO and is remembered for having developed chemistry into a separate empirical science that was distinct from its ancestors alchemy and medicine. His strong belief in the use of experiments to solve scientific problems and his writings on the SCIENTIFIC METHOD influenced subsequent practitioners, including ISAAC NEWTON (1642–1727).

Boyle was an accomplished experimenter himself. He concluded that, for a gas at constant temperature, the product of its pressure and its volume is a constant (Boyle's law). Boyle produced the modern definition of an element as a substance that cannot be divided into simpler constituents, and he conceived of matter as being composed of atoms. He was a founder of the world's first scientific organization, the ROYAL SOCIETY OF LONDON, and one of its most distinguished members. Boyle was interested in theology and saw no contradiction between religion and the mechanistic worldview of his chemistry.

See also BACON, FRANCIS; DALTON, JOHN; ROYAL SOCIETY OF LONDON; SCIENTIFIC METHOD.

bridge collapse and design flaws

Bridge construction often represents a large infrastructural investment and there have been a surprising number of bridge collapses in recent history, perhaps reflecting the increased length of modern bridges and the use of original designs. Many of these failures have been attributed to design flaws, including inadequate stability of foundations, use of substandard materials, and poor design either of elements of the bridge or of the structure as a whole. Many bridge collapses are thus due not to chance accidents but to ethically questionable lapses on the part of civil engineers.

One instructive example is the first Tacoma Narrows Bridge in Washington State that collapsed on November 7, 1940, shortly after its completion in July of the same year. The bridge was so unstable under windy conditions that crossing it was rather like taking a roller coaster ride. For this reason, it was nicknamed Galloping Gertie and was quite popular among locals. The demise of the bridge has been explained in terms of the elementary physical principle of resonance. The external force of wind gusts caused the bridge to vibrate at its natural frequency, just as a tuning fork causes the string of a guitar to vibrate. On November 7, 1940, there were unusually high winds that produced strong vibrations in the bridge literally caused Galloping Gertie to shake itself apart. A similar fate befell the Tay Bridge at Dundee, Scotland, in 1879. Bridge models are currently tested in wind tunnels to avoid this problem.

In principle, bridges can be designed to survive any weather conditions and to withstand earthquakes. While it may seem unreasonable to expect that no bridge should ever collapse, most collapses are avoidable and thus it has become an ethical issue for those who design a bridge, construct it, supply the materials, and are responsible for inspections and maintenance following construction.

See also EARTHQUAKES AND CONSTRUCTION CODES; ENGINEERING ETHICS.

Further Reading
Shepherd, R., and J. D. Frost, eds. *Failures in Civil Engineering.* New York: American Society of Civil Engineers, 1995.

Buddhist ethics

Buddha (Siddartha Gautama, c. 480–400 B.C.) and the early saints of Buddhism practiced an ascetic lifestyle because of the belief that the ultimate goal of Buddhism, attainment of nirvana (a permanent state of enlightenment), was impossible for individuals who were not morally perfect. Central to Buddhist ethics is the concept of KARMA which is shared by all the other major Indian religious traditions (Hinduism, Jainism, Sikhism). It involves the belief that moral and immoral actions have repercussions for a person's material and spiritual status. Moral actions bear good fruit in the present life and also determine one's spiritual well-being in future existences.

The doctrine of karma is thus very closely connected to belief in reincarnation, or rebirth. According to this belief, which preceded Buddha, the behavior of higher forms of life (gods, humans, animals) determines whether they move up or down in their next incarnation. Moral behav-

ior is rewarded by being reborn in better circumstances. Reincarnation means that existence extends indefinitely in time. Over the long haul, an individual might slowly climb up the moral scale from being a worm, for example, to being a higher animal to being human. Another might slowly move downward in progressive incarnations. Since all humans are thought to have the potential for attaining enlightenment, Buddhism encourages respect for all people and preaches tolerance and kindliness toward others.

The fact that the same being might appear as a human at one point in time and as an animal at another automatically implies that animals have a moral status that is not completely distinct from that of humans. Buddhism, and other Indian traditions, involve a high level of respect for animal life that resonates with modern conservationists. Even though the early Vedic religions used animal sacrifices, these were discontinued as cruel and uncivilized. Offerings of fruit, vegetables, and milk were used instead.

Respect for animal life was carried to extremes by Buddhist monks who took precautions to avoid treading on insects and even carried a strainer with them to ensure that they did not destroy small animals in their drinking water. In general though, Buddhism does not consider the taking of animal life to be wrong unless it is intentional.

See also ANIMAL RIGHTS; DHARMA; KARMA.

buffalo (bison), slaughter of

Commonly referred to as the buffalo, the American bison (*Bison bison*) is a noteworthy example of an animal that has been hunted to the verge of extinction. It is estimated that at the time the first European settlers arrived in North America there were some 30 million bison on the grasslands between the Mississippi River and the Rocky Mountains. These large animals could maintain their numbers against the simpler weapons of the Plains Indians but were no match for the repeater rifles used by professional buffalo hunters. Some, like the legendary "Buffalo Bill" Cody, made a living as providers of meat for the workers involved in the westward expansion of the railroads. In that sense, the destruction of the buffalo was not only brought about using modern technology but also served to promote technological expansion. Wholesale destruction of buffalo herds was not only unregulated by the government but also could have served as part of a deliberate policy for clearing the plains of indigenous people by destroying one of their principal resources.

See also BIODIVERSITY AND INDUSTRIALIZATION.

Further Reading
Stone, L. M. *Back from the Edge: The American Bison.* Vero Beach, Fla.: Rourke Book Company, 1991.

Burt, Sir Cyril Ludowic
(1883–1971)
English
Psychologist

Cyril Ludowic Burt was an influential English educational psychologist whose distinguished reputation has been posthumously destroyed by the recognition that some of his key research, involving a twin study of intelligence, was a complete fabrication. Although Burt's name will forever be linked to scientific fraud, he was nevertheless a capable and productive researcher and was the first psychologist to receive a knighthood for his academic contributions. Burt was an international figure and, in 1971, he received the Thorndike prize from the American Psychological Association, which was the first time that this honor had gone to a foreigner. The author of books such as *The Young Delinquent* (1925) and *The Backward Child* (1937), Burt was admired by peers for his contributions to statistical analysis of test data using factor analysis and analysis of variance.

BURT'S CAREER
In 1908 Burt received his first academic job, as a lecturer in psychology, in Liverpool. He was hired by the London County Council in 1913 as a psychologist for the school system. In this position, he adapted French and American intelligence tests for English use. He also developed tests of educational attainment. Burt was a professor of psychology at University College London from 1931 until his retirement in 1950.

Following World War II, Burt served as an influential consultant to a series of government-appointed committees charged with restructuring the English educational system. His contribution in this role was so important that he has been referred to as the father of English education. He was influential in devising the 11+ exam. This was based on the supposition that by the age of 11 years a child's educability can be fairly assessed. Children who did well on the 11+ were streamed for advanced academic education and entry to professions. Those who did not were streamed for developing practical skills, such as trades, and entry into blue-collar occupations. The philosophy was that there was no point in wasting money on an academic education for children who were not equipped to benefit from it.

The 11+ exam proved to be enormously divisive and was seen by political liberals as perpetuating a rigid class system. At the very least, it closed the doors of academic opportunity far too early in a child's life. Burt did not agree with this view because he was strongly hereditarian. He believed that intelligence is largely genetically determined and thus almost completely fixed at conception.

After Burt's retirement in the 1950s the 11+ exam began to come under heavy fire from political opponents (and was ultimately abandoned in 1969). He published a series of studies that appeared to shore up the hereditarian position. Supposedly based on evidence that had been collected in the 1920s and 1930s, when Burt worked in the London school system, the data ostensibly included IQ scores of identical twins reared apart. This is the gold dust of hereditarian research because it provides an ideal natural experiment in which the influence of heredity can be studied with the influence of a shared environment taken away. Since the twins were raised in different environments, the conclusion could be drawn that any similarity in intelligence was due solely to genetics. What is more, Burt's sample of identical twins raised apart purported to be larger than that available to any other researcher at the time.

Burt often referred to the assistance he received from coworkers, a Miss Margaret Howard and a Miss J. Conway, who helped him to update the IQ data. Burt edited the *British Journal of Statistical Psychology* for 16 years and, during his tenure, numerous articles appeared that praised Burt and heaped scorn on his environmentalist detractors. The rhetorical style of these articles has suggested to some modern scholars that they were written by Burt himself. Some of the articles were attributed to Miss Conway.

Burt's scientific fraud is believed by Leslie Hearnshaw, his sympathetic biographer, to have begun in 1943 but others believe that all of his scientific work is tainted by dishonesty. It is remarkable that he escaped detection for fully 30 years even though his results were not only widely known but actually took center stage in the promotion of a hereditarian position on intelligence in America by leading figures such as Arthur Jensen at Stanford University and Richard Herrnstein at Harvard. At a time when the British public was turning against both hereditarianism and Burt, his work was becoming more influential in the United States. Jensen's controversial 1969 article in the *Harvard Educational Review* pursued Burt's logic by arguing that intelligence is mainly due to genetic inheritance and that liberal programs designed to promote social equality of poor African American or white children through education were destined to fail. Writing two years later, hereditarian psychologist Richard Herrnstein claimed that IQ measurement was the most glorious achievement of psychology and that Burt's twin research was a shining example of accomplished work in this field.

BURT'S POSTHUMOUS UNMASKING
The most astonishing aspect of Burt's fraud is that it was so transparent. Credit with unmasking Burt is given to Leon Kamin, a psychologist at Princeton University. Oth-

ers certainly entertained doubts but no one said anything in public. They were awed, and silenced, by Burt's eminence. Kamin had never been interested in intelligence testing until a student encouraged him to read one of Burt's papers. Kamin claims that after reading the paper for 10 minutes, he was convinced that Burt was a fraud. If this is true, it suggests that other researchers of the day read Burt's work with a minimum of scientific skepticism, if they read it at all.

Kamin was struck by Burt's poor scholarship. His papers lacked precise details about how the data were collected, what tests were administered, who collected the data, where exactly it was collected, and when. Such vagueness is not normally tolerated in scientific papers. In reviewing Burt's twin studies over the years, Kamin spotted something even more unsettling. Between 1955 and 1966, Burt reported three sample sizes for identical twins reared apart, 21 pairs, "over 30" pairs, and 53 pairs. Despite the increasing sample size, the IQ correlation between identical twins reared apart was always reported as exactly .771. This uniformity of results is highly improbable. In general, if a researcher repeats a study of this type, he or she would feel lucky to repeat even the first digit of the correlation three times. If you make the assumption that the last two digits would vary randomly, the probability of getting the same result three times would be exactly one in a million ($1/100 \times 1/100 \times 1/100$). This is bad enough. Now Kamin discovered that the correlation for the identical twins raised together was always exactly .944. The probability of this coincidence using the above assumptions is one in a trillion, or a virtual impossibility. There is no chance that the results could be correct and the numbers are explainable only in terms of an implausible sequence of errors or deliberate fraud. In his book, Kamin stopped short of calling Burt a fraud but concluded that his work was undeserving of the attention of the scientific community.

Burt was publicly exposed in an article published in the *London Sunday Times,* on October 24, 1976, by medical correspondent Oliver Gillie. Gillie's angle was that he had been unable to discover any record of Burt's alleged coworkers, Miss Conway and Miss Howard. When this is combined with the extreme improbability of the published results, the thin facade of scholarship begins to crumble and it is clear that there were no coworkers, that no data was ever collected, and that Burt fabricated all aspects of his twin studies out of thin air.

Another smoking gun was produced by Leslie Hearnshaw, a professor of psychology at the University of Liverpool, who had given the eulogy at Burt's funeral and was commissioned by Burt's sister to write his biography. Hearnshaw concluded that since Burt had no means of collecting data following his retirement in 1950, the twin data reported in 1958 and 1966 could only be fraudulent.

Burt's personal papers revealed no evidence of meetings with his alleged coworkers. They did not exist and, during this period, Burt did no research.

CONCLUSIONS

The fact that complete fabrication of a program of research, like Burt's twin studies, could pass unnoticed for over a decade allows some fairly unsettling conclusions to be drawn about the ethics of scientific research and the ability of science to detect, and correct, errors. It would be easy to say that psychology is a "soft" science in which pure rhetoric may occasionally pass for scientific practice but psychology is as rigorous a science as most others with very high rejection rates for its top journals. Moreover, there are numerous examples of equally glaring data fabrication in the "hard" sciences, particularly in the biomedical area.

Contrary to the view that scientists are focused on data and carefully scrutinize the work of colleagues to identify errors, virtually no scholar actually took the trouble to read Burt's reports carefully. If they had, they could hardly have avoided the impression that something was wrong. One wonders at Burt's carelessness, or arrogance, in presenting data that were impossibly perfect. As a sophisticated statistician, he could hardly have failed to notice the extreme improbability of his "perfect" results. This is analogous to a criminal deliberately placing fingerprints all over a crime scene. Contemporary willingness to accept Burt's unlikely results at face value cannot be explained in terms of scientists being too willing to accept data that conform to their expectations (i.e., a confirmation bias) because his opponents were as slow to catch on to the fraud as his supporters were.

Another ethical implication is that Burt's work emerged from PEER REVIEW (a) without red flags being raised about the authenticity of the research and (b) without the author being asked to comply with basic aspects of editorial style, such as describing the procedure in enough detail to allow systematic replication. One wonders why. Is it possible that the editors were so pleased to publish Burt's work that they never even bothered to send it out for review? If it was reviewed, were the reviewers so overawed by Burt's name that they forgot to subject his work to the kind of critical analysis encountered by other authors in the field? It seems obvious that Burt must have manipulated the editorial review process in having his own articles published while he served as editor.

Whatever the answers to these questions, the Burt case is a clear example of the inability of the scientific community to police itself. Not only does science not emerge as the kind of impartial and objective process that it is supposed to be, but a forger of Burt's eminence can use the semblance of scientific method as a rhetorical device to further his own deep-seated prejudices and beliefs. These are uncomfortable truths for most scientists. Some would argue that the Burt case is an aberration, that fraud in science is rare, and that undetected data fabrication and other types of scientific dishonesty generally get by in dark shadows inhabited by obscure figures. Yet the list of probable cases of scientific fraud includes scientists of the caliber of PTOLEMY, GALILEO, NEWTON, DALTON, and MENDEL.

See also PEER REVIEW; PLAGIARISM; SCIENTIFIC DISHONESTY; SCIENTIFIC METHOD.

Further Reading

Broad, W., and N. Wade. *Betrayers of the Truth.* New York: Simon and Schuster, 1982.

Hearnshaw, L. *Cyril Burt: Psychologist.* London: Hodder and Stoughton, 1979.

Mackintosh, J., ed. *Cyril Burt: Fraud or Framed.* New York: Oxford University Press, 1995.

Cable Communications Policy Act of 1984

The Cable Communications Policy Act of 1984 was designed to deregulate the U.S. cable television industry. Regulation of rates paid by subscribers to cable companies, which had been governed by local authorities, was abolished. The expectation was that competition among cable companies would reduce rates. In fact, existing providers became established as monopolies and took the opportunity to raise rates. To correct this problem, Congress enacted the Cable Television Consumer Protection and Competition Act of 1992. One important measure designed to increase competition in the industry was a provision requiring cable programmers to provide their product to other delivery services, including satellite and wireless. Exclusive local franchises in which only one cable company was licensed to operate were also banned. Competition in electronic media is not only important for ensuring affordable services but also plays a role in improving the quality of television programming. Monopolies also stifle freedom of expression by controlling the content of television programs.

See also PORNOGRAPHY.

Callicott, J. Baird

(1941–)
American
Philosopher

A founder of environmental philosophy, J. Baird Callicott argued that modern environmental problems call for a radical reanalysis of the place of human beings in the natural world. He concluded that the natural world of evolved ecosystems has an inherent moral value that transcends the moral claims of our species. This position in ENVIRONMENTAL ETHICS is referred to as nonanthropocentric value theory. Other environmental ethicists have questioned whether a nonanthropocentric position can be logically coherent. Callicott's most influential book *In Defense of the Land Ethic* (1989) emphasized the need to protect the environment from the consequence of human industrial production.

See also ENVIRONMENTAL ETHICS.

Further Reading

Callicott, J. B. *In Defense of the Land Ethic.* Albany: State University of New York Press, 1989.

Carson, Rachel

(1907–1964)
American
Ecologist and Writer

Carson published both popular and scientific work on ecology and environmental issues. Her reputation as an author of accessible scientific books was established in

1951 with the bestseller *The Sea Around Us.* Her most influential book was *Silent Spring* (1962). At the time, it was the subject of virulent attacks by readers who could not accept its bleak message that widespread use of pesticides, particularly in America, was destroying ECOSYSTEMS.

See also DDT; ECOSYSTEMS; ENVIRONMENTAL ETHICS.

castration

Castration usually involves the surgical removal of the testicles of a male animal or human. A similar result can be obtained by destruction of the vas deferens through which the testicles transmit their secretions. Male farm animals are castrated to make them more docile. Pets may be castrated to stop them from reproducing and to prevent straying. Castration of females involves removal of the ovaries, an operation that is usually referred to as spaying.

Castration of human males is an ancient practice that was common in imperial China as well as the empires of Greece, Rome, and Turkey. Castrated men, or eunuchs, served in the courts of emperors and princes and their enforced sterility prevented them from being reproductive competitors of the rulers who often maintained harems containing several hundred wives. Eunuchs sometimes assumed considerable power and status within the courts.

In Europe, from the 16th to 18th centuries, male singers were sometimes castrated before puberty to prevent the normal thickening of the vocal cords and consequent deepening of the voice that occurs at sexual maturity. Some of these castrati were recognized as the finest opera singers of their day and enjoyed considerable prestige. They were also employed in Italian church choirs.

Today, castration is performed mainly for medical reasons. The operation may be performed either in the case of diseases of the testicle or because of prostate cancer. Early in the 20th century, many men were castrated and many women were sterilized for eugenic reasons. It is estimated that some 70,000 people in American institutions were sterilized for eugenic reasons. The timing of this unethical practice was determined both by the rise of SOCIAL DARWINISM and the development, at the end of the 19th century, of medically safe methods of sterilizing men and women.

The compulsory sterilization of criminals, the insane, and people judged to be of low intelligence is based on shaky assumptions concerning the genetic determination of these traits and the right of governments to curtail reproductive liberty and conduct harmful medical invasions of the bodies of citizens. Sterilization of institutional populations in the United States was undertaken not just to prevent reproduction of individuals considered "defective" but also to control their sexual motivation, particularly in the case of males. Castration not only sterilizes men but also interferes with their sexual function. Removing the testicles reduces circulating testosterone levels, which can reduce sexual desire and performance.

See also CASTRATION, CHEMICAL; EUGENICS; SOCIAL DARWINISM.

Further Reading

Areen, J. "Limiting Procreation." In *Medical Ethics,* ed. R. M. Veatch, 103–34. Sudbury, Mass.: Jones and Bartlett, 1997.

castration, chemical

Chemical castration is an inaccurate term because there is no permanent alteration of reproductive function as in the case of surgical castration. Chemical castration usually takes the form of a weekly injection of DEPO PROVERA, a birth-control drug ordinarily used on women, that reduces sexual desire in men and is used to decrease recidivism (reoffense) rates of convicted sex offenders. Depo Provera lowers the blood testosterone level of treated men and this tends to reduce the compulsive sexual fantasies that are acted out when sex crimes are committed by paraphiliacs such as exhibitionists and child molesters. Chemical castration has no permanent effects that persist if treatment is withdrawn. Side effects, including headaches and nausea, occur very rarely and are reversed by stopping the injections. Depo Provera treatment is a useful method for allowing some types of sex offenders to lead productive lives outside prison but it raises many ethical questions about the desirability of governments controlling the thoughts and actions of individuals through pharmacological interventions.

California was the first U.S. state to provide a legislative mandate for chemical castration. The 1996 California law mandated the use of chemical castration as a precondition for parole in the case of repeat sexual offenders of all types. In 1997, three other states passed similar laws so that a total of four states currently permit chemical castration. Research has shown that Depo Provera does reliably reduce sexual desire in men although it has no impact on sexual desire in women and therefore cannot be used in the case of female sex offenders. Chemical castration greatly reduces recidivism rates of paraphiliacs from over 90% to approximately 2%. This means that a chemically castrated parolee need not be considered as any more of a threat to victim populations, including children, than the male population in general.

Many ethical and legal issues are raised by chemical castration. Some legal scholars might argue that castration is a cruel and unusual form of punishment contrary to the Eighth Amendment to the U.S. Constitution. The most effective counterargument to this claim is that Depo Provera treatment is not a punishment at all but rather a form of medical therapy. The claim that it is a therapy rests on the finding of considerably reduced sexual reoffense rates.

It is true that the U.S. Constitution allows people to refuse medical treatment but this right is not undermined by the chemical castration laws because, in a narrow sense, the treatment is voluntary. Prisoners can reject the provisions of their parole, including chemical castration, and decide to serve out their sentences in prison.

Civil libertarians have been upset by the possibility that chemically castrated sex offenders are being denied the right to reproduce because of the sexual apathy that follows Depo Provera treatment. Yet, this concern may be misplaced because parolees receiving Depo Provera are capable of having normal sexual relations and fathering children. Their sexual desire may be considerably reduced but it is not eliminated. Another concern is that Depo Provera reduces the frequency of certain types of sexual fantasy and therefore infringes upon freedom of thought and expression as guaranteed in the First Amendment. One way around this problem is to argue that freedom of expression is not an absolute right but one that is constrained by considerations of the common good. By this reasoning, the compulsive sexual fantasies of a child molester would not constitute protected expression. A more ingenious, but less convincing, argument is that by reducing obsessive fantasy with Depo Provera treatment, the mental life of the sex offender is rendered freer to engage in other interests.

Chemical castration provides an instance in which the biological control of antisocial impulses can be ethically justified. This raises a more general question about whether it is desirable to control other types of antisocial impulse by similar means, even among the noncriminal population. For example, the high level of aggression and risk-taking among young men is arguably linked to their high testosterone levels. Should Depo Provera injections be mandatory for young men who want to drive? If not, why is testosterone-induced aggressive driving different from testosterone-induced sexual offenses? By a similar logic, could recidivism for crimes of violence be reduced using either Depo Provera or serotonin reuptake inhibitors, such as Prozac, since there is good reason to believe that each of these manipulations might control impulsive violence? If any of these measures ever becomes law, then the legalization of chemical castration might come to be seen as a slippery slope, which resulted in disturbing erosions of personal liberties.

Other concerns, such as the lack of privacy implied by having weekly Depo Provera injections, are less consequential because this is far less invasive than the sex-offender registration and notification practices that have been adopted in some states. It can also be argued that this is a small price to pay for the benefits of chemical castration. The major benefit to the public is that they are protected from the adverse effects of sexual offenses without having to bear the burden of costly prison incarceration. The main benefit to the offender is that he can overcome the compulsion to commit a serious crime and can therefore live a productive life outside prison.

The chief weakness of current laws is that they do not do enough to promote long-term rehabilitation of sex offenders. At present, with the end of the sentence, Depo Provera injection ends and the probability of reoffense increases. It is important that offenders should be given the option of continuing treatment. There is also a role for psychological counseling to help offenders overcome their compulsions before Depo Provera treatment ends.

See also CASTRATION; CRIME, AND BRAIN CHEMISTRY; DEPO PROVERA.

Further Reading
Meisenkothen, C. "Chemical Castration—Breaking the Cycle of Paraphiliac Recidivism." *Social Justice* 26 (1999): 139–54.

catalytic converter
Developed in the 1970s, catalytic converters promote the combustion of unburned hydrocarbons in car exhaust, and the chemical reduction of nitrous oxides, thereby reducing AIR POLLUTION. These chemical reactions are promoted by metal catalysts.

See also AIR POLLUTION; CLEAN AIR ACT.

catastrophism
Catastrophism is the view that geological features of the earth, including physical features like mountains, and the fossil record of life forms were affected by brief violent events, or cataclysms. Catastrophists were largely religious and were disposed to see cataclysms as having a supernatural origin. Since most accepted the biblical view that the earth was only a few thousand years old, it was difficult to understand how physical features like mountain peaks and canyons could have been formed by natural forces and their existence was thus rationalized as being due to cataclysms.

A leading light of catastrophism was French comparative anatomist George Cuvier (1769–1832). Cuvier studied fossils in the Paris basin. He produced two striking findings. The first was that animals had once lived that bore no relationship to anything currently alive. The second was that there are distinct breaks, or nonconformities, in rock layers. Each layer was associated with a marked difference in fossil content. Cuvier drew the conclusion that the life forms in each layer had been wiped out by a sudden cataclysmic event. Following their destruction, they had been replaced by either a new creation or the inward migration of species from unaffected regions.

Catastrophism only seemed reasonable if the age of the earth was equivalent to biblical indications. Once geologists realized the immense age of the earth's surface and the fossil life it contained, they realized that a gradualist explanation of the fossil record was possible. However weak the forces of natural change might be, given sufficient time, they could explain what seemed like the sudden transitions between geological strata. By the end of the 19th century the gradualist view, referred to as "uniformitarianism" was accepted by virtually all scientists and the catastrophist interpretations of the fossil record was abandoned.

Even though catastrophism was ultimately defeated in the 19th-century evolution debates, it presented some serious opposition to evolutionists like CHARLES LYELL and CHARLES DARWIN. One issue that was particularly troubling for the evolutionists was the argument in favor of a progressive series in the fossil record, that is, the sequence fish, reptiles, mammals, that suggests that the new life following cataclysmic extinctions was ascending from simpler to more complex forms. This evidence was so bothersome to Charles Lyell that he even denied the existence of any such progressive pattern. Evidence of straight-line progress is superficially contradictory to evolutionary change as described by Darwin's theory of random variation and natural selection. What may look like progress can also be attributed to a sequence of evolutionary changes in which a small number of relatively simple species gives rise to a large number of more specialized organisms (adaptive radiation). The evolutionary tree of speciation may look progressive but careful analysis of evolutionary sequences reveals that they can branch in any direction. Thus, having evolved from reptiles that evolved from fishes, some mammals, such as seals, and dolphins have returned to an aquatic life. Naked mole rats have not only lost their complex vision but have evolved a social system that is seen in no other mammals and resembles an ancient type of sociality found in insects like termites and wasps.

When Darwin wrote *On the Origin of Species* there were troubling gaps in the fossil record. Soon after its publication in 1859, some important fossil discoveries were made that supported uniformitarianism and cast doubt on catastrophism. One was the discovery of *Archaeopteryx*, a flying reptile that had feathers like a bird and thus supported the argument that modern birds evolved from reptiles. American paleontologists also suc-

ceeded in constructing an evolutionary series in the early evolution of horses that showed how modern grazing horses had evolved from leaf-browsing ancestors over tens of millions of years.

While catastrophism has been rejected as a general description of change over geological time, there are specific episodes for which a catastrophic account is plausible. Thus, the recent theory that the decline of the dinosaurs was hastened by the collision of an asteroid with the earth seems reasonable and is supported by converging lines of evidence, including the location of a crater in Mexico possibly caused by the collision and the detection of iridium (presumed to be of extraterrestrial origin) in the appropriate geological level at different points around the earth. Referred to as "neocatastrophism," these theories do not presume supernatural agency in the disappearance of species from the geological record or in the emergence of different ones to take their place.

See also CREATIONISM; DARWIN, CHARLES; DARWIN'S FINCHES AND SPEED OF EVOLUTION; EVOLUTIONARY ETHICS; EVOLUTIONARY THEORY AND EDUCATION CONTROVERSY; OWEN, RICHARD; WILBERFORCE, SAMUEL.

Further Reading
Ager, D. V. *The New Catastrophism.* New York: Cambridge University Press, 1993.
Huggett, R. *Catastrophism.* New York: Verso, 1997.
Ruse, M. *Darwinism Defended.* Reading, Mass.: Addison Wesley, 1984.

Cavalli-Sforza, Luigi, and human genetic variation   *See* SCIENTIFIC RACISM.

censorship   *See* PORNOGRAPHY.

Center for Democracy and Technology
The Center for Democracy and Technology is a nonprofit organization that seeks to enhance open communication on the Internet in the interest of democracy. The Washington, D.C.–based organization also aims to protect the privacy of Internet users.

See also ELECTRONIC PRIVACY; OPENNESS AND SCIENCE.

Center for Science in the Public Interest
The Center for Science in the Public Interest is a consumer watchdog group dedicated to improving food qual-

ity and safety. Founded in 1971, the Washington, D.C.–based organization has lobbied successfully for detailed labeling of food products that allows consumers to make more informed decisions about what they eat. It has also paid a great deal of attention to the quality of restaurant foods and has conducted tests that reveal a very high fat content of many fast foods, for example. These efforts have encouraged many restaurant chains to add more low-fat options to their menus. Misleading ads by food companies such as McDonald's, Campbell Soup, and Kraft have also been targeted with success. The center is dedicated to reducing the use of drugs, such as alcohol and tobacco, that are known to have adverse health consequences and has spoken out against research that seems to promote the health benefits of drinking red wine. Originally, the center was run primarily by scientists but the constitution of its board has shifted to favor nonscientists and it has become more of an advocacy group than a scientific organization.

See also FLUORIDATION; FOOD IRRADIATION.

Challenger accident

The space shuttle *Challenger* exploded soon after launch on January 28, 1986. Seven people, six astronauts and a teacher, died in the explosion. This tragic accident followed a long chain of technical errors associated with a faulty seal on the booster rocket, known as an O-ring. The O-ring had leaked, due to cold temperatures, allowing the booster's rocket fuel to ignite and explode. Morton Thiokol, the company that made the boosters, was aware of the problem according to the testimony of engineers before the presidential commission investigating the *Challenger* accident. It would have been possible to make new O-rings that were safer at low temperatures but this would have been very expensive. Since low-temperature launches were not anticipated, the O-ring problem did not appear to be a real threat. The day before the fatal launch, it was estimated that the O-rings would leak at 50 degrees Fahrenheit and Morton Thiokol engineers wanted to postpone the launch because of cold weather. According to their testimony, they were overruled by the company and by NASA (National Aeronautic and Space Administration) management. The *Challenger* accident illustrates the ethical implications of engineering problems. Partly as a result of such failures, there is a growing awareness of the ethical responsibilities of engineers and engineering ethics has emerged as a new subdiscipline.

See also ENGINEERING ETHICS; FORD PINTO.

Further Reading

Resnik, D. B. *The Ethics of Science.* New York: Routledge, 1998.
Schlossberger, E. *The Ethical Engineer.* Philadelphia: Temple University Press, 1993.

Westrum, R. *Technologies and Society.* Belmont, Calif.: Wadsworth, 1991.
Whitbeck, C. *Ethics in Engineering Practice and Research.* New York: Cambridge University Press, 1998.

Chem-Bio Corporation and pap smear controversy

Chem-Bio Corporation of Oak Creek, Wisconsin, was charged with reckless homicide in 1995 when two women died of cervical cancer after their pap smears were misread. Had the pap smears been correctly read, identifying the cancer in its early stages, the women would have had a 95% chance of survival, according to medical testimony. The case illustrates the importance of careful work by scientific technicians in the medical field. As applied scientists, they have the same obligation to doing careful research as academic researchers. Ironically, failure to detect disease can also mean that a fetus that normally would have been aborted is carried to term, as in the case of a pregnant Frenchwoman whose German measles went undetected, resulting in the birth of a severely disabled child, Nicholas Perruche, the subject of a successful 2001 damages suit against the responsible doctor and laboratory.

Indictment of Chem-Bio recognized that the quality of work by technicians is to a large extent determined by working conditions. In the Chem-Bio case, it seems that speed and cost effectiveness were put at a premium over scientific accuracy. The single Chem-Bio technician read an average of 179 slides per day even though the American Society for Cytology guidelines suggest that technicians should not read more than 100 slides per 24 hours. Retests showed that error rates were unusually high. The corporation pleaded no contest to the reckless homicide charges. In 1988, the Department of Health and Human Services introduced new regulations to promote tighter control of pap smear screening and other laboratory procedures known as Clinical Laboratory Improvement Amendments. Partly in response to the tighter regulations, new computerized screening technologies have been introduced that increase the reliability, and reduce the labor, of reading pap smears. These can be used either to isolate slides that are most likely to contain cancerous cells or to reread slides in search of human errors.

See also COMMERCIALISM IN SCIENTIFIC RESEARCH.

Further Reading

Resnik, D. B. *The Ethics of Science.* New York: Routledge, 1998.
Voelker, R. "Milwaukee Deaths Reignite Critical Issues in Cervical Cancer Screening." *JAMA* 273 (1995): 1559–60.

Chernobyl nuclear accident

On April 26, 1986, the Chernobyl nuclear reactor, located near the town of Pripyat, 80 miles north of Kiev, Ukraine, exploded, pouring eight tons of radioactive material into the atmosphere. This is estimated to be about 200 times as much radioactivity as was released in the HIROSHIMA and NAGASAKI bombs combined and roughly equivalent to the fallout produced in all tests of nuclear weapons. Immediate loss of life was much smaller than for the atomic bombs and was dwarfed by the BHOPAL, India, chemical accident. The immediate death toll at Chernobyl was approximately 31, according to official statistics, compared to some 2,500 at Bhopal.

Most of the ill effects of Chernobyl were delayed and took the form of increased incidence of cancers and other diseases. Wide dispersal of the radioactive dust around the globe means that the health and environmental effects are being detected in places as far distant as the United States and Britain. Given that some of the radioactive isotopes have long half lives, the impact will be felt for many decades to come. It is extremely difficult to estimate the number of premature deaths that will be due to the Chernobyl accident but many epidemiologists speculate that it will be of the order of hundreds of thousands.

The Chernobyl accident raises many thorny ethical questions:

Could adherence to improved safety protocols have prevented the meltdown?

The fact that the Chernobyl reactors were widely believed to be safe raises the issue of whether the Ukraine authorities had been fully informed by the engineers at the reactor site. If so, did the authorities inform the public about the extent of the threat to their lives and health?

Given the magnitude of the disaster, it is debatable whether the benefits of NUCLEAR POWER can ever justify this risk. Should poor countries like the Ukraine, which have a great need for energy because of long cold winters, be entitled to take a greater risk with the lives of citizens?

What about the ethics of American companies that profit from building reactors that turn out to have safety problems? Is it ethical to build new reactors in view of the demonstrated high level of risk at Chernobyl and in the case of the THREE MILE ISLAND NUCLEAR ACCIDENT?

After the meltdown, was it ethical to send in cleanup workers who were exposed to dangerously high levels of radiation?

Given that nuclear accidents have global consequences, can anyone be held morally or legally responsible for redressing this harm? If not, can the construction of nuclear power plants ever be ethically justified?

Given the global consequences of a nuclear accident, should nuclear power plants be regulated by the same kind of international treaty as applies to nuclear weapons?

These questions do not have easy answers but the student of applied ethics has much to gain by pondering them in the case of the Chernobyl accident.

WHAT WENT WRONG AT CHERNOBYL?

Advocates of nuclear power had long claimed that it was safe as well as sustainable. The fact that serious accidents have occurred at Chernobyl and Three Mile Island, in Pennsylvania, and that there have been many other safety problems at other reactors around the world does not necessarily mean that nuclear power cannot be safe. The safety of nuclear reactors is complex. To begin with, it is an issue for the engineers and scientists who design the plant. Are some designs currently in use flawed by safety problems that could have been prevented by a better design? Second, there is the question of whether the safety and maintenance procedures are adequate to prevent serious mishaps. Is the instrumentation adequate to allow engineers to spot problems in advance and are personnel sufficiently vigilant in looking out for signs of trouble? How much latitude should engineers have in modifying safety procedures when some system malfunctions or when there is an exceptionally high demand for power? Finally, there is the human side of the equation. How vulnerable is the operation of the reactor to human error? How open is it to the possibility of sabotage by a political terrorist, a common criminal, or a deranged person?

The disaster at the Chernobyl reactor has been variously attributed to human error and to design problems. It is probably safe to assume that both factors played a role. The accident occurred in the context of an experiment. The Chernobyl plant's No. 4 reactor was running with the emergency water-cooling system turned off. A sequence of misjudgments permitted neutron buildup in one area of the core. The nuclear reaction went out of control, which caused the fuel to disintegrate. Then there was a steam explosion that blew the lid off the reactor. The massive containment structure had not been designed to withstand this sort of pressure. A chemical explosion followed that rained burning fragments around the plant. The force of the nuclear reaction and fire propelled most of the radioactive material high up into the atmosphere.

The explosion was so powerful that the radioactive dust was dispersed over much of the Northern Hemisphere, finding its way all around the globe. A blanket of radioactive material spread out over Belarus, Russia, Scandinavia, Western Europe, and the United States. In America, radiation monitors were set off from New England to California. The fallout contaminated huge quantities of agricultural products in most European countries. In Finland, Lapps were told that it was unsafe to consume the meat of their reindeer. The amount of contaminated grass sheep were allowed to eat was restricted in Britain and consumption of the meat was also tightly regulated. In some areas, there were devastating consequences for wildlife. For example, the rate of bird hatching fell 62% at Point Reyes Station, north of San Francisco, reflecting adverse effects of the radioactive particles coming down with the spring rains. Even though much of the radioactive cloud was dispersed into the upper atmosphere, the region surrounding the nuclear reactor was the area most heavily contaminated, with devastating consequences for the local population and for cleanup workers.

THE AFTERMATH OF CHERNOBYL

Dispersal of radioactive dust in farmland in the vicinity of the Chernobyl plant produced a great deal of contamination of agricultural products. Farm animals were also adversely affected by the radiation, which produced widespread sickness and deaths and showed up in an epidemic of deformed newborns. One collective farm 30 miles from the plant, which maintained 400 cows and pigs, reported that there were 140 mutated births in the two years following the accident compared to only three in the previous five years. Some of the newborns lacked limbs, eyes, ribs, or even heads. Others were born with two heads. A similar rash of newborn deformities was observed in the case of the THREE MILE ISLAND partial meltdown in 1979 in Pennsylvania. Exposure of farm animals to radiation meant that some staple food products, such as milk, had to be avoided. Others, such as potatoes, had to be eaten because of a shortage of imported food.

The Chernobyl meltdown spewed radioactive dust over a vast area. The most severely affected region, with a radioactivity level in excess of one Curie per square kilometer was home to over four million people. The region included areas in Belarus (60%), the Ukraine (30%), and Russia (10%). The Soviet authorities are estimated to have evacuated some 116,000 people from the most heavily contaminated area within a 30-kilometer radius. Their policy, which was continued in the post-Soviet era, involved urgent evacuation of people in areas with greater than 15 curies of radiation and less urgent evacuation for areas with 5–15 curies. Evacuation of areas with 1–5 curies was not done unless it was impossible to obtain uncontaminated food and water.

Radioactive substances released in the blast included iodine 131, cesium 137, and strontium 90. Cesium and strontium stimulated the most initial concern because of their longer half lives (which means that their toxic effects will be present for many years to come) but iodine was the most widely dispersed and may have had the most serious consequences in terms of thyroid problems in children.

Health consequences for individuals were most severe for those close to the plant who experienced high levels of radioactivity. Most seriously affected were the "liquidators" who came in to stabilize the reactor following the meltdown. Most of these workers were army reservists who did not volunteer for the job. Estimates of the number of workers exposed to extremely high levels of radiation in the vicinity of the plant range from 600,000 to 800,000, a staggering number of people to do such dangerous work.

The cleanup workers were most exposed to radiation while attempting to seal the gaping hole in the roof of the reactor where the 2,000-ton lid had been blown clear. Wearing heavy shields, workers flung shovelfuls of graphite into the hole before retreating to relative safety. The whole area was so radioactive that Geiger counters went off the scale.

Perceptions of risk were much lower for other parts of the 30-kilometer exclusion zone surrounding the damaged reactor. Workers remained in the exclusion zone for as long as six months. There was a cavalier attitude to safety. A photographer for the German Magazine *Stern* snapped workers sitting on the contaminated ground eating their lunch. They did not even have protective clothing or headgear and the roofless reactor loomed behind them as they ate their sandwiches.

Estimation of the number of deaths of decontamination workers have ranged from 5,000 to 8,000. The National Committee for Radiation Protection of the Ukrainian Population estimated that there were 5,722 deaths of cleanup workers. In addition, local deaths of plant workers, Pripyat residents, farmers, and coal miners is estimated at approximately 100. Although not all of the deaths of cleanup workers are attributable to radiation sickness, because some would have died of other causes if the nuclear accident had never happened, most are clearly attributable to being exposed to dangerously high levels of radiation for several months.

The indirect effects of radiation dispersed through the atmosphere over long distances are much more difficult to pinpoint and quantify. Epidemiologists often resort to theoretical estimates of increased disease risk based on intensity of radioactive exposure. According to one such estimate, Chernobyl may have increased deaths from cancer by 30,000 over the following 50 years, approximately half of these deaths being expected outside the former

Soviet Union. Other estimates suggest that this many thyroid cancers may result for children alone in the most radiation-exposed country, Belarus.

The United States is believed to have suffered a four-month radiation epidemic during the summer of 1986. In the month of June, there was a 13% increase in infant mortality nationwide. Chernobyl has been linked to increases in thyroid cancer among children in Connecticut, Iowa, and Utah. The immature thyroid gland is particularly vulnerable to radioactive iodine, which has led to the suggestion that children exposed to radiation in Europe be given iodine tablets to minimize the chance of radioactive iodine being taken up by their thyroids. Radiation-linked hypothyroidism has been observed in babies born in the Pacific Northwest of the United States. Rates of thyroid cancer among children in some areas of Belarus are 200 times what they were before the accident.

Other types of cancer are elevated among those exposed to high doses of radiation. For example, among the 32,000 evacuees in Belarus, the lung cancer rate was four times higher than that of Kiev. The rate of congenital defects in Belarus increased by 24% from 1987 to 1992. In areas with high levels of cesium contamination (above 15 curies per square kilometer), the congenital defects increased 83%. These defects likely reflect damage to chromosomes caused by radiation. Chromosomal abnormalities have been observed in all the affected regions.

Alarming as these reports about direct effects of radiation exposure on specific diseases, some of the less specific effects on health are just as alarming. One study of 56,000 decontamination workers found that 80% of them had an impaired immune response, 40% had impaired hearing, and a third had experienced loss of libido. It is unclear whether any or all of these effects are attributable to radiation. They might also be related to psychological stress.

ETHICAL ISSUES OF LIABILITY AND SAFETY

The Chernobyl accident clearly caused devastating illnesses, deaths, and trauma for untold thousands, or even millions, of people that gives it the dubious distinction of being the world's worst industrial accident. Who is responsible for this damage? Are the responsible parties ethically, or legally, obliged to provide compensation to survivors?

It seems clear that the former Soviet Union bears much of the moral responsibility for the accident. Nevertheless, it is unclear how much the political authorities knew about the dangers posed by the plant. In the year preceding the accident, the Ukrainian authorities had boasted about how safe the plant was. The sincerity of this claim is questionable given the fact that three of the other reactors continued to be operated at the site even following the accident, which indicates that the need for energy may have been strong enough to override most safety considerations. It was only after several more incidents at the remaining reactors that the Ukrainian parliament finally decided to close the plant.

Immediately after the explosion, the authorities sought to minimize damage to the civilian population by evacuating those at greatest risk. Their failure to inform citizens of the levels of radiation at different locations and to provide adequate information about potential health problems can be considered a major ethical lapse of responsible government but it was entirely consistent with the secretive inclinations of the Soviet government. Even more regrettable was the decision to send hundreds of thousands of citizens into harm's way in the context of efforts to stabilize the reactor. It seems obvious that these workers not only had little choice about performing this dangerous work but also that they were poorly informed of the risks, had inadequate training, and in many cases did not even have rudimentary protective clothing. Since the Soviet Union no longer exists, it is difficult to see how it could be held legally liable for these abuses, although some might argue that the autonomous national governments that emerged with the collapse of the Soviet Union took on their local share of these responsibilities.

Assuming, for the purpose of argument, that the Soviet authorities believed Chernobyl to be safe, the primary responsibility for the accident would lie in the hands of the engineers running the plant. For whatever reasons, they had abandoned recognized safety protocols and were playing, not with fire, but with thermonuclear catastrophe. Even though they were putting their own lives on the line, they evidently took a calculated gamble with the lives of thousands of other people and lost. Many reactors in the former Soviet Union have been modified so that it is no longer possible for operators to override safety systems.

Apart from the bad judgments of on-site engineers, there is a more fundamental issue relating to the safety of nuclear reactor designs. The Chernobyl reactor was built to a defective design that would not have been approved in Europe or the United States and this view has not been disputed by Soviet engineers, or even by the Russian authorities who have cooperated with the United Nations in a project to improve the safety of Chernobyl-type reactors still operating in the former Soviet Union. Among other issues, it was vulnerable to overheating of the water resulting in a steam explosion that was the root cause of the accident. It also lacked an adequate containment building.

Advocates of nuclear power as a safe energy source claim that properly designed plants that use proper safety procedures need not pose a threat. Chernobyl has been the most serious nuclear accident in terms of the amount of radiation released and the continuing instability of the plant. It has had a decisive effect on public opinion in many industrialized countries, which has shifted against further development of nuclear power.

See also BHOPAL CHEMICAL ACCIDENT; THREE MILE ISLAND.

Further Reading

Balter, M. "Chernobyl's Thyroid Cancer Toll." *Science* 270 (1995): 1758–59.

Flavin, C. "The Case against Reviving Nuclear Power." In *Environmental Ethics,* ed. L. P. Pojman, 408–13. Belmont, Calif.: Wadsworth, 1994.

Marples, D. R. "The Decade of Despair." *Bulletin of the Atomic Scientists* 52, no. 3 (1996): 22–31.

Norman, C., and D. Dickson. "The Aftermath of Chernobyl." *Science* 233 (1986): 1141–43.

Wasserman, H. "In the Dead Zone." *The Nation,* April 29, 1996.

Wilson, R. "More on Chernobyl." *Environment* 38, no. 5 (1996): 3–5.

chimpanzee language-learning controversy

The 1970s was an exciting period in behavioral research on chimpanzees. Our closest evolutionary relatives were revealing astonishing capacities for abstract problem-solving, seemed to be self-aware, and evidently had the capacity to learn a version of American Sign Language according to the work of Beatrice and Allen Gardner with a female chimpanzee named Washoe. Raised in the Gardners' home, Washoe had apparently mastered 132 signs by the age of five years following four years of intensive training.

What looks like intelligent behavior on the part of laboratory animals sometimes turns out to be considerably less, as Oscar Pfungst discovered of CLEVER HANS, the horse who stunned European audiences by coming up with the correct answers to mathematical problems. Pfungst clearly showed that Hans had no mathematical ability whatever and simply responded to cues unconsciously given by his trainer.

The Oscar Pfungst of ape sign language was Herbert Terrace. Terrace raised a chimpanzee satirically named Nim Chimpsky (after Noam Chomsky, a linguist and proponent of the view that much of our language ability is built into our brains at birth and therefore unique to our species). Nim was a good sign language student, who acquired about 125 signs and apparently understand many more after four years of training. This good result was produced despite the fact that the labor of instruction was distributed among over 60 trainers, which inevitably disrupted the social environment of the young chimp. Terrace's critical nay-saying research came in the form of a videotaped analysis of interactions between Nim and her trainers. He discovered that about 40% of Nim's sign language utterances were straight repetitions of signs that had just been produced by the trainer. Only 12% of Nim's signs were spontaneous in the sense of not being directly provoked by a question from the trainer. Many of these were requests for food, suggesting that the signs were merely conditioned responses, which had been reinforced by receiving food. Terrace concluded that the performance of signing chimpanzees leaves a real question mark over whether they can construct meaningful sentences.

Although some ape language researchers have questioned the fine details of Terrace's training procedures, his findings had the same sort of ice-water effect on ape sign language as Oscar Pfungst's had on horse mathematics. Despite the setback, ape language researchers soldiered on, improving their procedures and ultimately vindicating their blind faith in the communicative abilities of chimpanzees.

One effective response to the Clever Hans problem is to use a computer, which takes the human trainer out of the training and testing situation and thus eliminates unconscious cueing. The first computer literate chimpanzee was Lana, a resident of the Yerkes Primate Center in Atlanta, Georgia. Lana pushed panels representing English words to construct simple sentences. The panels were lit up as a sentence was being written. An identical panel was illuminated in the experimenter's room that allowed progress to be monitored. Lana could request food or drink, which was automatically dispensed by the computer. The chimpanzee could also ask for her window shutters to be opened or closed. Lana also controlled the automatic delivery of toys, music, and even movies. She could send messages to the experimenter. The experimenter could communicate with Lana by having panels light up in her room.

Another advantage of the computer was that it only responded to requests for food that obeyed the rules of English grammar. Lana learned to be sensitive to word order in a way that signing apes were not. She constructed some fairly sophisticated sentences, although they were far from English prose: "Lana want drink milk eat bread." Moreover, the computer kept an objective record of all communication that took the guesswork out of interpreting videotapes that had been used as a scientific record for signing apes. Computers are also completely consistent (at least when they work). They operate around the clock as constant companions and recorders without ever getting tired. Under the more controlled conditions of this experiment, Lana produced a performance that was every bit as impressive as that of the signing chimpanzees. In addition to this more rigorous support for the view that chimpanzees can be trained to communicate using an artificial language, it was discovered that pygmy chimpanzees (or bonobos) may spontaneously learn to understand a human language.

In the 1980s, Sue Savage-Rumbaugh and Duane Rumbaugh were attempting to teach a female bonobo named Matata to communicate by pressing symbols on a keyboard. Matata's infant son Kanzi exceeded the language competence of his mother simply by observing efforts to train her. The experiments also had an uncanny feeling that Kanzi understood English. When someone asked for a light to be turned on, Kanzi was first to the switch and flipped it on. At the age of five years Kanzi understood 150 English words according to rigorous scientific tests, some using synthesized electronic voices. Kanzi not only recognized English words, he could respond correctly to requests such as: "Go to the refrigerator and take out the cabbage." Other pygmy chimps have accomplished similar feats.

Why bonobos have so much more facility than common chimps in learning language is an intriguing question for which there are only speculative answers. Bonobos are closer to humans in terms of their social organization. Thus, there are pair bonds between males and females as opposed to the segregation of the sexes typical of common chimps. Natural selection may have favored a capacity for communication between mates. Yet there is no evidence that bonobos use language in their natural habitat.

Another species that shows evidence of spontaneous ability to acquire English, the African gray parrot, is also a pair-bonded species. As pets, these birds are interesting but difficult. They tend to be deeply involved with their owners and often resent the intimate relationships that the owner might have with fellow humans. One parrot, named Alex, who has been trained by Irene Pepperberg, conducts simple English conversations about his toys. He responds correctly to questions such as "What color is the circle" by saying "blue." Linguists have trouble agreeing that these monosyllables can be considered language. Even if they are not, Alex's performance provides important evidence for the cognitive abilities of nonhuman animals.

Language ability of chimpanzees, coupled with intelligent problem-solving and self-awareness, puts them and other great apes, particularly orangutans, in a special ethical category that in the past was reserved exclusively for human beings. To say that humans deserve special consideration simply because they are human, and quite independently of their cognitive abilities, is to leave oneself open to the charge of SPECIESISM. Admitting great apes to the small club of highly intelligent primates creates a number of difficult ethical problems about how they should be treated by researchers. Is keeping apes in captivity morally analogous to enslaving human beings? If chimpanzees must be maintained in captivity, how should they be housed and managed by researchers to ensure their comfort and happiness? What should be done with them after their research lives are over?

See also ANIMAL RESEARCH, ETHICS OF; CHIMPANZEES, SELF-AWARENESS OF; CLEVER HANS EFFECT.

Further Reading

Pepperberg, I. M. "Cognition in an African Gray Parrot (*Psittacus erithacus*). Further Evidence for Cognition of Categories and Labels." *Journal of Comparative Psychology* 104 (1990): 41–52.

Pfungst, O. *Clever Hans (The Horse of Mr. Von Olsten).* New York: Holt, 1911.

Rumbaugh, D. M., ed. *Language Learning by a Chimpanzee: The Lana Project.* New York: Academic, 1977.

Savage-Rumbaugh, E.S. "Language Acquisition in a Nonhuman Species: Implications for the Innateness Debate." *Developmental Psychobiology* 23 (1990): 599–620.

Terrace, H. S. *Nim.* New York: Knopf, 1979.

chimpanzees, self-awareness of

Gordon Gallup reported in 1975 that chimpanzees can recognize themselves in mirrors, suggesting that they are self-aware. The finding was no surprise to people who were familiar with these intelligent animals. When exposed to mirrors, chimpanzees use them to examine visually inaccessible parts of their bodies, such as the inside of the mouth, and even to "dress up" where the chimp might place some object, such as a cabbage leaf, on its head and examine the effect.

These responses of apes to mirrors are very different from the responses of monkeys. The monkeys always see their own reflection as another monkey and often persist in threatening it. The ability to recognize oneself in a mirror requires a process of familiarization with reflective surfaces. We know this from the reactions of stone-age peoples on their first exposure to mirrors. The initial response of New Guineans tested by anthropologist Raymond Carpenter was to see the reflections as hostile spirits.

Before testing the chimpanzees, Gallup therefore allowed them to become familiar with the mirrors. He then placed them under general anesthetic. While the chimps were unconscious, he applied a nontangible dye mark over their brows. What would they do when they woke up and observed their changed appearance in the mirror? Each of the test animals gazed at the dye mark in the mirror, and just like a human, attempted to clean it off using the mirror to guide their movements. This elegant test has convinced most reasonable people that chimpanzees can recognize themselves in a mirror, which implies that they have visual self-awareness of a type not seen in most other animals. The mirror test has also been passed by orangutans, but, interestingly, has been failed by the relatively small-brained gorilla. Although David Premack has trained pigeons to peck at a mark on their feathers using a mirror, this performance differs greatly from the *spontaneous* use of mirrors by chimpanzees that have become familiar with them.

Self-awareness puts great apes in a special category and creates ethical problems about whether they should be maintained in captivity, how they should be housed and managed by researchers, and what should be done with them after their research lives are over. The line between apes and humans is further blurred by the fact that human infants are not self-aware, judging from the results of a modified version of the mirror test (without anesthetic). Children do not pass the mirror test of self-recognition until they are 12–24 months old according to developmental psychologist Michael Lewis, who also noted that the first signs of embarrassment emerge at around the same age.

See also ANIMAL RESEARCH, ETHICS OF; CHIMPANZEES, LANGUAGE-LEARNING CONTROVERSY.

Further Reading
Gallup, G. G. "Self-Awareness in Primates." *American Scientist* (1979) 67: 417–21.

Churchland, Paul

(1942–   )
Canadian-American
Philosopher

Paul Churchland and his wife Patricia are noted for their contributions to the mind-body problem using the findings of contemporary neuroscience. The Churchlands reject mentalism in favor of the view that the mind is fully explainable in terms of brain function (MONISM). This perspective has fundamental implications for ethical notions of free will and responsibility. For example, impulsive violence of the type that often leads to criminal incarceration could be explained in terms of brain chemistry and functional anatomy. This means that violent crimes could be predicted, and controlled, through advances in neuroscience. If so, responsibility for repeat violent offending might pass from the individual criminal to the criminal justice system.

See also CRIME, AND BRAIN CHEMISTRY; DESCARTES, RENÉ.

Further Reading
Churchland, P. M. *The Engine of Reason, the Seat of the Soul.* Cambridge, Mass.: MIT Press, 1995.
Churchland, P., and T. J. Sejnowski. *The Computational Brain.* Cambridge, Mass.: MIT Press, 1992.

Clean Air Act, 1970

The U.S. Clean Air Act of 1970 set standards of air quality in relation to six pollutants: sulfur oxides, nitrogen oxides, carbon monoxide, hydrocarbons, photochemical oxidants, and particulates. States were given responsibility for implementing these standards. In regions where the standards were not met, new industrial development, or expansion, was to be halted. Amendments to the Clean Air Act in 1990 generally tightened standards for automobile and industrial emissions. These regulations have helped to reduce air pollution. For example, emissions of sulfur dioxide, a key component of ACID RAIN, have been reduced by approximately one-third since passage of the act.

Such improvements have been achieved partly through the use of new technologies to reduce waste emissions. Thus CATALYTIC CONVERTERS have reduced the emission of nitrogen oxides, carbon monoxide, and hydrocarbons in vehicle exhausts. The catalytic converters promote oxidation of unburned gases. Emissions have also been brought down by reducing engine combustion temperatures. Areas with high ozone pollution (or SMOG) have switched to cleaner burning gasoline.

Efforts have also been made to reduce smokestack emissions from industry. Some of these have been quite simple and relatively inexpensive. For example, the electric utility industry, a major source of acid rain pollutants, has switched to a different form of coal that is lower in sulfur. The utilities were motivated to make this change by an innovative emissions trading system that allowed the electricity companies to lower costs by reducing emissions. Legislation modeled after the Clean Air Act has been passed in Europe and other countries around the world, producing measurable improvement in the air quality of large cities.

See also ACID RAIN; AIR POLLUTION; CLEAN WATER ACT; ENVIRONMENTAL PROTECTION AGENCY; HAZARDOUS WASTE.

Clean Water Act, 1977

The U.S. Clean Water Act of 1977 replaced the Federal Water Pollution Control Act of 1972. The 1977 act set an objective of national waters fit for swimming and fishing by 1973 and a halt to pollutant discharges by 1985. Although these goals have still not been met, there has been a tremendous improvement in water quality. In 1972, only about 35% of rivers were suitable for fishing and swimming compared to some 60% in 1998. These improvements were achieved by the end of the 1980s with a subsequent sharp decline in the rate of improvement.

The Marine Protection, Research, and Sanctuaries Act (1972) also gave the ENVIRONMENTAL PROTECTION AGENCY

(EPA) the authority to regulate the dumping of sewage and toxic chemicals in the oceans. Agreement by the states of New York and New Jersey to stop dumping sewage sludge and other solid wastes in the Atlantic Ocean by 1991 greatly improved the amenity value of beaches, which had regularly been closed after syringes and other medical wastes had washed up on the shores. Municipal wastewater treatment today involves skimming and suspension to remove most of the solids. This is followed by treatment with microorganisms to destroy organic materials that would encourage algal growth, which depletes water of oxygen and kills fish in polluted streams. Finally, the water is chlorinated to destroy any remaining microorganisms that would cause illness to humans. Some advanced wastewater treatment facilities produce an effluent that exceeds drinking water standards.

One of the main reasons that water cleanliness has not continued to improve is that framers of the legislation focused primarily on point sources of water pollution, such as sewers and industrial effluent, and ignored other sources, particularly farms and private septic systems that often run straight into streams. Agricultural effluents may seep into groundwater, contaminating sources of drinking water, or they may be washed into streams causing pollution, algal bloom, and fish kills. The EPA and the U.S. Department of Agriculture have agreed that farms with more than 100,000 chickens or 1,000 cows should be exposed to the same runoff controls as other industries are.

See also ACID RAIN; CLEAN AIR ACT; ENVIRONMENTAL PROTECTION AGENCY; HAZARDOUS WASTE.

Clever Hans effect

Named after a horse that was believed capable of counting and solving mathematical problems, this refers to the problem of human experimenters reading more into the performance of their animal subjects than is actually there. The horse had been trained by Wilhelm von Olsten, a retired German schoolteacher, to tap out numbers with his hoof and to stop tapping when he had reached the correct answer to a particular problem. Thus, when presented with the problem 6 x 4 written on a chalkboard, Hans would stop after the 24th tap of his hoof. His performances amazed and delighted audiences during the first decade of the 20th century and von Olsten was invited to bring his calculating horse to royal courts for the entertainment of monarchs. Hans's brilliant career as an equine mathematician came to an end when psychologist Oskar Pfungst demonstrated in a clever series of experiments that the horse was merely responding to unconscious movements of the trainer.

The Clever Hans effect has recently come to the fore in connection with the CHIMPANZEE LANGUAGE-LEARNING CONTROVERSY, wherein a whole generation of sign language research was destroyed by the discovery that signing chimpanzees often simply repeated signs already produced by the trainers. The Clever Hans effect forced ape language researchers to turn to computers, which do not provide the unconscious subliminal cues given by human trainers.

In the Clever Hans effect, people interpret a situation according to their expectations. Hans's audience expected him to solve mathematical problems. Ape language researchers expected to find that chimpanzees could communicate with them. Analogous expectancy effects played into the hands of spiritualists whose clients expected to hear from dead relatives. The spiritualists were mercilessly exposed as charlatans by HARRY HOUDINI.

See also CHIMPANZEE LANGUAGE-LEARNING CONTROVERSY; HOUDINI, HARRY.

Further Reading
Pfungst, O. *Clever Hans (The Horse of Mr. Von Olsten)*. New York: Holt, 1911.

Clipper chip

The Clipper chip is a technology for encrypting electronic communications. Worried that commercial encryption methods could prevent it from obtaining access to electronic messages, the U.S. government wanted to establish a national encryption system that would allow government access to all electronic communications when authorized by a warrant. The Clipper chip technology was opposed by computer professionals, partly because of the unwillingness of government agencies to divulge the encryption algorithms. The technical feasibility and cost effectiveness of the Clipper chip system have also been questioned. Moreover, the potential for government access to private communications evoked resistance from privacy rights advocates. The major advantage of Clipper chip technology is that it is more secure than the 56-bit encryption it was designed to replace because the data are passed through several different encryptions.

See also CRYPTOGRAPHIC RESEARCH; ELECTRONIC PRIVACY.

cloning research   *See* GENETIC ENGINEERING.

cold fusion controversy

The fusion bomb (or H-bomb) is one thousand times more powerful than the fission bomb (or A-bomb) used against the Japanese in World War II that was itself a million times more powerful than conventional explosives like TNT. The tremendous amounts of energy released when atoms fuse could theoretically satisfy all of the world's energy needs. In practice, the conditions under which fusion occurs are too extreme to allow fusion to be developed as a safe source of energy. Fusion is literally too hot to handle.

On March 23, 1989, it seemed that this scenario was about to be radically altered when two electrochemists announced that they had produced fusion at room temperature. Stanley Pons, chair of the chemistry department at the University of Utah, and Martin Fleischman, of Southampton University in England, held a press conference to announce that they had accomplished cold fusion using equipment available in most high school laboratories. Their press release was vague on the methodology but indicated that they had sent an electric current through a solution of lithium deuteroxide in heavy water causing the deuterium to get compressed at the cathode where the atoms fused into tritium with the release of large amounts of energy.

This story was eagerly accepted by the press but fusion researchers were more skeptical. Many nevertheless rushed to repeat the experiments insofar as they could understand what had been done based on press reports. Although not all of these experiments produced negative results, the consensus today is that cold fusion is not possible using currently available techniques. The positive findings of Pons and Fleischmann and others can be attributed to some combination of procedural sloppiness, wishful thinking, and a poor understanding of ordinary electrochemical phenomena. For example, it has been argued that the increased heat production measured near the cathode have been produced by ordinary chemical reactions if the solution was not mixed thoroughly.

Such controversies are seen by many as undermining the credibility of science, but this is hardly a fair conclusion. Pons and Fleischman's grasp of the potential commercial importance of their work distorted the normal procedures of science. Thus, the decision to announce the results in a press release, rather than as a scientific report, was evidently motivated by concern over receiving academic credit and protection of future PATENT rights. Had Pons and Fleischmann published a detailed account of their work, their "invention" could easily have been stolen by another laboratory. In order to receive a patent it is necessary to perfect an invention and describe in detail how it works. Had cold fusion been a real phenomenon and had Pons and Fleischmann published a detailed report that would have allowed other laboratories to repeat their experiment exactly, then they would have lost any advantage in the race to patent the process. Even though we may expect great feats of self-sacrifice from sci-

entists, it is more realistic to accept that their human needs for money and fame can occasionally interfere with the disinterested pursuit of scientific knowledge.

See also HIROSHIMA; N-RAYS; NUCLEAR WEAPONS.

Further Reading

Fleischman, M., and S. Pons. "Electrochemically Induced Nuclear Fusion of Deuterium." *Journal of Electroanalytic Chemistry* 261 (1989): 301.

Huizenga, J. *Cold Fusion.* Rochester, N.Y.: University of Rochester Press, 1992.

Resnik, D. B. *The Ethics of Science.* New York: Routledge, 1998.

cold war radiation experiments on human subjects

Following World War II, the rivalry between the United States and the Soviet Union took the form of a cold war that involved the buildup of long range nuclear weapons. The use of such weapons would have been irrational because it virtually guaranteed a response in kind leading to MUTUAL ASSURED DESTRUCTION. Many U.S. bureaucrats and scientists nevertheless believed that the prospect of a nuclear attack was likely enough to justify making preparations for such a catastrophe by learning about the effects of radiation on survivors. Even if an attack never materialized, the U.S. Army needed to know how exposure to radioactivity would affect military personnel involved in the building, storage, and maintenance of nuclear weapons.

When President Clinton ordered the Department of Energy to declassify cold war–era documents, in 1994, it became clear that the U.S. government had sponsored research in which citizens were deliberately exposed to harmful radiation without their knowledge. It was felt that the only way to survive a nuclear attack was to understand the effects of radiation on human tissues. This knowledge might be used to either increase survival rates in America or create even more devastating nuclear weapons to wipe out the enemy. The declassified documents indicate that the radiation experiments raised ethical doubts on the part of researchers and civil servants. Such qualms were silenced by a belief that the sacrifices of a few unsuspecting people were more than balanced by the goal of national survival in the event of nuclear attack. This ethical defense is similar to that of a commander ordering troops to advance in a battle. Soldiers will lose their lives as a result of the decision but this sacrifice is justified by the larger good of national security.

This utilitarian defense is unsatisfactory because it does not address the two most troubling aspects of the research. Participation was not voluntary and it was done surreptitiously so that people did not even know that they had been involved in an experiment. If the American public had found out about the experiments at the time they were performed, this could have been more destructive of national security than any information gained could have contributed to security because faith in the institutions of government could have been destroyed.

Among the most ethically objectionable experiments was one conducted at Vanderbilt University in which unsuspecting pregnant women were given iron supplements that were radioactive in an effort to study the effects of radiation on the fetus. Follow-up research revealed that the children of these pregnancies had an unusually high incidence of cancer. This experiment not only violated all rights to informed consent on the part of mothers but also involved research on individuals who were incapable of giving consent. Other experiments were conducted on unsuspecting terminally ill cancer patients who were injected with high doses of isotopes to study the absorption of radioactive substances into their tissues. These subjects had less to lose than the fetuses since they were already close to the end of their lives, but the research constituted an equally appalling violation by physicians of the HIPPOCRATIC OATH in which medical practitioners are enjoined to do no harm. Many of the patients survived for years and were surreptitiously monitored by the researchers in another illegal and immoral assault on their civil rights. In another horrifying experiment MIT researchers fed radioactive oatmeal to boys in a state school near Boston.

Not all of the participants in government-sponsored radiation experiments were involuntary. In one study, conducted at Oregon State Prison from 1963 to 1971, 67 prisoners were paid $200 to have their testicles irradiated with X-rays. The researchers wanted to study the effects of radiation on sperm function. This research might have been ethical if participants were informed of the increased risk of getting cancer but they were not. In 1979, nine of these prisoners were awarded $2 million in damages.

Other research was conducted on military personnel, who were not informed of the purpose of the experiments. In one study lasting two decades, approximately 1,500 men had radium capsules inserted in their nostrils for several minutes. The radiation exposure was severe enough for many to develop severe headaches.

Not content with observing the effects of radiation in laboratory studies, federal scientists released a cloud of radioactive iodine over eastern Washington in 1950. It is estimated that this cloud carried hundreds of times the radiation released during the THREE MILE ISLAND nuclear accident.

It is ironic that no sooner had the U.S. government formulated the NUREMBERG CODE to provide a basis for prosecuting Nazi researchers of abuses involving involuntary research than it began perpetrating atrocities on its

own citizens that were only somewhat less depraved. Much of the cold war radiation research conducted in America clearly violated the consent requirement of the Nuremberg Code. The U.S. Army, which played a key role in much of the research, had its own guidelines for the use of human subjects that were more stringent than the Nuremberg Code but these regulations were simply ignored in the zeal to discover the effects of radioactivity on human tissues that was stimulated by cold war contingencies. Most of the details of the involuntary radiation experiments were deliberately concealed from the U.S. public for almost half a century, virtually guaranteeing that those responsible would never be held accountable for their actions.

See also HIPPOCRATIC OATH; NAZI RESEARCH; NUREMBERG CODE; OPERATION WHITECOAT; RIGHTS OF HUMAN RESEARCH PARTICIPANTS; TUSKEGEE SYPHILIS STUDY.

Further Reading

Budiansky, S., E. Goode, and T. Gest. "The Cold War Experiments." *U.S. News and World Report,* February 28, 1994.

Pence, G., ed. *Classic Cases in Medical Ethics.* New York: McGraw-Hill, 1995.

Resnik, D. B. *The Ethics of Science.* New York: Routledge, 1998.

Welsome, E. *The Plutonium Files: America's Secret Medical Experiments in the Cold War.* New York: Dial Press, 1999.

commercialism in scientific research

Private companies, whose goal is to maximize profits, play an increasingly important role in scientific research both as employers of research scientists and as sponsors of studies conducted by scholars in an academic setting. The need to increase profits often conflicts with the objectives of science, and commercialism introduces many ethical problems into the practice of research. The two ethical principles that are most often compromised by commercial interests are objectivity of conclusions and openness of scientific communication.

Commercial drug trials are often of questionable objectivity. One well-known example of this is the drug SYNTHROID, a leading thyroid medication taken by eight million Americans every day. It costs considerably more than generic alternatives but the manufacturer, Boots Pharmaceutical (now Knoll Pharmaceutical), has been successful in convincing the medical profession for decades that Synthroid is better than its rivals and that the price premium is justified. Since there were no clinical trials showing that Synthroid was better, the phenomenal success of the drug was based exclusively on clever marketing. By the late 1980s, Boots Pharmaceutical had some reason to believe that its top-selling brand name was actually better because Betty Dong, a researcher at the University of California, San Francisco, published a small-scale study suggesting that Synthroid was more effective. Boots approached Dong, providing her with the $250,000 needed to complete a more rigorous and more reliable study.

After Dong completed her study, she found that there was no real difference between Synthroid and its three cheaper competitors. Dismayed by the results, Knoll used its contractual prerogatives to suppress publication of Dong's report. They also published the results preemptively coming to a conclusion that favored Synthroid. Dong's paper was finally published in 1997 and precipitated a class action lawsuit on behalf of Synthroid users, who felt defrauded by paying the extra cost of the drug. Knoll Pharmaceutical settled out of court for $100 million damages, a record amount in a suit of this type.

Synthroid is not an isolated case. There have been reports of other drug trials being suppressed, or distorted, such as unflattering results for Apotex's L1 drug and Sandoz's calcium-channel blockers. Yet, such incidents are liable to become public only when researchers stand up for their scientific integrity. A survey of published studies of new drugs has found that when the research is conducted without drug-company funding, 21% of the medications receive unfavorable evaluations. When the research is sponsored by the drug company, only 2% of the drugs get an unfavorable review. In other words, independent researchers are ten times more likely to criticize a drug than researchers who are in the pockets of the drug companies. This is a fairly depressing statistic, which suggests that researchers funded by drug companies lose most of their objectivity. Alternatively, the difference might be explained by drug companies vetoing publication of unfavorable studies. In either case, it is clear that published evaluations of drugs sponsored by the manufacturer do not discriminate between good and bad drugs and therefore have little value as scientific research.

Conflicts of interest between the obligations of researchers to scientific objectivity and the commercial interests of drug companies and other commercial sources of funding are common but are rarely disclosed. Many scientific journals do require scientists to disclose such conflicts as a condition of publication but researchers usually ignore this requirement. Thus, even though there is a commercial conflict of interest in over 20% of articles published in biomedical journals, and all the major journals have a disclosure requirement, most (68%) did not publish a single disclosure in 1997. In the case of researchers working on calcium-channel blockers, only 3% reported a conflict of interest even though 96% were being funded by a drug company.

Some scholars believe that FRAUDULENT RESEARCH is quite common in academic research where the rewards of dishonesty include benefits, such as renewed funding and

tenure. It might be supposed that scientific dishonesty would be more common in industrial research because the monetary rewards are greater and because the social pressures to obtain results that favor the success of a company might be intense. Similar pressures against objectivity are confronted by scientists who make a living giving EXPERT TESTIMONY in court cases.

A more subtle type of conflict occurs when the profit motive dictates that testing should be done more quickly than is compatible with good scientific practice. A clear example of this problem occurred in the case of speedy and inaccurate reading of pap smears that led to the indictment of CHEM-BIO on charges of homicide in relation to two women who died after their cervical cancers went undetected.

The social pressures of working for private companies does not excuse dishonest or low-quality research and there is no compelling reason that commercial researchers cannot maintain their scientific integrity if they are prepared to give up funding opportunities and jobs. (Givers of expert testimony in court are in a similar bind because they are hired to support one side or the other and if they do not perform as expected they will not be hired in the future.)

In addition to the threat to objectivity of results, commercial pressures also have implications for the open communication of results because the scientific ethic of free communication is opposed by the industrial ethic of guarding trade secrets.

The proper publication of scientific research can be threatening to proprietary information. An example of this sort of conflict emerged in the case of the COLD FUSION CONTROVERSY, in which researchers hoped to patent what they imagined was an important discovery. This inhibited them from publishing the fine details of their procedure, which caused many researchers to waste a lot of time in attempting to replicate results that turned out to be illusory.

Secrecy is necessary to protect the interests of commercial organizations from predatory rivals just as it may be required of researchers working in the military. In general, researchers who work in these settings can be said to have made a bargain with the devil in the sense that they enter into an implicit, or written, contract to protect their findings from the scrutiny of competitors, or the public. Many scientists are uncomfortable with these arrangements, but they accept that a bargain is a bargain. If they want to be paid, they are obliged to do what their employer wants. This might mean agreeing to suppress the results of their research if findings turn out to be antagonistic to the best interests of the corporation.

In general, the type of secrecy that protects the corporation's turf against rivals does not pose a major ethical problem for researchers. The worst that can happen is that publication of useful findings is delayed or that scientists waste money conducting research that would have been rendered obsolete. A more severe conflict emerges when the researcher discovers that the company's products, or actions, are causing harm to the public. In such cases, scientists have often been unwilling to come forward for the simple reason that blowing the whistle is actually illegal because it violates the secrecy clause in their employment contract, thereby exposing them to civil actions.

A particularly striking example of this kind of secrecy pressure is the case of work done by Victor deNobel and Paul Mele in the early 1980s on the harmful and addictive properties of nicotine in cigarettes. The researchers were working for Philip Morris, a major tobacco company, and their activities were conducted under a cloak of secrecy. When they attempted to publish results in the journal *Psychopharmacology*, Philip Morris forced them to retract the paper. Soon afterward, the research was terminated and they left the company. It was not until 1994 that the researchers agreed to discuss their research in public and then only after they had been released from their lifelong agreement with Philip Morris.

Not only can commercialism undermine the cardinal scientific virtues of objectivity, openness, and carefulness, but there is also a real sense in which commercial interests are taking control over academic research. Thus, leading universities are becoming more and more dependent on income derived from patents and licensing arrangements through which they (and usually the scientists themselves) capitalize on important discoveries. When a corporation funds a research facility on a university campus the laboratory may operate like a research-and-development arm of the corporation in the sense that its research agenda is driven more by the desire to establish lucrative patents, for example, than to advance scientific knowledge. Many ethicists feel that commercial sponsorship both debases the ethics of scientists and brings the scientific profession into disrepute. Others feel that the increasing collaboration between free enterprise and the universities, and the increasing entrepreneurial spirit that such collaboration stimulates in scientists, favors both prosperity and the advancement of knowledge.

See also BALTIMORE, DAVID; CHEM-BIO; EXPERT TESTIMONY; OPENNESS IN SCIENCE; PATENTS; SYNTHROID CONTROVERSY; TOBACCO COMPANY RESEARCH.

Further Reading

Broad, W., and N. Wade. *Betrayers of the Truth*. New York: Simon and Schuster, 1982.
"Money + Science = Ethics Problems on Campus." *The Nation* 268, no. 11 (1999): 11.
Resnik, D. B. *The Ethics of Science*. New York: Routledge, 1998.

compositionalism

Compositionalism is a perspective in the modern philosophy of conservation. It describes nature in terms of evolutionary ecology and considers human beings as distinct from the natural world.

See also DEEP ECOLOGY; ENVIRONMENTAL ETHICS.

computers and violence

The possibility that violent imagery on the Internet and in video games had a role to play in violent behavior, particularly of young people, was widely publicized in the context of a recent spate of shootings in American high schools. This followed several decades in which research on the association between viewing violent television and aggressive behavior did not provide compelling evidence of a causal relationship. The December 1, 1997, episode in which 14-year-old Michael Carneal opened fire on his classmates and friends at Heath High School in Paducah, Kentucky, illustrates the connection between computer games and violence.

Like most of his peers, Carneal had seen violent movies such as *Menace II Society* in which gratuitous violence in a school setting is depicted and glorified. The argument that his crime was an imitation of a violent movie is superficially plausible but does not explain why Carneal, rather than one of his peers who had also watched violent movies, was the perpetrator. Moreover, it does not explain one of the most striking aspects of the carnage, how a youth with no military training could engage in the cold-blooded execution of three young women who had been his friends. This mystery is partly resolved by familiarity with the world of violent video games with which Carneal was fascinated.

According to psychologist and retired army officer David Grossman of Arkansas State University, the natural aversion to killing other people in cold blood is a real problem confronted by military trainers. Thus, it is estimated that fewer than one-fifth of U.S. troops actually fired their guns in battle in World War II. Shooting games using man-shaped targets helped recruits overcome their deep inhibitions against killing soldiers. The basic principles of desensitization to violence have been taken to an extreme in violent video games. The games also reinforce players for good marksmanship, which makes them very useful in military training.

The families of the three victims have sued the publisher of some of the most violent games, charging that

Carneal's exposure to games like *Doom* played role in the high school students' deaths. The attorney for the families, Mike Breen, commented that "Michael Carneal clipped off nine shots in about a 10-second period. Eight of those shots were hits. Three were head and neck shots and were kills. This is way beyond the military standard for expert marksmanship. This was a kid who never fired a pistol in his life but because of his obsession with computer games had turned into an expert marksman." Breen was mistaken in one detail. Carneal had actually had the pistol in his possession for a couple of days before the shootings and apparently engaged in some real-world target practice.

Just as pilots train on flight simulators, so video games can be considered murder simulators. Some arcade games even feature realistic pistol grips complete with recoil. The Marine Corps is adapting a version of *Doom*, one of the most popular violent games, to train its own personnel for combat situations.

Practice in video-game marksmanship develops skills in accurate marksmanship and desensitizes players to the consequences of violence. Realistic touches include screams of agony, naturalistic wounds, pleas for mercy, and pools of blood. Desensitization occurs partly through repetition and familiarity but some of the games, such as *Doom*, have a progressive structure in which the player is rewarded for kills by receiving more powerful, and bloodier, weapons. The shotgun is replaced by an automatic weapon and eventually a chainsaw is used to rip virtual enemies apart.

A role for computer games in the rash of high school shootings is suggested by a number of unusual features of the murders, including inadequate motivation, and, in some cases, lack of a record of serious crimes. Many of the shooters were known to be obsessed with violent video games and spent time on Internet sites devoted to violent themes, such as bomb-making. These interests and pursuits probably desensitized them to the effects of violence and helped them to develop combat skills used in the murders. It is nevertheless worth pointing out that these violent influences are probably most important in the lives of children who lack a network of supportive parents, relatives, and friends that creates an alternative, nonviolent social reality. It has also been argued that parents can play an important role in limiting their children's access to violent games and activities.

If a causal relationship could be established between playing violent video games and increased propensity to commit murder, this might be seen as a compelling argument for removing these games from the market, or at least ensuring that they do not get into the hands of minors, as in the case of current controls on the sale of PORNOGRAPHY. Yet, it is unlikely that such controls could be successful if only because pirated versions of the games would inevitably be made available free over the Internet. As in the case of the debate over televised violence and violent movies, controlling the medium itself is difficult or impossible and a great deal of the responsibility falls to parents to regulate their children's access to violent materials.

See also ELECTRONIC TERRORISM; ELECTRONIC THEFT; TELEVISION AND VIOLENCE; V-CHIP.

Further Reading
Cannon, A., B. Streisand, D. McGrace, and D. Whitman, "Why?" *U.S. News and World Report,* May 3, 1999.
Grossman, D., G. Degaetano, and D. Grossman. *Stop Teaching Our Kids to Kill.* New York: Random House, 1999.
Hanson, G. M. B. "The Violent World of Video Games." *Insight on the News,* June 28, 1999.
Leo, J. "When Life Imitates Video." *U.S. News and World Report,* May 3, 1999.

confidentiality rights in research   *See* RIGHTS OF HUMAN RESEARCH PARTICIPANTS.

conflict of interest   *See* COMMERCIALISM; EXPERT TESTIMONY; PEER REVIEW.

consequentialism
Consequentialism in ethics implies that the morality of an action is determined solely by its consequences. Consequentialism is often placed in opposition to DEONTOLOGY, the view that morality is based on principles (or moral obligations). Consequentialism is often equated with UTILITARIANISM but it is a more general term. All utilitarians are consequentialists but all consequentialists are not utilitarian. For example, an egoist acts so as to produce consequences favorable to the self and is thus consequentialist but does not live up to the utilitarian ethic of maximizing the happiness of many people.

See also DEONTOLOGY; UTILITARIANISM.

contraception
Contraception includes all techniques used to prevent the fertilization or implantation of an egg in the case of otherwise fertile sexually active couples. The development of reliable contraceptive techniques in the 20th century has produced some profound changes in social behavior. For the first time, sexual behavior of women has been separated from reproduction. Moreover, the ability to plan families has allowed women to

participate more in businesses and careers. The small family size typical of developed nations has meant that populations in these countries have stopped growing and would decline if it were not for immigration. As more and more countries become economically developed, this will have an important impact on the world population crisis, which is expected to abate early in the 21st century.

Widespread use of modern contraceptive techniques has been accompanied by a dramatic change in attitudes to sexual behavior and reproduction. Their introduction, and legalization, was surrounded by a storm of ethical controversy based largely on the premise that people were tampering with the natural order via a dangerous separation of sexuality and reproduction. The intensity of these debates may seem difficult to understand today but it bears analogy with the more recent debates over artificial insemination, in vitro fertilization, cloning, and other REPRODUCTIVE TECHNOLOGIES that are producing profound revision in our basic social concepts of reproduction, marriage, and family and generating unprecedented ethical challenges in the process.

A BRIEF HISTORY OF MODERN CONTRACEPTIVE TECHNIQUES

Crude methods of birth control have been used for millennia. These sometimes involved the application of a viscous material, like honey, or oil of cedar, to the vagina with the intention of impeding the passage of semen to the uterus. Primitive barrier devices consisted of sponges or pieces of soft wool. These were often soaked with vinegar or lemon juice that was intended to function as a spermicide. Folk medicine also included the use of herbs believed to suppress conception.

Modern barrier techniques include a variety of substances such as foams, creams, jellies, and wax suppositories. Each of these contains a spermicide that is not harmful to vaginal tissues. None of these techniques is very reliable. The most effective barrier methods use an impermeable mechanical barrier that prevents sperm from entering the uterus and these include the diaphragm, the cervical cap, and the condom (worn by men). Another modern method is the sponge (worn by women and used with a spermicide, or sperm-inactivating chemical). The diaphragm is a shallow rubber cup that covers the cervix, or entry to the womb, during intercourse. The cervical cap is smaller but is also used to cover the cervix. The sponge retains its spermicidal effectiveness for 48 hours after insertion. None of these methods is highly reliable with an annual failure rate of around 15%.

The condom covers the penis and prevents semen from entering the vagina. It has a failure rate of approximately 16%. Problems may occur if the condom shifts, or even breaks during intercourse. The condom is popular as a barrier to sexually transmitted diseases, including AIDS, so that the terms "condom" and "prophylactic" are used as synonyms in America. Early condoms were made of linen and were therefore unlikely to have been effective as contraceptives. English diarist James Boswell describes his use of cloth condoms in encounters with prostitutes in the 18th century. The women evidently found them uncomfortable and objected to their use. Boswell complained of repeated reinfection with venereal disease. Since the 1840s, condoms have been made of vulcanized rubber. Latex condoms were developed in the 1930s and these are generally lubricated. Early condoms were designed for reuse and the modern disposable condom is a considerable advance so far as disease prevention is concerned.

Condoms are certainly important in disease prevention but they cannot prevent transmission of sexually transmitted diseases (STDs) unless used consistently and for all sexual acts. Since STDs are more easily transmitted from men to women than from women to men, women are generally at greater risk of contracting diseases as well as having the added risk of becoming pregnant. Possibly for these reasons women are more reliable in their use of contraception than men are. Since condoms play a very important role as a barrier to disease, a female variant of the condom, a pouch, has been developed to enable women to better protect themselves from sexually transmitted diseases.

Some of the most widely used contraceptive techniques take the form of a pharmacological intervention that interferes with a woman's reproductive physiology so as to prevent fertilization from occurring. One way of doing this has been to place a device in the womb to impede fertilization. Such intrauterine devices, or IUD's, are made of copper or plastic. Early IUD's were of a mechanical design. Their invention followed from the practice followed by camel herders of placing a stone in the uterus to prevent females from becoming pregnant. Why early IUDs worked has never been satisfactorily explained. Some more recent IUD's slowly release progesterone to prevent pregnancy. Modern IUDs are highly effective with a failure rate of only 4%.

However, IUD's have produced troubling side effects. They have caused bleeding and uterine infections and may also have played a role in the development of pelvic inflammatory disease. This blocks the fallopian tubes, trapping the egg and resulting in ectopic pregnancy. Their use can also cause sterility. The A. H. Robins company, producers of the Dalkon Shield IUD, were ordered in 1987 to pay $2.5 billion in damages to women injured by the device.

A popular method of birth control among American women is the pill, which is taken daily. The pill is a combination of estrogen and progesterone and it is highly effective, having a failure rate of only 3%, because it interferes with conception in different ways. High levels of estrogen block the development of egg follicles (i.e., the sheath of cells surrounding the egg) and prevent an egg being released. High levels of progesterone further guarantee that the egg will not be released.

The pill must be taken every day to prevent pregnancy and when women forget to take it, they may become pregnant. For this reason, long-acting contraceptives have recently been developed. In 1992, the U.S. Food and Drug Administration (FDA) approved Depo Provera, which is injected four times per year. It contains a synthetic progesterone. A year earlier, the FDA had approved Norplant, an even longer-lasting contraceptive that prevents conception for five years. Norplant is administered in the form of six flexible matchstick-sized tubes that are implanted in the upper arm and slowly release the hormone levonorgestrel, a progestin. Its annual failure rate is less than 1%.

Norplant has engendered its share of ethical controversy for several reasons. First, the implants have to be removed if the woman decides to have children. This surgical procedure has sometimes been botched with disfiguring results. Second, because of its long-term, yet reversible, effects, Norplant has attracted the notice of people who are interested in furthering a eugenics program without attempting the permanent sterilization carried out in such projects in the past. One example of such eugenic use of Norplant concerns the sentencing of a California woman convicted of child abuse in 1991. As a condition of her probation, the woman was ordered to use Norplant.

Court-ordered use of Norplant raises many ethical and legal problems and probably would not survive a challenge on constitutional grounds. In the specific case of the woman convicted of child abuse, Norplant could not be considered therapeutic so far as preventing the woman from abusing her children is concerned even though it could be argued that having fewer children reduces the social stresses that make child abuse more likely. A truly therapeutic approach would focus on parenting skills training, accessible day care, and other services to reduce the stressfulness of raising children. Children could be protected by close supervision of offending parents and the provision of foster homes to children who are at high risk of abuse by their parents. There is thus an important distinction between Norplant in this case and the Depo Provera injections received by child molesters, for example. Proponents of such CHEMICAL CASTRATION for sex offenders argue that it is genuinely therapeutic since it reduced the tendency to reoffend.

Forced use of contraception violates the right to procreate. This right was recognized by the U.S. Supreme Court in a 1942 decision striking down an Oklahoma law to the effect that repeat sex offenders should be sterilized. Yet, this is not an absolute right. Not all sterilization laws have been overturned. Seventeen states still allow involuntary sterilization of mentally incompetent persons who are deemed incapable of caring for children. In most cases, a court hearing is necessary at which it must be established that the sterilization is in the individual's best interest, not that of the state.

The only method of contraception accepted by the Roman Catholic Church is sexual abstinence. Abstaining from sexual intercourse during the fertile central period in a woman's monthly cycle can be moderately effective in preventing conceptions and has a failure rate of 19%. Although often described as a natural method, it is nothing of the sort. In fact scientific knowledge about the time of release of the ovum was acquired only in the 1930s. The rhythm method, or avoiding intercourse during the most fertile days of each month, is difficult to practice because cycles vary in length. Slight increases in body temperature provide one clue that a woman is ovulating and should avoid intercourse.

The pill is a very popular form of contraception (used by 28% of American women in 1992) but more American women rely on sterilization (35% used this method in 1992). In the case of married couples, or stable cohabitations, the woman may be sterilized by tying the fallopian tubes or the man may undergo a vasectomy, or cutting of the tube that carries sperm from the testicles. Of the two procedures, female tubal ligation is much more difficult to perform. It is categorized as major surgery requiring general anesthesia and a hospital stay. The vasectomy is minor surgery, requiring only local anesthetic, and can now be performed in a few minutes with a special forceps that punctures the scrotum.

ETHICAL AND LEGAL PROBLEMS IN THE USE OF CONTRACEPTION

Attitudes to contraception up until the 20th century were heavily influenced by the theology of St. Augustine, who wrote his treatises in the fourth century. St. Augustine distinguished between sexual intercourse for the purpose of producing children and sexual intercourse purely for the gratification of lust. Lust was always sinful, even in marriage, and St. Augustine felt that sexual intercourse was permissible only if engaged in for the purpose of procreation. He therefore rejected any effort at contraception as morally indefensible. Because Martin Luther's views on marriage were heavily influenced by Augustine, the Protestant Reformation brought no respite from Augustine's joyless prescription for relations between husbands and wives.

Modern legal regulation of contraception began in 1873 with the passage of the Comstock Act, which prohibited the use of the mails to distribute contraceptives, or even to forward information about contraceptive methods. The act also censored books dealing with sexual topics, which were described as "obscene." The timing of the Comstock Act may well have been related to an increased use of contraception in America. This has been inferred indirectly from fertility rates, which fell from 7.04 children per woman in 1800 to 4.55 in 1870. By 1940, rates had dropped to 2.10, reflecting widespread use of contraception rather than a decrease in sexual behavior.

The rationale behind the Comstock Act appears to have been generally religious. Two different themes can be distinguished. One was the fear of what unbridled lust might do to people. It was felt that contraception would allow an overindulgence in sex and that uncontrolled sexual urges would undermine morality. Another concern was that the use of contraception severed the connection between sexual intercourse and reproduction and was therefore an unwarranted interference in the natural order and hence with God's plan. This rationale is strikingly reminiscent of the concerns of St. Augustine 1,500 years earlier.

By the mid-19th century, after contraceptive use had become widespread, the women's rights movement in the United States promoted family planning, or "voluntary motherhood." Interestingly, women reformers advocated abstinence, rather than contraceptives, as the method of achieving this goal. It was felt that the use of contraceptives only made women more subject to the sexual tyranny of their husbands.

This view persisted into the 20th century. A leading feminist advocate for contraception was Margaret Sanger. Although she began with a sexual liberation agenda that owed something to the views of HAVELOCK ELLIS, Sanger ultimately couched her birth control agenda in terms of eugenics. Writing in 1919, she clearly put herself in the camp of SOCIAL DARWINISM by arguing that birth control helped to weed out the unfit by preventing the birth of "defectives." This strange alliance of radical feminism and extreme conservatism apparently reflected the unpalatability of a sexual liberation agenda at that time.

Contraception has become a very important issue because of the role it plays in addressing the problems of world population growth. Widespread adoption of contraception was delayed by a lack of reliable information. Although contraceptive techniques had been described in ancient Egyptian texts, as well as the Torah, the first detailed treatise on the subject was written by Soranus of Ephesus in the second century A.D. This information was not widely disseminated and those who wished to promote contraception encountered considerable resistance. Thus, American Charles Knowlton, who authored a book dealing with birth control titled *The Fruits of Philosophy* (1832), was prosecuted for obscenity. English distributors of the book encountered the same fate.

English economist THOMAS MALTHUS (1766–1834) was the first to call attention to the fact that human population was expanding faster than the food supply. He proposed that birth control should be used to avert the problem. Malthusian leagues were formed in Britain and Western Europe. The Dutch league founded the first birth control clinic in 1881, which was followed by Dr. Marie Stopes's clinic in England in 1921. Margaret Sanger's first clinic opened its doors in 1916 but it was quickly closed by the police. She opened another one in 1923. Sanger's National Birth Control League, which began in 1915, continues in the guise of the Planned Parenthood–World Population Organization.

The adoption of modern contraceptive techniques has been greatly affected by religious views and religious leaders. One influential development has been a change of heart on the part of the Protestant churches. In 1930, the Anglican bishops voted in favor of contraception by methods that did not call for sexual abstinence. In 1959, this position was endorsed by the World Council of Churches.

The Catholic Church has continued to oppose contraception by any method other than abstinence. Nevertheless, there have been some signs that the stranglehold of Augustinian sexual theology has been weakening. Thus, Pope Pius XI in 1930 approved marital intercourse at times when the wife could not conceive on the basis that such intercourse served two important social functions. First, it strengthened the bond of affection between husbands and wives. Second, it served to quieten lust, a perspective that was articulated memorably by St. Paul, who concluded that although chastity is preferable to marriage, it is better to marry than to burn. Consistent with this relaxation, a papal commission in the 1960s recommended the use of contraceptives by married people, only to be overruled by Pope Paul VI in 1968. The pontiff produced an Augustinian argument that allowing Catholics to use contraceptives would simply make it too easy for them to have sex without consequences. People, and especially young men, would give in to their sexual lust and respect for women would be lost. This official position of the Catholic Church has not changed. However, Catholics in some countries, particularly the United States, have rejected Papal authority in deciding such matters of private morality and have chosen to use contraceptives. This decision follows changes in the secular law. In 1936, a U.S. federal court ruled that the Comstock Act did not bar the distribution of contraceptives prescribed by a doctor. This decision prompted a backlash, with some states enacting their own laws that were even more restrictive than Comstock had been. In 1965, a Connecticut statute prohibiting the use of contraceptives was struck down by the Supreme Court in *Griswold v. Connecticut.* The Supreme Court decision was based on the tortuous argument that married couples have a right to privacy under the Constitution and that this right was compromised by the Connecticut law. What made this argument really remarkable is the fact that the word "privacy" is not even mentioned in the U.S. Constitution. Rather, a privacy right was inferred by the Supreme Court judges based on other guarantees in the constitution such as that preventing unreasonable searches and seizures of persons, houses, papers, and effects.

Since abortion is sometimes used to limit fertility after other methods of contraception have failed, abortion can be thought of as a form of birth control although the ethics of this practice has been a matter of vigorous debate. In *Roe v. Wade,* the U.S. Supreme Court in 1973 provided the same constitutional protection to abortion as had been given to contraception in the *Griswold* case. In striking down a Texas statute prohibiting abortions, the court once again appealed to the right of privacy that it had discovered in the *Griswold* case. The court ruled that the decision to have an abortion was to be made by the woman's physician. States were to permit abortions that occurred before the fetus became viable, or capable of existence outside the womb. Viability was defined to include fetuses that could be kept alive through medical interventions. This means that as medical technology advances and fetuses survive from ever more immature stages the period during which abortion is permitted will shrink.

Perhaps the most significant ethical aspect of the *Roe v. Wade* case was the conclusion that a nonviable fetus is not a "person" as far as this word is used in the U.S. Constitution. If states wished to protect the health and well-being of a fetus early in pregnancy (approximately during the first two trimesters before viability), they were free to do so but only on the basis that fetuses are not persons. Nonpersons, including animals and property, can be afforded legal protection.

The *Roe v. Wade* decision is seen by many feminists as an important assertion of the right of women to control their own bodies. In denying that the fetus is a person, the Supreme Court implied that, at least in early pregnancy, the reproductive and health interests of the mother are more important than the interests of the fetus because a person has more moral worth than a nonperson. This means that abortion, like contraception, is now largely a matter of individual decision in the context of medical advice. This has not satisfied many critics of the decision who do not accept that a fetus has less moral worth than the mother and who therefore see a decision

to have an abortion as equivalent to a decision to commit murder.

The most reliable and widely used contraceptive techniques, such as the pill and sterilization, represent advances in medical technology that are based on modern developments in reproductive biology. The most important social consequences of contraception have been a huge decrease in the birth rate in advanced countries and the separation of sexuality from reproduction. Ethical controversies about contraception have focused largely on its implications for sexual behavior and have paid less attention to its potential importance in controlling world population and hence relieving global problems of hunger. Contraceptive technology has had profound ethical implications because it has undermined a religious conception of sexuality as appropriate only in the context of reproduction.

See also DRUG USE DURING PREGNANCY; ELLIS, HAVELOCK; EUGENICS; MALTHUS, THOMAS; PORNOGRAPHY.

Further Reading
Areen, J. "Limiting Procreation." In *Medical Ethics,* ed. R. M. Veatch, 103–34. Sudbury, Mass.: Jones and Bartlett, 1997.
Brodie, J. F. *Contraception and Abortion in Nineteenth-Century America.* Ithaca, N.Y.: Cornell University Press, 1997.
Winikoff, B., and S. Wymelenberg. *The Whole Truth about Contraception: A Guide to Safe and Effective Choices.* Washington, D.C.: National Academy Press, 1997.

cookies and the Internet
Cookies are strings of text used by online companies to identify the computers used by visitors to their sites. The cookies are contained in a text file that is often placed in a folder labeled "cookies." They contain short strings of text attached to an Internet address like *Amazon.com.* The strings of text are unique identifiers that allow Amazon to identify a customer. This means, for example, that when a person has registered with a company to buy their products online, and when they return to buy again, they are likely to be greeted with a "welcome back" message. Marketers promote cookies as allowing for customization of advertising. If someone visits golfing sites, they can be told about golfing products in targeted advertisements.

Cookies have many sinister implications because they allow companies to scrutinize not just purchasing habits but also browsing behavior, including the amount of time spent at specific sites. Interpreting this as Big Business or Big Brother watching you is not just paranoia. Nothing that a person does online is truly private. Yet, the illusion of privacy may induce people to engage in very risky behavior.

Although cookies identify only the computer rather than the user, it is a small step from cookie identifiers to unveiling the real world identity of the user. A pioneer in this field is Abacus Direct, which was recently purchased by the electronic marketing firm DoubleClick despite earlier assurances that the company would not infringe upon the privacy of the individual in its online marketing practices. Cookies mean that the privacy of most on-line transactions is compromised unless computer users take steps to delete cookies from their hard drives and prevent them from being added in future. Cookies are not restricted by laws but Internet browsers can install software that alerts them whenever cookies are being sent and asks for their consent before accepting them. Cookies can also be automatically disabled by selecting that option in the Internet browser (e.g., Netscape, Internet Explorer) but disabling cookies in this way can make it impossible to use many e-commerce Web sites.

See also ELECTRONIC PRIVACY; ELECTRONIC SURVEILLANCE; ELECTRONIC TERRORISM.

Further Reading

Merion, L. "Internet Privacy." *Computer Life,* 5 no. 6 (1998) 59–61.

Rodger, W. "Cookies: How Sites Know What They Know." *USA Today,* January 25, 2000.

cooperation, evolution of   *See* ALTRUISM, EVOLUTION OF.

Copernicus, Nicolaus **(Nikolaj Koppernigk)**

(1473–1543)

Polish

Astronomer

Copernicus is considered one of the leading thinkers in the history of scientific thought. He studied astronomy, mathematics, and medicine in universities at Krakow, Bologna, and Ferrara; and he graduated in 1503 with a doctorate in canon law. He served as physician to his uncle, the bishop of Ermeland (1506–12). His uncle appointed him canon of Frauenburg in 1513, and he held this position until his death.

Copernicus's main claim to scientific importance is his rejection of the geocentric universe favored by PTOLEMY and ARISTOTLE. He replaced it with a heliocentric system in which the earth spun on its axis as it orbited the sun along with the other planets. Copernicus sketched out this revolutionary idea in his *Commentariolus* (1514), which was circulated in manuscript. It was only after an admirer, Rheticus, had published a book (*Narratio Prima,*

1540) advocating the geocentric cosmology that Copernicus was spurred to write his book-length treatment *De Revolutionibus Orbium Celestium* (On the revolutions of the heavenly spheres) and this was completed soon before his death. His delay seems to have been motivated by an awareness of the religious implications of his ideas and bears striking analogy to actions taken by CHARLES DARWIN three centuries later.

Copernicus's cosmology retained many classical features, including the Aristotelian notions of solid spheres and perfectly circular orbits, that were subsequently dispensed with but his deliberate break with classical authorities is seen as a turning point in the development of modern scientific thought. Copernicus demoted human beings from occupying the central position in the universe that stood at odds with the biblical view of humanity as the crowning glory of creation. He thus shattered the complacent anthropocentrism of previous cosmologies. Moreover, Copernicus envisioned a universe that was much larger than originally conceived, thereby ushering in the modern age of astronomy with its concepts of limitless space in which humans can be seen as inconsequential specks.

See also SCIENCE, HISTORY OF.

Further Reading

Moss, J. D. *Novelties in the Heavens.* Chicago: University of Chicago Press, 1993.

craniometry, as pseudoscience

Craniometry sought to explain individual differences in intelligence, criminality, and other traits in terms of measurements of the skull. Beginning as a scientific endeavor, it has fallen into disrepute as a PSEUDOSCIENCE. The Dutch anatomist Petrus Camper (1722–89) stimulated interest in the field by developing a method for measuring facial angles that was used in racial classification. Craniometry was a respectable discipline in the 19th century when SAMUEL GEORGE MORTON (1799–1851) studied collections of skulls to support prevailing preconceptions about European racial superiority. Italian criminologist Cesare Lombroso (1836–1909) believed that criminals were recognizable on the basis of their physical appearance, including the shapes of their skulls. Craniometry also developed into a craft analogous to palm reading in which personality was supposedly read from bulges in the skull reflecting development of underlying regions of the brain.

The overarching assumption that the moral qualities of individuals can be detected from the external appearance of skulls has been fairly thoroughly discredited and craniometry is seen as an archaic pseudoscience.

Nevertheless, skull measurement continues to play an important role in the study of hominid evolution. Skull size is used as a proxy in the study of brain development by obstetricians and pediatricians. Recent research has also demonstrated that brain size correlates with IQ score even though scholars have often dismissed this idea as part of the pseudoscience of craniometry.

See also INTELLIGENCE TESTS; MORTON, SAMUEL GEORGE; PSEUDOSCIENCE.

creationism

Creationism is the view that the account of creation presented in the Bible is literally true and that Darwin's account of evolution through natural selection is therefore false. Creationists have gone to great lengths to prevent the teaching of evolution in schools. In the United States, the state of Tennessee passed, in 1925, a law making it illegal to teach evolution. This precipitated the SCOPES TRIAL, a famous 1925 test case in which high school teacher John Thomas Scopes was prosecuted for teaching evolution in Dayton, Tennessee. Although Scopes was convicted (a result that was overturned on appeal), the trial turned into a circus and, thanks partly to the biting humor of journalist H. L. Mencken, the entire country was laughing at the backwardness of Tennessee.

Following this public relations disaster, creationism experienced a decline until the late 1960s. At this time activist creationists devoted their efforts not to banning evolution so much as to having creationism taught on an equal footing. Creationist legislation went into effect in several states in the 1970s and early 1980s but all these laws were struck down following legal challenges that culminated in a 1987 Supreme Court decision that teaching creationism in public schools is unconstitutional because it violates the principle of separation of church and state.

More recently, creationists have sought to advance their doctrine as a science. The Creation Research Society was founded in 1963 to foster the publication of scientific papers written from a creationist perspective. If creationist scholars were to publish in respectable scientific journals, then the discipline could be represented as a science and could be legally taught in schools. In general, such papers are not capable of passing the test of editorial review because they lack even the semblance of openness to scientific falsification of their hypotheses, fail to exhibit scientific coherence, and cannot provide objective data to substantiate their conclusions.

See also CATASTROPHISM; EVOLUTIONARY THEORY AND EDUCATION CONTROVERSY; OWEN, RICHARD; SCOPES TRIAL; WILBERFORCE, SAMUEL.

Further Reading

Numbers, R. L. *The Creationists*. New York: Knopf, 1992.
Ruse, M. *Darwinism Defended*. Reading, Mass.: Addison Wesley, 1982.

Crick, Francis Harry Compton

(1916–)
English
Physicist and Biochemist

Francis Crick's most important scientific achievement was deciphering the molecular structure of DNA (deoxyribonucleic acid) in collaboration with JAMES D. WATSON, in 1953. DNA encodes the hereditary information in genes. Crick and Watson discovered that the DNA molecule consists of a pair of helixes that are joined together by base pairs. The double helix could be unzipped into two strands that contained the same sequence of information, suggesting an elegant mechanism by which genetic information was encoded and could be duplicated. This work ushered in the modern age of biotechnology.

The collaboration of Crick and Watson is described in a fascinating memoir by Watson titled *The Double Helix*. The human side of scientific discovery as vividly portrayed by Watson is very different from the professional image projected by scientific publications. Among their many breaches of scientific ethics and etiquette, Crick and Watson surreptitiously gained access to the data of a colleague, Rosalind Franklin. They subsequently shared the 1962 Nobel Prize for physiology and medicine with Maurice Wilkins, who had let them see Franklin's X-ray diffraction studies apparently out of antifeminist spite against her. Most people agree that Franklin's work was of pivotal importance in solving the structure of DNA and that she should have received the Nobel Prize, but she was robbed of this distinction by an early death because the prize is never awarded posthumously.

Crick studied the process of genetic translation, through which sequences of bases code for the amino acids of a complex protein, before moving on to study neuroscience at the Salk Institute in San Diego. He has published several controversial books. *Life Itself* (1981) proposed that bacteria arrived on earth from other planets, a view known as panspermia. *The Astonishing Hypothesis* (1994) makes the case that conscious experience is purely a product of neural activity and attacks alternative mentalistic perspectives, including most religions, as scientifically naive.

See also DETERMINISM; WATSON, JAMES D.

Further Reading
Crick, F. H. C. *The Astonishing Hypothesis.* New York: Scribner's, 1994.
———. *Life Itself.* New York: Simon and Schuster, 1981.
Watson, J. D. *The Double Helix.* New York: Atheneum, 1968.

crime and brain chemistry

The ability to predict criminal tendencies from brain chemistry could be a major ethical problem if this was used as an excuse for compulsory testing and medication of potential offenders. Research conducted in the 1980s revealed that violent criminals can be distinguished by their brain chemistry. One difference that has received much attention is the low level of serotonin activity in the brains of violent criminals. This can be measured indirectly in terms of breakdown products in the blood, or cerebrospinal fluid. People with low serotonin turnover lack a normal sense of the potential negative consequences of their antisocial actions and may act with unusual bravery. Generalizing from animal studies, it is clear that brain serotonin level can be affected by genetics as well as by rearing environment (e.g., social isolation).

Violent impulsive behavior is strongly predicted by low serotonin levels of the human brain. Thus, blood levels of 5HIAA, a metabolite of serotonin, explain more than one-half of individual differences in aggressiveness scores on a personality test according to several studies. Measurement of serotonin metabolites in blood or cerebrospinal fluid can even be used to predict impulsive violence such as arson, suicide attempts, and other crimes of violence. In one follow-up study of killers and arsonists, serotonin levels before release from prison predicted reoffense with 84% accuracy.

These findings are strong enough to raise the possibility of ethically questionable applications. If a parole board was informed about a violent criminal's serotonin level, and they knew that this predicted future violence with such a high level of reliability, they would be very unlikely to recommend release. Indeed, they would be remiss in their duty if they did recommend parole. This raises the issue of whether chemical treatment (e.g., with Prozac, a serotonin agonist) could be as successfully used for repeat violent offenders as it has been for impulsive sexual offenders, whose crimes are related to the impact of high levels of circulating testosterone on the brain and can be prevented by Depo Provera treatment. In each case treatment would be ethically problematic only if it was involuntary. Circulating testosterone is not just important for sexual offenses but also plays a role in explaining violent crime of various types. Thus, younger men are both more criminally violent and have higher testosterone production than older men. Treating all young men with Depo Provera would likely reduce crimes of violence and other types of risk-taking, such as dangerous driving, but such treatment would be inconsistent with the cherished value of personal freedom in a democratic society.

See also CASTRATION, CHEMICAL.

Further Reading
Virkunen, M., J. DeJong, J. Bartko, F. K. Goodwin, and M. Linnoila. "Relationship of Psychobiological Variables to Recidivism in Violent Offenders and Impulsive Fire-Setters." *Archives of General Psychiatry* 46 (1989): 600–03.

cryonics

Cryonics is the practice of freezing people soon after death to preserve their tissues with a view to subsequent reanimation. Although considered a crazy idea by many, cryonics is now carried out by several private companies, including the Alcor Life Extension Foundation of Scottsdale, Arizona, that agree to preserve clients in liquid nitrogen until it is technologically possible to bring them back to life. At present, the freezing of bodily tissues causes irreversible damage to cells but the hope is that a superior future medical technology will be capable of restoring this damage.

Even though the resuscitation part of the cryonic procedure is still very much science fiction, it is difficult to be certain that it will never become a reality. Cryonics advocates argue that even a slim chance of reanimation is better than the certainty of permanent death associated with conventional burials. The practice of freezing people with a view to subsequent reanimation has a parallel in the mummification of the rulers of ancient Egypt in anticipation of an afterlife and might more appropriately be considered a religious rite rather than a medical or scientific procedure. Nevertheless, it raises a number of interesting problems that are worthy of the attention of students of ethics.

One issue has to do with the cost of the procedure. Preserving a full body carries a price tag of around $120,000 compared to $50,000 for the freezing of the head alone (which seems to rest on the shaky premise that individuals are defined by their brain alone). At present this cost is usually borne by a life insurance policy for which the cryonics provider is the sole beneficiary. This money covers only freezing. The cost of thawing and resuscitation will have to be borne at the other end.

The theoretical possibility of future resuscitation is obviously challenging to the definition of death. The cryonics procedure itself raises questions because the subject is placed on a heart-lung machine after their heart has stopped beating so that the circulatory system can be used to pump cooled blood around the body to initiate the cooling process. Are the clients alive or dead during this initial cooling? If the client wants to preserve the possibility of future function, then it is theoretically beneficial to initiate the cooling as rapidly as possible, before there has been a chance for serious brain damage to occur. If resuscitation were technically possible, then there would be a compelling ethical case for freezing people before they had reached any legal definition of death.

See also LIFE SUPPORT, TERMINATION OF; ORGAN TRANSPLANTATION.

Further Reading
Ettinger, R. *The Prospect of Immortality.* Garden City, N.Y.: Doubleday, 1964.
Trembly, A. C. "Cryonics Patients: Gone Today, Here (and Richer) Tomorrow." *National Underwriter Life and Health-Financial Services Edition* 103, no. 45 (1999): 7.

cryptographic research, government suppression of

Openness is one of the core values of scientific research. This means that researchers share their data, theories, ideas, and results. There are two contexts in which openness is necessarily compromised in research. One is if a researcher works in industry and is wary of divulging TRADE SECRETS. The other is when a researcher conducts military research and agrees not to divulge scientific information because of implications for national security. The U.S. government also has the power to suppress any civilian research that is perceived as threatening to national security. These powers are not often invoked but they have been used to suppress civilian research on encryption, or the making and breaking of codes, constituting a clear incursion into scientific freedom and openness.

Encryption assumed great military importance in World War II as each side attempted to intercept and decode the electronic communications of the other. In the Internet age, encryption has also assumed great significance for civilians as the security of electronic commerce, including credit card accounts, and bank and brokerage accounts is at stake.

Computer secrecy is also a major concern of the U.S. military, possibly reflecting the computerization of many weapons systems, including nuclear weapons. When computer scientist George Davida of the University of Wisconsin, Milwaukee, applied for a patent on an encryption device for computerized information in 1987 he received a notice warning him that he must not publish his invention or reveal any information about it. He

was informed that noncompliance with these instructions could lead to a two-year prison sentence and a $10,000 fine. Soon afterward, the Information Group of the Institute of Electrical and Electronic Engineers received a letter from the National Security Agency (NSA) warning it that a discussion of cryptography at an upcoming international conference it was planning would violate international treaties on the export of arms.

Such heavy-handed attempts to classify, or suppress, cryptographic research have met with strong opposition from the academic community. In a compromise, the National Science Foundation (NSF) has agreed to let the NSA examine all of its funding applications in cryptography. The idea is that if the NSA decides to sponsor a research project, it is automatically classified. The NSF also continues to sponsor unclassified research in cryptography.

See also CLIPPER CHIP; ELECTRONIC PRIVACY; NATIONAL SCIENCE FOUNDATION; TRADE SECRETS.

Further Reading
Dickson, D. *The New Politics of Science.* Chicago: University of Chicago Press, 1984.
Resnik. D. B. *The Ethics of Science.* New York: Routledge, 1998.
Stix, G. "Fighting Future Wars." *Scientific American* 276, no. 6 (1995): 92–101.

cultural relativism

Cultural relativism holds that people are shaped primarily by the societies in which they are raised. Articulated in the 20th century by Columbia University anthropologist Frans Boas and his students, notably Margaret Mead, relativism has been challenged by evolutionary psychologists, who present evidence of the existence of cross-cultural universals. These include universals of emotional expression, physical attractiveness phenomena, and sex roles.

See also EVOLUTIONARY PSYCHOLOGY AND SEX ROLES; MORAL RELATIVISM.

Curie, Marie (born Sklodowska)
(1876–1934)
Polish-born/French
Physicist and Chemist

Having studied at Cracow, Poland, Marie Curie (see photo) moved to Paris in 1891 and became the assistant of physicist Pierre Curie, whom she subsequently married in 1895. In 1898, the Curies discovered two new

The first woman to win a Nobel Prize, in 1903, Marie Curie has inspired subsequent generations of female scientists. (LIBRARY OF CONGRESS)

radioactive elements, plutonium and uranium, both of which were extracted from pitchblende, a dark-colored mineral. For this work, the Curies shared the 1903 Nobel Prize in physics with Henri Becquerel. After her husband's death, Marie became professor of physics in the University of Paris, a position previously held by Pierre. Her research focused on the radioactivity of radium and its possible medical applications. In 1911 she received the Nobel Prize in chemistry. In 1919, Curie was appointed professor of radiology at the University of Warsaw, a position she held for the rest of her life. She died of leukemia on July 4, 1934, which may have been caused by her exposure to radiation in the laboratory.

Marie Curie's daughter Irène Curie (1897–1956) worked as her assistant and shared the 1935 Nobel Prize for chemistry with Jean Frédéric Joliot, another assistant. These achievements of the Curie women were remarkable in an era in which there were few professional opportunities for women in science and they inspired female scientists by demonstrating that accomplishment at the highest level was possible for women.

See also AFFIRMATIVE ACTION IN SCIENCE AND TECHNOLOGY CAREERS; SCIENCE/TECHNOLOGY CAREERS, AND WOMEN.

Dalton, John
(1766–1844)
English
Chemist and Physicist

John Dalton is remembered as the scientist who established the modern scientific concept of the atom. The notion that matter is not continuous but broken up into discrete units is an archaic one that was current as an "armchair theory" in ancient Greece associated with the name of the philosopher DEMOCRITUS, among others. The idea of an atom, which has a characteristic weight for each element and which remains fundamentally unchanged in chemical combinations, is one of the foundational ideas of modern chemistry. Dalton used this concept to explain why chemical compounds are made of fixed proportions by weight of their constituent elements. When he published his ideas in *A New System of Chemical Philosophy* (1808) he included a table of atomic weights and a list of chemical symbols.

Dalton was an enthusiastic experimentalist and produced remarkably accurate results given the crudity of available equipment. Historians of science have concluded that his results were too good and that he must have fudged his data to fit the hypotheses. Attempts to replicate some of Dalton's key experiments on the oxides of nitrogen have not succeeded in producing anything close to the precision of his results. When historian J. R. Partington replicated these experiments, he found that Dalton's simple ratios could not be produced. This makes it unlikely that he merely engaged in selective reporting, publishing only those data that closely matched his predictions (a form of SCIENTIFIC DISHONESTY perpetrated by ROBERT MILLIKAN in his celebrated oil drop experiment).

Dalton spent most of his life working as a teacher and lecturer and considered himself primarily to be an educator. Nevertheless, he was a busy empirical scientist who is said to have made some 200,000 meteorological observations during his lifetime. He concluded that the aurora borealis (or northern lights) is a magnetic phenomenon. In accounting for the formation of dew, he produced a table of the vapor pressures of water at different temperatures. He published a law of partial pressures according to which equivalent increases in temperature of gases produced equivalent increases in volume. While most of Dalton's ideas about the structure of the atom have had to be changed in light of subsequent evidence, his contributions, flawed as they were by dishonesty, helped to establish chemistry and physics as modern sciences organized around experimental hypothesis testing.

See also DEMOCRITUS; MENDELEYEV, DMITRI IVANOVICH; SCIENCE, HISTORY OF; SCIENTIFIC DISHONESTY.

Further Reading
Broad W., and N. Wade. *Betrayers of the Truth*. New York: Simon and Schuster, 1982.

Nash, L. K. "The Origin of Dalton's Chemical Atomic Theory." *Isis* 47 (1956): 101–16.

Partington, J. R. "The Origins of the Atomic Theory." *Annals of Science* 4 (1939): 278.

Darwin, Charles Robert
(1809–1882)
English
Naturalist

Charles Darwin, author of the 1859 theory of evolution by natural selection, provoked religious and ethical debate by suggesting that human beings are descended from apes. (LIBRARY OF CONGRESS)

Charles Darwin's name is associated with the theory of evolution by natural selection. This theory implied that all living species, including human beings, are the result of descent with modification from earlier species, which thus conflicted with the biblical depiction of creation. Although Darwin (see illustration) was not personally antireligious, his evolutionary views of human origins brought him into conflict with orthodox religious views and he was attacked by religious polemicists such as BISHOP SAMUEL WILBERFORCE. A retiring man, who was plagued by health problems of unknown origin, Darwin steered clear of public debate and was championed by English biologists THOMAS HUXLEY and RICHARD OWEN. He preferred to devote his energies to scientific work and throughout his long career as a naturalist compiled a huge volume of detailed observations. For example, he devoted 10 years of his life to the study of barnacles, a topic on which he published four books.

Apart from the religious debate, Darwin's name is also implicated in the controversy over social Darwinism, a philosophical perspective that seeks to apply evolutionary principles to human economic and political behavior. Darwin is also important as the first person to write on the evolution of moral behavior and the emotions on which it is based.

DARWIN'S LIFE

Darwin was born into a privileged and distinguished family. One grandfather was Erasmus Darwin the naturalist, poet, and philosopher. The other was Josiah Wedgewood, a celebrated manufacturer of porcelainware. Darwin's cousin FRANCIS GALTON was one of the most brilliant scientists of his day who made important contributions to genetics and psychology. As a young man, Darwin showed little sign of brilliance and seems to have had minimal interest in his education at Shrewsbury School, which was followed by two years of medical training (1825–27) at Edinburgh University. Seeing surgery performed without anesthetic quickly convinced Darwin that he did not wish to follow his father into medicine. Sent to Cambridge University to study for the ministry in the Church of England he quickly discovered that he had

no vocation. Instead, he cultivated an interest in natural history. After receiving his B.A. degree in 1831, Darwin was recommended by John Stevens Henslow, a Cambridge professor and friend, for a position as unpaid naturalist aboard the HMS *Beagle*. The *Beagle's* five-year voyage, which began on December 27, 1831, was aimed at studying the Pacific coast of South America and some of the Pacific islands and setting up navigational stations. Darwin's job was to record geological formations and to study plant and animal life along the way. He had no training as a geologist but embarked on his geological research with enthusiasm, eventually producing three books in the field: *Coral Reefs* (1842), *Volcanic Islands* (1844) and *Geological Observations on South America* (1846). Darwin studied fossils and was impressed by the fact that the ancestral species preserved in the fossil record were not the same as those currently alive even though there was often an unmistakable resemblance. This did not square with the received wisdom that species had been created once and had remained the same ever since. Darwin became convinced that modern species are descended from different species that lived in earlier geological periods.

The *Beagle* voyage was a turning point in Darwin's life. He embarked as a young man with a vague ambition of making a mark in natural history and returned with a conviction about his theory of evolution by natural selection and a burning determination to compile a body of scientific work that would convince the scientific community of its validity.

In 1939, three years after his return, Darwin married his cousin Emma Wedgewood. In 1842, the couple moved to a quiet suburb of Downe, 16 miles from London. About this time he began to be plagued by the mysterious gastrointestinal symptoms and skin rashes that would bother him for the rest of his life. The origin of this illness is though to be at least partly psychological but other contributing factors might have been exposure to parasites during the *Beagle* voyage and attempts at self-medication. He coped with chronic illness by maintaining a rigid work schedule and felt best when he was most productive. Some biographers feel that Darwin used illness as a tool to keep worldly responsibilities at bay, so that he could focus on substantiating his evolutionary theory. Whether this is true or not, he was exceptionally fortunate in having a supportive wife who organized the household around facilitating Darwin's work.

Despite his illness, Darwin had an ideal environment for concentrating on his main problem. Yet his progress was notoriously slow. He let himself get sidetracked into largely descriptive work on barnacles and other projects that seem of trivial importance today. Six years after his return, he had still only produced a 35-page summary of his ideas on evolution by natural selection. This essay had been shared with a few close friends, including geologist Charles Lyell and biologists Thomas H. Huxley and Joseph D. Hooker, who happened to be some of the brightest natural scientists of their day. Darwin's friends urged him to expand on his ideas and publish them before someone else gained priority. By 1844, Darwin had expanded his essay into a 200-page paper but he was not satisfied with the result and refrained from publishing it. Instead, he spent the next 14 years obsessively poring over the details to make the theory as flawless as possible before publishing it to what he knew would be a hostile and skeptical readership.

Darwin was shocked into action by the arrival from the East Indies, in June 1858, of a letter from a young naturalist, Alfred Russell Wallace, that included a summary of a theory of evolution that was virtually identical to Darwin's, even in the terminology that was used. Ironically, Wallace had asked Darwin for help in getting his theory published. Darwin did help by reading Wallace's paper at a meeting of the Linnaean Society on July 1, 1858, along with excerpts from his own manuscript. Having published his theory in this way, Darwin worked quickly to bring out a book-length treatment. When it was published in 1859, *On the Origin of Species by Means of Natural Selection* was an immediate success and subsequently proved to be one of the most influential books ever written. Even though the work generated a great deal of controversy, Darwin succeeded in convincing his readers that evolution was a reasonable theory that needed to be taken seriously.

Although evolutionary theory conflicts with religious teachings concerning creation, its basic scientific ideas have never been controversial among biologists. Darwin pointed out that there is a struggle for existence (or survival). Some individuals are better equipped for survival than others and they pass on these favorable traits to their offspring. By the same token, individuals who do not have favorable traits are likely to die without reproducing, being thus weeded out by natural selection. As a consequence, over many generations, populations of a species come to have just those characteristics that help them to survive and reproduce. Once it has been stated in this way, evolutionary theory is as obviously true as a Euclidean theorem. The most serious gap, apart from evidence of evolutionary change in progress, was Darwin's complete ignorance of why offspring resemble parents. The young science of genetics had just been founded by an obscure monk named GREGOR MENDEL whose work on inheritance of traits in the garden pea was ignored by the scientific community until the dawn of the 20th century.

Among the other important books written by Darwin was *The Descent of Man* (1871), which explicitly applied evolutionary principles to human ancestry. In this work, Darwin argued, as he did in *The Expression of Emotions in Man and the Animals* (1872), that humans had evolved from nonhuman species. Both books stressed the continuity of mental life as well as biological ancestry. Darwin made the challenging case that humans and nonhumans not only have similar expressions of emotion but also that the mental life of nonhumans is continuous with our own, even to the extent of postulating the existence of moral sentiments in other species. Darwin died on April 19, 1882, at his home in Downe and was buried in Westminster Abbey.

DARWIN'S CONTRIBUTIONS TO EVOLUTIONARY ETHICS

Darwin did not have any pretension to being a moral philosopher. Yet he articulated some of the most important problems in evolutionary ethics in a way that is compelling to contemporary philosophers and other scholars. All of these problems are related to the continuity of human beings with other evolved species on this planet. Darwin himself had difficulty in believing that human beings were just another evolved species probably because this conclusion went so much against the grain of his religious background. This difficulty was resolved for him by an encounter with the Tierra Del Fuego Indians of Patagonia. To Darwin's eyes, here were a group of people so miserably lacking in foresight that they could not even clothe themselves to protect against the constant chilly winds of their miserable habitat. He was so repelled by these naked savages who seemed no better than animals that he could accept the continuity of species.

Even though Darwin shared the unabashed distaste for uncivilized peoples common to others of his time and milieu, he recognized that primitive peoples have a moral sense. He even speculated that a capacity for nurturance and self-sacrifice could have been favored by natural selection since tribes in which individuals had these morally admirable qualities would be better able to triumph over rivals. (Darwin frequently thought of selection acting on entire groups rather than individuals. This approach is dismissed by most contemporary biologists who see individual-level selection as much more important).

If morality has adaptive value for human beings, then it is reasonable to imagine that it would confer similar benefits on nonhuman animals. Darwin therefore predicted that moral sentiments, however rudimentary, should be found among other species and pointed to anecdotal accounts, from a zookeeper, concerning the capacity for empathy in CHIMPANZEES. When one chimpanzee was upset, another would embrace it in an apparent attempt to provide comfort. Modern research on the behavior of chimpanzees has provided some fairly compelling evidence that they are self-aware, capable of altruism in the context of food sharing, that they empathize with the distress of another, and that they form rudimentary social contracts involving reciprocal aid in aggressive encounters.

The notion of continuity of moral sentiments has some interesting and potentially troubling implications. One implication has to do with Kantian ethics, which espouses that the rights of individuals derive from their capacity for moral reasoning. If the moral community is defined by a capacity for morality, then it would have to include chimpanzees as well as humans! One practical ramification of this position is that chimpanzees and other great apes have rights implying they should not be held in captivity or subjected to invasive medical research. Of course, many ANIMAL RIGHTS activists maintain that these arguments apply to other laboratory animals also by virtue of their capacity to feel pain. Whether other species apart from great apes turn out to have primitive capacities for moral reasoning, as Darwin supposed, remains to be seen.

The whole direction of Darwin's influence in evolutionary ethics has been to break down barriers between humans and other evolved species. One aspect of this influence has been to promote the view that the ruthless forces of competition, without which evolution cannot occur, are also very much a part of human social life, including politics and economic affairs. This philosophy is referred to as SOCIAL DARWINISM.

Social Darwinism is greatly despised today as a theory that postulates no mercy for the weak and that elevates the interests of the powerful over the needs of the powerless. In the past, it was used as a justification for ruthless colonial expansion abroad and pitiless exploitation of workers and peasants at home. At an extreme, social Darwinists opposed VACCINATION of children because this allowed the biologically weak to survive and reproduce. For the same reason, any government programs designed to relieve the sufferings of the poor were resisted. Philosophers have argued that social Darwinism commits the NATURALISTIC FALLACY by assuming that just because ruthless competition is a natural feature of human social interaction, as it is in the interactions of other species, it should be considered either good or desirable.

There is some controversy about whether Darwin was a social Darwinist or not. This controversy is understandable. After all, how could a man who was so admirable in many ways have endorsed a philosophy that is so repellant to scholars today? The answer may be that Darwin was a man of his period and was not above its prejudices.

Social Darwinism has its roots in the writings of economist THOMAS MALTHUS (1766–1834) whose dismal pronouncements about the inevitability of the human population rising faster than the resources needed to sustain it greatly influenced Darwin's formulation of the struggle for survival. This was an essential ingredient of his theory of evolution. Strictly speaking, HERBERT SPENCER, a philosopher and founder figure in the discipline of sociology, was the first to extol the virtues of competition in weeding out unfit individuals and races and he did so before the 1859 publication of *On the Origin of Species*. Darwin's own views on the subject are clearly articulated in *The Descent of Man*. He describes the inevitability of superior tribes and races overcoming inferior ones and there is no question that he saw the expansion of the British Empire as an example of a superior group inevitably winning out over weaker ones.

See also DARWIN'S FINCHES; DARWIN/WALLACE AND CODISCOVERY OF EVOLUTION; EVOLUTIONARY ETHICS; SOCIAL DARWINISM.

Further Reading

Bowler, P. J. *Charles Darwin: The Man and His Influence.* New York: Cambridge University Press, 1996.

Browne, J. J. *Charles Darwin: Voyaging.* Princeton: Princeton University Press, 1996.

Darwin, C. R. *The Voyage of the Beagle.* Amherst, N.Y.: Prometheus, 1999.

Darwin, C. R., and F. R. Darwin, ed. *Autobiography of Charles Darwin and Selected Letters.* Mineola, N.Y.: Dover, 1991.

Hawkins, M. *Social Darwinism in European and American Thought, 1860–1945.* New York: Cambridge University Press, 1997.

Darwin's finches

These are a group of 14 related finch species (*Geospizinae*) distributed around the Galapagos Islands that were discovered by Charles Darwin in the context of his *Beagle* voyage. Darwin noted that the different species had a surprising variety of beak shapes ranging from the thick heavy bill of the large ground finch (*Geospiza magnirostris*) that fed on hard seeds to the delicate beak of the small insectivorous tree finch (*Camarhyncus parvulus*). He speculated that all of the finches had evolved from a single mainland species and speculated that their current differences could have emerged as adaptations to exploiting different food sources. These ideas influenced Darwin's formulation of the theory of evolution by natural selection.

Recent research has not only confirmed many of Darwin's intuitions about the finches but also has provided evidence that evolutionary change may occur within a single generation rather than the much more gradual change over hundreds of generations that Darwin had in mind when he formulated his theory of evolution. Peter R. Grant, a professor of zoology at Princeton University, studied changes in finch populations for two species, the medium ground finch (*Geospiza fortis*) and the cactus finch (*Geospiza scandens*) in the context of droughts, in 1977 and 1982 on Daphne Major, a small Galapagos islet. Grant found that the ground finch was particularly hard hit by the first drought with only 15% of the birds surviving. Interestingly, the surviving individuals were larger and had deeper, stronger, beaks. These traits favored survival for different reasons. The stronger individuals were better able to break open the hard seeds that were the only available food following the drought. In addition, their size and strength would have helped them to compete with other individuals over access to limited food sources. Since body size and bill thickness are highly heritable, the ground finch genotype had been modified in a single generation in a manner that made them more suited to surviving drought conditions.

Research on Darwin's finches has thus contributed to a substantial body of modern evidence. The research shows evolution in action that ranges from the emergence of drug-resistant bacterial strains to the phenomenon of INDUSTRIAL MELANISM, in which the blackening of trees by industrial soot favored the survival of dark-colored moths near English cities. Such findings have not only provided empirical backing for evolutionary theory among biologists but also have convinced many theologians previously opposed to evolutionary theory that to deny the reality of evolution by natural selection is equivalent to claiming that the earth is the center of the universe. The recent capitulation by Pope John Paul II accepting evolution as factual is an important milestone in the controversy between religion and evolutionary theory. It illustrates the weak hand held by modern creationists who continue to insist, despite the empirical evidence, that evolution is just a theory.

See also CREATIONISM; DARWIN, CHARLES ROBERT; INDUSTRIAL MELANISM.

Further Reading

Grant, P. R. *Ecology and Evolution of Darwin's Finches.* Princeton: Princeton University Press, 1986.

Lack, D. *Darwin's Finches.* Cambridge: Cambridge University Press, 1947.

Ruse, M. *Darwinism Defended.* Reading, Mass.: Addison-Wesley, 1982.

Weiner, J. *The Beak of the Finch.* New York: Vintage, 1994.

Darwin/Wallace and codiscovery of evolution

When scientists make original discoveries, they are entitled to recognition of their contributions. What happens if discoveries are simultaneous? How can priority be determined and credit apportioned? A particularly interesting case of codiscovery was the formulation of a theory of evolution by natural selection on the part of Alfred Russell Wallace and Charles Darwin. Darwin's ethical conduct in this matter was impeccable and provides a model for how to avoid unnecessary disputes between scientists over priority of discovery.

What makes the Darwin/Wallace story so intriguing is that from one perspective Darwin had clear priority having produced a detailed description of his theory of evolution in the early 1840s, almost 20 years before Wallace. Having come up with one of the most influential ideas of the 19th century, Darwin did nothing with it. Recognizing that the theory would be received as highly speculative, he shied away from publishing it and diverted his attention to a monumental four-volume study of barnacles. Meanwhile, he showed drafts of his ideas to friends who happened to be leading scholars and this limited publication established his priority over Wallace.

On the morning of June 18, 1858, Darwin received a nasty jolt in the form of a letter from one of his many correspondents, Alfred Russell Wallace, a young naturalist and collector who had been working in the Far East. Wallace had enclosed a short paper hoping that Darwin would help him to publish it. In the paper, Darwin recognized his own theory of evolution by natural selection and was struck by the fact that even the language used was similar to his own. In fact, the paper he received from this young naturalist whom he had never met might as well have come from Darwin's own pen. Some scholars interpret such coincidences as supporting the idea that most scientific discoveries are inevitable and will occur when the time is ripe rather than requiring unusual genius to bring them to light.

Deeply disturbed, Darwin consulted with his friends Charles Lyell the geologist and Joseph Hooker the botanist. His friends had been aware of Darwin's theory of evolution for some time and had urged his to publish before someone else anticipated him. It was agreed that Darwin should have his share of the credit. At the following meeting of the Linnaean Society, Darwin read Wallace's paper along with extracts from Darwin's own unpublished work. Significantly, Darwin's paper was read first.

Even though Wallace was treated ethically by his scientific contemporaries, few people today even remember Wallace's name. In modern life sciences, everything is Darwinian. There is no Wallacism. Darwin may have shared priority for the discovery of natural selection but he retains most of the credit. There are two good reasons why Darwin received the lion's share of the recognition. One is that he was a member of a privileged elite who had connections with many important people among the intellectual and political elite of his time. The other is that Darwin worked hard at developing evolutionary theory in book length treatises that were best-sellers of the day and made him a central figure in religiously motivated attacks on evolutionary theory.

See also DARWIN, CHARLES; DARWIN'S FINCHES; INVENTIONS, THEORIES OF; SIMULTANEOUS DISCOVERY.

Further Reading
Lamb, D., and S. M. Easton. *Multiple Discovery.* Trowbridge, England: Avebury, 1984.
Resnik, D. B. *The Ethics of Science.* New York: Routledge, 1998.
Ruse, M. *Darwinism Defended.* Reading, Mass.: Addison Wesley, 1982.

data cooking, fabrication, falsification, fudging, trimming   *See* SCIENTIFIC DISHONESTY.

DDT (dichloro-diphenyl-trichloroethane)

DDT is a pesticide that was widely used around the world until its potential for damage to ecosystems and to humans was recognized. Banned in most developed countries, DDT is still used in several developing nations to control human disease vectors such as mosquitoes. The story of DDT occupies a special place in the history of public concern over environmental issues because it marked an awakening to the ecological impact of distributing large quantities of persistent toxins into the environment.

DDT AS A MIRACLE CHEMICAL
First synthesized in 1874, DDT's effectiveness as an insecticide was recognized only in 1939. Swiss chemist Paul Muller discovered the effectiveness of DDT in controlling insect-carried diseases, including malaria, bubonic plague, river blindness, yellow fever, and typhus. He received a Nobel Prize in medicine for his work. DDT is a member of the chlorinated hydrocarbon (or organochlorine) group of compounds. During World War II, large quantities of DDT were used by the U.S. military to protect troops against malaria, typhus, and other insect-borne diseases to which they were exposed in conflicts far from home. Military historians recognize that DDT played an important role in saving the lives of thousands of soldiers and some claim that without its help the ability of American soldiers to wage war would have been greatly reduced, possibly affecting the outcome of the war.

Following the war, DDT was seen as a miracle chemical and American farmers applied large quantities of it to protect their crops from insect predation. In the 30 years prior to the banning of DDT in 1972, it is estimated that 1.35 billion pounds of the chemical were used in the United States. DDT was popular because it was cheap, versatile, effective in seeming to kill most insects, and long lasting, ruling out the need for frequent administration. Moreover, it seemed harmless to humans and people had no qualms about administering it to their domestic pets, for example.

DDT AS A BIOHAZARD
Public perceptions of DDT began to change in 1962 with the publication of RACHEL CARSON's book *Silent Spring.* Although Carson's book targeted DDT as one of the most widely used pesticides of the day, its logic applies to widespread use of other pesticides also. A science writer, with training in biology, Carson's critique of DDT was well researched and received favorable reviews from professional biologists. Yet it was received with incredulity and hostility by the general public and was publicly attacked by the chemicals industry.

The qualities that made DDT a successful and popular pesticide are the same as those that made it ecologically harmful. Chlorinated hydrocarbons such as DDT are highly toxic to insects. By some mechanism that has not been clearly established, these substances compromise the function of living cells and damage the nervous system resulting in death. DDT is a broad-spectrum insecticide, which means that it can kill most insects. It is fat soluble, which means that it accumulates in fatty tissues and cannot be excreted. For this reason, DDT tends to accumulate in the tissues of its victims.

Devastating as the effects of DDT were for insect populations, they had even more drastic effects on their natural predators, particularly birds, because of the effect of concentrating toxins as one ascends a particular food chain. Thus, if earthworms are exposed to DDT, their tissues acquire a certain concentration of the pesticide. Robins who eat the exposed earthworms may consume several times their own weight, which means that the robins will acquire a higher concentration of DDT than the worms. Similarly, hawks feeding on robins will acquire a much higher concentration of the toxin than their prey. By the time Carson's book had been published, scientists had observed several examples of bird populations being affected by accumulating levels of pesticide from their prey species. In 1957, the death of grebes in Clear Lake, California, had been attributed to DDD, a pesticide resembling DDT. In 1958, the use of DDT to control Dutch elm disease had wiped out a local population of robins at Urbana, Illinois, that had evidently consumed toxic prey. Carson pointed out that since humans are at the top of the food chain, we are exposed to the problem of concentrated toxicity.

Apart from the unintended devastation to bird populations, and the potential toxicity to humans, Carson was impressed by the ultimate futility of attempting to control pest populations through the use of dangerous chemicals. In particular, she recognized that killing off large numbers of insects using broad-spectrum insecticides sets up a scenario in which the evolution of pesticide resistance is guaranteed. Individuals that have partial resistance to the pesticide survive and reproduce passing on this resistance to their offspring. After a small number of generations, populations become established that are completely immune to the toxicity. When this happens, the farmer is likely to resort to a newer, usually more expensive insecticide, thereby initiating a new cycle in the evolution of insect resistance.

While this argument concerning the evolution of pesticide resistance may have seemed highly theoretical to Carson's readers, it is well established as an empirical phenomenon today. In 1938, only seven insect species were identified that were resistant to pesticides. By 1984, this number had climbed to 447. Some environmentalists argue that the loss of crops to insect pests today (around 15%) is similar to what it would have been before any pesticides were used. In summary, it appears that short-term and diminishing returns to farmers in terms of increased profitability are balanced by long-term damage to ecosystems.

EFFECTS OF DDT ON HUMANS AND OTHER MAMMALS

In the controversy that has raged over DDT ever since the publication of Carson's book, supporters of the pesticide have often pointed out that it has saved the lives of thousands of U.S. troops and that it has never been known to kill a single person. While this claim may still be technically true, a number of disturbing phenomena have been reported concerning the adverse health consequences of DDT for humans and other mammals. Many of these phenomena are recent, reflecting the continued widespread use of DDT in the developing world. Others reflect the continuing problems associated with the persistence of DDT in ecosystems where the pesticide has not been applied for decades.

The persistence of DDT in the environment is remarkable. The DDT molecule remains chemically stable for decades. It enters the atmosphere and is borne by wind all over the planet. It is more likely to condense in cold climates. For this reason, some of the highest concentrations of DDT have been found in penguins, polar bears, and among the Inuit people of Canada. DDT and related chemicals have a number of adverse health consequences. These pesticides can disrupt the functioning of vertebrate hormone systems producing at least two sorts of adverse effect. One is that immune function is suppressed. Thus, among the heavily exposed Inuit children, there is impairment of antibody production and an increased incidence of chronic infections. Another is that sexual development and reproduction can be affected. One example of this phenomenon is the unusually small reproductive organs of pesticide-exposed alligators in Florida's Lake Apopka. The relevance of this for humans is unclear.

Another disturbing problem is that pesticides like DDT can impair development of the central nervous system in the case of individuals exposed to high doses. Thus, learning problems and attention deficits have been observed in the case of mothers who consumed contaminated fish from Lake Michigan. Mexican children exposed to pesticides had diminished memory capacity (as well as reduced physical stamina) also suggesting impairment of central nervous system development. Such problems have convinced many world leaders that the time has come for a global ban on the use of DDT and other dangerous pesticides. Although experiments have shown that DDT can cause cancer in rodents, there is no direct evidence that it causes cancer in humans.

GOVERNMENT REGULATION OF DDT

Rachel Carson's book dealing with the toxic effects of DDT in ecosystems was highly controversial when published but it served to focus public interest on the problem of pesticides and probably accelerated regulatory action by the U.S. ENVIRONMENTAL PROTECTION AGENCY (EPA). Nevertheless, government regulation of DDT had begun five years before the publication of Carson's book. In 1957, the Forest Service of the U.S. Department of Agriculture (USDA) banned the spraying of DDT on strips of land around aquatic areas that came under its jurisdiction. The following year, the USDA began a general program of phasing out the use of DDT and after 10 years had reduced DDT use by about 98%. They also took a number of actions that restricted the domestic and agricultural use of DDT on food crops, agricultural animals, ornamental shrubs and turf, wood products, buildings, food processing plants, and in restaurants.

In 1970, the EPA was formed and took over responsibility for the regulation of pesticides. From the beginning, the EPA pursued a fairly aggressive course against DDT. On June 14, 1972, the EPA announced the final cancellation of all crop uses of DDT from December 31, 1972. An appeal launched by the chemicals industry was rejected by the U.S. Court of Appeals for the District of Columbia in the following year.

Under the Federal Environmental Pesticides Control Act, the EPA could authorize temporary use of DDT in situations in which there was a demonstrable economic necessity and in which no other pesticides were effective. Using these criteria, the EPA permitted DDT to be used in 1974 to control the pea leaf weevil in Idaho and Washington. In the same year DDT was used to control a Douglas fir tussock moth epidemic in the Northwest. However, the state of Louisiana was denied its request to use DDT for control of the tobacco budworm in 1975.

Although DDT has been banned in most of the industrialized countries of the world, its use is still legal in many developing countries and it is tolerated in other countries even where it is technically illegal. The most important reason for its continued use is that it is so effective against the mosquitoes that carry malaria. Countries in which DDT is still legal generally are located within regions where malaria is a potential problem. They include Bolivia, Bhutan, Ethiopia, India, Kenya, Malaysia, Mauritania, Mexico, Philippines, Nepal, Sri Lanka, Sudan, Tanzania, Thailand, Venezuela, and Vietnam.

The costs of DDT use in India are becoming apparent. Indiscriminate use of the pesticide in agriculture and for controlling malaria has wrought havoc with ecosystems as illustrated in the collapsing population of Indian vultures. India's residents now receive DDT in virtually all food products and have a body concentration of 22.8 PPM compared to 2.2 PPM for persons in the United States. A 1996 study of lactating mothers found that infants receive a dose of DDT in breast milk that is 40

times the acceptable daily intake according to World Health Association guidelines.

DDT use is therefore a demonstrable threat to the health and well-being of people and the integrity of ecosystems around the world, not just in the countries where it is used. As a result, representatives of over 100 nations gathered in Nairobi in 1999 to work out a global treaty that would eventually eliminate the use of DDT and 11 other dangerous chemicals, which together are nicknamed the "dirty dozen."

See also ENVIRONMENTAL PROTECTION AGENCY; LOVE CANAL.

Further Reading
Carson, R. *Silent Spring.* Boston: Houghton Mifflin, 1962.
Newton, L. H., and C. R. Dillingham. *Watersheds 2: Ten Cases in Environmental Ethics.* Belmont, Calif.: Wadsworth, 1997.

Declaration of Helsinki

The Declaration of Helsinki provides ethical guidelines for conducting experiments using human subjects. Originally formulated in 1964 by the World Medical Association, the declaration underwent revisions in 1975, 1983, 1989, and 1996. It reflects the same broad concerns as the NUREMBERG CODE in the sense that its primary concern is to protect the interests of subjects used in experiments.

A more detailed document than the Nuremberg Code, the Declaration of Helsinki goes beyond the older statement in several ways. While the Nuremberg Code guarantees protection against being subjected to research of dubious scientific value, Helsinki goes further in stating: "Concern for the interests of the subject must always prevail over the interests of science and society" (Basic Principle 5). In other words, the scientific merit of a piece of research can never be used to justify harm to the subject.

The Declaration of Helsinki recognizes a distinction between research that is conducted in a therapeutic context and that which is purely scientific in its objectives. In general, procedures that are potentially therapeutic are encouraged even if they incur risks. Thus, the physician is "free to use a new diagnostic and therapeutic measure, if in his or her judgment, it offers hope of saving life, reestablishing health, or alleviating suffering" (Part II, Principle 1). Withholding of effective treatment, as in the case of a PLACEBO-CONTROLLED STUDY, is prohibited except in cases where "no proven diagnostic or therapeutic method exists" (Part II, Principle 3). The word "proven" is subject to interpretation and might not rule out the replication of a placebo-controlled study if this is considered scientifically necessary to establish the therapeutic effectiveness of an experimental treatment.

The rights of human research subjects are more strongly asserted in the case of nontherapeutic research. For example, such subjects must be volunteers and the research must be discontinued if it poses a threat of harm to the individual. In therapeutic research, by contrast, research may be conducted without informed consent if it passes review by the relevant ethics committee. There is no explicit statement that the research has to be discontinued if it poses a threat to the subject presumably because many experimental therapies may pose such risks.

The declaration is influential in biomedical research since adherence to its ethical requirements is required by over 500 medical journals through the *Uniform Requirements for Manuscripts Submitted to Biomedical Journals.* The Declaration of Helsinki is regarded as a key foundational document in the ethics of biomedical research. Some ethicists argue, however, that its provisions are not strict enough in the case of therapeutic research that could be carried out on dying patients without their consent. Along with the Nuremberg Code, it has been highly influential in the formulation of the American *DHHS Regulations for the Protection of Human Subjects* that governs federally funded research.

See also BELMONT REPORT; DHHS REGULATIONS FOR THE PROTECTION OF HUMAN SUBJECTS; INFORMED CONSENT; MINIMAL RISK CRITERION; NUREMBERG CODE.

Further Reading

Woodward, B. "Challenges to Human Subject Protection in U.S. Medical Research." *JAMA* 282 (1999): 1947–52.

World Medical Association. "World Medical Association Declaration of Helsinki. Recommendations Guiding Physicians in Biomedical Research Involving Human Subjects." *JAMA* 277 (1997): 925–26.

deep ecology approach to environmental ethics

Ethical systems have always placed human beings at the center of the moral universe in the sense that good actions were defined in terms of what people are supposed to do to obey divine commands, increase human happiness, or promote virtue in the individual. The deep ecology approach to environmental ethics dethrones humans from their position at the center of the moral universe and asserts that the prime imperative is protection of ecosystems that are often seriously threatened by human actions. The term "deep ecology" was coined by Norwegian philosopher Arne Naess in 1973. Naess argued in favor of a deeper questioning and a deeper set of answers to environmental concerns.

Deep ecology is equivalent to ecological wisdom and is sometimes referred to as "ecosophy." According to Naess, the shallow approach to ecology is based on the self-interest of people in developed countries who want to avoid the problems of environmental pollution and resource depletion. Deep ecology is deep in a spiritual sense. It is not simply a way of solving immediate problems but a whole philosophy of life, and even a religion, that can be distinguished from widely held Western views about the exploitative relationship between humans and the rest of the natural world.

Deep ecologists see humans as an inseparable part of the natural world and they see the awakening of this sense of connectedness as a fundamental aspect of self-realization. Another key belief is that all forms of life are equal in their intrinsic worth because they all play an essential role in the ecosphere. This principle of biocentric equality applies not just to other complex animals but to microorganisms and plants. Thus, a vegetarian who claims that it is permissible to eat plants but not animals does not live up to the deep ecology belief system because they are suggesting that plants have less right to life than animals do. Deep ecology accepts the fact that species must use each other for food and shelter but argues that excessive interference by humans with other species is harmful to the ecosystem and, by implication, harmful to humans as part of that system.

The belief in biocentric equality suggests many principles that can guide our actions in ways that diverge from the prevailing technocratic and consumerist worldview. Instead of dominating nature and exploiting it to satisfy our own limitless needs, human beings should try to live in harmony with the ecosphere and to promote policies that are minimally destructive of ecosystems and use as little natural resources as possible. One of the most interesting positions taken by deep ecologists concerns the world population crisis. Whereas the conventional technocratic view has been that we should increase food production to feed the rapidly expanding human population, deep ecologists believe that unconstrained human population growth is harmful to other species on the planet and should therefore be resisted. In fact, the well-being of other species in the ecosystem calls for a reduction in the human population.

Deep ecologists generally do not favor the use of highly technological solutions to practical problems and have adopted the slogan "simple in means, rich in ends." This implies that traditional forms of agriculture are generally preferred over modern agribusiness. Consistent with this approach to economics, deep ecologists are in favor of decentralized government on the basis that centralization of power is most likely to promote environmental degradation in local areas, such as populous modern cities. Increased local autonomy also reduces pollution by reducing energy expenditure associated with the movement of goods to large centers of population. Together with decentralization goes increased respect for the autonomy and lifestyle of people living in traditional or subsistence economies. Deep ecologists are in favor of a classless society because a class hierarchy has the undesirable implication that one group is exploiting another contrary to the spirit of ecological egalitarianism.

Another ideological gulf between deep ecologists and the prevailing worldview involves the farmer's rejection of consumerism. Whereas industrial economies are predicated on a relentless pursuit of the profit motive, which calls for continual promotion of an affluent lifestyle and consumption of energy and natural resources, deep ecologists favor cutting back consumption as much as possible. This goal is incompatible with the ethos of a society in which consumption is increased and new needs are continually being created through aggressive advertisement techniques.

Deep ecology elevates the simple life in which the purpose of the individual is directed more toward spiritual growth than toward material wealth. The quality of life is separable from, and more important than, having a high material standard of living. Instead of craving for the latest consumer goods, people should attempt to consume less. Recycling plays an important role in reduced consumption.

Although deep ecologists believe in egalitarian biocentrism, there is a certain inconsistency of approach to different species that is not strictly egalitarian. In fact, some philosophers refer to it as "antianthropocentric biocentrism." Thus, human beings are expected to restrain their impulse to exploit and conquer the planet but no other species is expected to practice the same self-restraint. A skeptic might inquire why people should reverse the course of history and start behaving in an ecologically responsible fashion. The only compelling rationale for why this might happen is widespread realization that destruction of the environment is threatening to the well-being of our species and that self-interested remedial action is required to avert catastrophe. This is shallow ecology according to Naess's terminology.

See also ENVIRONMENTAL ETHICS.

Further Reading

Devall, B., and G. Sessions. "Deep Ecology." In *Environmental Ethics,* ed. L. P. Pojman, 144–46. Belmont, Calif.: Wadsworth, 1994.

Naess, A. "Deep versus Shallow Ecology." In *Environmental Ethics,* ed. L. P. Pojman, 137–44. Belmont, Calif.: Wadsworth, 1994.

Pojman, L. P., ed. *Environmental Ethics.* Belmont, Calif.: Wadsworth, 1994.

Watson, R. "A Critique of Anti-anthropocentric Biocentrism." In *Environmental Ethics,* ed. L. P. Pojman, 149–54. Belmont, Calif.: Wadsworth, 1994.

Democritus

(460–370 B.C.)
Greek
Philosopher

A native of Abdera in Thrace, Democritus developed classical atomism derived from his teacher Leucippus. Although he is understood to have written on every topic under the sun, only a few fragments of his work survive and most of what is known about him is through the work of other scholars.

According to atomism, the universe is made up of a combination of impenetrably hard atoms and empty space. Democritus believed that atoms had characteristic shapes that allowed them to join together resulting in the familiar shapes of everyday objects. His cosmology was based on constant change and he thought of the universe as containing many different worlds, each with a distinct beginning and end. Some worlds were growing, others decaying. Some had no sun or moon, others had multiple suns and moons. Worlds were created by the spinning of atoms in vortices and could be destroyed through collision with a larger world. Life develops naturally out of inanimate matter. In many respects, the atomism of Democritus is extraordinarily close to the philosophy underlying modern science.

Democritus was remarkably lacking in mysticism. He objected to any kind of supernatural explanation for the material world and despised the idea of an afterlife as a ridiculous fiction. He rejected any notion that there was an underlying purpose to human existence. His materialism extended to a description of the soul which he thought of as composed of round atoms. Thought was produced by the movement of these atoms. Although he is often criticized for believing in chance, Democritus, like Leucippus before him, was actually a determinist who saw all events as deriving from mechanical causes.

Democritus's ethics were based neither on moral obligations nor on consequentialism and can therefore be categorized as a virtue approach of sorts. He believed that cheerfulness was the only appropriate goal for a person's life and for this reason is often referred to as the "laughing philosopher." Democritus distinguished between happiness and pleasure. Like many other Greek philosophers, he disapproved of intense pleasures, such as sexual intercourse, but greatly valued the milder pleasures produced through intellectual cultivation and friendship. Democritus believed that everyone is responsible for their own happiness and he defined moral actions as those that are consistent with the conscience of the individual. In many respects his ideas resemble the secular philosophy of self-improvement that emerged in the 20th century, particularly in the United States. He was also an ardent democrat who believed that living in poverty in a democracy was preferable to living in wealth under a despot.

See also ARISTOTLE; MORAL PHILOSOPHY.

Further Reading
Russell, B. *A History of Western Philosophy.* New York: Simon and Schuster, 1972.

deontology

In normative ethics, deontological theories hold that the morality of an action is based on principle rather than on consequences. Deontology is derived from the Greek word *deon* meaning that which is obligatory. According to deontological theories, an action may be moral because it conforms to rules or principles, because it reflects good character, or because it is based on good motives. In the field of applied ethics, deontology arises frequently in connection with the rights of professional clients, research participants, the general public, and even, in the case of environmental ethics, future generations. The rights of clients, research participants, and the public impose corresponding obligations on scientists and technologists.

See also UTILITARIANISM; MORAL PHILOSOPHY.

Depo Provera   *See* CASTRATION, CHEMICAL; CONTRACEPTION.

Descartes, René
(1596–1650)
French
Philosopher and Mathematician

Descartes has the unusual distinction of being associated with the mind-body problem in philosophy, a problem which he neither originated nor solved. Descartes's legacy is that he presented mind-body dualism in a way that future philosophers found unsatisfactory but which they were unable to dismiss or ignore. His extreme skepticism, as reflected by the statement "I think, therefore I am," has also challenged subsequent philosophers. Descartes's most important contribution to mathematics was the unification of geometry with algebra through the use of Cartesian coordinates.

In Cartesian philosophy, the body was thought of as a machine. Thus, Descartes considered animals to be pure automatons that were incapable of either sensations or thoughts. Because animals could not feel pain, as he

believed, they could be dissected while still alive. Humans were unique in having a rational soul, or mind that was not made of physical substance and did not occupy space. Whereas most previous thinkers had seen the soul as controlling the body, Descartes's brand of interactionism involved a two-way process in which the mind received its sensations, for example, as a result of physiological mechanisms. In this scheme, the mind initiates and controls actions so that humans are capable of free will.

Descartes was a keen student of physiology and proposed a crude model, inspired by the technology of moving statues of his day, of how movement is controlled by nerves. In the process, he anticipated the stimulus-response association, or reflex that became a central idea of modern psychology. His view of animals as automatons was extended to humans by 20th century behaviorist psychologists who generally did not deny that the mind existed but argued that it had no control over action. According to this view, thoughts accompany actions in much the same way that noise accompanies a lawnmower. This epiphenomenalist position has been vehemently rejected by the majority of current ethical philosophers who argue that our ability to think is what is uniquely human and distinguishes us as ethical beings. Yet there are some who recognize that how we think is very much a function of the activities of neurons. Thus, modern philosophers like Paul and Patricia Churchland have chosen to study neuroscience as a means of understanding the human mind. Descartes never explained the critical issue of how a nonmaterial mind could communicate with a material body. Similarly, neuroscientists still cannot explain how neurons generate mental experiences except in the limited sense that damage to specific parts of the brain can result in specific losses of mental capacity.

See also CHURCHLAND, PAUL; WATSON, JOHN B.

Further Reading
Marshall, J. *Descartes' Moral Theory.* Ithaca, N.Y.: Cornell University Press, 1999.
Rozemond, M. *Descartes' Dualism.* Cambridge, Mass.: Harvard University Press, 1998.

"designer babies"   *See* EUGENICS.

determinism

This is the philosophical view that phenomena are produced by preceding events according to natural laws. Most scientists are determinists in their professional work but entertain notions of supernatural causation in their religious beliefs. In respect to human behavior, there are two kinds of determinism, genetic and environmental, that interact with each other. Many people, both religious and nonreligious, reject a strict determinism of human behavior in favor of free will. If people were capable of free will, as René Descartes claimed, then a science of human behavior would be impossible.

See also DESCARTES, RENÉ; GENETIC DETERMINISM.

dharma

In Buddhist thinking, dharma means the natural law. Dharma is all-encompassing. It includes both the physical laws that govern the movement of the stars, the changing of the seasons, and the moral laws (see KARMA) that distinguish right from wrong and define the duties of each individual. Since Buddhist ideology is believed to be grounded in natural law, the term dharma is extended to the entire body of Buddhist teachings. Dharma is also applied to the Buddhist Path by which enlightenment is approached.

Dharma is not controlled by a divine being. Rather, the gods themselves are controlled by its natural order. Buddhist ethics is not based on divine command in the way that Western religious ethics is. It is rational in the sense that moral action is seen as promoting the welfare of the individual and of the community. The highest ethical goal is happiness and personal well-being rather than simple obedience to a supernatural authority.

See also BUDDHIST ETHICS; KARMA; MORAL PHILOSOPHY.

DHHS regulations for the protection of human subjects

Since 1981, federally funded research in the United States that employs human subjects (or participants) must be conducted according to ethical guidelines promulgated by the Department of Health and Human Services (or DHHS). These regulations replaced the 1974 guidelines issued by the Department of Health, Education and Welfare. They are based on the conclusions of the BELMONT REPORT and revolve around its principles of beneficence, respect for persons, and justice. Ethical oversight of research is the responsibility of INSTITUTIONAL REVIEW BOARDS at the funded institutions, which must approve only research that is to be conducted in accordance with the regulations.

See also BELMONT REPORT; INSTITUTIONAL REVIEW BOARD; RIGHTS OF HUMAN RESEARCH PARTICIPANTS.

Further Reading

U.S. Department of Health and Human Services. *Code of Federal Regulations Pertaining to the Protection of Human Subjects.* Washington, D.C.: Government Printing Office, 1983.

Diagnostic and Statistical Manual of Mental Disorders (DSM)

This is the classification system of mental disorders used by most mental health professionals. The main purpose of the Diagnostic and Statistical Manual (DSM) is to provide a reliable system for categorizing mental disorders so that different professionals in different locations can agree on their diagnoses. With this aim in mind, the DSM provides an exhaustive list of behavioral descriptions, known as "diagnostic criteria," that can be used to determine the diagnosis. The DSM system collects information about patients in five categories, or "axes," that include: the primary diagnosis; personality and developmental disorders, if any; physical health problems; life stresses; and level of social functioning, such as ability to retain a job. All of this information is considered helpful in deciding on treatment and prognosis.

The DSM aims for objectivity and attempts to avoid inferences about the underlying causes of disorders, focusing instead on external manifestations. The system has gone through many editions (the most recent being DSM IV, 1994) and has been open to change and improvement. Thus, criticized for confusing moral judgments with medical diagnosis by including homosexuality as a mental disorder in earlier editions, DSM now considers sexual orientation to be a problem only if it is seen as such by the individual. Another controversial issue has been the emergence of personality disorders or long-standing patterns of maladaptive behavior. Many professionals feel that personality disorders include too much of what is arguably ordinary behavior.

See also BIOLOGICAL PSYCHIATRY; HOMOSEXUALITY, POSSIBLE BIOLOGICAL BASIS OF; SEXUAL ORIENTATION RESEARCH.

Further Reading

American Psychiatric Association. *Diagnostic and Statistical Manual of Mental Disorders: DSM IV.* Washington, D.C.: American Psychiatric Association, 1994.

dioxin   *See* AGENT ORANGE, TIMES BEACH.

DNA evidence

All individuals, except for identical twins and other multiples, are genetically unique. The differences among individuals can be analyzed in terms of the way that their DNA (deoxyribonucleic acid, the basic material of heredity) is cut by restriction enzymes. Restriction fragment length polymorphism (RFLP) patterns form a distinct "genetic fingerprint" that can have useful forensic applications.

Crime scenes that do not have actual fingerprints, perhaps because the perpetrator took the precaution of wearing gloves, almost always contain DNA evidence. Suspects can be placed at the scene of a crime with a very high degree of confidence using DNA analysis of blood stains, semen stains, hair roots, skin fragments, and so forth. There is controversy about the probability of error in DNA evidence because experts disagree on the underlying statistical techniques but in the absence of sample contamination, the error rate is usually far less than a million to one. DNA fingerprints are obtainable from samples that are several years old. This has enabled persons falsely convicted of crimes to be exonerated years after the event.

The use of DNA evidence in legal proceedings introduces two major ethical issues. One concerns the capacity of DNA evidence to, in effect, put the legal system itself on trial by ruling out the possibility that some individuals could have committed the crimes for which they were convicted and the other involves setting up DNA databases. Thus, the Innocence Project, run by well-known defense lawyer and DNA expert Barry Sheck has freed no fewer than 43 Americans who were sentenced to death by means of retrospective DNA tests on trial evidence. The fact that so many suspects can be wrongly convicted has strengthened the argument that the death penalty is unethical and ought to be abandoned. The pace at which old cases are being examined in the light of the new tests is very slow, however, due to shortage of forensic DNA labs.

If DNA tests are more widely used, they promise not only to minimize the likelihood of innocent people falling victim to false accusations but also to streamline criminal investigations by beginning with the name of the main suspect instead of conventionally identifying a large list of possible culprits and laboriously narrowing them down. Successful use of DNA evidence in the early stages of an investigation would rely on the creation of a genetic database that could be checked for matches. This idea has often met with resistance on the mistaken assumption that it would involve doing DNA tests on the entire population. All that would be required would be to test individuals who had been convicted of a felony.

See also EYEWITNESS TESTIMONY; FALSE MEMORY AND CRIMINAL CONVICTIONS; FBI CRIME LAB; GENETIC ENGINEERING.

Further Reading
Sheck, B., P. Neufeld, and J. Dwyer. *Actual Innocence.* New York: Doubleday, 2000.

Dolly, cloned sheep   *See* GENETIC ENGINEERING.

double effect, doctrine of
The doctrine of double effect is an ethical principle which recognizes that morally acceptable actions may have evil consequences. The evil must not have been intended, however, and the evil effect must not have been the means of producing the good effect. For example, if high doses of morphine are used to relieve suffering of a dying person, this may have the evil effect of hastening death. The intention behind such palliative care is to reduce pain and not to practice euthanasia. Hastening death is not the intended means of reducing pain. The doctrine of double effect can thus be invoked to justify dangerously high doses of morphine for dying people.

downwinders
Downwinders is a name applied to people who live downwind of nuclear weapons tests in the United States and thereby experience harmful effects of radiation. According to some authorities, the amount of radiation exposure due to nuclear testing in the United States has been far greater than that produced by the CHERNOBYL, NUCLEAR DISASTER, although the extent of nuclear contamination is controversial due in part to military secrecy. Some of the nuclear tests, particularly those conducted in the atmosphere, have revealed an astonishing lack of concern about health consequences, which is reminiscent of the COLD WAR RADIATION EXPERIMENTS ON HUMAN SUBJECTS in which people were subjected to dangerous doses of radiation in medical experiments without their knowledge or consent.

The dispersal of harmful radiation throughout the ecosystem in military tests raises many important ethical issues:

Can such military experiments ever be ethically justified?

If nuclear tests are carried out, who ensures that this is done in a manner that is minimally threatening to the health of local residents and downwinders?

After the tests have been conducted and radioactive contaminants have been dispersed in the atmosphere, what is the responsibility of governments to monitor radiation levels and to issue advisories and warnings that might influence where citizens chose to live, for example?

Given that radioactive contamination increases the incidence of cancers and other health problems, what are the ethical obligations of governments in respect to providing health care and monetary compensation for victims?

At present, the answers to these questions in the United States are greatly clouded by a lack of reliable information attributable to military secrecy and possibly to the desire to minimize government liability by revealing the extent of the effects on health and ecological damage produced by nuclear tests.

NUCLEAR TESTS IN THE CONTINENTAL UNITED STATES

The extent of radioactive fallout from American nuclear testing has only recently become clear. The health consequences to downwind residents and workers at 17 nuclear weapons plants around the country have received considerable publicity. In 1997, the National Cancer Institute released evidence that almost the entire continental United States was subjected to radiation from 1951 to 1962 when nuclear weapons were tested. All Americans of the period can thus be referred to as downwinders. The average exposure was equivalent to receiving 200 chest X-rays, but some people in highly exposed areas received over 10 times this much radiation.

The actual testing of the weapons posed the greatest amount of risk to military personnel and local residents. The Nevada desert was chosen as a location for most of the tests because it is sparsely populated. Testing by the Atomic Energy Commission began in the early 1950s. In a 40-year period, there were approximately 800 underground tests and 126 atmospheric explosions. Atmospheric tests are a great deal more hazardous because a huge cloud of radioactive dust is pushed up into the atmosphere and gets widely dispersed by the prevailing winds. Some of the atmospheric tests in Nevada were more powerful than the bombs that destroyed HIROSHIMA and NAGASAKI during World War II. Local residents often received skin burns from the blasts as they observed radioactive dust settling around them. Livestock died and the milk supply was contaminated with radioactive isotopes. Military personnel stationed only a mile-and-a-half from the blasts sometimes walked over ground zero where the sand had been melted to glass, thereby acquiring dangerously high doses of radiation that caused extensive bleeding.

The nuclear weapons industry would be "dirty" even if there were no atmospheric tests. One of the most heavily irradiated sites is the Hanford Nuclear Plant in Washington State, which produced over half of the plutonium used in nuclear weapons. It is estimated that 14,000 children living in the town of Hanford received at least 10 rads of radiation to their thyroids as a result of exposure to radioactive iodine 131. According to a study sponsored by the Centers for Disease Control, between 1944 and 1957 the Hanford plant released over 30,000 times as much radioactive iodine as the THREE MILE ISLAND plant did during its 1979 partial meltdown. The nuclear weapons program, along with NUCLEAR POWER, has also created a formidable HAZARDOUS WASTE disposal problem.

BIKINI ATOLL

This chain of coral islands, located in the Ralik chain of the Marshall Islands in the Pacific Ocean was a U.S.–administered territory during the cold war and was chosen as an atomic test site. Twenty-three nuclear devices were exploded there between 1946 and 1958. The residents, numbering about 200, were relocated to Kili Island before the testing began. In 1968, Bikini was declared habitable and the United States began bringing the residents back. In 1978, when the residents' blood levels of strontium 90 had reached dangerously high values, the Bikinians had to be evacuated again. The United States agreed to decontaminate the area in 10 to 15 years but radiation in the islets has remained dangerously high.

HEALTH CONSEQUENCES

Since Americans have been largely kept in the dark about the extent of their exposure to nuclear radiation as a result of military activities, the health consequences, with the obvious exception of journalistic accounts of high rates of cancers among downwind populations, are largely unknown. (Similar secrecy shrouds the effects of nuclear testing by other countries). According to a National Cancer Institute Study, the radioactive iodine released in nuclear tests could have caused as many as 75,000 case of thyroid cancers.

The kind of epidemiological studies that could quantify the actual increase in thyroid cancers, and other radiation-related illnesses for highly exposed populations such as those at Hanford and in the Nevada desert, remain to be done. As recently as 1998, Hanford residents were suing the Department of Energy (DOE) for funds necessary to initiate the Hanford Medical Monitoring Program. Ironically, this would cost only $12.9 million, a tiny fragment of the $10 billion the DOE has already spent in cleaning up the site under environmental regulations (the Superfund law) that were designed to protect communities from toxic sites.

MISINFORMATION AND SILENCE

While it might be argued that the unusual exigencies of the cold war era justified the taking of unusual environmental risks, many downwinders are angry that information that could have helped them avoid living in dangerously contaminated areas was not released by the U.S. government.

In 1982, after a 1978 lawsuit brought by residents living immediately downwind of the Nevada Test Site revealed health problems, Congress ordered the National Cancer Institute to conduct a study of radiation exposure and associated health consequences. Fifteen years later, their report was still pending. Whistle-blowers at the DOE leaked parts of the study to *USA Today* in 1997. The Cancer Institute admitted, in an August 1977 press conference, that the results of the study had been known for several years. One interesting piece of information from the National Cancer Institute study is that the average exposure of U.S residents to nuclear radiation was 200 times higher than the government said it was at the time of the tests. Such a divergence of estimates could be due to extraordinarily inept science or it might reflect a campaign of deliberate public misinformation.

The whole pattern of the U.S. nuclear testing program was one of excessive risk-taking. What seems like highly irrational conduct on the part of the military authorities is difficult to explain and probably has many causes. These include the unusually high level of paranoia encouraged by the nuclear arms race with the Soviet Union. Those responsible for the tests may well have had little appreciation of the long-term health consequences of exposure to nuclear fallout or they may even have considered that this was a small price to pay in the patriotic struggle for global domination. (One reason that the Mormon residents affected by testing in Nevada did not protest initially is that they believed their sacrifices were helping the country). In 1946, when J. ROBERT OPPENHEIMER, who worked on developing the first atomic bomb, voiced concern about nuclear testing in a letter to President Harry Truman, and advised postponement of the program in the interests of safety, Truman went on the offensive, referring to Oppenheimer as a crybaby scientist.

Before the first nuclear test was carried out at Alamogordo, New Mexico, in 1946, there was concern that some of the earth's atmosphere could be blown out but the test went ahead anyway in a breathtaking display of scientific recklessness. From this decision, it appears that the military scientists in charge of the tests acted without the restraint of common sense much less a well-developed ethical concern over the consequences of their actions for the health and well-being of downwinders. The test was conducted without any form of ethical review or other restraining device and provides an example of big science run amuck.

If the military scientists were behaving without ethical restraint, their actions were congruent with government policies and were supported by a deliberate campaign of official misinformation and silence. Thus, the public was repeatedly assured that the nuclear testing program was safe in spite of the fact that farm animals were dying of radiation sickness in the vicinity of test sites and that huge quantities of radioactive isotopes were being shot into the atmosphere and dispersed all over the country.

Not only did the U.S. government fail to protect its citizens in allowing so many extremely dangerous tests to proceed, a lapse which might be justified on the grounds of military exigency in the context of the Soviet threat, but they also failed to take due responsibility after the event. Monitoring of radiation levels and health effects was needlessly, and callously, postponed for decades. Not only were citizens harmed by the immediate consequences of the nuclear tests but those who chose to live in contaminated areas, having no information about the dangerously high radiation levels, were subsequently harmed by government secrecy. The tests were thus carried out in an ethical vacuum in which the government paid little heed to potential health consequences and was unwilling to take responsibility for those effects after the event.

See also CHERNOBYL NUCLEAR DISASTER; HAZARDOUS WASTE; HIROSHIMA AND NAGASAKI; NUCLEAR POWER; THREE MILE ISLAND NUCLEAR ACCIDENT.

Further Reading
Berger D. "We're All Downwinders." *The Nation,* October 13, 1997.
Skolnick, A. A. "'Downwinders' Angry about Medical Screening." *JAMA* 280 (1998): 408.
Ward, C. *Canaries on the Rim: Living Downwind in the West.* New York: Verso, 1999.

Draise test
The Draise test was once widely used in the cosmetics industry to determine the toxicity of chemicals. The test substance was placed directly on a rabbit's eye to see if it caused inflammation or other tissue damage. Widely regarded as an unethical infliction of pain and suffering on animal subjects, and used as propaganda by animal rights activists, the procedure has been dropped in favor of less controversial tests. It is worth noting that far more invasive and painful techniques continue to be used in biomedical research. For example, lethal doses of drugs are given to laboratory animals to measure toxicity. The goal of the cosmetics industry, namely, to make people look better, is evidently considered too frivolous to provide an ethical justification for inflicting pain on animals.

See also ANIMAL RESEARCH, ETHICS OF.

drug use during pregnancy
Recreational drugs used during pregnancy have the potential for producing permanent damage to the offspring. Common fetal effects of addictive drugs, particularly alcohol and cocaine, include growth retardation, permanent neurological impairment producing low IQ and attentional problems, and anomalies of physical appearance. In the United States, the unborn child has

little legal protection against such unwise maternal behavior because the fetus is not considered a person in the eyes of the law.

Legal attempts to protect the unborn have mainly taken the route of arresting mothers who use illegal substances, such as cocaine, amphetamines, and marijuana, but ignoring mothers who drink or smoke during pregnancy. There is no evidence that incarcerating pregnant mothers for drug use has improved the outcomes of their offspring and it can have damaging consequences for the psychological development of children. If mothers expect to be arrested they may also avoid prenatal care for fear of revealing their drug use. The hard-line incarceration approach loses an important opportunity for drug treatment because pregnancy gives women a strong motive for breaking addictions.

The criminalization approach also ignores the fact that most of the damage to the unborn is caused by legal drugs, particularly alcohol, which has been found to produce impairment of prenatally exposed infants, even at moderate doses. According to a 1992 National Institute on Drug Abuse Study, about 6% of U.S. women used an illegal drug during pregnancy compared to 19% who consumed alcohol and 20% who smoked cigarettes. A 1995 study by Patricia Shiono and others found that 2% of pregnant women had used cocaine, 11% had used marijuana, and 35% had smoked cigarettes.

Maternal drug use during pregnancy is a recent problem for several reasons. One is that addictions seem to be promoted by modern urban life and possibly reflect a relative lack of social integration in the anonymous conditions of cities. Another is that drug use among women has risen steadily as they have moved into the workforce in larger numbers and enjoy greater economic and social freedoms. Finally, the connection between prenatal drug exposure and infant health problems has been unambiguously established by modern scientific research that integrates the results of animal experiments with epidemiological evidence. Now that the problem is so well understood, legal and healthcare professionals are faced with the tough ethical dilemma of deciding what to do about it.

See also CONTRACEPTION; REPRODUCTIVE TECHNOLOGIES.

Further Reading
Marwick, C. "Challenging Report on Pregnancy and Drug Abuse." *JAMA* 280 (1998): 1039.
Schroedel, J. R. *Is the Fetus a Person? A Comparison of Policies across the 50 States.* Ithaca, N.Y.: Cornell University Press, 2000.
Shiono, P. H., M. A. Klebanoff, R. P. Nugent, et al. "The Impact of Cocaine and Marijuana Use on Low Birth Weight and Preterm Birth: A Multicenter Study." *American Journal of Obstetrics and Gynecology* 172 (1995): 19–27.

dualism

Dualism is the perspective in philosophy according to which body and mold are separate entities. Dualism is implied in most mainstream religions. The contrary view that there is only physical matter, referred to as MONISM, is typical of modern science.

See also DESCARTES, RENÉ.

duplicate publication

Duplicate publication means publishing exactly the same paper in different journals. This is prohibited by the ethical standards observed in some sciences. Authors engaging in duplicate publication are seen as dishonestly padding their publication list, an important, if unreliable, measure of academic productivity and success, while wasting valuable publication space that might have gone to original research. Duplicate publication can be justified, however, when there is a perceived need to bring a particular piece of research to a wider audience. Journal editors occasionally reprint articles from other journals that are of particular relevance to their readers. Duplicate publication also occurs ethically when journal articles are reprinted in book form. Authors avoid the charge of self-plagiarism by acknowledging any prior publication of their work whether in whole or in part.

See also PLAGIARISM; SCIENTIFIC DISHONESTY.

E

E. coli, as food contaminant   *See* FOOD IRRADIA-
TION.

earthquakes and construction codes
It often seems that earthquakes, tornadoes, and other natural disasters are more likely to strike poor countries and
poor people living in affluent countries. In reality, poor
people are often selected as victims because they inhabit
housing, such as shanty towns and trailer parks, that is
not built well enough to stand up to such cataclysmic
events. The importance of building codes is illustrated in
comparing the effects of earthquakes in California to
Turkey. On August 17, 1999, an earthquake that measured 7.4 on the Richter scale struck Izmit, a densely
inhabited suburb of Istanbul, destroying 40,000 buildings
and killing an estimated 15,000 people. A quake of comparable magnitude (6.7 on the Richter scale) struck
Northridge, a suburb of Los Angeles, in 1994, damaging
12,500 structures but killing only 57 people. Northridge
is one of the most earthquake-proof areas in the world
insofar as its building codes are concerned and there is
no question that the much lower mortality rate is largely
a function of sturdier buildings. The tragedy in Turkey
was made worse for families of the victims by allegations
that many of the buildings were constructed with minimal concern for safety, and the contractors responsible

for this shoddy work have been referred to as mass mur-
derers. As in the case of collapsing bridges, exploding
chemical factories, and melting down nuclear plants,
accidents are not so much random events as incidents
that are caused by inadequate safety protocols. A more
scientific approach to such phenomena compels us to see
them as largely preventable and therefore as an ethical
problem.

See also BHOPAL CHEMICAL ACCIDENT; BRIDGE COLLAPSE;
CHERNOBYL NUCLEAR DISASTER.

Further Reading
Neistat, V. "Earthquake Massacre." *Science World* 56, no. 5
   (1999): 16.

Echelon spy system   *See* ELECTRONIC PRIVACY.

ecosystem
An ecosystem consists of animals and plants, their physical environment, and the complex web of interrelation-
ships that determines how each species survives and
reproduces. The physical environment includes soil,
water, minerals, nonliving objects, sunlight, and climate.
An ecosystem can be analyzed in terms of the flow of
energy and the cycling of nutrients through its compo-

nents. In general, the energy that sustains an ecosystem is obtained from sunlight by plants through the process of photosynthesis. Other species obtain their energy from plants or by feeding on herbivores.

The concept of an ecosystem has been of central importance to environmental ethics ever since ecologist Rachel Carson documented the accumulation of DDT at ever higher concentrations as one moved up the food chain. Thus, robins feeding on DDT-exposed earthworms obtain a higher concentration of the toxin in their tissues than the worms, and hawks feeding on robins obtain a still higher concentration of DDT. It is clear that an understanding of the effects of environmental pollution must take ecosystems into account. Some environmental ethicists, such as J. BAIRD CALLICOTT, have argued that ecosystems merit our protection because the totality of co-evolved species in a biotic community has inherent moral value that is entirely independent of their usefulness to human beings.

See also DDT; DEEP ECOLOGY; GAIA HYPOTHESIS.

Edison, Thomas Alva
(1847–1931)
American
Inventor

With a record 1,039 patents held either singly or jointly, Edison was the most prolific of inventors. Trained as a telegraph operator, he produced, and patented, many improvements to telegraphic equipment. With the fortune made from these patents, he set up the world's first industrial research lab at Menlo Park, New Jersey, in 1876. Edison's most important inventions included the phonograph, the microphone, the electric light bulb, and a primitive movie camera and projector. His numerous improvements to existing devices included patents for electricity generation, electric storage batteries, and the telephone.

Perhaps the greatest mystery about Edison's life is why he was so extraordinarily productive as an inventor. He is often held up as an example of the inventor as genius. Yet most of his inventions were clearly determined by the context in which they occurred. Many of Edison's inventions took the form of modest improvements to existing technology that made it more effective or useful. Moreover, contemporaries were producing similar inventions to solve similar problems, resulting in legal conflicts over patents, as for example, in the case of the first commercial telephone. What is more, Edison was far from being a lone genius. Some of his successes were at least partly attributable to his entrepreneurial ability to exploit the efforts of other talented individuals who worked for him.

Edison learned quickly from the work of others, sometimes making a minor adjustment to an existing device and calling it his own.

Even if it is acknowledged that most of Edison's inventions could have been produced by other inventors of his day, contrary to the "genius" school of thought, Edison himself is something of an enigma with regard to his sheer energy and productivity. One clue to this puzzle is provided by his childhood education. Edison was deaf, which meant that he was unable to follow lessons at school and was something of a misfit. He compensated by developing a precocious interest in science and technology, which was fed by voracious independent reading.

Edison had other unusual personal attributes. He was unrelentingly practical and affected scorn for academic science. This scorn may have been more of a pose than anything else since he recruited academically trained researchers for his Menlo Park facility. Yet it is clear that Edison had no personal interest in making contributions to science and expended all of his energies in a focused attack on specific practical problems. Once he had identified a problem, he approached it with complete confidence that he could succeed and this optimistic attitude allowed him to attack obstacles that discouraged many other inventors of his day.

In summary, Edison's life story can be used to support both the contextual theory of invention and the genius theory. His accomplishments as an inventor were inextricably tied both to the period in which he lived and to his own individual qualities. He happened to live in an age when industry had become closely allied to science in the sense that scientific advances produced rapid changes in industry and the intellectual capital of inventors could be converted to economic gain associated with the new products to which they led. The intimate relationship between science and industry during Edison's life produced an interesting reversal in the direction of influence. Whereas scientific advances typically facilitate technological development, some of Edison's practical inventions led to the making of scientific discoveries that would influence academic research. One example is the "Edison effect," namely, the flow of electricity across the space between the filament of an electric light bulb and the metal plate positioned close to it. Scientists later attributed this effect to thermionic emission of electrons from the hotter filament to the colder metal plate.

See also INVENTIONS, THEORIES OF; SCIENCE, HISTORY OF; SCIENCE-TECHNOLOGY DISTINCTION; SIMULTANEOUS DISCOVERY; TECHNOLOGY, HISTORY OF.

Further Reading
Adair, G. *Thomas Alva Edison: Inventing the Electric Age.* New York: Oxford University Press, 1996.

Einstein, Albert
(1879–1955)
German-American
Physicist

Einstein's name is synonymous with scientific genius and this is a testament both to the complexity of his theory of relativity and its ability to challenge our everyday perceptions of the physical world. Einstein was not the first scientist to suggest that the dimensions of space and time are not absolutes as postulated by Newtonian physics. This distinction belongs to Irish physicist George Francis Fitzgerald (1851–1901), who suggested that when objects move at high speeds approaching the speed of light they shrink in size and that space and time are thus measurable relative to each other rather than standing alone as absolutes. Einstein's distinction was that he worked throughout his life to get all phenomena, from the movement of subatomic particles to the orbits of planets, to fit into the same relativity theory.

Scholars often distinguish between the special theory of relativity, which was published early in Einstein's career (1905) and which dealt with subatomic particles, and the general theory, which was published later (1915) and dealt with cosmology and astrophysics. The special theory received instant recognition from physicists probably because its predictions were more amenable to experimental tests. Even so, the general theory made some testable predictions. For example, Einstein predicted that light should be bent when it passed a gravitational field. This prediction was very publicly verified in 1919 in the context of an eclipse and the result brought Einstein considerable media attention that catapulted him to worldwide fame.

Einstein's career as a scientist was unorthodox. He was not a diligent student and following a mediocre academic performance, graduated from the Zurich Polytechnic as a secondary school mathematics and physics teacher. After two financially difficult years, he found a job as an examiner at the Swiss patent Office, where he remained for seven years (1902–09). This was an intellectually challenging job but it had little to do, at least directly, with his interests as a physicist. While working at the patent office, Einstein produced some remarkable papers that represent a triumph of amateur science. Not only was he not in personal contact with the great physicists of his time but he apparently spent little time doing library research.

The value of Einstein's contributions was recognized by the University of Zurich and he was appointed as an associate professor of physics there in 1909. This was followed by prestigious appointments at the Kaiser-Wilhelm Gesellschaft in Berlin (1914–33) and at the Institute for Advanced Study in Princeton, New Jersey (1933–55).

Despite such recognition, Einstein's theories remained controversial throughout his life. Thus, when he received the Nobel Prize for physics in 1921 he was cited for early (1905) work on the photoelectric effect (in which light strikes electrons off a negatively charged plate) rather than for his most important work on relativity.

Einstein had very mixed feelings about his native country. Having renounced his German citizenship after his family moved to Switzerland, he found himself back in Germany, working in Berlin in 1918. He was one of a small number of academics who retained to his pacifist principles and refused to support Germany's wartime expansionist ambitions. The fact that he was Jewish and supported Zionism occasioned further conflict with political conservatives in Germany. Einstein was denounced as a traitor and there were even attacks on his physics by "Aryan" physicists. His fortitude and ethics during this difficult period earned him considerable admiration outside Germany.

With the rise of fascism, Einstein was forced to give up his job and he moved to America to continue his work at the Institute for Advanced Study in Princeton (1933). He also abandoned pacifism. One aspect of Einstein's relativity theory involved the equivalence of matter and energy as expressed in the famous equation $E = mc^2$ (where E is energy, m is mass, and c is the speed of light). This equation suggests that matter contains a huge quantity of energy, which has two immediate practical implications. First, liberating the energy contained in atoms could provide a limitless source of power. Second, liberating this energy could have enormous destructive potential in warfare.

It was with the second possibility in mind that Einstein sent a letter to President Franklin D. Roosevelt in 1939 urging him to support U.S. development of atomic weapons before such weapons were developed by the Germans. Einstein's counsel in subsequent communications was influential in mobilizing the MANHATTAN PROJECT, which developed the first atomic bombs.

Following World War II, Einstein's pacifism re-emerged and he campaigned for nuclear disarmament. In the 1950s he lived through the period of McCarthyism and voiced his dissent to persecution before the House Un-American Activities Committee, just as he had resisted the forces of totalitarianism in Germany 30 years earlier.

Admirable as Einstein's public life was, his family life was not a happy one. His first child, a daughter, had to be given up for adoption apparently on account of the poverty of the parents. As in the case of many other people of great achievement in science, his single-minded devotion to physics detracted from his ability to place any great emotional investment either in his marriage, which broke up, or in his two sons, who ended up being bitterly estranged. Some feminists have criticized him for his treatment of his wife, a gifted mathematician. It has been claimed that he neither acknowledged her early collaboration in his work nor helped her to develop an academic career of her own. His neglect of his immediate family stood in striking contrast to his loyalty toward friends, such as mathematician Marcel Grossman, who shared his interests.

See also SCIENCE, HISTORY OF; MANHATTAN PROJECT; NUCLEAR WEAPONS.

Further Reading
Bernstein, J. *Albert Einstein.* New York: Oxford University Press, 1997.
Brian, D. *Einstein: A Life.* New York: John Wiley, 1997.

electroconvulsive therapy (ECT)

Originally developed to treat schizophrenia, electroconvulsive therapy (ECT) is used today to treat depressions, particularly those that do not respond to medication. In ECT the discharge of electricity through the brain produces a convulsive seizure analogous to that seen in epileptics.

Although ECT is quite effective as a treatment for depression, producing results that are slightly better than drug therapies, there is no particular scientific rationale explaining why it works. As a treatment, it is little more sophisticated than pounding the top of a TV set to improve the picture. Troubling side effects include confusion and loss of memory that can last for months following a course of ECT, which typically consists of three shocks per week for two to six weeks.

Ethical concerns about the use of ECT were aired in the novel *One Flew Over the Cuckoo's Nest,* which suggested that the procedure was being abused to cow mental patients into submission. Although the novel's fanciful plot may have had a tenuous relationship with reality, it seems that ECT was overused in the 1950s and that it did have the distressing side effect of causing severe confusion. Modern administration of ECT minimizes side effects by using the lowest possible level of electric current and sometimes by treating only one hemisphere of the brain.

See also PREFRONTAL LOBOTOMY.

electromagnetic radiation and cancer risk

In 1979, Nancy Wertheimer and Ed Leeper of the University of Colorado reported that children living near power lines were at increased risk of developing cancer. The study produced a great deal of sensationalist coverage in the mass media. Several other studies of children pro-

duced inconsistent results and failed to establish a dose-response relationship between the intensity of the magnetic fields to which children were exposed and the size of their increased risk for cancer. Study of electrical workers exposed to power lines also found an increased risk of cancer but this effect was fully explained by their exposure to solvents, like benzene, that are known to be carcinogenic and thus could not have been due to electromagnetic fields. However, large-scale epidemiological studies of adults residing for a long period close to power lines have failed to find any increased incidence of cancer. The whole controversy over electromagnetic fields and cancer is an interesting illustration of the uncertainty surrounding scientific findings and the popular hysteria to which this may lead.

The confusion may have been exacerbated by scientific fraud because biochemist Robert Liburdy, who purported to find that electromagnetic radiation affected calcium signaling in cells, was found guilty of falsification or fabrication of results by the OFFICE OF RESEARCH INTEGRITY in connection with a paper published in 1992 in the *Annals of the New York Academy of Sciences.* This work was of critical importance because it suggested a mechanism by which electromagnetic radiation could cause cancer.

See also ALAR SCARE.

Further Reading
Pool, R. "Is There an EMF-Cancer Connection?" *Science* 249 (1990): 1096–98.

electronic censorship   *See* PORNOGRAPHY.

Electronic Funds Transfer Act (1980)

This U.S. legislation was designed to protect the security of electronic accounts. It guarantees that customers are notified of third-party access to their accounts.

See also ELECTRONIC PRIVACY.

electronic privacy

Privacy can be defined as an area of inaccessibility that surrounds individuals and their social interactions in specific locations and contexts. For example, the casual conversation of people in their homes is normally considered private but what they say in a public street is not. Privacy has two main components. One aspect concerns preventing the movement of information out of the private sphere into the public one, as in the case of a home telephone being bugged. The other relates to the movement of potentially harmful, or unwanted, information from the public sphere into the private one. Thus, many people believe that the explicit depiction of sexual behavior on television threatens the privacy of families. This rationale has been used to promote censorship of such materials from television and the Internet in the United States. Such legislative attempts have generally been unsuccessful although commercial television is controlled by self-censorship because anything that alienates the viewership damages advertising revenues.

Even though there has been much controversy about what kinds of materials are appropriate for broadcast, this is not as pressing an ethical issue as the other threat to privacy, namely, the taking of information from the private sphere, because it is something of which we are aware and can control. If we do not like the content of a television or radio broadcast, we can turn it off. If someone detests commercial television, they can cancel their cable subscription or even remove receivers from their home. On the other hand, if their private information is being recorded, they are generally unaware that this is taking place and generally can do little to prevent it.

Electronic communications systems have tremendous advantages for their users but they also make them vulnerable to electronic invasion of privacy of various kinds. One conduit for the taking of information out of the private realm is the telephone line, which can be tapped, and taped, by law enforcement agencies. Another is the records kept by commercial organizations. Thus, video rental companies maintain records containing the titles of films rented to a particular account. Electronic databases make it ever easier to store and access information about individuals whether the files are kept by commercial organization or by governments.

The new medium of the Internet has got itself off to a very bad start by failing to protect the privacy of customer accounts. Many Internet companies are busily compiling extensive customer records without necessarily being aware of the purpose for which these will ultimately be used but with the conviction that they will be commercially valuable. The Internet has greatly fudged the conventional barriers between public and private spheres in several ways. Thus, e-mail communications, which have to some extent replaced letters, do not have the security of old-fashioned sealed correspondence, even though many users may be unaware of this. The Internet is actually a very public medium in which every click of a person's mouse can be, and is being, recorded by interested commercial entities. Moreover, companies are in the business of collecting, and selling, personal information over the Internet. This information makes people vulnerable to different types of criminals from stalkers to those who would steal their identity, run up credit card debts, and commit crimes in their name.

Electronic media, coupled with the power of super computers, have allowed an unprecedented degree of international espionage of ordinary electronic communications. Moreover, the miniaturization of video cameras has meant that privacy cannot be guaranteed to any person or in any setting. While it is foolish to exaggerate the immediate risk posed by all of these electronic assaults on privacy, it is also important to be concerned about them. Without such concern, the necessary legal protections against abuses will not be realized.

PRIVACY AND COMMERCE

Most people accept that when they go out into the public sphere to obtain a job, secure a loan, or make a purchase they must divulge personal information and thereby give up their privacy to some extent. This comparative loss of privacy generally does not cause problems because the information is given to trusted agencies who do not abuse it. Yet, in some cases, the confidence enjoyed by trusted commercial institutions, such as banks, has not been deserved. For decades, banks have been selling customer information to any companies willing to pay for it. This practice is controversial and many banks refuse to do it. Banks that do sell information defend themselves by claiming that they are not violating confidentiality because they do not reveal individual financial records. Instead, they sell lists of customers that are screened into various categories, such as frequent flier, mail-order shoppers, or individuals with the largest purchases in the previous year.

Commercial transactions today often lack privacy due to advances in electronic information technology. The use of electronic discount cards in the supermarket saves a person the bother of fumbling for cash but it means that every single item they purchase is electronically recorded. This is an extraordinarily rich dataset for the purposes of market researchers, which will allow them to study purchasing habits over time, responsiveness of individuals to sales promotions, and so forth. What is more, it allows profiles to be constructed for individual shoppers that can be very useful to marketing companies, allowing them to run precisely targeted advertising campaigns, for example. Similarly, the use of automatic payment of highway tolls allows the movement of individual vehicles to be tracked.

Everyday behavior may be recorded by video cameras that are used to scan urban streets. This system was introduced in England where it proved effective in inhibiting crime and has now been implemented in a number of other countries. Video is now recorded on disc rather than cumbersome tapes. Discs may be searched for a particular face, making it possible to track the movements of an individual over time.

People are watched not only electronically when they shop. A 1997 American Management Association survey of 900 large companies found that two-thirds admitted to electronic surveillance of employees. Bosses routinely listen in on the telephone conversations of employees, read their e-mail, and monitor their movements over the Internet.

PRIVACY AND THE INTERNET

Reasons for the lack of security and privacy on the Internet range from the hardware and software of this medium to the objectives of governments. In 1999, Intel and Microsoft were criticized when it was revealed that the chips and software of a personal computer are designed to transmit unique identifying numbers whenever that computer connects to the Internet. The companies hastily provided software to allow the identification numbers to be switched off. Such software fixes can be penetrated, however. There are two good practical reasons for the use of identifiers. One is that they help electronic devices and software to function more smoothly together. The other is that they inhibit computer crime. Electronic footprints are generally so easy to follow that few criminals are interested in attempting to steal money electronically from financial institutions.

Apart from the fundamental lack of anonymity in the design of personal computers, some companies store unique identifying strings, known as COOKIES, on the hard drives of computers that are used to visit their sites. Cookies allow Internet browsing habits to be electronically tracked. While cookies may seem like an ominous breach of privacy, they have never been used for anything more sinister than marketing. Nevertheless, many people were very upset when marketing company DoubleClick acquired a database that allowed them to connect the web habits of an individual with their personal information, including name, address, income, telephone number, interests, and so forth. This practice is known as "profiling." Profiling allows marketers to save a lot of money by targeting their advertising and sales efforts at individuals with a high probability of purchase. If you are selling golf equipment, for example, then it is enormously helpful to target your efforts at people who actually play golf.

Embarrassed by the adverse publicity that greeted their invasion of the privacy of Internet users, DoubleClick provided an option that allows individuals to opt out of having their every move electronically followed and recorded. This has not been enough to protect them from lawsuits and governmental regulatory action. A simple way of preventing such scrutiny is for computer users to delete the files in the cookies folder on their hard drive. Saving the cookies as empty read-only files prevents sites from adding information to them when they are revisited and thus frustrates efforts to track the visitor.

Such acts of resistance are encouraged by the online ELECTRONIC PRIVACY INFORMATION CENTER, a pressure group formed in response to the threats posed to privacy by technological innovations. Concerns over online privacy has also created a commercial opportunity. Some companies offer to strip e-mail of identifying information before remailing it. Anonymizer.com provides anonymous Internet browsing. Digital cash holds the potential for reducing the amount of personal information required for making online purchases.

Even though most online interactions are insecure and lack privacy, this is not an inevitable state of affairs. Good encryption programs are commercially available. With the exception of brokerage companies, very few commercial organizations use encryption on the Internet. This may partly reflect a general lack of interest in electronic privacy. After all, with the recent exception of cellular phones, telephone communications have generally not been scrambled either. Or it might reflect the view that in a free society it is difficult to prevent the flow of information. Moreover, the U.S. government has discouraged electronic encryption by thwarting research and criminalizing the export of encryption technology in the name of national security.

PRIVACY AND SURVEILLANCE

Even if companies encrypted all of their electronic communications, they would not be protected from industrial espionage, which can be carried on by planted employees. Even more alarming is the emergence of a miniaturized technology that threatens the security of interpersonal communications of all types. A miniature video camera installed in a light fixture can be used to record everything that transpires at a board meeting, for example. Microphones concealed in pens might pick up every whisper of conversation in the office of an industry leader.

Such devices used to exist only in the fictitious world of James Bond but are now not only widely available but also quite inexpensive. The worlds of international espionage and industrial espionage have come closer together in another way. A global electronic spying organization, run by the U.S. National Security Agency (NSA) with the assistance of allies (Britain, Canada, New Zealand, and Australia) has recently been accused by the European Union of industrial espionage targeted at European companies. Code-named Echelon, the electronic espionage system relies on 120 satellites that were originally launched to monitor the diplomatic communications of unfriendly nations. It also listens in to Internet traffic and even monitors conversations on undersea cables using a special submarine. The Echelon system is capable of intercepting millions of phone calls, e-mails, and faxes each day. It can access satellite transmissions as well as underground cables and the microwave, cellular, and fiber-optic channels. This vast quantity of information is searched by key words using computers that are believed to be located in 10 listening centers around the world. A small fraction of the data is then forwarded to the NSA's facility in Fort Meade, Maryland.

This information has been used to monitor money-laundering operations, study weapons proliferation, and fight corporate corruption according to NSA director Lieutenant General Michael Hayden. The extent to which it is used to spy on American citizens is unknown. Echelon's cooperative structure provides countries with a legal loophole to spy on their own citizens because the information can be technically collected by another country and then "shared." The NSA has been accused of passing intelligence on to U.S. companies that has helped them to win important international contracts but CIA director George Tenet insists that the United States intervenes only when the bidding process has been corrupted by bribery and that complaints are then taken up with the relevant governments.

The magnitude of the threat to personal freedom posed by Echelon is difficult to assess because the uses to which the information is put are largely unknown. With the capacity to listen in on virtually any electronic communication anywhere in the world, it is the closest to a practical realization of George Orwell's concept of a Big Brother that the world has yet seen.

Of more immediate and practical concern is the role of espionage in police work. Law enforcement in the United States has come to rely increasingly on electronic monitoring of suspects to produce arrests and convictions. The Communications Assistance for Law Enforcement Act of 1994 required telecommunications companies to install equipment that provided government access to all telephone and data communications. Technically, this information can be obtained only with a court order but judges almost never refuse to give police access to the communications of suspects in the context of a criminal investigation. Moreover, the FBI has insisted that wireless phones should be designed so as to allow location tracking. This means that law enforcement can identify the location and movements of a person who uses a cellular phone. There has been a steady increase in the number of court-ordered wiretaps from 564 in 1980 to 1149 in 1996. During the same period, the average duration of the wiretap increased from 21 days to 38 days. Needless to say, access to phone lines provides access to e-mail messages that are sent over these lines.

Given the many different channels through which electronic information about individuals can be collected, and the unprecedented ease with which this information can be retrieved from electronic databases, the progressive erosion of privacy with technological advancement has to be viewed with concern. The number of different kinds of information that can be collected and sold is staggering, ranging from driver photographs to purchasing patterns to DNA. Thus, Iceland's parliament recently agreed to sell the DNA information of its citizens to a medical-research company to promote Icelandic genetic medicine, in a move that is seen by many as a gross invasion of privacy.

Such extraordinary threats to privacy would appear to call for extraordinary legal protections. In general, these have not materialized. One exception is the European Union data protection directive of October 1998, which is designed to give individuals control over their own data. Consent is required for data to be processed and it can be used only for the purposes for which it was originally collected. Exporting data to countries with less stringent protections of privacy is also prohibited.

Recent legislation in the United States has been enacted to protect children who are thought to be particularly vulnerable to misuse of information gathered online. The Children's Online Privacy Protection Act of 1998 prohibits the collection of personal information from children under 12 years without the consent of their parents.

See also COOKIES, AND INTERNET; ELECTRONIC TERRORISM; ELECTRONIC THEFT; ENCRYPTION RESEARCH; PORNOGRAPHY; TROJAN HORSE (COMPUTER VIRUS).

Further Reading

Catan, T. "Secrets and Spies." *Financial Times,* May 30, 2000.
Dempsey, J. X. "Communications Privacy in the Digital Age." *Albany Law Journal of Science and Technology* 8, no. 1 (1997): 65–120.
Electronic Privacy Information Center. *Cryptography & Liberty 1999.* Washington: Author, 1999.
Gelman, R., S. McCandish, Electronic Frontier Foundation, and E. Dyson. *Protecting Yourself Online.* New York: Harper-Collins, 1998.
Hansell, S. "Getting to Know You." *Institutional Investor* 25, no. 6 (1991): 71–79.
Swire, P., and R. E. Litan. *None of Your Business: World Data Flows, Electronic Commerce, and the European Privacy Directive.* Washington, D.C.: Brookings Institution, 1998.

electronic publishing of scientific research

Print journals are still the most important mechanism through which scientists communicate with each other. This medium of communication has two important problems that can be addressed by publishing in electronic journals accessed over the Internet: producing print journals is extremely expensive whereas electronic publishing is much cheaper; and there are so many journals in some fields, particularly the biomedical area, that it is difficult for careful scholars to keep abreast of their field. Large volumes of information can be accessed quickly using electronic search engines such as Medline. The disadvantage of doing research in this way is that often only the abstracts, or summaries of articles, are available online. If scholars want to avoid doing superficial research, they may be compelled to go back to the print publications.

Electronic publication is generally a great deal more democratic than elitist journals that often publish less than 10% of the submissions they receive. Thus, virtually anyone can become an Internet publisher by creating a home page. The main problem with the explosion of online publication is that a substantial fraction of the information is of poor quality. Some of the research available on the worldwide web is unreliable because it is advocacy, deliberate misinformation, or pseudoscience. There is no PEER REVIEW PROCESS to screen out junk science and propaganda. In fact, insofar as web pages are concerned, there is no kind of scientific quality control. This means that scientists and researchers may waste their time pursuing faulty leads and that the public can be misled concerning the state of scientific knowledge on some subject.

See also PEER REVIEW PROCESS IN ACADEMIC RESEARCH.

electronic surveillance *See* ELECTRONIC PRIVACY.

electronic terrorism

The electronic age rests on two critical aspects of computers: the capacity to store information and the capacity to process it rapidly. Both of these capabilities are threatened by cyber terrorism, which may either destroy stored computer files by unleashing destructive viruses or cause computer networks to crash as in the recent denial-of-service attacks against e-commerce companies and other well-known sites like Yahoo. Such attacks are relatively easy to launch and can have devastating economic costs. Determined law enforcement efforts against perpetrators are likely to make such crimes less common in future. Their main motive seems to be the thrill of wreaking havoc. However cleverly designed, they generally leave electronic footprints that make it fairly easy to identify the perpetrator.

Another type of electronic terrorism consists of attacks on electronic infrastructure, such as that which controls the power grid of a city. Such potential attacks are often referred to as information warfare and they are expected to play an important role in the warfare of the future. Some experts believe that the United States used information warfare to disable Iraq's missile defense sys-

tem, which proved to be strangely silent during the air raids of Operation Desert Fox in 1991, although this is officially denied.

See also ELECTRONIC PRIVACY; ELECTRONIC THEFT; ELECTRONIC VIRUS.

Further Reading
Alexander, Y., and M. Swetnam, eds. *Cyber Terrorism and Information Warfare.* Dobbs Ferry, N.Y.: Oceana, 1999.
Wilson, J. "Information Warfare." *Popular Mechanics* 176, no. 3 (1999): 58.

electronic theft

Early computer hackers of the early 1970s amused themselves by electronically shifting money from the accounts of figures such as President Richard Nixon to left-wing organizations. There was virtually no security in electronic financial transactions. Since then, many financial institutions have moved to protect their clients using encryption that prevents electronic communications from being deciphered. Nevertheless, most accounts are still accessible using a fairly simple personal identification number. Despite the apparent vulnerability of financial institutions, such as banks and brokerages, electronic theft has not turned out to be a major problem. The real issue in electronic theft today is the piracy of electronic products. Another problem is that thieves may use the Internet to obtain personal information, including credit card numbers, that is used to commit fraud, and in some cases even to steal a person's identity.

A highly publicized instance of electronic theft involved the stealing of $10 million by Russian hacker Vladimir Levin from Citibank, a crime that was revealed in the summer of 1995. Still, such crimes are not as clever as they might appear because they leave electronic footprints that make it comparatively easy to detect the culprit. In fact, Levin was being observed as part of a sting operation. He and his accomplices were arrested and most of the money was recovered.

Producers of software, electronic games, recorded music, and even videos have long been concerned about the issue of piracy on the Internet, which is estimated to cost manufacturers large amounts of revenue. Thus, the software industry alone is estimated to have lost some $16 billion to piracy in 1998. The American Society of Industrial Security estimates that American business loses $250 billion in intellectual property annually (a speculative figure that includes the fruits of organized industrial espionage as well as piracy by individuals), much of this being passed over the Internet. In 1997, the U.S. Congress passed the No Electronic Theft (NET) Act making it illegal to copy pirated material even though no financial gain was realized from the theft. This legislation was prompted by the 1994 case against student at MIT who placed popular software on the web to be downloaded for free by anyone who wanted it. The copyright infringement case had been dismissed because the student had not made any profit.

There is little evidence that the NET law has been enforced. One problem is that Internet Service Providers (ISPs) automatically make copies of materials that their users publish online. If these are pirated, then the ISPs are technically guilty of piracy. There is a flourishing cyber underground that operates primarily on a Hotline network that is separate from the worldwide web. Unlike conventional browsers, the *Hotline Connect* software used to access Hotline sites is very good at transferring large files. Hundreds of Hotline sites offer pirated software that can be downloaded with the click of a mouse. Another center of piracy is *Site0* (Site Zero) where providers (many are employees of software companies or computer stores) upload the latest versions of commercial software so that "crackers" can break the protection codes allowing them to be copied at will. The cracked programs are liable to end up on Hotline sites. Finally, there are a half-dozen sites like *Napster* that specialize in the swapping of pirated music. It has yet to be determined by legislative and judicial authorities what will be the ultimate future of such sites, now that Napster is fee-based.

The problem of infringement of intellectual property rights on the Internet highlights the fact that new media are always difficult to regulate and none more so than cyberspace. Advocates of the Internet argue that too much regulation is bad in that it can choke off development of a medium that can have revolutionary, and beneficial, consequences for our lifestyles. One intrinsic aspect of the Internet is that it tends to promote the free transfer of information. Such transfer is obviously inimical to the laws protecting intellectual property. Inevitably, some form of compromise will have to be reached since software companies, writers, musicians, and others who cannot sell their intellectual property may be inhibited from producing it.

See also ELECTRONIC TERRORISM; PATENTS; TRADE SECRETS.

Further Reading

Hawaleshka, D., and R. Scott. "Beware the Internet Underground: A Secret Passage Is Leading Kids to Stolen Software and Hardcore Porn." *Macleans,* November 8, 1999.
Radcliff, D. "Invisible Loot." *Industry Week,* November 2, 1998.
Schut, J. H. "Stop Cyberthief!" *International Investor* 30, no. 4 (1996): 95.

electronic virus

Like biological viruses, electronic viruses have the capacity to spread, or infect, and to reproduce themselves, thus harming their electronic hosts. One common type of electronic virus infiltrates e-mail address books and sends itself to all those addresses to begin a new cycle of infection. Another type, known as a TROJAN HORSE, allows distant programmers to observe every keystroke of a host computer with a complete loss of security of passwords and other confidential information. Although some viruses may be accidentally generated, or released, most are deliberately created by sophisticated programmers.

See also ELECTRONIC TERRORISM; TROJAN HORSE COMPUTER VIRUS.

Ellis, Havelock
(1859–1939)
English
Physician, Writer

The wall of silence around human sexuality posed a considerable obstacle to those researchers who wished to investigate it. The earliest serious scientific investigator of human sexual behavior was English physician and writer Havelock Ellis. His seven-volume *Studies in the Psychology of Sex* (1897–1928), which was banned on charges of obscenity, anticipated some of the major findings of American researchers concerning the actual sexual behavior and sexual problems of ordinary people.

Ellis reported the omnipresence of masturbation. He theorized that homosexuality and heterosexuality are a matter of degree rather than absolutes, thereby anticipating the Kinsey scale of sexual orientation. Ellis also appreciated the role of psychological factors in sexual dysfunctions like impotence and frigidity.

See also KINSEY, ALFRED; MASTERS AND JOHNSON.

Further Reading

Draznin, Y., ed. *My Other Self.* New York: Peter Lang, 1993.
Grosskurth, P. *Havelock Ellis.* New York: Knopf, 1980.
Robinson, P. *The Modernization of Sex.* New York: Harper and Row, 1988.

embryonic/fetal tissue research   *See* FETAL CELL GRAFTS; HUMAN CLONING RESEARCH.

encryption and Internet security   *See* CRYPTOGRAPHIC RESEARCH.

Endangered Species Act, 1973

An endangered species is defined as one whose population has been reduced to the point that extinction is possible. A species can be endangered for natural reasons, such as climatic change, or as a result of human activities. Human activities can threaten species because of ecological destruction, as in the felling of forest habitats, environmental pollution, or overuse, as in the hunting of the American bison (or BUFFALO) to the verge of extinction. The International Union for the Conservation of Nature and Natural Resources in Morges, Switzerland, maintains a list of endangered species of birds, mammals, reptiles, and amphibians and, in 1973, 80 countries agreed to halt international trade in these endangered species. In the same year, the United States passed the Endangered Species Act (ESA) legislation that has been reauthorized in five-year intervals.

The ESA operates under the premise that all species are of inestimable value but it does not mandate the limitless supply of government funding for recovery efforts. Between 1991 and 1997, it is estimated that the ESA cost state and federal governments $529 million per year. As of 1999, 1,138 species had been listed as threatened or endangered with a backlog of several thousand species that are candidates for listing because their numbers are believed to be declining. The illustration shows the stilt bird that was sacred to the old Hawaiians but is now on the endangered list. Consistent with the philosophy that it is important to preserve all species, not just the kind of "charismatic megafauna" that people like to visit in zoos, most of the listed species are now insects and other invertebrates.

Once a species is listed as threatened or endangered, the Fish and Wildlife Service creates critical habitat designations and develops a recovery plan. The idea is to save a species by protecting its habitat and then, by implementing the recovery plan, to raise its population to levels at which it no longer requires protection and can be delisted.

Despite some widely publicized cases of conflict between the ESA and economic activity, such as the conflict between the spotted owl and logging interests in Oregon, the act is regarded as having made an important contribution to restoring species such as the bald eagle, wolf, and grizzly bear. Of 20 species that had been delisted by 1994, eight had become extinct, and eight more were delisted because the data used to list them in the first place was inaccurate.

Only four species were delisted because their populations had increased. The gray whale, the American alligator, and the brown pelican were said to have recovered but were maintained in a special listing category. Whether the ESA actually helped any of these species or not is arguable. In each case a logic can be presented

Hawaiian stilt birds, long protected and considered sacred, are now an endangered species. (NATIONAL ARCHIVES)

showing that these species might have recovered without the ESA. For example, the rebound of pelican populations may be due to the ban on DDT. The alligator may have been listed due to misunderstanding of its population dynamics. Still, the prognosis for most of the listed species is not good. Only one-tenth have increasing populations. It is clear that the ESA has not been very effective at saving endangered species and that the Fish and Wildlife Service is not adequately funded to carry out its provisions.

Some of the bitterest objections to the ESA have come from landowners whose property values have decreased as a result of being designated a critical habitat for some species or whose occupation has been otherwise affected. Thus farmers have been prosecuted for protecting their livestock from attack by endangered species, such as grizzly bears and wolves.

The costs of the ESA to individual landowners can be severe. In North Carolina, Ben Cone had 1,560 acres of his land designated as habitat for the red-cockaded woodpecker. Since no timber can be harvested, among other provisions of the protection plan, Cone's land lost 96% of its value, or $73,914 for each of the 29 resident woodpeckers. This taking of private property by the government, without compensation, would appear to be contrary to the Fifth Amendment of the U.S. Constitution, which stipulates: "nor shall private property be taken for public use without just compensation."

In this case the appropriation of Cone's land had consequences that were not necessarily favorable to the protected species. In fact, local landowners were provided with an incentive to clear-cut their forests to ensure that the land was not attractive to red-cockaded woodpeckers. In this way, they could defend their property interests by restricting the potential habitat of the endangered species.

Preventing residential and commercial building projects constitutes perhaps the biggest economic cost of the ESA. It is estimated that protection of around a quarter of listed species conflicts with development projects or other economic activities. Since the ESA assigns all endangered species an inestimable value, it does not include a mechanism for weighing the benefits to an endangered species against the economic costs to local residents and the U.S. economy. Pervasive costs of habitat protection under the ESA include reroutings of highways and prohibitions on expansions of dangerously congested airports.

Such costs can be substantial and their distribution across communities is inevitably unequal. Thus, it has been estimated that property values in Travis County, Texas, fell by $359 million after two local birds, the black-capped vireo and the golden-cheeked warbler, were put on the endangered list. It would be difficult to justify such an unequal distribution of the costs of protecting these species even if the protection measures were going to be effective. The limited data thus far do not inspire confidence on this score. Great as the economic costs of the ESA are to private landowners, there is little persuasive evidence that it is providing a commensurate benefit to the endangered species.

See also CLEAN AIR ACT; CLEAN WATER ACT.

Further Reading
Simmons, R. T. "The Endangered Species Act: Who's Saving What"? *Independent Review* 5 (1999): 309–26.

endocrine disrupters   *See* HORMONE MIMICS.

engineering ethics
Increased reliance on machines and infrastructure in the modern world has meant that people are more vulnerable than ever before to the consequences of faulty engineer-

ing. This awareness has been fostered by a number of spectacular failures of mechanical and structural engineering products in the 20th century. They include problems with collapsing bridges, exploding cars, the Space Shuttle CHALLENGER ACCIDENT, the sinking of the TITANIC due to structural design flaws, the initial failure of the HUBBLE SPACE TELESCOPE, and the construction of industrial plants, including chemical factories and nuclear power stations that were dangerous, polluting, or both.

Engineers involved in the construction of dangerous infrastructure and machines often have unique access to information relating to safety and this gives them an ethical responsibility to bring any problems to light even though this is harmful to the profitability of the company for which they work. This responsibility also applies to routine operation and maintenance of industrial plants by which public health and safety are threatened. Although ethics has historically received less emphasis in the training of engineers than was true for scientists, this is changing as we become more aware that industrial accidents are often caused by avoidable design flaws and by faulty safety procedures for which engineers are directly responsible.

See also BHOPAL CHEMICAL ACCIDENT; BRIDGE COLLAPSE AND DESIGN FLAWS; CHALLENGER ACCIDENT; CHERNOBYL NUCLEAR DISASTER; EARTHQUAKES AND CONSTRUCTION CODES; FORD PINTO; HUBBLE SPACE TELESCOPE; SEVESO CHEMICAL ACCIDENT; THREE MILE ISLAND NUCLEAR ACCIDENT; TITANIC, DESIGN FLAWS OF.

Further Reading
Schlossberger, E. *The Ethical Engineer.* Philadelphia: Temple University Press, 1993.
Westrum, R. *Technologies and Society.* Belmont, Calif.: Wadsworth, 1991.
Whitbeck, C. *Ethics in Engineering Practice and Research.* New York: Cambridge University Press, 1998.

Enlightenment, the
The Enlightenment was an intellectual movement that prevailed in the West during the 18th century. It emphasized the use of reason and rejected established philosophical, religious, and political views. Enlightenment philosophy emphasized the dignity and worth of the individual and the importance of education and the pursuit of personal fulfillment. Leading figures of the Enlightenment included Voltaire (1694–1778) in France, David Hume (1711–76) in England, and IMMANUEL KANT (1724–1804) in Germany. Intellectuals of the period felt a desire to educate the masses and produced a variety of encyclopedias and dictionaries that made large quantities of technical information available to the average reader. In religion, the Enlightenment promoted free thought

and agnosticism as opposed to unquestioning acceptance of religious beliefs. In politics, there was an emphasis on freedom of expression and equality before the law. The ideals of rationalism, liberty, and equality found extreme expression in the French Revolution (1789), whose aftermath marked the beginning of a more conservative period.

See also SCIENCE, HISTORY OF.

environmental ethics

Environmental ethics emerged in the 1960s as a response to a growing awareness that pollution and other consequences of human economic activities were causing irreparable damage to the environment and to ECOSYSTEMS. Given that all life on the planet can be changed by processes such as GLOBAL WARMING produced by the greenhouse effects of industrial gases, it is essential for ethicists to ponder these effects. To the extent that the health and quality of life of the entire human population are threatened, it is a matter of practical necessity to deal with these problems before they become unmanageable.

Given that large numbers of species are threatened with extinction as a direct result of human economic activity and that fragile ecosystems are being destroyed or degraded, ethicists are preoccupied with determining the value of what is being lost and hence determining what we must attempt to save. These are two opposing points of view. According to the anthropocentric perspective, we should not pollute the air because this will eventually damage our health. Acid rain is to be avoided because it degrades infrastructure and destroys the amenity value of lakes by killing fish. According to this point of view, the passing of the dodo and the woolly mammoth attributable to human predation is no more regrettable than the passing of the dinosaurs due to nonhuman influences. Since the loss of the dodo has no important practical consequence for most people its loss is not important.

The opposite perspective claims that each species has an intrinsic value that must be protected. This point of view is explicitly expressed in the U.S. ENDANGERED SPECIES ACT (1973), which ascribes a limitless value to each and every species. This is mainly a rhetorical device, however, since the claim is not backed up by a limitless (or even substantial) appropriation of government funding to protect each species. Many biologists affirm that biological diversity is valuable in itself and that the loss of any species that reduces this diversity makes our planet poorer. From this perspective, there is inherent value in genetic information although this value must be expressed in living individuals. Thus, the saving of genetic information concerning an extinct species, which

will be technically possible in the near future, is unlikely to satisfy most conservationists.

Many environmental ethicists believe that the inherent value of living creatures is not to be found in a particular species but rather resides in the complex system of co-evolved relationships between different species in an ecosystem. This perspective takes into account the complex pattern of interdependencies between species in a biotic community. For example, many plants rely on insects to pollinate them. Many seeds cannot germinate until they have first passed through the digestive system of an animal that feeds on the fruit. J. Baird Callicott, a pioneer in environmental ethics, maintains that ecosystems have intrinsic moral value because of such complex information that is stored in them and not because they are of use to humans.

Callicott's position has been criticized on logical grounds. Many philosophers do not accept that it is possible to step outside our skin, as it were, and take a nonanthropocentric viewpoint. Critics argue that it is possible to protect the environment using more conventional ethical principles. Thus, many ethicists accept that we have obligations to future generations. This means that we should not do anything today that can have the foreseeable, but distant, consequence of making life on this planet less pleasant. Obligation to future generations can be a useful and flexible principle, which, for example, suggests that we should be careful not to accumulate toxic wastes. Moreover, it can be used to provide a rationale for protecting endangered ecosystems, such as the rain forest that contains many thousands of unknown species whose biochemistry could potentially be of enormous benefit to medicine.

Intimately related to the central issue about how to define moral value in environmental ethics is the practical issue of industrial development and its potential for harm to the environment. Many environmental ethicists recognize that the benefits of living in an affluent society are counterbalanced by environmental damage due to increased industrial production. How should ethical individuals behave in an affluent society if they wish to minimize environmental damage? One answer is that they must consume as little as possible. Reduced consumption can be accomplished through a radical simplification of lifestyle, such as the use of a bicycle rather than a car. Recycling is also an important strategy for reducing both energy use and solid waste.

In general, if most of the environmental problems we are experiencing today are a consequence of the INDUSTRIAL REVOLUTION, then the solution to these problems could involve a return to a lifestyle that more closely resembles that which existed prior to industrialization. Since it seems unlikely that enough people will favor reduced industrial growth to make this feasible, it is more likely that ethical concerns will create the impetus for new technologies that minimize industrial damage to the environment. One example is the recent development of "green" cars that produce virtually no exhaust, thereby potentially removing an important source of environmental pollution.

See also AIR POLLUTION; CALLICOTT, J. BAIRD; DEEP ECOLOGY; SOLID WASTE DISPOSAL; WATER POLLUTION.

Further Reading
Pojman, L. P., ed. *Environmental Ethics.* Belmont, Calif.: Wadsworth, 1994.

EPA (Environmental Protection Agency)

Founded in 1970, the Environmental Protection Agency (EPA), a part of the executive branch of the U.S. government, administers legislation designed to protect the environment. Examples of its responsibilities include dealing with pollution of the air, earth, and water by pesticides, solid wastes, radiation, toxic chemicals, chimney emissions, vehicle exhausts, acid rain, and even by heat and noise. Operating out of 10 regional offices, the agency coordinates the pollution-fighting activities of state and local governments.

See also CLEAN AIR ACT; CLEAN WATER ACT; ENDANGERED SPECIES ACT; LOVE CANAL.

Epicurus

(341–270 B.C.)
Greek
Philosopher

Epicurus was the son of a poor Athenian colonist on the Greek island of Samos. He studied philosophy and established his own school, at Mytilene, on the Greek Island of Lesbos, in 311 B.C. In 307 B.C., he moved to Athens and operated his school in the garden of his home. Although he is believed to have written hundreds of philosophical treatises, only a few fragments remain and what survives of his work has come down to us through his disciple the poet Lucretius (99–55 B.C.) and the biographical work of Diogenes Laertius, who lived in the third century A.D.

Epicurus is one of the most misunderstood of philosophers. Technically, his philosophy revolved around hedonism as reflected in his statement, "Pleasure is the beginning and end of the blessed life." Such statements were as much misunderstood by his contemporaries as they are today. The fact that his school accepted women and slaves led to accusations that it was a center of debauchery. These allegations, which may have been maliciously cultivated by his philosophical rivals, the Stoics, evidently had no basis in reality.

Most of the confusion revolves around defining what is meant by pleasure. Epicurus thought of pleasure as freedom from pain. The lifestyle that he advocated, and practiced, was monastically austere. His diet consisted mainly of bread and water and he ate small quantities to avoid indigestion. If he wanted to have a feast, he consumed a little cheese. Sexual intercourse was seen as having no value and marriage and children were considered a distraction from the pursuit of wisdom. Epicurus favored passive pleasures over active ones and therefore preferred friendship to sexual love. Far from encouraging indulgence in the pleasures of the flesh, Epicurus defined pleasure as the absence of pain which could be achieved by prudent living and by training the mind to contemplate pleasant thoughts. All pleasure depended on sensory experiences because the pleasure of the mind was nothing more or less than contemplation of pleasures of the body.

Epicurus thought of philosophy as having the pragmatic goal of securing peace of mind and had little time for the pursuit of knowledge for its own sake. He detested religious superstition and felt that knowledge and science could be usefully deployed to dispel such mistaken ideas.

Although he was a materialist, Epicurus was not a determinist like DEMOCRITUS, who was a major influence on his philosophy. He adapted Democritus's idea of the soul as consisting of atoms but, unlike the laughing philosopher, he claimed that soul atoms were capable of swerving off course. This meant that human actions were not predetermined and that free will was possible.

See also DESCARTES, RENÉ; HEDONISM.

Further Reading
Russell, B. *A History of Western Philosophy.* New York: Simon and Schuster, 1972.

epiphenomenalism

Epiphenomenalism is one solution to the philosophical question of the relationship between mind and body (the mind-body problem). Epiphenomenalists minimize the importance of the mind in controlling actions. According to this view, thoughts accompany actions in much the same way that noise accompanies a lawnmower. They accompany behavior but do not determine it. Epiphenomenalism became influential among 20th century behaviorist philosophers such as JOHN B. WATSON and BURRHUS FREDERIC SKINNER, who were motivated to remove the mind from scientific psychology because it could not be scientifically observed. The epiphenomenalist position has been vehemently rejected by most living ethical philosophers who argue that our ability to think is what is uniquely human and distinguishes us as ethical beings.

See also CHIMPANZEES, SELF-AWARENESS OF; DESCARTES, RENÉ; SKINNER, BURRHUS FREDERIC; WATSON, JOHN B.

ether, the

The ether was a hypothetical medium that 19th-century physicists supposed to be necessary for the transmission of light over otherwise empty space. In addition, the concept of the ether was used to account for the phenomena of heat, magnetism, and electricity. The ether is an interesting example of the ability of scientists to believe in phenomena that subsequently turn out to be illusory. In the end, the ether was rejected because of the failure of experiments (particularly the Michelson-Morley experiments of 1881 and 1887) to detect any evidence of its influence. Rejection of the ether by ALBERT EINSTEIN and others marked an important departure in modern physics.

See also N-RAYS; PHLOGISTON.

ethic of care   *See* GILLIGAN, CAROL; MORAL PHILOSOPHY.

ethics   *See* MORAL PHILOSOPHY.

Euclid

(c. 330–260 B.C.)
Greek
Mathematician

Euclid can be described as the most successful textbook author who ever lived. His *Elements,* an introduction to geometry, established a complete monopoly in the field for over two millennia after his death and remains the most printed book after the Bible. Topics covered included plane geometry, number theory, irrational numbers, and solid geometry. Although Euclid probably did compose some of the proofs in his *Elements,* his main contribution consists in organizing the impressive mathematical knowledge of his contemporaries at Athens, where he studied, and Alexandria, where he spent most of his adult life.

Mathematicians are still delighted by the rigor and creativity of the Euclidean proofs. Moreover, Euclid organized his textbook cleverly so that what follows in sequence is seen to proceed logically from what went

before. Euclidean mathematics is elegant, making the minimum of assumptions, and its formulation satisfies a high criterion of clarity. An interesting feature of mathematical proofs is that the work of Euclid is as true today as when it was composed, whereas scientific theories are subject to constant change. Although Euclid was a scientist, contributing to astronomy, musical harmonics, and optics, much of this work has been lost. Euclid's geometry not only preserves the finest flowering of ancient Greek mathematics but also influences modern intellectual life by providing a model for logical proof.

See also PYTHAGORAS.

eugenics

The term eugenics, derived from the Greek word *eugenes,* or wellborn, refers to the application of hereditarian principles to understanding, and fostering, desirable human traits. Eugenics first emerged in the late 19th century, in England, as an offshoot of SOCIAL DARWINISM, which applied evolutionary notions of a struggle for existence and survival of the fittest to human society.

HISTORICAL EUGENICS

FRANCIS GALTON, the English biologist and psychologist, is considered to be the founder of eugenics. He coined the term in 1883. Galton had long been interested in the role of heredity in accounting for individual differences in mental abilities. In 1869, he published *Hereditary Genius,* a study of the pedigrees of eminent men. Galton's data showed that leading scientists, lawyers, authors, and so forth, often had intellectually distinguished ancestors. Moreover, the form of genius that was inherited tended to be specific. Eminent men of law had prominent jurists in their family trees. Distinguished authors had ancestors who were also writers. Galton assumed, along with others of his period, that dominant classes and races were intrinsically superior to dominated ones for hereditary reasons. His eugenics program advocated the elimination of the unfit (or negative eugenics) and the systematic improvement of human heredity by encouraging reproduction of desirable individuals (positive eugenics).

During the first half of the 20th century, eugenic programs became popular in many countries in Europe, as well as in North and South America, China, and Japan. One of the most influential such movements was in the United States where eugenic theories focused primarily on perceived racial differences. Two of the most important practical effects were seen in immigration policy and in the compulsory sterilization of institutionalized individuals. The U.S. Immigration Restriction Act of 1924 expounded eugenics in promoting the immigration of people from Northern Europe in preference to all others who were seen as "biologically inferior." By 1937, 32 states had passed compulsory sterilization laws, which were designed to prevent the reproduction of undesirables defined to include the mentally ill, mentally retarded, and those convicted of sexual crimes and drug offenses. In practice, sterilizations were carried out only in the case of those individuals who were unfortunate enough to inhabit state institutions. The laws thus tended to discriminate against poor people and ethnic minorities. In California, which had the most active sterilization program, blacks and foreign immigrants were twice as likely to be sterilized as the rest of the population.

The logic behind eugenic sterilization was clearly stated by U.S. Supreme Court Justice Oliver Wendell Holmes in the infamous *Buck v. Bell* case, which upheld the state of Virginia's decision to sterilize Carrie Buck. Although of normal intelligence, Buck was a resident of a state home for "mental defectives." The daughter of a "feeble-minded" woman, she had produced an illegitimate child as a consequence of being raped. Holmes argued that if the state could ask desirable people to sacrifice themselves in battle, then it should be entitled to ask for a lesser sacrifice from undesirable individuals, who were already sapping government resources. He also argued that if compulsory vaccination were permissible then compulsory cutting of the fallopian tubes could be allowed. His famous punch line was: "Three generations of imbeciles are enough."

The emergence of the eugenics movement in Germany has many eerie parallels to that in the United States and it is by no means obvious why a eugenic ideology produced genocide in Germany rather than America. In each case there was belief in white North European racial superiority and a willingness to sterilize members of unfit classes. In 1933, under the Nazis, a Law for the Prevention of Congenitally Ill Progeny called for compulsory sterilization of schizophrenics, manic depressives, epileptics, and alcoholics. Sterilizations were carried out in the case of some 350,000 mentally ill people. Sterilization was also used in the Nazi racial hygiene program with the compulsory sterilization of some 30,000 German Gypsies as well as a few hundred blacks. The racial hygiene program moved beyond sterilizations to the "final solution" of genocide. By the time of the Allied liberation in 1945, the Holocaust had resulted in the deaths of 70,000 German psychiatric patients, 750,000 Gypsies, and 6 million Jews.

MODERN EUGENICS

The horrors of the Holocaust have convinced most reasonable people of the great dangers of eugenic thinking but it would be naive to imagine that the ideal of genetic betterment of the population has disappeared. In fact, a recent Chinese law, euphemistically titled the Law on Maternal and Infant Health Care, which came into effect on June 1, 1995, calls for compulsory sterilization of unfit individuals as a precondition for marriage. Couples undergo a premarital screening to find out if either is a carrier of serious genetic disease, has an infectious disease (gonorrhea, syphilis, AIDS, leprosy), or suffers from a mental illness. The motivation behind this legislation was practical. The government wanted to reduce the number of disabled persons that had to be cared for in state institutions. It is interesting that this legislation has widespread support in China and is endorsed by many in the medical professions. The same is true of other nations in which there has not been a vigorous tradition of individual rights and for whom the collective good weighs heavier in the moral balance than reproductive freedom.

Popular enthusiasm for eugenics in the West was sapped by the Nazi Holocaust but recent advances in GENETIC TESTING, as well as ARTIFICIAL INSEMINATION and other REPRODUCTIVE TECHNOLOGIES, have ushered in a new interest in, and acceptance of, eugenic notions. The important difference is that modern eugenics, rather than being imposed on individuals by the state, is practiced voluntarily by parents who wish to prevent having children with genetic diseases or to produce children having what are seen as desirable qualities.

AMNIOCENTESIS can be used to detect serious genetic diseases, which can be prevented through selective abortion. (In future, genetic diseases may be prevented through permanent alterations of the germ line.) The underlying ethical rationale for selective abortion of fetuses with genetic disorders takes two forms. One is that some genetic diseases make for such a miserable quality of life that the affected individuals wish they had never been born. The other is that caring for seriously incapacitated children destroys the quality of life of parents and impairs their capacity to care for other healthy children that they might have. Opponents of this type of eugenics may be opposed to abortion in principle. Otherwise, they must explain what good is served by compelling parents to raise a child whose quality of life is likely to be very poor insofar as this is determined by health and vigor.

A different kind of eugenics is made possible by advances in reproductive technologies. Instead of the "negative" eugenics of preventing the development of children suffering from serious disorders, desirable traits may be selected for offspring in the practice of positive eugenics. In the future, positive eugenics may take the form of modification of the germ line to make offspring taller, stronger, or more intelligent, for example. At present such eugenics is practiced through the selection of sperm and eggs. Women who choose to reproduce through donor sperm obtained from a sperm bank can select the donor based on desirable traits such as height, good looks, intelligence, and economic status.

There is also a trend for egg donors to be selected on the basis of similar qualities. One example of this phenomenon is the appearance, in 1999, of an advertisement in the newspapers of several Ivy League universities. The ad specified an egg donor who should be tall, athletic, and highly intelligent. A fee of $50,000 was offered, which was a considerable increase over the market rate of $1,000 to $5,000 paid to donors who receive somewhat risky injections before having their eggs aspirated. Many commentators saw in this advertisement an unseemly example of positive eugenics in which a market value is placed upon desirable genetic qualities. They also rejected its elitist implication that wealthy people can purchase good genes for their offspring. Apart from the unusually high price offered, it is clear that there is no important ethical distinction between selecting egg donors in this manner and the positive eugenics practiced by women who use sperm banks to conceive offspring.

Positive eugenics may not be much more important in the future than it has been in the past. There are two fundamental problems. First, what may seem like good genes may turn out to be disappointing. A very bright sperm donor may produce children that are of below average intelligence. This reflects the obvious fact that offspring only acquire 50% of the genotype of the father. It is possible that the genes that have the strongest impact on intelligence are obtained mainly from the mother. Just as important is the other fact that a genetic potential for high intelligence is most likely to emerge in an appropriately stimulating environment. For young children this means having access to toys, being read to by parents, and so forth. By the same token, none of the other desirable traits is guaranteed by having good genes. For example, a person may inherit an excellent immune system but become ill as a result of continued exposure to environmental toxins or cigarette smoke.

The HUMAN GENOME PROJECT and its commercial competitors, which are sequencing the entire human genome, have led to speculations about how these data will be used. One goal of this work is to acquire a better understanding of the genetic basis of disease. This information can be applied to developing more sophisticated genetically engineered therapies. One possibility is direct modification of germ lines so that serious genetic diseases might be completely eliminated. Another possibility is that the information could be used to promote positive eugenics. People of the future might be genetically engineered to be more intelligent, to see or hear better, to be more athletic, or to live for 200 years. Although few people would deny that these objectives are intrinsically desirable, many are appalled at the prospect of "playing God" with the human genome. Whether this sense of eugenic horror prevents scientific manipulation of the human genome to cure diseases or to produce "designer babies" remains to be seen.

See also AMNIOCENTESIS; ARTIFICIAL INSEMINATION; GENETIC ENGINEERING; REPRODUCTIVE TECHNOLOGIES.

Further Reading

Caplan, A. L., G. McGee, and D. Magnus. "What Is Immoral about Eugenics"? *British Medical Journal* 319 (1999): 1284–85.

Dikotter, F. *Imperfect Conceptions: Medical Knowledge, Birth Defects, and Eugenics in China.* New York: Columbia University Press, 1998.

Kevles, D. J. *In the Name of Eugenics.* New York: Knopf, 1985.

Kevles, D. J. "Eugenics and Human Rights." *British Medical Journal* 319 (1999): 345–48.

Muller-Hill, B. *Murderous Science: Elimination by Scientific Selection of Jews, Gypsies, and Others, Germany 1933–1945.* Oxford: Oxford University Press, 1988.

Shannon, T. A. "Eggs No Longer Cheaper by the Dozen." *America*, May 1, 1999, p. 6.

European Bioethics Convention

Signed by 20 European countries in 1997, the European Bioethics Convention provides ethical guidelines that will regulate medical research and practice once it has been ratified. The ethical provisions concerning treatment of research participants are broadly similar to those of the NUREMBERG CODE. In addition, a number of articles are specifically designed to address potential abuses in the modern fields of ORGAN TRANSPLANTATION and GENE THERAPY.

The convention bans research on human embryos and outlaws trade in human organs or tissues. It also prohibits in vitro fertilization for the purpose of sex selection of offspring. It bans the involuntary removal of organs or tissue from any person but makes an exception in the case of regenerative tissue, such as bone marrow that has the potential for saving the life of the donor's brother or sister, for example. Consistent with a strong European distaste for GENETIC ENGINEERING in other contexts, including agriculture, the convention prohibits human germ line modification even to cure heritable genetic disorders.

See also AMNIOCENTESIS; GENE THERAPY; GENETIC ENGINEERING; ORGAN TRANSPLANTATION.

Further Reading

Watson, R. "European Bioethics Convention Signed." *British Medical Journal* 314 (1997): 1066.

euthanasia

The term "euthanasia" often serves as a synonym for mercy killing, although the word can also refer to deliberate nonmerciful killing of the sick, elderly, or socially disfavored, as practiced by German Nazis. "Euthanasia" is also used to describe the disposal of stray pets—acts, however, whose motive may stem less from concern for the animal's welfare than for the convenience of shelter administrators. The killing of animals can never be considered voluntary, of course. Involuntary euthanasia, such as that practiced by the Nazis against humans, is seen by most ethicists as morally indistinguishable from murder, and therefore as ethically indefensible. Debate concerning euthanasia almost always assumes that the practice is voluntary and that it is carried out with the intention of preventing unnecessary suffering.

Euthanasia has a long tradition of disapproval by the medical community that dates back to the HIPPOCRATIC OATH, which explicitly forbade the taking of life either by hastening death of the infirm or elderly or by abortion. Even though medical assistance in the taking of life has always been considered a violation of professional ethics, some authors have argued that it is actually a common practice even in the vast majority of countries where it is also illegal. This view is supported by a great deal of questionnaire data showing that a large number of medical professionals (and possibly the majority) favor euthanasia under some circumstances even though it is contrary to the Hippocratic Oath they have taken.

Contemporary interest in euthanasia is related to several convergent developments in medical technology. The recent development of lethal drugs can take life without causing undue pain and suffering unlike many of the poisons used to kill people in the past. Knowledge of these methods has increased following development of the lethal injection as a form of execution in the United States, which is viewed as a more humane "alternative" to the electric chair. In lethal injection executions, the victim is anesthetized (often with sodium pentothal) before receiving drugs (such as pentobarbital, potassium bromide, and curate) that shut down breathing and heart function thereby causing death. Knowledge of painless techniques for causing death allows a stronger case to be made for euthanasia as promoting a patient's best interest by helping them to avoid pain. Interestingly, the concurrent development of more effective forms of pharmacological palliative care tends to undermine the case for euthanasia as not being required in the overwhelming majority of cases.

Another reason for contemporary interest in euthanasia centers around the development of effective LIFE SUPPORT systems that can sustain life well beyond the point that a person would have died without such support. Such innovations raise difficult ethical questions about whether terminally ill patients should be placed on life support. Even more problematic is the issue surrounding the question of if and when life support should be discontinued once it has started. What is more, the possibility of artificial support of heart and lung function considerably muddies the definition of death. People may be kept alive for an indefinite period in a vegetative state without any reasonable prospect of recovery of normal brain function. Physicians are much more interested in euthanasia today because advances in medical technology force them to make difficult decisions that were not faced by preceding generations of doctors.

In the debate over whether participation by physicians in the taking of life is ever permissible, there has been a great deal of discussion over whether euthanasia is active (as in the withdrawal of life support) or passive (as in the failure to provide life support or other heroic medical procedures). Even though professional ethicists cannot pinpoint any meaningful moral distinction between withdrawing and failing to provide medical services for a dying person, it seems that passive procedures are more acceptable to many doctors, suggesting that there is a diminished sense of responsibility compared to taking an action that results in the death of a patient.

Another frequently mentioned distinction is between physician-assisted suicide (PAS) and euthanasia. The only meaningful difference between these two is whether it is the patient who administers the lethal dose obtained from the physician (PAS) or whether it is the physician who administers the poisonous drugs (euthanasia). While it might be argued that self-administration of the lethal dose clarifies the issue of voluntariness, or intent, there are some cases in which the patient is so physically incapacitated that they literally cannot take the drugs without assistance, even when prepared as an oral dose. Physician-assisted suicide is more acceptable to medical professionals, and to the public, than euthanasia. Yet the distinction involves a question of semantics more than ethics. It is clear that both procedures are voluntary from the patient's perspective. Moreover, the question of responsibility for the patient's death is not materially different. In each case, the patient and physician collaborate to bring about the patient's demise.

Even though there is no ethical distinction between PAS and euthanasia, there may well be a legal distinction. Thus, the state of Michigan could not convict Dr. Jack Kevorkian of murder for attending, and facilitating, suicides. This changed when Kevorkian filmed himself actually killing a woman who could not commit suicide because she suffered from a severe neuromuscular disorder.

PHYSICIAN-ASSISTED SUICIDE IN OREGON

There are only two civil jurisdictions in the world in which euthanasia is legal when carried out by physicians:

the Netherlands, where euthanasia (including PAS) is legal, and the state of Oregon, where PAS is legal but euthanasia is not. Under Oregon's law, 24 patients received lethal prescriptions in 1998, of which only 16 actually committed suicide. In 1999, 33 lethal prescriptions were written and 27 people committed suicide. Most of these people were suffering from terminal cancer. The small numbers indicate that Oregon's physicians are conservative about writing lethal prescriptions, complying with only about one in six requests for lethal prescriptions. Since almost all of the patients were suffering from incurable illnesses, the Oregon law has had the effect of producing a modest reduction in the length of life of a small number of people, which was presumably the intent of the legislators. It is interesting to speculate as to how many of these might have had their lives shortened in any case by the use of standard palliative therapies, such as morphine, that is administered to relieve pain but often hastens death.

THE NETHERLANDS AND THE SLIPPERY SLOPE CONTROVERSY

The Netherlands is the only country in which euthanasia is legal. Dutch law had explicitly prohibited euthanasia and physician assisted suicide since 1886. Several legal cases in the 1960s and 1970s resulted in the evolution of tolerance for PAS and euthanasia for terminally ill patients who were mentally competent. Significantly, the very first such case involved the new technology of life support systems. A 21-year-old woman named Mia Versluis suffered cardiac arrest while under anesthesia and entered a persistent coma. After the young woman had spent four months on life support, the anesthesiologist independently determined that life support should be withdrawn. The father lodged a complaint with the Medical Disciplinary Tribunal. In the ensuing court case, the anesthesiologist was fined and the judgment against him was made public. These penalties were unusually harsh by Dutch standards because physicians enjoy a great deal of respect in the Netherlands and medical malpractice lawsuits are extremely rare. Yet, the court ruling established guidelines under which termination of life support was permissible. These called for consultation with family members as well as among medical colleagues. Because the withdrawal of life support inevitably results in the patient's death, this procedure is morally equivalent to euthanasia, although there may be important legal distinctions.

The second pivotal case involved a more active form of euthanasia in which a physician, named Postma, hastened her paralyzed mother's death with a lethal injection of morphine. In the course of the 1973 trial it was revealed that most Dutch doctors considered it appropriate to give high doses of morphine, or other pain medica-tion, to relieve the pain of a dying patient even though this had the effect of shortening life. Although Postma was convicted, and received a one-week jail term together with a year of probation, the court ruled that euthanasia was permissible under certain conditions.

The Royal Dutch Medical Association later formulated guidelines under which euthanasia could be carried out. The request had to come from the patient and be both well considered and stable over time. The patient's suffering had to be both unbearable and without prospect for improvement. The physician had to obtain the approval of a colleague before acting. There should be a fully documented record of the events leading up to and including the euthanasia, a requirement that most physicians never complied with for a variety of practical and legal reasons. One rather astonishing omission was the lack of any requirement that next-of-kin should be consulted.

In the Dutch example, it is clear that legal permission for PAS and euthanasia is a logical extension of other practices that have long been considered good medical practice. These include withdrawing care at the request of the patient, terminating treatment when it was clear that the patient's life could not be saved, and using palliative treatments that hastened death. There is controversy about whether the Dutch medical system has produced a uniquely supportive environment for dying or whether the Netherlands has started down a slippery slope in which the rights of patients are inadequately protected and abuses are inevitable.

Proponents of the latter view can point to the sheer numbers of people dying with medical help. In 1990, some 3,700 people died as a result of PAS or euthanasia. In addition, the 1986 guidelines of the Royal Dutch Medical Association are vague and ambiguous and grant considerable discretion to the individual physician. One result is that euthanasia has spread from the voluntary type condoned by the courts to the involuntary killing of terminally ill patients. Thus, in 1990, there were over 1,000 cases in which physicians caused the death of their patients without the patients' consent. It is estimated that one-quarter of these patients were mentally competent but the physicians did not consult them before taking their lives. The fact that involuntary euthanasia occurs is shocking to most people and the large number of these cases is also startling even though the overwhelming majority of these persons were dying anyway and had their lives shortened only by a matter of hours or days.

Further evidence in favor of the slippery slope argument comes from the broadening of the categories of people who can be killed. Perhaps most significant here is the broadening of "unbearable pain" criteria to include unbearable psychological as well as physical pain. In 1984, the Dutch Supreme Court exonerated psychiatrist Boudewijn Chabot who assisted the suicide of a 50-year-old depressed, but otherwise healthy, woman who refused treatment for depression and wanted to die. Earlier, in 1982, the euthanasia of a 95-year-old woman who was not terminally ill had been upheld by the Supreme Court on the basis that her life was unbearable due to the chronic deterioration of old age. In 1999, the Dutch ministers of justice and health introduced a bill that would extend euthanasia and assisted suicide to children aged 12 to 16 years even over the objections of their parents.

In summary, even though it can be presumed that the majority of physicians carrying out euthanasia in the Netherlands do so because they are motivated by what they perceive to be the best interests of the patient, a line has been crossed into involuntary euthanasia that opens the door for abuses. The fact that so many patients are killed and that so few of their deaths are documented according to legal regulations should also be a cause for alarm because there is an inadequate record to enable prosecution of doctors who break euthanasia laws. The lack of documentation is evidently tolerated by the government. Surveys of elderly people in the Netherlands reveal that most are worried about the prospect of involuntary euthanasia.

See also EUGENICS; LIFE SUPPORT, TERMINATION OF.

Further Reading

Dworkin, G., S. Bok, and R. G. Frey. *Euthanasia and Physician Assisted Suicide: For and Against.* New York: Cambridge University Press, 1999.
Foubister, V. "Doctors in Oregon Aided 27 Suicides Last Year." *American Medical News,* March 13, 2000.
Grassi, L., M. Agostini, and K. Magnani. "Attitudes of Italian Doctors to Euthanasia and Assisted Suicide for Terminally Ill Patients." *The Lancet* 354 (1999): 1876–77.
Griffiths, J., A. Bood, and H. Weyers. *Euthanasia and the Law in the Netherlands.* Amsterdam: Amsterdam University Press, 1998.
Hendin, H. *Seduced by Death: Doctors, Patients and Assisted Suicide.* New York: Norton, 1998.
Spanjer, M. "Dutch Bill to Change Euthanasia Laws." *The Lancet* 354 (1999): 660.

evolutionary ethics

Attempts to derive ethical principles from what is known about biological evolution originate with HERBERT SPENCER (1820–1903). Spencer did not have a good understanding of evolutionary ideas as presented by CHARLES DARWIN in *On the Origin of Species by Means of Natural Selection* (1859) and many of his ideas on evolutionary ethics predated Darwin's book. Spencer incorrectly believed that biological evolution is progressive with newer and more complex creatures inevitably emerging from simpler ones. This is an understandable mistake because, as a general principle, older animals, like snails, are simpler in

design than more recent ones like mammals. Yet the premise that evolution is intrinsically progressive is false. Thus, its direction may reverse, as in the return of mammals, such as dolphins and seals, to an aquatic way of life that had been abandoned by their ancestors. Natural selection produces adaptation to environmental niches rather than making organisms better, as Spencer imagined. In many cases evolution is regressive, resulting in the loss of a complex trait. Thus, in the case of naked mole rats, which spend their entire lives under ground in complete darkness, the complex mammalian eye has completely lost its function.

In addition to the optimistic, but false, assumption that evolution drives a species ever onward and upward, Spencer acquired from evolutionary theory a few catch phrases such as "the struggle for existence" and "survival of the fittest." He equated economic and political competition within, and among, human societies with the struggle for existence described by Charles Darwin. He saw this struggle as being good because it allowed the best sort of people to prevail. Spencer did not deceive himself into thinking that natural selection was favoring the traits of wealthy industrialists of his day, for example, but he felt that competition in nature was good and that we should use it as a model for how human affairs should be run.

THOMAS H. HUXLEY (1825–95) rejected Spencer's approach of finding moral inspiration in evolution. Huxley accepted that evolution might be helpful in explaining why we all have both good and bad tendencies but it does not help us to understand why some of our tendencies are labeled "good" and others are labeled "bad." In the parlance of philosophy, he rejected the NATURALISTIC FALLACY. We are not entitled to derive moral statements from factual ones. As the Scottish philosopher David Hume first pointed out, we cannot legitimately switch from statements about how things are to conclusions about how they should be.

Charles Darwin was less pessimistic about evolutionary ethics than was Huxley, his friend and champion. Darwin not only accepted that moral impulses could have been favored by natural selection but also suggested an evolutionary rationale for why some behaviors are considered good and others bad. Darwin speculated that helping others could be favored by natural selection if it facilitated the survival of the whole group. Based on this rationale he accepted that moral sentiments could have been shaped by natural selection. He further speculated, based on some anecdotes about apes in captivity, that moral sentiments, such as sympathy and the desire to help others, are present in nonhuman animals. These anecdotes have received reasonably good support from modern observational study of aid-giving among captive and wild primates.

Darwin's sketchy ideas on evolutionary ethics have received little attention from philosophers. One field of study in which they have been taken seriously is economics. Robert Frank, in his book *Passions within Reason* (1989), has attempted to explain some paradoxes in human economic behavior in terms of evolutionary theory. For example, the practice of tipping a waiter when one visits a distant city is economically irrational because the tip will not influence future service. Frank concludes that such behavior is only superficially irrational. It is based on a deep-seated concern over reputation that would have helped our ancestors to survive and reproduce in the small groups in which they lived. In such an environment, obtaining a reputation for not living up to one's obligations could have been literally fatal. In hard times the cheat would not receive the support and protection of the group and might die.

Consistent with the view that human morality has been shaped by natural selection, Frank argues that we carry emotional baggage that helps us to stay honest. In particular, there is the emotion of embarrassment, as expressed by blushing and other nonverbal signs, such as bowing the head. Frank argues that such body language serves to keep us honest. Even if we wanted to deceive a friend, for example, our body language would give us away. Why do we betray ourselves in this fashion? According to Frank, the trait of sincerity, as expressed by blushing, and other signs of candor make individuals particularly attractive as friends because their reliability is assured. This is a clear example of how moral behavior could be affected by natural selection.

Most of the recent work in evolutionary ethics is connected in some way to the topic of altruism. Modern gene selectionism has provided a mechanistic explanation for the emergence of kin-selected ALTRUISM in an otherwise selfish world. Reciprocal altruism, in which a favor is done on the basis that it will be reciprocated, can also evolve through natural selection and has been documented for species as diverse as vampire bats and chimpanzees. Reciprocal altruism is exceedingly rare, however, because of the vulnerability of this system to cheating. Reciprocal altruism is particularly common among humans, forming the basis, for example, of the monetary economy. Any kind of reciprocal altruism can be interpreted as involving a social contract. At a minimum this requires not only a stable relationship between at least two individuals but also some sort of knowledge of the contract and also a means of enforcing it. The last requirement might explain why people are likely to get very upset when they receive a bad check or when they discover that their spouse has been having an affair. Moralistic anger works on the enforcement side of a contract just as shame and embarrassment can be seen as promoting compliance. The central role of contractual

ideas in ethics might reflect psychological adaptations underlying reciprocal altruism.

See also ALTRUISM, EVOLUTION OF; DARWIN, CHARLES ROBERT; NATURALISTIC FALLACY.

Further Reading
Frank, R. *Passions within Reason*. New York: Norton, 1989.
Ruse, M. *Darwinism Defended*. Reading, Mass.: Addison Wesley, 1982.
Skyrms, B. *Evolution of the Social Contract*. New York: Cambridge University Press, 1996.

evolutionary psychology and sex roles
The new academic specialization of evolutionary psychology has spent a lot of effort investigating predictable sex differences in behavior that are amenable to evolutionary theorization. For example the greater size, strength, and proneness to violence of men than women can be attributed to greater masculine competition over status and participation in warfare. Similarly, the greater interest of most men than most women in casual sexual relationships could reflect an evolutionary past in which men's reproductive success was increased by having several partners. One of the most interesting and successful areas in evolutionary psychology has been the study of sex differences in physical attractiveness phenomena. Thus women are attracted by bodily indicators of social dominance in men more than men are attracted by such indicators in women, whereas men are more preoccupied with indicators of health and fertility, such as narrow waists, than women are. Such preferences would have increased the reproductive success of our ancestors by promoting adaptive mate choice.

By emphasizing the ancient biological basis of sex differences in motivation and behavior evolutionary psychologists have often been criticized as endorsing the sex role stereotypes they are ostensibly describing and explaining. They are accused of perpetuating the NATURALISTIC FALLACY by assuming that a state of affairs that is influenced by sex differences in brain biology is (a) desirable and (b) unchangeable. In reality, most evolutionary psychologists recognize that sex roles are changing rapidly and most are in favor of a more sexually egalitarian society. The existence of sex-typed adaptations in the human brain, however, implies that the ideal of a society in which gender does not matter is impracticable and contrary to the happiness of both sexes.

Some of the most compelling evidence that sex roles are, to some extent, present in the brain at birth comes from the harrowing life stories of people whose sex was misassigned, or reassigned, in early life causing them to play the role of a woman while possessing the brain of a man or vice versa. This has happened in the case of accidental removal of the penis of a boy during a botched circumcision and in the case of Dominicans suffering from a rare disorder that caused penises to develop at puberty in the case of individuals raised as girls. Despite often-repeated false claims to the contrary, such individuals have rarely been happy and many have reverted to the sex role consistent with their biological sex.

There is abundant evidence that the brains of men and women process information differently, much of it reviewed recently by brain researcher Doreen Kimura:

- The left cerebral cortex is thicker in women while the right cerebral cortex is thicker in men, reflecting a female advantage in verbal skills and a male advantage in three-dimensional spatial ability.

- Men have faster reactions than women. Women are faster at recognizing complex visual patterns, like faces.

- Women are much better than men at remembering the location of objects in an unfamiliar setting. This difference apparently represents an inherited adaptation of women to finding food by foraging.

- When asked to spell a word, most men use the left hemisphere but most women use both sides of the brain according to brain-imaging of energy use (PET scans).

- Women have thicker connections between the left and right side of the brain (although this observation remains controversial due to inaccurate measurement).

- Psychologists have long recognized that women are more emotional than men. Thus, they are twice as vulnerable to depression. This difference reflects greater activity of the limbic system, which regulates emotions.

- Women excel in nonverbal communication. They send more nonverbal signals. They are more sensitive to the emotional content of a photograph and can detect the emotion conveyed by tone of voice in experiments where the meaning of speech is garbled. They are better than men at using body language cues to determine when someone is lying.

These neurobehavioral differences are clearly relevant to sex roles in subsistence societies. For example, women's nurturant roles as care givers and socializers of children are favored by communicative skills and emotional sensitivity. Men's lower emotionality and better three-dimensional skills may favor hunting, aggressive competition with peers, and warfare.

Just as women excel at remembering the location of objects, a skill that is directly relevant to foraging, men have better motor skills, at least those that are relevant to hunting. There is a large sex difference favoring men in targeting tasks. On a dart-throwing task for example, the average man performs better than 84% of the women. This difference remains after the greater sports experience of men is statistically controlled.

Conversely, women perform much better than men on tasks that require them to control fingers individually, such as bending a finger at the joint without moving any of the other fingers. This advantage in fine manual dexterity is consistent with the kind of work women perform most in subsistence societies, such as gathering fruits and nuts, weaving baskets, and making thread and clothing.

Although sex roles do vary considerably in different societies, the often-repeated assertion that there are societies in which sex roles are completely reversed is not correct. Thus, anthropologist Margaret Mead claimed that among the Tchambuli in New Guinea, men act like women and women act like men. Tchambuli men were alleged to be submissive and anxious while Tchambuli women were bold and domineering. Mead was mistaken. In fact, the Tchambuli were a polygynous society in which wives were purchased by husbands and in which men felt free to beat their wives whenever they felt dissatisfied.

Evolutionary psychologists claim that men are politically more powerful and more physically aggressive than women in all societies. Women are universally more nurturant and spend more time and effort than men in caring for children and the elderly. These differences can be modified by rearing experiences but their universality points to evolved sex differences in the brain and they cannot be explained away in terms of different socialization experiences.

It is now accepted that boys and girls have clear toy preferences that emerge in spite of socialization pressures. Most boys prefer to play with vehicles and construction toys and most girls prefer to play with dolls and stuffed animals. These preferences are not taught by parents as many socialization theorists have claimed and this conclusion is highlighted by the toy preferences of CAH females whose brains are masculinized by prenatal exposure to high levels of testosterone. They show a distinct preference for "boy" toys over "girl" toys.

Further Reading

Geary, D. C. *Male, Female: The Evolution of Human Sex Differences.* Washington, D.C.: American Psychological Association, 1998.

Kimura, D. *Sex and Cognition.* Cambridge, Mass.: MIT Press, 1999.

Mealey, L. *Sex Differences.* New York: Academic, 2000.

Townsend, J. *What Women Want—What Men Want.* New York: Oxford University Press, 1998.

evolutionary theory and education   *See* CRE-ATIONISM, SCOPES TRIAL.

expert testimony

Scientists are often called upon to give expert testimony in court cases and their evidence can be of critical importance in determining the outcome. In adversarial judicial systems, such as that of the United States, experts for the defense may be called to rebut the testimony of experts for the prosecution. Since experts are paid for their testimony, they are analogous to mercenaries in a war.

This situation is very different from the environment in which scientists customarily operate and is therefore quite challenging to their code of ethical conduct. In conducting scientific research, scientists are expected to be honest, even-handed, open, and objective. Since experts in the courtroom are testifying on one side or the other, they may experience subtle, or not so subtle, pressures to bias their testimony in one direction. Can they bias their testimony by selective reporting of facts, deliberate obfuscation, or misleading presentation of test results? To do so would clearly be unethical. Moreover, to the extent that experts engage in deceptive practices, they not only betray the judicial system that they are ostensibly serving but also bring the profession that they represent into disrepute.

While it is very easy to argue that a scientist in the courtroom can be expected to live up to the same very high professional standards that are expected of scientists in other settings, this may be naively optimistic. The mere fact of being paid to testify for one side may not be unethical in itself but when this monetary relationship between lawyers and scientific experts is viewed in terms of long-term relationships, it can be quite compromising. Experts that produce good outcomes for the lawyers who hired them are much more likely to be called again than careful scientists who presented the evidence in a more objective fashion. This means that there is a conflict of interest between scientific honesty and obtaining future work as an expert witness.

This conflict is not a trivial matter because some scientists make so much money as expert witnesses that they do not need any other employment. The polarized, and biased, nature of expert testimony is well illustrated in the case of DNA EVIDENCE, which is one of the most powerful kinds of scientific evidence that can be presented in court. In a typical case, a DNA expert for the prosecution presents evidence based on analysis of biological material allegedly left by the defendant at the crime scene or, in rape cases, in the victim. DNA evidence is far more reliable than EYEWITNESS TESTIMONY and experts testifying on its behalf like to brag about its accuracy. The probability of a false positive, in which the defendant is wrongly implicated, is estimated at anywhere from one in millions to one in billions. In other words, the odds are almost always long enough to virtually ensure a determination of guilt.

Experts for the defense are called to whittle away at the aura of certainty that large numbers often bring. Their options are limited and therefore fairly predictable. To begin with, the defense expert might quibble with the mathematical calculations by arguing that the exact probability of a false positive cannot be accurately calculated given the present state of the science. Moreover, if there is a possibility that a biological relative of the defendant could be a suspect, the probability of a false positive shoots up. The most effective defense gambit might be to attack the integrity of the evidence itself by showing that it was not appropriately sealed, that it could have been contaminated after the event, or even that the evidence had been planted by the prosecution. Such attacks were notoriously successful in the defense of O. J. Simpson, a former football star who was accused of murdering his wife. In that case, they were facilitated by evidence of astonishingly sloppy and unethical practices of the FBI CRIME LAB.

Given the importance of scientific testimony in court cases, it may not make much difference to point out that experts often do not live up to their professional ethical codes. This outcome seems inevitable in a system that favors the unethical over the scrupulous. In the final analysis, the ethics of conducting a court case has to be seen more as a legal issue than a scientific one. Lawyers hire experts and judges determine whether they can testify and supervise the actual delivery of the testimony. Even though the testimony may not live up to scientific standards of honesty and objectivity, it can be argued that the differing experts tend to cancel each other out, allowing juries to determine the truth.

See also DNA EVIDENCE; EYEWITNESS TESTIMONY; FBI CRIME LAB; SCIENTIFIC DISHONESTY.

Further Reading
Capron, A. "Facts Values and Expert Testimony." *Hastings Center Report* 23, no. 5 (1993) (5): 26–28.
Hubbard, R., and E. Wald. *Exploding the Gene Myth.* Boston: Beacon Press, 1993.
Huber, P. *Galileo's Revenge: Junk Science in the Courtroom.* New York: Basic, 1991.
Resnik, D. B. *The Ethics of Science.* New York: Routledge, 1998.

extrasensory perception

A century ago, scientists believed that there are only around five different senses and that the same sensory modalities are used by different species. This view has changed with the discovery of exotic sensory systems in other species. We now know that bats locate the moths on which they feed by bouncing high frequency sounds off them and listening for the echo. The ears of moths are tuned exclusively to very high frequency sounds such as those produced by the bats. To call them "ears" is rather misleading since there is no overlap between what the moth hears and the much lower frequencies detected by human ears. Other exotic senses that have been discovered in the last century include the infrared (or heat sensitive) system used by snakes to "picture" prey animals; the ultraviolet "vision" of insects; the electro location of fish living in muddy rivers; and the apparent magnetic sense of birds that allows them to navigate using the earth's magnetic field. Given the unexpected richness of these sensory systems, it hardly seems unscientific to investigate the possibility of an exotic sense in humans. To the extent that study of extrasensory perception (ESP, also known as psi) is a scientific endeavor, it can be seen as a search for hitherto unrecognized human sensory systems. The most unethical outcome that can occur is that scientists will have wasted their time in an unproductive endeavor.

Even though parapsychology is now being pursued as a scientific subdiscipline, mainstream psychologists have been unwilling to consider the possibility of ESP because many of its alleged phenomena are the stock and trade of psychic healers, mediums, palm readers, and fortune tellers who unscrupulously separate clients from their cash. Spiritualists often claim to be telepathic, to be able to tune in to the thoughts of others. Psychic healers may claim the ability to see a tumor lying deep in a person's body. Many also claim to be able to predict the future, albeit in frustratingly vague terms from the point of view of gamblers and investors.

Scientific parapsychology struggles to distinguish itself from such trickery and duplicity. Its most important organ, *The Journal of Parapsychology,* publishes methodologically sound articles. Parapsychology has had a slow beginning with many false starts in which promising findings failed to replicate. Despite this, there has been an accumulation of apparently replicable positive findings in the field. The most promising procedure so far has been the ganzfeld method, a controlled experimental test for telepathy, or thought transference from one person, the sender, to another, the receiver. The sender typically concentrates on a picture and tries to convey it to the receiver who is isolated in another room and who provides verbal commentary on his or her thoughts that is available to the sender. The sender tries to telepathically influence the receiver to think about the target picture. The receiver, who is unaware of participating in an ESP study, is kept under conditions of perceptual isolation to ensure there is no leakage of information through con-

ventional sensory channels. "Ganzfeld" means "total field" in German and refers to the monotonous sensory input of white noise and diffuse light that the receiver experiences.

At the end of the session, the receiver is shown four pictures, one of which is the one that was being "sent," and rates how closely each one matches the imagery experienced during the session. If the correct picture is given the highest rating, this is referred to as a direct hit. A 1994 analysis by social psychologist Daryl Bem of Cornell University and the late Charles Honorton, a University of Edinburgh parapsychologist, of 28 different ganzfeld studies, in which 835 people were tested, found that the correct picture was detected 38% of the time compared to the 25%, or one-in-four, probability of being correct due to chance. This result was highly statistically significant and would have arisen by chance less than one time in a billion. This seemed like a stunning confirmation of the ganzfeld phenomenon until the 1999 publication of a study by Julie Milton of the University of Edinburgh and Richard Wiseman of the University of Hertfordshire that analyzed 30 recent ganzfeld studies not covered by Bem and Honorton's work. Milton and Wiseman's study concluded that the pooled results of the studies were not appreciably different from chance.

Even if ESP had been demonstrated satisfactorily by the conventional standards of science, and there is ongoing controversy as new findings come in, it would be a weak phenomenon, for which no practical consequences have yet been demonstrated. Ultimately, parapsychologists may tire of attempting to replicate a phenomenon that may turn out to have been a scientific illusion analogous to N-RAYS.

Scientists are expected to be highly conservative and skeptical about accepting new ideas. In the case of ESP research, the public is actually more skeptical than scientists are. Surveys have found that only 50% of the American public believes in the reality of ESP, far fewer than the number who accept supernatural phenomena in a religious context, e.g., by believing in God. This number rises to two-thirds for people with a college education, apparently because they have read about parapsychology experiments in newspapers and magazines. Among professional academics, belief in ESP also hovers around two-thirds of the population. Intriguingly, only 34% of psychologists accept the reality of ESP, compared to 55% of people in the natural sciences and 77% of academics in the humanities and education.

For ESP to gain much respect from psychologists, it would be necessary not merely to demonstrate that it may exist but to show that it can be manipulated in a scientifically interesting way. For example, if the ganzfeld effect disappeared when the sender was surrounded by copper shields, or even if it decayed systematically with transmission distance, mainstream psychologists would be more enthusiastic.

See also FAITH HEALING; HOUDINI, HARRY; N-RAYS.

Further Reading
Bem, D. J., and C. Honorton. "Does Psi Exist? Replicable Evidence for an Anomalous Process of Information Transfer." *Psychological Bulletin* 115 (1994): 4–19.
Honorton, C. "Meta-analysis of Psi Ganzfeld Research: A Response to Hyman." *Journal of Parapsychology* 49 (1985): 51–91.
Hyman, R. "The Ganzfeld Psi Experiment: A Critical Appraisal." *Journal of Parapsychology* 49 (1985): 3–49.
Milton, J., and R. Wiseman. "Does Psi Exist? Lack of Replication of an Anomalous Process of Information Transfer." *Psychological Bulletin* 125 (1999): 387–91.

eyewitness testimony, unreliability of

The recent development of DNA EVIDENCE has shown quite unambiguously that innocent people are often convicted of crimes. The reasons for this are complex, including factors such as a presumption of guilt in the case of individuals with a criminal record and the ineffectual defense cases presented by court-appointed lawyers representing poor defendants. Innocent people are often convicted on the basis of eyewitness testimony. Recent research in cognitive psychology has shown that although such evidence is highly convincing to jurors, it is often woefully unreliable.

There are few pieces of evidence that can sway a jury like eyewitness testimony. For example, in an experiment by Elizabeth Loftus, mock jurors evaluated a case of a gro-cery store robbery in which two people were killed. When told that there was an eyewitness, 72% of the students voted for the guilty verdict compared to 18% otherwise. When the eyewitness was discredited by evidence he had not been wearing his glasses and could not have seen the defendant clearly, 68% still voted guilty. In other words, half of jurors may be swayed by the testimony of an eyewitness, even when he is totally lacking in credibility!

A major problem with eyewitness testimony is that it tends to change to incorporate new information. Elizabeth Loftus has found that the reliability of witness identification of faces can be destroyed by subsequent incorporation of misleading information. For example, if subjects see a picture of a man who is clean-shaven and are subsequently told he is bearded, they tend to pick a bearded face as the one they have seen.

This suggests that people can, in effect, be tried in the press. For example, if a witness has seen a picture of the suspect in a newspaper, it is possible that this picture can influence their memory of the person they saw committing the crime. Such interference effects have been repeatedly demonstrated in experiments by memory researchers who have also shown that memory of events can be contaminated by subsequent suggestion. The research findings are so clear that lawyers have begun to accept that convictions based on eyewitness testimony alone are unethical.

See also DNA EVIDENCE; FALSE MEMORY AND CRIMINAL CONVICTIONS.

Further Reading
Loftus, E. F. *Eyewitness Testimony.* Cambridge, Mass.: Harvard University Press, 1996.

F

fact-value distinction

This refers to the view that the world of scientific fact is distinct from the philosophical realm of values and that to draw ethical conclusions from scientific observations is fallacious. First identified by Scottish philosopher David Hume in 1740, it is synonymous with the is/ought barrier. To violate the distinction is to engage in the NATURALISTIC FALLACY.

See also NATURALISTIC FALLACY.

faith healing

Modern medicine is an applied science that uses research findings to improve patient outcomes. Yet there are phenomena in medicine that are not easily consistent with scientific analysis and might even appear to invalidate it. In particular, the placebo effect, in which patients improve after receiving a pharmacologically inactive treatment, such as a sugar pill, still defies adequate scientific explanation. Placebo effects can be as large as the effect of a scientific medicine and have been reported in connection with treatments as diverse as cell grafts for Parkinson's disease and drug treatments for depression and hair loss. Many forms of faith healing seem to rely on phenomena that are either equivalent to, or analogous to, placebo effects.

If placebos can improve health in clinical trials, then the psychological conditions provided by religious experiences might also boost health. A variety of evidence suggests that people who attend religious services regularly have better health and recover more quickly from illness. The strongest evidence comes from longevity data. It has been found that people who practice a religion regularly live some seven years longer than those who do not.

Such data have received a variety of explanations. Skeptics argue that they are largely due to statistical artifacts. Most of the research failed to control for relevant personal characteristics including sex, education, ethnicity, marital status of parents, social support, medical history, and socioeconomic status. When such variables are controlled the health benefits of religious observance are small and unreliable. Another explanation focuses on the physiological correlates of religious participation. Religious services and individual prayer tend to lower blood pressure, promote immune function, and reduce depression. These phenomena mirror the health benefits of secular meditation. Few physicians endorse a supernatural explanation for the health benefits associated with faith and prayer.

Given that religious observance could be as effective as some medications, and has no harmful side effects, can physicians ethically recommend to patients that they should go to church more often? Some medical ethicists argue that it is appropriate to support patients in their religious practice. Others argue that religious observance is none of the physician's business. Thus, married people are known to experience better health than singles, but doctors would not be justified in encouraging patients to marry.

Even if many doctors may feel uncomfortable about dispensing unscientific medicine, there is a growing acceptance of the importance of holistic approaches to healing, which consider a variety of lifestyle issues as well as conventional therapies. This trend is reflected by the fact that half of American medical schools now offer courses on religion, although some commentators see this as evidence of the curricular influence of the religious right wing.

Telling patients to attend church more often may go against the grain of scientific medicine but it is unlikely to do the patient any harm. A much more severe ethical problem is posed by those religious groups, such as Christian Scientists, who see scientific medicine as intrinsically evil and refuse life-saving medication and surgery in favor of the healing power of prayer. Christian Scientists see illness as an illusion that can be cured through prayer alone. Legal responses to the faith healing preferences of the Christian Scientists within the United States have been strangely mixed. Parents of children who died after being denied life-saving treatments have been convicted of child neglect. On the other hand, Christian Science sanatoria have been allowed to receive medical insurance payments as though healing exclusively through prayer were a legitimate medical practice.

See also EXTRASENSORY PERCEPTION; HOUDINI, HARRY; PLACEBO-CONTROLLED DRUG TRIALS.

Further Reading
Benson, H., and M. Stark. *Timeless Healing: The Power and Biology of Belief.* New York: Charles Scribner's, 1996.
Koenig, G., J. C. Hays, D. B. Larson, L. K. George, et al. "Does Religious Attendance Prolong Survival?" *The Journals of Gerontology, Series A* 54 (1999): 370–77.
Mitka, M. "Getting Religion Seen As Help in Being Well." *JAMA* 280 (1998): 1896–97.

false memory and criminal convictions

Many criminal proceedings revolve around the testimony of witnesses. Such testimony can be unhelpful in the case of crimes of violence, such as murder, rape, and child abuse, because the trauma associated with the attack impairs the ability of victims to recall details of the crime. In extreme cases, they might have no recollection of the event whatever. This possibility has led to the development of controversial psychological techniques for facilitating memory of traumatic events.

When George Franklin, a retired Foster City, California, firefighter was convicted in 1990 of raping and murdering an eight-year-old girl, the prosecution's case hinged on the memories of his daughter Eileen who "remembered" witnessing the crime almost 20 years after the event. Following this precedent, memory retrieval has been used to secure convictions in many other cases despite expert testimony that false memories can be induced by subtly suggesting that some event has occurred. The use of hypnosis in memory retrieval makes subjects particularly susceptible to suggestion and there is no reliable method of distinguishing between real memories, memories concocted due to the retrieval process, and products of pure imagination. The False Memory Syndrome Foundation was established in 1992 to help people who believe that they have been falsely accused using retrieved memories.

See also DNA EVIDENCE; EYEWITNESS TESTIMONY, UNRELIABILITY OF.

Further Reading
Loftus, E. "Remembering Dangerously." *Skeptical Enquirer* 19, no. 2 (1995): 20–30.
Ofshe, R., and E. Waters. *Making Monsters.* New York: Charles Scribner's, 1994.
Resnik, D. B. *The Ethics of Science.* New York: Routledge, 1998.

Faraday, Michael
(1791–1867)
English
Chemist and Physicist

Largely self-taught, Michael Faraday impressed Sir Humphry Davy, a leading chemist of the day, sufficiently to receive a job as a laboratory assistant at the Royal Institution in London. He was promoted to professor of chemistry in 1833 and remained at the Royal Institution for the rest of his life. Faraday's important discoveries in electromagnetism included the induction of electricity by a magnetic field, the laws of electrolysis (known as Faraday's Laws), and the fact that the plane of polarized light is rotated by passage through a magnetic field (the Faraday Effect). His discoveries led directly to the invention of the dynamo for generating electricity and the electric motor for converting electrical energy into work. He described electromagnetic fields as consisting of lines of force. In chemistry, Faraday discovered benzene and two new chlorides of carbon, developed colloidal gold, and was the first to liquify chlorine. In addition to his achievements as a researcher, Faraday was a popular lecturer at the Royal Institution and even gave lectures for children that were published as *The Chemical History of a Candle* (1860), a delightful instance of scientific popularization.

See also DALTON, JOHN; POPULARIZATION OF SCIENTIFIC FINDINGS; SCIENCE, HISTORY OF; TECHNOLOGY, HISTORY OF.

FBI crime lab and scientific inadequacy

Ethical scientists are committed to doing careful objective work. A number of high-profile criminal trials, including the UNABOMBER, O. J. Simpson, Oklahoma City, and World Trade Center cases, have provided evidence of some extremely careless work in which elementary procedures, such as containing evidence samples in sealed plastic bags, were not followed and therefore prevented cross-contamination of evidence during collection and storage. Another embarrassing incident was the false identification of Richard Jewell as the chief suspect in the Atlanta Olympics bombing. Even more serious has been the charge that FBI scientists are affected by illegal biases. Thus, the lab has been accused of a conspiracy to withhold evidence concerning the FBI assault on Ruby Ridge and of generally suppressing evidence that might be useful to defendants. Long considered a benchmark of objective forensic science, the credibility of the FBI crime lab has been seriously damaged by a welter of accusations in highly publicized cases.

See also DNA EVIDENCE; EYEWITNESS TESTIMONY, UNRELIABILITY OF.

Further Reading

Kelly, J., and P. Wearne. *Tainting Evidence: Behind the Scandals at the FBI Crime Lab.* New York: Free Press, 1998.

Fechner, Gustav Theodor

(1801–1887)
German
Physicist, Philosopher, and Psychologist

Although his academic training was in physics, and he was professor of physics at Leipzig from 1834 to 1839, Gustav Fechner is remembered mainly for his contributions to scientific psychology. In addition to his independent research, he had made a name for himself as the translator of many physics texts from French to German. The rest of his life was devoted to research, writing, and lecturing. Fechner is the father of psychophysics, a branch of psychological research that looks for quantitative relationships between physical stimuli and their psychological impact.

Fechner was plagued by ill health and his illness was associated with a loss of appetite accompanied by unusual sensitivity to light. He spent much of his convalescence lying in bed in a dark room being read to by his mother through a slit in the doorway. His recovery began when a friend dreamed that she had made him a dish of spiced ham soaked in wine and lemon juice. She convinced Fechner to eat some of it and be began to feel better, his spirits buoyed by a dream of his own indicating that he would be completely recovered in 77 days. He felt so good that he began to imagine God has chosen him to solve all of the world's mysteries.

Fechner experienced a great moment of insight on October 22, 1850, when he realized that the mind-body problem could be investigated using scientific techniques. This was exciting for Fechner not only because it drew a connection between his interest in physics and his interest in philosophy but it suggested that a scientific study of the mind was possible, which promised to illuminate the ancient, seemingly intractable, philosophical problem of the mind-body relationship. The idea was to study changes in the perceived intensity of a stimulus (or the sensation) produced by changes in the intensity of a physical stimulus. While this idea was a simple one, the practical question of measuring changes in the sensation remained. Fechner solved this problem in an ingenious way. Beginning with the smallest stimulus that could be detected (the absolute threshold), he increased the intensity until the difference could be reliably detected. Then he increased the intensity again until the next just noticeable difference was reached. These just noticeable differences (JNDs) comprised units in the psychological scale.

Fechner discovered that successive steps in the JND scale required progressively larger jumps in the physical intensity of the stimulus. Mathematically stated, the sensation is proportional to the logarithm of the stimulus intensity. This generalization is known as Fechner's Law. It has had many practical applications, forming the theory behind the decibel scale of loudness and the Richter scale for earthquake intensity, for example. Although 20th-century research has demonstrated that Fechner's Law does not provide a good description of our response to very low intensity or very high intensity stimuli, it can nevertheless be seen as an impressive achievement, which related the internal world of sensation with the physical world by means of a mathematical relationship that applies to most sensory modalities that have been studied.

It is interesting that Fechner's ideas were anticipated by Ernst Weber (1795–1878), who published studies on the subject four years before Fechner had his moment of insight. Weber was also interested in the JND and discovered that the size of the JND was proportional to the size of the reference stimulus (a generalization known as Weber's Law). Mathematically speaking, this means that there is a logarithmic relationship between sensation and stimulus intensities, exactly what Fechner had discovered!

Fechner denied that he had been aware of Weber's work. If so, then the uncanny similarity between Weber's Law and Fechner's represents a remarkable instance of SIMULTANEOUS DISCOVERY that occurred not just around the same time but actually in the same city. Either this was a unique coincidence or Fechner was unconsciously influenced by some vague knowledge of what Weber had done. In any case, history has decided that Fechner is the father of psychophysics and has allowed Weber priority in relation to his law. Weber is not seen today as the founder figure of psychophysics for much the same reason that we no longer associate Alfred Wallace with the discovery of evolution by natural selection. Fechner, like Darwin, invested more energy and enthusiasm in compiling empirical support for the discovery. Fechner performed extensive investigations, and he devised many of the techniques used in classical psychophysics. Even though Darwin held actual priority in respect to the theory of evolution and Fechner did not have priority for the most basic generalization of psychophysics, both scientists worked hard to amass the evidence that would make their names synonymous with a whole field of investigation.

Despite the rigor of his scientific work, Fechner was every bit as much a mystic as a scientist. Thus he wrote on life after death, star life, and the soul of plants. He also believed that dreams predicted the future. If a modern scientist entertained such ideas, he or she would be unlikely to publish them for fear of losing scientific credibility but they did not harm Fechner's reputation as a pioneer in psychophysics.

See also DARWIN/WALLACE AND CODISCOVERY OF EVOLUTION; HELMHOLZ, HERMANN; SIMULTANEOUS DISCOVERY.

Further Reading
Schultz, D. P., and S. E. Schulz, *A History of Modern Psychology.* San Diego: Harcourt Brace Jovanovich, 1987.

feminist ethics

Feminist ethics is defined by two key assumptions. One is that women's moral reasoning is qualitatively different from that of men, focusing more on concrete interpersonal relationships than on abstract moral principles and social contracts. The other is that women have a different life experience from men, generally enjoying less economic and political power and lower social status and that this difference effects ethical decisions and situations. For example, the historical lack of economic power by women left them vulnerable to sexual harassment in the workplace. The greater vulnerability of women to this sort of attack means that sexual misconduct by men in the workplace has more serious potential consequences for the victim than equivalent behavior directed by women against men.

Feminist ethics began as a revolt against mainstream moral philosophy that had been conducted almost exclusively by men and therefore arguably contained male biases. These biases were most clearly described by CAROL GILLIGAN in her study of the developmental psychology of moral development. Gilligan's work constituted a critique of the work of Lawrence Kohlberg, who developed a moral scale based on how people resolved moral problems presented in the form of vignettes, or scenarios. According to Kohlberg, the highest stage of moral reasoning was reached when people showed evidence of being able to construct their own abstract moral principles that sometimes conflicted with social expectations.

Kohlberg found that women were less likely to attain what he considered to be the highest level of moral reasoning. Gilligan argued that the Kohlberg scale was biased in favor of men in that it arbitrarily elevated male-type concerns for abstract notions like justice and fairness over female-type preoccupation with the welfare of individuals and the integrity of interpersonal relationships. If this criticism applies to developmental psychology, then precisely the same kind of argument could be made against moral philosophy in which the masculine concern with what came to called an ethic of justice prevailed over the feminine ethics of care.

Feminist ethics of moral philosophy have contended that moral philosophy reflects the interests and experiences of white middle-class men and omits the experiences of women. Thus, ethics concerns itself with realms of human activity in which men predominate, such as politics and business, and is relatively silent about the ethics of conduct in spheres that are of more central relevance to women, such as the family.

At first glance, it might appear that feminist ethics constitutes an attack on the possibility of an objective ethics, and therefore a kind of MORAL RELATIVISM, but most feminists appreciate that moral relativism is fundamentally threatening to their agenda. According to moral relativism, in some societies the subordination of women to men in seen as desirable and is therefore permissible in those societies. Feminists argue to the contrary that a lack of power for women is always morally objectionable.

The objective of feminist ethics to develop a theory of moral action that is based on women's experiences has turned out to be extremely difficult. The problems have been articulated most often by feminists themselves. Thus, Virginia Held has argued that the mother-child relationship should be seen as an alternative paradigm for ethics. Yet many women are not, and will never be, mothers. In a similar vein, different subgroups of feminist philosophers have questioned whether it makes any sense to discuss the experiences of women as a single group. If ethics is to reflect women's experiences, then it must be diverse, reflecting the differing experiences of African-American women, lesbians, disabled women, elderly women, and so forth. Feminist ethicists have failed to devise any compelling resolution to this ethical relativity problem.

Even though feminist ethics began as a challenge to mainstream ethical philosophy, it has ended up being incorporated as a subdiscipline of ethics. There are many different reasons why this has happened. Male philosophers have responded to criticisms of mainstream ethics by incorporating feminist ideas in their own work. Feminist philosophers who began their career from a feminist perspective have developed an interest in more mainstream ethical topics. Finally, some feminist philosophers have recognized that many of the ideas of rights-based ethics can be reformulated in terms of relationships. Rights need not be seen as tied to an individualistic, self-assertive, and competitive conception of the self as earlier feminist critics of traditional ethical theories imagined. Thus political rights can be redefined in terms of opportunities to create desirable interpersonal relationships. If so, then disputes about rights can be settled by considering the sort of relationships that are being fostered.

The acceptance of some feminist concepts in moral philosophy may not be an isolated instance and it probably reflects a growing recognition, attributable to the work of evolutionary psychologists, that there are predictable differences between the thought patterns of men and women. Most of these differences are comparatively small, however, and the same applies to differences in moral reasoning. If the reasonable assumption is made that most women do care about rights and that most men do care about the quality of their interpersonal relationships, then it is clear that feminist ethics and traditional ethics are each applicable across gender boundaries. Feminist ethics can be seen as a useful correction to biases in moral philosophy that developed when it was practiced exclusively by men.

See also ETHICS OF CARE; EVOLUTIONARY PSYCHOLOGY AND SEX ROLES; GILLIGAN, CAROL; MORAL PHILOSOPHY.

Further Reading

Brown, S. "Recent Work in Feminist Ethics." *Ethics*, 109 (1999): 858–93.

Card, C., ed. *Feminist Ethics.* Lawrence: University Press of Kansas, 1991.

Clement, G. *Care, Autonomy, and Justice: Feminism and the Ethic of Care.* Boulder, Colo.: Westview, 1999.

Koehn, D. *Rethinking Feminist Ethics: Care, Trust, and Empathy.* New York: Routledge, 1998.

Nelson, H. L. *Feminism and Families.* New York: Routledge, 1997.

feminist ethics and the philosophy of science

Feminist philosophers of science have launched a critique of traditional male-dominated science that has many parallels with the objections of feminist ethicists to mainstream moral philosophy (see FEMINIST ETHICS). Central questions that have been raised are whether science is objective, gender neutral, and value free. While feminist philosophers of science began with the postmodernist perspective that science is extremely male-biased, and that science is inevitably contaminated by the values of its practitioners, making true objectivity impossible even in principle, current feminist scholars have changed tack considerably. The current view is that even though science is flawed by masculine biases, objectivity is possible and female scientists have an important role to play in helping to identify and correct these biases. Instead of railing at science as a male-biased "narrative," many people argue that women should take up scientific careers in larger numbers and that their efforts should be supported by affirmative action programs specifically targeted to science.

The early feminist attack on science took the form of a central claim that science represented a masculine way of knowing the world. As a masculine activity science was dominating, analytical, and competitive. Women saw the world more holistically, empathized with their subject matter, and favored social cooperation. Fewer females succeeded in science careers because the whole epistemic basis of the endeavor was hostile to women's ways of knowing. This argument has not found favor among feminist scientists because it is hostile to science. If it were really possible to have a feminist science operating in parallel with the masculine one, then science could not be a universal or objective body of knowledge. Science as commonly understood would not be possible.

Feminist scientists generally do not believe that men and women scientists are assembling diverse universes of knowledge, or even that they conduct their research in fundamentally different ways. What is different is their characteristic choice of subject matter. Female scientists are often drawn to questions that have been ignored by their male counterparts. One example is the study of symbiosis in ecology. Symbiosis involves mutually beneficial relationships between species, such as the role of small cleaner fish in clearing larger fish of parasites. Male ecologists tended to steer clear of such phenomena because they were incompatible with the view that competition and hostility are the ruling principle of interspecific relationships. Modern ecologists recognize that stable ecosystems are comprised of a finite number of plant and animal species that help each other to survive and reproduce. Many of the ecologists doing this work have been women.

An even more obvious example of how gender influences scientists' choice of subject matter concerns disciplines that study relationships between the sexes, including animal behavior, primatology, psychology, anthropology, and archaeology. Male primatologists, for example, have tended to focus on the role of dominant aggressive males in studying mating competition and social organization. In the case of baboons, the larger males, with their impressively large canine teeth, lead a harem of females and are aggressively dominant. This encouraged the view that male aggression was the basis of social organization in baboons. Subsequent research, by feminist primatologists, who examined interactions more from the perspective of females, revealed not only that female baboons often actively solicited copulations with males outside the harem but that older females played an important role in leadership of larger groups by determining the direction of their daily foraging movements.

In keeping with their selection of different topics in the social sciences, women often emphasize the role of different explanatory concepts. For example, feminist anthropologists have challenged the man-the-hunter explanation for the evolution of increased hominid brain size, noting that large brains also facilitate language communication and social integration, capacities in which women excel. By the same token, feminist archaeologists have argued that exclusive reliance on stone tools as a clue to subsistence economy is a mistake because the tools used by women in existing subsistence societies, such as nets used to gather food and gourds used to collect water, leave no trace in the archaeological record.

Feminists have also been able to point to male biases in the biomedical field. Even though the egg plays an active role in fertilization by sending out microvilli to draw in a sperm, a role that was discovered over a century ago, this fact was largely ignored until the 1960s. Older textbooks concentrated on the heroic competition among sperm to reach and penetrate the egg. While rediscovery of this phenomenon preceded feminist attacks on science, it seems likely that the finding would have received more recognition sooner if there had been more women scientists working in embryology.

There has been a perception of gender bias in the conduct of biomedical research in which the bulk of research supposedly focused on male subjects. According to this view, research on women's illnesses has been excessively preoccupied with reproductive diseases reflecting societal norms concerning the role of women and there has been insufficient attention to leading causes of death, such as heart disease. Another complaint has been that there was a failure to include women in drug trials even though they actually consume the great bulk of most medications. Despite the well-publicized case of the Harvard

Medical School study of the effects of aspirin on heart disease in male physicians, published in 1989, there is no evidence of a systematic neglect of women in medical research. Recent U.S. legislation has mandated the inclusion of women in drug trials, which may have implied that there was a real bias against women. In reality, the health of women receives more attention than that of men. Twice as much money is spent on health care for women than men because women are more likely to seek medical care than men are. The National Institutes of Health (NIH) spends twice as much money on research into diseases that are specific to women as it does on diseases that are specific to men.

It is hard to disagree that a diversity of perspectives among scientists can correct distortion and improve the objectivity of a scientific discipline. There is considerable controversy, however, about whether gender bias can afflict the "hard" sciences, such as chemistry and physics, and whether it is present in mathematics.

It is easy to claim that a male-dominated science is one that has blind spots that can be redressed by the greater participation of women in scientific careers. A more challenging question to answer is why there are comparatively few women who work as leading researchers in science, engineering, and mathematics. There are two plausible explanations that are often seen as rivals but may be complementary. One is that women are not drawn to these fields because they find them intrinsically cold and uninteresting. The other is that women have a similar level of interest as men do but that social norms, including the sexist social environment in which science is taught and conducted, are hostile to the entry of women to these professions.

Average sex differences in cognitive aptitudes are generally quite small but in the specific case of the tiny minority of individuals who score high on mathematical abilities there is a large overrepresentation of males. Mathematical ability is predictive of interest in and success at careers in science and engineering as well as mathematics. If follows that the interest in, and success of, women in these careers may never reach parity with that of men. Note that this argument does not claim that women cannot achieve at the same level as men but only that such high achieving women will be outnumbered by similarly high achieving men. The fact that women can contribute to science at the highest level is supported by the emergence of female Nobel Laureates in all of the major branches of science in the 20th century. While it might be argued that the dearth of women in mathematics and engineering is entirely due to a lack of equal opportunity, this view is not supported by the fact that social changes favorable to women in the second half of the 20th century did not produce an influx of women into these careers. For example, male engineers in America still outnumber females by approximately nine to one even though engineering schools have made great efforts to recruit women.

Even if sex differences in mathematical aptitude play a role in explaining the underrepresentation of women in science, engineering, and mathematics, they are unlikely to constitute the whole story. Feminist philosophers have made the case that these disciplines are, in various ways, unfriendly to women. One way of increasing the number of female scientists would be to reduce the very high attrition rates in the first two years of college. It has been suggested that this could be done through a pedagogy that involves increased emphasis on collaborative learning, practical experiences, and group projects. Whether these pedagogical reforms would be of more benefit to women than men is not clear, although they do reflect feminist insights about the importance of pluralistic pedagogies.

Even if women achieve parity in respect to the number of graduates from science, engineering, and mathematics programs, their problems are not over. The world of professional science is a tough competitive environment in which few women rise to the top. A recent study by Cambridge University found that while 40% of the researchers in chemistry were women, only 2% of the tenured faculty in that discipline were female.

How do we begin to explain the huge difference in success between men and women of science? One view is that science remains one of the most harshly competitive fields in which individuals struggle to be published, achieve fame and status in the eyes of peers, and receive funding. Men supposedly enjoy this challenging environment and are stimulated to be highly productive. Women supposedly are less interested in competing for priority and status. Consistent with this view is the fact that tenured women are generally much less academically productive than tenured men. Feminists argue that women's productivity is undermined by competing obligations of relationships and families, which they often set as an ethical priority. Sex bias also plays a role. Even in Sweden, a country that is noted for its ethos of gender egalitarianism, male-dominated review committees do not give women fair treatment. Female Swedish academics have to publish twice as much as their male counterparts to be deemed worthy of a fellowship in medicine.

There is also evidence of bias against women in academic careers in the United States. Thus, in one study of medical school faculty hired in 1980, only 5% of women had become professors after 11 years compared to 23% of men. Women also received lower salaries than men in comparable positions in academic medicine. In prestigious institutions such as Stanford University, women are underrepresented on faculties. Thus, women comprise just 19% of the faculty compared to 28% at doctoral

degree–granting universities nationally. Female academics at Stanford University recently brought their complaint of sex bias at Stanford to the U.S. Department of Labor. Internal studies at some institutions have established evidence of gender bias leading to voluntary programs designed to eliminate barriers to female promotion. Thus, at Johns Hopkins University School of Medicine, after a five-year program initiated in 1990, the number of women at the rank of associate professor increased by a staggering 550% (from 4 to 26). Barriers to female promotion may include reduced mentoring, reduced access to funds, equipment, and personnel, as well as outright gender discrimination.

There is a long history of women receiving lower academic salaries but this finding has been complicated by numerous factors, including seniority, field of specialization, time devoted to academic work, and research productivity, all of which worked in favor of men in the past. Women in British universities are paid 15% less than men and are still underpaid when factors including age, rank, tenure, and field of specialization are statistically controlled. In the United States, however, at least one study, dealing with the University of Alabama, found that women were actually better paid than men when the effects of research productivity were also statistically controlled. In summary, discrimination against women in academic life appears to be more pronounced in some fields of study, such as medicine, and it seems to be more prevalent at some institutions than others. Bias against women in academic careers may be removed by deliberate administrative interventions.

Attempts to improve the status of women in science take two different, and apparently contradictory, positions. The first is that AFFIRMATIVE ACTION should be used to improve both women's representation among tenured faculty and women's receipt of government funds. Responding to legislation produced by pressure from THE ASSOCIATION FOR WOMEN IN SCIENCE, the National Science Foundation (NSF) conducts an affirmative action program for women in science that involves continuous monitoring and biennial publication of statistics. This program has funding for career development awards earmarked for women.

The second approach is designed not so much to help women compete in the harshly competitive male-dominated world of professional science as to change that climate to make it more friendly to women. The NSF conducts site visits designed to improve the climate for women in academic departments that have traditionally excluded them. This will presumably mean that refusing tenure to women will become more difficult than it has been in the past. Women may also receive more opportunities to develop their research programs through relief from teaching and other obligations. Such programs may help women cross the threshold in some academic departments that currently have very few tenured females. The deeper philosophical question about whether the scientific enterprise itself can be changed to make it more friendly to women remains open. Is it possible to change the incentive structure of scientific investigation such that cooperation is placed at a premium over egocentric competition? This would indeed constitute a profound change in the way that science is done. Such a change is unlikely if women do not have more administrative control over research laboratories.

See also AFFIRMATIVE ACTION; ASSOCIATION FOR WOMEN IN SCIENCE; FEMINIST ETHICS; MORAL PHILOSOPHY.

Further Reading

Benbow, C. P. "Sex Differences in Mathematical Reasoning Ability in Intellectually Talented Preadolescents." *Behavioral and Brain Sciences* 11 (1988): 169–232.
Burrelli, J., C. Arena., C. Shettle, and D. Fort. *Women, Minorities, and Persons with Disabilities in Science and Engineering.* Upland, Pa.: Diane Publishing, 1998.
Fried, L. P., C. A. Francomano, S. M. McDonald, et al. "Career Development for Women in Academic Medicine," *JAMA* 276 (1996): 898–905.
Kadar, A. G. "The Sex-Bias Myth in Medicine." *The Atlantic Monthly* 274, no. 2 (1994): 66–70.
Lindley, J. T., M. Fish, and J. Jackson. "Gender Differences in Salaries: An Application to Academe." *Southern Economic Journal* 59 (1992): 241–59.
McNabb, R., and V. Wass. "Male-Female Salary Differentials in British Universities." *Oxford Economic Papers* 49 (1997): 328–43.
Nelson, H. L., and J. Nelson, eds. *Feminism, Science and the Philosophy of Science.* Boston: Kluwer, 1996.
Rosser, S. *Re-engineering Female-Friendly Science.* New York: Teacher's College Press, 1997.
Schiebinger, L. L. *Has Feminism Changed Science?* Cambridge, Mass.: Harvard University Press, 1999.
Wenneras, C., and W. Wold. "Nepotism and Sexism in Peer Review." *Nature* 387 (1997): 341–43.

fetal cell grafts

Some diseases of the brain, such as Parkinson's disease (a movement disorder with tremors and difficulty initiating actions), are caused by the loss of cells in a specific brain structure. If the cells are to be replaced, grafted fetal brain cells are ideal because they can mature and differentiate to serve the required function. Up to this point, fetal cell grafts in patients with Parkinson's disease have produced modest improvements in motor function but are still far from being accepted as a cure for the disorder.

The fact that the cell grafts were derived from aborted fetuses has produced ethical problems and governmental roadblocks for this kind of research. Responding to pressure from his antiabortion constituents, President Ronald Reagan denied NIH (National Institutes of Health) funding for fetal tissue research. The moral objection to this work was not that human tissues were being used in the grafts. After all, transplanted organs from cadavers are now routinely used with the prior consent of the deceased without any moral outcry. Rather, the research could be seen as "benefiting" from abortion, which was itself seen as immoral. President Bill Clinton restored funding for fetal grafts soon after he came into office.

See also HUMAN CLONING; ORGAN TRANSPLANTATION.

Further Reading

Resnik, D. B. *The Ethics of Science.* New York: Routledge, 1998.

fetish, conditioning of in lab

A fetish is defined as an object or body part that assumes a focal role in the sexual life of an individual to the extent that they cannot achieve sexual gratification without the real or imagined presence of the fetish object. For example, a foot fetishist might be unable to have intercourse with his wife without caressing her feet or fantasizing about shoes. Although fetishes are classified as abnormal either when they interfere with normal sexual expressions or get the individual into trouble, as in the case of peeping Toms, exhibitionists, or pedophiles, the tendency to focus erotic interest on specific body parts, items of clothing, or situations, appears to be characteristic of the sexual lives of many people.

Case histories suggest that many fetishes can be traced to specific sexual experiences and that virtually any inanimate object associated with sexual arousal could become a fetish object that is capable of evoking sexual excitement itself. Fetishes are not entirely arbitrary, however. Most relate directly to women's bodies or to articles of women's clothing that would ordinarily be removed prior to intercourse.

The notion of a man being excited by women's clothing is analogous to Pavlov's dog salivating when the buzzer is sounded. The idea is that both the buzzer and the clothing suggest that an appetite is about to be satisfied. Fetishes may be an example of Pavlovian association learning. The most convincing method of establishing how fetishes develop is to create a fetish in the laboratory. The project of turning normal young men into sexual deviants is clearly an ethical problem that would frighten many researchers.

The 1966 study by S. Jack Rachman and Ray J. Hodgson of the Institute of Psychiatry in London was modeled on Pavlov's experiments with salivating dogs. Pavlov's idea was to make the dogs salivate to something completely unrelated to food (a buzzer) by presenting the buzzer just before the food.

Instead of measuring drops of saliva, the researchers measured sexual arousal among a selected sample of men directly using a penile strain gauge. This ingenious device consists of a mercury-filled rubber cuff placed around the shaft of the penis. As the subject becomes aroused, the diameter of the penis increases. This stretches the cuff and makes the column of mercury inside it thinner thereby decreasing the electrical conductance of the mercury. Changes in the mercury column are continuously recorded as changes in electrical conductance allowing for an unobtrusive measure of sexual arousal.

In the experiment, the researchers first presented a slide of the intended fetish object, a pair of black fur-lined boots (chosen for the obvious reasons that many fetishes involve articles of feminine clothing). They wanted to be sure that none of the participants in the experiment were already erotically interested in footwear. No one became aroused.

Next, a slide of the black fur-lined boots was repeatedly followed by a slide of an attractive nude from a pornographic magazine. After 20 such pairings, most of the men were sexually aroused by the slide of the boot. The normal participants has been turned into sexual deviants!

The researchers were sensitive to the ethical implications of what they had done. What if the lives of the subjects were to be overtaken by kinkiness, to the extent that they could not have sex without the aid of articles of footwear? Suppose that this destroyed their prospects for a happy marriage? Using Pavlov's work as an inspiration, they devised a solution. Pavlov had repeatedly sounded the buzzer without giving his dogs food until they eventually stopped salivating at the sound. Rachman and Hodgson presented the boots slide repeatedly, without the nude. They found, to their presumed relief, that the fetish slowly went away.

This would have been a gratifying story of responsible researchers expunging the possible harm caused to participants had the researchers not decided to contact the same participants again a year later to pursue their studies. When they showed them the same black fur-lined boots, the reaction was unanimous—strong sexual arousal. Pavlov noted the same phenomenon in his salivating dogs and referred to it as "spontaneous recovery," a process by which extinguished conditioned responses naturally regain strength over time. Participants underwent the same procedure for getting rid of the fetish a second time. The extinction procedure was a success the second time also. Had they finally solved the problem? We may never know.

The Rachman and Hodgson experiment falls into the unusual category of research that is designed to create a behavior problem. As such, it clearly operates in dangerous territory from an ethical standpoint and would be unlikely to receive funding, or pass ethical review, in the contemporary ethical environment. In their own defense, the researchers could say that they had taken all reasonable measures to return participants to their original state; that the benefits to our knowledge outweigh the minor harm suffered by participants; or even that the risk of developing a fetish in the experiment was little greater than what might be expected in their ordinary lives (the MINIMAL RISK CRITERION).

The benefits of scientific knowledge about the causes of abnormal sexual interests are potentially great. Knowing the important role played by classical conditioning in the erotic lives of men allows programs to be developed that are designed to alter sexual behaviors which have been acquired by association learning and are considered problematic by the individual or by the legal system.

For example, a man who is bothered by a clothing fetish can get rid of it through a procedure in which an unpleasant association is built up for items of women's clothing without affecting attraction to women's bodies. In this treatment (known as aversive counterconditioning) slides of items of feminine clothing are repeatedly followed by painful electric shocks. Eventually, clothing loses its sexually arousing properties. Similar procedures have been used to treat pedophiles. In this case, pictures of children are paired with electric shock. While pedophilia is very difficult to eradicate, the recidivist rate is substantially reduced by counterconditioning.

See also ELLIS, HAVELOCK; KINSEY, ALFRED; MASTERS AND JOHNSON; SEXUAL ORIENTATION RESEARCH.

Further Reading
American Psychiatric Association. *Diagnostic and Statistical Manual of Mental Disorders: DSM IV.* Washington, D.C.: American Psychiatric Association, 1994.
Bootzin, R. R. *Abnormal Psychology.* New York: Random House, 1988.
Rachman, S., and J. Hodgson, "Experimentally Induced "Sexual Fetishism": Replication and Development." *The Psychological Record* 18 (1968): 25–27.
Seligman, M. E. P. *What You Can Change and What You Can't.* New York: Knopf, 1993.

file drawer problem

In the publication of scientific research, journal editors have a distinct bias in favor of papers with positive findings. Submissions that fail to establish significant results are seen as contributing nothing new to knowledge and are generally not allowed to take up valuable space in journals. For this reason, authors of studies that lack statistically significant effects usually do not bother submitting them for publication. They accumulate in file drawers. If only studies with positive results get selected for publication, then effects that are entirely due to chance can seem real. The resulting error is called the file drawer problem.

See also EXTRASENSORY PERCEPTION.

Fitzgerald, George Francis
(1851–1901)
Irish
Physicist

A leading theoretical physicist, Fitzgerald anticipated some key aspects of ALBERT EINSTEIN's theory of relativity, particularly the notion that objects moving at high speeds shrink in size. This concept is new referred to as the Fitzgerald-Lorentz contraction.

See also LORENTZ, HENDRIK ANTOON; SIMULTANEOUS DISCOVERY.

fluoridation of water supply
Fluoridation of the public water supply to reduce tooth decay has been controversial in the United States and was rejected in some areas as a form of involuntary medication of the population by government. Fluoride was first added to public water in 1945 and since that time has been credited with halving the incidence of dental cavities and tooth loss. Fluoride occurs naturally in some waters but usually at a lower level than the one part per million that is necessary for optimal reduction of dental caries.

Some 60% of Americans live in areas where the public water supply is fluoridated, compared to only around 10% of the inhabitants of Britain. In Britain, physicians and researchers have expressed grave concerns about the possible toxicity of fluoride in the water supply, although there is little convincing scientific evidence of this, and they have argued that it is an inappropriate substitute for better dental hygiene and better dental services. Despite these concerns, fluoridation has been endorsed by the World Health Organization and is considered safe in America, where it has been used most. Resistance to water fluoridation may be based on irrational fear of technology, which has been identified in other contexts in which the fear of ALAR, FOOD IRRADIATION, and GENETICALLY MODIFIED FOOD, for example, seemed out of proportion to any identifiable risk.

See also ALAR SCARE; FOOD IRRADIATION; GENETICALLY MODIFIED FOOD; PASTEURIZATION.

Further Reading
Martin, B. *Scientific Knowledge in Controversy: The Social Dynamics of the Fluoridation Debate.* Albany: State University of New York Press, 1991.
"The Facts about Fluoride." *Chemist and Druggist,* March 13, 1999.

food irradiation
Exposure of food to ionizing radiation produced by radioactive isotopes can kill pathogenic bacteria and fungi and may prevent the reproduction of larger organisms, such as *trichinella spiralis,* the parasitic worm in pork that is responsible for trichinosis. This is a significant advantage over simple refrigeration that slows down food spoilage by merely retarding the growth of microorganisms. Food irradiation is used in the United States and some 20 other countries but its use is restricted to a few categories of foods that are believed to hold a particular public health threat. Irradiated foods include fish, shellfish, and poultry for which salmonella poisoning is a particular threat. Other irradiated products are potatoes, tropical fruits in South Africa, and Dutch spices.

Even though wider use of irradiation could save many lives attributable to food poisoning, the process is viewed with extreme suspicion in the United States and elsewhere. Thus, even though irradiation of food and poultry is permitted by the U.S. Food and Drug Administration, demand for these irradiated products has not developed. Consumer resistance is based partly on a controversy about the potential for adverse health effects of irradiated foods and partly on the irrational fear of radiation. This fear is irrational because even though the food is irradiated, it does not itself acquire radioactive properties and poses no radiation threat to the consumer. There is no scientific evidence that irradiated food causes any harm to consumers. The worst that can be expected is that lipids in the food are damaged and may acquire a somewhat rancid taste. This problem is minimized by vacuum packing of food and refrigeration during irradiation.

One of the biggest health threats from contaminated food is posed by hamburger meat, which is estimated to cause some 10,000 cases of *E. coli* illness in the United States alone each year. A small minority of these cases are fatal. Such tragedies can be averted by food irradiation. In February 1999, the U.S. Department of Agriculture proposed allowing irradiation of uncooked meat and meat products but only on a voluntary basis and with labeling that clearly tells consumers that the product has been irradiated. It therefore seems unlikely that the problem of contaminated hamburger meat will be solved in the near future. Preventable deaths will continue to occur because the public is not ready to accept as safe a technology that could prevent them. Similar reluctance has caused an irrational delay in the implementation of other technologies for making the food supply safer, such as resistance to pasteurization of milk.

See also ALAR SCARE; FLUORIDATION; GENETICALLY MODI-FIED FOOD; PASTEURIZATION.

Further Reading

"Food irradiation—An Update." *Frozen Food Digest,* December, 1999.

Satin, M. *Food Irradiation: A Guidebook.* Lancaster, Pa.: Technomic, 1996.

Ford Pinto

The Ford Pinto was notorious for fires that resulted from low-speed rear end collisions. A subcompact car designed to be inexpensive and light, it was vulnerable to rear end impacts that ruptured the gas tank causing fuel spillage that often turned minor accidents into deadly infernos. The Ford Pinto case is often held up as an example of the harmful effects of the profit motive on design safety. Once interpreted in terms of the immoral decisions of leading executives, contemporary applied ethicists take a broader systems approach to the development of tragically unsafe products, placing them in the context of distributed decision-making processes within corporations and the industrywide constraints under which these operate.

PINTO MADNESS

Developed between 1967 and 1970, and first sold in September 1970, the Pinto received widespread public attention as a safety problem in 1977 with the publication of a Pulitzer Prize–winning article by Mark Dowie in *Mother Jones* magazine entitled "Pinto Madness." The punch line of Dowie's article was that even though Ford was aware of the gas tank problem and could have fixed it for $11 per vehicle, it chose not to do so because the cost of the change would have exceeded company liability in the case of death and injuries attributable to Pinto fires. A 1973 Ford memo obtained by Dowie calculated that the benefit associated with installation of a modified fuel tank would amount to $49.5 million. This assumed 180 deaths at a cost of $200,000 each (the value of a human life according to calculations by the National Highway Traffic Safety Administration rather than company liability as Dowie presumed), 180 serious burns at a cost of $67,000 each, and the loss of 2,100 vehicles at $700 each. The cost of the safety modification was estimated at $11 each for 12.5 million vehicles produced by all American car manufacturers in a year, or a total of $137 million. Since the dollar benefit of making the Pinto safer far exceeded the cost, the conclusion was drawn that the modification was not cost effective. The Ford memo cited by Dowie (often referred to as the Grush/Saunby Report) was written in 1973 several years after the Pinto had been designed and therefore could not possibly have affected decisions made by the design engineers as Dowie claimed.

Ignoring the fact that the Ford memo was dealing with all fire-related automobile accident deaths, Dowie disputed the estimate of 180 deaths and concluded that over 500 people had actually died due to Pinto fires. Yet the National Highway Traffic Safety Administration (NHTSA) could document only 27 Pinto fire-related deaths. In response to Dowie, Ford claimed that the Pinto actually had a better fire safety record than the average car on the road, accounting for 2% of vehicles and only 1.2% of fatal fires.

Dowie's article did for the Pinto what RALPH NADER had done for GM's Corvair in a 1965 book titled *Unsafe at Any Speed.* Nader claimed that design flaws caused the Corvair to turn over at high speeds. Nader not only criticized this one model but argued that it was symptomatic of an industry ruled by greed in which the best interests of consumers were sacrificed to increase profits. Nader's book was a best-seller and the public outrage it stimulated was influential in passing the Highway Safety Act in 1966. This was the first serious attempt at governmental regulation of safety issues in the automobile industry and it led to the establishment of the NHTSA, which was responsible for developing safety codes as the Pinto was being designed. During the design process, the Pinto engineers were guided by internal company standards and industry norms. Government regulations had not yet been implemented. The automobile industry had largely escaped safety regulation by promoting the view that traffic fatalities were due primarily to unsafe driving rather than to cars that were not crashworthy.

REVISIONIST ACCOUNTS OF PINTO MADNESS

The linchpin argument of Dowie's exposé of the Ford Pinto is that design engineers were aware of the safety problem created by fuel spillage during rear collisions but that they failed to fix the problem because this would not have been cost effective. Yet, there is no evidence that the engineers believed the Pinto to be particularly unsafe because of the fuel tank, although they certainly did consider the problem. The engineers simply did not expect inexpensive cars to live up to the same safety criteria as larger more expensive vehicles. Moreover, the fixed barrier tests in which Pintos were towed backward into walls at speeds of 20–30 miles per hour, and in which the gas tanks were ruptured, did not set off the same kind of alarm bells as they would today because engineers at that time were skeptical of the extent to which a fixed barrier test mimics a real-world crash.

During the design phase, several different possible gas tanks were considered for the Pinto but they were rejected because their consequences were uncertain or because they were anticipated to fail in real-world situations. When Pintos were equipped with foam-like gas tanks, they did not leak fuel during 30-mile-per-hour rear collisions. Engineers rejected this tank because they were afraid it would melt. The best protection for the gas tank in the event of a collision is to place it over the axle where it is unlikely to be crushed. This solution was rejected because the tank was closer to the passenger compartment and was also more likely to be ruptured by pointed objects placed in the trunk. Another design that was considered involved the use of rubber bladders inside the metal gasoline tanks. These were successful in preventing spillage, even when the tank was ruptured, but they have never been adopted in any car, suggesting some serious problem.

The design engineers were not callous in their disregard of the fuel tank safety issue and were not generally complacent in respect to safety issues. It is worth pointing out that corporate safety decisions are ultimately made by higher management using the information provided by engineers, rather than by the engineers themselves. In fact, some had joined the company with the specific intention of improving car safety. They did not see the fuel tank issue as being particularly important and were far more concerned about the windshield, which was not made of safety glass and had a possible retention problem. A Pinto transmission problem that produced exactly the same number of deaths as the fuel tank problem never received any publicity.

Contrary to the Pinto Madness story, there is no clear evidence that the Pinto was any more unsafe than the average car of its size class. The reason that the Pinto was targeted by Dowie seems to be that Ford had submitted its cost-benefit analysis of the modified fuel tank to the NHTSA, thus bringing it into the public domain and making it available for journalistic scrutiny. Ironically, General Motors had produced a similar report on rear-end collisions of its own cars that used a similar cost-benefit analysis using the NHTSA's $200,000 value for a life. General Motors did not pass their report on to the NHTSA. The poor safety record revealed in this report might have made General Motors a more appropriate target for Dowie's criticism.

Bowing to adverse publicity concerning the Pinto, the NHTSA launched an investigation of the Pinto gas tank and announced, in 1978, that the car had a safety defect after two of the vehicles burst into flames during a crash test. It made no difference that other small cars would not have passed the same test or that the Pinto had always complied with NHSTA standards for fuel system integrity.

FORD'S LIABILITY

Following the NHSTA investigation, Ford voluntarily agreed to recall all 1971–76 Pintos, or 1.5 million cars,

making this the largest product recall in history. Two plastic shields were added to prevent the fuel tank from being ruptured by bolts in the differential housing. A longer fuel tank pipe and an improved sealing cap were installed to prevent gasoline leakage following collisions.

In addition to the cost of recalling the vehicles, Ford also settled numerous lawsuits at a cost of some $50 million. The most famous of these was the case of *Grimshaw v. Ford Motor Company* that was filed in connection with the death of driver Lilly Gray and serious injury of her passenger, 13-year-old Richard Grimshaw, in a rear-end collision after which their Pinto hatchback burst into flames. The jury awarded $2.5 million in compensatory damages and a record $125 million in punitive damages. By filing a motion for retrial, Ford succeeded in having the punitive damages thrown out.

In addition to the civil suits, Ford was the subject of a sensational and unprecedented criminal trial. On September 13, 1978, the company was indicted on three counts of reckless homicide by an Indiana grand jury after three girls burned to death in a Pinto following a rear-end collision. The company was found innocent on a legal technically. This was not an important trial in terms of the potential legal consequences, which might have amounted to a tiny fine, but it did provide undesirable publicity that might have been much worse if the company was found to be guilty of a crime.

CONCLUSIONS

The Ford Pinto case was once considered a notorious example of a company continuing to manufacture and distribute an unsafe product, long after it had become aware of problems in the fuel tank of the vehicle, because it saw profit as more important than the lives and health of customers. This "Pinto madness" view was developed by Mark Dowie in his *Mother Jones* article. Careful analysis of the process by which the Pinto was developed shows that this scenario is highly inaccurate in several respects and quite misleading. Despite Ford's legal liability for the problems of the Pinto's fuel tank, there is no compelling evidence that this was actually worse than the fuel tanks of other cars in the same class at that period. More important, there was no cost-benefit analysis according to which Ford design engineers decided to release an unsafe vehicle because it would be cheaper to pay lawsuits for death and injury liability than to produce a safer vehicle. The engineers did not see the Pinto as unsafe by the standards of the day (which were lax, particularly for inexpensive vehicles) and the cost-benefit analysis was made several years after the release of the Pinto by a separate division of the company whose primary function was to influence safety regulations.

See also ENGINEERING ETHICS; HYBRID-POWERED CARS.

Further Reading

Birsch, D, and J. H. Fielder, eds. *The Ford Pinto Case.* Albany: State University of New York Press, 1994.

Dowie, M. "Pinto Madness." *Mother Jones,* September–October 1977, pp. 18–32.

Lee, T. M., and M. D. Ermann, "Pinto 'Madness' As a Flawed Landmark Narrative." *Social Problems* 46 (1999): 30–45.

Nader, R. *Unsafe at Any Speed.* New York: Grossman, 1965.

fraudulent research   *See* SCIENTIFIC DISHONESTY.

Freon   *See* OZONE DEPLETERS.

Friends of the Earth

Friends of the Earth (FOE) is an environmental organization dedicated to preserving the health and biodiversity of the planet for the benefit of future generations. Based in Washington, D.C., FOE was founded more than 30 years ago. It acts as a political pressure group that favors government incentives promoting sustainable development and seeks to eliminate government support for companies and industries that harm the environment, for example, by using energy sources that pollute the air. Its periodical *Atmosphere* first appeared in 2000 and focuses on issues such as GLOBAL WARMING and ozone depletion.

See also GREENPEACE.

frontal lobotomy   *See* PREFRONTAL LOBOTOMY.

fur trade and endangered species

The use of exotic animal skins and furs in high fashion has long been a favorite target of animal rights activists and others concerned about the impact of trade in such materials on the well-being of wild animals, particularly those in danger of extinction. Species allegedly threatened by the fashion trade include crocodiles, alligators, lizards, snakes, ostriches, and sables. In reality, the fashion trade often provides an economic incentive to raise rare animals in captivity, which can buffer the wild population against extinction. Thus the Florida alligator population rebounded after landowners were allowed to take some skins in return for protecting the eggs. Since international trade in endangered species is illegal, most fur is now produced by farming.

Nevertheless, activists have threatened to throw paint at women wearing fur coats. Many celebrities, including actresses Daryl Hannah, Kim Basinger, Candice Bergen, and Betty White and talk show host Bob Barker have contributed to the antifur campaign in less militant ways, such as promoting the use of synthetic substitutes. Barker walked down Fifth Avenue in New York carrying signs and shouting "Shame!" at women wearing fur coats. He withdrew from hosting the Miss Universe pageant because contestants wore furs. Long intimidated by the highly publicized antics of organizations like People for the Ethical Treatment of Animals (PETA) fashion designers in the mid-1990s sensed that consumers had grown weary of environmental activism and some returned to using furs in their designs.

See also ENDANGERED SPECIES ACT.

Gaia hypothesis

The Gaia hypothesis is the notion that all life on earth comprises a single living system. Just as the body of a multicellular being is composed of single cells, so all living organisms on his planet can be thought of as contributing to the functioning of a single superorganism. Named after the Greek earth goddess Gaea, the hypothesis was proposed by English biologist James Lovelock in 1969. The idea was further elaborated through the joint efforts of American cell biologist Lynn Margulis culminating in Lovelock's (1979) book *Gaia: A New Look at Life on Earth.*

The Gaia hypothesis has had a mixed reception. To begin with, it received a bad review from scientists. Contemporary biologists were very much of the mind-set that there is a Darwinian struggle for survival that pitted rival organisms against each other. The notion that organisms may work together for the common good was associated with the much despised theory of GROUP SELECTION in biology. Many scientists were offended by the mystical and religious elements in Gaia and were put off by its cultish following. Long dismissed as New Age mysticism, the Gaia hypothesis has again been taken up serious scientists thanks to new insights into ecosystems and the mathematical theory of complexity.

Central to the Gaia hypothesis is the notion of the superorganism in which the collective behavior of individuals subserves group needs. Thus, the temperature of a termite colony is maintained within a very narrow range due to the combined activities of thousands of individuals, which provides an analogy with temperature homeostasis in mammals due to the combined activities of millions of cells.

Biologists once believed that the coordinated activities of superorganisms were produced by natural selection operating on individuals. Now, with the conceptual aid of a new branch of mathematics, known as complexity theory, they are leaning toward the view that order can emerge from the system itself. Illustrative of this idea is the brood care of some ant species of the genus *Leptothorax.* When colonies are very small, the individual ants move around randomly caring for the young in a highly erratic fashion. After the number of ants passes a threshold value, the ants begin to synchronize their movements and adopt a rest-activity cycle with a 25-minute period. The conclusion that this cyclicity rests in the system itself rather than natural selection serving to shape individuals is supported by computer models in which simulated ants following very simple rules of interaction exhibit a similar rhythmicity.

The concept of the superorganism can also be applied to whole ecosystems, and, by extension, to the entire biotic system of the planet. Computer simulations of ecosystems in which species are added in random order from a pool of 100 have found that new species enter the system quite easily when the existing number of species is low. After the number of species exceeds about 15, entry becomes increasingly difficult even if the new entrant has some clear competitive advantages over already established species. It thus appears that the ordered relationships between species in an ecosystem are an emergent property of the system that makes it quite resistant to new species. These ideas are consistent with empirical research on peatland ecosystems, which are remarkably similar the world over.

Even if the superorganism metaphor is useful in describing specific ecological systems, it remains to be seen whether it will be as useful in describing the ecological dynamics of the entire planet. Problems of particular interest here are whether the climate and atmosphere of the earth are combined with living organisms in a complex system that tends to promote ecological stability.

The Gaia hypothesis is of particular interest to environmental ethicists. To begin with, if the Gaia principle is taken at face value it means that traditional views of the role of human beings in nature need to be revised. We are not standing apart from, and using, nature. Instead, we are an integral element of the earth's biotic systems. This may seem to be too metaphorical and abstract to have practical implications but one key insight follows, namely, by harming the environment we inevitably injure ourselves.

Some environmental ethicists have argued that ecosystems should be thought of as having inherent moral value, which obliges us to protect them. This perspective can have important practical effects on environmental policies because it suggests that evaluation of the effects of industrialization should focus on the extent to which whole ecosystems, rather than particular species, are threatened. From this perspective, the destruction of rain forest to create agricultural land may be more damaging to the earth's ecosystems than industrial pollution, even though most people would view pollution as a worse threat because it affects the quality of life of urban residents. To the extent that the Gaia hypothesis encourages environmental ethicists to think globally, it will foster global solutions to the earth's ecological problems.

See also ENDANGERED SPECIES ACT; ENVIRONMENTAL ETHICS; RAIN FOREST, DEPLETION OF.

Further Reading

Joseph, L. *Gaia: The Growth of an Idea.* New York: St. Martin's Press, 1991.
Lewin, R. "The Comeback of Gaia." *New Scientist,* December 14, 1996.
Lovelock, J. *Gaia: A New Look at Life on Earth.* New York: Oxford University Press, 1979.
———. *The Ages of Gaia.* New York: W. W. Norton, 1995.

Galen

(130–200)
Greek
Physician

Galen's writings provided the single most important source of authoritative medical information until his influence waned with the rise of modern scientific medicine in the 16th and 17th centuries. Although restricted to dissecting animals such as goats, dogs, and pigs, Galen made important discoveries in anatomy and physiology. He was the first to describe many muscles and his experiments in severing the spine of animals at different levels demonstrated the key role of the spinal cord in controlling movement. Galen's theoretical work concerned issues such as digestion, blood formation, and neural function. Even though his work contained many errors, these did not become obvious until the 16th century when ANDREAS VESALIUS (1514–64) began conducting dissections.

See also AVICENNA; HARVEY, WILLIAM; VESALIUS, ANDREAS.

Galileo Galilei

(1564–1642)
Italian
Physicist, Astronomer

Known to history for his conflict with the Roman Catholic Church as much as for his impressive contributions to physics and astronomy, Galileo's reputation as a champion of scientific objectivity has been tarnished by claims that he, like many other distinguished figures in the history of science, fabricated the results of his experiments. His contemporary, Père Marsenne, tried to replicate one of Galileo's experiments on falling bodies in which the time taken for a brass ball to roll down the groove in a long board was measured. Marsenne's results were so different that he suspected Galileo had never carried out the experiment in the first place. It is ironic that today Galileo is considered an important champion of the importance of testing theories by experiment instead of relying on the world of authorities. He was certainly sincere in wanting to depose authorities, but his empiricism was at least partly rhetorical. His books reveal a fondness for describing thought experiments and imagining the results. If these were logically compelling, Galileo did not see the point of actually conducting the experiment.

Ethical controversy aside, there is no question that Galileo made many important discoveries. He was unusually predisposed to find order in everyday experiences and expanded the range of scientific observation by inventing or improving instruments, including the telescope (which he greatly improved in 1609, soon after its invention in Holland), the thermometer, and the sector, a mechanical calculating device. At the age of 18, Galileo was observing a swinging candelabrum in the Pisa Cathedral when he noticed that the time taken by the swing was always the same regardless of the distance traveled. He is said to have used his own pulse to count the time. Near the end of his life, he discovered that the arc of a projectile thrown upward is in the shape of a parabola, a discovery that has influenced the design of artillery.

Galileo's academic career was marked by indecision, as he abandoned his medical studies at the University of Pisa in favor of mathematics and left without a degree. Following a few years as a private tutor, he was appointed professor of mathematics at the University of Pisa in 1589. While there, he discovered that the speed with which dropped objects fall is independent of their size and weight. Whether he ever actually tested this hypothesis by dropping various objects off the Leaning Tower of Pisa and timing their descent relative to each other, as legend claims, is questionable. This discovery turned out to be highly controversial because it falsified a key claim of ARISTOTLE's theory of motion, namely, that heavier objects should fall faster. The controversy was so intense that Galileo had to leave Pisa for Florence in 1591. In the following year, he was appointed professor of mathematics in Padua.

Galileo's problems with the Catholic Church began with his astronomical observations using a telescope he had built, which was then the most powerful in the world. This opened up a wealth of information that only he could observe and consequently could expect to receive no support from other scientists. Among the astronomical novelties he recorded were the mountains and valleys of the moon, the satellites of Jupiter, the rings of Saturn, and sunspots (which he used to deduce the sun's rotation). A 1613 treatise on sunspots openly advocated the heliocentric worldview of COPERNICUS, which was considered a heresy. Following ecclesiastical intimidation, Galileo agreed to renounce the Copernican theory and to refrain from teaching or writing about it. This promise was very pointedly broken in 1632 with the publication of *A Dialog on the Two Chief Systems of the World,* which supported Copernicus and rejected the older Ptolemaic worldview propounded by the Church. Found guilty of heresy, Galileo was sentenced to life imprisonment and all of his books were banned. Threatened with torture, he again recanted and was allowed to serve out his sentence under house arrest in his own home.

Galileo's life illustrates the importance of academic freedom for scientists, not just because of the closed minds of church leaders but because of the general reluctance of people, including philosophers and scientists, to accept scientific observations and ideas that are at odds with their perception of reality. It is interesting that the Catholic Church has acknowledged the error of persecuting Galileo but it did not do so until 1992. The same pope who vindicated Galileo, John Paul II, has also acknowledged that it was a mistake to oppose Darwin's theory of evolution by natural selection.

See also NEWTON, ISAAC; SCIENCE, HISTORY OF; SCOPES TRIAL; TECHNOLOGY, HISTORY OF.

Further Reading
Broad, W., and N. Wade. *Betrayers of the Truth.* New York: Simon and Schuster, 1982.
Cohen, J. B. *Lives in Science.* New York: Simon and Schuster, 1957.
Sharatt, M. *Galileo: Decisive Innovator.* New York: Cambridge University Press, 1996.

Galton, Sir Francis
(1822–1911)
English
Biologist, Psychologist

Although Francis Galton made his primary contributions to the study of heredity in biology and to the study of intelligence in psychology, his range of scientific interests and achievements was extremely wide. Thus, he made contributions to meteorology, statistics, and physical anthropology. He was also a keen explorer.

Deeply influenced by the evolutionary theory of his cousin, Charles Darwin, Galton overestimated the role of heredity in human characteristics such as criminality and intelligence. In his study of the 100 men he judged to be the most distinguished, he found that other family members also tended to be eminent and that the probability of eminence increased with closeness of relationship to the reference individual. Galton interpreted this as supporting the conclusion that "genius" is inherited. He was aware that the evidence could also be interpreted in terms of environmental advantages such as wealth and good education. Against this view, he produced many arguments that seemed compelling to him. To begin with, particular families excelled in particular fields, such as law, science, or writing. Galton understood this as evidence for the biological inheritance of specific aptitudes. Moreover, some very eminent men had humble beginnings, suggesting that affluence is not a necessary precondition for achievement. Galton also argued that more widespread education, such as occurred in America, did not increase the number of eminent individuals being produced. Finally, he argued that conferring wealth and privilege on individuals, such as happened in the case of favored kinsmen of the Roman Catholic popes, did not make them more likely to become eminent in their own right.

Galton was the first to conduct a twin study. He was aware of the distinction between identical (or monozygotic) and fraternal (or dizygotic) twins but he was not aware of the genetic explanation for this difference. (Galton's contemporaries knew almost nothing about genetics, as such, because the important early work of GREGOR MENDEL was ignored for almost 40 years). He collected biographical and autobiographical information on 55 pairs of twins (35 sets of identicals and 20 pairs of fraternals). Twins that were said to be similar at birth remained similar but the dissimilar (fraternal) twins did not become more similar over time despite living in the same environment. Galton concluded that heredity was more important than the shared environment in making twins similar to each other. This result, in 1883, anticipated the MINNESOTA TWIN STUDY by a century.

Galton's contributions to data collection in psychology and physical anthropology were impressive. He developed apparatus and procedures for measuring visual acuity and auditory sensitivity judgment of length, weight discrimination, reaction time, and memory span. He was the first to use photographic composites in psychological research. He developed a quantitative method of studying mental imagery and thus anticipated the work of early experimental psychologists. He was the first to use the questionnaire as a research tool. Between 1884 and 1892, Galton arranged for the collection of bodily measurements and psychological data on some 17,000 individuals. Interestingly, these individuals actually paid for the privilege of having their measurements taken, which gives some idea of the novelty of what Galton was doing.

Galton was the father of EUGENICS, which seeks to improve human heredity both by encouraging reproduction of the most valuable individuals in a society (positive eugenics) and by discouraging reproduction on the part of less valued individuals (negative eugenics). His basic argument stipulated that we could either let natural selection follow its blind course or we could intervene so as to improve the health and well-being of the human population by improving its hereditary characteristics.

There is no doubt that Galton shared the prejudices of his contemporaries about which people were valuable to society. He assumed, as did many others, that those individuals who constituted eminent members of a dominant society owed their prominence to hereditary superiority. This unfortunate doctrine was brought to its logical conclusion in the racist immigration laws and sterilization practices carried out in the United States, and in the event that marked the darkest hour of eugenics—the Nazi Holocaust—perpetrated on European Jews, Gypsies, and mentally ill people. It would be unfair to judge Galton as responsible for these abuses of eugenics. He simply did not have the benefit of subsequent research showing that genetic differences within ethnic groups are more important than genetic differences between groups.

See also EUGENICS; GENETIC DETERMINISM; INTELLIGENCE TESTS; MINNESOTA TWIN STUDY.

Further Reading
Blacker, C. P. *Eugenics Galton and After*. New York: Hyperion, 1986.
Murphy, G., and J. K. Kovach. *Historical Introduction to Modern Psychology*. New York: Harcourt Brave Jovanovich, 1972.
Plomin, R., J. C. DeFries, and G. E. McClearn. *Behavioral Genetics: A Primer*. San Francisco: W. H. Freeman, 1980.

gene therapy

Many genetic diseases are caused by the inability of an individual to produce a normal protein. The goal of gene therapy is to insert normal genes for such proteins into cells so that they can be incorporated into the cell's DNA, permitting the missing protein to be produced. Gene therapy thus involves two critical stages. The first is to identify the faulty gene. The second is to transport the normal gene inside cells. This has proven to be technically challenging. One of the most effective approaches has been to use retroviruses. A copy of the beneficial gene is incorporated in the retrovirus, which then carries it inside the cell and attaches it to the cellular DNA. Initiated in 1980, gene therapy has achieved some exciting successes but has not had a happy beginning due to unforeseen side effects and to a cavalier attitude by some researchers toward research protocols. Gene therapies that show promise include treatments for heart disease, hemophilia, cancer, and severe combined immunodeficiency (SCID).

The idea of gene therapy has always been controversial because researchers are seen as playing God with the human genome. From the beginning, there have been serious ethical problems about the way in which clinical trials have been conducted. In 1980, when Martin Cline at UCLA introduced normal hemoglobin genes into the bone marrow of two patients suffering from beta-zero thalassemia, a type of severe heritable anemia, he did not have approval from the relevant INSTITUTIONAL REVIEW BOARD. He lost his NATIONAL INSTITUTES OF HEALTH grant. The treatment was unsuccessful. Ten years would pass before gene therapy was used successfully on a human. A four-year-old girl, Ashi DaSilva, was treated for SCID by having normal genes inserted into her cells.

Many other promising results were produced. Then, in September 1999, tragedy struck at the University of Pennsylvania when 18-year-old Jessie Gelsinger died after a dose of genetic material had been injected into his liver in a clinical trial. According to the Food and Drug

Administration (FDA), Gelsinger's liver had not been functioning well enough for him to have been included in the trial. A number of other protocol violations were also noted by the FDA. Gelsinger's parents filed suit against the University of Pennsylvania in September 2000, complaining that he had not been adequately informed of the risks of the gene treatment.

In the aftermath of the Gelsinger tragedy, the NIH wrote to gene therapy researchers asking if they had observed any adverse reactions in their trials that could shed light on Gelsinger's death. They received no fewer than 691 reports of adverse effects, only 39 of which had previously been reported to the NIH. Even though the University of Pennsylvania study was not allowed to recruit any new participants, the FDA has continued to approve trials in human gene therapy. Promising results have been reported for gene therapy in relation to a number of disorders, including heart disease, squamous cell cancer, Alzheimer's disease, diabetes, and hemophilia.

See also GENETIC ENGINEERING; GENETIC TESTING.

Further Reading
Meager, A. ed. *Gene Therapy Technology, Applications, and Regulations.* New York: John Wiley, 1999.

genetic determinism

Genetic determinism is a theory that holds that individual differences in a particular trait (such as height, weight, intelligence, or sociability) are at least partly the product of genetic differences. Modern behavior genetics research has produced clear evidence supporting genetic determination of human intelligence and personality traits. Although genetic determinism is often linked to the phrase "biology is destiny," this leap is not warranted by the evidence. For example, even though INTELLIGENCE TEST scores are strongly influenced by genetics, they are also strongly influenced by schooling and by living in a technologically advanced society. Even in the case of a phenomenon that is entirely explained by genetics, such as the mental retardation produced by phenylketonuria (PKU), biology is not destiny. Thus, removal of phenylalanine from the diet allows PKU children to develop normally.

See also INTELLIGENCE TESTS; MINNESOTA TWIN STUDY.

genetic engineering

Genetic engineering is an applied science in which knowledge of genetics is used to solve practical problems as diverse as food production, medicine synthesis, waste disposal, treatment of genetic disorders, and crime solving. Genetic engineering techniques have always been controversial because of the potential for unanticipated ill effects in relation to social consequences and health outcomes. These ethical concerns reached a crescendo with the widespread publicity surrounding the successful cloning of a sheep named Dolly in Scotland in 1997. While genetic engineering logically encompasses the technique of SELECTIVE BREEDING through which animals and plants have been domesticated, the term is more commonly used to refer to molecular techniques. The direct manipulation of genetic material is referred to as recombinant DNA technology or gene splicing.

LABORATORY TECHNIQUES
In recombinant DNA technology, genetic material from one organism is inserted into another. The key to this technique is the use of restriction enzymes that split DNA strands wherever a specific sequence of nucleotides (the basic structural units of DNA) occurs. Donor DNA fragments thus formed can combine with similarly produced fragments from other organisms. For example, if human DNA and bacterial DNA are cut with the same restriction enzyme, both have matching "sticky" ends that make it possible for the human genetic material to be inserted into the bacterium. The sticky end of the human DNA then joins up with the sticky ends of the bacterial DNA.

Many genetic engineering experiments use viruses and plasmids (rings of bacterial DNA) as vectors, or carriers, for getting the donor DNA into a host cell. Once the vector gets inside a cell, the recombinant DNA fragment from the host is replicated every time the cell divides. The resulting daughter cells are identical and can produce the protein coded by a donated gene.

When large quantities of gene product are required, as in the case of drug production, waiting for cells to reproduce may be inefficient. In 1985, a new procedure was developed that allowed for a much more rapid replication of the genetic material. This is known as polymerase chain reaction (PCR) and it was discovered by Kary Mullis working at the Cetus Corporation of California. Mullis won the 1993 Nobel Prize in chemistry for this work. In PCR, the DNA fragment is first heated to separate the two strands. In the second stage, segments of DNA called primers are attached to each strand to identify them for the final stage, copying. Copying is facilitated by adding the natural enzyme DNA polymerase. Each repetition of the procedure results in a doubling of the genetic material. Within a few hours, it is possible to multiply the original sample by a factor of one million. Genetic engineering procedures have been used successfully in addressing many practical problems.

PRACTICAL APPLICATIONS

- Waste Disposal. Many waste products of agriculture and industry either do not break down naturally or break down so slowly that they can cause problems for humans and degrade the habitat of many species of plant and animal. These problems can be addressed through the use of bacteria with altered genes. For example, the yeast species *Saccharomyces cerevisiae* has been genetically engineered to dispose of whey, a waste product of cheese-making, by converting the lactose it contains into alcohol, which is used in medicine and industry. Many agricultural waste products, like corn husks, contain cellulose that normally decomposes slowly. Such waste can be converted into sugar by the enzyme cellulase. Bacteria that naturally contain the gene for cellulase are not effective waste disposal agents. *Escherichia coli* in which the cellulase gene has been inserted can be used in waste disposal programs. Similarly, bacteria have been genetically engineered that are capable of breaking down oil and other organic wastes.

- Genetic Disorders. Recombinant DNA technology has been useful for mapping the chromosomal location of genetic disorders. When a particular restriction enzyme cuts DNA, it does so by recognizing a sequence of six nucleotide bases. If an individual does not have this sequence at a particular location, then their DNA will not be broken at that point. Instead of having two shorter fragments of DNA, the individual will have one longer one. This difference between individuals is referred to as a restriction fragment length polymorphism (RFLP, or "riff-lip"). Over 1,000 human RFLP's have been identified and their chromosomal locations have been established, allowing them to be used in the construction of gene maps. Diseases caused by a single gene can be linked to one of these RFLP markers allowing their chromosomal location to be determined. Moreover, the gene itself can be taken out and inserted into a bacterium. Study of the resulting gene product is helpful in understanding how the gene produces the disorder. RFLP markers not only allow people to be screened for genetic diseases but also are fundamental to the development of GENE THERAPIES, in which the genome of human somatic cells is altered.

- Forensic genetics. Because all individuals, except for identicals, are genetically unique, their RFLP patterns form a distinct "genetic fingerprint" that can have useful forensic applications. For example, it is possible to place a suspect at the scene of a crime with a very high degree of confidence using DNA analysis of blood stains, semen stains, hair roots, and so forth. DNA fingerprints are obtainable from samples that are several years old. The relia-

bility of the technique also allows suspects to be cleared of crimes, thereby preventing miscarriages of justice.

- Food plant improvement. Gene-altered plants can be made more resistant to pests and diseases or acquire other qualities that enhance their commercial value. For example, the corn borer *Pyrausta nubilalis,* which bores into the stems of corn plants, eventually killing them, can be controlled using a genetically altered bacterium *Clavibacter xyli* that lives on the stems of the corn plant. *C. xyli* is normally not harmful to the corn borer but the genetically modified strain received the capacity to produce an insect-killing endotoxin from another species, *Bacillus thuringiensis,* that does not live on corn plants. Corn borers that ingest the genetically altered bacteria along with their food are killed, thus protecting the corn plants from a serious pest and increasing corn yields.

- Another problem in food production to which recombinant DNA technology has been applied is the spoilage of delicate products between harvesting and sale. This problem is particularly serious in the case of tomatoes, which, if ripened on the vine, are too soft for shipping. In the past, tomatoes were harvested while green and hard, refrigerated while shipping, and artificially ripened in ethylene gas. The softening of tomatoes is caused by the enzyme polygalacturonase. The polygalacturonase gene has been inserted, in reverse order, into tomato plants. This inactivates the original enzyme-producing gene and produces a dramatic reduction in softening of the tomatoes. The sale of genetically modified tomatoes was approved by the United States Food and Drug Administration but there has been more resistance to GENETICALLY MODIFIED FOOD in Europe. The genetically altered tomatoes can be shipped without expensive refrigeration, which allows them to be produced more cheaply and sold to consumers at a lower price.

- Medicine production. Many genetic diseases result from a deficiency of some chemical that is in short supply in nature. Examples include primary diabetes resulting from insulin deficiency, and hemophilia caused by a deficiency of clotting factor VIII. These valuable substances are synthesized by inserting human genes for the chemicals into bacteria using recombinant DNA technology.

Another approach to the manufacture of human gene products has been to transfer human genes into animals (e.g., cows, sheep, goats), so that they can yield the gene product in their milk. In this technique, the human gene is attached to a section of animal DNA that ensures the gene will be activated only in the mammary gland. The combined human-animal donor fragment is inserted into the fertilized egg of the host species where it may join up with one of the host's chromosomes. After the transgenic animal is reimplanted, born, and matures, it produces the gene product in its milk. Moreover, the altered gene is transmitted to offspring via the gene line allowing for the development of transgenic herds that function as specialized pharmaceutical factories.

The production of vaccines is another important contribution of genetic engineering to medicine. Vaccines are created by transferring the genes that determine a pathogen's surface configuration to a harmless microorganism. Metaphorically speaking, a sheep in wolf's clothing is produced. When the genetically modified microorganism is used in a vaccine, its surface stimulates the production of antibodies. The presence of these antibodies protects an individual against the pathogen. Using this method, successful vaccines have been produced for influenza, cold sores, and hepatitis B, among others.

Apart from actual medicine production, genetic engineering may play an important role in earlier stages of disease research. Animals may be genetically altered to be vulnerable to a disease so that the mechanism of the disease can be investigated and potential therapeutic agents tested. Thus, mice have received a cancer-predisposing gene. These mice have been used to test the consequences of exposure to various potential carcinogens and to test the efficacy of preventive drugs.

CLONING RESEARCH

Cloning refers to the production of genetically identical cells or complete organisms as a result of descent from a single individual. Cloning is extremely common at the level of cells. Thus growth in tissues normally relies on the production of new cells. When a cell divides, two genetically identical daughter cells are produced. Among asexually reproducing species, such as bacteria, yeast, and algae, population consist of clones. Even in more complex species of plants and animals, such as dandelions and flatworms, vegetative reproduction produces populations of clones. Cloning occurs naturally in humans in the case of identical multiple births. In these cases, a single fertilized egg divides, creating several genetically identical individuals. Although they are genetically identical, human multiples can be quite different in terms of their biology, as reflected, for example, in disease vulnerability, and in terms of their behavior.

Artificial cloning of farm animals has been conducted for over a decade using embryonic cells. The cloning of a sheep named Dolly from a mammary gland cell of an adult sheep elicited astonishment in the scientific community as well as generating an ethical frenzy over potential human applications. As the first example of a mammal cloned from an adult cell, Dolly contradicted the dogma that DNA from a specialized adult cell cannot start over and guide the egg's development into a complex organism.

The technique used by Ian Wilmut at the Roslin Institute in Scotland to clone Dolly consisted of inserting the nucleus of an adult mammary gland cell into an egg from which the nucleus had been removed. The Roslin team got the donor nucleus to fuse with the egg's cytoplasm by using an electrical pulse. The same electrical pulse triggered the fused cell to develop into an embryo. After the egg cell had been activated, it was implanted in a surrogate mother and experienced a normal gestation.

In the year following reports about Dolly, some researchers entertained doubts concerning the limited genetic analysis carried out by Wilmut's team to demonstrate that Dolly was not the natural offspring of the surrogate mother. More detailed tests using both RFLP fingerprinting and detailed analysis of short stretches of DNA (microsatellites) have confirmed that Dolly is indeed genetically identical to the cultured mammary gland cells from which she was created.

Since Dolly's cloning became public, a Japanese team produced two calf clones by fusing oviduct cells from one cow with the enucleated egg cells of another. Careful DNA testing has verified that these calf clones are also what they are supposed to be. Leading researcher Yukio Tsunoda reported that four additional cows were pregnant with cloned embryos.

The repeatability of cloning of mature mammalian cells was most clearly demonstrated in 1998 at the University of Hawaii, Honolulu. Researchers at the John A. Burns School of Medicine cloned more than 50 mice. The Honolulu team developed a new technique for activating the fused eggs. The newly fused egg is allowed to sit for six hours so that the egg cell alters the donated DNA, allowing its developmental genes (which had been turned off during maturation of the donor cell) to be expressed again. Development is then triggered by placing the eggs into a culture medium containing strontium. The strontium stimulates the release of calcium from the egg's internal stores, which is the normal signal that induces fertilized eggs to begin dividing.

ETHICAL CONTROVERSIES

Genetic engineering is a brave new world in which our understanding of the basic principles of biology is constantly being challenged by new discoveries. Not only is it possible, and practicable, to transfer genetic material across species boundaries, thereby undermining the definition of a species, there is even some evidence that animal genes can be transferred to plants. In one spectacular demonstration of this possibility, researchers inserted the genes responsible for producing the enzymes that allow fireflies to glow into a tobacco plant. When the roots of the plant were immersed in a solution having the appropriate ingredients, it began to emit light.

Some ethicists object to alteration of the genetic code of humans, or nonhuman animals, in principle, seeing it as an act of intolerable human arrogance often expressed as playing God. A secular version of the same argument is to be found in modern *environmental ethics* in which alteration of the genetic code of animals and plants is seen as an unwarranted interference with, and degradation of, the inherent value built into species and ecosystems through the process of evolution by natural selection. Cloning of herds of identical farm animals to produce pharmaceuticals in their milk would be seen as an unwarranted act of interference with the course of evolution from this point of view. Such use of animals is clearly within the tradition of domestication of plants and animals practiced in all societies since the development of agriculture.

Most of the ethical debate associated with genetic engineering focuses on whether the beneficial consequences to humans outweigh the costs. The costs to animals are weighed more heavily by PETER SINGER, an ethical pioneer in the animal rights movement, who would question a program of genetically altering pigs so that their organs would be less likely to be rejected by human recipients as intrinsically "speciesist." Many of the concerns about genetic engineering have to do with the possibility of inadvertent harmful consequences. For example, the *E. coli* bacterium that is a subject of much genetic engineering research is also a normal resident of the human intestines. There is a fear that the accidental release of a genetically modified strain of *E. coli* could cause a dangerous epidemic in the human population. The recent opposition to genetically modified foods in Europe is also based largely on the premise that such foods might have long-term deleterious consequences for the health of people who eat them. Such concerns are far more muted in the United States, which might reflect the fact that genetically modified food plants have been approved as safe by the Food and Drug Administration.

In addition to issues of health consequences, biotechnology has raised a number of social issues, many of which have focused on new REPRODUCTIVE TECHNOLOGIES, including the production of test tube babies, artificial insemination, surrogate parenting, and cloning research. Such technologies generally have the effect of fragmenting the roles of parenthood thereby creating the potential for ethical disagreements and legal conflicts concerning the rights and duties of biological parents, surrogate parents, and custodial parents.

Cloning research has raised two distinct ethical problems. The first revolves around the culturing of embryonic cell clones for research purposes. The question arises as to whether such tissue should be treated differently from that used in research on nonhuman animals. The second problem relates to the potential use of human cloning as an alternative to sexual reproduction. Although viable human embryos have not yet been cloned, it now seems likely that this is technically feasible. Assuming that human cloning is technically possible and that it is legally permitted, the resulting offspring would have only one parent instead of the customary two. A change to asexual reproduction would produce a dramatic alteration in family structure since the sexual relationship of husbands and wives would no longer be of central importance to reproduction. This raises the issue of whether children can be thought of as having an inherent right to two parents who can potentially look out for their best interests.

In addition to these larger social questions, cloning raises a number of technical issues that have ethical ramifications. One relates to mitochondrial DNA, an unsung portion of the genotype that resides outside the nucleus and is normally inherited from the maternal line. In the procedure of nuclear transfer, in which the nucleus of one egg is transferred to another, this portion of the genetic material is lost, which would artificially alter the heredity of a child. This procedure has been performed successfully using rhesus monkeys and is being contemplated to prevent genetic disorders carried by human mitochondrial DNA.

Another unresolved technical issue relates to the biological age of clones. Early work on Dolly has found that her chromosomal telomeres are shortened, a phenomenon that is associated with cellular aging. If clones are the same biological age as the parent cell, then cloning of humans might produce a reduction in lifespan, which could be interpreted as an unwarranted reduction of the quality of life to be expected for individuals produced by sexual reproduction. *See also* GENE THERAPY; GENETIC DETERMINISM; GENETIC TESTING; GENETICALLY MODIFIED FOOD; REPRODUCTIVE TECHNOLOGIES; SELECTIVE BREEDING.

Further Reading

Chet, I. *Biotechnology in Plant Disease Control.* New York: Wiley-Liss, 1993.

Drlica, K. *Understanding DNA and Gene Cloning.* New York: Wiley, 1992.

Humber, J. M., and R. Almeder, eds. *Human Cloning,* Totowa, N.J.: Humana Press, 1998.

Kass, L. R., and J. Q. Wilson. *The Ethics of Human Cloning.* Washington, D.C.: AEI Press, 1998.

Levin, M. A., and H. S. Strauss. *Risk Assessment in Genetic Engineering.* New York: McGraw-Hill, 1990.

Marteau, T., and M. Richards. *The Troubled Helix.* New York: Cambridge University Press, 1999.

Nussbaum, M. C., and C. R. Sunstein, eds. *Clones and Clones: Facts and Fantasies about Human Cloning.* New York: W. W. Norton, 1999.

Plomin, R. *Nature and Nurture*. Pacific Grove, Calif.: Brooks/Cole, 1990.

Resnik, D. B. *The Ethics of Science*. New York: Routledge, 1998.

genetic testing

Approximately one person in 20 suffers from a genetic disorder or birth defect. Genetic testing helps people to avoid producing children with serious genetic disorders. In some cases, diagnosis of a genetic disease can allow parents to engage in effective treatment strategies. Thus the treatment of PKU (phenylketonuria), a genetically determined form of mental retardation, consists simply of withdrawing phenylalanine from the diet. Hence the compulsory screening of infants in most U.S. states and the labeling of foods containing phenylalanine.

In the case of diseases like cystic fibrosis, which does not have a cure, the benefits of screening are more controversial, although it does detect carriers. When two carriers marry, there is a 25% chance that each child will be affected by the disease. Carrier testing is thus often conducted where there is a family history of diseases such as cystic fibrosis and muscular dystrophy.

Genetic testing is also conducted for purposes of prenatal diagnosis. Fetal cells are obtained for testing by AMNIOCENTESIS, chorionic villus sampling, or fetal blood sampling. Prenatal diagnosis is ethically controversial because the main reason for such early diagnosis is that mothers expecting genetically disordered offspring may choose to have an abortion. Opponents of abortion generally do not make exceptions in these cases. Even among those who accept the principle of selective abortion based on prenatal tests, many are opposed to the use of prenatal testing for sex selection. This practice is common in many Asian countries where there is a strong preference for male children.

Not all genetic disorders are evident early in life. Some, like Huntington's disease, a devastating progressive neuromuscular disorder, emerge in middle age (30–45 years). Genetic testing now reveals which members of an affected family are destined to develop this disorder. Such information may help affected individuals to plan their lives appropriately but it can also be a devastating blow that produces severe anxiety and depression. As more information accumulates about genetic vulnerability to cancer, heart disease, and strokes, etc., new ethical issues will be raised concerning how much individuals should find out about their own vulnerabilities. Another problem relates to the possibility that this information could fall into the wrong hands (e.g., employers, insurance companies) and be used to the detriment of the individual.

See also AMNIOCENTESIS; EUGENICS; GENETIC DETERMINISM.

genetically modified food

A number of agricultural plants have been altered through GENETIC ENGINEERING. They include tomatoes, potatoes, corn (maize), tobacco, and cotton. Genes inserted in these plants have either improved the product, protected the plants against viruses, pests, or herbicides, or turned them into pharmaceutical "factories." Thus, genetically modified (GM) tomatoes are firmer and easier to transport long distances. Genetically modified corn has been used to produce tryptophan, which can be synthesized into a medicine for Parkinson's disease (L-dopa). Other pharmaceuticals produced by plants include anticlotting hormones, such as erythropoietin, and alpha interferon used to treat hepatitis C. Between 1986 and 1997, field tests were conducted on over 60 different crops.

Despite recent protests against testing of GM foods, spearheaded by GREENPEACE in Britain, this new food technology is remarkably safe. (Greenpeace activists uprooted GM maize that was being grown at a test site in Lyng, Norfolk, to protest the creation of GM food plants). American consumers prefer GM potatoes because of their better taste and firmer texture. GM soya products have been consumed by hundreds of millions of people in America and Europe for several years with no adverse consequences. GM food has never been identified as the cause of any human illness, making it one of the safest new technologies ever introduced.

While there are no laws against GM food in Europe, the European Union (EU) bowed to populist concerns and, in 1998, stopped approving new GM foods for use in Europe. This has resulted in a ban on all corn imports from the United States because bulk shipments are likely to contain a variety that has not been improved. The 15 member states are now obliged to label all packages containing GM corn or soy. American companies selling food in Europe, such as McDonald's and Burger King, have pledged not to use GM ingredients in order to forestall boycotts and protests. Other large international food companies, including Gerber, H. J. Heinz, and Frito Lay, have also stopped using GM products.

Although there are some legitimate concerns about the ecological effects of genetically engineered food plants, the evidence that they cause damage to insect populations living around cultivated land remains unconvincing. The current hysteria about the potential danger of GM foods suggests a disturbing lack of public trust in the scientific profession and in the protective function of government. Thus, Europeans have virtually ignored the report of the NUFFIELD COUNCIL ON BIOETHICS dealing with GM food. This report was released in May 1999, following 18 months of study by a panel of experts, and drew the unreserved conclusion that GM food currently on the market is safe for consumption. Some commentators believe that public trust in the ability of scientists and government to guarantee the safety of foods has been badly weakened by the recent outbreak of "MAD COW DISEASE" in Europe.

Recent fears have focused on the effects of GM potatoes on experimental rats in England. Arpad Pusztai of the Rowett Research Institute claimed that a diet of modified potatoes had harmed rats in an experiment. Instead of waiting for the study to undergo peer review, he immediately took his results to the press. He was criticized by his employers for doing so and the Rowett Institute also refused to support his conclusions. When the *Lancet* published Pusztai's paper, the editor added a note to the effect that the paper was being published out of a duty to bring the work out into the open, implying that the journal was not endorsing it. An investigation by the Royal Society found that the study was so badly conducted that no legitimate conclusions could be drawn from it. The rats may have become sick by consuming a large quantity of potatoes—an effect independent of the genetic modification of the plant.

See also BOVINE SOMATOTROPIN; GENETIC ENGINEERING; GREENPEACE.

Further Reading
Paarlberg, R. "The Global Food Fight." *Foreign Affairs* 79 no. 3 (2000): 24–28.

germ warfare   *See* BIOLOGICAL AND CHEMICAL WARFARE.

Gilligan, Carol
(1936–)
American
Psychologist

Carol Gilligan is a developmental psychologist whose research has focused on female moral and emotional development. Although her own research was primarily concerned with how adolescents' self-confidence is related to social expectations, Gilligan is best known for her feminist critique of theories of moral development in psychology, particularly that of Kohlberg, and these ideas have had far-reaching implications for ethics.

American psychologist Lawrence Kohlberg extended Jean Piaget's work on moral development in young children to later development. Kohlberg presented his research participants with scenarios involving ethical conflicts and used their solutions to these conflicts to study moral development. Kohlberg found that people tend to reach higher levels of moral reasoning as they mature. At the highest level (postconventional morality) action is guided by abstract moral reasoning rather than concern about the opinions and feelings of others (conventional morality). Research using Kohlberg's methods often found that more men than women solved ethical dilemmas using postconventional morality. This led to the implication that women do not have the capacity to separate their moral judgments from specific contexts, from their own interpersonal relationships, and from feelings and emotions.

Faced with such evidence, Gilligan suggested that Kohlberg's theory is flawed. Instead of seeing women as deficient in moral reasoning, it is more appropriate to recognize that they see the world in a different way and make moral judgments using different criteria. Thus, in one of Kohlberg's dilemmas, participants were asked whether a man should steal a loaf of bread to feed his starving family. Men who said that he should not steal the bread were likely to cite rules of justice. Women who said that he should steal the bread were moved by compassion for the family. Gilligan concluded that there are two different ways of talking about moral problems: the ethic of justice and the ethic of care. The ethic of care, or compassion is more likely to affect women's moral judgement. When this "different voice" of women is brought into a discussion of moral development, it changes how we interpret the results of studies using Kohlberg-type scenarios.

Gilligan's 1982 book *In a Different Voice* has been enormously influential in generating hundreds of academic articles and books dealing with her feminist approach to ethics. In addition to its influence in developmental psychology, it has provided moral philosophers with alternative formulations to traditional ethical theories that were dominated by contractual thinking. Gilligan's ideas have been particularly attractive to postmodernist feminist scholars, who see them as evidence of the failure of scientific objectivity in the sense that before Gilligan male scholars working in moral development were constructing a male-biased narrative.

See also POSTMODERNISM AND THE PHILOSOPHY OF SCIENCE; SOKAL, ALAN.

Further Reading
Gilligan, C. *In a Different Voice.* Cambridge, Mass.: Harvard University Press, 1982.
Taylor, R. "The Ethic of Care versus the Ethic of Justice." *The Journal of Socio-Economics* 27 (1998): 479–94.

global warming
There has been considerable scientific controversy about whether the emission of industrial gases into the atmosphere is producing a warming effect for the entire planet.

Few climatologists dispute that the planet is getting warmer at a rapid pace but there is less consensus about whether the warming tend is due to human activities.

Global warming can be produced by a "greenhouse effect" according to which the sun's heat is trapped by atmospheric gases, including carbon dioxide, water vapor, and methane, through a mechanism analogous to the trapping of solar energy by the panes of a greenhouse. The greenhouse effect as a natural phenomenon is not controversial. It is estimated that in the absence of a natural greenhouse effect the earth's surface would cool from an average of 57 degrees Fahrenheit to -4 degrees Fahrenheit, making it uninhabitable by humans or most other familiar life forms. What is controversial is the extent to which human activity has increased the concentration of greenhouse gases and thereby contributed to global warming.

The most important human contribution to greenhouse gases is through the burning of wood, coal, oil, and gas, which emits gases such as carbon dioxide, carbon monoxide, and sulphur dioxide into the atmosphere. Since plants use up carbon dioxide, the main contributor to global warming, global deforestation tends to increase the greenhouse effect and the cultivation of agricultural crops tends to reduce it.

CLIMATE CHANGE IN HISTORICAL PERSPECTIVE

That the atmosphere traps some of the sun's energy was appreciated as early as 1827 when Jean-Baptiste-Joseph Fourier described the atmosphere as analogous to a greenhouse. An even more explicit statement of the greenhouse effect was presented by Svante Arrhenius in 1896. Arrhenius recognized the important role of carbon dioxide in warming the earth and predicted that a doubling of this gas in the atmosphere would warm the earth by approximately 5 degrees Centigrade (9 degrees Fahrenheit). Although it is difficult to evaluate this prediction with any precision, modern evidence suggests that it is too extreme.

Precise recording of atmospheric concentrations of carbon dioxide were begun only in 1957 and earlier values rely on statistical inference. Between 1957 and 1988, the carbon dioxide concentration measured at Mauna Loa, Hawaii, increased from 315 parts per million (ppm) to 350 ppm, an increase of 11%. The 1988 value is 25% higher than the estimated concentration of carbon dioxide in 1800. During the 20th century, carbon dioxide concentrations increased 30% to 365 ppm while average global temperature increased by approximately one degree Fahrenheit.

It is thus clear that carbon dioxide increases and global warming have occurred simultaneously during the 20th century. Correlation does not imply causation, however, and climatologists were for a long time unwilling to

make this leap. One nagging problem has been the inability to demonstrate an orderly relationship between the time course of atmospheric change and the time course of temperature. If global warming was being driven largely by human-generated increases in greenhouse gases, then the inexorable increase in atmospheric carbon dioxide should be accompanied by a steady rise in global temperature. Yet this pattern was not obvious. In fact, at some periods the graphs moved in opposite directions with global temperature falling as atmospheric carbon dioxide rose. This phenomenon might appear to rule out a causal connection between carbon dioxide concentration and average world temperature but climate is such a complex system that this is not necessarily true. Thus, volcanic activity, which sends clouds of dust into the atmosphere, has a cooling effect that is independent of greenhouse gases.

Such uncertainties inhibited most scientists from making any strong declarations about the contribution of industrialization to global warming. Yet, it was hard to deny that there was something unique about global atmosphere and temperature changes in the 20th century. First, the increase in carbon dioxide is distinctively modern. We know this from study of ancient air bubbles trapped in ice. Samples of prehistoric air from ice cores in Antarctica and Greenland go back 400,000 years. During this period, carbon dioxide concentrations remained below 300 ppm until the dawn of the Industrial Revolution. The current value is above 365 ppm. The one-degree rise in global temperature might seem modest until it is compared with the last 1,000 years of temperature data based on study of ice cores and annual tree-ring diameter (which turns out to be a reliable index of temperature change). In the previous nine centuries, global temperatures declined by an average of 0.04 degrees.

Perhaps the single most important development in the global warming debate was the application of complex computer models to the problem. Despite anomalies, such as volcanic eruptions and climatic disturbances due to changing ocean currents of the El Niño type, computer models have been quite successful at using increases in atmospheric carbon dioxide to predict increases in temperature with good accuracy.

In 1988, James Hansen, director of the Goddard Institute for Space Studies, decided that the time for academic waffling had come to an end and informed a U.S. Senate Committee on Energy and Natural Resources that the scientific evidence is strong enough to conclude that human activities are responsible for global warming. While many scientists resented his bluntness, none has come forward with a compelling demonstration that this conclusion was incorrect. In fact, the Intergovernmental Panel on Climate Change (IPCC), an alliance of 2,500 scientists studying climate change on behalf of the United Nations, came to a similar conclusion in 1995. Hansen's testimony both raised public consciousness of the potential harmful consequences of the greenhouse effect and stimulated international efforts to mitigate these problems.

CONSEQUENCES OF GLOBAL WARMING

Current predictions indicate that without changes in the rate of emissions of industrial gases, average global temperature will rise between 2 and 6 degrees Fahrenheit during the 21st century. While such temperature changes may seem trivial in the context of day-to-day changes in weather, from the perspective of global change, they can have profound consequences. For example, during the last Ice Age, some 18,000 years ago, the average temperature was only about 7 degrees Fahrenheit colder than it is today.

The most immediate concern about global warming is that it will melt the polar icecaps producing a rise in sea level. Since 25% of the world's population lives less than 3.7 feet above sea level, even minor rises in sea level can have profound economic and lifestyle consequences. Many cities are located on low-lying coasts, where flooding could force the evacuation of large numbers of people. Much agricultural land around coasts and deltas will be inundated with salt water making it unsuitable for food production. It is estimated that a three-foot rise in sea level would displace 72 million people in China, 11 million people in Bangladesh, and 8 million people in Egypt. In the process, some of the most productive agricultural lands would be lost in countries that have difficulty feeding themselves at present. Rising seas would also cover beaches, an important amenity that draws many people to seashores with a corresponding loss of property values and tourism revenues to affected areas. The upper movement of the snow line would also put many ski resorts out of business.

Global warming is likely to have an important impact on climate and ecology. Some climatologists consider that these changes are already in evidence. When a huge iceberg broke loose from the Larsen Ice Shelf in Antarctica in 1995 the event, although of little practical consequences, called attention to the fact that temperatures in this continent have been rising more rapidly than elsewhere on the planet. Since Adelie penguins live on the ice, their shrinking habitat has been associated with a one-third decline in the population. (The penguins have also been adversely affected by pesticides like DDT).

Increasing global temperature is expected to produce major changes in weather. Weather events are forecast to become more severe. Along with anticipated increases in droughts, expansion of deserts, and the emergence of longer, hotter, summers, the IPCC predicts that global warming will increase the severity of rainstorms and flooding as well as the amount of snow falling in some regions during winter.

As warmer conditions extend to larger regions of the planet, tropical diseases are expected to rise. This effect will be engendered by increases in the number of insects that serve as vectors for diseases such as yellow fever, dengue, malaria, and viral encephalitis. The expansion of insect populations is likely to be exacerbated by the expansion of the human population so far as disease transmission is concerned.

Some of these changes have already been observed. For example, in the Colombian Andes mosquitoes that carry yellow fever and dengue used to be found only at elevations below 3,300 feet but were recently identified at 7,200 feet. Similarly, in the Irian Jaya province of Indonesia, malaria recently appeared for the first time at an altitude of 6,900 feet. Malaria has also recently spread to the highlands of Kenya.

Needless to say, the consequences of global warming will not be uniformly negative. For example, large regions of Siberia that are currently too cold for agriculture may become suitable for raising food crops. Residents of North America and Northern Europe may benefit from milder winters and a protracted growing season. Despite such likely benefits, most governments see global warming as a major threat and have responded to this perceived threat by formulating agreements and regulations to control emission of greenhouse gases. Among the more vocal supporters of political action have been the leaders of small island states whose territories are threatened with substantial reduction by encroaching sea water. The IPCC predicts a 1.5 foot rise in sea level by 2100.

POLITICAL RESPONSES TO GLOBAL WARMING

Acting on the implications of the IPCC findings, the world's political leaders met at Kyoto, Japan, in December 1997 to work out an agreement for reduction of greenhouse gas emissions. This agreement focused mainly on carbon dioxide produced by the burning of fossil fuels. The Kyoto Protocol committed nations of the industrialized world to reduce greenhouse gas emissions to 5.2% below their 1990 level, by the year 2012. The United States agreed that its contribution would be to reduce emissions to 7% below their 1990 level. The Kyoto agreement may turn out to be more of a theoretical than a practical accomplishment. It is clearly the kind of international agreement, setting definitive targets, that is necessary to reduce greenhouse gases, but it has generally not received much political support among nations that are currently responsible for producing most of the world's greenhouse gases.

For many business leaders, reducing carbon emissions implies reduced energy use, which has implications of slowing economic growth. This is not necessarily true because switching to cleaner-burning fuels and alternative energy sources could theoretically result in reduced emissions without cramping economic growth. As of 1999, no European Union state had ratified Kyoto. In the United States, the Senate voted unanimously to reject Kyoto, even before the agreement had been signed, unless it imposed restrictions on developing nations. President Clinton has demanded that such restrictions be imposed before he sends the treaty to the Senate for ratification. It thus appears that some time will pass before Kyoto has any practical effect on greenhouse gases.

If nothing else, the Kyoto agreement reveals some of the complex social and ethical implications of reducing carbon dioxide emissions. It demonstrates that increasing levels of greenhouse gases in the 20th century were closely tied to increased world population. On the positive side, the velocity of population increase is slowing and some experts estimate that world population may stabilize early in the 21st century. On the negative side, countries that already have large populations, such as China, are inevitably going to produce large quantities of greenhouse gases as their economies develop. Thus China is expected to be the world's largest producer of carbon dioxide by 2030. Underlying this conclusion is the assumption that the amount of coal burned by the Chinese will double in the next 20 years.

Clearly, any agreement that fails to impose some restrictions on developing nations is doomed to failure. Meanwhile, the poorer nations of the world are arguing that the problem of greenhouse gases has been caused by wealthy industrialized countries and that these should therefore bear the brunt of cleaning it up. This may mean that developing countries will balk at applying emissions controls unless they feel adequately compensated for the expense of doing this. Many leaders of developing nations see emissions controls as a political lever that can be used to promote some kind of global economic redistribution that will be in their favor.

Political conservatives in the United States have rejected Kyoto as implying a loss of economic sovereignty. If emissions controls are internationally dictated, then the economic development of the United States could be choked off in favor of economic development of poorer countries. Conversely, the failure of the Kyoto accord to control greenhouse gas emissions in developing countries can be seen as the result of a consequentialist calculation to the effect that reducing greenhouse gases in developing countries is ethically indefensible if it causes them to lapse into deeper poverty relative to the developed world. It can therefore be argued that there is a fundamental connection between controlling world pollution and dealing with poverty attributable to industrial underdevelopment. This conflict could be resolved through the development and global implementation of cleaner more efficient energy sources. Examples of such improvements include the "decarbonization" of fossil fuels, that is, burning them without releasing much carbon into the air. Another promising innovation is the development of HYBRID-POWERED CARS that are capable of producing efficiencies of over 100 miles to the gallon of gasoline and emit no greenhouse gases.

See also AIR POLLUTION; HYBRID-POWERED CARS.

Further Reading

Easterbrook, G. "The Real Evidence for the Greenhouse Effect: Warming Up." *The New Republic,* November 8, 1999.

Gantenbein, D. "The Heat Is On." *Popular Science* 255, no. 2 (1999): 54–59.

Newton, L. H., and C. K. Dillingham. *Watersheds 2: Ten Cases in Environmental Ethics.* Belmont, Calif.: Wadsworth, 1997.

White, R. M. "Kyoto and Beyond." *Issues in Science and Technology* 14, no. 3 (1998): 59–66.

GreaterGood.com

Greater Good (http://www.greatergood.com/cgi-bin/Web Objects /GreaterGood) is an E-commerce portal that donates at least 5% of purchases made through its site to charitable causes that may be selected by the shopper. Causes to which GreaterGood contributes include World Wildlife Fund, The Nature Conservancy, Special Olympics, Save the Children, and the United Nations World Food Programme. GreaterGood offers access to more than 80 well-known retail sites and brands such as Amazon.com, Lands' End, J. C. Penney, Neiman Marcus, e Toys, and OfficeMax. These companies sponsor the philanthropy of GreaterGood not just because they want to do good but because they see it as a favorable marketing opportunity. Market research has found that the majority of Americans prefers to buy from firms associated with charitable causes.

It is estimated that at least 20 other sites emerged between 1997 and 2000 that claim to help nonprofit organizations benefit from the growth of Internet shopping. They receive merchant rebates that vary from 3% to 25% of the retail price of items purchased through their site. Most of the charity sites retain a portion of the merchant rebates to cover their expenses. An exception is 4charity.com, which says that it donates the entire rebate to charity and covers its expenses with money from investors who are drawn to profit-making services that are planned for the site.

Apart from uncertainty about how much of their donation is going to charity, shoppers have no guarantee that online charities they contribute to are legitimate. For example, a *U.S. News and World Report* reporter listed a fictitious Save the Peacocks fund on iGive.com and the fakery was not detected. Another problem is that online giving of this sort is not considered charitable by the U.S. Internal Revenue Service because shoppers do not pay over the retail value and cannot use their contribution as a tax deduction.

Online shopping rebates make it very easy for shoppers to contribute to charitable causes without themselves feeling economic pain. In this respect, the services of GreaterGood.com resemble those of THE HUNGER SITE (which it now manages) that allows visitors to contribute food to the hungry that is paid for by advertisers whose logos are presented on a thank-you screen. Even though it is highly convenient and economically painless, sponsored giving from online retailers is not a large contributor to conventional charities, although its role is liable to grow with E-commerce itself. For example, in the two months after it signed up with GreaterGood.com, the Humane Society received $1,200 in donations, which is scarcely a drop in the bucket of its annual $48 million budget. Nevertheless, GreaterGood.com is a useful example of how Internet technology can allow for painless collection of charitable donations. In the process, it empowers people who may not be wealthy themselves to contribute to causes about which they care.

See also HUNGER SITE, THE.

Further Reading
The Hunger Site *About GreaterGood.com.* www.thehungersite. com (accessed 25 April 2000).
Perry, J. "Buy Online, Save the Whales." *U.S. News and World Report,* October 11, 1999.

green architecture

Green architecture refers to building design that is sensitive to global as well as local environmental issues. The guiding principle is to construct buildings that do the least possible amount of harm to the environment. For this reason "green" building plans are sometimes referred to as "sustainable design."

It is estimated that building construction uses more than 40% of raw materials, demolition waste accounts for one-quarter of landfill volume, and buildings are responsible for over one-third of energy use. How buildings are designed can therefore have enormous practical implications for environmental quality. Sustainable design is a guiding philosophy rather than any particular type of construction. The following are examples of practical features of green architecture:

- Energy use is minimized by taking advantage of natural heating and cooling phenomena. Thus, the use of deciduous shade trees allows for shade in summer without blocking sunlight in winter. Solar energy is preferred to burning wood or oil because it does not pollute the air or rely on dwindling resources of solid fuel. Energy conservation is facilitated by using software to control heating appliances.

- Construction calls for the conservation of existing resources by using materials from demolished buildings as well as recycled materials.

- In the ideal situation, sewage is handled on-site through the use of waterless composting toilets and the construction of artificial wetlands to recycle waste water. Water from roofs is collected to irrigate gardens. Residential designs sometimes make it possible for residents to grow food for themselves.

- Choice of construction materials involves avoidance of many paints, sealers, adhesives, insulators, etc., that are potentially toxic and could be harmful either to the interior, the interior air quality, or the external environment. Ignoring such principles in the past has been unwise since it is estimated that some 30% of commercial buildings suffer from SICK BUILDING SYNDROME, i.e., have such bad air quality that it undermines worker health. Green principles may also be applied to building maintenance that eschews toxic cleaners and polishes, for example.

- Site development is planned in ways that minimize destruction of the local ecology. Thus, existing trees may be preserved where possible and the amount of paving is reduced to a minimum. Architects want to design their buildings in harmony with the natural site and they aim for unobtrusiveness in appearance.

Green buildings are more expensive than conventional ones but the cost increase is often surprisingly modest, approximately of the order of 10%. This premium may be more apparent than real because the energy savings achieved alone can far outstrip the increased cost over the lifetime of the building. Green buildings are complicated to build and maintain and most architects have not been trained in this area. They also suffer from the perception that they are a passing fad designed to appeal to a minority of environmental enthusiasts.

Perhaps the most important aspect of green buildings is that they create pleasant environments in which to live and work. Employees in a green office with good air quality are not only happier and possibly willing to work for a lower salary, they are also healthier, lose fewer days to illness, and are thus more productive. Employers who do not have a strong environmental agenda have begun to recognize that increased productivity and reduced maintenance costs in green buildings can more than compensate for the increased costs of construction.

See also ARCHITECTURE AND CRIME; SICK BUILDING SYNDROME.

Further Reading
Wilson, A., J. Uncapher, and L. H. Lovens. *Green Development: Integrating Ecology and Real Estate.* New York: John Wiley, 1998.

green revolution, the

The green revolution is a term referring to substantial increases in yields of agricultural crops, particularly cereals, due to the development of genetically improved varieties. As early as the 1940s, researchers in Mexico developed a short-stemmed variety of wheat that resisted stem rust, produced high yields, and proved highly efficient at converting fertilizer and water into food. The improved wheat strain boosted wheat production in Mexico and helped to fight off famine in India and Pakistan. Research team leader American Norman E. Borlaug of the International Maize and Wheat Center in Mexico (CIMMYT) won the 1970 Nobel Peace Prize for this work. These successes stimulated a worldwide research effort to improve the yields of staple food crops in the developing world so as to avert famines. Despite initial optimism about the green revolution, the problem of world hunger never went away. Even though food crops have been improved, the green revolution is predicated on intensive agriculture using expensive fertilizers and irrigation systems that are often beyond the reach of poor farmers. Moreover, it has been argued that green agriculture is damaging to the environment.

The green revolution has been particularly important because of soaring world population. Increases in agricultural yields have staved off what might otherwise have been a severe worldwide shortage of food. This point can be appreciated by looking at the case of India, which produced 12 million tons of wheat annually in the early 1960s but produces approximately 70 million tons annually today. This astonishing rate of increase in production of a staple food crop is partly attributable to the development of a cross-bred wheat that is part-Mexican, part-Japanese, and part-Indian and that is both high-yielding and sturdy. The first year that this crop was grown, the yield was three times what it had been in the previous year. Increased wheat production in India has also been affected by an aggressive government drive to achieve self-sufficiency in food.

Another staple crop for which production has soared due to the green revolution is rice. From 1967 to 1984, Asian rice production increased at an average rate of 3.2% per year (or 70% for the entire period). According to UN estimates 56% of the world's population lived in countries with average per capita food production below 2,200 calories per day in 1969 compared to less than 10% by 1994. Without such colossal increases in food production, a dreadful Malthusian scenario would have played out in which the world's soaring population would have quickly outstripped the available food supply. As it is, approximately one-third of the world's population is estimated to be chronically malnourished, which reflects inefficient food distribution systems as well as underproduction.

The green revolution has spread to all the major regions of the world, most recently reaching Africa. Ethiopia is a particularly interesting example because this country, which has traditionally been associated with food shortages and famines, has now become a food exporter. The use of improved cereal varieties boosted grain production from 6 million tons in the 1994–1995 harvest to 11.7 million tons in 1996–1997.

The green revolution has not been without its critics. From an economic perspective, it has been argued that green agriculture is capital intensive because of the high cost of seeds, fertilizer, pesticides, and irrigation systems. The necessity for high capital investment to produce a crop tends to exclude small farmers and convert agriculture from a traditional enterprise managed by family farms to a business controlled by a small number of large producers. Even when small farmers stay in production, they are often compelled to take on a large burden of debt, which makes them extremely vulnerable to crop failures and price swings.

Environmentalists have argued that the high-yielding green revolution crops encourage farming practices that promote environmental degradation. For one thing, high yields require the use of high levels of nitrogen fertilizer. The production of these fertilizers requires the use of large quantities of energy, which increases the emission of greenhouse gases. For another, the high-yielding crops are often highly vulnerable to weeds, pests, and diseases. To protect their valuable harvest, farmers are obliged to use large quantities of herbicides and insecticides that leave lingering toxic effects in the environment and can cause illnesses in humans. This problem may be worsened by the use of genetically modified food plants that have vulnerabilities to pests, weeds, or diseases.

See also BOVINE SOMATOTROPIN; GENETIC ENGINEERING; GENETICALLY MODIFIED FOOD; GREENHOUSE EFFECT.

Further Reading
Dawe, D. "Reenergizing the Green Revolution in Rice." *American Journal of Agricultural Economics* 80 (1998): 948–53.
Ehrenfeld, D. "A Techno-pox upon the Land.' *Harper's Magazine,* October 1997, pp. 13–17.
Hazell, P. B. R. *The Green Revolution Reconsidered.* Baltimore, Md.: Johns Hopkins University Press, 1991.

Mann, C. "Reseeding the Green Revolution." *Science* 277 (1997): 1038–42.

Greenpeace

Greenpeace is an international organization dedicated to protecting endangered species and the environment. It is militantly activist and promotes awareness of environmental problems through direct confrontation of polluters and others who are felt to be causing ecological harm. The organization was founded in 1971, in British Columbia, in opposition to U.S. nuclear tests at the Alaskan island of Amchitka. Its actions have included sealing up pipes that discharged toxic substances into bodies of water and destroying experimental plots of English genetically modified (GM) food in Lyng, Norfolk, in 1999.

Greenpeace has been active in preventing seal hunting and whaling and trade in endangered species. Its tactics included actions such as steering inflatable crafts between the whales and the harpoon guns of the whalers.

The Greenpeace movement received unprecedented publicity in 1985 following a July 10 incident in which their craft, the *Rainbow Warrior,* was sunk by two bombs while berthed in a New Zealand harbor (at Auckland). The vessel was scheduled to sail for Moruroa Atoll to protest French nuclear tests conducted there. It appears that the craft was sunk by French intelligence agents and the international incident that followed culminated in the resignation of the French minister of defense.

Apart from calling attention to the unethical behavior of those who would cause damage to the environment, Greenpeace's own actions have often received ethical scrutiny. The most salient question is whether activists are entitled to break laws and endanger their own lives and those of other people in unlawful protests. Activists would argue that the costs of protecting animals and their habitat is more than offset by the benefits of promoting this cause.

See also AIR POLLUTION; FUR TRADE AND ENDANGERED SPECIES; GENETICALLY MODIFIED FOOD; WATER POLLUTION.

group selection

Group selection is the belief that natural selection promotes the interests of the species, or group, rather than the individual. This argument was advanced by V. C. Wynne-Edwards in his 1962 book *Animal Dispersion in Relation to Social Behavior.* He argued that much of animal behavior is designed to restrict population growth and that species without effective population regulation mechanisms became extinct because they destroyed their food base. Group selection is theoretically interesting because it provides a rationale for altruistic, and even self-sacrificial, behavior. This rationale is seriously flawed but it has been historically important because it stimulated biologists to provide alternative explanations for altruistic behavior in the animal world, thus initiating the modern evolutionary study of behavior. The basic problem with group selection is that it is likely to be undermined by the effects of selection at the individual level. Thus, individuals that refrain from reproducing in the interests of conserving the food supply of the species will leave no offspring behind to inherit, and practice, their altruistic tendencies. Modern evolutionists generally reject group selection but accept two other mechanisms through which altruistic behavior can evolve, namely, kin selection and reciprocal altruism.

See also ALTRUISM, EVOLUTION OF; GAIA HYPOTHESIS.

Gutenberg, Johan
(c. 1398–1468)
German
Printer

Born in Mainz, Germany, Johan Gutenberg began his working life as a goldsmith. He developed the first European printing press with movable type (see illustration). This achievement allowed books to be printed faster and more cheaply. The modern printing industry has placed huge amounts of information in the hands of ordinary people for the first time in history, which has facilitated social changes that include universal education and democracy. The printing industry thus marks the beginning of the information age. In recognition of his important role, Gutenberg is sometimes referred to as the "father of printing." The modern age of mass production of books and other print media is sometimes referred to as the "Gutenberg Revolution."

Gutenberg spent approximately 10 years perfecting his printing technique (1440–50). He did for printing in the 15th century what THOMAS EDISON would do for electricity in the 20th. His printing press incorporated an astonishing series of innovations that covered virtually all aspects of the printing process. He invented replica casting in which single letters were engraved in relief and then stamped into a brass plate to produce a mold that was used to cast replicas made with molten metal. The individual letters were combined to make a flat printing surface.

Drawing on his experience as a goldsmith, Gutenberg developed an alloy suitable for casting letters. He also developed a special oil-based ink that would stick to the metal type. The printing press itself was a modified version of a winepress.

Having received financial backing from a rich goldsmith named Johann Fust, Gutenberg went into business

Johan Gutenberg stands beside his 1450 printing press, the first in Europe to use movable type. (LIBRARY OF CONGRESS)

as a commercial printer around 1452. Following a quarrel in 1455, Fust foreclosed on a mortgage and took over the printing press. Fust then went into the printing business for himself. Gutenberg set up a second press. He retired from printing in 1465, possibly due to vision problems. When he died, in 1468, he was comparatively poor. Unlike Edison, he could not benefit from his inventions through PATENT laws because intellectual property had no legal protection in his day.

Since dates and printers' names did not go on books at this period, it is difficult to be sure which books were actually produced by Gutenberg. We do know from its distinctive type that he produced the Gutenberg Bible, also known as the 42-line Bible because there were 42 lines per printed page. The ornate type resembled the gothic writing style of the period.

See also EDISON, THOMAS ALVA; PATENTS; TECHNOLOGY, HISTORY OF.

H

Harvey, William

(1578–1657)
English
Physician

William Harvey has claimed a place in the history of science by discovering the circulation of the blood in humans and other mammals. Harvey received his first medical training at Caius College, Cambridge, although this was inadequate, even by the standards of his day. He took a two-and-one-half year course at the University of Padua, which was thought to have the best medical school in Europe. He studied under Hieronymus Fabricius ab Aquapendente, a noted anatomist. During this period, Harvey became interested in the problems of cardiovascular function about which very little was then known despite Fabricius's detailed work on the anatomy of veins.

Harvey returned to England in 1602 and received a license to practice medicine in 1604. In 1607, he received a fellowship of the College of Physicians and an appointment to the hospital of St. Bartholomew's and St Thomas's which he held until 1643, when he was displaced for political reasons due to the rise of Oliver Cromwell to power. He developed a large private practice that included many distinguished citizens, such as SIR FRANCIS BACON. In 1618, he was appointed personal physician to King James I.

Despite his flourishing medical practice, Harvey was active in research. He not only established that blood is pumped through the circulatory system but performed careful experiments capable of quantifying the rate of flow. Harvey recognized the difference between arterial blood leaving the heart and venous blood (i.e., that which is contained in veins) returning to it. These accomplishments mark a turning point in circulatory system knowledge because his research convincingly discredited the widely accepted views of GALEN (c. 130–201) and ushered in a new age of medicine based on scientific physiology. After the Royalist side lost the English Civil War, Harvey retired from work as a physician and devoted all of his energies to medical research, which included embryology as well as physiology. He died of a stroke probably at the house of his brother Eliab in Roehampton outside London.

See also GALEN; SCIENCE, HISTORY OF; VESALIUS, ANDREAS.

Hastings Center

The Hastings Center is an independent interdisciplinary research institute that studies APPLIED ETHICS in the fields of biomedicine and the environment. Located in Garrison, New York, it is the oldest institute of its kind, having been founded in 1969 (see www.thehastingscenter.org). The center was created in response to ethical problems arising from rapid changes in medicine and biology. It publishes two periodicals, *The Hastings Center Report* and *IRB: A Review of Human Subjects Research*.

See also KENNEDY INSTITUTE OF ETHICS; NUFFIELD COUNCIL ON BIOETHICS.

Hawking, Stephen
(1942–)
English
Theoretical Physicist

A leading figure in modern cosmology, Hawking has worked at connecting quantum physics and gravitation, a field now known as quantum mechanics. Some of his key speculations include black holes that contain immense amounts of matter but are no larger than subatomic particles and multiple universes connected by "wormholes." In addition to his successful academic career, Hawking has popularized physics in books like *A Brief History of Time* (1988) and *Black Holes and Baby Universes and Other Essays* (1993) that have been best-sellers. Hawking has also been the subject of much attention because of his physical disability due to motor neuron disease (or Lou Gehrig's disease) and the determination with which he has faced it.

The field of cosmology inevitably invites religious controversy as GALILEO and others found to their cost. Modern cosmology, as represented by the work of EDWIN HUBBLE, tends to represent human beings as inconsequential specks and Hawking is definitely of this school of thought. Some readers of *A Brief History of Time* have come away with the impression that Hawking was attempting to prove the nonexistence of God. While there was little theological discussion as such, Hawking did speculate that the universe is finite and self-contained and that it lacks beginning or end. If so, there is no necessity to infer a creator.

See also GALILEO GALILEI; HUBBLE, EDWIN; SAGAN, CARL; SCIENCE, HISTORY OF.

Further Reading
Filkin, D. *Stephen Hawking's Universe.* New York: Basic, 1997.
Hawking, S. *A Brief History of Time.* New York: Bantam, 1988.
Hawking, S. *Black Holes and Baby Universes and Other Essays.* New York: Bantam, 1993.

hazardous waste
Hazardous wastes are those that pose special threats to public health and therefore must be either neutralized before disposal, stored safely, or dumped in a secure landfill that is designed to prevent leaching of toxins into groundwater. The various types of hazardous waste include toxins (encompassing carcinogens and mutagens), radioactive isotopes, highly flammable or explosive substances, chemically reactive agents such as acids, and infectious biological material.

Hazardous waste is most often transported over public highways, which leads to the potential for toxic spills and endangerment of the motoring public and local inhabitants. The practice of "midnight dumping," in which toxic wastes were intentionally discharged at unapproved locations, has been greatly curtailed by tighter regulations in the United States and other countries. One effective strategy for preventing unauthorized dumping is a manifest system that requires companies to account for toxic wastes from the moment that they are created.

If it is impractical to inactivate hazardous wastes by other means, they may be incinerated. Incineration is useful for destroying many chemical toxins and organic wastes but it has the adverse effect of increasing air pollution. Even though the absolute quantity of air pollution from incinerators may be small, they frequently discharge very highly toxic substances, of which dioxin, a potent carcinogen, has caused most concern.

Some organic wastes can be destroyed by applying microorganisms that digest them. For example, genetically engineered bacteria are capable of breaking down oil. Another technique of managing hazardous wastes is solidification. The material is concentrated and mixed into a solid substance, such as concrete or plastic, that prevents leaching. In deepwelling, liquid wastes are injected into porous rock that lies below a layer of impermeable rock, using high pressure. Although this system is intended to be permanent, leaks may occur. Groundwater must therefore be constantly monitored. Deepwelling is inexpensive but public concern about the possibility of leaks has forced major chemical companies in the United States to give up the practice.

Insecure dumping of hazardous wastes first came to the fore as a major public health problem in 1970 with the evacuation of residents from LOVE CANAL, a housing development at Niagara Falls, New York, that had been built directly over a chemical dump. THE ENVIRONMENTAL PROTECTION AGENCY (EPA) did not have the legal authority to deal with the problem and this resulted in the Comprehensive Environmental Response, Compensation, and Liability Act of 1980 (or CERCLA, also known as Superfund). A special tax on chemicals and oil producers was used to create a fund that could be used by the EPA to act quickly following discovery of dangerously toxic sites.

In addition to such emergency actions, the Superfund is used to effect long-term remediation of toxic sites that are identified on a National Priorities List. As of 1977, 599 Superfund sites had been remediated, 464 were undergoing cleanup, and a further 208 had a low level of activity. Contingent on continued funding, it is anticipated that 85% of the sites will have been remediated by 2005. Over 30,000 other sites are listed on a national registry as possibly containing hazardous chemicals, which gives some idea of the extent of the problem. Between 1980 and 1990, the EPA spent $6 billion of Superfund money in cleaning up priority sites.

The Superfund is not a public works project but aims to recover the costs of remediation from the parties considered primarily responsible for the dumping. In the past, the cleanup has tended to proceed followed by protracted legal action to recover the costs. These tactics were often viewed as heavy-handed and unfair, especially if all of the burden was placed on smaller companies. In recent years, the EPA has shifted to negotiated agreements with responsible parties in advance of remedial work.

In addition to the work of sovereign states in cleaning up hazardous waste sites, there has been some international effort. For example, countries affected by pollution of the Rhine River have signed an agreement to limit release of pollutants into the waterway. The same is true for countries bordering the shallow and heavily polluted Mediterranean Sea. The United Nations has also played a role in monitoring worldwide distribution of hazardous chemicals.

See also ENVIRONMENTAL PROTECTION AGENCY; LOVE CANAL.

Further Reading

Freeze, A. R. *The Environmental Pendulum: A Quest for the Truth about Toxic Chemicals, Human Health, and Environmental Protection.* Berkeley: University of California Press, 2000.

Woodside, G. *Hazardous Materials and Hazardous Waste Management.* New York: John Wiley, 1999.

hedonism

Hedonism is the philosophical position that pleasure is the most important goal of a person's life. It is a much misunderstood concept because different people, and different philosophers, define pleasure differently. EPICURUS (341–270 B.C.) is a central figure in the hedonist philosophy of ancient Greece but his life is better characterized by the cautious avoidance of pain than by the reckless pursuit of pleasure that the term "hedonist" suggests to modern ears. It is interesting that Epicurus's extremely monastic lifestyle was misunderstood even in his own time because he accepted female students and slaves and his school was therefore falsely presumed to be a den of iniquity. Hedonism is extremely important in modern ethics for at least two different reasons. First, the influential branch of MORAL PHILOSOPHY known as utilitarianism is based on maximizing pleasure and minimizing pain for whole groups of people. Second, the modern ethics of self-improvement and self-realization, as reflected in the motto "be all that you can be," is based on the notion that accomplishment and enlightenment contribute to the happiness of the individual and are worth pursuing for this reason.

See also EPICURUS; MORAL PHILOSOPHY; UTILITARIANISM.

Helmholz, Hermann Ludwig Ferdinand von
(1821–1894)
German
Physiologist

Hermann Helmholz was a leading 19th-century scientist who contributed to fields as diverse as meteorology, electromagnetism, and optics but is most remembered for his work on sensory physiology. In his day, the controversy over whether mental phenomena could be reduced to mechanical events, the mind-body problem, posed in its modern form by RENÉ DESCARTES (1596–1650), still raged. Contrary to Descartes's dualist position, Helmholz was dedicated to the view that mental experiences are the result of physiological events, and he provided support for this hypothesis by investigating the physical explanation of sensory perception, including the role played by nerve impulses.

One of Helmholz's most important achievements was to measure the speed of conduction of the nerve impulse. He did this using a subtraction method in which a person's time to respond to a stimulus at the ankle by moving their hand was compared to the reaction time for a stimulus delivered to the shoulder. By estimating the extra distance the impulse had to travel to the brain from the ankle compared to the shoulder, Helmholz could calculate the speed of the impulse as it traveled this additional distance. Although he was disappointed by the variability in his human data, his estimate of around 50 meters per second was reasonably accurate. The use of frogs produced more reliable results with a conduction speed of some 30 meters per second. These speeds were much lower than would have been expected. They suggested that actions are not simultaneous with thoughts as many contemporaries believed. The finding had important philosophical, and even theological, ramifications because it indicated that movement is controlled subject to mechanical laws and therefore cannot be explained purely in terms of volition, or spirit.

Helmholz's work on the physiology of vision and hearing were original in their day and are still cited in psychology textbooks. He investigated the way that internal muscles of the eye focus the lens and worked on a trichromatic theory of color vision. Helmholz was interested in the different sound qualities of musical instruments when they produce the same fundamental note, or pitch. He believed that the different timbre of various instruments was attributable to overtones or sounds at a higher frequency than the fundamental one and demonstrated this by using resonators to produce a synthetic version of the characteristic sounds of different instruments. His resonance theory of pitch perception was based on the idea that different parts of the inner ear act as resonators for different sound frequencies. Modern research supports this theory.

The general philosophical significance of such work is that it suggests the possibility of accounting for mental experiences in terms of mechanical events, thus supporting the theory of materialism, or MONISM, that is accepted by a majority of scientists today, at least in principle. Progress in our contemporary understanding of how sensory experiences are received and decoded by the brain has not been matched by substantial progress in understanding how all of this information is assembled to produce our experience of the world in which we live. *See also* CRICK, FRANCIS H.; DESCARTES, RENÉ, FECHNER, GUSTAV; WEBER, ERNST.

Further Reading
Schultz, D. P., and S. E. Schultz. *A History of Modern Psychology.* San Diego: Harcourt Brace Jovanovich, 1987.

Hippocratic oath

The Hippocratic oath was probably not authored by Hippocrates (ca. 460–377 B.C.) but is believed to have been produced by a Greek school of medicine of which he was the head. Hippocrates' school was located on the island of Cos and it produced a large volume of scientific and ethical writings that were subsequently gathered into various collections known collectively as the Hippocratic corpus. The Hippocratic oath is important as the first clearly articulated statement of the need for ethical conduct by physicians and therefore as a statement of the rights of patients. The Hippocratic oath is ethically far more complex than is often realized. For example, it not only protects patients but also shields the professional interests of physicians. Some aspects of the oath apply only to the social context in which it was written and these are often simply ignored by ethicists discussing its contemporary relevance.

The Hippocratic oath has two parts. The first is an oath of allegiance by a physician to his teacher. The teacher was to assume the status of a parent even to the extent that the physician was to treat his teacher's children as though they were siblings. Moreover, the oath-taker pledges not to share medical knowledge with anyone else unless they are also pledged to secrecy. These notions of protecting professional territory may well be descriptive of some aspects of the history of medical practice but they are diametrically opposed to some mod-

ern ethical principles. Thus, it runs counter to the scientific obligation to share information openly. In addition, it violates modern notions concerning the patient's right to know. The latter issue has always constituted a nagging problem of the Hippocratic tradition and to medical ethics more generally.

The second part of the Hippocratic oath is commonly referred to as the code of ethics. In the section on dietetics, which was one of three branches of contemporary medicine, the others being pharmacology and surgery, the physician agrees to act "for the benefit of the sick according to my ability and judgment." In the minds of many, this phrase captures the essence of the Hippocratic description of the doctor-patient relationship. Some more explicit ethical protections are also offered to the patient. Sexual intercourse with patients is proscribed and the physician undertakes to shield the patient's privacy with some exceptions.

Acting for the benefit of the sick is held up as an ideal in modern medical practice and any suspicion that other interests may come first is likely to produce heated controversy. For example, the emergence of managed care organizations, or HMOs, has raised the specter of a conflict between physicians serving the financial needs of health insurance companies and the health needs of patients. Even in the oath itself, some conflicts lie close to the surface. Thus, the guarantee of privacy is ambiguous stating only that the physician should not divulge "those things that ought not to be spread abroad." This is far from being a seal of absolute privacy. It can be interpreted to mean that the physician should use discretion in passing on privileged information, not that he should refrain from doing it. It might even be concluded that there are situations in which the physician has an ethical obligation to divulge patient confidences. For example, if a young woman is being treated for a venereal disease, the physician might feel a paternalistic obligation to inform her parents. If a man has paranoid delusions, the physician might feel a social obligation to inform the persons being threatened.

Such discretionary looseness has often led to criticism that the oath is paternalistic. Similarly, the injunction to benefit the sick according to the physician's ability and judgment seems to open the door for judgment calls that may not always be what the patient would wish. For example, it seems to imply that the physician can treat patients against their will. Moreover, it has been interpreted by some modern ethicists to mean that if a patient is terminally ill they should not be informed of this because the knowledge can only increase their distress.

In the section of the oath dealing with pharmacology, a prolife position is taken in respect to both abortion and euthanasia. The physician pledges never to administer a deadly poison nor to do anything that causes an abortion.

The Hippocratic ban on EUTHANASIA is still in effect. In modern medical ethics, physicians may cause death by turning off life support systems but generally do not take active measures to end a life. Exceptions arise in countries where euthanasia or medically assisted suicide is legal (the Netherlands and the United States). The medical practice of abortions today clearly goes against the Hippocratic tradition.

One of the strangest aspects of the Hippocratic oath to modern minds is the pledge never to conduct surgery. This seems to have been based on a simple division of labor. Hippocratic physicians practiced only in dietetics and pharmacology. Other physicians were more skilled in the use of the scalpel and it was in the best interest of the patient if the Hippocratic physician stood aside and let a surgical specialist do the work.

The Hippocratic oath has stood the test of time even though many other Greek ethical systems have been ignored. One reason for this is that it has many points of similarity to Christian ethics. Physicians still see it as a defining statement of their ethical obligations to patients even though some aspects are anachronistic. Technological advances in medicine, such as life support machines, have also presented physicians with ethical problems that those who crafted the Hippocratic oath could not have been aware of.

See also EUTHANASIA; NUREMBERG CODE.

Further Reading

Smith, W. D., ed. *Hippocrates*. Cambridge, Mass.: Harvard University Press, 1994.

Veatch, R. M. *Medical Ethics*. Sudbury, Mass.: Jones and Bartlett, 1997.

Hiroshima, Japan

Hiroshima was the site of the first atomic bomb dropped by the United States on August 6, 1945. It was followed by another at Nagasaki, Japan, three days later that brought World War II to a sudden end (see photo). Both

The explosion of the second atomic bomb, at Nagasaki, Japan, on August 9, 1945, with the distinctive mushroom-shaped cloud. (LIBRARY OF CONGRESS)

cities were largely destroyed with a death toll estimated at 75,000 men, women, and children in each case. Subsequent exposure to radioactive fallout in the regions produced an increase in the rate of birth defects and cancer that has lasted for more than half a century.

The atomic bombs were produced by scientists working in the MANHATTAN PROJECT, many of whom detested bloodshed. They felt that it was necessary to liberate the awesome destructive power contained within the atom to combat the great evil of fascism that was on the verge of world domination. Dropping the bombs on civilian targets was clearly unethical as a matter of principle but it has often been justified on utilitarian grounds. Large numbers of civilians had already been killed either as a consequence of air raids on military targets located in cities or as a result of saturation bombing designed to undermine civilian morale. A good case can be made that by bringing the war to a speedy conclusion, the atomic bombs greatly reduced the number of civilians that would otherwise have died. Using this logic, it can be argued that use of this new weapon was not only ethically defensible but also ethically necessary. A weakness in this logic is that at the time of the decision to bomb, the surrender of the Japanese could not have been predicted. Another troubling implication is that no form of warfare, however, horrifying, is beyond the ethical pale.

See also CHEMICAL AND BIOLOGICAL WEAPONS; CHERNOBYL NUCLEAR DISASTER; DOWNWINDERS; MANHATTAN PROJECT.

Hobbes, Thomas
(1588–1679)
English
Philosopher

Associated with the royalist camp during England's Civil War (1642–51), Thomas Hobbes worked as a tutor for the Cavendish family and traveled with them to Europe on several occasions, which provided him with an opportunity to meet GALILEO, DESCARTES, and other leading intellectuals of his day. As a person who had supported the absolute rights of kings in print, Hobbes was obliged to leave England once the Civil War had broken out. In Paris, he became tutor in 1646 to the exiled Prince of Wales and future king Charles II. While there, he also penned *Leviathan*, an influential statement of his conservative political philosophy.

Hobbes is well known for expressing an extremely bleak view of human nature. He believed that humans are essentially motivated by egoistic self-gratification and that if they were left to themselves the resulting conflicts would make life "nasty, brutish, and short." According to Hobbes version of the social contract theory of government (also supported by JOHN LOCKE, 1632–1704) people give up their natural right to self-government in the interest of obtaining security against the adverse effects of lawlessness. The main difference between Hobbes and Locke is that Hobbes was an absolutist believing that absolute power is given by people to the monarch, or republic, and that the contract cannot be renegotiated. Hobbes believed that rebellion against the state was always immoral because it violated the social contract. Locke argued that the continuance of the social contract must be contingent on the government living up to its obligations.

Hobbes's dark view of human nature is disturbing, even today, and most of his work was quite unpopular among contemporaries. Not only did he see human beings as craven animals, but he denied that any supernatural or religious force could rescue us from our depravity. Only the absolute power of the government could do that. Even after the restoration of the English monarchy, in 1660, Hobbes was out of favor and his description of the Civil War, *Behemoth* (1680), was actually banned. The very nihilism that repelled his contemporaries makes him seem interestingly modern today. For example, many of his insights about human nature were repeated in the nonscientific psychological ideas of Sigmund Freud in the 20th century.

See also LOCKE, JOHN; ROUSSEAU, JEAN JACQUES; SOCIAL CONTRACT THEORY.

Further Reading
Martinich, A. *Hobbes: A Biography*. New York: Cambridge University Press, 1999.

homosexuality, possible biological basis of
It is useful to distinguish between homosexual behavior and orientation. Homosexual behavior occurs in many societies, including our own, among heterosexual people whose primary sexual interest is in the opposite sex. Societies vary greatly in the extent to which such sexual contact is socially, or legally, sanctioned. Homosexual behavior is particularly common among prison inmates due to the unavailability of individuals of the opposite sex. These inmates generally do not develop a lifelong homosexual orientation or preference. People with an exclusive homosexual orientation are sexually attracted to members of their own sex and are sexually uninterested in the other sex. (Bisexuals may be sexually attracted to individuals of both sexes). Many leading sexologists, including ALFRED KINSEY, have concluded that people do not fall neatly into the categories of homosexuality or heterosexuality but that there is a continuous range of variation from exclusive homosexuality through

bisexuality (or about equal preference for both sexes) to exclusive heterosexuality.

When we talk about homosexuals, we are usually referring to people whose preference for members of their own sex is consistent and clearly established. Such exclusive homosexuality is comparatively rare, encompassing less than 4% of the adult male population according to the best available evidence. Many readers may be accustomed to seeing higher figures based on the data of Alfred Kinsey that include any experience in a person's lifetime (including adolescent sexual experimentation) rather than the more stringent criterion of recent sexual behavior for adults. Homosexuality is more common in men than women.

Exclusive homosexuality is puzzling to biologists since it undermines reproduction, one of the primary imperatives built into organisms by natural selection. Homosexuality occurs in many animals, as various as rats, pigs, and zebra finches, and thus cannot be a unique product of human civilization.

Despite reducing reproductive rates, homosexuality in men and women is affected by genes, having a heritability of approximately 25% to 50% according to several twin and family studies. How could this be? If homosexuals have far fewer children, then genes predisposing individuals to homosexuality should become less frequent from generation to generation until they disappear! Genes predisposing individuals to exclusive homosexuality must provide some other benefits to their carriers that make up for the lost reproduction. Geneticists have come up with many ingenious possibilities. Perhaps carriers of a single copy of the hypothetical gene have an advantage in health, survival, or reproduction but are not exclusive homosexuals. Perhaps it is an X-linked trait that provides a benefit for female carriers.

Much public fanfare attended the reported discovery of a "gay gene" on the X-chromosome by Dean Hamer and his associates as published in *Science* in 1993. This report was based on a study of the banding patterns of chromosomes, a technique that is notoriously unreliable. The results could not be replicated by an independent team of investigators whose work was published three years later in the same journal. Although the controversy over discovery of a gene locus for homosexuality has shaken the faith of many in the quality of the science and called the motives of the researchers into question, this reaction has been overdone. Thus, many features of human personality and behavior are heritable, even though the locations of genes responsible for these influences are almost completely unknown.

Moreover, there is abundant additional evidence for a biological basis for homosexual orientation. Perhaps most relevant is the 1991 finding of Simon Le Vay that the brains of homosexual men differed from those of hetero-

sexuals. The difference was discovered in a nucleus of the hypothalamus that is known to regulate sexual behavior. This nucleus, known as INAH3 (or the third interstitial nucleus of the anterior hypothalamus), was smaller in homosexuals, suggesting low testosterone exposure before birth. Heterosexual men also differ from homosexual men in other aspects of brain anatomy, including the anterior commissure and the suprachiasmatic nucleus. The fact that homosexual men are unusually good-looking according to one scientific report also suggests reduced testosterone exposure early in life.

There is a variety of fairly compelling evidence that the prenatal hormonal environment plays a role in sexual orientation. Women who are unaware that they are pregnant have continued to take certain birth control pills (particularly those containing diethylstilbesterol, DES), thereby exposing their unborn daughters to high levels of sex hormones that masculinize their brains. When they mature, these daughters are more likely to be sexually attracted to women. The same conclusion has been drawn concerning women who were exposed to high levels of sex hormones prenatally because of congenital adrenal hyperplasia (also known as adrenogenital syndrome).

Animal research provides valuable evidence that sexual orientation can be altered by the effects of prenatal hormones, although many scholars insist on pointing out the distinctions between human sexuality and that of nonhuman mammals in terms of neural organization, capacity for voluntary control, and the role of social norms among humans. When homosexuality is demonstrated in laboratory rats, this means that male rats (a) ignore sexually receptive females and (b) direct sexual behavior toward males.

Exposing pregnant rats to intense stressors (bright light and immobilization) causes them to secrete stress hormones (endorphins) that cross the placenta and produce antitestosterone effects in the hypothalamus of a fetus. When the male offspring are born and mature, their sexual behavior is altered although the precise outcome is affected by rearing conditions. If they are raised in isolation or with other males, they exhibit a sexual responsivity to other males. If they are raised in mixed-sex groups, they become sexually responsive to both males and females.

Although the prenatally stressed rats look male in all respects and behave like males otherwise, if they are raised without females they become sexually responsive only to other males. This is an interesting analog for human homosexuality but there are important differences. Homosexual male rats are feminized in their behavior. They exhibit the female-typical pattern of lordosis in which a sexually receptive posture is adopted by arching the back. Human male homosexuals do not necessarily have a preference for stereotypically feminine postures or actions and are characterized purely by a preference for male sexual partners. The evidence that mothers of human homosexuals might experience stressful pregnancies, thus providing an exact parallel with the animal research, is weak and unreliable.

The importance of biological factors in homosexuality is further implicated by the early emergence in childhood of gender-atypical behavior (such as boys being more interested in cookery and needlework than sports, and girls being more interested in trucks as playthings than in dolls). There is also evidence that homosexuals have different patterns of cognitive abilities reflecting early differences in exposure of the brain to sex hormones. Thus according to the research of Doreen Kimura, homosexual men have better manual dexterity but are not as good as heterosexuals at throwing an object to hit a target.

The weight of evidence for a biological influence on the development of exclusive homosexuality in humans is thus fairly compelling. Equally telling, perhaps, is the fact that there is almost no persuasive evidence that the home environment plays any role in the development of sexual orientation. Extensive surveys, such as the one carried out in the Bay Area of San Francisco, failed to find important experiential differences between homosexuals and heterosexuals. The only striking exception was evidence of strained or distant relationships with fathers. It is more plausible that this tension was a response to early gender nonconformity on the part of the child than to assume the relationship with the father played a causal role in the development of a homosexual preference in men and women.

Faced with the mountain of evidence implicating biological factors in homosexuality, conservatives continue to assert that homosexual behavior, like any other action, is a moral choice. This may be true but it does not explain why some people are sexually attracted to persons of the same sex. In conclusion, the moral choice view does not explain differences in the size of the INAH3 nucleus, or the fact that homosexuals are more likely to have older brothers and homosexual relatives. Indeed, if choice played a role in sexual orientation this would make the phenomenon even more puzzling. Why would anyone choose to be a member of what has most often been a despised minority?

See also KINSEY, ALFRED; MASTERS AND JOHNSON; SEXUAL ORIENTATION RESEARCH.

Further Reading

Hamer, D., S. Hu, V. A. Magnuson, N. Hu, and A. M. L. Pattatucci. "A Linkage between DNA Markers on the X Chromosome and Male Sexual Orientation." Science 261 (1993): 321–27.

Kalat, J. Biological Psychology. Pacific Grove, Calif.: Wadsworth, 1995.

Kimura, D. Sex and Cognition. Cambridge, Mass.: MIT Press, 1999.

LeVay, S. "A Difference in Hypothalamic Structure between Heterosexual and Homosexual Men." Science 253 (1991): 1034–37.

Murphy, T. F. Gay Science. New York: Columbia University Press, 1997.

Zucker, K. J., J. Wild, S. J. Bradley, and C. B. Lowry. "Physical Attractiveness of Boys with Gender Identity Disorder." Archives of Sexual Behavior 22 (1993): 23–36.

Hooke, Robert
(1635–1703)
English
Physicist and Biologist

Hooke was a central figure in the ROYAL SOCIETY OF LONDON and served as the curator of experiments for this organization of scientists from 1662. Beginning his scientific career as an assistant to ROBERT BOYLE (1627–91), Hooke is believed to have played an important role in facilitating the scientific achievements of his boss. He had a talent for devising, or improving, scientific equipment. Hooke's pioneering use of the microscope in biology marks the beginning of microbiology. He was the first to use the word "cell" with its modern biological meaning and was the founder of insect anatomy. In conjunction with many of the other men of science of his day, he had a breadth of interests that is rarely seen in contemporary academic life. He presented influential theories of light, combustion, gravitation, and fossils.

Whereas his contemporary, FRANCIS BACON, is often seen as a leading advocate for the kind of serious professional science represented by the Royal Society, Hooke can be seen as the consummate practitioner. His scientific practice convinced him that nature, at whatever scale we contemplate it, from the microscopic to the astronomical, is a great machine that can be understood in terms of mechanical laws.

See also BACON, FRANCIS; ROYAL SOCIETY OF LONDON.

hormone mimics
Also known as endocrine disruptors, hormone mimics are chemicals in the environment (of natural as well as synthetic origin) that act on hormone receptors with adverse consequences for biological development and health. Most concern has focused on synthetic substances that mimic sex hormones and are believed by some biologists, physicians, and environmental activists to impede normal sexual development, sexual function, and reproduction in humans as well as wildlife and also to increase

the risk of various cancers. These conclusions are highly controversial among scientists. This controversy is important because activists argue that biological systems may be exquisitely sensitive to small quantities of endocrine disruptors at critical periods during development. If correct, this means that the standard assumptions of toxicologists about what is a permissible amount of a specific toxin, assumptions that guide the regulatory activities of governments, may have to be revised downward to much more conservative levels.

Recent concern about hormone mimics was raised by biologist Theo Colborn of the World Wildlife Fund and others in *Our Stolen Future* (1996), a book that is hailed by some as an eye-opening publication whose importance is equivalent to *Silent Spring* (1962), in which Rachel Carson first alerted the public to the dangers of pesticides like DDT. Hormone mimics can have adverse effects on development because they influence gene expression by interfering with the normal pattern of interaction between sex hormones and genes. For example, sex determination occurs because sex hormones either turn on the genes promoting development of male reproductive structures or turn on those promoting female reproductive development.

The many substances known to mimic hormones in animals and in cell cultures encompass products that people are exposed to on a daily basis. They include: organochlorine pesticides such as DDT, DIOXIN; surfactants used in cleaning products, paints, and other products; resins used in dental fillings; PCBs (polychlorinated biphenyls); and plasticizers used in food packaging. Many of these substances have been found to affect sexual development and function in rats and other lab animals whose mothers had been exposed during pregnancy. Typical results included changes in reproductive organs, disrupted sexual behavior, diminished sperm production, and reduced fertility. The relevance of these results to humans is unclear both because of species differences and because the level of exposure in experiments is typically much higher than people might expect to encounter under ordinary circumstances. Human offspring may be exposed in utero and there is evidence that breast milk in some populations may contain surprisingly high levels of endocrine disruptors.

The laboratory work has been verified by natural experiments, including spillage of pesticide in Lake Apopka, Florida. Male alligators were observed that had unusually small penises. A variety of other abnormalities concerning sexual development in wild animals have been attributed to chemical spills but it is not impossible that the changes could have been produced by natural hormone mimics. In general the effects of endocrine disruption tend to masculinize females and feminize males. Masculinization has been observed in female fish exposed to pollution from paper plants and in female shellfish exposed to tin pollution from ship paint. Whatever their cause, some of these changes have been quite extreme, with ovarian tissue appearing in testes and male reproductive organs occurring in females.

The evidence that human development may be adversely affected by endocrine disruptors derives from two main types of study: examination of the correlates of exposure to various chemicals and epidemiological studies that have found increases in disorders associated with hormone mimics as the industrial use of such substances increased beginning in the 1940s.

The offspring of women who took the drug diethylstilbesterol (DES), which was used to prevent miscarriages, suffered from a number of abnormalities in sexual development and had increased rates of cancer. Thus, some males had undescended testes, or unusually small penises, and females had higher rates of vaginal cancer and were more likely to report homosexual behavior. Further concrete evidence comes from the finding that maternal exposure to PCBs is associated with reduced IQ of children.

Epidemiological studies have found increased rates of undescended testes, hypospadias (abnormal penises), and increased rates of testicular and prostate cancer among men, as well as increased rates of endometriosis (thickening of the womb) and breast cancer among women. Although these phenomena are consistent with the effects of endocrine disruptors, each can also be explained according to a variety of different theories. Clearly, a great deal more research is needed. To this end, the U.S. Environmental Protection Agency has begun the process of screening tens of thousands of industrial chemicals to determine which need to be regulated as potential endocrine disruptors. Such research constitutes a preliminary step to discovering whether the age of chemicals really has stolen our future, as Theo Colburn claims.

See also DDT; DIOXIN.

Further Reading
Colborn, T., D. Dumanoski, and J. P. Myers. *Our Stolen Future: Are We Threatening Our Fertility, Intelligence, and Survival? A Scientific Detective Story.* New York: Dutton, 1996.

hormone replacement therapy

When women go through menopause (or cessation of menstrual cycles) there is a decline in estrogen production which has adverse health consequences, including increased incidence of osteoporosis (loss of bone density and strength) and cardiovascular disease. Replacement therapy, usually with a combination of estrogen and progesterone, alleviates these problems.

Hormone replacement therapy involves complex ethical issues, including the definition of a disease. From one perspective, menopause is a natural phenomenon and not a disease and is therefore no more appropriate for medical intervention than age itself. Many physicians would respond that medicine is concerned not just with treatment of disease but with improving the quality of life, which would be the primary rationale for cosmetic surgery, for example. Moreover, as life expectancy continues to increase, improving the quality of life of the older population will become an ever more important role of medicine.

Hormone replacement is usually continued indefinitely once it starts. This has two distinct ethical ramifications. First, can the economic burden of chronic medication for a natural condition of aging be justified if it takes resources away from other branches of medicine? In practice, though, treatment is rarely denied because there is a more "worthy" condition to be treated. The economic issue boils down to who is willing to bear the costs and this problem is solved by health insurance companies' willingness to cover much of the expense. Second, the possible adverse consequences of long-term use (i.e., decades rather than years) are unknown.

An even more pressing ethical problem than either of the above surrounds the finding that hormone replacement increases the chances of developing breast cancer and clotting of venous blood. Such complications can be troubling to physicians because they violate the "do no harm" commitment of the HIPPOCRATIC OATH. Recent research has found that the increased cancer risk is minimal, however, and most physicians accept the utilitarian view that virtually all medications have the potential for adverse effects in some individuals but that are more than compensated for by their net beneficial effects for most patients. Moreover, the recent development of lower-dose treatment capable of combating osteoporosis will likely reduce the incidence of severe adverse effects.

See also HUMAN GROWTH HORMONE.

Further Reading
Hope, S., J. Brockie, and M. Rees, eds. *Hormone Replacement Therapy*. New York: Oxford University Press, 1999.

Houdini, Harry
(1874–1926)
Hungarian American
Magician

Harry Houdini was the stage name of Ehrich Weiss. The son of Hungarian immigrants to America, he became an

In his 1925 stage show at the New York Hippodrome, Houdini demonstrates techniques used by dishonest mediums. Seated to the left, Houdini rings a bell using his toes after slipping off his shoe. (LIBRARY OF CONGRESS)

aviator and worked in movies before developing his own magic show. Houdini's remarkable ability to escape from straitjackets, shackles, locked boxes, sealed containers, submerged packing crates, and so forth brought him worldwide celebrity. He invented many magic tricks based on clever devices that he put together in his basement laboratory. The most famous of these was the Chinese water torture cell in which Houdini's ankles were clamped in stocks and he was lowered, upside down, into a tank of water that was then sealed.

Some of Houdini's tricks were so successful that spectators were convinced he had supernatural powers. Even his friend Arthur Conan Doyle, author of the Sherlock Holmes mystery stories, believed that Houdini became "dematerialized" during his escapes. Doyle was a believer in spiritualism and himself became the victim of unscrupulous mediums who pretended to contact the spirits of the dead for a hefty fee.

As a professional illusionist Houdini quickly saw through the tricks of the spiritualists and incorporated them in his stage act to expose the fraud and provide entertainment. The illustration shows how a "spirit" could make its presence known by ringing a bell during seances. Houdini is pulling the bell cord with his toes, having slipped off his shoe underneath the table. His 1924 book, *A Magician among the Spirits,* chronicles his exposure of well-known mediums of the day, which also provided popular stories in contemporary newspapers. His crusade against spiritualists illustrates how vulnerable people may be to irrational beliefs when their experiences do not fit into a commonsense pattern. If a person at a seance does not know why the bell is ringing, it is easy to jump to the conclusion that their dead child is ringing it.

Even though Houdini believed that all of the professional mediums he visited were complete charlatans, he continued to believe in the reality of the spirit world and arranged that he and his wife would attempt to communicate after death. This story illustrates the remarkable persistence of supernatural beliefs even in highly skeptical people.

See also EXTRASENSORY PERCEPTION.

Hubble, Edwin Powell
(1889–1953)
American
Astronomer

Edwin Hubble was a pioneer in the study of galaxies. Using the 100-inch telescope at Mount Wilson Observatory, he discovered that the Andromeda "nebula" is actually in a different galaxy from our own Milky Way

galaxy. An important generalization he made, known as Hubble's Law, states that the more distance there is between galaxies the faster they are moving away from each other. Hubble's Law indicates that the universe is expanding and the rate of expansion is given by Hubble's constant.

Hubble's research revealed that there are many other galaxies in the universe. Instead of seeing ourselves as the center of the universe, as in the Ptolemaic worldview, modern astronomy reveals the earth to be an insignificant speck in a vast cosmic expanse. One interesting possibility that this raises is that our planet may not be unique in supporting life. Some scientists have even begun to anticipate the discovery of intelligent life in other galaxies and to speculate about how we might communicate with it.

See also SAGAN, CARL; SCIENCE, HISTORY OF; TECHNOLOGY, HISTORY OF.

Hubble Space Telescope
The Hubble Space Telescope (HST), named for astronomer Edwin Hubble, has allowed us to peer farther into the universe than ever before.

By orbiting 613 kilometers above the earth, the solar-powered HST's view is not obscured by atmospheric particles. Even though the orbiting telescope is now seen as a success, its faulty engineering has received a great deal of criticism as an unethical waste of public funds.

When the HST was first launched from the Space Shuttle Discovery, on April 15, 1990, it had already experienced substantial delays and cost overruns. Due to an error in constructing the main mirror of the telescope, it could not be focused properly. A successful Shuttle repair mission was launched in December 1993 and the HST has since transmitted a remarkable stream of images taken from the visible, ultraviolet, and infrared regions of the electromagnetic spectrum. Detailed images, such as glowing rings of hot gas surrounding the remains of a dying star in the Aquila constellation, have provided astronomers with a more concrete explanation of such events than was ever previously possible. HST observations have confirmed the existence of some 50 billion more galaxies than were previously supposed to exist.

See also CHALLENGER ACCIDENT; ENGINEERING ETHICS; FORD PINTO.

human cloning research
Long the substance of science fiction, artificial cloning of viable human beings has not yet been carried out, or at least it has never been reported, although two Italian

doctors, Severino Antinori and Panos Zavos, told a National Academy of Sciences panel, on August 7, 2001, that they would attempt to do this. The first report of successful cloning of human *embryos* was a paper presented by Jerry Hall, Robert Stillman, and others, to the October 1993 meeting of the American Fertility Society. Working with embryos that had been fertilized by more than one sperm, and were therefore incapable of developing into babies, Hall and Stillman reported that when the embryos had developed to the eight-cell stage, they separated out the cells, which began to divide again. In this way, a single embryo yielded eight new embryos, each genetically identical to the others. Human clones had been created.

Hall and Stillman's work yielded very different responses among their scientific peers compared to the general public. Their paper was awarded the General Program Prize at the American Fertility Society meeting. Colleagues recognized that the ability to multiply embryos could provide a major advantage in attempts to help infertile couples produce babies. From a technical standpoint, there is no reason that these procedures could not be repeated with viable embryos.

The public reaction to Hall and Stillman's work was overwhelmingly negative. The research was seen as horrifying and as extremely unethical. This judgment was based both on what the researchers had actually done and on what most people foresaw as possible future applications. President Clinton issued an executive order prohibiting the use of federal funds to support human cloning research. Expressing the popular distaste for cloning, Clinton stated that it undermines the uniqueness and sacredness of human life. He set up a federal bioethics commission to examine the ethical and legal issues surrounding cloning and appealed to corporations to put off any research planned in this field.

The public outrage over Hall and Stillman's work was nourished by *Brave New World* scenarios in which science ruled a world peopled by thousands of cloned Saddam Husseins, Albert Einsteins, or Bill Gates, and many leaped ahead to eugenics programs and manipulation of the genetic code. These fears were largely misplaced given the technology of the day but the successful demonstration of cloning an adult sheep in Scotland has revived many of the same fears. While it is not yet possible to modify the genetic program in embryos so as to produce "designer" human beings, the fact that the entire human genome will soon be decoded does open the door for deliberate genetic modification designed to eliminate genetic disorders.

With 93% of Americans opposed to human cloning, according to a Time/CNN poll, this kind of research is obviously politically unpopular. Sentiment in Britain may be even more strongly negative. Human cloning there is now illegal, as it is in other countries. There has also been a political attempt to prevent human cloning by discouraging cloning research on nonhuman mammals. Thus, the funding of Ian Wilmut, who cloned the sheep named Dolly, was cut off by the British Ministry of Agriculture in April 1998.

See also GENETIC ENGINEERING; HUMAN GENOME PROJECT; REPRODUCTIVE TECHNOLOGIES.

Further Reading

Drlica, K. *Understanding DNA and Gene Cloning.* New York: Wiley, 1992.
Humber, J. M., and R. Almeder, eds. *Human Cloning.* Totowa, N.J.: Humana Press, 1998.
Kass, L. R., and J. Q. Wilson. *The Ethics of Human Cloning.* Washington, D.C.: AEI Press, 1998.
Nussbaum, M. C., and C. R. Sunstein, eds. *Clones and Clones: Facts and Fantasies about Human Cloning.* New York: W. W. Norton, 1999.

human factors research

Human factors research attempts to improve the interface between humans and technology (broadly conceived) and is thus an interdisciplinary endeavor involving psychologists and engineers. Its problems might be as simple as determining the correct location of brake lights to prevent rear-end collisions or as complex as designing the living quarters of a space station.

Human factors research has contributed greatly to the safety of air flight and research is often conducted using flight simulators. Thus, it has been found that when pilots approach runways built on sloped terrain they misjudge distances and are prone to land beyond the tarmac. This perceptual error led to otherwise unexplainable accidents and can be corrected by appropriate training.

Human factors research deals with physical as well as psychological aspects of human-machine interface, and such research investigates whether machine controls are designed to maximize comfort and performance as well as safety. Many machines have been poorly designed from this perspective. Thus, it has been estimated that traditional lathes (which are still in use) were built for optimal performance by a person who is 4 1/2 feet tall and has an 8-foot shoulder span!

An issue of ever-increasing importance to human factors specialists as more people reach old age is the extent to which homes and automobiles are designed to reduce accidents among this age group. Thus, simple modifications, such as changing floor coverings or making stair steps shallower, could reduce dangerous falls that are a frequent cause of injury to the aged. Better design of automobile controls and dashboards, improved highway lighting, and more legible signs could reduce the rate of car accidents among the elderly.

Human factors issues have been raised in attempts to explain many major industrial accidents of recent history, such as the nuclear reactor accidents at CHERNOBYL and THREE MILE ISLAND and the chemical factory mishaps at BHOPAL and SEVESO. Some of the issues were very simple, as in the case of unreliable or unreadable gauges. Others were more complex and involved management decisions that ignored, or overruled, safety procedures. From a human factors perspective, many of these incidents were not really accidents at all so much as the inevitable consequence of insufficient attention to safety procedures in the interface between human operators and technology.

See also BHOPAL CHEMICAL ACCIDENT; CHERNOBYL NUCLEAR DISASTER; THREE MILE ISLAND NUCLEAR ACCIDENT.

Further Reading
Kantowitz, B. H., and R. D. Sorkin. *Human Factors: Understanding People-System Relationships*. New York: Wiley, 1983.

Human Genome Project

Efforts are currently under way to decode the sequence of bases in the DNA of all the chromosomes that make up the human genome. The Human Genome Project, which began in 1990 and is run by the NATIONAL INSTITUTES OF HEALTH (NIH), operates in competition with the biotechnology company Celera Genomics. Both groups published preliminary reports, based on sequencing of over 90% of the human genome, in February 2001. One interesting finding is that there are only between 30,000 and 40,000 human genes, approximately one-third of earlier estimates.

There has been considerable speculation about how these data will be used. One goal is clearly to acquire a better understanding of the genetic basis of disease. It is anticipated that the information can be applied to developing more sophisticated genetically engineered therapies. Thus, direct modification of germ lines may be possible so that serious genetic diseases could be eliminated, just as serious infectious diseases like smallpox were wiped out by VACCINATION.

A more ethically troubling possibility is that the information could be used to promote positive EUGENICS. People of the future might be genetically engineered to be more intelligent, more athletic, or to live much longer. Although few people question that these objectives are intrinsically desirable, many are appalled at the prospect of "playing God" with the human DNA. Deciphering the human genome takes positive eugenics a step closer to technical feasibility. Whether such manipulation of human DNA will be permitted on ethical grounds remains to be seen.

Another major ethical concern involves the potential discovery of genes that are predictive of a variety of personal attributes, such as alcoholism, vulnerability to heart disease or cancer, and potential to commit violent crimes. Many fear that such information could be misused by employers, governments, or insurance companies. Arguably, such fears are overblown for several reasons. First, such information is already contained in medical records and generally does not get abused. Second, where such personal qualities are not revealed by the individual's behavior or appearance they are difficult to hide from coworkers, for example, with whom a certain degree of intimacy accrues over time.

A more technical issue has to do with the implied assumption of genetic determinism. Just because an individual has a hypothetical gene that is statistically associated with alcoholism, this does not mean they will become alcoholics. They may, for example, decide to abstain from alcoholic drink because of the damage it caused to a family member, or they may happen to live in a Muslim country in which no alcohol is produced. Even when it can be established that behavioral tendencies are highly heritable, there are usually many genes involved, which makes it exceedingly difficult to predict behavior from genotype. If genes do not provide a good prediction of behavior then knowledge of a person's genotype may not be as threatening as many currently fear.

See also CRIME, AND BRAIN CHEMISTRY; EUGENICS; GENETIC DETERMINISM; GENETIC TESTING.

Further Reading
Cantor, C. R., C. L. Smith, and Human Genome Project. *Genomics: The Science and Technology behind the Human Genome Project*. New York: John Wiley, 1999.
Sternberg, S. "Human Genome Makes Mind-Boggling Reading." *USA Today*, February 13, 2001.

human growth hormone

Human growth hormone is produced by the anterior pituitary gland. A deficiency of growth hormone during childhood results in low adult stature (sometimes referred to as dwarfism) whereas an excess produces abnormal height, or gigantism. Children suffering from growth hormone deficiency can have their supply boosted by injections (usually around three per week) to stimulate growth. Growth hormone therapy has been controversial for a number of reasons, including viral contamination, side effects, the high cost of the treatment, and the basic rationale for who should be treated.

In the past, human growth hormone could be obtained only from the pituitaries of cadavers. This source was limited and unreliable. Moreover, it was subject to contamination by the Creutzfeldt-Jakob disease virus, which, in the early 1980s, killed several people who had received hormone supplementation. For this reason, human growth hormone was banned in the United States in 1984. The following year, genetically engineered growth hormone based on recombinant DNA technology became available. This synthetic hormone alleviated the supply problem and solved the problem of viral contamination but it did not reduce, and arguably increased, the cost of the treatment. In 1990, this was estimated at $5,000 to $30,000 depending on the weight of a child.

Given the great cost of growth hormone treatment, it is important to know how effective it is and if the benefits to the treated individual outweigh the economic cost. The expensiveness of the therapy raises the problem of whether all children with growth hormone deficiency should be treated. In the United States, it is estimated that there are over 14,000 such children.

One interesting aspect of growth hormone treatment is that its effectiveness is largely an unknown quantity. Ideally, children receiving the hormone would be compared to a matched group receiving no hormone and differences between the two groups could be attributed to the treatment. In practice, the effectiveness of growth hormone is assessed by using the individual as their own control, that is, by comparing growth rate before intervention with that after intervention. Therapy is considered effective if there is a 2 cm increase in the patient's growth rate during the first year of injections. Using this criterion, over half of the treatments are successful. Yet there is a nagging uncertainty about whether these "successful" treatments really produce individuals who are taller or whether their growth is merely accelerated at the time of treatment and they end up being the same height as they would have been without hormone supplementation.

Apart from the issues of expensiveness and effectiveness, growth hormone treatment produces some troubling side effects. The most worrying of these are (a) formation of antibodies against growth hormone that develop in 30% to 40% of treated children and (b) the development of glucose intolerance. For children developing antibodies to growth hormone, there is a distinct possibility that a child will end up shorter than he or she would have been if they had received no treatment. Although there are slight increases in insulin and blood glucose in the case of children treated with growth hormone, whether these are of any clinical importance is unknown. Other problems include swellings from the

repeated injections and, in unusual cases, slippage of the growth area of the thigh bone (femur).

Physicians find themselves in an uncomfortable position when they recommend growth hormone injection as a treatment for short stature because shortness is not an illness. While it is true that some children have growth hormone deficiency that can be demonstrated in clinical tests, endocrinologists cannot agree on useful objective definitions of growth hormone deficiency and many believe that accelerated growth rates in response to hormone treatment is the only useful criterion. This brings the ethical debate back to the question of stature.

Even though shortness is not itself an illness, it can be argued that children of low stature are at risk of developing serious problems due to stigmatization and social isolation, which put them at risk for developing clinical depression. If these problems can be averted by treatment, then growth hormone could be construed as preventive health care. The main problem with this approach is that there is insufficient evidence (a) that short children necessarily have more psychological problems or (b) that these problems are improved by growth hormone treatment.

If neither growth hormone deficiency nor low stature can be considered an illness, then it follows that many children are receiving growth hormone for cosmetic reasons. From this perspective, growth hormone as a medical intervention is analogous to the fitting of dental braces. Neither of these manipulations is going to improve health but each is designed to benefit a child by improving their appearance and social acceptability. Such a cosmetic approach is possibly unethical for at least two different reasons. The first is that growth hormone, even if uncontaminated, can do harm with at least the potential for producing diabetes and cardiovascular problems. The mere fact of repeated injections produces some level of risk that is questionable if used only for cosmetic purposes. The second issue has to do with the high cost of growth hormone, which can potentially divert economic resources away from more important areas of child health.

See also BST; GENETIC ENGINEERING.

Further Reading

Bercu, B. D., ed. *Basic and Clinical Aspects of Growth Hormone.* New York: Plenum, 1988.

Lantos, J., M. Siegler, and L. Cutler. "Ethical Issues in Growth Hormone Therapy." *JAMA* 261 (1989): 1020–24.

Shulman, L., J. L. Miller, and L. I. Rose. "Growth Hormone Therapy." *American Family Physician* 41 (1990): 1541–46.

Thomas, P. "Hormone Stirs Hype and Hassle." *Medical World News,* October 10, 1988.

Humphreys, Laud

(1930–1988)
American
Sociologist

In the past, laws and customs prevented men from expressing their desire for sexual variety with the exception of rich and powerful figures like King Solomon, or the Emperor Ismail the Bloodthirsty, who maintained huge harems for their exclusive sexual use. Some men such as Italian nobleman Giovanni Casanova (1725–98) have flouted the laws and customs of their period by engaging in the dangerous pursuit of extramarital partners. Others, like English diarist James Boswell (1740–95) have endangered their health, and that of their wives, by consorting with prostitutes. The craving for sexual variety may be so great that men who are married and live ordinary humdrum lives engage in homosexual activities for the sheer sexual gratification this provides.

This conclusion was drawn by sociologist Laud Humphreys who described his ethically controversial study of homosexual practices conducted in public toilets in his 1975 book *Tearoom Trade*. Within this subculture, restrooms are referred to as "tearooms" and homosexual activities are referred to as "trade." This slang term is somewhat misleading since it usually has no commercial connotation of making a profit from prostitution, for example.

As a participant observer, Humphreys played the role of voyeur and lookout. As "watch queen," he was relied on as someone to alert the participants in homoerotic activity at the approach of unknown individuals, particularly police. Oral sex (fellatio) was the most common form of sexual gratification. As a lookout, Humphreys observed hundreds of acts of impersonal sex.

Sex in public lavatories was the occasion of many arrests in the United States for "homosexuality" and indecent exposure and Humphreys could therefore be considered an accomplice in illegal activities. Prosecution of homosexual activity in public places has been less vigorously pursued in recent decades despite a highly publicized 1999 case in which 18 were arrested in a sting operation conducted in Wasena Park in Roanoke, Virginia. Virginia is one of a handful of states that retains archaic antisodomy laws, in which sodomy is defined as use of any bodily orifice except the vagina. Humphrey's greatest breach of professional ethics involved invasion of privacy. Participants did not know they were being observed in a scientific study or that the "watch queen" was really a professional sociologist.

Humphreys wanted to explode what he felt were dangerous myths held by the public, and by authorities, about men who seek impersonal sexual gratification in public restrooms. He believed that the best way to dispel negative stereotypes about these men was to collect objective evidence. Humphreys gained the confidence of a few of his subjects and explained that he was a sociologist intent on studying their sexual behavior. By this means, he elicited information about their personal lives and motivations.

It was not possible to get to know most of his subjects sufficiently well to establish a relationship of trust. Humphreys surreptitiously recorded their license plate numbers. He then used an elaborate cover story to trick the Motor Vehicles Bureau into releasing the men's home addresses. Humphreys approached the men in their homes one year later while wearing a disguise and pretending to be a health service interviewer. In this fictitious role, he administered a survey asking the men about their personal lives.

Humphreys's data certainly exploded some stereotypes about impersonal sex in public restrooms. The majority of the men were not only highly respectable but were also happily married "heterosexuals." Of the 50 men he interviewed, 54% were married and 8% were separated or divorced. Of the single individuals, a minority (14%) could be considered a part of the gay subculture. More than two-thirds of the interviewees had military experience.

What surprised Humphreys most about his subjects was the aura of respectability with which they were surrounded. They lived in the neatest houses, drove the newest cars, were well dressed and immaculately groomed. Moreover, most were deeply religious. They saw devoting time to their families and homes as top priorities. Humphreys gained the impression that the Bible on the table and the flag on the wall could be indicative of secret deviance. He concluded that they used a "breastplate of righteousness" or a glistening shield of respectability to divert attention from their deviant sexual behavior.

The controversy surrounding publication of Humphreys's research was not primarily focused on the seamy subject matter but reflected the feeling of many colleagues that the rights of the participants had been flagrantly violated. Of paramount concern was the issue of invasion of privacy. Humphreys had connected their tearoom personae with their home lives in a way that could have destroyed their marriages. For example, a subject might have recognized Humphreys and his spouse might have wanted to know where they had met. The right not to be studied without consent had been denied. Many ethical critics were also upset by the tissue of lies, ruses, roles, and disguises that Humphreys employed to connect tearoom activity with the home lives of the subjects. In the view of many opponents, these were not only unethical as research practices but actually broke the law in different ways.

Despite the ethical reservations, Humphreys received a prestigious award, the C. Wright Mills Award for the Study of Social Problems, in 1970. His colleagues felt that the work asserted the right of social scientists to investigate any, and all, areas of human experience. His research is often discussed today in professional ethics courses as an example of the ethical difficulty of studying phenomena, such as human sexuality, that most people see as inherently private even if it occurs in public places. Some ethicists would argue that the insights into human sexuality derived from Humphreys's research justified the potential risks to participants and the generally dishonest and marginally illegal methods used to obtain the data. Others would argue that the same information could have been obtained using less invasive procedures and that the use of unethical and illegal techniques is damaging to the professional credibility of researchers in the social sciences.

See also KINSEY, ALFRED; SEXUAL ORIENTATION RESEARCH; SYMONS, DONALD.

Further Reading

Humphreys, L. *Tearoom Trade: Impersonal Sex in Public Places.* Chicago: Aldine, 1970.
Nardi, P. L. "'The Breastplate of Righteousness' Twenty-five Years after Laud Humphreys' *Tearoom Trade.*" Journal of Homosexuality 30, no. 2 (1995): 1–10.
Symons, D. *The Evolution of Human Sexuality.* New York: Oxford University Press, 1979.

Hunger Site, the

The hunger site (www.thehungersite.com) is an example of creative use of the Internet for charitable purposes. By clicking on an icon, visitors to the site can donate food to feed the starving people of the world. The donation is paid for by sponsors who underwrite the cost of one-sixth of a cup of food in return for having their logo presented on a "thank you" screen. This not only familiarizes consumers with the brand names but gives them a point of entry to the web sites of the sponsors. Founded by Indiana software engineer John Breen in June 1999, the Hunger Site is now run by GREATERGOOD.COM, a charitable organization that receives donations when its site is used by online shoppers.

When visitors click on the "donate free food" button, they are taken to the "thank you" screen containing the sponsor banner ads. Sponsors, including companies such as Sprint, ProFlowers.com, InsWeb.com, iVillage, and Novica (a seller of crafts from around the world), each donate one-sixth of a cup of food per visitor and pay their fees to the United Nations World Food Programme, which distributes food through a wide variety of programs in 80 countries. Less than a year after its creation, the Hunger Site was receiving approximately a million visitors per month and generating some 2.5 million pounds of food donations, enough to make a dent in world hunger. As the Internet grows, the form of donation is likely to become much more important. Technological innovations, such as the Worldwide Web, can thus be harnessed to address world hunger and other seemingly intractable moral problems. In this way, people in wealthy technologically advanced countries can contribute to those living in less fortunate circumstances.

See also GREATERGOOD.COM.

Further Reading

The Hunger Site. *About us.* http://www.thehungersite.com (accessed April 25, 2000).
Roha, R. R. "A Mouse Roars to Fight Hunger." *Kiplinger's Personal Finance Magazine,* December 1999, p. 66, 1999.

Huxley, Thomas Henry

1825–1895
English
Biologist

Thomas Henry Huxley was one of the most distinguished biologists of his day. His early reputation was earned for his observation of jellyfish (*Medusae*) while working as a physician on board the H.M.S. *Rattlesnake* on its exploratory voyage to Australia and New Zealand. He is remembered today for his staunch defense of CHARLES DARWIN's book *On the Origin of Species by Means of Natural Selection* (1859). For this reason, contemporaries nicknamed him "Darwin's bulldog." Huxley's own book *Man's Place in Nature* (1863) presented the view that humans evolved from apes and was bitterly attacked by religious leaders.

See also DARWIN, CHARLES; WILBERFORCE, SAMUEL.

hybrid-powered cars

Hybrid cars are powered by both a gasoline (or diesel) engine and an electric motor. The gasoline engine charges the electric batteries and adds acceleration for passing, for example. Hybrids can be thought of as a compromise between the emission-free, but highly inconvenient, electric car that is sluggish and has a range of only about 100 miles and the highly convenient, but highly polluting, gasoline engine. Most of the major car manufacturers have introduced these environment friendly vehicles at prices similar to those of conventional vehicles. Hybrid vehicles are much more fuel efficient and this provides consumers with a real economic incentive to buy them.

This incentive will prove to be more important in countries where gasoline is more expensive than in the United States where the increased efficiencies are expected to have more appeal for cost-conscious public transportation companies. Ultimately the hybrid vehicle is likely to be replaced by the emission-free fuel cell car that burns hydrogen and emits only water. This is an ideal solution to the problem of polluting car exhausts but it is still prohibitively expensive.

See also AIR POLLUTION; CLEAN AIR ACT, 1970.

Further Reading

Motavalli, J. *Forward Drive: The Race to Build the Car of the Future.* San Francisco: Sierra Club, 2000.

I

industrial melanism

Industrial melanism refers to the darkening of English pepper moths (*Biston betularia*) as an adaptation to make them invisible to predators on soot-blackened trees associated with the rise of industry. It is one of the best-known and best-documented examples of evolution in progress and has been influential in convincing many people involved in the theological debate over evolution versus creationism that evolution by natural selection is an empirical fact and not just a theoretical speculation.

Darwin himself could point to no such example of evolution in progress and this lack of solid evidence was probably one reason that he delayed publication of his ideas on evolution by natural selection for some 20 years. Late in the 19th century, pepper moths had light wings with dark spots. During the past century, a completely dark variant has evolved because it is less conspicuous resting on a soot-blackened surface than the light-colored type. Since it is better camouflaged, the dark, or melanic, type is less vulnerable to predation by birds and has thus increased in numbers in polluted areas.

This interpretation is not only consistent with the records of entomologists over the decades, it is also verified by experimental work. The release of hundreds of moths has shown that the light type is much more vulnerable to predators in polluted areas, such as Birming-ham, whereas the dark type is more vulnerable in unpolluted regions, such as Dorset. Control of air pollution, a kind of natural experiment, has also been shown to promote a resurgence of the light variety, which becomes less conspicuous as the amount of soot on trees and other surfaces is reduced.

The pepper moth is not an isolated case; over 100 other moth species exhibit industrial melanism. Evolutionary biologists have recorded scores of other compelling examples of evolution in progress from the emergence of drug-resistant bacterial strains to the evolution of beak depth in DARWIN'S FINCHES and the adaptation of body size among giant tortoises in the Galapagos to food availability. Many of these examples show that evolution can be much more rapid than the gradualistic multigeneration process conceived by Darwin. If evolution were really so slow, it would be extremely difficult to observe.

Highly speculative when first formulated, Darwin's theory of evolution by natural selection is on much firmer ground today because of concrete evidence of natural selection at work.

See also CREATIONISM; DARWIN, CHARLES ROBERT; DARWIN'S FINCHES.

Further Reading

Kettlewell, H. B. D. *The Evolution of Melanism.* Oxford: Clarendon, 1973.

Industrial Revolution, the

The Industrial Revolution refers to a broad pattern of economic changes associated with the shift from an agrarian society to an urban one. This transition, which began in 18th-century Britain, and is still happening throughout the world, is associated with greatly increased affluence. Although the increased wealth made possible by industrialization generally improves health and increases longevity, the increased burning of fossil fuels by both industry and modern transportation has resulted in environmental pollution that threatens ecosystems and poses health problems.

Why industrialization first emerged in Britain cannot be easily explained but historians see several factors as being important. They include: population growth; increased efficiency of agriculture; mechanization of traditional handcrafts like cloth production; better transportation infrastructure due to road improvement and canal building; and availability of raw materials, such as coal and iron ore; and a climate of intellectual freedom that promoted inventiveness and social mobility.

Britain's growing population (which nearly doubled in the 18th century) increased the demand for manufactured goods. The increased population was fed by an increase in agricultural production. This was made possible by dividing up common lands into private holdings, which were farmed more intensively using crop rotations that increased productivity.

The increased demand for goods, particularly clothing, led to the abandonment of traditional methods of weaving in favor of a mechanized textiles industry. This change meant that traditional crafts workers were put out of business leading to the violent protests of the LUDDITES who wrecked machinery beginning in Nottinghamshire in 1811 and ending in 1816. The process of mechanization had begun 50 years earlier in the cotton industry. Many of the critical inventions either were developed or became widely used in the 1760s. They included the flying shuttle that allowed broad pieces of cloth to be woven by a single worker and that was invented in 1733 by John Kay. James Hargreaves's spinning jenny (invented around 1765) allowed cotton thread to be produced much more rapidly. Richard Arkwright's water frame was a much heavier spinning machine originally designed to be driven by horses rather than water. The first water-driven cotton mill opened in 1770, at Cromford, and this marked the true beginning of the Industrial Revolution because workers had to be present at the source of power and could no longer work from their homes. JAMES WATT's improved Newcomen steam engine (developed in 1769) was widely used from the 1780s onward, which meant that factories no longer had to rely on the limited power produced by water-driven mill wheels.

These developments allowed for a huge increase in production of cotton goods, mostly for exports. Between 1751 and 1861, British cotton exports increased by a factor of one thousand. During the 19th century, the overall gross national product of the industrialized economy surged 400%.

Industrialization spread slowly across Europe, occurring first in Belgium (1820s), then France (1830s), and Prussia (1840s). An English textile worker, Samuel Slater established the first cotton mill in America, in Rhode Island, before the end of the 18th century.

The social consequences of the Industrial Revolution were initially quite bleak. Factory labor was often both tedious and dangerous, involving long hours of repetitive activities in which people were reduced to the role of functional appendages of the machines they tended. The urban environment in which workers lived was often crowded, with highly polluted air, inadequate sewage systems, and a shortage of water for washing. Such conditions favored the spread of infectious diseases and death rates were high. Such dire conditions, as portrayed in Charles Dickens's novels *Hard Times* and *Our Mutual Friend* were gradually improved through the introduction of public water and sewer systems, the passage of social welfare legislation, and the organization of workers in trade unions to bargain for better conditions. Despite its initial adverse social consequences, industrialization has had the effect of creating unprecedented wealth. This meant that the populations of many countries could continue to grow without experiencing the catastrophic increase in starvation predicted by population theorist THOMAS MALTHUS (1766–1843) whose life coincided with the onset of the Industrial Revolution in England.

See also AIR POLLUTION; BIODIVERSITY AND INDUSTRIALIZATION; GLOBAL WARMING; LUDDITES.

Further Reading
Sterns, P. N. *The Industrial Revolution in World History.* Boulder, Colo.: Westview, 1998.

informed consent

In ethical research, human participation is voluntary. To make a rational decision about whether to participate, subjects must be informed of any potentially harmful or unpleasant consequences they might experience from participation. When they agree to take part in the research after they have been provided with all of the information that might influence their decision, informed consent has been given.

See also NUREMBERG CODE; RIGHTS OF RESEARCH PARTICIPANTS.

institutional review board (IRB)

Institutional review boards (IRBs) are charged with overseeing the ethical quality of research conducted at institutions receiving U.S. government funds. Such ethical quality control formerly rested entirely in the hands of individual researchers, which allowed some notorious lapses to occur such as the TUSKEGEE SYPHILIS STUDY and COLD WAR RADIATION EXPERIMENTS ON HUMAN SUBJECTS. The U.S. government had originally considered regulating all research involving human subjects but it would have been difficult to do so without compromising academic freedom. Ethical regulation by IRBs thus applies only to institutions receiving federal funding for research. Researchers at nonfunded institutions are guided by the ethical code of conduct of their disciplines. In addition, many of these researchers put their research proposals through a voluntary review process designed to ensure ethical conduct of research, protect institutions against legal liability, and educate students in the ethics of research.

The IRB consists of at least five members drawn from nonscientific as well as scientific disciplines. To provide further independence of opinion, at least one member of the IRB must be a member of the local community who has no other ties to the institution. The IRB reviews research proposals to determine if the protocol is ethically acceptable. If not, the proposal can be rejected. Alternatively, the IRB may request changes to bring the research into line with ethical requirements and review the proposal again. IRBs are also responsible for monitoring ongoing research to ensure that it is carried out ethically according to federal regulations.

With a steady increase in the volume of research, IRBs find themselves so overworked in some institutions that they may devote only a minute or two to each study. This has raised reasonable concern about the quality of the review process. Under the federal regulations, expedited review is conducted for research that is identified as MINIMAL RISK (i.e., poses no more risk to subjects than ordinary life) and also in the case of minor modifications to previously approved protocols.

Apart from problems of work overload, there has been some controversy about whether IRBs understand what they are supposed to be doing. Research by social psychologists has investigated the effectiveness of IRBs by conducting an experiment in which research proposals of nine types were reviewed. The proposals either had no ethical violation, were ethically questionable, or contained clear ethical violations. They also varied in terms of the social sensitivity of their hypotheses. Two-thirds of the proposals were socially sensitive, including one-third testing liberal hypotheses and one-third having a politically conservative slant. One-third of the proposals were politically neutral.

Results of this study were quite disturbing because there was little agreement between the various IRBs as to which study was accepted and which was rejected. In the case of politically neutral studies, those with ethical problems were less likely to be passed. Politically sensitive studies were less likely to be accepted than politically neutral ones. What is more important is that in the case of politically sensitive studies, those that were in clear violation of ethical principles were just as likely to be passed as those that had no violation. The disturbing implication of this finding is that IRBs behave rather like censors who screen out studies that are politically sensitive regardless of the ethical properties of their procedures. When there is a distracting influence of political controversy, IRBs completely ignore their primary role as watchdogs protecting the interests of human subjects.

Federally funded research using animal subjects is also subject to ethical oversight but this is the responsibility of an Animal Care and Use committee rather than an IRB. The animal committee not only conducts ethical reviews of experimental protocols but also carries the added responsibility of ensuring that experimental animals are housed and maintained according to very detailed and exacting standards.

See also ANIMAL RESEARCH, ETHICS OF; COLD WAR RADIATION EXPERIMENTS ON HUMAN SUBJECTS; DHHS REGULATIONS FOR THE PROTECTION OF HUMAN SUBJECTS; MINIMAL RISK CRITERION; TUSKEGEE SYPHILIS STUDY.

Further Reading

Ceci, S. J., D. Peters, and J. Plotkin. "Human Subjects Review, Personal Values, and the Regulation of Social Science." *American Psychologist* 40 (1985): 994–1002.

U.S. Department of Health and Human Services. *Code of Federal Regulations Pertaining to the Protection of Human Subjects.* Washington, D.C.: Government Printing Office, 1983.

intellectual property   *See* PATENTS; TRADE SECRETS.

intelligence tests, ethics of

Many people angrily reject the notion that an individual's intelligence can be reduced to a single number such as an IQ score. Others object to the narrowness of the tests in current use that elevate a few academic skills and virtually ignore other abilities, such as the ability to grasp a tune, play sports, or put others at ease in a tense social situation. The greatest ethical issue in the use of intelligence tests in the United States has been their connection to a history of SCIENTIFIC RACISM.

Alfred Binet and others developed intelligence tests in an educational context at the end of the 19th century in

France. He wanted to assess which children were too dull to benefit from a regular public education. Group-administered tests developed in a military context in the United States. The belated entry of the United States into World War I created a personnel crisis for the army. Large numbers of recruits had to be trained for action but before this could be done an officer corps had to be created and a decision had to be made about whether individuals comprised officer material or not. Robert Yerkes, the president of the American Psychological Association, was determined that the fledgling discipline of psychology would make an important contribution to solving this problem.

Yerkes assumed the rank of major and assembled a team of 40 psychologists charged with the task of developing intelligence tests suitable for administering to large groups. The Stanford Binet intelligence test, developed in 1916 by Louis Terman, could be given to only one person at a time and had to be administered by a highly trained tester. Yerkes's group reviewed a number of little used intelligence tests that had been designed for group administration. They were particularly impressed by the test of Arthur S. Otis, which was the first to use the recently devised multiple choice question, one of the most important labor-saving devices of educational testing. Otis's work was used as a model for the army's Alpha and Beta intelligence tests. The Alpha test was the standard version. The Beta was designed for use with people who did not speak English or who were illiterate.

Recruits taking the Beta test received a demonstration or pantomime of what they were to do. Yerkes' team used the test results to make recommendations about which recruits were officer material, which were capable of skilled labor, and which should be discharged as intellectually incapable of military service. Whether these recommendations were ever actually followed is unknown. Even if they were not, the genie was out of the bottle. After the war, a mania for similar testing for civilian personnel recruitment purposes began. Americans found themselves facing batteries of tests that determined their occupational futures. Some disturbing implications emerged from the results of the army Alpha and Beta tests, released in 1921. Recruits were measured according to their mental age, a concept introduced by ALFRED BINET. If a person scores at the same level as a typical 10-year-old child on an intelligence test, then they are said to have a mental age of 10. The average mental age of the draftees was only 13. Since men were drafted primarily on the basis of age criteria, they formed a reasonably representative sample of the American population as a whole. The obvious implication was that the majority of Americans fell below the level of intellectual competence expected of adults (which had been standardized by Terman as a mental age of 16 years). Using a

mental age of 13 years as his cutoff point, Yerkes concluded that 37% of American whites and 89% of blacks were "morons"!

In view of the disappointingly low performance of the population as a whole, it is surprising that the army tests should have been interpreted as confirming the racist biases of the day. Yet the army intelligence tests were used not only to make the case for white racial superiority but also formed the basis for a fine-tuned theory of racial differences that claimed to distinguish between Germans and Italians, English and Poles. Respectable psychologists, such as Robert Yerkes and Henry Goddard, concluded, based on the test results, not only that blacks (with an average mental age of only 10 years) had a lower level of intelligence than whites but also that Russians, Italians, and Poles (all having an average mental age of 11) were racially inferior. Goddard used the test results to form a ranking of different groups according to intellectual ability. Luckily, members of his own group, people of North European ancestry, were at the top of the listing. Members of groups originating from progressively more southerly locales scored progressively worse. Inhabitants from Africa scored at the bottom.

Scientifically flimsy though such racial categorizations may seem today, in their time they had a mesmerizing influence because it seemed that the worst suspicions about Africans and other immigrant groups had been confirmed by the best techniques of scientific measurement. Racist interpretation of the army tests followed a long tradition of biased race science. The results were used for decades in America to support all kinds of appalling political agendas from controlling immigration by race and ethnicity to compulsory sterilization of the mentally retarded population in state institutions to banning interracial marriages and racial segregation of schools. In Europe, they were adapted to the needs of Nazi theorists.

With time, it was noticed that people scored higher on the Stanford Binet than would be predicted from their Alpha test scores. It was clear that the army tests were normed incorrectly.

Such technical problems of intelligence tests has been largely resolved over the decades although the theoretical issues of what constitutes intelligence, or what intelligence tests really measure, are just as puzzling as they have ever been. For example, the 20th century saw a rise in IQ scores of some 20 points in industrialized countries, known as the Flynn effect. This change has not been adequately explained but might reflect increasing complexity of the world in which we live, and increased duration of formal education.

James Herrnstein and Charles Murray (authors of *The Bell Curve,* 1994) presented a more sophisticated modern version of the concerns of Henry Goddard about the idiocy of inferior races and painted a nightmare scenario in which successive generations become more and more idiotic as the stupid out-reproduce the intelligent. Such concerns are based on a pessimistic view of intelligence as being largely determined by genes and ignores a plethora of evidence that intelligence increases with education level, income, nurturant parenting, economic opportunity, and level of economic development in a country.

Intelligence tests can be defended as useful predictors of educational achievement, although many educators believe that school grades should bear more weight in college admissions procedures. The problem with grading is that high schools vary greatly in the rigor of their assessment techniques. Intelligence tests have been largely replaced by aptitude tests, such as the SAT (Scholastic Aptitude Test), that are used to screen students for college admission in the United States. Such tests can be considered as intelligence tests that also assess academic preparation for third-level education. Their use has resulted in disproportionate exclusion of African Americans and others. Although it has often been claimed that the tests discriminated against racial minorities, no convincing scientific evidence has been brought forward to substantiate these claims. Nevertheless, affirmative action quotas were introduced that permitted African Americans and Hispanics to enter college with lower SAT scores than Europeans or Asians. The quota system has been successfully challenged on constitutional grounds in California and affirmative action in college admissions is being abandoned in many other states.

See also MORTON, SAMUEL GEORGE; SCIENTIFIC RACISM.

Further Reading

Flynn, J. R. "Massive I.Q. Gains in 14 Nations: What Do IQ Tests Really Measure?" *Psychological Bulletin* 101 (1987): 171–91.

Hanson F. A. *Testing Testing: Social Consequences of the Examined Life.* Berkeley: UCLA Press, 1993.

Hernstein, J. J., and C. Murray. *The Bell Curve.* New York: Free Press, 1994.

Shipman, P. *The Evolution of Racism.* New York: Simon and Schuster, 1994.

Zenderland, L. *Measuring Minds: Henry Goddard and the Origins of American Intelligence Testing.* New York: Cambridge University Press, 1998.

interactionism

Interactionism is the philosophical position that mind and body are distinct from each other but interact in controlling behavior. It is more compatible with a religious worldview than with modern science, which favors the view that mind and brain are equivalent and that mental experiences are the product of communication between brain cells (MONISM).

See also DESCARTES, RENÉ; MONISM.

Internet and information ethics   *See* ELECTRONIC PRIVACY.

inventions, theories of

An invention consists of the practical working out of an idea in a way that improves subsequent technology. Many people have good ideas for inventions but few succeed in making them a practical reality. Conversely, a person who takes an idea and constructs a device is not considered the inventor unless they were the first to do so. Thus Galileo did not invent the telescope even though he built it without ever having seen the Dutch original.

Historians of science have often been struck by the clustering of inventions in a particular time and place with little innovation occurring in most populations for most of human history. This suggests a situational, or deterministic, explanation of inventions. Study of the life of inventors such as THOMAS EDISON, who produced far more inventions than any individual before or since, suggests that inventors may be unusual geniuses. Such conclusions form the heart of the individualistic theory of inventions.

According to the deterministic theory, inventions occur when the time is ripe. The particular individual who actually comes up with a particular invention is of no consequence. The deterministic theory is supported by the phenomena of simultaneous invention and independent invention. Simultaneous invention occurs when the same device is produced by two or more inventors working independently. For example, Elisha Gray and Alexander Graham Bell developed the telephone simultaneously. Bell submitted a patent application on the same day that Gray filed a caveat (or statement of intent to perfect) for a device to transmit voice.

The frequency of simultaneous invention has given rise to many PATENT disputes. Independent invention occurs when an inventor perfects a device unaware that it has already been produced by someone else. Thus, many inventions developed in the West, including guns and the printing press, had long been in use in China.

The deterministic theory of invention helps us to understand why the pace of technological innovation has picked up in recent centuries. Inventions are highly prized and inventors are rewarded through patent royalties. Current patent laws are modeled largely on the English Statute of Patents and Monopolies (1623). In the past century, scientific research and development laboratories funded by industry have become an important source of inventions, which supports the contextual account of inventions and tends to contradict the individualistic theory.

Edison's career is interesting in this respect since he began life as an individualistic inventor and went on to found a large research laboratory, the first of its kind, in Menlo Park, New Jersey. If the deterministic theory were correct, then Edison's productivity as an inventor should have risen after he had established his state-of-the-art laboratory and hired many talented assistants. Even though the productivity of the whole Edison enterprise may well have increased (and this is controversial) there is little question that his personal productivity went down. Bogged down with the responsibilities of running a large enterprise, he had less time and effort to devote to his personal work as an inventor. Thus, what seems to be a natural experiment testing one theory against the other turns out to be inconclusive. The subsequent proliferation of research and development laboratories has not solved the problem. Although many important inventions have emerged from group laboratories (e.g., nylon from the Dupont lab in 1938 and transistors from Bell Labs in 1948), some of the most important inventions (e.g., television, the jet engine, the ballpoint pen) continue to be made by individuals. In summary, it is clear that some contexts are favorable to innovation. Within those favorable environments, some individuals are much more capable of producing important inventions than others are. Both theories of invention would appear to be partly correct.

See also EDISON, THOMAS ALVA; PATENTS; TECHNOLOGY, HISTORY OF.

Further Reading

Lamb, D., and S. M. Easton. *Multiple Discovery.* Trowbridge, England: Avebury, 1984.
Williams, T. *The History of Invention.* New York: Facts On File, 1987.

IUD (intrauterine device)   *See* CONTRACEPTION.

Johnson, Virginia   *See* MASTERS, WILLIAM.

junk e-mail

Junk, or unsolicited and unwanted, e-mail clutters up in-boxes and wastes the time of employees and individuals in their private life. Also referred to as spam, it is the electronic equivalent of junk mail that differs in two important ways. First, it is much cheaper to disseminate. Second, it can harbor viruses that damage electronic files or cause computer systems to crash. Viruses can be avoided by deleting junk e-mail without opening attachments.

   See also ELECTRONIC PRIVACY; ELECTRONIC VIRUS.

junk science   *See* PSEUDOSCIENCE.

jury selection, scientific

Scientific jury selection uses science to bias the outcome of a trial in favor of a client. Lawyers first use surveys to establish correlations between demographic characteristics (age, sex, education, religion, etc.) of local inhabitants and attitudes to the trial. The survey results allow a profile to be developed of the ideal juror whose attitudes to the trial are favorable to the client (e.g., young men with a college education and no religious preference). During jury selection peremptory challenges are used to exclude jurors that do not fit the profile and the trial outcome is thus more likely to be in favor of the client.

   Scientific jury selection is legal (provided prospective jurors are not contacted) although ethically questionable. Lawyers who use this technique have a remarkable winning percentage but it cannot be determined whether this success is due to the use of surveys or other aspects of an effective representation. To the extent that scientific jury selection is an ethical problem, it is one that is produced by advances in social science research that allow juror decisions to be reliably predicted.

   See also DNA EVIDENCE; EXPERT TESTIMONY; EYEWITNESS TESTIMONY.

K

Kaczynski, Theodore *See* UNABOMBER.

kamikaze

A kamikaze was a member of a Japanese air corps in World War II who was assigned to make a suicidal attack on a U.S. target. The term *kamikaze* is Japanese for "divine wind" and it refers to a typhoon that destroyed a huge armada of invading Mongols in 1281, saving the country from being overrun.

Kamikaze bombers sank more than 30 American ships and damaged several hundred others resulting in a great loss of life. They also had important effects on the morale of combatants. The suicide missions were deployed late in the war, as an act of desperation, and played a significant role in the Japanese defense of Okinawa in April 1945 and the Battle of Leyte Gulf in October 1944.

The technology of the kamikaze attacks comprised a missile delivery system consisting of the vehicle, usually a battered old aircraft, and the guidance system, which was the role of the suicide pilot. The pilot would train his missile on an enemy vessel and fire the missile's rocket engine thereby sending the plane and missile on a steep descent to collide with the target ship. The collision detonated the missile resulting in certain death for the pilot, who had no way of escaping.

The use of kamikaze attacks bears witness to the primitive quality of mechanical guidance systems in 1944. Advances in missile guidance systems have relegated kamikaze attacks to being a footnote in history.

Of course, the deficiency in World War II missile guidance systems does not provide a complete description of the motivation behind kamikaze attacks. Thus, they were never used by the Allies, who rejected such tactics as offensive to Western concepts of freedom and individuality. For the Japanese, the kamikaze fit in with an ancient samurai tradition. The samurai were an aristocratic warrior class who followed an exacting chivalric code. This code called for suicide rather than surrender. Hence the emergence of the kamikaze pilots when the tide of the war turned against the Japanese and defeat and surrender were imminent.

Similar military codes of honor applied among the ancient Greeks. In the Roman Empire they are exemplified by the Jewish defense of Masada against the Romans in the year 70, when the defenders committed suicide rather than surrender after a two-year siege.

All military codes emphasize the necessity for sacrificing one's life, under extreme circumstances, in the name of country and family but the samurai tradition is unusual in asking the individual to give up his life voluntarily. Such codes can be extremely destructive in warfare when combined with modern technology.

See also BIOLOGICAL AND CHEMICAL WARFARE; NUCLEAR WEAPONS.

Kammerer, Paul
(1880–1926)
Austrian
Biologist

Paul Kammerer rose to prominence on the basis of claims that he had conducted experiments on amphibians (salamanders and toads) that demonstrated the genetic propagation of traits that had been acquired during an individual's lifetime. Such experiments it appeared to contradict the prevailing opinion among biologists that Lamarckian inheritance of this type was impossible. Kammerer worked at the Institute for Experimental Biology in Vienna. The evidence was presented in a series of technical papers published between 1904 and 1928 (the last paper being posthumous).

Some of Kammerer's key evidence involved midwife toads that lack the thick pigmented thumb pads characteristic of other toads. Most toads mate in the water and males have thick pads on their feet that they use to cling onto females during mating. Midwife toads normally mate on land and do not have nuptial pads. Kammerer claimed to have induced pads in midwife toads by forcing them to mate in water for several generations. He also claimed that once the pads appeared, they could be passed on to subsequent generations by genetic means. Kammerer's claims received much publicity and were highly controversial. They elicited extreme skepticism among contemporary biologists. In 1926, it was discovered that key evidence in the experiment had been tampered with. Thumb pads on preserved toads had been colored with India ink. There is a striking similarity between the details of this forgery and WILLIAM SUMMERLIN's use of ink to simulate the grafting of black skin onto white mice in his Sloan-Kettering Institute laboratory.

In an account that was sympathetic to Kammerer, writer Arthur Koestler suggests in his book *The Case of the Midwife Toad* that Kammerer's research might have been conducted with integrity. Koestler believes that the pads were real but that the pigment faded with time and that either one of Kammerer's assistants tampered with the evidence to re-create their original appearance or an enemy perpetrated the crude forgery to discredit him. Kammerer himself claimed to have no knowledge of the falsification. His suicide, which quickly followed publication of the fraud, is suggestive of guilt, however.

Even though the true facts of Kammerer's research may never be known, there clearly was scientific fraud and it seems unlikely that Kammerer could have remained unaware of it. What is particularly fascinating about Kammerer is that his earlier career, involving a string of publications in respected scientific journals, looks completely trustworthy. His credibility in the scientific community was reduced, however, by the publication of a 1919 book *Das Gesetz der Serie* (The law of seriality) in which he claimed that coincidences were governed by a natural law that was independent of physical causation. Kammerer's Lamarckian ideas appealed to Russian communists and he had accepted an appointment as professor of biology at Moscow University shortly before his death.

See also BURT, SIR CYRIL; LAMARCK, JEAN-BAPTISTE DE MONET DE; SCIENTIFIC DISHONESTY; SUMMERLIN, WILLIAM.

Further Reading
Koestler, A. *The Case of the Midwife Toad.* New York: Vintage, 1973.

Kant, Immanuel
(1724–1804)
German
Philosopher

Immanuel Kant (see illustration) studied theology, natural science, and philosophy (in that order) at the University of Königsberg, left for nine years to work as a private tutor (1746–55), and spent the rest of his life there after being appointed lecturer in philosophy in 1756. His life in Königsberg was devoted mainly to intellectual pursuits that included physics and mathematics as well as philosophy. In philosophy he made important contributions to ethics (see MORAL PHILOSOPHY) and epistemology. Kantian ethics is built around the concept of the categorical imperative. Thus, actions are moral if the principle on which they are based is desirable as a universal law. Thus it is immoral to steal from other people because we would not wish to live in a society in which no one could be trusted with property. His ethics placed an unusual degree of emphasis on the moral value, and rights, of the individual.

Kant's best-known work, the *Critique of Pure Reason*, which was first published (in German) in 1781, was devoted to epistemology, or the theory of knowledge. According to Kant, our knowledge always derives from two sources: the inputs of our sensory organs and the order (or "categories") that the mind imposes on them, such as ideas of space, time, and causality. It is as though the mind provides spectacles that modify what we perceive and we always look at the world through the same spectacles.

Immanuel Kant in 1791. Kant's ethical philosophy has influenced modern ideas of the rights of individuals. (LIBRARY OF CONGRESS)

Kant felt that his redefinition of the philosophical problem of the relation between the mind and external reality was of great importance. Previous philosophers had conceived the operation of the mind as being constrained by sensory experience but Kant felt that our phenomenal experience is structured by the categories imposed by the mind. Kant is one of the most influential philosophers and the impact of his work is most clearly seen in the writings of other leading German idealist philosophers, including Arthur Schopenhauer (1788–1860) and Georg Hegel (1770–1831), much as these differed from him. Kant was also a key influence for JOHN RAWLS, the 20th century's leading political philosopher. Rawls's influential ideas on political fairness toward the less privileged members of a society are based on his version of Kant's categorical imperative, in which we design a society so that we could be happy living out any possible role in it.

See also MORAL PHILOSOPHY.

Further Reading

Paton, H. J. *The Categorical Imperative: A Study in Kant's Moral Philosophy.* Philadelphia: University of Pennsylvania Press, 1999.

karma

Karma is the Eastern ethical concept that moral and immoral actions have consequences in the future for both the worldly fortune of the individual and their spiritual status.

See also BUDDHIST ETHICS.

Kennedy Institute of Ethics

Established in 1971 at Georgetown University in Washington, D.C., the Joseph P. and Rose F. Kennedy Institute of Ethics focuses on ethical perspectives in public policy. A teaching and research institute, it houses the largest ethics library in the world. It also maintains *BIOETHICS-LINE®*, an online database of medical ethics.

See also HASTINGS CENTER.

Kepler, Johannes

(1571–1630)
German
Astronomer

Johannes Kepler not only advocated the Copernican world system but also worked out laws governing the movement of the planets around the sun. His first book *Mysterium Cosmographicum* (Cosmographic mystery, 1596) defended the heliocentric (or sun-centered) planetary system of Copernicus. It impressed Tycho Brahe (1546–1601) so much that Kepler became his assistant and succeeded him as mathematician to Emperor Rudolf II. Kepler worked in Prague.

Kepler was convinced that God had created a beautifully harmonious universe and he set out to discover this order with a religious zeal. His efforts culminated in the formulation of three laws of planetary motion. The first stated that planets moved in elliptical orbits with the sun as one focus of the ellipse, which was contrary to the Aristotelian insistence on circular perfection in the heavens. According to the second law, as planets orbit around the sun they sweep out equal areas in equal time. The third law stated that the square of the time taken for a complete orbit by a planet was proportional to the cube of its distance from the sun. The first two laws were published in *Astronomia Nova* (New astronomy, 1609) and the third in *Harmonices Mundi* (Harmonies of the world, 1619).

Kepler wrote many well-received books on Copernican astronomy and also contributed to optics, particularly the optics of telescope lenses. A contemporary of Galileo (1564–1641), he wrote a 1610 commentary approving Galileo's novel telescope observations. His *Epitome Astronomiae Copernicae* (Introduction to Copernican astronomy,

1618–21) was one of the most widely read works in contemporary astronomy. Kepler's *Rudolphine Tables* (1627) contained accurate tables of planetary positions.

It is interesting that Kepler's advocacy of the Copernican system did not arouse the same religious controversy as had greeted Galileo's work in a similar vein. The most likely reason for this is that Kepler did not emphasize the potential for conflict between theology and science, as Galileo had done. He was also a Lutheran, whereas Galileo was a Roman Catholic who was expected to follow the authority of the Vatican.

See also COPERNICUS; GALILEO GALILEI; NEWTON, ISAAC; PTOLEMY; SCIENCE, HISTORY OF.

Further Reading

McClellan, J. E., and H, Dorn. *Science and Technology in World History.* Baltimore, Md.: Johns Hopkins University Press, 1999.

Kinsey, Alfred Charles
(1894–1956)
American
Biologist

Alfred Kinsey was America's leading sex researcher (see photo). His scientific career as a biologist initially focused on the gall wasp and the two monographs he published on this subject, as well as a biology textbook, established his scientific reputation. He coauthored the influential Kinsey reports while a professor of zoology at the University of Indiana. This work began when Kinsey was asked to teach a course on marriage and was impressed by the lack of reliable scientific information on the subject. He resolved to collect as much information on human sexual behavior as possible and interviewed people concerning their sexual histories. This work received many grants and, in 1947, he established the Institute for Sex Research at Indiana University. *Sexual Behavior in the Human Male* (1948) by Alfred C. Kinsey, Wardell B. Pomeroy, and Clyde E. Martin was followed by *Sexual Behavior in the Human Female* (1953) by the same authors and Paul H. Gebhard.

These reports were based on interviews with 5,300 white males and 5,490 white females. Sexual practices about which participants were questioned included masturbation, petting, marital and extramarital intercourse, oral sex, and female orgasm. This work provided the first reliable statistical norms for sexual behavior and is still considered authoritative despite procedural criticisms. The work of Kinsey's group met with a storm of ethical protest and had to endure attempts to stop the study or prevent publication of the findings.

One reason for the controversy was the conclusion that violation of the socially accepted norms of sexual

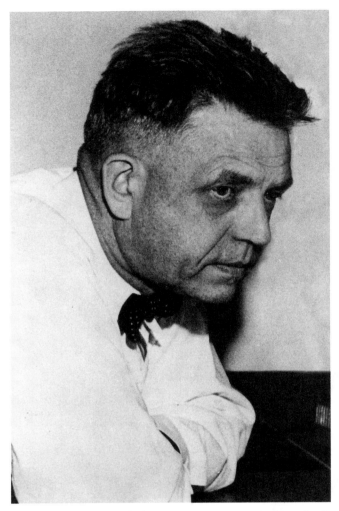

Alfred Kinsey, photographed in 1954, was a pioneer in the study of human sexual behavior whose survey data challenged existing ideas of normal sexuality. (LIBRARY OF CONGRESS)

behavior was extremely common in respect to masturbation, homosexual activity, and extramarital sexual behavior. Kinsey was criticized by Margaret Mead for "allowing" his studies of human sexuality to become best-sellers. Mead argued that by letting young people know just how often sexual behavior fails to confirm to social norms, Kinsey was making it harder for them to resist their nonconforming sexual impulses. Kinsey argued the contrary case that attempts to cover up, or reject, natural sexual impulses caused a great deal of unnecessary psychological distress.

One conclusion from Kinsey's data that surprised many people concerned the frequency of homosexual behavior, which was reported by some 37% of his male respondents. More recent studies have found that the incidence of homosexuality and bisexuality is less than

3% for men. This difference can be explained in terms of the way that questions were framed. Kinsey's numbers involved any sexual experience during the individual's lifetime, including sexual experimentation occurring only during adolescence. Contemporary researchers use the stricter criterion of recent sexual behavior. Another problem with Kinsey's data relates to the problem of sampling. Interviews were conducted with volunteers instead of using a random sample that would be truly representative of the American population. At the time when Kinsey conducted his pioneering research, it is hard to imagine that he would have made much progress by contacting people at random and expecting them to divulge the intimate details of their sex lives to a stranger.

Kinsey is also criticized for seeking out informants whose sex lives were bizarre or deviant because he wanted to make the case that such behavior is extremely common. By showing that such behaviors are common, Kinsey evidently hoped to promote greater tolerance of a variety of sexual behaviors. These criticisms were made in a 1997 biography by James H. Jones. Jones also claims that even though Kinsey maintained a dignified public persona as the leading scientific authority in his field, his private life was different. According to Jones, Kinsey was a homosexual with masochistic tendencies who enjoyed exhibitionism and had other unusual sexual tastes. He created an environment in which sexual experimentation was encouraged and expected his research assistants to serve as subjects in informal studies of sexual responses conducted in the evenings in his own home. Such claims add up to the conclusion that rather than being an objective researcher concerned with unearthing the scientific truth about human sexuality, Kinsey was really a crusader determined to free Americans from sexual repression. Kinsey died in Bloomington hospital of an embolism that resulted from a bruise received while gardening.

See also ELLIS, HAVELOCK; MASTERS, WILLIAM, AND JOHNSON, VIRGINIA; WATSON, JOHN B.

Further Reading

Gathorne-Hardy, J. *Sex the Measure of All Things: A Life of Alfred C. Kinsey.* Bloomington: Indiana University Press, 2000.

Jones, J. H. *Alfred C. Kinsey: A Public/Private Life.* New York: Norton, 1997.

Robinson, P. *The Modernization of Sex: Havelock Ellis, Alfred Kinsey, William Masters, and Virginia Johnson.* New York: Harper and Row, 1976.

Symons, D. *The Evolution of Human Sexuality.* New York: Oxford University Press, 1979.

Krafft-Ebing, Richard von

(1840–1920)
German
Neurologist

Krafft-Ebing performed a brilliant but ethically questionable experiment that produced a breakthrough in the understanding and treatment of syphilis. His research also established that mental disorders result from brain pathology. In 1884, when he performed his experiment, the asylums of Europe were crammed with men suffering from general paresis, the final stage of syphilis.

According to the prevailing view of the medical establishment, general paresis was a product of bad habits. Leading German psychiatrist Wilhelm Griesenger felt that it was due primarily to inhaling the smoke of cheap cigars. A small minority, represented by neurologist Krafft-Ebing, believed that general paresis was the end result of contracting syphilis. This was a controversial idea, since many of those struck down with general paresis claimed never to have had a venereal disease. Moreover, it was believed that some individuals had not been sexually active for decades. Of those who did admit to having syphilis, this was often reported as a minor inconvenience, which produced a sore on the penis that lasted only a matter of days.

Nevertheless, the evidence for a brain disorder of some kind was fairly compelling. For example, the pupils of a general paresis patient did not respond by contracting when a light was shone on them. In addition, the crude autopsy information available indicated that their brains were quite shrunken, suggesting the work of a brain-wasting disease. Krafft-Ebing finally decided to settle the issue by means of a horribly unethical experiment.

Like other streetwise men of his day, he knew that syphilis could only be contracted once. As in the case of measles, it produced skin eruptions sometimes combined with a high fever. In addition to genital sores and high temperature, urination might sting for a few weeks. Then the infected man could indulge in sexual pleasure with the busiest of prostitutes without ever contracting syphilis again.

Suppose that the men with general paresis had previously been infected with syphilis. If so, then reinfection could not occur. Krafft-Ebing obtained infected material by scraping the penile sores of men recently infected with syphilis. This material was injected into nine general paresis patients thereby violating the HIPPOCRATIC OATH by deliberately exposing them to the risk of infection. None developed a penile sore, however. This meant that all must have had syphilis previously and therefore supported the hypothesis that general paresis is an organic disorder that is due to the ravages of syphilis on the brain. This view was subsequently established when *Treponema pallidum* was identified as the bacterium that causes syphilis and was discovered to be present in the brains of general paresis patients. This discovery led to the development of a blood test, and an effective drug treatment, for syphilis. Krafft-Ebing is therefore viewed as a heroic figure in early neurology who expanded the frontiers of what is known about the relationship between insanity, previously seen as a character disorder, and brain pathology, thereby ushering in a medical solution to the epidemic of general paresis.

See also NUREMBERG CODE; RIGHTS OF HUMAN RESEARCH PARTICIPANTS.

Kropotkin, Prince Peter (Pyotr) Alekseyevich
(1842–1921)
Russian
Political Philosopher

Peter Kropotkin was an aristocrat who become well known initially as an explorer of Siberia. He became an anarchist and renounced his title. Kropotkin's iconoclastic ideas landed him in jail for two years (1874–76) which ended with his escape to England. Having been expelled by Switzerland, he was again imprisoned in France, this time for five years (1881–86). Kropotkin then returned to England, where he lived until 1917. Following the Russian Revolution (1917) he was allowed to return to Russia but quickly came to dislike the autocratic Bolshevist government.

Even though Kropotkin was much persecuted during his life, he practiced nonviolence and was targeted mainly for his antiauthoritarian views. His study of animal behavior convinced him that the natural world is characterized more by cooperation between members of a species than by destructive competition. These ideas are presented in his book *Mutual Aid* (1920). Similarly, in the realm of human affairs, he rejected SOCIAL DARWINISM with its claims that ruthless competition was both inevitable and desirable. In Russia, his attacks on social Darwinism were admired and seen as a potent objection to capitalism. Yet, Kropotkin never approved of communism and stuck to his anarchist principles as expounded in a book titled *Ethics* (1920).

See also GROUP SELECTION; LYSENKOISM; SOCIAL DARWINISM.

Kuhn, Thomas Samuel
(1922–1996)
American
Historian of Science

Kuhn's highly influential book *The Structure of Scientific Revolutions* (1962) presented a provocative new theory of progress in science. Instead of the conventional view of science as a cumulatively building enterprise, Kuhn proposed that scientists operate within "paradigms" that are periodically scrapped. Thus, the Ptolemaic worldview was discarded with the emergence of the Copernican system and Kuhn sees this change as a "revolution." The world of Newtonian physics to which the Copernican revolution gave rise was itself subsequently discarded in favor of quantum mechanics and general relativity.

A paradigm is a scientific worldview that consists of shared theories, defining experiments, and methods of collecting data. In short, it is a complete intellectual tool kit that scientists in a particular field bring to their work. Scientists continue to conduct research within their paradigm until they accumulate unexplained facts that are inconsistent with it. At a certain point, the anomalies reach a critical mass and there is a revolt against the old paradigm. This usually happens when a genius comes along who develops a new paradigm that gets around some of the problems of the old one. For example, ALBERT EINSTEIN's theory of relativity overcame problems that could not be explained by Newtonian physics. Although historians of science accept Kuhn's argument that scientific progress is not a straight line, many feel that the word "revolution" is an overstatement. While physicists operated under the Newtonian paradigm, many were aware that it had problems. Moreover, when relativity was published most did not completely abandon Newtonian ideas.

The notion of the scientific paradigm contributes to the modern view that science should be seen as a human activity. Paradigms impose blinkers on scientists constraining the kinds of questions they ask, the types of data they collect, and the experiments they perform. Science is no longer viewed as an entirely objective pursuit but rather as one that suffers from the relativity imposed by the paradigm. A scientific paradigm is essentially a belief system shared by scientists that has a tenuous connection to the real world it is intended to describe.

This relativistic view of the scientific enterprise has made scientists vulnerable to charges of subjective biases in research. Postmodernist philosophers and feminists have often suggested that instead of a high-minded and objective pursuit of knowledge scientists really construct a "narrative" that is subjective and self-serving and that their hypotheses and experiments are really no more than rhetorical devices. Such notions are clearly extreme and forget that the objectivity of science is constantly being reaffirmed by the success with which it is used to solve practical problems.

See also FEMINIST ETHICS AND PHILOSOPHY OF SCIENCE; POSTMODERNISM AND THE PHILOSOPHY OF SCIENCE; SCIENCE, HISTORY OF; SCIENTIFIC METHOD; SOKAL, ALAN.

Further Reading
Kuhn, T. S. *The Structure of Scientific Revolutions*. Chicago: University of Chicago Press, 1962.

L

Lamarck, Jean-Baptiste Pierre Antoine de Monet, chevalier de
(1744–1829)
French
Naturalist

Jean-Baptiste de Lamarck studied both botany and invertebrate zoology and produced distinguished textbooks in each of these fields. After three years in the army and a brief career in banking, he devoted himself to natural history. In 1778, he published *French Flora* and, in the following year, was appointed to run the royal garden in Paris. In 1793, when *Jardin des Plantes* (essentially a museum of natural history) was founded in Paris, Lamarck was put in charge of the invertebrates and lectured on this topic. Forced by failing eyesight to retire in 1818, Lamarck died blind and in poverty. He is best known today for his theory of evolution via inheritance of acquired traits, which used to be considered a rival to Charles Darwin's theory of evolution by natural selection but is generally discredited today. Lamarck believed that animals and plants changed in response to the environment they experienced and that this change could be passed on to offspring. For example, he believed that giraffes developed long necks by craning to reach at leaves on the tops of trees and that the benefit of this experience was passed on to offspring in the form of longer necks.

The notion of inheritance of acquired traits has never been demonstrated to the satisfaction of modern biologists, although PAUL KAMMERER, an Austrian biologist, and Trofim Lysenko, a Russian geneticist, both claimed to have produced such evidence in the 20th century. Kammerer's research was found to involve fraudulent data, however. Lysenko's ideas were accepted by the Communist Party, which initiated a disastrous episode in Russian agriculture and genetics. The resulting crop failures once again demonstrated the error of the Lamarckian theory.

See also DARWIN, CHARLES; KAMMERER, PAUL; LYSENKOISM.

land mines
A land mine is a defensive weapon that is designed to protect territory by exploding beneath enemy troops. Anti-tank mines are detonated by heavy vehicles and smaller antipersonnel mines are set off by the presence of an individual soldier. Antipersonnel mines normally contain no more than 0.5 kg of explosive but this may be enough to kill, or dismember, soldiers. The major ethical problem with the use of land mines is that although they are usually targeted at soldiers, as much as 80% of the casualties they produce are civilians. The injuries may occur long after hostilities have ceased and the victims are likely to be children.

The United Nations estimates that there are some 110 million land mines buried around the world with three countries (Egypt, Iran, and Angola) accounting for almost half of this number. A land-mine ban was approved by 135 countries in Ottawa, Canada, in 1997. Signatory countries undertook not just to plant no new mines but to remove all of their existing mines. Within 15 months of signing the treaty, unfortunately, at least 13 countries had broken the treaty by planting new mine fields. In fact, mines are being planted at a far more rapid rate today than they are being removed. Approximately 26,000 people are killed by land mines each year and a third of this number are women and children.

Some of the world's major countries, including China, the United States, and Russia, refused to sign the Ottawa treaty, which further weakens the ban. U.S. leaders say that land mines are needed to protect the demilitarized zone between North and South Korea. This region was mined at the end of the Korean War as part of an American commitment to protect South Korea from attack by the Communist North.

Further Reading

Winslow, P. C. *Sowing the Dragon's Teeth: Land Mines and the Global Legacy of War.* Boston: Beacon, 1998.

lead poisoning and neurological problems

Lead poisoning in children today is unlikely to be the life-threatening condition that used to be associated with the ingestion of peeling lead paint. Rather, lead from the environment produces chronically high levels in the bloodstream that have an insidious effect on health, especially of the nervous system. The developing nervous system is particularly vulnerable and children who are exposed to low levels of environmental lead may sustain neural damage resulting in permanent cognitive impairment and behavior problems. Lead exposure is a factor in poor school performance, attention deficit disorder, and poor impulse control.

Some 900,000 U.S. children between one and five years of age have dangerously high levels of lead in the bloodstream. In the late 1970s, as many as 88% of children had elevated blood lead levels. With the gradual ENVIRONMENTAL PROTECTION AGENCY (EPA) phaseout of lead additives in gasoline, beginning in 1973 and ending in 1986, the proportion of lead-intoxicated children fell to 4.4% by the early 1990s. (Europe did not ban leaded gasoline until 2000 and it is still used in many of the world's developing countries). Lead has also been eliminated from paints, food cans, and window blinds. Even though lead-based paint was banned in 1971, children living in older homes are still exposed to it.

Lead may enter the air from car exhaust or from industrial sources. Once it gets into the air, it is deposited on soil and in dust. Children are probably more exposed to lead from putting contaminated objects in their mouth than by direct inhalation. Researchers found that children living within 110 feet of a major roadway have higher blood lead levels than those living farther away. Even though cars in the United States no longer emit lead, the toxin has accumulated in the soil around major roadways and is thus a problem in inner cities. African Americans, who remain the main occupants of inner cities, have unusually high levels of lead poisoning. Approximately 12% of black children under five have dangerously elevated lead levels of 30 micrograms per decaliter of blood or more compared to the national incidence of 4%. Poor white children also have significantly higher blood lead, reflecting their residence in undesirable inner-city locations. Household dust, which contains lead from paint, is a major source of contamination and thorough weekly home cleaning can be an effective abatement strategy.

Lead exposure during childhood produces mental retardation, growth reduction, elevated blood pressure, and other health problems. The toxic metal concentrates in the central nervous system and other organs. Its effects on the brain are still poorly understood but lead is believed to take the place of calcium, which plays a key role in the process by which cells release neurotransmitters (the chemicals used by brain cells to communicate with each other).

It is difficult to obtain a reliable estimate of how much lead toxicity affects academic performance and IQ scores because there are so many confounding variables, such as amount of intellectual stimulation available to children in the home. The average IQ deficit associated with a rise in blood lead levels from a low value of 10 micrograms per decaliter to an elevated value of 20 micrograms per decaliter varies in different studies from approximately 1 to 4 IQ points. This means that lead toxicity may play a nontrivial role in explaining socioeconomic and ethnic group differences in IQ scores. To complicate matters further, maternal lead exposure is a factor in low birth weight, which is itself a risk factor for low IQ.

The history of lead additives in gasoline is a depressing story of unethical conduct by industrialists and by the engineers and scientists they employed or funded. In 1921, Thomas Midgley (1889–1944), a mechanical engineer and chemist, discovered that tetraethyl lead (TEL) added to gasoline prevents engine "knock" (i.e., helps the engine to run more smoothly). The commercial development of TEL as a gasoline additive was pioneered by the DuPont Corporation, which opened the first TEL plant in Deepwater, New Jersey, in 1923. In the following year,

General Motors and Standard Oil entered into a partnership to produce their leaded gasoline known as Ethyl. From the beginning of production, workers became ill and some even died of lead poisoning. For example, in 1925, five workers died of lead poisoning at Standard Oil's Bayway TEL plant.

Even though the U.S. public health service warned of the dangers of leaded gasoline as early as 1922, and Ethyl was voluntarily withdrawn from the market between 1925 and 1926, it was fully 50 years before the EPA introduced its phaseout of lead additives in 1973. It had never been necessary to use lead as an additive. Before leaded gasoline had been introduced, alcohol-gasoline blends had been developed that prevented engine knock just as well. These were not exploited commercially because there was less profit in producing them.

One reason that leaded gasoline remained on the market so long is that industrial scientists like Thomas Midgely and Robert Kehoe, a toxicologist employed by the Ethyl Gasoline Corporation were so effective as advocates for the lead additive. Robert Kehoe (1893–1992) perpetrated some extraordinary works of scientific obfuscation designed to prove that leaded gasoline was safe. He advanced the notion that all animals on the planet carry a naturally high lead load. Moreover, he produced evidence that residents of underdeveloped countries had the same lead levels as people living in the U.S. cities who were exposed to the exhaust from leaded gasoline. One of his "control" populations, located in Mexico, turned out to have been exposed to high levels of lead from the manufacture and use of ceramic vessels that were finished with a lead glaze. From his conclusion that high lead levels were "natural" it was a short leap for Kehoe to claim that any further increases from leaded gasoline would not be harmful to health. Yet, he must have known that workers in the additive plants were dying from lead poisoning, a condition that had been known to physicians for two thousand years. Kehoe evidently abandoned the scientific ethic of pursuing the truth and sacrificed his scientific integrity to the commercial interests of his employer.

See also AIR POLLUTION; CLEAN AIR ACT; COMMERCIALISM IN SCIENTIFIC RESEARCH; FORD PINTO.

Further Reading

Kessel, I., J. T. O'Connor, and J. W. Graef. *Getting the Lead Out: The Complete Resource on How to Prevent and Cope with Lead Poisoning.* New York: Perseus, 1997.

Needleman, H. L., ed. *Human Lead Exposure.* Boca Raton, Fla.: CRC Press, 1992.

Nriagu, J. O. "Clair Patterson and Robert Kehoe's Paradigm of 'Show Me the Data' on Environmental Lead Poisoning." *Environmental Research* 78 (1998): 71–78.

Rice, D. C. "Behavioral Effects of Lead." *Environmental Health Perspectives* 104, suppl. 2 (1996): 337–51.

learned helplessness and animal research

Clinical depression is the commonest form of severe mental disorder, afflicting over 10% of the U.S. population at some point in their lives. One way of developing a scientific understanding of depression is to study the phenomenon in nonhuman animals. When experimental animals, such as rats, dogs, or monkeys, are exposed to painful electric shocks that they cannot avoid, they are initially distressed and attempt to escape. After repeated exposure to the painful experience, they become apathetic and stop resisting as though they have learned that they are helpless to control the situation. They lose the ability to learn adaptive responses in new situations and exhibit other signs of depression, such as a loss of interest in food and in members of their own species.

Learned helplessness research is extremely distressing for animal subjects and is rarely performed today because of the ethical questions it raises. Shock generators used to be standard equipment in behavioral laboratories but these are rarely used today because researchers who inflict pain on their animal subjects must provide a clear ethical justification for doing so. Researchers justify such techniques in arguing that the pain inflicted on research animals is balanced by the potential relief of human suffering. Animal rights activists would argue that even if every animal serving in a learned helplessness experiment saved one person from depression, the research would not be justified because the suffering of animals is just as important as that of humans. To assert otherwise is to be guilty of SPECIESISM. In practice, research that causes pain to animal subjects is permitted by scientific research guidelines if it has any reasonable potential for relieving human suffering.

See also ANIMAL RESEARCH, ETHICS OF.

Leary, Timothy Francis
(1920–1996)
American
Clinical Psychologist

Timothy Francis Leary, who is best known as an advocate for the use of psychedelic drugs, was an academic psychologist with appointments at the University of California, Berkeley (1950–55), the Kaiser Foundation (1955–58), and Harvard University (1959–63), before coming to national prominence as a seer of the drug counterculture.

During his Harvard years, Leary began to experiment with psychedelic drugs, including psilocybin and LSD (lysergic acid diethylamide) and collected data on the psychological impact of these drugs. In 1963, he and a collaborator, Richard Alpert, were dismissed from Harvard for using student volunteers in LSD experiments, which was considered an unethical abuse of their position as scientific researchers. Not only had they encouraged illegal behavior but they had also exposed their research participants to unjustifiable risks. Together, they founded the International Foundation for Internal Freedom, whose purpose was to study LSD and promote its use as a "consciousness-expanding" agent. Its Castilia Center, based at a donated estate near Poughkeepsie, New York, continued Leary's research and became a mecca for leading intellectuals of the psychedelic movement, including Aldous Huxley, Allen Ginsburg, William Burroughs, and Thomas Pynchon. Leary authored slogans that summarized the aims of this movement, such as "Turn on, tune in, drop out." To some, he was a visionary prophet who exposed the spiritual barrenness of American life. To others, he was a self-serving corrupter of youth.

From about 1965 to 1976, Leary encountered severe legal problems following frequent charges of drug possession and related crimes. He spent several years in jail punctuated by a spectacular escape to Algeria, where he was granted political asylum, in 1970.

In addition to his messianic attitude to psychedelic drugs, an attitude he never lost despite the fact that some young people, unconnected to him, committed suicide during LSD trips, Leary promoted many other causes during his life. These often united science with broadly religious objectives. He wanted to use the electronic revolution to increase human intelligence. He believed science could be applied to extending the human lifespan and making people happier. He was even interested in setting up space colonies and formed a cooperative (known as Starseed) for this purpose. Critics have often dismissed his enthusiasms as vehicles for cynical self-promotion. Much that Leary said and did was apparently designed to shock the middle-class America that he professed to hate. It is thus hardly surprising that he very quickly exceeded the bounds of scientific ethics at Harvard and moved on to the more colorful theater of life outside the academy.

See also PSYCHEDELIC DRUG RESEARCH.

Further Reading

Leary, T., R. Metzer, and R. Alpert. *The Psychedelic Experience: A Manual Based on the Tibetan Book of the Dead.* Guelph, Ontario: University Books, 1964.

Lemon Law

A lemon is a defective item, such as a car, so named because it leaves a bitter taste in the mouth, figuratively speaking. In the past, individuals who purchased new vehicles and discovered that they had irreparable flaws had little legal recourse apart from suing the car dealership, or the car manufacturer, or both. California's 1982 Lemon Law was a landmark in consumer protection legislation because it entitled owners to a full or partial refund for a car or truck that qualified as a lemon. A lemon was defined in the law as a car that in its first year (or first 12,000 miles) either spent 30 days being repaired or had to be fixed four times due to the same problem. About 5,000 vehicles per year satisfied this definition and were bought back by car companies.

California's Lemon Law was copied by most other states. Although laws vary from state to state, most agree that a lemon is a vehicle that, in its first year, has been off the road for 30 days and cannot be fixed after four attempts by a manufacturer-authorized facility. States vary in the period covered (from one to two years), out-of-service days (15–30), mileage covered (12,000–18,000 miles), vehicle classes covered (cars, trucks, motorcycles, etc.), whether the vehicles can be secondhand, and classes of owners (private only or small businesses also).

Lemon laws require aggrieved owners to notify the dealer of the problem and provide an opportunity for the defect to be fixed. If this fails, the owner is obliged to enter a mediation process sponsored by the vehicle manufacturing industry. If arbitration fails, as happens in a small minority of cases, legal action may be taken against the dealers and manufacturers. Such cases rarely come to trial, most being settled out of court to the satisfaction of the consumer.

The lemon laws mark an important break with the *caveat emptor* (let the buyer beware) tradition in business law according to which the purchaser of an item with a problem assumed full responsibility for the problem at the time of purchase. Previous legislation regulating the vehicle manufacturing industry had focused on safety issues and allowed purchasers to assume full responsibility for purchasing a lemon. Now that car manufacturers are being legally compelled to pay a steep price for putting defective vehicles on the roads, they have an economic incentive for improving quality control and reducing compensation paid to lemon owners. Vehicles that are merely mechanically defective do not generate the intense interest in ENGINEERING ETHICS as do those that are dangerous and cause deaths on the road; still, they are an ethical issue because they reduce the quality of life for a large number of people and because they evoke deep feelings of injustice. Lemon laws constitute a response to each of these issues.

See also ENGINEERING ETHICS; FORD PINTO.

Further Reading

Freedman, E. "Virginia Court Says Lemon Law Covers Used Vehicles." *Automotive News,* July 27, 1998.

Mouchard, A. "California's Lemon Law Could Be in for a Major Overhaul." *Knight-Ridder/Tribune Business News*, April 23, 1996.

"These Laws Put the Squeeze on Lemons." *Changing Times* 40, no. 1 (1986): 57–60.

life support, termination of

The development of many artificial life support systems that extend life beyond the point where it would have naturally ceased has created a whole category of new and difficult ethical problems. In the case of a mentally competent patient, is the decision to forgo, or to withdraw, life support different from committing suicide? For mentally incompetent patients, how can the decision by a physician to turn off life support systems be distinguished from homicide?

Life support systems have necessitated a revision of the concept of death. In the past, death was defined by the cessation of heartbeat and breathing but the perfection of heart-lung machines allowed life to continue after these vital signs had stopped. Death is now defined in terms of cessation of brain function. Even this definition can be problematic because the loss of all cortical function can be accompanied by maintenance of neural function by the respiratory control systems at the base of the brain. The person is alive in a technical sense but has little or no prospect of retaining consciousness.

Patients who are in a persistent coma and are kept alive through artificial life support present a special ethical problem. Although there have been exceptional cases of individuals recovering from a coma after several months, or even years, physicians are generally justified in assuming that there is no prospect of full recovery after months lying in a coma or persistent vegetative state (in which the eyes periodically open and there is some evidence of primitive responsiveness to stimulation). In 1976, the New Jersey Supreme Court helped to clarify the situation of physicians contemplating termination of life support by ruling that doctors could unplug a respirator that was keeping a comatose patient, Karen Ann Quinlan, alive. The rationale given was that since there was no reasonable hope of recovery, the respirator was preventing the patient from dying with decency and dignity.

This decision was followed by the passing of "right-to-die" laws by many U.S. states. In 1990, the U.S. Supreme Court ruled that people have a constitutional right to have life support discontinued if that is their wish. This allowed the family of 33-year-old comatose Missouri woman Nancy Cruzan to have the feeding tube removed that was keeping her alive. In the case of comatose patients, their previously expressed wishes are to be respected. If they have not made their wishes known, or if they are mentally incompetent, physicians may consult with surrogates such as close relatives, spouses, friends, and legal guardians. The Supreme Court ruled that the request of family members to have life support terminated for a comatose patient can be overridden by the state in cases where the person left no clear wishes. This ruling encouraged the making of living wills in which people clarify what should be done in such situations.

However the decisions are made, it is clear that there is a strong trend toward allowing patients to die in intensive care units (ICUs). According to the results of one study conducted at two different ICUs, 51% of patients died following a decision to withhold, or withdraw, life support in 1987–88. By 1992–93, this number had increased to 90%, indicating that the overwhelming majority of deaths at ICUs follow decisions to limit therapy. Most of these decisions are based on a consensus between physicians and patients or their surrogates, although physicians may sometimes refuse requests from surrogates to withdraw life support.

Another important practical problem concerns which life support systems are to be terminated. If several systems are involved, what is the order in which they should be removed? A survey of American internists revealed a pattern in the preferred form of life support to be withdrawn in the case of critically ill patients for whom the decision to withdraw life support has been made. The physicians generally preferred to withdraw forms of support that produced speedy rather than lingering death. Their preferences, ranked from most likely to least likely, were: blood, dialysis, intravenous vasopressors, total parenteral nutrition, antibiotics, mechanical ventilation, feeding tubes, and intravenous fluids.

Withdrawal of life support is difficult for many physicians because it violates the HIPPOCRATIC OATH that requires prolonging life. The legal gloss on this situation is that discontinuing, or withholding, life support does not cause death. Instead, it allows the disease, or injury, that is the fundamental cause of death to have its effect. Such legal language is necessary to protect physicians from charges of assisting suicide or homicide. Yet, it does not satisfy most ethicists who tend to evaluate actions in terms of intentions.

There is no clear ethical distinction between an individual causing his or her own death by refusing life support and an individual causing his or her death by refusing food. In fact, there are situations, as in the Nancy Cruzan case, in which the critical support system being withdrawn was food, albeit delivered via intranasal tubes. When food is delivered intranasally, it is categorized as a medical treatment, which allows this form of life support to be terminated without inadvertently legalizing the starvation of patients in hospitals and nursing homes. Such legal distinctions may be necessary to resolve many of the practical problems surrounding the termination of life support but it is important to realize that they have little basis in ethics.

See also EUTHANASIA; HIPPOCRATIC OATH.

Further Reading
Brock, D. W. "Death and Dying." In *Medical ethics,* ed. R. M. Veatch, 363–94. Sudbury, Mass.: Jones and Bartlett, 1997.
Christakis, N. A., and D. A. Asch. "Biases in How Physicians Choose to Withdraw Life Support." *The Lancet* 342 (September, 1993): 642–47.
Prendergast, T. J., and Luce, J. M. "Increasing incidence of withdrawal of life support from the critically ill". *JAMA 279*: 814, 1997.
"Withdrawal of life-support from patients in a persistent vegetative state". *The Lancet* 337 (January): 96–99, 1991.

lifespan enhancement research

Animal research has begun to illuminate the biological mysteries of lifespan and aging. There are several good reasons for thinking that lifespans can be enhanced and that people in the future will live much longer than they do today. This prospect raises a number of practical and ethical problems that we are currently not well equipped to solve, such as when, if ever, people should retire, when they should stop having children, who should carry the burden of increased healthcare costs associated with an aging population, and so on.

Selective breeding experiments using fruit flies and other species have shown that it is possible to increase average lifespan by allowing only long-lived individuals to reproduce. A different line of research has shown that melatonin acts as an elixir in mice, allowing them to live about 50% longer than normal without any signs of aging or ill health. Melatonin is produced by the pineal gland and old mice given the pineals of younger ones have lived twice as long as normal. Whether this hormone will have the same effects for humans is unknown but it is reasonable to look at the pineal gland and its hormones as a possible means of enhancing lifespan.

Scientists have also developed a good understanding of aging at the cellular level. Normal cells can divide only about 50 times because of the loss of DNA from the telomeres at each division. Genetically engineered cells that express the enzyme telomerase that prevents telomeres from fraying can divide more than 100 times before succumbing to senescence.

Another promising lead is provided by evidence that mice consuming a low-calorie diet live much longer and enjoy better health than those eating as much as they wanted. While this effect is so far unexplained, it is consistent with research on people with exceptionally long lifespans. Given the number of promising leads available

to aging researchers, it seems almost inevitable that the future will bring important revelations concerning human aging mechanisms and that this will lead to the development of manipulations capable of doubling our current lifespans.

If people routinely lived to be 120 years old, then it seems obvious that they could not retire at the age of 60 years because the burden on any such retirement system would be too great as it would have far more beneficiaries than contributors. As heavier-than-average users of healthcare services, the large number of elderly people would also create an enormous strain on healthcare delivery systems. Longer lifespans would also have many interesting social consequences with children often reaching senescence at approximately the same time as their parents. There would thus be a great lag in the cross-generational transmission of wealth. Some elderly people would become extremely wealthy due to compounding over time whereas others, perhaps those having heavy medical expenses, might accumulate huge burdens of debt. These phenomena, already present as trends, would become more pronounced with further increases in lifespan.

See also ORGAN TRANSPLANTATION.

Further Reading
Bova, B. *Immortality: How Science Is Extending Your Life Span and Changing the World.* New York: Avon, 2000.

Linnaeus, Carolus (Carl von Linné)
(1707–1778)
Swedish
Naturalist

As a young man, Carolus Linnaeus embarked on a 4,000-mile journey around Lapland during which he discovered many plants unknown to science. He went on to establish a career as professor of botany at Uppsala. Linnaeus was interested in classification and divided up plant species according to their number of pistils and stamens (female and male reproductive structures, respectively). He also devised the binomial nomenclature system in biology that identifies each organism in terms of a pair of Latin names. The first name gives the genus, or category, in which the species fits and the second gives the individual species (e.g., Canis *lupus*, the wolf; Canis *familiaris*, the dog). His classification work was inspired by a religious desire to celebrate God's creation (see NATURAL THEOLOGY).

Initially, a classification system for plants, the mature form of Linnaeus's system, as contained in his book *Systema Naturae* (1758), contained over 4,000 animals. Humans were listed as *Homo sapiens*. Linnaeus was knighted by the Swedish Government in 1758 in recognition of his contributions to natural science, and he promptly changed his name to Carl von Linne to reflect his new status. It is difficult to exaggerate the importance of Linnaeus's classification system for stimulating research in the field. It was to biology what the periodic table was to chemistry. After Linnaeus's death in 1778, following a stroke in 1776, his personal collections of plant specimens and books passed to an English biological society that subsequently took the name *The Linnaean Society.*

See also NATURAL THEOLOGY; RELIGIOUS IDEOLOGY AND SCIENCE; SCIENCE, HISTORY OF.

Little Albert and fear conditioning   *See* WATSON, JOHN B.

Locke, John
(1632–1704)
English
Philosopher

John Locke was a key figure in the ENLIGHTENMENT in England and France. His works promote the Enlightenment values of reason, learning, and political freedom. He was interested in scientific investigations and collaborated with his friend ROBERT BOYLE, a leading pioneer of modern chemistry.

Locke's philosophy of knowledge was very much a product of the scientific age in which he lived with its core belief in the importance of sensory data rather than the authority of traditional ideas. In *An Essay Concerning Human Understanding* (1690), he made the case that all knowledge is acquired through the senses and used the metaphor of a young mind being a blank slate, or *tabula rasa,* on which the lessons of experience are written. Locke's philosophy of ideas involved the formation of associations, or connections, among sensory impressions. Complex ideas consisted of connections being made between simpler ones. Associationism proved to be extremely influential both in philosophy and in 20th-century associationist psychology, or behaviorism.

Locke was influential as a political philosopher and is considered to be the father of political liberalism. He articulated the view that the purpose of government is not to control people but rather to create conditions in which they are free to pursue happiness. He assumed that humans were basically good and that the pursuit of individual happiness would promote the common good. Government was a contract between the ruler and the ruled

that could be dissolved by the latter if they were dissatisfied. This optimistic perspective stands in marked contrast to the views of English philosopher THOMAS HOBBES (1588–1679) who believed that people had evil impulses that needed to be restrained by the absolute power of government. Locke was consulted by the leading English liberal politicians of his day and his political philosophy greatly influenced the drafting of the U.S. Constitution. He died at Oates in Sussex in 1704.

See also ENLIGHTENMENT, THE; HOBBES, THOMAS.

logical positivism   *See* POSITIVISM.

Lorentz, Hendrik Antoon
(1853–1928)
Dutch
Theoretical Physicist

Hendrik Antoon Lorentz was professor of mathematical physics at Leiden from 1877. He shared the 1902 Nobel Prize in physics with Pieter Zeeman for work on the theory of electromagnetic radiation. He also anticipated some basic concepts in the theory of relativity that are often attributed to ALBERT EINSTEIN. In particular, he stated that motion has the effect of compressing molecules along the direction of movement and observed that this contraction could not be detected since any ruler used to measure the length of an object would be similarly compressed. Lorentz died in Haarlem, Netherlands, on February 4, 1928.

Lorentz was himself anticipated by Irish physicist GEORGE FRANCIS FITZGERALD (1851–1901) and this is accurately reflected in the term "Fitzgerald-Lorentz contraction" that is used today to describe the hypothetical phenomenon. Even though many scientists may entertain similar ideas, the retrospective process of assigning priority is often quite selective. There are two good reasons why this should be so. First, it is easier to remember a theory, or invention, as the product of a single mind rather than giving partial credit to many individuals. Second, once a scientific idea is "in the air" some scientists work harder at developing it than others do and the fruits of their labor may bring a disproportionate share of the credit for originality. This would explain why Einstein, rather than Lorentz (or Fitzgerald), is given the credit for the theory of relativity. Similarly, CHARLES DARWIN, rather than Alfred Wallace, is remembered as the author of the theory of evolution by natural selection even though Wallace worked out a theory independently that was almost identical to Darwin's down to the terminology used. Some historians of science believe that such MULTI-PLE DISCOVERY is the rule rather than the exception as the well-prepared minds of many scientists contemplate the current state of knowledge and conceive of similar advances.

See also EINSTEIN, ALBERT; DARWIN/WALLACE AND CODISCOVERY OF EVOLUTION; FITZGERALD, GEORGE FRANCIS; HAWKING, STEPHEN; MULTIPLE DISCOVERY.

Further Reading
Lamb, D., and S. M. Easton. *Multiple Discovery*. Trowbridge, England: Avebury, 1984.

Lorenz, Konrad
(1903–1989)
Austrian
Animal Behaviorist

Konrad Lorenz was a pioneer in the study of animal behavior from an evolutionary perspective, or *ethology*. He shared the 1973 Nobel Prize for physiology with Austrian zoologist Karl von Frisch (1886–1992) and Dutch zoologist Nikolaas Tinbergen (1907–88). Lorenz's extrapolation of the principles of animal behavior to human beings has been highly controversial.

Initially pursuing a career in medicine, Lorenz received an M.D. from the University of Vienna in 1928. Medicine gave way to an interest in animal behavior and he obtained a Ph.D. in zoology in 1933. Lorenz loved animals and began to conduct detailed observations of pets such as jackdaws, ducks, and geese. These observations formed the basis of scientific reports as well as popular books such as *King Solomon's Ring* (1952).

Primarily interested in species-typical behavior, Lorenz showed that closely related species had similar behavior patterns just as they had similar bodily traits. This derives from the action of natural selection, which produces adaptive patterns of behavior just as it shapes morphology and physiology. He was also interested in how natural selection shapes learning, and he studied imprinting, the process by which young animals, like ducklings, learn to follow their mothers. The survival value of such learning is obvious because the mother protects chicks from predators, keeps them warm, and helps them to feed.

In 1937, Lorenz became editor of the new journal *Zeitschrift für Tierpsychologie*, which became a leading publication in the study of animal behavior from an evolutionary perspective. In the same year, he was appointed lecturer in comparative anatomy and animal psychology at the University of Vienna. His academic work was interrupted by World War II in which he served as a physician

in the German army. Following the war, he began a long association with the Max Planck Institute (1950–73) in which he served as a director of the Seewiesen Institute for Behavior Physiology (1961–73).

Lorenz developed a drive theory to explain why particular behavior sequences get switched on at a particular time. This theory has been elaborated by his collaborator Nikolaas Tinbergen in studies of herring gulls. Lorenz applied drive theory to human aggression with indifferent results. He concluded that humans have an inborn aggressive drive and that if they do not have channels for allowing this aggression to run off harmlessly, such as playing sports, they will engage in destructive aggression, including warfare. This analysis has two major flaws. The first is that expressing aggression in any form does not reduce anger, as Lorenz would predict, but actually increases it. The second is that there is a very tenuous connection between individual aggressiveness and warfare. Most soldiers at the battlefront are frightened rather than angry. Lorenz had limited credentials in psychology and these, arguably irresponsible, claims about human aggression were made in a popular work (*On Aggression*, 1966).

Throughout his academic career, Lorenz maintained a fascinating feud with American animal behaviorists, which stood in marked contrast to the dispassionate pursuit of knowledge that many people expect of scientists. The academic basis of this dispute was the feeling of Lorenz that American psychologists paid too little attention to evolution and genetic influences on behavior and the equally strong feeling of American comparative psychologists that Lorenz and his allies paid far too little attention to environmental influences, such as reinforcement and classical conditioning. Some idea of the strength of Lorenz's feelings on this subject can be obtained from his book *Evolution and Modification of Behavior* (1965). Modern psychology has moved beyond this nature versus nurture debate by accepting that both influences are important.

See also EVOLUTIONARY PSYCHOLOGY AND SEX ROLES; RESEARCH ETHICS; VIDEO GAMES AND VIOLENCE.

Further Reading
Lorenz, K. *Evolution and Modification of Behavior.* Chicago: University of Chicago Press, 1965.
———. *King Solomon's Ring.* New York: Crowell, 1952.
———. *On Aggression.* New York: Harcourt Brace Jovanovich, 1966.

Love Canal
Love Canal, in Niagara Falls, New York, is the site of one of the most notorious toxic chemical dumps in the country and marks a turning point in American governmental

regulation of the chemicals industry. Although Love Canal is often held up as an example of the harm that can be caused to people by irresponsible dumping of toxins, Hooker Chemical Company, which dumped most of the hazardous materials, neither broke any laws nor did anything that can be considered blatantly unethical.

THE UNFOLDING OF AN ECOLOGICAL TRAGEDY

Love Canal was dug in the first decade of the 20th century as part of a dream community planned by industrialist William T. Love. Love's idea was to dig a canal between the upper and lower Niagara River that could be used to navigate around the falls and provide inexpensive electric power for the homes and industries located in his planned community that would take advantage of the beautiful local scenery. Love's dream died due to an economic slowdown and the discovery by Nikola Tesla (1856–1943) of a method to transmit electric current over long distances using alternating current. This meant that housing could benefit from electric power without being built close to a generator. By 1910, the partially dug canal had been abandoned as Love decided to cut his losses. In the 1920s, it was used as a municipal and industrial chemical dump site. From the 1930s to 1952, the site was owned by Hooker Chemical Company, which gave it to the city of Niagara Falls in 1953. Over the years, Hooker dumped approximately 20,000 tons of chemicals contained in metal drums into the abandoned canal. The drums contained waste solvents, process slurries, and pesticide residues. In 1953, Hooker covered the site with an impermeable clay cap, designed to prevent chemicals from leaking to the surface, as the law of the day required. The site was then sold by Hooker to the Niagara Falls School Board for one dollar. When Hooker deeded the land to the city, the document specifically admonished that the clay cap should never be pierced, no structures were to be built on the site, and the surface was not to be disturbed by digging.

Sadly, these warnings were ignored because, during the postwar baby boom, growing demand for housing confronted a scarcity of building land. In the late 1950s, approximately 100 homes and a school were constructed at the dump site. By the time the issue of the site's toxicity arose, in the late 1970s, over 200 homes had been built in the vicinity of the landfill. The nature of the environmental catastrophe was revealed in 1978 when unusually heavy rains seeped into the corroding chemical drums, which caused their contents to percolate to the surface in a toxic brew. Puddles of poisonous chemicals were seen in the yards and basements of many of the homes and the school. Vegetation around homes turned black and died. Rusted chemical drums poked through the surface and the air had a faint but pungent odor. The dump is believed to have contained at least 82 different compounds, 11 of them suspected to be carcinogens, including DIOXIN and benzene.

HEALTH CONSEQUENCES

At the same time that the presence of the chemicals became obvious, disturbing health problems began to show up. Children returned from play with chemical burns on their hands and feet. Residents complained of dizziness and difficulty in breathing that they attributed to chemical fumes. Birth defects (such as cleft palates and an extra row of teeth) were common. Moreover, there was an unusually high rate of miscarriages. A large number of residents were found to have high white blood–cell counts, a possible harbinger of leukemia. Beginning in May 1978, the *Niagara Gazette* collected the stories of individuals who blamed their illnesses on exposure to the chemicals and this led to national media coverage.

New York State Health Commissioner Dr. Robert Whalen declared a health emergency in August 1978. Families were ordered to relocate and the state agreed to purchase the chemically contaminated homes. In all, 238 families moved and their homes were bought by the state. In 1980, a second evacuation included 792 families from the periphery of the original danger zone. This action was provoked by a study, conducted by the Biogenic Corporation of Houston, indicating chromosomal damage. This finding could not be replicated by the U.S. Department of Health and Human Services (DHHS) in a follow-up study and seems to have been in error. The second evacuation from a much less contaminated area was thus probably unnecessary. Follow-up studies sponsored by the DHHS supported claims of health problems, particularly for infants whose gestation had occurred at Love Canal. They were twice as likely to have low birth weight, and elevated rates of birth defects were found in 19 of 174 children studied. Cancer registry data did not find elevated rates of liver cancer, lymphoma, or leukemia in a census tract containing the Love Canal area. Lung cancer rates were higher (25 cases observed versus 15 predicted) but this could have been due to high air pollution levels rather than to breathing volatile chemicals from the dump site.

LEGISLATIVE RESPONSE AND LEGAL LIABILITY

Public outcry over the Love Canal catastrophe resulted in the enactment of legislation designed to prevent similar problems in future. Congress passed the Comprehensive Environmental Response Compensation and Liability Act (CERCLA) in 1980. This legislation empowered the Environmental Protection Agency (EPA) to prevent emission of hazardous substances into the environment and to identify and clean up all old toxic waste sites. One of the most remarkable aspects of this legislation was that the chemical companies were required to pay for the cleanup.

Business leaders saw CERCLA as unfair to the chemicals industry because companies who would be held liable under the act had not necessarily broken any laws. Moreover, they felt that government should contribute as much as businesses to clean up toxic dump sites.

The legal side of liability for the cleanup in Love Canal was convoluted. Hooker Chemical had been acquired by OxyChem, a subsidiary of Occidental Chemical. In 1996, after a 17-year legal battle over who should pay to clean up Love Canal, Occidental agreed to pay $129 million to the federal government to cover the EPA's cleanup costs for the site. This followed a 1994 settlement with New York State for $98 million. Residents received $20 million in damages, which was paid by the chemical company and the city of Niagara Falls. Legal closure came in 1999 when the last four cases were settled at a further cost of $7.1 million.

Following a major cleanup effort, state authorities declared that two-thirds of the chemically contaminated area of Love Canal was safe for repopulation and set aside the remainder for commercial use. By 1999, 250 families had resettled the area. In a deliberate attempt to distance the community from its tragic history, it has been renamed Black Creek Village.

Love Canal occupies a unique position in the history of HAZARDOUS WASTE management for several reasons. It is one of the most alarming and shocking stories of domestic chemical contamination encountered by Americans, who were exposed to a great deal of sensational media coverage of the issue. From that perspective, it is the story that more than any other has sensitized people to the necessity for safe disposal of hazardous materials. The really unethical decision was not to have dumped chemicals at Love Canal but rather to have built over the landfill. This distinction was often lost in journalistic coverage of the event, which demonized chemical companies, possibly because it is natural to associate the harmful effects of chemicals with the companies that produce them. As toxic polluters, the chemical companies have had more than their fair share of adverse publicity, which has succeeded in convincing many executives of the advisability of working harder to prevent environmental damage. If chemical companies do not take more responsibility for toxic emissions, they are inviting ever more restrictive governmental regulation. This philosophy was formally adopted by the Chemical Manufacturer's Association in 1985 (see BHOPAL, INDIA, CHEMICAL ACCIDENT).

Perhaps the most important practical consequence of the Love Canal fiasco was the deliberate attempt to avert the same sort of tragedy from occurring in other communities by aggressive EPA efforts to clean up all hazardous dump sites. In this context, holding chemical companies responsible for the damaging consequences of their dumping activities, even when these activities were per-fectly legal, is an important departure insofar as protecting the environment from pollution is concerned. This precedent may go beyond strictly environmental issues. For example, a conceptually similar ethical issue is involved in asking tobacco companies to pay for the cost of medical care attributable to use of their product. Love Canal not only awakened the U.S. government and American citizens to the problem of toxic chemical waste but also stimulated vigorous remedial actions.

See also BHOPAL, INDIA CHEMICAL ACCIDENT; DIOXIN; ENVIRONMENTAL ETHICS; HAZARDOUS WASTE; TOBACCO COMPANY LITIGATION.

Further Reading
"Love Canal." *Mother Jones* 24, no. 5 (1999): 4–9.
"Love Canal Closure." *Chemical Week,* July 28, 1999.
"Love Canal Settlement Highlights Superfund Debate." *Chemical Marketing Reporter* 249, no. 1 (1996): 3–4.
Newton, L. H., and C. R. Dillingham. *Watersheds 2: Ten Cases in Environmental Ethics.* Belmont, Calif.: Wadsworth, 1997.
Stringer, J. "OxyChem Fights for a Clean Name." *Chemical Week,* May 29, 1996.

Luddites

The Luddites were a group of English craftsmen who protested against mechanization of the textiles industry by destroying the machines that were displacing them. Possibly named after Ludlum, a character of legend who destroyed a knitting frame to spite his father, the Luddites signed their proclamations "King Ludd," "General Ludd," or "Ned Ludd." The movement began in the lace and hosiery industries in Nottinghamshire in 1811 and spread to the cotton and woolen industries in Lancashire and Yorkshire in 1812. The Luddites avoided violence against persons but a number of protesters were shot down in 1812 by an employer named Horsfall whose factory was being threatened. Horsfall's subsequent murder initiated harsh government repression. Fourteen Luddites were hanged in York in 1813. After some scattered episodes of violence, the Luddite movement finally disappeared in 1816.

The motivation of the Luddites was primarily economic. They disliked mechanization because it was ruining their livelihoods. Unlike antitechnological movements of the 20th century, there was no ideological objection to technological advances as such. The attack on industrial plants has parallels with the activities of GREENPEACE today although that organization is attempting to prevent damage to the environment resulting from industrial pollution rather than protesting economic conditions. An ideological objection to electronic advances was what motivated the violence against people perpetrated by the UNABOMBER. Although most modern environmentalists disapprove of the Unabomber's violent methods, many share his view that technological development erodes the quality of life.

See also GREENPEACE; INDUSTRIAL REVOLUTION; UNABOMBER, THE.

Lyell, Sir Charles
(1797–1875)
English
Geologist

Charles Lyell was the most influential geologist of the 19th century. His three-volume text *Principles of Geology* defined the field for contemporaries. Lyell rejected the theory of CATASTROPHISM, which attempted to explain away the richness of the fossil record of animal life by speculating that there had been numerous mass extinctions followed by the emergence of new species. Catastrophism was an attempt to fit the great diversity of geological phenomena, and the richness of the fossil record of animal life, into the limited time period prescribed by the age of the earth as represented in the Bible. Lyell's alternative, known as uniformitarianism, posited that geological time was much longer than previously supposed and that with sufficient time all of the observed geological phenomena could have been produced by gradual change. CHARLES DARWIN was greatly influenced by Lyell's work, which he read before departing on the *Beagle* voyage and which prepared him to think of organic evolution in terms of the accumulation of gradual changes. Upon Darwin's return the two men became friends and correspondents and Lyell played a key role in arranging for the simultaneous publication of the evolutionary theories of Darwin and Alfred Russell Wallace. Lyell died in London on February 22, 1875, after months of failing health. Lyell's many honors included knighthood and baronetcy. He received the Copley medal and the Royal Medal of the ROYAL SOCIETY. In 1864, he was elected president of the British Association for the Advancement of Science.

See also CATASTROPHISM; DARWIN, CHARLES; DARWIN/WALLACE AND CODISCOVERY OF EVOLUTION.

Lysenkoism

Lysenkoism is the name applied to Russian genetics, which was conducted according to approved Soviet ideology under the leadership of T. D. Lysenko (see photo composite). It is often held up as an example of the dangers to science of being too closely tied to politics. In that sense, Lysenkoism resembles accounts of the deleterious effects of the Nazi movement on German science. Lysenkoism was influential in the Soviet Union from the 1930s until the 1960s.

T. D. Lysenko was a Lamarckian. Contrary to the received opinion of the scientifically orthodox Mendelian genetics of his day, he believed that acquired characteristics are inherited. If true, this meant, for example, that a plant that had adjusted to growing in a dry environment during its lifetime could pass on the acquired drought resistance to its progeny. According to Mendelian genetics, plants can pass on drought resistance only if their genes encode for drought resistant traits to begin with.

The quarrel between Soviet communism and Mendelian genetics is founded on the socialist view that Mendelian genetics offers too much resistance to the political desire to control nature. Self-copying genes that are changed only by random mutations offer little hope of producing a speedy improvement in the quality of life. Selective breeding, the only reliable method of improving crop quality in Lysenko's day, seemed too slow a method of increasing agricultural productivity.

Mendelism also had ideologically unappealing implications for changing humans. If some inequalities in the population of the U.S.S.R. were due to unchangeable Mendelian characters, then the desire to promote equality by changing the social order might not work. How much better it would be if a few generations of living under the Soviet system were sufficient to level out the genetic quality of the population! By contrast, Mendelism seemed to imply persistent human inequalities and helplessness to change genetic destiny.

The son of a peasant, Trofim Denisovich Lysenko was born in 1898. He received a doctorate in agricultural science from the Kiev Agricultural Institute. Lysenko first obtained publicity through the announcement in 1929 of a technique known as "vernalization," which he claimed to have invented but was actually a traditional treatment of soaking and chilling winter wheat to improve its yield. Treated in this way, the seed was actually planted in the spring (hence vernalization) rather than in the fall. Lysenko's claims were greeted with skepticism by agricultural experts but he silenced them by doing an experiment on his father's farm in which a plot of vernalized winter wheat was planted alongside ordinary spring wheat and was said to have produced a better yield.

This result was seized on by politicians as a way of increasing the productivity of Russian agriculture and allowed Lysenko to promote himself as a man whose science made a practical difference. Initially, Lysenko's ideas were dismissed as eccentric and lacking in scientific rigor. Even though Lenin paid lip service to the independence of science from politics, Lysenkoism appealed to him. Russian geneticists noticed which way the wind was blowing and generally refrained from criticizing Lysenko

Огромная армия ученых, специалистов сельского хозяйства, мастеров колхозного производства, вооруженная мичуринским учением, плодотворно работает под руководством талантливого продолжателя дела Мичурина академика Т. Д. Лысенко

В Москве с 31 июля по 7 августа 1948 года проходила очередная сессия Всесоюзной академии сельскохозяйственных наук им. В. И. Ленина

Выступление Т. Д. Лысенко на сессии

Около 700 научных работников и специалистов сельского хозяйства, собравшихся на сессию, заслушали доклад Т. Д. Лысенко „О положении в биологической науке"

Участники сессии на опытном поле академии в Горках-Ленинских

T. D. Lysenko combined science and politics. Lysenko holds up onions before a crowd of people at a meeting in Moscow, and a group of people examining grain in a field, all in 1949. (Library of Congress)

in public. By 1935, the Communist Party was supporting Lysenko not because they were persuaded by his science, or blinded by his rhetoric, but because his work seemed to offer some hope of immediate improvement in agricultural yields, something that was not forthcoming from the Mendelian scientific establishment.

Lysenko also launched a political campaign against his scientific opponents. During the 1930s, he demonized Mendelian geneticists as opposed to materialism ("idealistic") and as standing in the way of change. By 1939, Lysenko had gained political control over Russian science. His approach to biology had the full support of the Soviet Academy of Sciences as indicated by a letter of support members sent to Joseph Stalin. In the letter, they wholeheartedly endorsed Lysenko's genetics as embodying the principle of the use of science to change nature

for the common good and explicitly rejected Mendelism for assuming that nature could not be altered.

Lysenko applied his politically approved ideology to agriculture but with little practical success. When faced with the choice of believing in his ideology or believing the results of his own experiments, he apparently chose ideology over the scientific method. Out of touch with scientific reality, he nevertheless wielded enormous political power. Whether Lysenko played a direct role in the assassination of his scientific opponents through Stalinist purges is controversial. Critics of this interpretation point out that scientists on the Lysenkoist side were occasionally purged as well as the Mendelians. In this hostile climate, many geneticists left the country just as German scientists fled during the Nazi era. Those who stayed renounced their Mendelian views and supported those of

Lysenko. Genetics as a useful science disappeared from the Soviet Union for two decades.

By 1964, implementation of Lysenko's policies had produced disastrous crop failures. With the downfall of Nikita Khrushchev, who had supported him, Lysenko fell from grace and Mendelian genetics returned to set about repairing the damage he had caused to Soviet agronomy.

The Lysenko affair is quite unusual because few scientists become so enamored of politics as he evidently was. Politics and science can make for a volatile mix and Lysenko engaged in many tactics that are far more acceptable to politicians than they are to scientists. Among his other lapses of scientific ethics, Lysenko chose to attack his opponents as individuals instead of focusing on their scientific work. Instead of engaging in detailed scientific discussion in scientific journals, he took his case to the popular press. His message focused on simplistic slogans that had popular appeal rather than an honest discussion of scientific differences. He also sidestepped the scientific method, telling his research staff that if they wanted to obtain a particular result strongly enough then they would obtain it. There can be no more compelling illustration of his personal renunciation of scientific objectivity.

Lysenko was not the first scientist to believe more strongly in the correctness of his ideas than in the empirical data. Perhaps a more disturbing side of the Lysenko affair lies in the ease with which the Russian scientific establishment was subverted. This is a far cry from the checks and balances to bad science that should characterize scientific method and which elevates science above other types of human discourse.

See also NAZI SCIENCE; SCIENTIFIC DISHONESTY; SCIENTIFIC METHOD.

Further Reading

Broad, W., and N. Wade. *Betrayers of the Truth*. New York: Simon and Schuster, 1982.

Huxley, J. *Heredity East and West: Lysenko and World Science*. New York: Schuman, 1949.

Medvedev, Z. A. *The Rise and Fall of T. D. Lysenko*. New York: Columbia University Press, 1969.

Plomin, R., J. C. DeFries, and G. E. McClearn. *Behavioral Genetics*. San Francisco: W. H. Freeman, 1980.

M

Machiavelli, Niccolò
(1469–1527)
Italian
Political Theorist, Diplomat

In his most famous book *The Prince* (1513), Niccolò Machiavelli (see portrait) argued that the conduct of rulers should be exempt from ordinary standards of morality. Given that the primary purpose of a head of state was to remain in power, he had to be cunning, treacherous, and bloodthirsty to undermine the efforts of his enemies. That meant being willing to betray friends when it was in his political interest to do so. Being religious was generally not an advantage for leaders but it was easier to secure popular support if they *pretended* to be religious. This scheming approach to politics, and selfish manipulation of others more generally, is referred to as Machiavellianism. Machiavelli has often been interpreted as having no personal ethics but Jean Jacques Rousseau, among others, has claimed that *The Prince* was intended as a satire on the vices of rulers. Machiavelli's personal political views, as described in *Discourses on the First Ten Books of Titus Livius* (1513–21), were apparently a great deal more democratic and liberal than those espoused in *The Prince*, which was written in an unsuccessful attempt to obtain the patronage of the Medici family who ruled Florence.

    See also SOCIAL CONTRACT THEORY; SOCIAL DARWINISM.

Italian Renaissance author Niccolò Machiavelli from frontispiece in Opere di Niccoló Machiavelli. *His name is associated with lack of moral scruples.* (LIBRARY OF CONGRESS)

MacLean, Paul D.

(1913–   )
American
Physician and Neuroscientist

Paul MacLean introduced the idea that the human brain has three main parts corresponding to evolutionary age: reptilian, old-mammalian, and new-mammalian. The reptilian brain at the anatomical center controls breathing, walking, copulation, fighting, and other stereotyped and reflex actions necessary for survival and reproduction. The limbic system (or old-mammalian) brain provides emotions such as fear, anger, sexual love, and parental feelings, according to MacLean's triune (or three-part) brain model. The new-mammalian brain, or neocortex, lies above the limbic system, was last to evolve, and is particularly well developed in humans to permit complex processing of sensory information and abstract reasoning. Hence MacLean's witticism that a client in psychotherapy brings along a crocodile (reptile) and an anteater (old-mammal) in addition to themselves.

The triune brain is one of the most influential generalizations in neuroscience. Yet, it has been ignored by the majority of researchers in the field as too large and "fuzzy" to be a useful guide for research. It is also vulnerable to detailed scrutiny. Thus, the entire human brain has been subject to natural selection at all times during our evolution as a species. It is therefore unlikely that the "reptilian brain" has remained exactly as it was for our reptilian ancestors even if it has changed more slowly than the forebrain. Despite such problems, the triune brain is useful in accounting for psychological conflicts in terms of the different levels of brain function. Thus, the difficulty of young children in controlling their emotions of sadness and anger is explainable in terms of the late maturation of the neocortex relative to other brain structures. As they get older, the neocortex matures and inhibits the limbic system, making them less likely to cry and to throw tantrums. Recognizing such complexity in brain function helps us to understand some of the psychological conflicts we experience and explains why people should count to 10 (a neocortical function) if they have lost their temper (a limbic function).

The triune brain is of clear relevance to moral philosophy. If ethical behavior is thought of as controlling antisocial impulses, then the triune brain provides a brain explanation for moral behavior. Consistent with this view, there is some evidence that problems of impulse control can, at lease in some unusual cases, be attributed to brain abnormalities. One example of this phenomenon is the story of Phineas Gage, an American railroad foreman who survived an 1848 explosion that damaged his brain but who lost the ability to restrain his emotional impulses, making him seem childlike and irresponsible. Recent attempts to link impulse problems, such as criminal behavior and attention deficit disorder, to brain trauma have not been convincing, however. In general, explanations of crime in terms of brain chemistry have been more successful than explanations in terms of anatomy.

If there is a neuroanatomical basis to ethical behavior, then some aspects of ethical decision making are open to scientific analysis. This would violate the is/ought distinction by combining scientific and ethical discourse. Yet it does so without committing any fallacy, naturalistic or otherwise.

See also CHURCHLAND, PATRICIA; CRIME, AND BRAIN CHEMISTRY; NATURALISTIC FALLACY.

Further Reading

Kalat, J. *Biological Psychology*. Belmont, Calif.: Brooks/Cole, 1988.

mad cow disease (bovine spongiform encephalopathy)

Bovine spongiform encephalopathy (BSE) is a degenerative brain disease of cattle that has recently received a considerable attention because of an outbreak of the disease in Britain beginning in 1985 and peaking in 1993 when 170,000 cattle died of the disease. BSE belongs in a category of diseases known as transmissible spongiform encephalopathies (TSEs) that may take a new form in passing from one species to another. Thus, a probable cause of the BSE outbreak is the feeding of meat and bone meal to cows that was made of scrapie-infected sheep. (Scrapie is a TSE affecting sheep). The major concern about BSE is that people who eat beef from infected livestock may develop a fatal human TSE known as new variant Creutzfeldt-Jakob disease (nvCJD). By 2001, 80 Europeans, mostly Britons, had died of the disease.

Apart from the fatalities, BSE has had enormous economic consequences for beef farmers, with many countries refusing to accept beef from affected regions. Even in the United States, where the incidence of BSE has been extremely low, the disease has produced new regulations to protect the blood supply. On the untested assumption that people who are infected with nvCJD can transmit the disorder in their blood donations, blood is no longer being accepted from Americans who have recently visited Britain.

Perhaps the most important political consequence of mad cow disease in Europe is that it has eroded public confidence in the ability of governments to protect the safety of the food supply. The British government has been accused of late response in admitting that there was a crisis, failure to appropriate adequate research funding for nvCJD, slowness to develop a good BSE test, and incompetent inspection of meat and meat-handling facilities. Critics have charged that it has failed to provide convincing rationale for its slowness to act. The resulting erosion of public trust in the safety of food has grown to include a general debate over GENETICALLY MODIFIED FOOD, which has been vehemently rejected in Britain and Europe despite the absence of any scientific evidence that it poses a health risk.

See also GENETICALLY MODIFIED FOOD.

Further Reading

Brown, P. "On the Origins of BSE." *The Lancet* 351 (1988): 252.

Matza, M. "The Stealthy Assassin That Is Mad Cow Disease Survives Heat and Cold." *Knight-Ridder/Tribune News Service*, February 1, 2001.

Tan, L., M. A. Williams, M. K. Khan, H. C. Champion, and N. H. Nielson. "Risk of Transmission of Bovine Spongiform Encephalopathy to Humans in the United States: Report of the Council on Scientific Affairs." *JAMA* (1999) 281: 2330–39.

Malthus, Thomas Robert

(1766–1834)
English
Economist

Thomas Robert Malthus was trained as a parson but spent most of his adult life as a professor of history and political economy at the East India College close to London (1805–34). His name is synonymous with the problem of human population growth as presented in *An Essay on the Principle of Population As It Affects the Future Improvement of Society* (1798). Malthus argued that population increases geometrically whereas food supply increases only arithmetically. From these premises, he drew the dismal conclusion that the starvation of large numbers of people was a practical inevitability. Otherwise, the explosive increase of population could be checked only by disease and warfare.

These depressing ideas went very much against the grain of contemporary ENLIGHTENMENT thinking with its strong belief in the perfectibility of society, and they were bitterly attacked by many leading thinkers of the period, including philosopher William Godwin (1756–1836), historian Thomas Carlyle (1795–1881), and poet Samuel Taylor Coleridge (1772–1834). What bothered social reformers most about Malthus's ideas was the suggestion that nothing could be done to relieve the miseries of the poor. Malthus was bothered by this himself and he suggested policies to reduce their suffering, including an increased emphasis on agriculture and land reform as a means of increasing food production and delayed marriage to reduce population growth.

Malthus's essay has been extremely influential. Not only did it make a mark in economic theory but it was a potent influence in formulation of the theory of evolution by natural selection by both Charles Darwin and Alfred Russell Wallace. Eugenic attempts to limit reproduction of the poor led to the founding of the English Malthusian League (1877–1927), which advocated birth control.

Contemporary scholars are still debating the merits of the Malthusian thesis as the population of the world has exploded and they examine the implications for world hunger. In some respects, Malthus's dismal forecast has been justified. With the population of the world having increased approximately sixfold in the past 200 years, there are currently almost as many people starving as the entire world population in Malthus's time. In other respects, Malthus was clearly wrong in his predictions. First, the food supply has increased at a far greater rate than he could have imagined. In recent decades, the GREEN REVOLUTION has helped some countries to escape from starvation. Second, food distribution systems in a country may be as important as actual food production in causing famine. Third, modern contraceptive technology has greatly limited fertility in developed countries and the aging of the population has further slowed the reproduction rate. Malthus died near Bath, England, on December 23, 1834.

See also CONTRACEPTION; DARWIN, CHARLES ROBERT; ENLIGHTENMENT, THE; EUGENICS; GREEN REVOLUTION, THE.

Further Reading

Brown, L. R. 1999. *Beyond Malthus*. New York: W. W. Norton.

mammograms, reliability of

Accuracy of measurement is an ethical goal of all scientists, including applied scientists, such as physicians. This is particularly true in the case of tests such as breast mammograms that are used to make potentially life-saving determinations such as whether an individual has a cancerous tumor. Mammograms are used to detect tumors early while they are still treatable. In summarizing decades of empirical study, researchers have found that mammograms for women over 50 years of age play an important role in reducing mortality from breast cancer (by around 30%). The use of mammography in younger women is much more controversial.

Due to age effects on the density of breast tissue, mammograms are far less reliable in women under the age of 50. For women in this age group, about 5% of mammograms produce positive results but about 87% of these are false positives. Moreover, of women who develop breast cancer about 12% have had false negative mammograms. Mammograms are so unreliable for this age group that researchers cannot agree on whether there is actually any benefit from the procedure. Yet young women continue to receive mammograms because the prospect of developing breast cancer is anxiety provoking and they have been led to believe that regular mammography is a responsible preventative measure. To the extent that mammography for young women diverts huge amounts of economic resources away from other medical needs, it is difficult to defend as an ethical procedure. The costs of the procedure itself are compounded by the unnecessary medical procedures for a large number of false positives, not to mention the psychological trauma of being told that one may have a life-threatening illness.

Given the overwhelming problems with mammograms for young women, it is surprising that they continue to be recommended by medical experts and are routinely performed for women in their forties. There are many possible reasons. One is that the benefits of mammography have been consistently either misunderstood or misrepresented. Another is that modern medicine is prone to doing too many tests because (a) these defend physicians against malpractice suits alleging negligence and (b) remuneration of medical practices is often related to the number of procedures performed. There is an added problem of explaining why mammograms are advisable in the case of older women but not for younger ones. For these reasons, mammograms are likely to continue for younger women even though they lack scientific support and are ethically questionable.

See also CHEM-BIO AND PAP SMEAR CONTROVERSY.

Further Reading
Kopans, D. B. *Breast Imaging.* Philadelphia, Pa.: Lippincott-Raven, 1997.
Wright, C. J., and C. B. Mueller. "Screening Mammography and Public Health Policy." *The Lancet* 346 (1995): 29–32.

Manhattan Project

This was the code name for the U.S. effort to develop an atomic bomb during World War II. The Manhattan Project was prompted by the fear that Hitler would develop a fission bomb following the 1938 discovery of atomic fission by German scientists. It coordinated the efforts of many civilian and military scientists under the direction first of Vannevar Bush, head of the National Defense Research Committee, and then of General Lesley Groves of the Army Corps of Engineers. The project was conducted under tight security at several sites, including the Oak Ridge, Tennessee, facility at which scientists mastered the difficult task of enriching uranium ore to weapons-grade material and the Los Alamos weapons

laboratory, directed by physicist ROBERT OPPENHEIMER, at which the first fission bomb was designed and built.

The first atomic bomb had a gun-type design in which one piece of uranium-235 was fired at another, in a barrel, so that they reached critical mass allowing an uncontrollable chain reaction to develop. This weapon was deployed at HIROSHIMA, Japan, on August 6, 1945, (without having been tested) and resulted in a huge loss of civilian lives. Following the work of physicist Enrico Fermi of the University of Chicago on the synthetic radioactive element plutonium-239, five reactors were set up at Hanford, Washington, to produce this fuel, which was then shipped to Los Alamos for bomb construction. A plutonium bomb was tested at Alamogordo, New Mexico, on July 16, 1945. In this design, a core of noncritical plutonium was compressed to critical mass using conventional high explosives. The plutonium bomb was used on Nagasaki, Japan, on August 9, 1945. Its devastating consequences, coming so soon after the Hiroshima bomb, led to Japanese surrender and the end of the war.

The Manhattan Project is one of the most controversial military research projects ever conducted, primarily because it launched the age of nuclear weapons and involved the use of weapons of mass destruction against largely civilian targets. A major ethical issue for scientists involved in the Manhattan Project was the requirement to work under conditions of complete secrecy. Many were required to move their families to remote locations in which freedom of action and expression was constrained by the requirements of military security. Such extreme secrecy is incompatible with the scientific ethic of open communication of ideas.

Scientific support for the development of newer and better nuclear weapons (based on atomic fusion) waned after the defeat Germany and Japan in World War II. This phenomenon is reflected in the subsequent fate of Robert Oppenheimer, who lost his influential position as chairman of the Atomic Energy Commission because he opposed the development of fusion bombs and favored worldwide arms control, and also because he was considered a political liberal who would not be tough on communism. Like many other civilian physicists who had favored the development of nuclear weapons to forestall Hitler, he took the humanitarian position that even more destructive weapons were unnecessary and ought to be resisted.

See also DOWNWINDERS; HIROSHIMA; MUTUAL ASSURED DESTRUCTION; OPERATION WHITECOAT; OPPENHEIMER, ROBERT; NUCLEAR WEAPONS.

Further Reading
Hales, P. B. *Atomic Spaces: Living on the Manhattan Project.* Urbana: University of Illinois Press, 1997.

Masters, William H.

(1915–2001)
American
Gynecologist

and

Johnson, Virginia E.

(1925–   )
American
Psychologist

At the forefront of attempts to demystify sexual response as a physiological process were William Masters and Virginia Johnson who dared to expose this most private facet of human experience to objective scientific analysis. They pushed open a door that had previously been held tightly closed by rigid codes of sexual morality constraining the efforts of scientists and physicians alike. They did so in the mid-1960s at the height of the sexual revolution (a period during which the norms regulating sexual behavior were liberalized). Their book *Human Sexual Response* (1966) was the culmination of over a decade of scientific investigation of human sexual physiology. As a gynecologist, Masters saw himself as researching in the field of medical physiology, but his work had been anticipated by the clandestine investigations of psychologist JOHN B. WATSON. Virginia Johnson, his colleague, is a psychologist. They married in 1971 and divorced in 1993. The researchers have trained thousands of sex therapists. Even though their lives were surrounded by controversy related to the exaggerated claims that they made for their therapies, few scholars question the value of their early scientific work.

RESEARCH PROCEDURES

It is hard to imagine any research topic that is more ethically sensitive than study of the physiology of human sexuality. How precisely the study was conducted is not recorded in detail by the researchers, and the story has never been told by participants, whose complete anonymity was assured as a precondition of participation. At least some of the important facts related to the design and administration of their study were presented, however tersely, by Masters and Johnson in their published work. They restricted themselves primarily to detailed descriptions of the typical physiological reactions of their study population to genital stimulation.

Who were the participants? How were they recruited? What effects did the study have on their normal sexual behavior? What effects did the study have on the *researchers* as professional voyeurs? The information that Masters and Johnson provide concerning these questions is scanty. One motivation of the participants was stated to be financial need, which implies that they were paid for their time. Another was sexual need.

One young woman, whose sexual history is presented, wanted what Masters and Johnson refer to as release from sexual tension. She had been referred to the study by her physician, as were many of the other participants and was attracted by the anonymous nature of the study situation, which allowed her to have sexual intercourse with male participants without jeopardizing her chances of remarriage. Her research experiences also included sequences of masturbation and artificial intercourse with a mechanical device.

Many of the participants were married people who participated in the study because they had experienced sexual difficulties or felt that their marital sexual expression lacked intensity. Perhaps the most surprising aspect of participant motivation was that the commonest reason for participation was the desire to contribute to greater knowledge and understanding of the sexual response.

Since ethical researchers are always concerned about the long-term consequences of their procedures, Master and Johnson conducted follow-up evaluations of participants at five-year intervals. One of their major concerns was that the study situation would create a concern about adequacy of sexual performance, which might carry over to sexuality outside the lab. The study population (which was not typical of the general population for various reasons, including high level of education) experienced relatively few failures to reach orgasm in the study, for either male or female participants. When this happened, researchers gave advice as to how sexual function could be enhanced.

There was no reason to believe that the study experience caused future sexual difficulties. If anything, the study tended to increase sexual interest and satisfaction. This conclusion was drawn from the willingness of many married couples to continue in the study after their specific project had ended. Following participation, it appears that subjects used what they had learned to enhance their marital satisfaction. There was no indication that it led to extramarital sexual experimentation.

Masters and Johnson acquired their research participants in two ways. Most were recruited directly from the academic community on the campus that housed the Reproductive Biology Research Foundation in St. Louis. Others were recruited from specific projects, such as those dealing with geriatric sexuality or sexual responsiveness in pregnancy. The remainder were referred for treatment of infertility or problems of sexual function, such as impotence, frigidity, and those associated with reconstructive surgery. All of the participants were volunteers who agreed to be observed during sexual intercourse. This followed an extensive orientation process that helped them to accept, and even feel comfortable with, the unusual context. All of the subjects were sexually experienced in terms of masturbation and intercourse.

Each individual first provided detailed medical, social, and sexual histories. This information was collected by both a male and a female interviewer working together because previous research suggested that this procedure helped participants to divulge intimate sexual information and to present it more accurately. An even more important motive for using men and women as interviewers was to habituate subjects to the fact that all subsequent procedures, including natural intercourse, masturbation, and artificial coition (using a machine) would be observed by investigators of both sexes.

Masters and Johnson have probably done more than any other researchers to demystify the female sexual response. In order to observe responses within the vagina, they employed sophisticated recording equipment contained within a mechanical copulating machine. This equipment was devised by radio physicists. It included a penis fashioned from perfectly transparent plastic that allowed for continuous cold-light observation and recording. Movement of the artificial penis was powered by electricity. The responding woman could control both the depth of penetration and the speed of thrusting, generally increasing both as she neared orgasm.

Participants were first familiarized with the physical surroundings in which the research was conducted. The equipment was shown and its function explained. Initially, participants engaged in sexual activity in private in the research facility. Once they had gained confidence in their ability to perform, they did so in the presence of investigators while attached to a variety of physiological recording instruments.

It takes unusual courage to pursue a field of research that breaks such powerful social taboos as those against sexual behavior in public. There is certainly no quick reward for such courage except for the satisfaction that comes in knowing that one is devoting one's efforts to a project that is worthwhile and which has the potential for making a lasting contribution to scientific knowledge and human happiness.

Research that is intrinsically so invasive of privacy might be expected to encounter a host of ethical and legal problems. The fact that no such problems ever came to the surface, either as legal problems or in any other form, is a testament to the cooperativeness of the participants and their completely professional treatment by researchers. Quite apart from breaking sexual taboos (which were very weak in the 1960s in America), Masters and Johnson also exploded a number of myths about human sexual responses including:

- The myth that a woman's sexual satisfaction is related to the size of a man's penis.

- The myth that some women are physiologically incapable of orgasm.

- The myth that men and women differ greatly in the overall pattern of their response to sexual stimulation.

Breaking taboos regarding study of sexual responsiveness can be ethically justified because it contributed to treatment of sexual dysfunction. Demystification of sexual responses can do a great deal to alleviate the misery of sexual difficulty in marriages—difficulties which can be upsetting enough to result in divorce. Even though some ethicists might argue that the study was immoral because it induced participants to violate ethical codes of sexual conduct, this argument does not implicate the researchers in unethical conduct because participants were fully informed about the nature of the study before agreeing to participate.

Masters and Johnson readily acknowledged that sexual experience is much more than a physiological reaction. The physiological aspect of human sexuality is also of great importance. It is not just a matter of sexual stimulation, but rather how that stimulation is perceived. For example, women are just as physiologically responsive to pornography as men are but they have little interest in pornographic material and generally do not purchase it. Sociological influences are also important. For example, women are less likely to experience orgasm in societies and family environments where female sexual expression is discouraged. The importance of psychological influences in sexual expression is also highlighted by the issue of sexual orientation wherein it is not just sexual stimulation that matters but rather the object of desire about which sexual needs are focused and expressed.

In *Homosexuality in Perspective* (1979) Masters and Johnson described the sexual responses of homosexual men and lesbians. Their claim that they could change the sexual preference of homosexuals who wanted to change it is seen as a gross exaggeration today and created intense controversy at the time. Masters and Johnson have also been accused of overstating their success rate in treatment of sexual dysfunction among heterosexuals. In their 1988 publication, *Crisis: Heterosexual Behavior in the Age of AIDS,* they forecast that the spread of AIDS among heterosexuals would reach epidemic proportions.

See also ELLIS, HAVELOCK; KINSEY, ALFRED CHARLES; RIGHTS OF HUMAN RESEARCH PARTICIPANTS; WATSON, JOHN B.

Further Reading
Robinson, P. *The Modernization of Sex: Havelock Ellis, Alfred Kinsey, William Masters and Virginia Johnson.* Ithaca, N.Y.: Cornell University Press, 1989.

Symons, D. *The Evolution of Human Sexuality.* New York: Oxford University Press, 1979.

Maxwell, James Clerk

(1831–1879)
Scottish
Physicist

James Clerk Maxwell was one of the leading physicists of the 19th century. His most important contributions included seminal work on electromagnetism and the electromagnetic spectrum. He extended the work of MICHAEL FARADAY on lines of force in electromagnetic fields and introduced mathematical formulae, now known as "Maxwell's equations," that describe these phenomena. Maxwell calculated that electric fields move at approximately the speed of light and this prompted him to suggest that light is an electromagnetic phenomenon. Another triumph of inference was his conclusion that the rings of Saturn must be composed of small solid particles.

Maxwell's physics made heavy use of the concept of a hypothetical ETHER, which was understood to carry light without being changed by it, analogous to a ripple passing over the surface of water. The concept of the ether has been abandoned by modern physicists as unnecessary and unhelpful. One of the first to dismiss the ether was ALBERT EINSTEIN, who demolished it in introducing his theory of relativity. Interestingly, Maxwell's major conclusions, including his equations, were generally unaffected by the loss of the ether as a theoretical construct. He died of stomach cancer in Cambridge, England, on November 5, 1879.

See also EINSTEIN, ALBERT; ETHER, THE; FARADAY, MICHAEL; SCIENCE, HISTORY OF.

Mendel, Gregor Johann

(1822–1884)
Austrian
Geneticist and Monk

Gregor Mendel founded the science of genetics in the course of conducting crossbreeding experiments on pea plants. This research illuminated the way that the traits of two parent plants are combined in the offspring. Mendel presented his results at a meeting of the Brunn Natural History Society and they were published in the society's journal the following year, in 1866. Ignored by the contemporary scientific community, Mendel's revolutionary work was unearthed by three scientists, working independently, in 1900 (Carl Correns, Erich Tschermak, and Hugo DeVries), long after his death. Mendel died in Brno, Bohemia, then in the Austro-Hungarian Empire, on January 6, 1884, of kidney inflammation, edema, uremia, and cardiac hypertrophy. It is possible that even Mendel himself forgot about his research, which he abandoned in 1868 after promotion to the position of abbot of the monastery.

Apart from the lack of credit for his seminal research, Mendel's life has received ethical scrutiny from an altogether different perspective. In a close examination of the published results, British statistician R. A. Fisher concluded that the results of Mendel's experiments were far too perfect to constitute an accurate reflection of his true findings.

Mendel's experiments involved the crossing of garden peas and the study of the plants grown from their seeds. He found that when tall strains were crossed with dwarf ones, all of the first generation offspring were tall, leading to the conclusion that tallness is dominant and shortness recessive. His more interesting result came from self-pollinating these tall hybrids and studying the offspring grown from their seeds. The result was that a quarter of the plants were like the ancestral dwarfs, a quarter were like ancestral talls, and a half were talls that did not breed true (i.e., did not have offspring all of the same height as themselves).

Mathematically, Mendel's experiment resembles an exercise in probability. It is as if one were to write "tall" and "dwarf" alternately on 200 disks of paper and shuffled them in a hat, carefully extracting two at a time. The *average* probability is that you will get 25 tall-tall pairs, 50 tall-dwarf pairs, and 25 dwarf-dwarf pairs. However, it is extremely unlikely that one would get exactly this outcome on any one experiment.

Fisher estimated that the odds are 10,000 to 1 against Mendel having given an accurate account of his research. In other words, Mendel is suspected of fudging the data to bring the results in line with theoretical predictions. From the point of view of genetics, this controversy is unimportant since the basic laws of genetics that Mendel established have been verified in countless subsequent experiments. The more fundamental question concerns the extent to which science is truly empirical.

In principle, when scientists test a hypothesis, they collect observations that determine whether the hypothesis is true or false. The data are thus "sacred" and accurate collection and reporting of results is the linchpin of scientific practice. In Mendel's case, it seems as though he used his raw data to infer the general principles underlying the results and then "cleaned up" the data to fit those principles. To modern scientists, that would be considered scientific fraud. It is difficult to imagine how Mendel, a self-educated scientist, would have viewed the issue because he did not leave behind any record of his private reflections.

See also GALILEO GALILEI; MILLIKAN, ROBERT; SCIENCE, HISTORY OF; SCIENTIFIC DISHONESTY.

Further Reading

Broad, W., and N. Wade. *Betrayers of the Truth*. New York: Simon and Schuster, 1982.
Fisher, R. A. "Has Mendel's Work Been Rediscovered?" *Annals of Science* 1 (1936): 115–37.
Henig, R. M. *The Monk in the Garden*. Boston: Houghton Mifflin, 2000.
Stern, C., and C. R. Sherwood, eds. *The Origin of Genetics: A Mendel Sourcebook*. San Francisco: W. H. Freeman, 1996.

Mendeleyev, Dmitry Ivanovich

(1834–1907)
Russian
Chemist

Mendeleyev created the periodic table of elements that allowed elements to be arranged in ascending series according to atomic weights. Elements were arranged in eight columns and the family of elements within a column possessed similar chemical properties. This arrangement suggested that chemistry was much more orderly than had previously been thought and it excited considerable skepticism among contemporary scientists. Perhaps the most provocative aspect of the periodic table is that it contained blanks, or spaces to which no known element corresponded. Discovery of three of these elements (gallium, scandium, and germanium) forced chemists to accept the periodic table. Mendeleyev died at St. Petersburg in Russia, on February 2, 1907.

Even though Mendeleyev is today credited with discovery of the periodicity of elements, he may have been anticipated by Lothar Meyer, a German scientist who produced his first table in 1868. Meyer delayed publication until 1870, however, and thus lost the race for priority to Mendeleyev, who published in 1869.

See also SCIENCE, HISTORY OF; SIMULTANEOUS DISCOVERY.

Mengele, Josef

(1911–1979)
German
Physician, Anthropologist

Josef Mengele was the chief medical officer at the Nazi concentration camp in Auschwitz, Poland. Known as "the angel of death," Mengele participated in the deaths of 400,000 people. Trained as a physician, he exploited his

position to conduct unethical biomedical experiments. Mengele's complete lack of regard for any moral worth of his research victims has made him notorious.

Raised as a conservative Catholic, Mengele was the oldest of three sons of a wealthy farm equipment manufacturer. He was ambitious and chose a career in medicine over the limitations of the family business. His medical studies at Munich between 1930 and 1936 focused on anthropological genetics, a specialization that had grown up in the context of a EUGENICS movement that was influential in America as well as Germany.

Munich was an important center for the Nazi Party and its "racial purity" agenda. Anxious to succeed, Mengele cultivated professors who were Nazi sympathizers and chose a politically relevant research topic. His dissertation on racial jaw morphologies was published. Mengele further ingratiated himself with the academic establishment by marrying a professor's daughter. He also became a Brown Shirt, or Nazi storm trooper.

Mengele's career of appallingly unethical human experimentation began in 1943 after he had been appointed as women's physician in Birkenau concentration camp. His experiments were actually conducted at the Auschwitz concentration camp even though he had no official appointment there. Mengele's first order of business at Birkenau was to clear the camp of typhus. This was quickly accomplished by the simple expedient of identifying and gassing some 1,000 sick prisoners. His ruthless efficiency and unsentimental attitude toward the prisoners earned him the immediate respect of his superiors.

Mengele then turned his attention to research dealing with eugenic themes. He was evidently motivated by academic ambition and his incomprehensibly brutal experiments were designed to improve his chances of a professorship at Munich. Interestingly, Mengele wanted to develop methods of changing bodily traits (or phenotype) in spite of genetic characteristics (or genotype). This objective seems strangely inconsistent with Hitler's Final Solution according to which "inferior" races and "defective" individuals were to be exterminated. In pursuing this research agenda, Mengele combined a very weak understanding of biology with a complete lack of regard for the suffering and death of his victims. His attitude seems to have been that the people were going to die in any case. Meanwhile, they might as well be used to advance medical research and his personal objective of a full professorship.

Mengele wanted to devise ways of producing what were considered desirable phenotypes, such as blonde hair, blue eyes, and freedom from genetic disease. In his attempts to change the effects of genotype, Mengele used identical twins as controls. He selected his victims by standing at the train station as boxcars came in filled with

Jews about to be slaughtered. In this way, he chose twins and any other people who interested him, giving those whom he selected an effective stay of execution as they participated in his ghastly experiments.

An example of the childlike crudity of Mengele's work is furnished by his project of changing eye colors to blue by injecting a blue dye. When the experiment was over, he cut out the eyes, preserved them, and displayed them in his office along with other human organs. Also exhibited were the skeletons of two hunchbacks, a father and son, which reflected his interest in body types. A family of seven dwarfs from a Romanian circus were kept alive for the amusement of visiting physicians.

Mengele appears to have been interested in immunology at some level. He experimented by interchanging blood among and between twin pairs to see what would happen. One of his most appalling projects was perpetrated on a pair of fraternal (or nonidentical) twin children, one of whom had a hunchback. In an effort to create conjoined twins, Mengele sewed the back of the normal twin onto the hunchback of the other. As if for dramatic effect, he also sewed the backs of their wrists together. The wounds were dirty and gangrene quickly set in. An observer was impressed by the horrible odor of the gangrene and the fact that the children cried every night after they had been returned to the barracks.

It is estimated that Mengele was directly responsible for the deaths of 150 to 200 twins. He dispatched many of them personally with the cool detachment for which he was noted. On one occasion, when 300 children escaped from the gas chamber due to an accident, Mengele had a gasoline fire lit in a large trench and watched as the children were thrown into the flames. When some of the burning children desperately clawed their way over the side of the pit, Mengele helped the SS men to kick them back in.

Although some commentators see Mengele as an example of the banality of evil, that is, a very ordinary man caught up in an extraordinary situation, it is hard to believe that he did not derive pleasure from the sufferings of his victims. Not content with his eugenics "research" he also investigated the limits of human endurance by subjecting 75 prisoners to severe electric shocks. He found that a third of them died immediately. Other prisoners were subjected to high levels of radiation that caused severe burns.

Although painful to contemplate, Mengele's activities provide a useful reference point for ethicists who are interested in defining ethical research. When Nazi war criminals were tried at Nuremberg, there was a special trial, known as the Doctor's Trial, held to judge Mengele's colleagues, many of whom were engaged in parallel horrifying experiments. These were so shocking to the judges that they resulted in the promulgation of the NUREMBERG CODE, designed to protect the welfare of human research participants. In general, the major ethical problems raised by Nazi research included that it was conducted without the consent of the participants and that it harmed the participants. Moreover, the academic quality of the research was so low that it had virtually no potential for producing scientifically useful information.

Mengele escaped capture and was, therefore, never tried for war crimes in connection with his human experimentation. He appears to have spent the rest of his life in Brazil under the alias Wolfgang Gerhard. A man with this name died in a drowning accident in 1979. When the remains were tested in 1985, they provided a DNA match to Mengele's son, indicating without any doubt that this was indeed the remains of the angel of death. Mengele never expressed any remorse for his activities in telephone conversations with his son and seems to have been unwilling to admit that he had done anything wrong.

See also NUREMBERG CODE.

Further Reading
Pence, G. E. *Classic Cases in Medical Ethics*. New York: McGraw-Hill, 1995.
Posner, G., and J. Ware. *Mengele: The Complete Story*. New York: McGraw-Hill, 1986.

mercury poisoning

Mercury is a toxic heavy metal that accumulates in bodily tissues, particularly the kidneys and liver. It also accumulates in the central nervous system where it may produce mental confusion, convulsions, and other symptoms. At high doses, mercury is fatal. Mercury poisoning is a factor in mental retardation of children and also plays an important role in producing neurofibrillary tangles in brain cells that are a feature of Alzheimer's disease.

The main sources of mercury poisoning are dental fillings; mercury in food, particularly fish; environmental sources, including broken thermometers and medical equipment; and industrial exposure particularly on the part of independent gold miners who use mercury to extract the precious metal from ore. Of the nonoccupational sources of mercury, dental fillings are estimated to contribute more than all other sources combined. Silver amalgam fillings are actually an amalgam of silver and mercury and they have been widely used in tooth restoration for over 180 years. The use of amalgam fillings has continued despite a German controversy, in the 1920s, concerning potential toxicity. Although many dentists maintain that the mercury is locked in the amalgam, many careful scientific studies have shown that dental mercury is continuously being released. It is estimated that a person with eight biting-surface amalgam fillings receives a daily dose of 10 micrograms of mercury that is released mostly during chewing. Autopsy data have shown that brain mercury concentrations are correlated with the number of mercury fillings. Moreover, experiments on sheep have clearly shown that mercury fillings quickly compromise kidney function (as reflected in reduced albumin excretion). Experiments on monkeys have found that mercury from amalgam fillings promotes resistance to antibacterial drugs, which is a serious problem in modern medicine.

There is considerable uncertainty about whether the level of mercury to which people are exposed through their tooth fillings is sufficient to cause health problems. Some dentists have established lucrative practices that focus on removal of amalgam fillings, but whether this actually improves health remains controversial. Nevertheless, it is clear that the risk posed by toxic fillings is not balanced by any particular benefit derived from using them. Several European countries, such as Sweden, Germany, and Austria, have banned amalgam fillings or are in the process of phasing them out. Most industrialized countries have chosen to ignore the issue, however, for reasons that are unclear.

Apart from dental fillings, the second most important source of contamination may be methylmercury from fish. This has been a cause of concern because of evidence that fish have been accumulating in their tissues increased levels of mercury found in salt- and freshwater attributable to industrial pollution. Several recent studies have concluded that dietary exposure to sea fish is not predictive of tissue mercury levels but one study, in French Guiana, found that consumption of freshwater fish and livers from game animals was associated with a doubling of tissue mercury concentrations. Forty-one U.S. states currently issue advisories concerning amounts of freshwater fish that may be safely eaten.

In conclusion, there is evidence that dental fillings are a major source of easily preventable exposure to mercury poisoning. Dentists have been unwilling to face up to this problem, apparently because they do not want to accept that a product that has been used throughout the history of modern dentistry could be harmful to patients. Instead of confronting the scientific evidence, they have fallen back on anecdotal defense of their traditional procedures. This is not only bad science, if dentists are considered as applied scientists, but it is unethical medicine since it clearly contravenes the "do not harm" precept of the HIPPOCRATIC OATH. The continued use of mercury in dental fillings could be justified only if it were established that the level of exposure from amalgam did not cause health problems. Recent research with sheep and monkeys does not inspire confidence on this score.

See also FLUORIDATION OF WATER SUPPLY; LEAD POISONING.

Further Reading

Murray, J. V. "Toxic Teeth: The Chronic Mercury Poisoning of Modern Man." *Chemistry and Industry,* January 2, 1995.
Smith, J. C., P. V. Allen, and R. Von Burg. "Hair Methylmercury Levels in U.S. Women." *Archives of Environmental Health* 52 (1997): 476–80.

meta-ethics   *See* MORAL PHILOSOPHY

metal detectors

Electronic metal detectors are used to spot concealed metal objects, such as weapons, and thus to enhance security at airports, prisons, public buildings, and schools. A common type of detector is based on the principle that metal objects distort an electrical field thereby altering electrical inductance. The use of metal detectors can be problematic in the case of individuals with electronic pacemakers but the main ethical objection to their use in schools, for example, is that they create a prison-like atmosphere that is oppressive and demeaning.

See also ELECTRONIC PRIVACY.

Milgram's obedience research

American social psychologist Stanley Milgram (1933–84) carried out important research on the human capacity to obey authority figures that has evoked considerable ethical controversy and has shaped professional guidelines for conducting ethical research in psychology. Milgram's based his point of departure on an attempt to understand the Nazi Holocaust. He was intrigued by the fact that when Nazi officers were tried as war criminals they often continued to insist that they had merely been obeying orders even though they had very high rank, such as S. S. leader Adolf Eichmann, or had no chance of being acquitted and therefore little motivation for lying. Moreover, the Nazi officers who were accused of appalling crimes against humanity did not look particularly evil. Many observers were struck by their sheer banality.

Could S. S. officers who perpetrated the Holocaust have been ordinary people caught up in an unusual situation who merely obeyed the orders of superiors? This defense was ineffectual at Nuremberg and did not save Eichmann's life at his trial in Jerusalem. Many observers dismissed it as self-serving nonsense.

Milgram devised an experimental test that sought to distinguish between the role of the individual and the role of the situation in producing atrocities. His logic held that if ordinary people can be induced to follow orders to commit atrocious acts, atrocities are produced by social situations rather than by evil individuals.

MILGRAM'S EXPERIMENTAL PROTOCOL

Milgram's experimental setup required subjects to play the role of teachers. Under the watchful eye of an experimenter, they were ostensibly teaching another person arbitrary connections between pairs of words. The learner responded with the second word in a pair when the first word was given as a prompt. For the first error, the teacher was instructed to deliver a 15-volt shock as punishment. For each subsequent error, the shock was to be increased by 15 volts, up to a maximum of 450 volts. Milgram wondered how much the shock level would have to be incremented before subjects began abandoning their role as instructor. Would ordinary people be likely to raise the shock above the life-threatening level of 300 volts?

Creating an atrocity in the psychology laboratory calls for theatrical flair. Milgram set up an impressive looking bank of electronic devices that produced a series of loud clicking noises followed by an ominous hiss to stimulate the effects of delivering electric shocks. In reality, the machine only made noises and did not generate shocks. The illusion of shocking another person was enhanced through the services of a skilled actor who played the part of the shock victim. He responded to stimulated shocks by grunting in pain and asking to be let out of the experiment. His groaning and pleading continued until the 330-volt level, at which point the actor fell silent, evidently unconscious.

Subjects became very upset. Milgram described the reaction of one "instructor":

> I observed a mature and initially poised businessman enter the laboratory smiling and confident. Within 20 minutes, he was reduced to a nervous stuttering wreck, who was rapidly approaching a point of nervous collapse. He constantly pulled on his earlobe and twisted his hands. At one point, he pushed his fist into his forehead and muttered: "Oh God, let's stop it." And yet he continued to respond to every word of his experimenter and obeyed to the end.

Instructions from the experimenter followed a strict protocol. Beginning with a mild "Please continue," they progressed to the brutal "You have no other choice, you must go on." Not one of Milgram's subjects stopped giving "shocks" before reaching the 300-volt level. Upset though they were, two-thirds of the subjects continued to press switches even after the "learner" had stopped responding.

The findings were not just a freak of the psychology lab. One real-world experiment found that 95% of nurses were prepared to give excessive doses of an unlisted drug to their patients when ordered to do so, by telephone, by a "Dr. Smith in Psychiatry" whom they recognized as being on the hospital staff but had never met. Their unthinking impulse to obey someone they accepted as an authority figure overrode several of their written instructions for drug administration, including one which forbade taking telephone instructions.

Contrary to the views of most moralists, Milgram concluded that ordinary people may commit atrocities when ordered to do so by an authority figure. Atrocities are thus a feature of social situations rather than evil individuals. Milgram speculated that a strong tendency to obey leaders is built into us by natural selection. This would have enhanced the prospects for survival and reproduction of individuals in the hunter-gatherer bands of our ancestors even though it may have horrendous consequences in modern states with centralized authority.

ETHICAL ISSUES RAISED BY MILGRAM'S RESEARCH

Milgram has been accused of several ethical lapses in conducting his research, but it is important to realize that the outcome of such original research can be difficult to predict. It is reasonable to hold researchers responsible for unethical outcomes of their research only if these can be predicted. Another important point to make is that at the time of the research, in the early 1960s, few professional ethical guidelines were available that helped researchers to treat their subjects in an ethical manner.

The primary ethical objections is that Milgram exposed participants to an extremely stressful experience without alerting them to this possibility in advance. Subjects were also deceived in various ways. They were told that the experiment was concerned with the effects of shock on learning and its true purpose was not disclosed. They were misled into believing that the "victim" was another subject rather than a confederate of the researcher. Most important, they were deceived into thinking that they were administering dangerous levels of shock to another human being. Subjects were also prevented from withdrawing from the experiment (by the verbal instructions of the experimenter), which compromised the voluntariness of their participation.

In Milgram's defense, he could not have predicted that so many subjects would have persisted for so long in the teacher role making their experience extremely stressful. Yet, there were more than 1,000 participants in the research. The harrowing nature of the experience was obvious to Milgram soon after the experiments began. Interestingly, when subjects were interviewed following their participation, most (84%) were actually glad that they had the opportunity of participating in the experiment (15% were neutral and 1% regretted participation). Three-quarters of the subjects said that they had learned something of personal importance from the experience.

Deception has been justified in social psychological research on the basis that the research could not have been carried out effectively in any other way. Ethical researchers who use deception are obligated to conduct a debriefing in which the adverse consequences of lying are redressed by establishing the true purposes and circumstances of the experiment. Milgram carried out an extensive debriefing and this may have been partially responsible for the subsequent lack of bad feeling from participants.

When subjects were told that the experiment must continue, it is arguable that their right of voluntary participation, as specified in the NUREMBERG CODE, was being compromised. They could, of course, have got up and left at any time but they were restrained by social pressures (which was the point of the experiment). Modern ethical researchers avoid this ambiguity by having participants read, and sign, consent forms that reassure them that they have the right to withdraw participation at any time without a penalty of any sort.

See also NUREMBERG CODE; RIGHTS OF HUMAN RESEARCH PARTICIPANTS.

Further Reading

Milgram, S. *Obedience to Authority.* New York: Harper and Row, 1974.

Miller, A. G. *The Obedience Experiments.* New York: Praeger, 1986.

military research

The history of the connection between science and warfare is an ancient one. For example, ARCHIMEDES (298–212 B.C.), one of the earliest scientists, constructed burning mirrors to defend his native city, Syracuse in Sicily, from attacking Romans.

Many modern scientists see the objectives of death and destruction produced by military technology as a misuse of science, although this sentiment often changes along with historical events. Thus, the realization that German scientists had been working on nuclear fission, and might soon produce an atomic weapon, prompted ALBERT EINSTEIN, an avowed pacifist, to participate in a political campaign to develop the atomic bomb in America. The thought of Hitler with atomic weaponry may have been enough to shake the most ardent of pacifists and this situation clearly illustrates an ethical defense of military science. Not only did the Nazis threaten the liberty and autonomy of countries around the world but their ideology, like that of other totalitarian regimes, including the Soviet Union, was inimical to freedom of scientific enquiry and greatly degraded the quality of scientific practice. The interests of science may thus coincide with military objectives.

However, the extensive cooperation between civilian scientists and the military in the context of nuclear weapons development (see MANHATTAN PROJECT) masks inevitable conflicts between the goals of science and military objectives. The most obvious of these relates to the issue of secrecy. The modern scientific enterprise revolves around publication of results, allowing for a relatively free sharing of conclusions and ideas. Scientists conducting military research generally work under a binding code of strict secrecy. Many scientists choose not to work under such intellectually stifling conditions. Many others make this compromise, however. Approximately 500,000 scientists and engineers are employed in military research around the world. Moreover, about a quarter of all research and development expenditures in the world are allocated to military science.

Most reasonable people accept that secrecy is necessary to accomplish military objectives, namely, national security, and that military innovations freely shared with potential enemies would end up doing far more harm than good. Secrecy produces two major kinds of ethical problem for military scientists. The first is that military secrecy can be a way of concealing clearly unethical research that denies the rights of human research participants. The most egregious example of such research conducted in America involved the testing of radioactivity effects on unknowing and involuntary participants in COLD WAR RADIATION EXPERIMENTS. Similar, if less extreme, ethical problems were raised by the use of soldiers in PSYCHEDELIC DRUG RESEARCH that was aimed at developing new chemical weapons. Had the public been aware of any of these experiments, they could not have been conducted. These episodes in U.S. military science are painful to contemplate and difficult to reconcile with the values of a democracy. Military secrecy not only permitted these atrocities to be committed in the name of scientific research but also provided shelter for the perpetrators who were never brought to justice.

Secrecy in military research also can create pressure toward inaccurate, or fraudulent, reporting of results to the public, who are entitled to transparency because they fund the research through payment of taxes. The pressures to produce a particular outcome in a military test are analogous to the distorting effect of COMMERCIALISM IN SCIENTIFIC RESEARCH. For example, drug tests that are sponsored by pharmaceutical companies almost never criticize the drug being tested either because the researcher feels obliged to report positive outcomes of the drug or, more likely, because the research contract allows drug companies to suppress publications of findings that they do not like.

Military scientists also have to be concerned about the financial implications of their data. A spectacular example of funding motivated scientific fraud occurred in the context of research on the strategic defense initiative, or "Star Wars" program initiated by President Ronald Reagan. Researchers faked a critical test of an antimissile missile. They were worried that a string of dismal failures in real tests would result in cancellation of the whole program as a huge waste of public funds. The 1994 test was rigged by placing a transmitter in the target missile making it easy to locate and destroy and this deception came to light in August of the same year. Needless to say, a real attacking missile would not broadcast its location and identity in this way. Congress was fooled by the rigged test and funding for the Star Wars program was continued.

The ethical pressures facing scientists in the military are in many ways quite similar to those confronting scientists who work in commercial settings in that most of the issues in both settings relate to the need for secrecy and the distorting effects of having to come up with the required answer. Scientists who have ethical reservations about their work may have difficulty in bringing their concerns to the attention of the public because they are legally bound to preserve state secrets. Similar issues of secrecy apply in commercial settings where scientists sign legally binding agreements not to divulge their results without prior approval of the employer. There is no easy answer to many of these conflicts except to say that scientists who are uncomfortable with such compromises should probably avoid research in a military setting.

See also BIOLOGICAL AND CHEMICAL WARFARE; COLD WAR RADIATION EXPERIMENTS; MANHATTAN PROJECT; OPERATION WHITECOAT; NUCLEAR WEAPONS.

Further Reading

Fotion, M., and G. Elfstrom. *Military Ethics.* Boston: Routledge and Kegan Paul, 1986.
Resnik, D. B. *The Ethics of Science.* New York: Routledge, 1998.
Weiner, T. "Inquiry Finds 'Star Wars' Plan Tried to Exaggerate Test Results." *New York Times,* July 22, 1994.

military research on humans *See* COLD WAR RADIATION EXPERIMENTS, OPERATION WHITECOAT.

Mill, John Stuart
(1806–1873)
English
Philosopher

John Stuart Mill (see portrait) was the most influential English philosopher of his day and his utilitarian ethics continues to be widely discussed. Mill defended individual liberty against the power of the state and was an early

John Stuart Mill was a leading figure in utilitarian ethics of the 19th century. (LIBRARY OF CONGRESS)

proponent of equal rights for women. A philosophical heir to Jeremy Bentham, he was groomed by his father, philosopher James Mill, to assume leadership of the Philosophical Radicals. This group was organized around the utilitarian philosophy of Bentham and campaigned for social and political reforms.

While Mill contributed to several areas of academic philosophy, he is remembered primarily for his books on utilitarian ethics and political philosophy. *Utilitarianism* (1863) promoted the Benthamite principle that individuals and states should strive to promote the greatest good of the largest number of people. For utilitarian philosophers, the good of the individual is equivalent to happiness, which is equivalent to pleasure or freedom from pain. Whereas Bentham did not distinguish between the value of different kinds of pleasure, Mill followed many classical Greek philosophers in elevating the pleasures of the mind over those of the body. Mill's book *On Liberty* (1859) is still read as a classic statement of the liberal view that people should enjoy the maximum amount of freedom of action, thought, and expression, and that these freedoms should be respected, and defended, by the state. His view that the liberty of the individual should be constrained only by the need to protect the freedom of others is memorably expressed in the aphorism that a man's right to swing his fist ends with his neighbor's nose. Mill collaborated with his wife Harriet (formerly Taylor) to produce *The Subjection of Women* (1869), a foundational book on women's liberation. He died in Avignon, France, on May 8, 1873.

See also BENTHAM, JEREMY; MORAL PHILOSOPHY; SOCIAL CONTRACT.

Further Reading
Donner, W. *The Liberal Self: John Stuart Mill's Moral and Political Philosophy*. Ithaca, N.Y.: Cornell University Press, 1992.

Millikan, Robert Andrews
(1868–1953)
American
Physicist

Robert Millikan was one of the most gifted and distinguished scientists of his day whose work on measuring the charge of the electron (in his celebrated oil-drop experiment) and on photoelectricity, won him the 1923 Nobel Prize in physics. He went on to do important work on the extraterrestrial radiation he christened "cosmic rays." Millikan's *First Course in Physics* (1906), coauthored by Henry Gale, was a successful textbook. Millikan conducted research at the University of Chicago (1896–1921). In 1921, he moved from the University of Chicago to California Institute of Technology (Caltech). His efforts as chairman of Caltech's executive council helped turn the institution into a top-notch research facility.

Despite this life of dedication to science, Millikan's scientific integrity has been called into question in relation to the oil-drop experiment that played a pivotal role in his career. As in the case of many other luminaries of science, including GALILEO, ISAAC NEWTON, JOHN DALTON, and GREGOR MENDEL, the published results turned out to be too good to be true. In the oil-drop experiment, Millikan observed the movement of a charged oil droplet between two electrically charged plates. Previous research had used water drops, but Harvey Fletcher, one of Millikan's graduate students, suggested that oil drops would be an improvement because oil does not evaporate. Fletcher's contribution was never acknowledged in Millikan's publications. Millikan's technique involved adjusting the electrical charge on the droplet using an X-ray beam and studying the altered speed of movement of the charged drop against gravity in the electrical field. From changes in the speed of the electron he could easily calculate changes in the charge on the drop. It was Millikan's hypothesis that these speeds would change by a series of unit steps corresponding to the addition of one, two, three, of four electrons to the drop, thus supporting his view that the electron is the irreducible unit of electric charge.

In practice, the measurements were difficult and subject to considerable variation. In his first published paper, in 1910, Millikan reported the results for 38 different drops. The results were graded from "best" to "fair" and he revealed that data for seven tests had been thrown out. Felix Ehrenhaft, at the University of Vienna, who was conducting similar research, believed that there were charges less than the charge on an electron (fractional charges or subelectrons). He quickly pointed out that the variability in Millikan's data was consistent with his theory of subelectrons carrying fractional charges. The possibility of fractional charges became a burning issue in physics circles around the world.

In a 1913 paper, which was clearly motivated by the desire to rebut Ehrenhaft, Millikan published new data that made a clearer case for his unit-charge hypothesis. He claimed that the data had not been selected but represented all of his results from 60 days of experimentation. When historian Gerald Holton went back to check Millikan's laboratory records on which the 1913 paper was based, he found that the 58 observations reported had actually been culled from a total of at least 107. The laboratory notebooks contained a private notation that showed Millikan was excising data for trials that did not match his predictions. Opposite the data he made telltale comments such as "very low, something wrong" and "beauty, publish this surely, beautiful."

Since the test of Millikan's hypothesis depended on the distribution of his data, rather than any single calculation, it is clear that the practice of reporting only selected data is scientifically indefensible although some scientists have championed Millikan's case arguing that he knew more about the experiment than many of his critics. Millikan believed strongly in the electron as the irreducible minimum of electric charge and was not open to falsification of the hypothesis. According to philosopher of science KARL POPPER, openness to falsification of hypotheses by data is the central defining characteristic of science. If so, then Millikan's oil-drop experiment, as reported by him, was not science but a circular exercise of working backward from what he knew to be the correct answer.

Dishonest though he was, Millikan clearly prevailed in the scientific battle with Ehrenhaft, perhaps because heavyweights such as MAX PLANCK and ALBERT EINSTEIN were also enamored with the quantum. Ehrenhaft, who had more accurate equipment, continued to publish all of his data and this did not fit in with the notion of the electron as an indivisible unit of electric charge. Subsequent research at Stanford University, using similar techniques, has supported Ehrenhaft's conclusion that there is a subelectronic charge. Millikan received the Nobel Prize, became the president of the American Association for the Advancement of Science, and was an adviser to U.S. presidents while Ehrenhaft sank into obscurity and became quite disillusioned with physics.

Millikan died in Pasadena, California, on December 19, 1953.

See also SCIENTIFIC DISHONESTY.

Further Reading

Broad, W., and N. Wade. *Betrayers of the Truth.* New York: Simon and Schuster, 1982.

Franklin, A. "Millikan's Published and Unpublished Data on Oil Drops." *Historical Studies in the Physical Sciences* 11 (1981): 185–201.

Holton, G. "Subelectrons, Presuppositions, and the Millikan-Ehrenhaft Dispute." *Historical Studies in the Physical Sciences* 9 (1978): 166–224.

Resnik, D. B. *The Ethics of Science.* New York: Routledge, 1998.

minimal risk criterion

In making ethical decisions about projected research, researchers and INSTITUTIONAL REVIEW BOARDS (IRBs) are required to evaluate the potential for harm to participants. If research is categorized as minimal risk, this means that participants are no more likely to experience ill effects of the research than if they were going about their normal daily activities, including receiving routine medical and psychological examinations.

See also RIGHTS OF HUMAN RESEARCH PARTICIPANTS.

Minnesota Twin Study

The Minnesota Twin Study has compared the personality test scores of more than 56 pairs (as of 1990) of identical twins who were reared apart due to adoption. The twins resembled each other strongly, which provided convincing evidence for the importance of genetics in personality development and intelligence. Perhaps the most striking conclusion of the Minnesota Twin Study was that identical twins raised apart were just as similar as identical twins reared together, indicating that being raised in the same home does not make the twins more alike. The Minnesota Twin Study indicates that about half of the differences between individuals in personality traits can be attributed to genetic differences. The study has often been misinterpreted as showing that the personalities and life outcomes of children are almost entirely determined by genetics and that parents have little influence on the development of their children. Such misconceptions could be used to justify an unethical neglect of children by parents.

See also BURT, SIR CYRIL; GENETIC DETERMINISM.

monism

Whereas DUALISM sees mind and matter as inhabiting distinct realms, monism is the philosophical view that only one of the two is real. Most scientists are materialistic monists, which means that they deny the existence of a mental realm independent of matter. From this point of view, thought is nothing more or less than the pattern of communication between cells in the nervous system.

A more unusual version of monism, idealism, holds that only the mind is real and that matter is an illusion. Philosophers espousing variants of this position include the Dutch Jew Spinoza (1634–77) and Irishman George Berkeley (1685–1753). For Spinoza, only God was real. For Berkeley, we could not know material objects directly but only through our own mental impressions of them. Idealism is unpopular because it defies common sense. As one of Berkeley's critics pointed out, according to his solipsistic (or literally self-centered) philosophy trains only have wheels when they are in the station because that is the only point at which passengers see the wheels!

Materialism is much more compatible with the scientific point of view than dualism is because the existence of an independent realm of the mind ushers in problems of free will and indeterminacy that most scientists prefer to avoid. Yet such avoidance can be difficult, particularly in medicine, where scientists are confronted with unexplained mentalistic phenomena such as the PLACEBO EFFECT, FAITH HEALING, and VOODOO DEATH.

See also DUALISM; DESCARTES, RENÉ; MOORE, GEORGE EDWARD.

monosodium glutamate (MSG) food additive

Monosodium glutamate (MSG), a sodium salt of the amino acid glutamic acid, is added to food as a flavor enhancer. Manufactured from seaweed, MSG is an important element in the cuisines of Japan and China. MSG was first identified as a flavor enhancer in Japan, in 1908, by Kihunae Ikeda, who made soups using seaweed. It is also a natural constituent of tomatoes and Parmesan cheese. MSG produces a distinct taste sensation known as *umami* (Japanese for delicious). Umami is a savory, meaty flavor. The use of MSG in restaurants, and in packaged foods, has created an ethical problem because it may produce an adverse reaction characterized by burning at the back of the neck, facial tightness, and tingling sensations. This reaction is known as Chinese restaurant syndrome because of the liberal use of MSG in many of these establishments.

There has been a great deal of controversy about the safety of using MSG in food but scientific reports have found that in normal quantities, of less than three grams per meal, it produces no adverse reaction in the general population regardless of age. However, a small number of people can be adversely affected by the flavor enhancer. They include a small proportion of asthmatics, and those who are vulnerable to migraine headaches. According to the U.S. Food and Drug Administration MSG is safe for use in foods.

See also PRESERVATIVES IN FOOD.

Further Reading

Geha, R. S., A. Beiser, R. Patterson, et al. "Review of Alleged Reaction to Monosodium Glutamate and Outcome of a Multicenter, Double Blind, Placebo-Controlled Study," *Journal of Nutrition* 130 (2000): 1058S–62S.

Moore, George Edward
(1873–1958)
English
Philosopher

George Edward Moore was elected to a fellowship at Cambridge University in 1898 where he subsequently taught moral science and became a professor of philosophy (1925–39). He also edited the philosophical journal *Mind* (1921–47). Moore was a member of the Bloomsbury group, an intellectual coterie that included philosopher Bertrand Russell, economist John Maynard Keynes, and writers E. M. Forster and Virginia Woolf. He died in Cambridge, England, in 1958.

G. E. Moore was a realist who took issue with the commonplace assumption of turn-of-the-century philosophers that the reality of the material world could not be accepted. Thus he made fun of idealist philosophers by pointing out that, according to them, trains only have wheels when they reach the station (because that is the only point at which passengers see the wheels). In *A Defense of Common Sense* (1925), he argued that commonsense beliefs were superior to philosophical ones. Sometimes referred to as the father of analytic philosophy, Moore insisted on a rigorous definition of terms so as to avoid the linguistic confusions that have always dogged philosophers.

Moore was well known as an ethicist. His *Principia Ethica* (Elements of ethics, 1903) was widely read outside the philosophical community as well as by philosophers. One of his main ethical contentions was that moral good could not be satisfactorily defined. It had to be grasped directly, through intuition. Those who defined goodness in terms of social approval or desirable outcomes committed the NATURALISTIC FALLACY (i.e., inferred moral conclusions from nonmoral factual premises).

See also MONISM; MORAL PHILOSOPHY.

moral objectivism
Moral objectivism is the opposite of MORAL RELATIVISM and thus claims that there are universal moral standards that apply to all people at all times and in all places. Moral objectivists vary in how binding these moral standards are seen to be. The weaker approach holds that moral standards, although universal, are not binding, but merely serve as useful guides as to how one should behave. Moral objectivists account for the presumed cross-cultural universality of some ethical standards in three different ways. Naturalists claim that morality is naturally acquired either through human biology or social experiences. Rationalists argue that moral standards are based on reasoning. We obey moral standards, such as the prohibition on murder, because any rational being can see that this is the correct way to behave. Supernaturalists derive moral standards from the will of God.

See also MORAL PHILOSOPHY; MORAL RELATIVISM.

moral philosophy
Moral philosophy, or ethics, is the branch of philosophy that is concerned with conduct and character and more specifically with ways of distinguishing between right and wrong actions and good and bad outcomes. There are three main branches of ethical discourse: normative ethics, meta-ethics, and applied, or practical, ethics (see chart). Meta-ethics deals with the philosophical underpinnings of ethical theories, principles, values, and codes, by asking such fundamental questions as: "Is there a valid difference between right and wrong?" and "Is morality objective?" Important and interesting though such questions might be, the focus of a recent upsurge of interest in moral philosophy is normative ethics, the study of moral theories and their constituent concepts, values, principles, and standards. Moral theories have excited increasing interest because of the emergence of ethical dilemmas due to social change and the need for professions to develop standards of ethical conduct in a changing world.

The increased interest in ethical decision making is a product partly of social change and partly of technological development (insofar as these are separable). Professions that were not explicitly concerned about ethics in the past have been forced to take more of an interest because their activities are more open to public scrutiny, for example, by ruthless investigative journalists, and because they feel more vulnerable to legal actions brought by clients. Journalists, lawyers, physicians, scientists, and engineers are more educated about ethical issues than ever before and many professions have recently developed detailed codes of ethical conduct

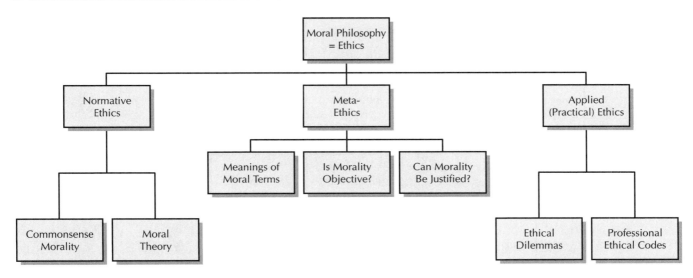

The major branches of ethical philosophy.

that help individuals to traverse the ethical minefields of modern life. One important source of new ethical problems has been technological innovation. These changes have generated ethical dilemmas that are entirely new. Recent developments in biotechnology, for example, have forced scientists and the public to ponder the ethics of genetically modified food, cloning, and alteration of the human genome. Similarly issues of invasion of privacy due to electronic monitoring of the Internet by marketing companies pose a new and, to many, a troubling problem. Now that we are aware of the adverse effects of economic expansion on the environment, and on the diversity of life forms remaining on our planet, there is a new sense of obligation to future generations to conserve and protect existing flora and fauna. In the field of medicine, the development of sophisticated life support systems has muddied the definition of death and raises unprecedented problems as to when such support systems should be disconnected. The emergence of successful organ transplant operations has raised often harrowing questions as to who should receive the limited supply of donated organs.

MAJOR ETHICAL THEORIES

Normative ethics, or prescriptive ethics, provides a basis for making moral decisions. Ethicists like to distinguish between commonsense ethics and formal ethical theories. Commonsense ethics refers to the norms of conduct that most people acquire from a wide variety of authority figures in their social environment and practice without any formal theory or philosophical analysis. Commonsense morality involves moral principles such as "love your enemy," "neither a borrower nor a lender be," "honesty is the best policy," and "do not do unto others as you would not have them do unto you." Commonsense morality revolves around ethical values including generosity, honesty, courage, freedom, wisdom, integrity, and love. Ethical theories are designed to provide a reasoned justification for moral action. Western ethics has produced a large number of ethical theories (see chart), the most influential of which are sketched here.

DIVINE COMMAND

For most religious people, the morality of an action is determined by whether it corresponds to God's will. According to this perspective, a person behaves morally when their actions are what God wants them to do and they behave immorally when their actions go against divine command. In the Judeo-Christian tradition, Moses is supposed to have received the ten commandments directly from God hewn on a stone tablet. This is a list of moral prescriptions that has deeply influenced Western legal codes. Religious fundamentalists believe that the will of God is represented in the entire Bible despite the fact that it is riddled with inconsistencies. For example, revenge is presented as a moral duty in the Old Testament but is seen as immoral in the New Testament. Such textual inconsistencies, as well as problems concerning which texts to include in the religious cannon, and problems with textual provenance and translation, have encouraged some religions to advance the view that divine will is being promulgated through church leaders.

Divine command theories claim that ethical standards can be founded only on religion but most philosophers dispute this. Applied ethicists usually attempt to accommodate the wishes of those whose ethical choices are

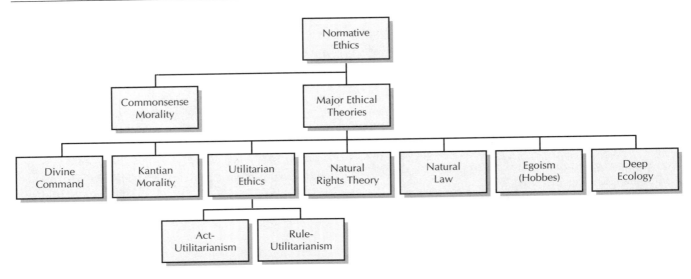

The major theories of normative ethics.

based on religious belief but generally prefer to derive ethical principles from secular theories.

KANTIAN MORALITY

According to the views of IMMANUEL KANT (1724–1804), a German ENLIGHTENMENT philosopher, rational beings are set apart by having intrinsic moral worth. This means that each individual must be treated as an end unto themselves rather than as an external object or an instrumentality by which certain goals can be achieved. This theory of intrinsic moral worth provides an ethical basis for democracy and the rights of the individual and it is influential in the modern ethics of science and technology.

Kant held that the morality of an action is determined not by divine command nor by consequences, but by the intention of the actor. He introduced the concept of the categorical imperative according to which an action is justified only if you behave according to a rule that you would want to be universally observed. For example, Kant believed that borrowing money was immoral because if everyone tried to borrow there would be no money left to lend. The categorical imperative can be used to condemn actions that are proscribed by many moral systems, such as theft, murder, adultery, and deceit.

Kantian ethics provides an important basis for protecting the rights of individuals but it is not particularly helpful in determining ways that conflicts of interest between the sovereign wills of two individuals can be resolved. Another weakness of Kantian ethics is that the categorical imperative does not always bring the behavior of the individual into line with the needs of the larger community. If I am a smoker, I might wish that smoking be permitted in all locations. Yet this would cause universal harm.

UTILITARIAN ETHICS

The possibility that actions can have good or bad consequences and that the morality of an action is determined by its consequences is the central idea of utilitarian ethics. In utilitarian ethics the goal of ethical action is to produce the maximum amount of beneficial consequences, or utility, for the maximum number of people. According to *act* utilitarianism individual actions should maximize utility while *rule* utilitarianism is based on the view that actions should follow a system of rules that maximize utility. Utility is normally measured in terms of happiness or a preponderance of pleasure over pain. This means that utilitarianism has some overlap with the ancient Greek philosophy of HEDONISM.

At face value, utilitarianism seems to imply that any method is permissible if the outcome is good, namely that the end justifies the means. It also has the unappealing implication that the good of minorities must always be sacrificed to the good of the majority. Sophisticated versions of utilitarianism have sought to counter these problems.

The essential feature of ethical action for a utilitarian is that it is determined by a judicious weighing up of possible good and bad consequences. Although the ethical codes employed in science and technology often focus on the rights of people, whether as clients, research participants, or residents of a particular place, ethical practice generally calls for a weighing up of potential good and bad consequences of some course of action that is utilitarian in spirit.

NATURAL RIGHTS THEORY

Natural rights theory holds some similarities with Kantian ethics in the sense that it emphasizes intrinsic rights of the individual. These rights include the right to life, political and intellectual freedom, and the right to hold property. Individuals may do whatever they wish so long as their actions do not violate the rights of others. Rights are negatively defined. For example, the right to life means that no one does anything that causes death. It does not have the implication that anything must be done to save a person from dying. For this reason, natural rights theory is sometimes characterized as "minimal morality." We have an obligation not to infringe on the rights of others but we have no moral obligation to help those who are less fortunate than others.

NATURAL LAW

Natural law theories hold that we are behaving morally when we follow our natural tendencies. Immoral action is based on resistance to natural impulses, emotions, and social relationships. Natural law theories are likely to become more influential as the impact of evolutionary thinking in the social sciences increases and there is greater awareness of how human motivation is affected by evolved predispositions.

DEEP ECOLOGY

The deep ecology approach to ethics is inspired by the environmentalist movement. Instead of assuming that the moral of human individuals is the central issue in ethics, this approach is based on the intrinsic moral value of ecosystems. According to this perspective, pollution is not wrong because of its potential for damaging the health of human beings but because it threatens the survival of nonhuman species by destroying their habitat. Deep ecologists argue that a human-centered ethics cannot begin to address problems that are larger than our species, such as the threat of industrial development to ecosystems, and they focus on values in nature that are independent of human needs and rights.

ETHICS OF CARE

Psychological studies of the development of ethical reasoning discovered sex differences that seemed to indicate females were less likely to reach higher stages of morality. In reaction to this conclusion, feminist scholars argued that when you look at sex differences in behavior an opposite pattern can be observed. Women spend much more time and effort helping out relatives and friends than men do. This led to the criticism that formal ethical theories are cold and legalistic and distant from the real world in which people carry out the moral imperative of taking care of each other. Psychologist CAROL GILLIGAN has been a leading advocate for this position. The ethics of care has been compared to Jesus' prescription that we should love our neighbor and to his criticism of the more rationalistic and legalistic morality of the Pharisees.

VIRTUE APPROACHES

Virtue approaches to ethics see the development of good character as the central goal of ethical conduct. Dating to Aristotle, who stated, for example, that the unexamined life is not worth living, virtue theories have never been very popular among philosophers (as distinct from religious mystics like St. Thomas Aquinas) and do not play an important role in applied ethics.

EGOISM

Egoism is not so much an ethical theory as a challenge to ethics. Egoists argue that if there is no advantage to being moral you cannot expect people to behave in moral ways. This perspective was developed by THOMAS HOBBES (1588–1679) who used the conviction that people always act out of their own brutish self-interest to argue in favor of the need for government. Hobbes was a monarchist and derived the absolute power of the monarch from divine right.

SCIENCE AND TECHNOLOGY ETHICS

Although the variety of ethical theories is quite bewildering, there is actually a great deal of overlap in the values that they support and the standards of conduct they recommend. Many ethical theories have similar practical ramifications. For this reason, many applied ethicists avoid restricting themselves to any one ethical theory. Instead of focusing on theory, such pluralists direct their attention to core moral principles. Pluralists may eschew the complexity of controversial theories and concentrate on a small number of principles with which most people can agree.

Applied ethics, including scientific ethics, and professional ethics in general, can operate effectively on the basis of a few fundamental principles drawn from widely held values. Thus, the NUREMBERG CODE, which stipulates the rights of human research participants, has three major principles (rights to informed consent, freedom from harm, and termination of participation) that are based on the ethical value of autonomy or personal liberty. The code also states that scientific research using human participants should have the potential for benefiting humanity, reflecting the value of beneficence. Other values that form the basis for applied ethical principles include privacy, honesty, fidelity, and fairness. Thus, professionals are expected to respect the privacy of their clients, avoid defrauding them, live up to promises they have made, and treat people with the same consideration regardless of differences in age, ethnicity, religion, or income.

Applied ethicists are guided by fairly straightforward ethical principles. Ethical dilemmas arise when these principles are in conflict with each other. When that happens, the ethicist might elevate one principle over another. For example, an ethicist on an INSTITUTIONAL

REVIEW BOARD might decide that the scientific benefits of an experiment in social psychology were more important than the participants' right to informed consent.

See also DEONTOLOGY; GILLIGAN, CAROL; HEDONISM; HOBBES, THOMAS; KANT, IMMANUEL; NUREMBERG CODE; UTILITARIAN ETHICS.

Further Reading
Gilligan, C. *In a Different Voice.* Cambridge, Mass.: Harvard University Press, 1982.
Naess, A. *Ecology, Community and Lifestyle,* trans. D. Rothenberg. New York: Cambridge University Press, 1989.
Nozick, R. *Anarchy, State, and Utopia.* New York: Basic Books, 1974.
Pojman, L. *Ethics.* Belmont, Calif.: Wadsworth, 1995.
Resnik, D. B. *The Ethics of Science.* New York: Routledge, 1998.
Russell, B. *A History of Western Philosophy.* New York: Simon and Schuster, 1972.

moral relativism

Moral relativism is the view that standards of right and wrong are relative to a particular society at a particular point in its history. Thus, cannibalism, which is viewed with such horror in most Western societies, may be permitted or even considered a moral obligation in others. Relativism is an important challenge to the credibility of ethical standards in the sense that it portrays them as quite arbitrary. If ethical standards are so variable and so capricious, why should much weight be attached to them?

Moral relativism falls into two categories. General ethical relativism focuses on differences between societies. Special ethical relativism recognizes that the standards vary between different social institutions and professions. Special ethical relativism is less controversial. Most people recognize that there are essential differences between the codes followed in various occupations. The gap between scientific ethics and the ethical standards of other occupations becomes clear when scientists work in applied settings in collaboration with other professionals. When scientists give expert testimony in court as a way of making their living, for example, the scientific ideal of objective presentation of evidence may get left by the wayside. Similarly, when scientists work in industry, the need to protect trade secrets and other proprietary information may get in the way of openness in scientific communication. Special ethical relativism is real and its problems are not easily overcome for scientists who make their living in applied settings. In general, scientists who leave the protected environment of academic life can expect to make ethical compromises. If they find it difficult to make such compromises, then it might be argued that they should attempt to make a living in a university setting where it is easier to follow professional ethical standards.

General ethical relativism is historically recent and its emergence can be attributed to a number of coinciding trends. One is the CULTURAL RELATIVISM disseminated by influential anthropologists such as Margaret Mead who claimed that cultural conditioning makes people completely different in different societies. This would explain why their ethical proclivities might be so different. Another relevant trend is the decline in Western religious belief. If morality is interpreted as following divine commands, then these would be the same in different places. A decline in religious belief thus creates a favorable environment for general ethical relativism.

Relativism has been fostered by revulsion at the excesses of colonial domination. When European powers conquered the indigenous people of Africa, Asia, and the Americas, their stated aim was to civilize and convert the heathen. In the process, they destroyed indigenous civilizations and exploited, or enslaved, the inhabitants. The barbarity of this process caused many to question its moral underpinnings and thus to adopt a relativist stance.

Some modern scientific ideas, such as the big bang theory in cosmology and the theory of evolution by natural selection in biology, have arguably undermined belief in objective standards of morality because they suggest that the earth and the creatures living on it are products of random events. The same is true of modern quantum physics, which can be interpreted as undermining all notions of objectivity.

A cultural relativist stance is supported by the scientific fact of diversity of moral standards in different civilizations. Most of the world's societies permit polygamy but marrying several wives in the West is not only a crime but is considered to be morally wrong. In India, people worship cows. In America cows are turned into hamburgers.

Critics of moral relativism attempt to minimize the importance of cross-cultural ethical diversity in several ways. One argument is that even though there are indeed societal differences, these may mask an important shared core of moral standards. Thus, most societies condemn murder, rape, incest, theft, dishonesty, and sadism. It is true that not all people are treated similarly. In some societies, one's enemies are seen to fall outside the moral community so that it might be considered acceptable to torture men and rape women when the opportunity presents itself.

The universality of moral standards has been explained in at least two ways. First, if societies do not adhere to certain moral standards, they tend to disintegrate. For example, if a society condoned impulsive homicide, there would be a high death rate and this would make a particular group vulnerable to domination by its neighbors. Second, it can be argued that some aspects of morality are supported by emotional adaptations. Thus the anger that people feel when they find out that they have been cheated is universal and may play a role in the expectation that individuals should behave honestly. Moreover, economist Robert Frank has argued that the emotion of embarrassment that is expressed by blushing is a mechanism that has evolved to keep people honest in their social contracts.

The fact of cross-cultural moral diversity may not be so troubling to moral objectivists if they can provide an environmental explanation for the differences. For example, it has been argued that the Jewish moral prohibition against eating pork is a defense against trichinosis, a parasite of pig meat. According to this view, trichinosis would have been a particularly serious health problem in the historical environment in which the Jewish dietary laws arose. Such arguments are rarely well developed, or sophisticated enough, to be very compelling. Yet, they could be improved by being stated as testable scientific hypotheses.

Perhaps the weakest aspect of relativism is logical in nature. Opponents of relativism accuse it of jumping from "is" statements to "ought" statements. Because there is ethical diversity among the societies of the world does not mean that this is desirable. To make this inference is to perpetrate the NATURALISTIC FALLACY. Because societies differ in their ethical standards, it does not mean that they would not benefit from following a universal moral code in which societal differences were lost.

See also CULTURAL RELATIVISM; MORAL OBJECTIVISM; NATURALISTIC FALLACY.

Further Reading
Frank, R. *Passions within Reason.* New York: W. W. Norton, 1989.
Frankena, W. *Ethics.* Englewood Cliffs, N.J.: Prentice Hall, 1973.
Pojman, L., ed. *Philosophy.* Belmont, Calif.: Wadsworth, 1990.
Resnik, D. B. *The Ethics of Science.* New York: Routledge, 1998.

morning-after pill *See* RU-486, ABORTION PILL.

Morton, Samuel George
(1799–1851)
American
Anthropologist

The view that races are fundamentally different received its most substantial scientific backing in the mid-19th century from the work of Samuel Morton, whose research, focused on racial differences in head size. When

Morton died in Philadelphia, on May 15, 1851, his *New York Times* obituary declared him to be the single most eminent American scientist in terms of his international impact.

Morton flatly rejected the biblical image of a single Adam (the monogenic view). He believed that the different races could be considered separate species, each descended from a different Adam (the polygenic view). Instead of relying on travelers' tales and other kinds of anecdotal evidence that fueled the popular, and scientific, debates about race differences, he set out to collect objective data in support of his view that the races are distinct.

According to Morton, Europeans were intellectually and morally superior to other races. If so, they ought to have larger brains. (While Morton never *explicitly* connected intelligence with brain size, it is quite clear that his interest in brain size derived from the knowledge that the brain is the seat of intellect). He set out to test the hypothesis of European superiority by measuring skulls. Since the skull had encased the living brain, measuring cranial capacity is a good proxy for the size of the brain.

Having collected hundreds of complete human skulls from different racial groups, including a large collection of Native American skulls and crania from Egyptians and Negroes buried in the Pyramids, Morton set out to measure their cubic capacity. His first method consisted of filling the skulls with sifted white mustard seed and then pouring the seed into a graduated cylinder to measure its volume. Morton, like all scientists, was concerned with accurate measurement. It bothered him that the mustard seed method produced inconsistent results. Remeasuring the same skull could produce a value that differed by as much as 4 cubic inches from the previous measure. The problem was that the mustard seed did not pack well and varied in size despite sifting. (The imprecision of this early work has stimulated the charge of unconscious racist bias leveled by biologist Stephen Jay Gould, who remeasured some of the skulls. However, a further reanalysis of the skulls by anthropologist John S. Michael concluded that Morton's work had been conducted with integrity and called Gould's own academic integrity into question.) When Morton switched to one-eighth inch lead shot, he found that the error value decreased to 1 cubic inch. This level of accuracy finally allowed for meaningful comparison of the cranial capacity of different racial groups.

Morton produced two major publications. His 1839 book *Crania Americana* was based on the mustard seed method. The lead shot data were published in 1849 in the *Proceedings of the Academy of Natural Sciences.* The more reliable lead shot method produced disappointing results, from Morton's perspective. He found that Caucasians, with an average cranial capacity of 92 cubic inches, had larger brains than Mongolians (82 cubic inches), Malays

(83 cubic inches), Native Americans (79 cubic inches), and Negroes (82 cubic inches). This may have tended to support the idea of European superiority, but it was actually rather damaging to the more general hypothesis that different human groups could be ranked in terms of their brain size with implications for intellectual and behavioral differences. Excluding the European data, the other four groups in Morton's classification were virtually identical in brain size.

Morton's real scientific problem was neither measurement error nor unconscious bias but a poor grasp of the importance of sampling techniques. His samples differed in size and there are several reasons why they might not have been representative of the groups from which they were drawn. One is body size. Another is age at the time of death. Perhaps the most important issue was Morton's failure to take account of the role of gender in brain size. Since male brains are larger than female ones, the more males in the sample the larger the group average will seem and the more females in the sample the smaller the group average will become. Marred by such ambiguity, his results are scientifically worthless although they have alerted modern students of craniometry to the pitfalls of sampling bias. Morton's work was used by unscrupulous popularizers to make the case that there were superior and inferior human races.

See also SCIENTIFIC RACISM.

Further Reading
Gould, S. J. *The Mismeasure of Man.* New York: W. W. Norton, 1996.
Michael, J. S. "A New Look at Morton's Craniological Research." *Current Anthropology* 29 (1988): 349–54.

mountaintop removal

In this little-used technique of coal mining, entire mountains are blown up as the coal is extracted. Unused rubble is simply tipped into valleys where it clogs streams. The natural landscape is completely destroyed and local communities must endure the noise of blasting, dust in the air, and the real threat of substantial mudslides during wet weather. Mountaintop removal (MTR) apparently violates the Surface Mining Control and Reclamation Act of 1977, which requires mine operators to restore mined terrain to its approximate original contours. However, the law suffers from a loophole. MTR is allowed if the mining companies can demonstrate that it is important for the long-term economic prospects of a region. MTR variances have been awarded in poor regions of West Virginia and Kentucky even though this heavily mechanized method of coal extraction provides few jobs. Despite increases in the rate of coal extraction, these regions have remained among the poorest in the country. It is likely that MTR will come under increasingly strict control as environmental groups challenge mining companies and the local political representatives that are often biased in their favor.

See also STRIP MINING.

Further Reading
Ward, K., Jr. "West Virginia Congressmen Fight Ban on Mountaintop Removal Mining." *Knight-Ridder/Tribune Business News,* November 1, 1999.

multiple discovery   *See* SIMULTANEOUS DISCOVERY.

mutual assured destruction

Mutual assured destruction (or mutually assured destruction) was the philosophy underlying the buildup of NUCLEAR WEAPONS during the cold war period. The idea was to build up enough of an arsenal so that if the enemy struck, first wiping out a certain amount of your offensive capability, you still had enough weapons left to launch a devastating counterattack that wiped out the population of the country launching the initial strike. If an enemy expected such a catastrophic consequence following their initial strike, they would be suicidally foolish to launch an attack. In this way, the build up of nuclear weapons guaranteed that they would never be used.

See also NUCLEAR WEAPONS.

N

Nader, Ralph
(1934–   )
American
Lawyer, Consumer Advocate

Ralph Nader has devoted his life to defending the rights of individuals against the power of corporations. He first became interested in car safety while representing the plaintiffs in accident damage suits. Horrified by the lack of interest of manufacturers in automobile safety, he wrote *Unsafe at Any Speed* (1965). His careful detailing of safety problems of American cars in this book proved influential in passage of more stringent car safety laws and Nader served as a government consultant in formulating the new standards. Other public issues in which Nader has played an important role include: meat safety, care of the elderly and mentally ill, air and water pollution, ELECTRONIC PRIVACY, and NUCLEAR POWER. He was an independent candidate in the 2000 presidential election.

See also FORD PINTO.

Naess, Arne   *See* DEEP ECOLOGY.

Nagasaki, Japan   *See* HIROSHIMA.

NASA (National Aeronautics and Space Administration)

NASA is the U.S. federal agency that coordinates research in the development of aviation and space technologies. Formed by the National Aeronautics and Space Act of 1958, NASA replaced the National Advisory Committee for Aeronautics that had been founded in 1915. During the 1960s NASA commanded a huge budget as the United States, which had lost in the race to be first in space to the Russian *Sputnik* satellite in 1957, set its sights on sending an American to the moon in the Apollo Project. This cost approximately $25 billion and, at its height, employed over 400,000 people, including contractors.

Following the success of the moon landing, NASA's budget was greatly reduced but it continued an ambitious program of space exploration. A reusable space shuttle was also developed to make access to space more reliable, easier, and cheaper. The 1986 explosion of the shuttle *Challenger* during launch was a devastating political blow to the space mission and it was followed by further budget reductions. The space shuttle program has had important commercial uses, including the launching of communications satellites. Although NASA has lost considerable prestige since the 1960s, it has played a central role in propelling the United States to the forefront in electronic technologies and thus has arguably more than repaid large budget expenditures over the years.

See also CHALLENGER ACCIDENT; HUBBLE TELESCOPE.

National Academy of Sciences (NAS)

Established by the U.S. Congress in 1863, the National Academy of Sciences (NAS) is a private nonprofit society of scholars that promotes scientific research and serves in an advisory capacity to the federal government on issues in science and technology. Including its offshoot, the National Academy of Engineering (NAE), and the Institute of Medicine, the National Academies have over 5,000 members. Membership is honorific and is by election of existing members. Both the NAS and the NAE are operated by the National Research Council, which was founded in 1916 and reorganized in 1974 to include more scientific disciplines. The National Academies promote research both nationally and internationally by organizing conferences, administering funds for research and fellowships, and sponsoring scientific publications.

See also NASA; NATIONAL INSTITUTES OF HEALTH; NATIONAL SCIENCE FOUNDATION.

National Institutes of Health (NIH)

A branch of the Public Health Service, which is itself a division of the U.S. Department of Health and Human Services, the National Institutes of Health (NIH) are responsible for conducting and supporting research into the causes, cures, and prevention of diseases. The NIH comprises national institutes for a variety of medical specialties such as aging, cancer, diabetes, environmental health, neurological disorders, etc. The overarching mission of all of these institutes is to improve the health of all Americans.

See also NASA; NATIONAL ACADEMY OF SCIENCES; NATIONAL SCIENCE FOUNDATION.

National Science Foundation (NSF)

The National Science Foundation was established in 1950 as an independent agency of the U.S. government. Its mission is to promote scientific progress while enhancing national health, welfare, and prosperity. It has a board of 24 part-time members and a director all of whom are appointed by the president with the consent of the U.S. Senate. The NSF has an annual budget of $3.3 billion that is used primarily to support research and educational projects through grants and contracts to universities and scientific organizations. In addition, the NSF supports a number of ground-based astronomic observatories, the Ocean Drilling Programs, and the U.S. Antarctic Program. The NSF provides scientific information and advice to other branches of government. It also promotes affirmative action policies in science and technology careers.

See also AFFIRMATIVE ACTION IN SCIENCE AND TECHNOLOGY CAREERS; NASA; NATIONAL ACADEMY OF SCIENCES; NATIONAL INSTITUTES OF HEALTH.

National Security Agency (NSA)   *See* CRYPTOGRAPHIC RESEARCH, ELECTRONIC PRIVACY.

natural law theory   *See* MORAL PHILOSOPHY.

natural rights theory   *See* MORAL PHILOSOPHY.

natural theology

Natural theology refers to knowledge of a divine will or presence that is obtained directly from the natural environment instead of through religious texts or sermons. It played an important role in the emergence of natural history, and hence modern biology, as a field of study. Some of the leading pioneers in the systematic description of nature such as Swede CAROLUS LINNAEUS (1707–78) and his predecessor Englishman John Ray (1627–1705) were motivated to describe animals and plants to prove that the earth was created by God as a habitat for human beings. Evidence of God's handiwork was to be found in the wonderful design of animals that enabled them to survive and reproduce in their natural environment, a theme developed in John Ray's book *The Wisdom of God Manifested in the Work of Creation* (1692).

Ironically, study of the natural world quickly revealed inconsistencies with the biblical story of creation, including the discovery of fossils of large numbers of extinct species that necessitated a revisionist theory of multiple creations, known as CATASTROPHISM. Natural theology is a direct ancestor of evolutionary theory and John Ray's "divine wisdom" differs from CHARLES DARWIN's "adaptation" primarily in the underlying explanatory theory (i.e., divine creation versus natural selection, respectively).

See also CATASTROPHISM; DARWIN, CHARLES; LINNAEUS, CAROLUS; RELIGIOUS IDEOLOGY AND SCIENCE.

naturalism

Naturalism is the philosophical position that denies there is any supernatural reality. Reductionist naturalism, which prevailed in the 17th through 19th centuries, held that all natural phenomena were reducible to physical objects. This meant that human beings were bounded by determinism. According to contemporary naturalism, physical science is capable of explaining only a small fraction of the phenomena we experience. Physical science cannot account for the diversity, richness, and value we attribute to our surroundings or explain our sense of freedom and spontaneity.

See also GENETIC DETERMINISM; POSITIVISM.

naturalistic fallacy

The naturalistic fallacy states that we are not entitled to derive moral statements from factual ones. As the Scottish philosopher David Hume (1711–76) first pointed out, we cannot legitimately switch from statements about how things are to conclusions about how they should be. By making this distinction, he erected a barrier between ethics and science that few scholars have been willing to cross. This intellectual boundary is sometimes referred to as the is/ought distinction.

SOCIAL DARWINISM is often held up as an example of the naturalistic fallacy in which the natural state of competition for survival and reproduction among nonhuman animals was inappropriately used to justify the unrestrained economic competition, and exploitation, of laissez-faire capitalism. Both ethical naturalists (who define morality in terms of natural impulses) and evolutionary ethicists (who seek to explain the emergence of moral sentiments in humans, and potentially in other species), provide a possible exception to the naturalistic fallacy because they combine ethical statements and scientific ones in a way that is not obviously fallacious. Another possible exception is the study of impulse control, and hence moral behavior, from the perspective of neuroscience. Similarly, descriptive ethics (the scientific study of moral standards in different societies) crosses the is/ought barrier.

See also EVOLUTIONARY ETHICS; SOCIAL DARWINISM.

Nazi research   *See* MENGELE, JOSEF; NUREMBERG CODE.

Newton, Sir Isaac
(1642–1727)
English
Physicist, Mathematician

Isaac Newton's name is synonymous with scientific genius in the same way that ALBERT EINSTEIN's is. Both men made revolutionary discoveries while separated from the academic mainstream. The young Einstein labored in the Swiss Patent Office in Berne while making some of his most important advances in the theory of relativity as an amateur researcher. The context of Newton's early scientific breakthroughs was even odder. By the age of 22, he had attracted little attention at Cambridge University. In 1666, Cambridge was closed by the plague and Newton returned to his mother's farm in Lincolnshire where he spent the next eighteen months initiating revolutionary advances in physics and mathematics. During this period, he developed calculus, which was independently discovered by German mathematician and philosopher Gottfried Leibniz several years later. His other accomplishments included formulation of the three laws of motion and the experimental demonstration by prism experiments that white light consists of light of several different colors. Perhaps the most important fundamental discovery made by Newton in this period was universal gravitation, the notion that all objects are attracted to all other objects. Using a clever synthesis of his own laws of motion and Kepler's laws of planetary motion, he showed that gravitational force is proportional to the inverse square of the distance between two heavenly bodies.

In 1669, Newton obtained the Lucasian chair in mathematics at Trinity College. In 1687, he published *Philosophiae Naturalis Principia Mathematica* (or The mathematical principles of natural philosophy), a book that provided a synthesis of his mathematical work on gravitation and the orbits of planets. The *Principia,* as it is called, established Newton as the leading scientist of his day. Within a few years after its publication he suffered a nervous breakdown and abandoned research altogether for a career at the Royal Mint. Some scholars believe that his psychological problems were aggravated by contact with poisonous chemicals in the context of his enthusiastic pursuit of the esoteric *pseudoscience* of alchemy. In 1703, he was elected president of the Royal Society of London and, in 1708, he was knighted by Queen Anne, the first scientist to receive this honor. Newton died in London on March 20, 1727, having been ill with gout and inflamed lungs.

Like Galileo (1564–1641), Newton was committed to empiricism as a means of establishing the truth. Also like Galileo, he has been accused of failing to practice what he preached. Galileo is believed to have conducted some key experiments in his head! Similarly, Newton is accused of fudging the data to make it match more closely with his theoretical predictions.

Newton's fudging took place in the context of a rivalry with German philosopher Gottfried Leibniz (1646–1716), who disagreed with Newton's theory of universal gravitation. In later editions of the *Principia*, Newton adjusted his calculations to bring the results more closely in line with his theoretical predictions and then used the numerical precision of his findings as a rhetorical weapon against Leibniz. In the final edition of his masterpiece, he boldly claimed an accuracy of better than 1 part in 1,000. It is astonishing that Newton's contemporaries were oblivious to his finagling and this bears testimony to the skill and aplomb with which he faked the numbers. In fact, over 250 years would pass before his dishonesty was exposed (see SCIENTIFIC DISHONESTY).

Newton played to win and this is nowhere clearer than the abuse of his position as president of the Royal Society to further his case for priority over Leibniz in the development of calculus. In 1712, the Royal Society issued a report that backed up Newton's claim to have been first to devise calculus. Ostensibly the work of an impartial scientific committee, the entire report appears to have been written by Newton himself. The unctuous preface emphasized the importance of objectivity and fair play in adjudicating the dispute, pointing out that a fair judge would never admit a witness in his own cause.

Why distinguished scientists with such an obvious stake in defending scientific objectivity should indulge in this kind of scientific arrogance is difficult to explain. Newton, like Galileo, was a highly argumentative man who cultivated bitter scientific rivalries. The desire to overwhelm his intellectual enemies, particularly Leibniz and ROBERT HOOKE, may have motivated him to cross the line into SCIENTIFIC DISHONESTY.

See also EINSTEIN, ALBERT; POSTMODERNISM AND THE PHILOSOPHY OF SCIENCE; SCIENCE, HISTORY OF; SCIENTIFIC DISHONESTY.

Further Reading

Broad, W., and N. Wade. *Betrayers of the Truth.* New York: Simon and Schuster, 1982.
Christianson, G. E. *Isaac Newton and the Scientific Revolution.* New York: Oxford University Press, 1998.
McClellan, J. E., and H. Dorn. *Science and Technology in World History.* Baltimore, MD: Johns Hopkins University Press, 1999.
Westfall, R. S. "Newton and the Fudge Factor." *Science* 179 (1973): 751–58.
———. *The Life of Isaac Newton.* Cambridge: Cambridge University Press, 1993.

Nietzsche, Friedrich
(1844–1900)
German
Writer, Philosopher

Friedrich Nietzsche is usually described as a philosopher but his academic background was in ancient Greek and Roman literature and his philosophical ideas were transmitted through the literary forms of his day, including books of aphorisms and works of fiction. He contributed no formal academic theories to philosophy and his primary interest was in ethics.

After his 10-year career as a professor of philology at the University of Basel (1869–79) was brought to an end by ill health, Nietzsche spent the next decade writing books while he lived in Switzerland and Italy as well as Germany. His health had been severely damaged by an 1870 experience as a volunteer medical orderly in the Franco-Prussian War when he suffered from dysentery and diphtheria. In 1899, while living in Turin, Italy, Nietzsche suffered a severe mental breakdown, probably caused by the advanced stages of syphilis, and he died the following year.

Nietzsche was essentially a romantic who identified with his close friend the composer Richard Wagner and admired the aristocratic nihilism of English poet Lord Byron. Like Byron, he was fond of announcing that conventional morality had become irrelevant. For Nietzsche, God was dead, having been killed off by the rise of scientific knowledge due to the ENLIGHTENMENT. He neither accepted religion nor the scientific worldview that was replacing it and sought refuge in the construction of a pagan mythology that could give ethical meaning to his life.

In *Thus Spake Zarathustra* (four volumes, 1883–85) Nietzsche's most important work, he attacked both Christianity and democracy as mass ideologies of the weak and obedient. In their place, he developed the aristocratic ideal of the superman who is driven by the will to power, celebrates the moment by living dangerously, and develops his capacity for the creative use of passion. Nietzsche's superman bears more than a passing resemblance to Siegfried, the hero of Wagner's *Ring* cycle of operas.

Nietzsche is often compared with MACHIAVELLI (1469–1527) because he developed an ethic of conduct that ran directly counter to the religious views of his day. Just as the bloodthirsty Cesar Borgia represented an ideal of princely conduct for Machiavelli, Napoléon was an embodiment of Nietzsche's superman. As far as Nietzsche was concerned, the fact that Napoléon had come out of the French Revolution was a complete justification for the murder and anarchy that prevailed during that period. Contrary to the democratic ideals of his day, Nietzsche did not think that the loss of life of ordinary people was of any account. It was his custom to refer to the ordinary run of humanity as the "bungled and botched."

Nietzsche's ideas were borrowed by the Nazis to bolster their conception of the master race as analogous to the superman concept, but this was clearly a distortion. Nietzsche was not a nationalist. In fact, his contempt for the smugness and mediocrity of German society produced a falling out with Richard Wagner. Moreover, he was an elitist who had no time for populist mass movements of any stripe.

There is much in Nietzsche's writing that was designed to shock. In *Beyond Good and Evil* (1886), he tells us that evil is preferable to good. He has a stoical quality that is at odds with the utilitarian philosophers of his time and views pain as good because it can strengthen the will. The democratic formulae of the utilitarian philosophers, as represented by JOHN STUART MILL (1806–73), whom he referred to as a "blockhead," were singled out for particular scorn. Nietzsche detested the implication that there is an equivalence between the actions and prerogatives of one individual and any other individual as represented in Mill's dictum, "Do not do to others that which you would not that they should do unto you."

Nietzsche's rejection of the utilitarian ideal of promoting the greatest happiness of the greatest number of people is just as unambiguous, and just as provocatively stated. As far as he was concerned, the suffering of a heroic single individual can weigh more heavily in the ethical balance than the misery of an entire nation. "The misfortunes of all these small folk do not together constitute a sum total except in the feelings of mighty men."

Nietzsche's spirited attacks on the prevailing ethical ideas were an important influence on the existentialist generation of philosophers who followed him. He was a great writer whose books are exhilarating to read despite the unpalatable conclusions that they contain. In all this, he was essentially a romantic describing an escapist mental realm in which we are allowed to experience more grandiose ideas than the real world permits us.

See also ENLIGHTENMENT, THE; MACHIAVELLI, NICCOLÓ; SOCIAL DARWINISM.

Further Reading
Conway, D. W. *Nietzsche's Dangerous Game.* New York: Cambridge University Press, 1997.
Russell, B. *A History of Western Philosophy.* New York: Simon and Schuster, 1972.
Thiele, L. P. *Friedrich Nietzsche and the Politics of the Soul.* Princeton, N.J.: Princeton University Press, 1990.

noise and cognitive development

Noise is an ever-present reminder of industrial development and there is good reason for concluding that the noise produced by cars, trucks, and airplanes is just as much a pollutant as their exhaust. Noise has pervasive health consequences quite apart from industrial deafness and it is particularly deleterious to the developing nervous systems of children. Thus, children living in an apartment building built directly on top of the entrance to the George Washington Bridge in New York City were found to be doing badly in school. This was most likely due to the noise because children on the lower floors, who were exposed to louder traffic noise, did worse in reading than children who lived higher up in relative silence.

An even more compelling case for the adverse effects of long-term exposure to noise comes from study of academic performance in a school built next to the Los Angeles International Airport. Children attending this school had higher blood pressure than children in quieter schools in the city, which indicated that the prolonged noise was stressful. They were also worse at solving problems and their performance declined with the amount of time they had spent in the noisy schools. Instead of learning ways of coping with disturbance from the airliners, the children actually became more distractible with continued exposure. They developed a marked tendency to give up instead of persisting at difficult problems until a solution was achieved. The results were so striking that the school authorities decided to soundproof the classrooms.

Adverse effects of noise for the cognitive development of children place an ethical responsibility on those who design both machines and buildings to protect children from noisy environments. Given this knowledge, it is no longer acceptable to locate schools near airports or beneath the flight paths of large planes. Residential housing should also be protected from highway noise through appropriate zoning decisions and the use of sound-attenuating barriers such as belts of trees. More attention should also be paid to soundproofing homes, particularly in inner-city neighborhoods having high exposure to traffic noise where so many other environmental influences, including LEAD POISONING, tend to undermine academic performance.

See also LEAD POISONING; NOISE POLLUTION.

Further Reading
Cohen, S., G. W. Evans, D. S. Krantz, and D. Stokols. "Physiological, Motivational, and Cognitive Effects of Noise on Children." *American Psychologist* 35 (1980): 231–43.
Cohen, S., G. W. Evans, D. S. Krantz, D. Stokols, and S. Kelly. "Aircraft Noise and Children." *Journal of Personality and Social Psychology* 40 (1981): 331–45.
Cohen, S., D. C. Glass, and J. E. Singer. "Apartment Noise, Auditory Discrimination, and Reading Ability in Children." *Journal of Experimental Social Psychology* 9 (1973): 407–22.
Fay, T. H., ed. *Noise and Health.* New York: New York Academy of Sciences, 1991.

noise pollution

Noise pollution, mainly from traffic, is a constant of city life and even rural havens are increasingly invaded by noises from pleasure boats and jet skis. The ordinary noise level in most cities is far greater than anything our ancestors would have experienced moving about in the wilderness gathering and hunting. Can we make the adjustment with complete success, or is there a price to pay?

Modern scientific interest in the effects of noise began with the premise that loud noise is stressful. Like other stressors, it ratchets up bodily arousal in terms of increased heartbeat, breathing, blood pressure, muscle tone, and so forth. These bodily responses are part of a coordinated set of physiological reactions to an emergency situation (known as the fight-or-flight response). Noise also impairs our ability to concentrate and solve problems. Repeated exposure to noises of 90 decibels or higher damages the hearing receptors in the inner ear resulting in diminished ability to hear, particularly high-pitched sounds. Such levels are routinely exceeded in many industrial workshops necessitating the wearing of ear protectors. Industrial deafness is common among rock musicians exposed to the high sound amplification used at concerts.

When people were exposed to repeated bursts of loud noise in the laboratory an interesting pattern emerged. At first, the noise produced a strong bodily reaction and undermined the ability to perform arithmetic and solve word problems. However, participants in the experiment adapted very well. After 25 minutes, their physiological response to the noise was much lower and their problem-solving ability completely recovered. Despite this remarkable adjustment, exposure to the noise had a delayed effect. In the second part of the experiment, subjects who had been exposed to the noise performed poorly in solving problems and also became frustrated very quickly. Noise is thus a psychological stressor. Prolonged exposure to a psychological stressor wears down our ability to adapt to it. Eventually, it can undermine well-being and health.

The potential for adverse health effects of repeated exposure to noise places an ethical responsibility on those who design machines and buildings, as well as on industrial managers. Industrial workers exposed to high noise levels should be encouraged to wear ear protectors. Residential housing should be protected from highway noise through appropriate zoning decisions and the use of physical barriers such as fences and belts of trees. More attention should also be paid to soundproofing homes, particularly in inner-city neighborhoods having high exposure to traffic noise. Noise pollution has been greatly reduced by the development of automobiles with quieter running engines and the use of sound-attenuating mufflers in exhaust systems. As machines of the future become more energy efficient (to reduce AIR POLLUTION) they will also inevitably become quieter since noise is a form of energy and thus represents inefficiency in design.

See also NOISE AND COGNITIVE DEVELOPMENT.

Further Reading

Fay, T. H., ed. *Noise and Health.* New York: New York Academy of Sciences, 1991.

Glass, D., and J. Singer. *Urban Stress: Experiments on Noise and Social Stressors.* New York: Academic, 1972.

Hansen, C. *Understanding Active Noise Control.* London: E and F N Spon, 2000.

Smith, B. J., R. J. Peters, and S. Owen. *Acoustics and Noise Control.* Reading, Mass.: Addison-Wesley, 1996.

normative ethics

Normative ethics is one of the three main branches of MORAL PHILOSOPHY, (the others being meta-ethics, or the philosophical underpinnings of ethical theories, and applied, or practical, ethics). It focuses on the study of moral theories and their constituent concepts, values, principles, and standards. The main branches of normative ethics are UTILITARIANISM (or ethical judgement of actions on the basis of consequences) and DEONTOLOGY (ethical decisions made on the basis of obligations and rights) with the virtue (or character-building) approach to ethics remaining a minority interest. Normative ethics is prescriptive (in the sense of telling people how they ought to behave) and is thus contrasted with descriptive ethics (the scientific study of moral standards in different societies). Moral theories have excited increasing interest because of the emergence of ethical dilemmas due to social change and the need to refine standards of ethical conduct in a changing and increasingly complex world.

See also DEONTOLOGY; MORAL PHILOSOPHY; UTILITARIANISM.

Nozick, Robert

(1938–)
American
Philosopher

Robert Nozick is an ethicist who bases his ethical theory on rights. He argues that our rights to life, liberty, and property ownership are absolute and cannot be legitimately violated by the state or by anyone else. According to Nozick, our primary ethical obligation is to avoid infringing on the rights of others. We do not have any obligation to protect the interests of others. Thus poor people may appeal to the compassion of the wealthy but governments have no right to tax the rich to provide for the poor. Although Nozick's ideas appeal to individuals having right-wing political views, many ethicists are repelled by their harshness. Politically and ethically, Nozick is often seen as being a polar opposite to John Rawls, whose conception of justice called for a radical redistribution of property.

See also DEONTOLOGY; MORAL PHILOSOPHY; RAWLS, JOHN.

Further Reading
Nozick, R. *Anarchy, State, and Utopia.* New York: Basic, 1974.

N-rays, as shared illusion of scientists

Although science, unlike most other forms of enquiry, has an effective mechanism for rejecting errors and for that reason steadily increases its accumulation of reliable knowledge, this mechanism is far from perfect. Practicing scientists are exposed to a great deal of uncertainty and for that reason are sometimes vulnerable to a bias in confirming what they believe to be true. As a result of this bias, scientists may waste a great deal of time and effort in exploring phenomena that turn out to be illusory. In general these biases affect the work of individual scientists pursuing their individual intuitions but there are episodes in the history of science in which belief in illusory phenomena is widely shared among scientists. Examples include the belief that there is an "ether" through which light travels and the view that substances can burn only if they contain "PHLOGISTON." A 20th-century example concerned N-rays.

At the end of the 19th century, scientists were busily exploring the electromagnetic spectrum and had recently discovered many new kinds of radiation, such as X-rays, cathode rays, and radio waves. In 1903, French physicist Rene Blondlot claimed the discovery of N-rays, which were observed as an increase in brightness that was visible to the naked eye. During the next few years, around a hundred scientists produced over 300 papers on N-rays, which were supposedly observed in chemicals, gases, magnetic fields, and even in the human brain. Many of the scientists were highly respected and Blondlot's priority in the field was recognized when he received the Leconte Prize from the French Academy of Sciences.

The illusion was punctured when American physicist R. W. Wood visited Blondlot's laboratory. Blondlot claimed that he could observe the N-rays splitting into different wavelengths in a darkened room after they had passed through a prism. Unfortunately for him, he continued to make the same claim even after Wood had removed the prism.

Following Wood's debunking of N-rays, most of the world's scientists lost interest in them although they continued to have a following in France for several years. As in the case of the more recent COLD FUSION CONTROVERSY, bad scientific conclusions can be seen to follow from sloppy, inadequate, scientific practices spurred on by an unscientific capacity for self-deception and a lack of appropriate skepticism. There is some controversy over whether such episodes ought to be seen as aberrations or whether they reflect imperfections that are inherent to the scientific process. One important source of bias stems from the failure of scientific journals generally to publish reports of failures to establish some phenomenon. For this reason, scientists usually do not write such reports. This means that positive findings from badly conducted studies are much more likely to be published than negative results from impeccably conducted ones. The resulting bias is often referred to as the FILE DRAWER PROBLEM because unpublishable negative findings are hidden away in files.

See also CLEVER HANS; "PLANARIA SOUP" EXPERIMENTS.

Further Reading
Broad, W., and N. Wade. *Betrayers of the Truth.* New York: Simon and Schuster, 1982.
Huizenga, J. *Cold Fusion.* Rochester, N.Y.: University of Rochester Press, 1992.
Resnik, D. B. *The Ethics of Science.* New York: Routledge, 1998.

nuclear power

About 5% of the world's electricity is produced in some 400 nuclear reactors. Nuclear power is politically unpopular and there is good evidence that the 1986 explosion at the CHERNOBYL nuclear power plant has been influential in this respect. Before Chernobyl, the populations of developed nations were about evenly divided between those in which a majority favored, and those in which a majority opposed, nuclear power. After Chernobyl, the majority in all developed countries surveyed expressed opposition to nuclear power. This suggests that safety of operation is a major concern in the nuclear power debate. Another major concern about nuclear energy is the generation of large amounts of radioactive waste that is difficult to dispose of safely. In addition, the establishment of nuclear generators in politically unstable countries may contribute to proliferation of nuclear weapons, although it is technically very demanding to enrich reactor fuel to the point that it can be used for weapons.

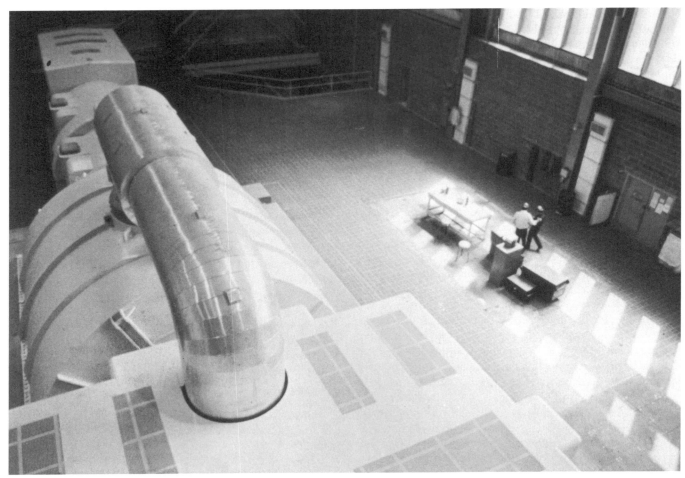

Turbines at St. Urain Nuclear Power Plant, near Denver, Colorado. (NATIONAL ARCHIVES)

Fears about the danger to local residents of having a nuclear plant (see photo) located close to them may be overblown. First, normal operation would appear to be completely harmless. People residing near nuclear power plants receive an increased exposure to ionizing radiation that is less than 1% of the normal background exposure to such radiation (from radon gas, cosmic rays, minerals, food and water, and medical and industrial sources). There is no evidence that such low levels of exposure can have any deleterious health consequences.

The risk of reactor meltdown, or partial meltdown, as in the case of Chernobyl and THREE MILE ISLAND accidents, respectively, about which the public has expressed grave concern, need not pose a major risk either, except to the extent that psychological stress undermines health. The Three Mile Island accident was caused by a series of human errors. Despite these errors, the plant proved to be reasonably stable in the sense that no one was killed in the accident. The amount of radiation released into the atmosphere was small. It is esti-

mated that the extra dose of radiation for residents within 10 miles of the plant was equivalent to that from cosmic rays associated with a return flight from Dallas to London. In summary, it is clear that the public health effects of nuclear power in the United States are far less severe than those of fossil fuels, which fill the air with pollution contributing to respiratory diseases as well as GLOBAL WARMING.

The Chernobyl accident, distressing though it was, can be considered an aberration for several reasons. First, the accident occurred during an experiment rather than in the course of normal operation. Second, the operators displayed an astonishingly casual attitude to safety that is unlikely to occur again if only because the Chernobyl accident illustrated the dangers. Third, some of the design problems of the Chernobyl plant are unlikely to be repeated in the future and are not characteristic of Western reactors. These include vulnerability to steam explosions and the lack of an adequate containment building.

Concerning the operation of power plants, it is clear that nuclear power is safe. In terms of deaths from air pollution and accidents, nuclear power is far safer than coal- or oil-burning plants and is even safer than hydroelectricity. Only natural gas is safer and this method of electricity generation contributes to the global warming problem via carbon dioxide emission.

An enlightened evaluation of fuels used in electricity generation must include consideration of not only the health consequences of using the fuel but also the health costs of producing it. Coal mining is notoriously dangerous and coal-burning plants expend huge amounts of this fuel. By comparison, URANIUM MINING is much safer. Even though uranium ore is radioactive, miners receive a small dose of radiation because the material that is mined must be greatly concentrated, or enriched, before it can be used as reactor fuel.

Perhaps the most vexing safety question involves the disposal of hazardous waste in the form of spent reactor cores and radioactive isotopes produced during power generation. Safe methods of disposing nuclear wastes, such as containment in stainless steel or glass followed by burial beneath mountains, is possible although there can be no absolute guarantee that none of these materials will ever reach the surface during the many centuries necessary for some of the isotopes to decay. Radioactive waste is thus a problem that is bequeathed to the future, but the same can be said of the greenhouse gases and other air pollutants associated with the burning of fossil fuels.

Radioisotopes with either very long or very short half lives (the time taken for the amount of radiation emitted to decrease by one-half) do not pose a major disposal problem. Those with very long half lives (greater than a million years) are not a threat because if they decay so slowly, they are producing very low levels of radiation. Examples include uranium-235 and neptunium-237. Isotopes with short half lives (less than 10 years) decay rapidly so that even though they are very hazardous to begin with, they decay quickly and do not need to be placed in a permanent safe containment. Examples include strontium-90 and cesium-137.

The most worrisome radioisotopes are those such as plutonium-239 that have half lives of the order of thousands of years. Such isotopes can be dangerous if present in large enough quantities and their comparatively long half life means that they must be contained for tens of thousands of years before they decay to levels that are considered safe. The length of such time periods has caused many people to think twice about the desirability of nuclear power. Yet, disposal is not the only option. Spent fuel that contains most of the problematic isotopes can be reprocessed so that 99% of the uranium and plutonium is recovered along with other isotopes. This greatly extends (by a factor of 100) the useful life of the fuel.

Reprocessed waste generally has a much shorter half life, which means that it needs to be contained for around 1,000 years instead of 10,000 years. One problem with reprocessing radioactive waste is that it generates large amounts of chemical wastes that must also be disposed of.

In summary, the declining popularity of nuclear power following the Three Mile Island and Chernobyl accidents may be somewhat irrational. When all aspects of the power generation process, including mining of fuel and disposal of waste, are taken into consideration, nuclear fuel is safer and cleaner than most of the widely used alternatives (the only real exception being natural gas). Meltdown has never occurred in a properly constructed plant and the probability of such an event in the future is small. Nuclear power generation may be necessary to avert the global warming arguably produced by burning fossil fuels. Many environmentalists want to develop other energy sources, including wind, solar, hydroelectricity, and geothermal energy because of the problems of safe nuclear waste disposal. Yet, it is questionable whether these alternatives could ever satisfy the world's huge energy needs.

See also CHERNOBYL NUCLEAR DISASTER; HAZARDOUS WASTE; NUCLEAR WEAPONS; THREE MILE ISLAND NUCLEAR ACCIDENT.

Further Reading

Flavin, C. "No: The Case against Reviving Nuclear Power." In *Environmental Ethics,* ed. L. P. Pojman, 467–72. Belmont, Calif.: Wadsworth, 1994.

Jagger, J. "The Natural World of Radiation." In *Environmental Ethics,* ed. L. P. Pojman, 463–66. Belmont, Calif.: Wadsworth, 1994.

Lennsen, N. "Nuclear Waste: The Problem that Won't Go Away." In *Environmental Ethics,* ed. L. P. Pojman, 485–92. Belmont, Calif.: Wadsworth, 1994.

"Yes: Nuclear Power Is Safe, and We Need It." In *Environmental Ethics,* ed. L. P. Pojman, 474–84. Belmont, Calif.: Wadsworth, 1994.

nuclear weapons

Nuclear weapons can be categorized as those that produce nuclear fusion (like the more recent hydrogen bomb) and those that rely solely on atomic fission (like the atom bombs used at HIROSHIMA and Nagasaki). The are many thousands of times more destructive than conventional explosives and the building of thousands of nuclear weapons in the United States and the Soviet Union has threatened the entire global population with annihilation.

The sheer destructiveness of nuclear weapons has prevented their use by the superpowers for fear of retaliation in kind. This perspective is known as the doctrine of MUTUAL ASSURED DESTRUCTION. Its inevitable logic has been to promote the signing of treaties reducing these expensive and unusable weapons, and a similar logic has attended biological and chemical warfare. The breakup of the former Soviet Union and technological advances in space exploration and laser weapon development may allow a shift from the philosophy of mutual assured destruction to one that admits the possibility of defensive strategies against long-range nuclear weapons. This philosophy underlies the recent resurgence of "STAR WARS" initiatives in the Unites States. Some strategists also feel that some limited defensive program is necessary to counter the threat posed by small nations such as North Korea and terrorist individuals and groups that are developing or may develop nuclear capabilities.

The first atomic weapons were developed in the United States through a collaboration between civilian scientists and the military, known as the MANHATTAN PROJECT. This program was initiated in response to the fear that Germany, led by Adolf Hitler (1889–1945), would develop a fission bomb following the 1938 discovery of atomic fission by German scientists. The project was conducted under tight security at several sites, including the Oak Ridge, Tennessee, facility at which scientists mastered the difficult task of enriching uranium ore to weapons-grade material, and the Los Alamos weapons laboratory, directed by physicist ROBERT OPPENHEIMER, at which the first fission bomb was designed.

The fission bomb was based on the concept of critical mass. Radioactive materials are constantly producing radiation as atoms decay. This radiation can destabilize other atoms, which then decay producing more radiation. If enough radioactive material is packed in a small space, the amount of radiation produced increases rapidly until the energy becomes so great that an explosion is produced. Achieving critical mass depends on the purity of the radioactive material and the quantity of it present. In the Manhattan Project, scientists solved the difficult problem of enriching uranium to the point that it could be used in weapons. The basic design of the atomic bomb involved bringing together two subcritical (and therefore stable) masses so as to reach the critical mass at which an uncontrollable chain reaction developed.

The first atomic bomb had a gun-type design in which one piece of uranium-235 was fired at another, in a barrel, so that they reached critical mass at which an uncontrollable chain reaction developed. This weapon was deployed at HIROSHIMA, Japan, on August 6, 1945, (without having been tested) and resulted in a huge loss of civilian lives. Following the work of physicist Enrico Fermi of the University of Chicago on the synthetic radioactive element plutonium-239, five reactors were set up at Hanford, Washington, to produce this fuel that was then shipped to Los Alamos for bomb construction. A plutonium bomb was tested at Alamogordo, New Mexico, on July 16, 1945. In this design, a core of noncritical

plutonium was compressed to critical mass using conventional high explosives. The plutonium bomb was used on Nagasaki, Japan, on August 9, 1945. Its devastating consequences, coming so soon after the Hiroshima bomb, led to Japanese surrender and the end of the war.

The fusion bomb can be vastly more destructive than the fission bomb and has never been used except in the context of testing. Although several different workable designs have been developed, all use a fission bomb to initiate a fusion reaction that burns hydrogen fuel. (This is the same type of reaction that produces the sun's heat energy. The first fusion bomb was tested by the United States on November 1, 1952, at Eniwetok in the Pacific Ocean. It used deuterium (an isotope of hydrogen with an atomic weight of 2) as the fuel. The blast had a destructive force the equivalent of 10 million tons of TNT (or 10 megatons). The booster bomb uses a large fission bomb to initiate the fusion of a small quantity of hydrogen. Neutrons produced in the hydrogen burn, which increases the intensity of the ongoing fission reaction. A U.S. booster bomb was tested at Bikini Atoll on March 1, 1954 (the Bravo Test). It produced a 15-megaton blast. The Soviet Union had tested a small booster device in 1953 and matched the Bravo test in 1955.

Whereas the energy produced by a fission bomb is limited by the flying apart of the fissionable material in the early stages of the explosion, the size of a fusion reaction is limited primarily by the amount of fuel present. The largest fusion test involved a 60-megaton bomb exploded by the Soviet Union in 1961. Smaller fusion weapons have been developed that can be delivered accurately from aircraft, submarines, and intercontinental ballistic missiles (ICBMs).

At the time of the first use of nuclear weapons against Japan in 1945, the United States was the only country with nuclear weapons. Several other countries developed nuclear capabilities: the USSR in 1949; Britain in France in 1960; China in 1964; and India in 1974. Since that time, a number of smaller countries have begun to develop nuclear weapons. They include North Korea, Iran, Pakistan, and Israel. The problem of nuclear proliferation has created a great deal of uncertainty and has threatened the cold war stasis produced by the deterrent effect of MUTUAL ASSURED DESTRUCTION. The recent threat from smaller nations has been used as a rationale for developing a limited land-based missile defense system in the United States that has developed out of a more grandiose "STAR WARS" concept.

The creation of horrifyingly destructive forms of warfare such as nuclear, biological, and chemical weapons produces a strange paradox. Such weapons can almost never be used. If they are almost never used, then it is illogical to divert a huge proportion of national wealth to building and stockpiling these weapons of mass destruction. In fact, the only logic for doing so is a scenario in which one large group of people is willing to risk its own annihilation to wipe out another large group of people. Such a logic arguably existed during the cold war. Even then, any reduction in the nuclear arms buildup was in the profound self-interest of both sides.

Thus, even when the cold war was at its chilliest, treaties were signed to reduce the pace of nuclear arms development and stockpiling. The first treaty, the Limited Test Ban Treaty (1963) bound the United States, the Soviet Union, and Britain to conduct no more nuclear tests above ground, under water, or in space. Underground tests continued. The Nuclear Non-proliferation Treaty (NPT, 1968) sponsored by the United Nations sought to reduce the number of nations developing nuclear weapons. In return for their pledge not to develop nuclear weapons, small countries received assurances from the nuclear "club" that their arsenals would be reduced.

The NPT played an important role in setting the stage for bilateral treaties between the United States and the Soviet Union, which allowed for substantial reductions in nuclear arsenals. The first stage was the Strategic Arms Limitation Talks (SALT), which produced two agreements in 1972. The Antiballistic Missile Treaty (ABM) prevented either side from developing national defensive systems. The logic was that if either side viewed itself as impervious to counterattack, it would be more likely to launch a nuclear war. The ABM treaty is threatened by the Strategic Defense Initiative of 1983 (and its successor, known as National Missile Defense) and is likely to either undergo revision or be abandoned if the United States pursues its defensive policies. The other treaty, known as the Interim Agreement on Offensive Weapons, prohibited the development of new long-range missile launchers.

Several other agreements followed including the Threshold Test Ban Treaty, which limits the size of underground tests that are permitted. An important obstacle was reached in 1979 when the U.S. Senate failed to ratify the SALT II treaty, which was basically an extension and refinement of its predecessor, due to a deterioration in the relationship between the two countries. Despite this problem, both sides continued to observe the provisions of the SALT II agreement.

During Ronald Reagan's presidency (1981–89), Strategic Arms Reduction Talks (START) achieved an agreement in principle to reduce long-range nuclear weapons by 50%, and to have an inspection system to ensure compliance. Progress on these objectives was stymied by Reagan's enthusiasm for the Star Wars program, which the Soviets viewed as contrary to the ABM treaty signed by President Nixon in 1972.

Further progress in arms reduction was made when Mikhail Gorbachev came to power. Gorbachev was strongly committed to reducing military tensions between the Soviet Union and the West. Under Gorbachev, the Intermediate Range Nuclear Forces (INF) Treaty (1987) was signed. This banned all land-based nuclear weapons having a range between 500 and 5,500 kilometers (310–3,415 miles). In July 1991, the START treaty was finally signed. Shortly thereafter the Soviet Union fell apart. When this happened, Russia became the sole nuclear power in the region. By 1994, all of the other former Soviet republics had agreed to give up their nuclear weapons and joined the NPT as nonnuclear countries.

The collapse of the Soviet Union has created an environment in which it is easier for nuclear powers to negotiate cooperative agreements. For example, in 1992, Russia and the United States negotiated an agreement to send Russian weapons-grade uranium to the United States for use in electricity generation. Unfortunately, there has been a rise in ethnic and religious conflicts both within the former Soviet Union and around the world. Many states that were previously allied with the Soviet Union are now without a military champion and may feel the need to develop their own nuclear weapons. One example is North Korea. The threat of nuclear proliferation has fueled interest in defensive technologies in the United States and this may increase tension among the existing nuclear powers.

See also BIOLOGICAL AND CHEMICAL WARFARE; MANHATTAN PROJECT; MILITARY RESEARCH; MUTUAL ASSURED DESTRUCTION; NUCLEAR POWER; "STAR WARS."

Further Reading
Baylis, J., and R. J. O'Neill. *Alternative Nuclear Futures.* New York: Oxford University Press, 2000.
Hagerty, D. T. *The Consequences of Nuclear Proliferation.* Cambridge, Mass.: MIT Press, 1998.
Kaplan, F., and M. J. Sherwin. *The Wizards of Armageddon.* Stanford, Calif.: Stanford University Press, 1991.
Weeramantry, C. G. *Nuclear Weapons and Scientific Responsibility.* New York: Kluwer, 1999.

Nuffield Council on Bioethics

The Nuffield Foundation was formed by car manufacturer William Morris, Lord Nuffield, in 1943, to promote wide educational and social welfare goals. In 1991, the foundation established the Nuffield Council on Bioethics. Its mission is to promote public understanding and discussion of ethical issues raised by recent advances in biological and medical research. To this end it issues reports compiled by multidisciplinary panels on topically important questions. As of 2000, five reports had been issued dealing with genetically modified crops, genetic screening, research on human tissue, animal-to-human organ transplants, and the genetics of mental disorders.

See also GENETIC ENGINEERING; GENETIC TESTING; GENETI-
CALLY MODIFIED FOOD; ORGAN TRANSPLANTATION.

Nuremberg Code

Human experimentation was essentially unregulated
before World War II. The immediate impetus for the
development of the Nuremberg Code was to provide a
rationale according to which Nazi scientists could be
prosecuted as war criminals for their horribly unethical
treatment of human research subjects. The code was
developed in 1947 during the Nuremberg Trials (see
photo).

The Nuremberg Trials were conducted by the U.S.
Army and Nazi war criminals were prosecuted before a
military tribunal presided over by an international panel
of judges. At the doctors trial, which was the first of 12
cases, defendants were accused of murder and torture
conducted in the name of medical experimentation. Of
the 23 defendants (of whom 20 were actually physicians),
16 were found guilty, and 7 of the guilty were executed.
The final judgment, delivered in August 1947, contained
the Nuremberg Code, a 10-point statement articulating
principles that protected the welfare of human research
participants. This internationally recognized declaration
provides the foundation for currently used codes of ethical conduct by scientists using human participants.

It has two principal themes. First, research must be
scientifically valid and have the potential to yield beneficial results for the society in which it is conducted.
Attempts by Nazi scientists to modify the color of the
human iris exemplifies a project that had no scientific
value. Moreover, if this project had been a success, its
likely consequences would have been harmful, that is, a
program to make Germans look more "Aryan."

The second theme is that human research participants
have certain key rights that must be protected. They have

*Herman Göring (standing) is accused, at Nuremberg, in 1946, of organizing the Nazi extermination of Jews while head of the German
secret police, the Gestapo.* (LIBRARY OF CONGRESS)

the right to participate only on a voluntary basis after they have been fully informed about the nature of the research (the right of informed consent). The participant has the right to discontinue participation at any time during an experiment for any reason (the right of termination). Researchers have an obligation to terminate the experiment if it is likely to cause serious injury or death. When a person agrees to participate in research they have the right to expect that they will not be seriously harmed in any way by the procedure (the right of nonmaleficence). Researchers have the obligation to design procedures that are safe and should take steps to minimize risks of pain and discomfort. The rights of nonmaleficence, informed consent, and termination are the cornerstone of ethical treatment of human participants although several other more minor principles, such as the right to confidentiality, have gained currency since the Nuremberg Trials.

It is interesting that although the Nuremberg principles are of central importance in conducting research, they derive from deontological ethics whereas professional ethical codes that have developed since then are largely utilitarian. The Nuremberg Code is an example of Kantian conceptions of the inherent moral value and dignity of the individual. Professional ethical guidelines are less absolute calling on individual researchers to weigh the potential benefits of their research against possible harm to subjects. That is, they are asked to make a utilitarian (consequentialist) calculation. This means that there are situations in which the Nuremburg Code may not be strictly observed. The most obvious examples involve experiments in social psychology in which fully informed consent would undermine the purposes of the research (e.g., MILGRAM'S OBEDIENCE RESEARCH).

Even though the Nuremberg Code provides a useful statement of key rights that should be afforded to participants in ethical research, it has no legal force since it was never formally adopted, even by the Western Allies. There have been many cases in which the Nuremberg Code has not been observed in medical research in the United States. The clearest violations occurred in the context of COLD WAR RADIATION EXPERIMENTS.

See also COLD WAR RADIATION EXPERIMENTS; DECLARATION OF HELSINKI; DHHS REGULATIONS FOR THE PROTECTION OF HUMAN SUBJECTS; INFORMED CONSENT; MENGELE, JOSEF; MILGRAM'S OBEDIENCE RESEARCH; RIGHTS OF HUMAN RESEARCH PARTICIPANTS.

Further Reading
Faden, R. R., S. E. Lederer, and J. D. Moreno. "U.S. Medical Researchers, the Nuremberg Doctors Trial and the Nuremberg Code." *JAMA* 276 (1996): 1667–71.
Grodin, M. A., and G. J. Annas. "Legacies of Nuremberg: On Medical Ethics and Human Rights." *JAMA* 276 (1996): 1682–83.

"Nuremberg Code (1947)." *British Medical Journal,* December 7, 1996.
Resnik, D. B. *The Ethics of Science.* New York Routledge, 1998.
Veatch, R., ed. *Medical Ethics.* Sudbury, Mass.: Jones and Bartlett, 1997.

Nutrasweet (artificial sweetener)

Artificial sweeteners have been developed so that people can indulge their sweet tooth without weight gain or other undesirable biochemical effects of sugar (e.g., in the case of diabetics). Nutrasweet (aspartame), produced by a branch of Monsanto Company, contains amino acids that are much sweeter than sugar and was approved as safe by the U.S. Food and Drug Administration (FDA) in 1981. It is used in many low-calorie soft drinks and foods.

Nutrasweet became a focus of public controversy in 1996, when psychiatrist John Olney of the University of Washington published an analysis of brain tumors from 1975 to 1992 in the *Journal of Neuropathology and Experimental Neurology.* Olney reported a 10% rise in glioblastoma three years after Nutrasweet was approved. He admitted that the rise could be due to any number of environmental carcinogens (such as pesticides, air pollution, industrial solvents, etc.) but argued that the evidence pointed to Nutrasweet. The manufacturers objected that Nutrasweet could not cause brain tumors because it does not enter the bloodstream. Moreover, the incidence of tumors was falling from 1991 to 1993 when Nutrasweet use would have been rising. Despite these facts, British scientists launched a three-year study in 1999 of the possible link between asparatame and cancer.

The Nutrasweet incident illustrates the ease with which groundless scares can develop over food products, particularly those that are known to be "artificial." When people use such products for health reasons, they are disturbed to learn that their health is actually being undermined. Such products are thus a convenient target for sensational journalism. This underscores the need for public trust in the work of scientists, such as those assessing the safety of foods for the FDA. Unfortunately these scientists have not always behaved in ways that fostered trust. For example, the analogous ALAR SCARE revealed serious conflicts of interest among FDA scientists that undermined public confidence in the safety of apples.

See also ALAR SCARE.

Further Reading
Henderson, C. W. "Scientists Study Sweetener's Link to Cancer." *Cancer Weekly Plus,* August 16, 1999.

obligations to future generations

Environmental ethicists are divided over whether the protection of endangered ecosystems should be based on the interests of our own species or whether the value of non-human life forms, which are threatened by human industrial activities, should take precedence. Critics of the second, nonanthropocentric view question whether it is logically possible to elevate the value of other species over our own and suggest that the environment can be protected using more conventional ethical principles. Thus, many ethicists accept that we have obligations to future generations. This means that we should not do anything today which can have the foreseeable, but distant, consequence of making life on this planet less pleasant. Obligation to future generations can be a useful and flexible principle. It suggests, for example, that we should be careful not to accumulate toxic wastes. Moreover, it can be used to provide a rationale for protecting endangered ecosystems, such as the rain forest, which is believed to contain biochemical resources that could be of enormous benefit to pharmacology in the future. Although the term is superficially deontological (i.e., based on moral obligation) the notion of obligations to future generations is really a type of utilitarianism that promotes the happiness of the greatest number of people living in the future.

See also DEEP ECOLOGY; DEONTOLOGY; ENVIRONMENTAL ETHICS; UTILITARIANISM.

Office of Research Integrity

This branch of the U.S. Department of Health and Human Services (DHHS) is responsible for investigating allegations of scientific misconduct. Out of some 1,500 allegations of misconduct, the Office of Research Integrity (ORI) has issued 74 guilty findings, six of which were reversed on appeal to the DHHS Department Appeals Board. Some of the reversals were high-profile cases, such as that against Theresa Imanishi-Kari in the BALTIMORE AFFAIR. The ORI has received a barrage of criticism for mishandling cases, slow case resolution, and a more deliberate pursuit of "small fry" (students and technicians) rather than important researchers. Such criticisms have led to a shift in policy toward greater reliance on scientific misconduct hearings conducted by the relevant institutions, which generally have more familiarity with the cases they handle and produce findings that are less vulnerable to being overturned on appeal.

See also BALTIMORE AFFAIR, THE.

Further Reading

Brainard, J. "Federal Office of Research Integrity Adjusts to a New Role." *Chronicle of Higher Education* 46, no. 13 (1999): A44.

Friedly, J. "Ori's Self-assessment: A batting average of .920"? *Science* 275: 1255, 1997.

Official IRB Guidebook

This publication, which first appeared in 1986, gave researchers and INSTITUTIONAL REVIEW BOARD members assistance in understanding and applying U.S. federal regulations for the protection of human participants in research. These DHHS REGULATIONS were first drafted in 1981 with revisions in 1983 and 1991. A revised version of the *Official IRB Guidebook* (titled *Institutional Review Board Guidebook*) was published in 1993 by the Office for Human Research Protections (formerly known as Office for Protection from Research Risks) to reflect revisions in the federal regulations.

See also BELMONT REPORT, THE; DHHS REGULATIONS FOR THE PROTECTION OF HUMAN SUBJECTS; INSTITUTIONAL REVIEW BOARD.

Further Reading

Penslar, R. L. *Institutional Review Board Guidebook*. Washington, D.C.: Office for Human Research Protections of the Department of Health and Human Services, 1993.

President's Commission for the Study of Ethical Problems in Medicine and Biomedical and Behavioral Research. *Official IRB Guidebook*. Washington, D.C.: Government Printing Office, 1986.

openness in science

Modern science is characterized by an ethic of openness in the transmission of information. The importance of scientific publication for furthering both the advancement of science and social progress was emphasized by FRANCIS BACON, a leading proponent of the scientific revolution in England. During the 18th century ENLIGHTENMENT, this ideal was democratized as encyclopedists set about the task of making a large amount of information, including scientific information, easily available to ordinary people who could read but did not have specialized academic training.

The free flow of information in science has not always been permitted for various reasons. For example, medical initiates in the Hippocratic tradition were not allowed to pass on professional secrets. Similarly, commercial interests have often militated against the free exchange of scientific knowledge, as illustrated in the phenomena of PATENTS and TRADE SECRETS. The existence of such forms of proprietary scientific information can be seen as sometimes creating a drag on scientific progress. For example, the belief of researchers Pons and Fleischman that they could obtain an enormously lucrative patent for COLD FUSION prevented them from freely divulging the fine details of their experimental protocol. As a consequence, many scientists wasted their time in a futile attempt to replicate the findings. On the other hand the potential for taking out patents can be a stimulus to scientific innovation. This is illustrated by the focus of current research in biotechnology in areas that are producing lucrative patents. Unfortunately research that is irrelevant to patents tends to be abandoned.

Another motive for secrecy of scientific information is national security in the context of military research. In the MANHATTAN PROJECT, U.S. civilian scientists working on the development of the first NUCLEAR WEAPONS had to operate under conditions of complete secrecy that applied not only to their own social contacts but also those of their spouses and children. Military secrecy has sometimes been used to conceal bad, or even fraudulent, science as in the case of the "STAR WARS" program in which scientists faked a critical 1994 test to obtain continued funding for their work.

Most scientists are ethically and temperamentally opposed to such restriction on scientific openness. They feel that scientists have a duty to propagate scientific information and that doing so both strengthens science by exposing it to possible criticisms and increases public trust in scientific work. Academic scientists are rewarded for the volume of information they share with peers in the form of academic publication. Many of the world's most distinguished scientists have labored to transmit their ideas, and findings, to as large an audience as possible, not just because this promotes their careers but because of the feeling that knowledge is beneficial in itself and therefore that it should be shared with nonscientists as well as colleagues. The sharing of information between scientific colleagues is a cooperative aspect of the scientific process that facilitates the making of important advances.

See also BACON, FRANCIS; COMMERCIALISM IN SCIENCE; ENLIGHTENMENT, THE; PATENTS; POPULARIZATION OF SCIENTIFIC FINDINGS; TRADE SECRETS.

Further Reading

Bird, S., and E. Houseman. "Trust and the Collection, Selection, Analysis, and Interpretation of Data." *Science and Engineering Ethics* (1995) 1: 371–82.

Grinnell, F. *The Scientific Attitude*. New York: Guilford, 1992.

Marshall, E. "Publishing Sensitive Data: Who Calls the Shots." *Science* (1997) 276: 523–25.

Resnik, D. *The Ethics of Science*. New York: Routledge, 1998.

open-pit mining

Also known as strip mining this technique is used to extract materials that lie close to the surface. Waste materials are first removed from the top in an operation known as "stripping." The site is drilled and loaded with bulk explosives. Following blasting, the ore, or coal, is removed, often using huge mechanical shovels. Open-pit mining typically produces enormous devastation of the

Strip mining for coal in southeastern Ohio, in 1974, disrupts the natural landscape. (NATIONAL ARCHIVES)

landscape (see photo) and produces large heaps of spoil with a danger that hazardous substances will leach (i.e., percolate) out of the material thus polluting groundwater. Surface mining is subject to stringent government regulations in the United States. The Surface Mining Control and Reclamation Act of 1977 requires mine operators to restore mined terrain to its approximate original contours.

See also MOUNTAINTOP REMOVAL.

Operation Whitecoat

This biomedical research program on human subjects grew out of an arrangement between the U.S. Army and heads of the Seventh-Day Adventist Church. Beginning in 1954, and continuing for almost 20 years, soldiers who were Seventh-Day Adventists avoided combat and participated, strictly on a volunteer basis, in biological experiments, some of which involved exposure to infectious diseases. The Adventists supported community service but were ethically opposed to combat. Some 150 experiments were performed using a total of 2,300 volunteers.

See also COLD WAR RADIATION EXPERIMENTS; NUREMBERG CODE; RIGHTS OF HUMAN RESEARCH PARTICIPANTS.

Oppenheimer, Julius Robert
(1904–1967)
American
Physicist

Robert Oppenheimer (see photo on next page) is remembered as the "father of the atomic bomb" because he directed the Los Alamos, New Mexico, laboratory at which this was developed during World War II. The atomic bomb was based on the comparatively simple idea that radioactivity can destabilize atoms causing them to become unstable and emit radiation themselves. If enough of a fissionable material such as uranium-235, or plutonium, is present, an explosive chain reaction develops. The dropping of atomic bombs on HIROSHIMA and Nagasaki, in Japan, caused huge loss of civilian life and immense suffering from radiation sicknesses but engineered a speedy end to hostilities, thereby arguably pro-

Atomic physicist J. Robert Oppenheimer was responsible for construction of the atomic bombs used against Japan in World War II but resisted development of even more destructive fusion bombs. (NATIONAL ARCHIVES)

ducing a net reduction of civilian casualties that would have occurred had the war continued indefinitely.

The ethical defensibility of the Hiroshima and Nagasaki attacks is likely to be debated far into the future as is the role of civilian scientists in bringing military objectives to fruition. Even though Oppenheimer and other distinguished physicists, such as ALBERT EINSTEIN, were clearly in favor of the atom bomb, many drew the line at developing and stockpiling even more frightening weapons such as the hydrogen bomb that was based on atomic fusion. As chairman of the board of scientific advisors of the Atomic Energy Commission, Oppenheimer opposed development of the hydrogen bomb, which placed him at the center of a heated political controversy. During the era of McCarthyism in the early 1950s, Oppenheimer was suspected of being disloyal to his country and was denied security clearance in 1953. He died of throat cancer, at Princeton, New Jersey, on February 18, 1967. Before losing political favor, he had been one of the most influential scientists in the world.

See also EINSTEIN, ALBERT; NUCLEAR WEAPONS.

Further Reading
Holloway, R. L. *In the Matter of J. Robert Oppenheimer: Politics, Rhetoric, and Self-Defense.* New York: Praeger, 1993.
Rival, M. *Robert Oppenheimer.* New York: Flammarion, 1995.

organ donation

The emergence of ORGAN TRANSPLANTATION as a standard medical procedure has meant that there is a much greater demand for organs than is sufficient to meet current needs. Since it is difficult to obtain consent from relatives when a healthy person, having good transplantable organs, dies unexpectedly, the consent of a donor is usually obtained when they are in good health. Donated organs are matched to potential recipients in the United States using a national computerized system that maximizes the probability that the organ will be immunologically compatible with the recipient. Most people do not consent to donate organs after death even though they do not anticipate further personal use of them. Evidently, the thought of surgical invasion of the body is repellent even when it occurs following death.

Some kinds of donated organs, such as kidneys and bone marrow, may be obtained from living persons. Since close relatives are most likely to be immunologically compatible, they may experience considerable social pressure to provide a donation in order to save the life of an ill relative. In such cases, people with few close relatives are more likely to receive a needed organ than those with many relatives, contrary to the mathematical probabilities. If there is only one potential donor in the family, he or she is likely to feel much greater responsibility than if there are several. Conversely, if there are many potential donors, each individual is liable to pass on the responsibility to others.

In some cases, the probability of obtaining a donor may be so small that parents have resorted to having an additional child to serve as a donor for an older child. In at least one case, the Nash family of Colorado, this was done to provide a bone marrow transplant. Adam Nash was born as a result of in vitro fertilization and implantation of an embryo selected for immunocompatibility with his sister Mollie who suffered from a genetic bone marrow disease. In many other cases, pregnancies have been planned to provide umbilical cord stem cells to treat sick children. Such cases raise serious questions about voluntary consent, particularly if the act of donation entails potentially harmful health consequences. Similar issues have been raised in cases where parents consented to providing part of a lung or liver to save the life of a sick child. In general, organ donations are not accepted if they would result in loss of life of the donor, since this would constitute suicide or murder. Moreover, donations that have the potential for undermining health, such as a person giving their only good kidney to save a relative, are considered unethical for the same reason.

Considerable ethical controversy surrounds the possibility of using the organs and tissues of aborted fetuses, reflecting the ambiguous moral and legal status of such individuals. Aborted fetuses do not have the legal rights of full-term infants and they are certainly not capable of providing consent, but most ethicists feel they should be treated with the same kind of respect that is accorded the bodies of deceased adults whose organs cannot be harvested without their prior consent. Similar problems surround the potential use of organs from nonviable infants with serious birth defects such as anencephaly.

See also FETAL CELL GRAFTS; ORGAN TRANSPLANTATION; STEM CELL RESEARCH.

Further Reading

Simmons, R. G., S. D. Klein, and R. L. Simmons. *Gift of Life: The Social and Psychological Impact of Organ Transplantation.* New York: Wiley, 1997.
Veatch, R. M. *Transplantation Ethics.* Washington, D.C.: Georgetown University Press, 2000.

organ transplantation

The field of modern organ transplantation began in 1954 with the first successful kidney transplant by Joseph Murray and other surgeons at Boston's Bent Brigham Hospital. The first successful heart transplant was by Christian Barnard, in 1967, at Schur Groot Hospital in Cape Town, South Africa. The major technical barrier to successful organ transplantation has been the immune response to foreign tissue but this problem has yielded somewhat to the development of effective immunosuppressant drugs that have allowed patients to survive for years after their transplants rather than days. Another important technical advance was the development of the heart-lung machine that kept patients alive during the often lengthy operation. Over 20,000 organ transplants are performed each year (mostly kidneys, livers, and hearts, in that order).

The routine performance of organ transplantations has raised many questions for medical ethicists. The most fundamental issue is whether it is morally acceptable to use the organs of one person to keep another alive. The harvesting of organs from cadavers raises a delicate issue about determination of death since organs are likely to function better if removed quickly. One of the most harrowing ethical issues involves decisions as to who should receive scarce organs and who should not. The scarcity of organs raises the further issue of whether, in a rational society, people who could theoretically save many lives with their donated organs should be free not to donate them. If they are, how can the number of organs ever match the demand? Should the sale of organs ever be permitted? Is it ethical to raise, or to genetically engineer, animals whose sole purpose is to provide organs for transplantation into humans?

CONSENT OF THE DONOR

From one point of view, the ethics of organ donation from a deceased person, who had consented, to another in need of the organ is relatively simple. There is generally no cost to the donor and there is a great benefit to the recipient in the sense that their life may be prolonged. Yet, some religious traditions have strong taboos against dissecting cadavers. Jewish religious law forbids mutilation of the dead but modern Jews have made some accommodation by accepting removal of an organ after death if a life can thereby be saved. Belief in reincarnation, or resurrection, as in the case of Buddhists and Muslims, respectively, means that these traditions are generally opposed to dissecting cadavers. Yet, in most of the relevant countries, the right of the individual to provide consent for removal of their organs after death is respected and transplants can therefore be ethically performed.

One of the most fundamental ethical consequences of organ transplantation has been a need to redefine death. The traditional definition of death was irreversible cessation of heartbeat, pulse, and breathing. This was a problem for the harvesting of organs because the loss of oxygenated blood flow produced rapid damage that quickly rendered the organs useless for transplantation. A new definition of death introduced by the President's Commission for the Study of Ethical Problems in Medicine in 1981 states that death occurs either with irreversible loss of respiration or irreversible loss of brain function. This has meant that organs can be harvested from people after their brains had died while their bodies were still "alive" thanks to the use of heart-lung machines to sustain respiration.

The biggest practical problem in organ transplantation is the scarcity of donated organs and this has led to considerable ethical discussion about whether the taking of organs from cadavers requires consent. Some ethicists feel that consent must come from the donor before death. The U.S. Uniform Anatomical Gift Act of 1968 allows potential donors to sign a contract that authorizes the taking of their organs after death. Moreover, in the case of deceased individuals who had not expressed opposition to being a donor, family members may authorize the taking of organs.

Strictly utilitarian ethicists take the point of view that all organs of cadavers should be harvestable. The organs of dead people will simply decay and be of no use to humanity. Meanwhile great human misery is created by an artificial scarcity of transplantable organs. One rational way to produce the greatest happiness of the greatest number of people is compulsory universal donation. A more moderate position recognizes that taking organs without consent can produce societal harm because it could undermine the inviolability of the body before death. For example, the needs of a dying person might take second place to the desire to spare his, or her, organs. Moreover, people belonging to religions that credit reincarnation might have strong feelings of oppression and outrage creating social unrest. The legal situation in France is more nearly utilitarian than that in the United States because French law assumes consent to organ donation in the absence of dissent either from the donor while still alive or from the family after death.

SELLING ORGANS AND NONHUMAN SOURCES

The continued scarcity of organs suggests that there is some deep-seated objection to donation that may reflect limitations to human altruism. Just as few people bequeath their property to complete strangers, there may be a disinclination to give organs to unknown individuals. The most obvious way around this obstacle is to have a market in which transplantable organs are bought and sold. If it is possible to sell blood, semen, and human eggs, according to this view, what possible objection could there be to selling organs? One objection, articulated most often in Britain where commercial blood donation is not allowed, is that the opportunity to engage in a truly altruistic act is a valuable institution in a modern anonymous industrial society.

This virtue argument has an obverse side, namely, that where the sale of organs is permitted it is generally degrading. Thus, under pressures of poverty, individuals may decide to "cash in" one of their disposable organs. In India, the sale of organs is permitted and most of these organs are believed to end up in wealthier countries. The problem with selling life-saving organs is that they tend to be very expensive. This introduces what many see as an unacceptable element of unfairness because wealthy people can always afford to purchase organs but poor people cannot. Moreover, there is a strong economic motive for poor people to sell their organs but none for wealthy people to do so. The inevitable consequence of a free market in organs would be that they would be purchased by wealthy people and countries from poor individuals and nations. Many people reject the prospect of such unfairness in the practice of medicine.

In the United States, sale of organs was prohibited by the National Organ Transplantation Act of 1984. The very high costs of organ transplantation are borne mainly by medical insurance companies and are thus distributed in the form of increased health insurance premiums paid by businesses and individuals. These expensive operations add at least 4% (and potentially as much as 24%) to the national health care budget, which means that the small number of transplant recipients are draining over 700 times as much out of the system as the average individual. Whether this huge expense is merited by the benefits to organ recipients or ought to be spent elsewhere in the health care system merits careful analysis.

Given the pressing issue of organ scarcity, medical researchers have investigated alternatives, such as artificial hearts and organs from nonhuman animals. Neither of these approaches has worked well due to a variety of technical problems. It is possible that these difficulties will be overcome with time. For example, the development of better immunosuppressant drugs may allow for widespread use of animal organs. Animal rights activists object to the use of animal organs as speciesist, namely, unfairly elevating the interests of humans over other species, but medical researchers reject this view and physicians generally place a higher value on human than animal life.

In addition to artificial organs and animal grafts, there are at least two other technical approaches that could be important for increasing the supply of transplantable organs in the future. One is the development of genetically engineered animal strains whose organs produce less of an immunological reaction in human recipients. Another is the genetic engineering of isolated organs that can be designed for compatibility with different kinds of recipients. The perfection of such techniques would raise a whole range of new ethical problems. Thus, if it were possible to replace organs at will, life might be greatly extended raising the issue of when heroic medical interventions to prolong life should be ended.

SELECTING RECIPIENTS

The scarcity of transplantable organs has not only raised the ethical question of how to increase the supply of organs but has also highlighted the issue of who should receive them. The decision as to who on a long waiting list should receive an available organ is difficult because nonselection usually means early death. Such decisions often rest on medical criteria predicting success of the transplant operation and are affected by other criteria of utility of the organ such as how long it is likely to be used for (which favors young and healthy recipients).

Transplants are thought of as an unusual privilege rather than a right. Some medical providers have qualms about giving a liver transplant to an alcoholic suffering from cirrhosis of the liver. There are two problems. One is that the liver has been destroyed as a result of an abusive lifestyle. The other is the fear that the transplanted organ will be destroyed by alcohol abuse. The introduction of such concerns about moral worthiness into the determination of who receives a medical treatment is seen by many ethicists as troubling and irrelevant. From the point of view of practical outcomes, former alcoholics are just as likely as others to survive liver transplantation and apparently have good records of abstinence following this gift of life.

Concern over inequities in recipient selection prompted the U.S. Department of Health and Human Services to issue guidelines promoting consistent medical criteria for organ allocation in 2000. The ruling calls for sharing of organs over wider geographical regions than previously practiced but not for a single national list. An independent advisory committee was formed to oversee organ allocation procedures in transplant operations.

See also EUTHANASIA; LIFE SUPPORT, TERMINATION OF; ORGAN DONATION.

Further Reading
Jonsen, R. A. "Ethical Issues in Organ Transplantation." In *Medical Ethics*, ed. R. M. Veatch, 239–74. Sudbury, Mass.: Jones and Bartlett, 1997.
Simmons, R. G., S. D. Klein, and R. L. Simmons. *Gift of Life: The Social and Psychological Impact of Organ Transplantation.* New York: Wiley, 1997.
Veatch, R. M. *Transplantation Ethics.* Washington, D.C.: Georgetown University Press, 2000.

ozone depleters

These are volatile chemicals that cause thinning of the ozone layer in the earth's atmosphere. In 1985, a hole in the ozone layer over Antarctica was discovered. Thinning of the ozone layer has been accompanied by a worldwide rise in skin cancers attributable to increased exposure to harmful radiation penetrating the ozone. American chemists F. Sherwood Rowland and Mario Molina proposed that the ozone hole was caused by chlorofluorocarbons (CFCs) used as aerosol propellants. Their hypothesis received additional support from Dutch scientist Paul Crutzen (with whom they shared the 1995 Nobel Prize).

The view that aerosol propellants were responsible for the ozone layer was vigorously attacked by the manufacturers and distributors of hundreds of cosmetics and other products that used aerosols. The industry argued that Rowland and Molina's claims lacked scientific credibility and funded scientists who showed signs of hostility to the theory. They even boycotted the trade journal *Drug and Cosmetic Industry*, which supported the Rowland-Molina thesis, by canceling advertisements.

Despite, these efforts, the case against aerosol propellants has been accepted by political leaders around the world and a 1987 agreement signed by the United States and 22 other countries in Montreal agreed to phase out CFC production. In a 1992 meeting, in Copenhagen, the date for phaseout of CFCs, along with other ozone-depleting chemicals (carbon tetrachloride and methyl chloroform), was accelerated to January 1, 1996. This agreement was signed by 107 countries. CFCs are also used in air conditioning systems and these are now being recycled under Environmental Protection Agency regulations promulgated in 1992 and 1993.

Even though the ozone hole has not been mended, global action to address the problem has been rapid and has prevented further deterioration of the ozone layer. The seriousness of the problem and the necessity for a speedy global response is analogous to the issue of ACID RAIN.

See also ACID RAIN; AIR POLLUTION; GLOBAL WARMING.

Further Reading
Davis, D. A. "Saving the Earth." *Drugs and Cosmetic Industry* 157, no. 5 (1995): 20.

P

pantheism

Derived from the Greek words *pan* (all) and *Theos* (God), pantheism refers to a religious belief system in which God is equated with the universe. It is often used in reference to the worship, or admiration, of nature. Pantheism can take many different forms and may be philosophical, poetic, or scientific, as well as religious. Religious pantheists include Christian mystics like German Dominican theologian Johannes Eckehart (1260–1327) and German mystic Jakob Boehme (1575– 1624). In Hinduism, the only reality is Brahman, the supreme unity. Among philosophers, Spinoza developed a monistic system in which God is everywhere. For many of the English romantic poets of the early 19th century, the contemplation of natural beauty allowed the observer access to a sense of the infinite and the immortal. Many scientists have expressed a similar feeling of awe and inspiration when contemplating the order inherent in nature. CARL SAGAN, host of the Public Broadcasting Service series *The Cosmos,* offers a good example of scientific pantheism. A similar appreciation of the intrinsic value of nature is to be found in modern ENVIRONMENTAL ETHICS.

See also DEEP ECOLOGY; MONISM.

pap smear controversy   *See* CHEM-BIO CORPORATION AND PAP SEMAR CONTROVERSY.

paranoia, experimental induction of

Psychological stress produced in psychology experiments is generally mild and does not affect participants more than their normal daily experiences (thus satisfying the MINIMAL RISK CRITERION). In rare cases of ethically questionable research, psychologists have deliberately set out to induce a highly unpleasant mental state. One striking example is the late 1970s experiment of Phillip Zimbardo, a professor of social psychology at Stanford University, in which participants were rendered paranoid.

Elderly people who suffer from paranoia, or irrational suspicion of other people, are likely to have hearing problems. Zimbardo felt that hearing loss with age may be so gradual that it goes unnoticed. When people become deaf without being aware of it, they might imagine that acquaintances are acting strangely, for example, whispering together in their presence. Could it be that the unrecognized deafness is making them paranoid? Zimbardo felt that this was an important question. If his hypothesis was correct, then much needless suffering could be avoided. Paranoia could be treated through the simple provision of a hearing aid, thereby avoiding costly and ineffective psychiatric treatments.

Zimbardo devised a convoluted and improbable experiment to test his hypothesis. Hypnosis was used to cause deafness in traditional-age students. Responding to a posthypnotic suggestion, the six experimental participants became deaf without knowing why; six control subjects became deaf with knowledge of the cause; and, in a second control group, subjects experienced a posthypnotic suggestion to scratch an itchy ear without remembering that they had been told to do so by the hypnotist.

The experiment, like other deceptive research in the field, was intricately theatrical. A cover story, or fictional explanation of the purpose of the experiment, threw participants off the scent of the real hypothesis being tested. They were told that the experiment was concerned with the effects of hypnosis on creative problem solving. Another type of deception involved the use of confederates, or experimenters posing as subjects. The main purpose of the confederates was to set a social context in which paranoia could be induced but they also recorded the behavior of the real subjects.

The experiment followed an elaborate protocol. Participants received instructions from timed slides. Two confederates engaged in a well-rehearsed conversation designed to establish a previous connection between them. They recalled a fictitious party they had attended, described a hilarious incident, made a funny face, and decided to work together. During their conversation, deafness was induced in the subject by means of a posthypnotic cue (the word "focus" presented on a screen). Since the conversation was no longer audible, it could have been perceived as hostile.

Zimbardo and his coworkers found that people who lost their hearing without knowledge of the cause became paranoid according to a number of different measures, including self-report questionnaires. Control subjects who either knew why they were deaf, or scratched an ear without knowing why, did not become paranoid. The findings were published in *Science* but the unethical aspects of the research were criticized by *New York Times* reporter Morton Hunt, who based his story on an interview with Steve Kaufman, one of the participants.

Kaufman had been selected for the study because he was highly susceptible to hypnosis as revealed in a preliminary training session. He was interested in learning more about hypnosis because he was a fencer and wanted to control pain using hypnosis to improve his fencing. In the second session, Zimbardo showed the volunteers how to put themselves into deep hypnotic trances and taught them how to use autosuggestion to counter pain. The experiment followed. Each of the three students was given a set of earphones and Zimbardo signaled a woman assistant to start the tapes. Kaufman heard Zimbardo's voice explaining to him that he was to put himself in a deep hypnotic trance with the help of the accompanying music, a series of flute-like arpeggios. The voice later returned, informing him that when he saw the word "focus" on the screen in the laboratory, sounds would become very low and difficult to hear. He would not remember the instruction until a hand was placed on his

shoulder and he was told "that's all for now," at which point he would recall everything that had happened.

As soon as the word "focus" appeared on the screen, Kaufman began having a very strange and stressful experience. As the confederates talked hilariously about the fictitious party they had attended at which one of the participants threw up all over a TV, he wasn't sure whether he couldn't hear them or couldn't understand what they were saying. He wondered who these people were and why they were sitting there laughing at him. In a panic, he asked himself why they were mumbling or talking in a foreign language. He became extremely agitated, rating himself at the top of scales measuring paranoia.

The slide projector clicked, and an instruction appeared: "As you view the next slide, please write a brief description of what the characters portrayed in the slide are doing, who they are, and what the mood of the scene is." The first picture depicted a man looking down on a woman lying in a bed. Hands trembling and sweat running down his sides, he started to work.

When researcher Susan Andersen came in, put her hand on his shoulder, and said, "That's all for now," his hearing returned. As the researcher reminded him of the procedures he had been through, he suddenly remembered everything. Greatly relieved to be able to hear, he was also shocked and upset about how he had been used. He could not believe that the distinguished researcher whom he admired so much could have deceived him in this way.

Then Zimbardo himself entered the room, clearly delighted by how the experiment had gone. He had been observing from an adjacent room. Slowly and carefully he explained the real purpose of the experiment. Kaufman was fascinated. He saw how the experiment made sense, how deafness really does make a person paranoid, and how the experiment could not have been performed correctly without concealing the real point of the study at the beginning. By the time the debriefing was over, he felt calm and in control of himself. However, as soon as he left, the traumatic events kept racing around in his thoughts and it was several days before he calmed down.

Kaufman returned after a week for a check on his welfare. The follow-up revealed that there were no lingering paranoid effects of the experiment. Kaufman was proud to have had the privilege of participating in what he perceived as an important piece of research. Participants in other stressful social psychology experiments, such as Milgram's obedience study, have also generally been glad that they participated.

The major ethical problem with the paranoia experiment was that it took unjustifiable risks with the welfare of participants. Thus, traumatic experiences have the potential for producing long-term psychological harm, as in the case of posttraumatic stress disorder. It might be argued that the increase of knowledge made possible by the study justifies its cavalier approach to inflicting emotional distress on subjects with the potential for lasting psychological harm. Zimbardo maintained that looking for situational explanations of psychopathology was a whole new frontier that marked an important departure from seeing all psychopathologies as due to brain disorders. Yet, the series of studies that he envisioned along these lines never materialized. It is also arguable that Zimbardo's results could have been produced without inducing such trauma to subjects. Why did Zimbardo not simply provide hearing aids to elderly people to test if their paranoid ideation declined?

Other important ethical issues involved the use of deception and surreptitious surveillance. Misled about the true purpose of the research, and unaware that they were about to undergo a traumatic experience, subjects did not know what to expect and could therefore not give informed consent to participate, contrary to the Nuremberg Code. Even though Zimbardo demonstrated ethical awareness in providing a full debriefing to correct for the deception, and in conducting a follow-up to check for aftereffects of trauma, he may not have been sufficiently sensitive to the rights of his research participants in planning the study.

See also MINIMAL RISK CRITERION; NUREMBERG CODE; RIGHTS OF HUMAN RESEARCH PARTICIPANTS.

Further Reading
Hunt, M. "Research through Deception." *New York Times* September 12, 1982.
Zimbardo, P. G., S. M. Andersen, and L. G. Kabat. "Induced Hearing Deficit Generates Experimental Paranoia." *Science* 212 (1980): 1529–31.

Parkinson's disease and agricultural chemicals

Parkinson's disease is a movement disorder that mainly afflicts elderly people. It is a rare disease, resulting in the death of one person in 1,000 who is over the age of 75. The symptoms include slowness of movements, difficulty in initiating actions, muscle tremors, rigidity, particularly of the facial muscles, which form an expressionless "mask." Parkinson's disease is sometimes associated with depression and may cause intellectual impairment. Its immediate cause is degeneration of nerve cells in the substantia nigra (SN) region of the midbrain. Parkinsonian damage to the SN may be caused by exposure to chemicals. In 1982, several young people in northern California who took a heroin substitute developed Parkinson's disease that was similar to that seen in the elderly. Since one of the chemicals in the designer drug, MPTP, is similar to common herbicides, researchers began to investigate whether exposure to agricultural chemicals might *affect* risk of developing Parkinson's disease.

Many epidemiological studies have found that residence on a farm, or in the country, increases the risk of Parkinson's disease by a factor of at least three. Exposure to agricultural herbicides and pesticides that contain MPTP-like molecules may play a role in onset of the disease. This impression is supported by findings that gardeners, and people who frequently use insecticides around the home, also have an increased risk of Parkinson's. Herbicides and pesticides clearly play an important role in the development of Parkinson's disease, but most people who are occupationally exposed to them do not develop the disease. This has led researchers to conclude that the disease is caused by the combination of exposure to environmental toxins with a genetic vulnerability to them. Family history is at least as important a predictor of Parkinson's as exposure to toxins is and researchers have begun to zero in on the biochemical basis of vulnerability to toxins. One attractive hypothesis is that it resides in the glutathione system for detoxifying cellular mitochondria.

The connection between environmental toxins and Parkinson's disease provides yet another powerful argument that challenges the widespread use of agricultural chemicals. Not only are we poisoning the environment, as RACHEL CARSON first pointed out in connection with the use of the insecticide DDT, but we are poisoning ourselves. While it might be argued that Parkinson's still affects a small minority of people, it is also true that this disease is an extreme manifestation of brain toxicity. Thus, of the many young Californians exposed to MPTP in designer drugs only a few developed Parkinson's but all of those exposed to the toxin had evidence of substantia nigra damage. Moreover, it is possible that other parts of the brain, not to mention other organ systems, are adversely affected by agricultural pesticides and herbicides. The ethical, and prudent, course is for farmers, gardeners, and householders to use such toxic chemicals as little as possible.

See also DDT; HORMONE MIMICS.

Further Reading
Golbe, L. I. "Parkinson's Disease: Nature Meets Nurture." *The Lancet,* October 24, 1998.
Pearce, J. M. S. *Parkinson's Disease and Its Management.* New York: Oxford University Press, 1992.
Stern, G. M., ed. *Parkinson's Disease.* Philadelphia: Lippincott, Williams and Wilkins, 1999.

pasteurization

Pasteurization is a method of food preservation that uses heat to destroy microorganisms thus preventing spoilage.

Originally devised by French chemist Louis Pasteur (1822–95) the process was first used, in the early 1860s, to protect wine from being spoiled by a yeast strain that produced lactic acid. In the United States milk, as well as milk and egg products, must be pasteurized according to federal laws. Milk may be pasteurized either by heating it to 145 degrees Fahrenheit for 30 minutes or by heating it to 160 degrees Fahrenheit for 15 seconds (the flash method).

The discovery of pasteurization marked an important advance in microbiology in the sense that it stimulated Pasteur to investigate microorganisms as the cause of diseases. This work culminated in his development of vaccines for anthrax and rabies (see VACCINATION). It also led to the development of antiseptic systems to prevent infection during surgery, an important contribution of English surgeon Joseph Lister (1827–1912).

French wine producers quickly adopted pasteurization because spoilage of their product had created an economic burden. It is interesting that widespread pasteurization of milk was delayed for almost a century after its discovery apparently reflecting a widespread phobia about the application of poorly understood scientific procedures to the food supply. A similar irrational reluctance to adopt new technologies has been observed in the case of water fluoridation, food irradiation, and genetically modified food. It appears that new food technologies are quite unlikely to be adopted unless there is a pressing need for them.

See also FLUORIDATION; FOOD IRRADIATION; GENETICALLY MODIFIED FOOD; VACCINATION.

patents

Most Western nations have laws that recognize rights to intellectual property, including original works of art, trademarks, trade secrets, and inventions. The rights of an author to control reproduction of a book, picture, or work of art is protected by copyright law. Similarly the right of an inventor to control the reproduction of an invention is protected by patent law. Patent laws are comparatively recent and reflect the value attached to innovation in modern society. The first patent law was enacted in Venice in 1474. The English Statute of Patents and Monopolies (1623) is the model for most subsequent patent laws around the world. It gave the right of patent exclusively to the first developer of a new device. The device itself is patented rather than the idea which gave rise to it.

In the United States, patent applications are submitted to the Patent and Trademark Office and inventions receive a patent only if they (a) are described in sufficient detail to be reproducible, (b) are original and useful, (c)

are not designed to break the law, and (d) do not constitute a threat to national security. Patents must relate to concrete processes or objects. Just as it is impossible to copyright an idea, so it is impossible to patent one.

They are many examples of the potential value of a patent interfering with scientific progress. One of the clearest relates to the COLD FUSION controversy. When researchers Pons and Fleischman falsely believed that they had discovered how to produce atomic fusion at low temperatures, they went public in a press conference instead of first submitting their work for scientific review. A likely reason for not going the usual route is that they were afraid that if they presented their work in sufficient detail someone else could have beaten them to get a patent. The lack of clarity and detail in describing what they had done impeded the efforts of fellow scientists to replicate their work. This is an example of the more general principle that commercialism in science tends to subvert professional ethical standards. In this case, the researchers not only failed to conduct their research with the care and precision that is expected but they also violated the principles of openness and objectivity in reporting their results.

See also COLD FUSION; COMMERCIALISM; RESEARCH ETHICS; TRADE SECRETS.

Further Reading

Foster, F., and R. Shook. *Patents, Copyrights, and Trademarks.* New York: John Wiley, 1993.
Resnik, D. *The Ethics of Science.* New York: Routledge, 1998.

Pavlov, Ivan Petrovich

(1849–1936)
Russian
Physiologist, Psychologist

The first-born of 11 children of a poor rural priest, Ivan Pavlov (see photo on next page) seemed destined to follow in his father's footsteps until he read CHARLES DARWIN and changed his mind. He walked several hundred miles from his native village of Ryazan to attend the University of St. Petersburg where he specialized in animal physiology. Pavlov's single-minded devotion to science is legendary and he was assisted in this by his wife Sara, who relieved him of most of the daily responsibilities of running a household. The Pavlovs were wretchedly poor from their marriage, in 1881, until he obtained a job as professor of pharmacology at the Military Medical College in St. Petersburg, in 1890. This was followed by successively more prestigious appointments at the Institute of Experimental Medicine and the Academy of Sciences.

Pavlov first worked on the physiology of the cardiovascular system before moving on to digestive physiology of the dog. His work on digestion was distinguished and would ultimately bring him a Nobel Prize for physiology and medicine (1904). His decision to change, in mid career, to the study of conditioned reflexes thus involved a certain amount of courage. In the course of his work on digestion in dogs, Pavlov surgically inserted a fine tube in the dog's salivary duct so that the salivary response to food could be studied outside the animal's body. As soon as meat powder was placed in the dog's mouth, it began salivating copiously. After repeated experiments with the same dog, Pavlov could not help noticing that the dog would begin salivating before any food was put in its mouth. Like others who observed the phenomenon, he assumed that it was caused by the animal's thoughts and referred to it as "psychic secretions."

Instead of dismissing this as an annoyance that introduced error into his digestion experiments, Pavlov quickly realized its theoretical importance. Here was an objective, if indirect, method of studying what was going on inside the dog's mind and brain. From 1902 onward he devoted himself to the study of conditioned reflexes, as they came to be called.

In a basic, experiment, a buzzer (serving as a conditioned stimulus) was sounded and the number of drops of saliva recorded. When the buzzer was accompanied by placing food in the dog's mouth, the animal salivated (the unconditioned response). Then in the experimental test, the buzzer was sounded without food being presented. The dog continued to salivate (the conditioned response), indicating that it had learned to make a connection between the buzzer and receiving food.

Pavlov's devotion to precise measurement was remarkable and paid off in the form of precise repeatability of results. He developed laboratory cubicles in which an experiment could be operated remotely without ever allowing the dog to see the experimenter, thus influencing B. F. SKINNER's work on automatic running of operant conditioning research in the United States. When he received the necessary financial support, he even designed a laboratory building that blocked out background noises from the dog's sensitive hearing system. Known as the Tower of Silence, the three-story laboratory was supported by steel girders that were surrounded by sand. A deep trench filled with straw encircled the building. Windows were made of unusually thick glass and the double steel doors had airtight seals to shut out extraneous sounds.

Pavlov's conditioning experiments made him a worldwide celebrity. His experiment on conditioned salivation in dogs may be the most widely known piece of empirical research ever carried out. Pavlov played a critical role in

Ivan Pavlov united philosophy and physiology by demonstrating that his experimental dogs could learn to associate two events, such as a buzzer and food, that previously had no connection. (LIBRARY OF CONGRESS)

the acceptance of psychology as a rigorous science. Later in his career, he avoided any assumption that dogs were thinking about the experimental situation much as a human might and, as far as possible, dropped mentalistic references (such as "psychic secretion") from the published descriptions of his work, thus influencing the development of American behaviorism under the leadership of JOHN B. WATSON.

Among the many basic phenomena of conditioning described by Pavlov was discrimination. In discrimination learning, an animal learned to salivate to a stimulus predicting food but not to a stimulus that was never accompanied by food. In one experiment, conducted by Pavlov and a colleague named Shenger-Krestovnikova, dogs learned to discriminate between a circle that predicted food and an ellipse which was never accompanied by food. The ellipse was made more and more like the circle in successive series of trials until the dog had great difficulty in telling the stimuli apart. Instead of simply ignoring the stimuli, the dogs became very upset. When led into the apparatus, the animals would bark, whine, and attempt to bite whatever parts of the apparatus were within reach even though they had previously been calm and cooperative. Moreover, even when the simple discrimination was brought back, the dogs took twice as long to relearn it as they had taken to master it the first time around. By analogy with people suffering nervous breakdowns under highly stressful conditions, Pavlov christened this phenomenon "experimental neurosis." Experimental neurosis bears many similarities to the phenomenon of LEARNED HELPLESSNESS studied a half century later in American animal learning laboratories. Learned helplessness experiments are considered to be excessively cruel to animals and are rarely conducted today on ethical grounds.

Pavlov was extremely fortunate in that his research was favored by the Communist government. Apparently due to his international reputation, he was seen as a poster boy of Russian science and received generous government support for his work. Despite this favorable treatment, Pavlov was openly hostile to the Soviet regime. Famously straightforward in all of his interpersonal relationships, he wrote blunt and angry letters of protest to Soviet leader Joseph Stalin (1879–1953). He also boycotted Russian scientific meetings to express his opposition to the Communists. His bravery, or foolhardiness, recalls the principled stand of ALBERT EINSTEIN against totalitarianism in Germany. Just how big a risk Pavlov was taking is illustrated by the subsequent episode of LYSENKOISM in which Russian geneticists were murdered because they accepted principles of Mendelian inheritance. Pavlov died in Leningrad, Russia, on February 27, 1936. On that morning he had intended to get up to go to his laboratory.

The study of conditioned physiological reflexes has been overshadowed in America and the rest of the world by Skinner's work on the learning of operant responses that deal with visible voluntary movements such as a rat pressing a lever. Conditioned responses play an important role in the functioning of the immune system, comprise a critical aspect of human sexuality, form the basis of phobias, and have inspired the design of advertisements. Scientific knowledge in each of these areas has been greatly advanced by Pavlov's fundamental research on conditioning.

See also ADVERTISING, AND SCIENCE; EINSTEIN, ALBERT; FECHNER, GUSTAVE; FETISH, CONDITIONING OF; LEARNED HELPLESSNESS; LYSENKOISM; WATSON, JOHN B.

Further Reading

Todes, D. P. *Ivan Pavlov: Exploring the Animal Machine.* New York: Oxford University Press, 2000.

PCBs   *See* HORMONE MIMICS.

peer review process in academic research

Imperfect though science may be, one of its proudest claims is that it is a self-correcting system of knowledge. Scientists, it is true, sometimes publish papers that are badly in error. These errors usually become obvious when other researchers try to replicate the work. The peer review process is designed, in part, to ensure that really bad science never sees the light of day. Peer reviewers try to identify errors of fact, inference, methodology, calculation, and even typing, in papers they review. They are also concerned that any literature review contained in the paper does not omit important information or sources and is unbiased in its presentation of the state of the art in research knowledge on the topic of the paper. In general, reviewers must decide if a paper is appropriate for a particular journal in terms of its subject matter, meets criteria of methodological rigor such that the procedures described in the paper support its conclusions, and makes enough of an original contribution to merit page space in the journal.

Peer review, or at least peer criticism, was discussed by Francis Bacon early in the 17th century but was not formally practiced for two centuries and did not become standard procedure until the middle of the 20th century. When the editor receives a manuscript from the author, a preliminary decision is made about whether the paper is suitable for the journal and good enough to merit review. It may then be sent to several readers, or reviewers, who may be either members of the editorial board of the journal or outside reviewers.

The reviewers generally either accept or reject the paper for publication. If the paper is deemed acceptable, reviewers often recommend changes that may be either superficial or profound. It is the editor's job to act on the recommendations of the various reviewers (usually there are three), which can be a difficult task if, as often happens, the opinions of the reviewers are divided and their comments contradict each other.

ETHICAL ISSUES IN PEER REVIEW

Most reasonable scientists agree that peer review has an important role to play in improving the quality of scientific communication and increasing the fairness of editorial decisions. This is only true if the review process is conducted in an unbiased manner and many scholars have concluded that the peer review process is degraded by the biases of reviewers and editors. Reviewers may unfairly criticize a paper merely because they happen to have a personal dislike of the author or disagree on purely ideological grounds. Similarly, editors who dislike a particular author may deliberately send out the manuscript to like-minded reviewers who are likely to reject it. What is more, editors outrank reviewers and they can go against the majority opinion. This rarely happens but when it does it undermines the whole point of peer review and generally convinces authors that their work will not be fairly reviewed by this journal in the future.

Most journals practice single-blind review, which means that the reviewers are protected by anonymity. Some scientists object to this lack of transparency but it can be argued that anonymity helps reviewers to do their work impartially without having to fear reprisals, which might possibly take the form of hostile reviews of their own work. In double-blind reviews, the name of the author is withheld from the reviewers. Double-blind review has the laudable goal of improving objectivity in the review process by helping to ensure that the work receives the same treatment regardless of the academic status or other personal attributes of the author, including ethnicity and gender. Most scientists agree that double-blind review is a sham. Reviewers cannot help detecting the identity of the author using a very simple heuristic: scientists are more likely to cite their own work than others are!

There is considerable controversy about how effective the editorial process in scientific publishing is. The quality of editorial decisions may be adversely affected by the quantity of submissions. Thus, in fields such as social psychology in which the vast majority of submissions are rejected by editors, the average paper that is accepted may be little better than the average paper that is rejected. This is an inevitable consequence of large rejection rates. If most submitted papers have some merit and if most of the submitted papers are rejected, then editors are turning down large amounts of good research. Where the volume of submissions is lower, editors can probably give more thought to individual papers and to the comments of reviewers and make higher-quality decisions.

If authors in a particular field have little trust in the objectivity of the review process, they may tend to ignore reviewers' suggestions for improvement and send a rejected paper to the next journal on their list. In this way, a large volume of second-rate papers may clog the review process making it more and more dysfunctional.

Journals like *Nature* and *Science* that receive a high volume of submissions protect themselves from overload through the use of an initial screening process. Only a small fraction of the submitted papers actually get into the review process.

One important bias in scientific review is a tendency to reject new or unfamiliar ideas. Scientists are conservative people and they tend to screen out innovative ideas and controversial findings, particularly those that do not square with their preconceptions. This bias against innovation slows down the pace of scientific progress and could encourage frustrated authors to sidestep the peer review process and announce discoveries directly in books or through the press.

Some journals make the peer review process completely open by publishing reviewer comments along with the target article. This format is too wasteful of space to be effective for mainstream journals but it is useful for dealing with controversial issues. An example of an open peer review journal is *Behavioral and Brain Sciences*. *Current Anthropology* is another journal that uses peer review but limits this to only some of its articles.

The limited research literature dealing with the effectiveness of scientific peer review does not promote confidence in its quality. One study deliberately introduced eight serious errors into a paper that had already been accepted for publication and sent it to more than 200 reviewers. Of these, no reviewer spotted all of the errors and the average reviewer spotted only about two of them. This has several implications. Reviewers may need to be trained. If this is not possible, then editors should try to send more of their manuscripts out to high-quality reviewers. It also suggests that a large number of reviewers might be needed if all of the serious errors in a paper are to be detected.

The peer review system in science is analogous to the jury system in law. Each may have its flaws and biases but each also makes an important contribution to the fairness of important decisions. Many people believe that no system promotes fairness more effectively than peer review whether in science or in law. The very strength of this faith may have prevented experimentation with alternative systems, such as reliance on paid professional junior editors who could be trained in effective reviewing.

See also RESEARCH ETHICS; SCIENTIFIC METHOD.

Further Reading

Chubin, D., and E. Hackett. *Peerless Science*. Albany, N.Y.: State University of New York Press, 1990.
Fletcher, S., and R. Fletcher. "Evidence for the Effectiveness of Peer Review." *Science and Engineering Ethics* 3 (1997): 35–50.
Resnik, D. B. *The Ethics of Science*. New York: Routledge, 1998.

People for the Ethical Treatment of Animals (PETA)

Based in Norfolk, Virginia, PETA is one of the largest animal rights organizations in the world, claiming a worldwide membership of over 600,000. Founded in 1980, the organization received considerable publicity in the following year when cofounder Alex Pacheco infiltrated the laboratory of neuroscientist Edward Taub and notified the authorities of what was seen as improper treatment of the SILVER SPRING MONKEYS. As a consequence, Taub was arrested on charges of cruelty to animals and the monkeys were confiscated. Since then, a number of prominent animal research facilities have been targeted for violations of animal welfare legislation. They include the Carolina Biological Supply House (the largest in the country), Ohio's Wright State University, where painful scabies experiments were conducted on dogs and rabbits, and the University of Pennsylvania head trauma lab that used primates as research subjects. As a result of these actions, it is probable that animal care and use committees have become wary of approving research that could make their institution a target of such undesirable publicity. This inhibits research that inflicts pain on animals but it also has a chilling effect on academic freedom and possibly dissuades bright young scientists from entering the biomedical field.

Other favorite PETA issues include the abusive treatment of animals on fur farms and in entertainment, and the use of animals in General Motors crash tests. In addition to such activist agendas, PETA has an educational goal designed to make people aware of the suffering involved in many animal products and distributes a large volume of literature in schools. PETA has received the help of celebrities to promote its campaign against the use of animal furs in clothing and in so doing has made itself a fashionable cause.

See also ANIMAL RESEARCH, ETHICS OF; BATTERY CHICKENS; GREENPEACE; SILVER SPRING MONKEYS; FUR TRADE AND ENDANGERED SPECIES.

pesticides   *See* DDT.

phlogiston
Phlogiston is a hypothetical principle of fire that early scientists believed was contained in all combustible substances. Although the phlogiston idea owes much to ancient metaphysics, it was actually a scientific theory. German chemist Georg Ernst Stahl first used the word *phlogiston* early in the 18th century. Stahl proposed the rusting of iron and other metals was a slow form of burning in which phlogiston was released leaving a sort of ash. By 1800, scientists realized that there was no such phenomenon as phlogiston but that oxygen performed a similar role. The realization that oxygen was necessary for burning and rusting had been established by the experiments of French chemist Antoine Lavoisier, among others. The phlogiston theory is thus a fascinating example of the evolution of scientific ideas from metaphysical precursors. It is paralleled by the belief, in physics, in a hypothetical ETHER through which light rays were propagated.

See also ETHER, THE; N-RAYS.

physician-assisted suicide   *See* EUTHANASIA.

Piltdown man
The Piltdown man refers to fake fossils planted in a gravel pit evidently with the intention of establishing that the first humans were natives of Britain. A crude forgery consisting of a human skull and an ape's jaw with molar teeth roughly filed, the Piltdown remains were accepted as the missing link between humans and apes from the time of their discovery in 1908–12 until the mid-1920s when the discovery of hominid fossils in Africa, which had an apelike brain and a humanoid jaw, suggested that the Piltdown remains were anomalous.

The Piltdown man was discovered by Charles Dawson, a lawyer and amateur geologist. Dawson had asked a laborer working in gravel pits at Piltdown Common, close to Lewes in Sussex, England, to bring him any artifacts he might find. In 1908, the laborer found a piece of bone that turned out to be part of a thick human skull. During the next three years, other bits of the skull emerged and these were carefully put together. Then, in 1912, when Dawson was digging along with his friend Arthur Smith Woodward, an expert on fossil fishes at the British Museum of Natural History, he unearthed a complete jaw that seemed to belong to the reconstructed skull. Smith Woodward took the skull and cranium back to the British Museum where they were assembled.

Kept secret at first, Piltdown man was revealed at a meeting of the Geological Society in December 1912. The unveiling created a sensation. At the same time, some skeptics pointed out that the human skull did not really belong with the ape-like jaw. Yet most were so delighted by the find, which established Britain as the cradle of humanity, that the objections were quickly brushed aside. Moreover, another Piltdown man was discovered a few years later that seemed to clinch the evidence.

The perpetrator of the Piltdown hoax was never revealed although Charles Dawson, the discoverer of the first ape jaw, would have to be a candidate. Yet it is doubtful whether Dawson had the necessary access to ape bones. What is remarkable about the whole Piltdown episode is the leap of faith that scientists were willing to make without considering the obvious deficiencies of the evidence before them. The thickness of this mental fog is nicely illustrated by the counterhoaxing activities of Martin A. C. Hinton, a young zoologist at the British Museum who visited the site in 1913 and quickly came to the conclusion that Piltdown man was a fake.

Hinton decided to expose the hoaxers by himself planting an obvious fake, a crudely filed ape's tooth, and waiting for it to be "discovered" and turned into a laughing stock. To his intense annoyance, the tooth was accepted as a new find that backed up the earlier evidence. Realizing that the scientists were not inspecting the evidence very carefully, he decided to aim his satire at a broader audience. Taking an antique elephant leg bone, he carved it into a cricket bat. When the bat was discovered, Smith Woodward saw it as a unique, and therefore important, example of a paleolithic tool. Smith Woodward and Dawson wrote up a technical report on the find which described it in detail but avoided drawing the obvious conclusion—that it was a cricket bat. Defeated by the gullibility of the scientists, the counterhoaxers simply gave up in their attempts to expose the Piltdown man as a fraud.

Hinton's use of satire in an attempt to unmask phony scholarship is reminiscent of ALAN SOKAL's *Social Text* parody. Sokal was fighting the academic vacuousness of postmodernist critiques of science and made his point by having a worthless piece of verbiage accepted in a top journal in the field. Yet no one in the academic community recognized Sokal's hoax. He had to write a detailed account of how he had constructed the fake paper. Perhaps that is where Hinton's scheme failed.

See also ARCHAEOLOGICAL ETHICS; SCIENTIFIC DISHONESTY; SOKAL, ALAN.

Further Reading
Broad, W., and N. Wade. *Betrayers of the Truth.* New York: Simon and Schuster, 1982.
Weiner, J. S. *The Piltdown Forgery.* London: Oxford University Press, 1955.

PKU (phenylketonuria)

Phenylketonuria (PKU) is a hereditary metabolic disorder that involves inability to metabolize the amino acid phenylalanine. It used to be associated with severe mental retardation. The unmetabolized phenylalanine builds up in the blood, which evidently prevents the developing brain from acquiring other amino acids and impairs normal brain development. Infants are routinely screened to detect PKU. If they test positive for the condition, they are kept on a diet free of phenylalanine and do not suffer mental retardation. Maintaining such diets has been facilitated by the compulsory labeling of foods containing phenylalanine in the United States.

PKU is often used as an example of the principle that high heritability does not imply that an outcome is independent of environmental influences. This error has often been made in discussing the heritability of IQ scores, for example. PKU is a genetically determined trait but the mental retardation to which it would lead can be prevented by environmental intervention.

See also GENETIC DETERMINISM; GENETIC TESTING; IQ TESTS.

placebo-controlled drug trials

Properly conducted drug trials must include a placebo control group, that is, a group of participants who believe that they are receiving the active drug but actually receive a pill containing a pharmacologically inactive substance such as sugar. Placebo-treated individuals typically improve and this raises the bar for declaring that a medicine is effective. The effect of the active drug must be significantly greater than the effect for the placebo control group.

Ideally, placebo-controlled drug trials should be double-blinded. That is, neither the researcher administering the medicine nor the recipient should know whether the active medicine or the placebo has been administered. Otherwise, expectancy effects could contaminate the outcome of the study. In practice, participants often know which group they are in because pharmacologically active drugs usually produce side effects whereas placebos do not. This means that there may be a larger placebo effect for the drug group than for the placebo control group, which would tend to exaggerate the potency of the drug. Researchers can solve this problem through statistical control for correct guessing of group assignment.

The requirement for a placebo control group may be ethically questionable in the case of drug tests for life-threatening illnesses, such as AIDS. By participating in a study of this sort, terminally ill patients may be deprived of a life-saving medication. Yet drugs must be tested in placebo-controlled studies before their effectiveness can be accepted. Once the effectiveness of a drug has been experimentally established, it is no longer ethically defensible to withhold treatment from terminally ill people according to the DECLARATION OF HELSINKI. Thus, a large-scale study testing the protective effects of aspirin in relation to heart attacks was cut short after the effect had been established at a high level of statistical reliability. Researchers felt that it would be unethical to deny the placebo group the preventative effects of aspirin administration after they discovered that this reduced the probability of heart attacks.

See also COMMERCIALISM IN SCIENTIFIC RESEARCH; DECLARATION OF HELSINKI; HIPPOCRATIC OATH; PLACEBO EFFECT.

placebo effect

A placebo is a pharmacologically inactive substance, such as a sugar pill, that is administered as a medicine. Placebos often produce a substantial therapeutic effect that may rival that of a pharmacologically active drug. Such placebo effects have been observed in connection with testing of drugs as diverse as antidepressants and hair loss treatments. Physicians have often resorted to the use of placebos to calm hypochondriacs and in situations where the cause of a complaint is obscure. Such use of placebos is paternalistic and contrary to the spirit of modern scientific medicine. Placebos have a long history as a profitable business of quacks, charlatans, and frauds. While placebo effects are obviously based on a patient's expectation that they will improve, the underlying physiological mechanisms are poorly understood.

See also FAITH HEALING; PLACEBO-CONTROLLED DRUG TRIALS; VOODOO DEATH.

plagiarism

Plagiarism is defined as the use of another person's work without due acknowledgment. Plagiarism can be regarded as a type of theft, or dishonesty, in which the work of another person is passed off as the work of the plagiarist. Technically, plagiarism can refer to any kind of intellectual property but in academic life it arises most often in relation to written work.

Obvious academic plagiarism is very common among students but is quite rare among professional academics due to the high likelihood of exposure and the very damaging consequences of detection. Although plagiarists often break copyright laws, there is an important distinction between plagiarism, which is an ethical failing, and breach of copyright, which is a legal transgression. Viewed in legalistic terms, copyright does not cover ideas but only the particular form in which the ideas are expressed. Plagiarism is a broader term that refers not just to the unacknowledged use of sentences and phrases composed by another but also to the unacknowledged use of the ideas that are being expressed.

Plagiarism can be very easily spotted or it can be a difficult judgment call, depending on how blatant the plagiarist is. The clearest example of plagiarism occurs when a student finds that they are a few hours away from the final deadline for a term paper and have done no work. At this point, rather than fail the course, the student might borrow a paper written by a friend and substitute their own name for the friend's on the title page. This might solve the practical problem in a large college where the paper is unlikely to go to the same reader twice. If no friend has recently taken the relevant course, the desperate student might visit an online site where term papers can be selected by topic. The paper is then pasted into a word processor, allowing the plagiarist to add their name and print the paper just on time to beat the deadline. This is the purest example of plagiarism because the offender contributes only their name and passes off both the ideas and the form of expression of another person as their own. It might seem that professors would be helpless in the face of such organized dishonesty but the technically sophisticated have struck back by compiling a database of plagiarized papers that can be searched by long word strings thereby allowing suspect papers to be traced back to their source.

Most students probably do not want to engage in such blatant cheating. After all, if they are going to do something that is so clearly immoral, why spend several years in college to obtain a degree. It would be more efficient to fake their diploma. Most student plagiarism is not so blatant. Instead of stealing a whole paper, the student may steal sentences or paragraphs without using quotation marks and acknowledgments. This kind of plagiarism is often easy to detect. It may have exactly the same intent as stealing an entire paper or it might result more innocently when the student is unversed in the correct method of acknowledging sources. Correct use of quotation and acknowledgment does not necessarily detract from the contribution of the student and may actually enhance a reader's perception of the paper by making it more intelligible.

A much grayer area in plagiarism occurs when the student uses their own words but steals the ideas without due acknowledgment. This is sometimes difficult to detect. Students who engage in this form of plagiarism often cite their sources. Even if they do cite sources, this practice can still be seen as plagiarism because the reader of the paper may be given the false impression that the ostensible author has contributed more of the ideas in the paper than they actually did.

Most professional academics avoid blatant plagiarism but many are guilty of excessive paraphrasing, which is not discouraged by the editorial policies of many journals. Moreover, academics often plagiarize themselves. This can occur quite easily when a researcher repeatedly produces theoretically similar papers. This may be permissible with two qualifications. One is that the original source must be acknowledged. The other is that the practice may be in breach of copyright law and, if it is, the author is legally compelled to seek permission from the copyright holder to reuse his or her own work.

Wholesale plagiarism may occur without detection in fields such as biomedicine where there are thousands of professional journals in many different languages. The most prolific scientific plagiarist may be the Iraqi medical student Elias A. K. Alsabti who is believed to have plagiarized some 60 complete papers between 1977 and 1980. Alsabti's method was simple, he retyped whole papers changing only the title and author's names and occasionally adding some fictitious coauthors. His dubious success provides evidence of a conspicuous failure of the editorial review system. Some journals even failed to publish retractions after the plagiarism had been pointed out to them. Alsabti himself disappeared soon after he had been exposed as a cheat.

If an author publishes exactly the same paper in different journals, this is referred to as DUPLICATE PUBLICATION and it is prohibited by the ethical standards observed in some disciplines. The author is felt to be dishonestly padding the publication list in their curriculum vitae while wasting valuable publication space that might have gone to original research. Despite this, duplicate publication can be justified when there is a perceived need to bring a particular piece of research to a wider audience. Moreover, journal editors may decide to reprint articles from other journals that are of particular relevance to their readers. Duplicate publication also occurs ethically when journal articles are reprinted in book devoted to a specific research topic or when the previous publication of the article is clearly acknowledged prior to publication.

See also DUPLICATE PUBLICATION; PATENTS.

Further Reading

Broad, W., and N. Wade. *Betrayers of the Truth*. New York: Simon and Schuster, 1982.

LaFollette, M. *Stealing into Print: Fraud, Plagiarism and Misconduct in Academic Publishing*. Berkeley: University of California Press, 1992.

Resnik, D. B. *The Ethics of Science*. New York: Routledge, 1998.

"planaria soup" research

Planaria are flatworms and these were used in the 1960s in experiments on the molecular basis of learning that produced initially positive findings but were subsequently abandoned as a waste of time. Ideally, science is supposed to be empirical. That is, scientists are expected to arrive at truth via the collection of data. Yet there are many episodes in the history of science in which scientists paid more attention to what their beliefs and intuitions were telling them than to the data. Modern examples include the persistent widespread belief of physicists in illusory phenomena such as N-RAYS and COLD FUSION. There is nothing unique about physics in this respect. All sciences encounter considerable uncertainty, particularly in new fields of research. Examples in psychology include research on the molecular basis of learning in *planaria* and other species and on EXTRASENSORY PERCEPTION.

During the 1960s, many researchers investigating the brain basis of learning and memory believed that each memory was encoded in the structure of a molecule of RNA or a protein. In 1962, J. V. McConnell reported that he had trained *planaria* to respond to a light and fed the trained individuals, ground up in a "soup," to other *planaria*. Thus fed, the cannibal flatworms seemed to have acquired some of the training because they learned the light discrimination more rapidly than usual.

Conceptually similar experiments were performed on rats in which subjects consumed brain extracts from trained animals prior to being trained themselves. In one study, rats that had received brain extracts from other individuals trained to approach a clicking sound for food, learned to approach the clicking sound more quickly than control animals. This positive finding stimulated many attempts to replicate the transfer of training via brain extracts. About half of these studies produced positive findings and half did not. There was a distressing lack of consistency in the results, even within a single laboratory where exactly the same procedures were repeated.

Studies of transfer of training via brain extracts persisted for 20 years. This line of research was not abandoned because of disconfirming evidence but because investigators stopped believing that such transfer of training was possible despite many published studies indicating that it had occurred. These positive results were outweighed by the unreliability of the phenomenon. Researchers (and grant proposal reviewers) came to the conclusion that their time would be better spent on other projects and moved on to more promising ventures.

Abandonment of a field of scientific research is rather unusual. In general, scientists do not abandon a research idea unless they receive clear evidence disconfirming their hypothesis. Such evidence was provided in the case of the shared scientific illusion of N-rays but even then scientists in France, where the illusion originated, continued with this line of research. Abandonment of research on transfer of training via brain extracts means that neuroscientists do not know whether this was also based on a scientific illusion or whether they are ignoring a potentially important phenomenon.

See also COLD FUSION CONTROVERSY; EXTRASENSORY PERCEPTION; N-RAYS; SCIENTIFIC METHOD.

Further Reading

Kalat, J. *Biological Psychology*. Belmont, Calif.: Wadsworth, 1988.

Planck, Max

(1858–1947)
German
Physicist

Max Planck won the 1918 Nobel Prize in physics for his discovery that energy is transferred in quanta (or small bundles) rather than as a continuous stream. Planck was one of the most respected scientists of his day. In 1930, four years after he retired from his professorship at the University of Berlin, he was invited to head the Kaiser

Wilhelm Society in Berlin, which was renamed the Max Planck Society to honor him. His outspoken opposition to the Nazi regime, however, resulted in his removal from this position in 1937.

Apart from his specialized research interests, Planck wrote about the relationship between physics and philosophy, religion, and society. A deeply religious man, he agreed with EINSTEIN's view that God does not play dice with the universe and thus rejected indeterminacy, a key concept in the quantum physics that he and Einstein had been instrumental in creating. Planck died in Gottingen, Germany, on October 4, 1947.

See also EINSTEIN, ALBERT; HAWKING, STEPHEN; SCIENCE, HISTORY OF.

Plato
(428–348/7 B.C.)
Greek
Philosopher

Plato is probably the most influential philosopher who ever lived, the closest contender being ARISTOTLE, his student. Although Aristotle and Plato endorse a similar virtue approach to ethics and share similar political views, their attitudes to scientific knowledge were quite different. Plato was essentially a mystic who believed that knowledge could not be derived from the senses. His other-worldly approach to knowledge has appealed to Christian scholars through the ages whereas Aristotle came to prominence only after the rise of medieval universities and the associated scholastic tradition of learning in the 12th century. Aristotle not only developed the logic that has been used by scientists but was an influential scientist in his own right whose work in zoology was influential right up to the lifetime of CHARLES DARWIN.

Plato's strange mix of rationality and mysticism is partly attributable to his association with SOCRATES, who was not only his teacher but also a close personal friend. Born into a privileged Athenian family, Plato might well have entered a political career but the excesses of the Thirty Tyrants (404–403 B.C.) and of the restored democracy that followed them seem to have discouraged him. Deeply affected by the execution of Socrates, he spent several years traveling in Italy and Sicily. In 387 B.C., he returned to Athens to found the Academy, a school of higher learning similar in its goals and activities to a modern university. Students at the Academy, including Aristotle, conducted research in many different branches of learning. Most of Plato's working life was spent teaching and managing the Academy. He died in Athens in 348 or 347 B.C.

In addition to Socrates, Plato was influenced by other Greek philosophers. He acquired from Parmenides the notion that reality is timeless and unchanging and that everything that changes is an illusion. From Heraclitus he extracted the idea that there is nothing permanent in the external world of the senses. Combining these mystical concepts, he came to the conclusion that knowledge cannot be acquired from the world of the senses and can be approached only through the intellect. These views are typified by Plato's theory of forms, which postulated that the world of our senses consists of imperfect and transitory copies of the real world that is ideal and eternal. It would be difficult to imagine a philosophy that is more antagonistic to scientific investigation. Despite these views, Plato apparently encouraged his students to cultivate learning in all branches of knowledge, including science (see ARISTOTLE).

Plato's ideas were expressed in 26 dramatic dialogues dealing with various philosophical topics, all of which have survived. The dialogues feature Socrates as a leading character and depict the Socratic method, according to which knowledge is elicited from a student by asking a long series of probing questions. The most influential dialogue is the *Republic,* which describes an ideal form of government and constitutes the first in a long series of literary Utopias.

The *Republic* is a great deal more aristocratic than the name implies to modern ears. Plato describes rulers who possess an unusual degree of intellectual enlightenment. Thus, rulers must learn mathematics, not because they will apply this knowledge in levying taxes or developing military strategy, for example, but because this kind of knowledge is essential for developing their intellectual character so that they can know the Good. Rulers must be the best kind of people in the sense of being capable of intellectual and moral discipline. In order to be wise, they would also have to be leisured. Plato, like most of the other philosophers of his time, believed that wisdom was impossible without leisure. In short, the wise and good rulers he envisaged could only be aristocrats.

Plato's ethics is essentially a refinement of the virtue approach of Socrates. Both men agree that virtue comes from understanding and that to live the good life people must develop their intellectual faculties. Plato also saw philosophical education as essential for developing internal harmony between reason and passion. The resulting self-mastery allowed the good life to be lived.

See also ARISTOTLE; MORAL PHILOSOPHY; PYTHAGORAS; SOCRATES.

Further Reading
Melling, D. J. *Understanding Plato.* New York: Oxford University Press, 1988.
Russell, B. *A History of Western Philosophy.* New York: Simon and Schuster, 1972.

polygraph ("lie-detector") test

Taking a "lie-detector" test involves being hooked up to physiological recording devices. Heart rate and breathing are recorded using different automatic inked styluses. (Hence the term "polygraph.") Sweating is cleverly measured as changes in the electrical conductivity of a patch of skin when it is moist (the galvanic skin response). Ethical controversy concerning the use of polygraph tests has focused on their basic unreliability and on the deceptive practices resorted to by test administrators to convince subjects that the test is more powerful than it actually is.

The polygraph measures bodily arousal produced by anxiety. The underlying rationale is as follows. Lying produces anxiety. Anxiety stimulates the sympathetic nervous system thereby producing increased heartbeat, breathing rate, and sweating. If there are marked changes in the polygraph traces, the subject is anxious and must therefore be lying.

The polygraph may be wrapped up in scientific trappings but it is actually a crude and primitive test. Thus, a traditional Arab test of lying employs exactly the same physiological principles. In the Arab lie-detector test, a knife blade was heated and pressed briefly against the tongue of the suspect. If the flesh burned, the person was guilty of lying! This bears exact analogy with the modern polygraph because if a person is nervous, they stop salivating, which allows the hot blade to burn them.

The assumptions underlying the measurement of anxiety as an indirect measure of deception are faulty in several respects. People are not always anxious when they lie. Compulsive liars may be just as relaxed when lying as when telling the truth. Sociopaths (e.g., many career criminals) are also very good at controlling their level of bodily arousal in stressful situations. Moreover, some honest people may be quite anxious when they are telling the truth, if only because they are worried about the outcome of the test.

Such flawed assumptions help to explain why the polygraph produces poor results in controlled studies. When someone is lying, the test catches them about three times out of four. However, when they are telling the truth, the polygraph incorrectly identifies them as liars half of the time. Overall, the polygraph is correct about two times out of three (65%) compared to the 50% correct answers obtainable by tossing a coin. Some critics believe that it is even more unreliable in the hustle and bustle of real-world settings such as police stations.

New techniques are being developed such as voice stress analysis. The underlying principle is that when people are tense, voice tremor is suppressed by the activity of the sympathetic nervous system. The accuracy of this technique has not yet been satisfactorily established, but there is no good reason for believing that it can improve on the polygraph.

Given the problems and unreliability of the polygraph test, it is rather disappointing that many U.S. states permit its use as courtroom evidence even though it is not admissible in federal cases unless both the defense and the prosecution agree to its use. The American Polygraph Association claims that polygraph accuracy is far higher (90%) than independent researchers have determined.

In conclusion, the so-called lie-detector test is a faulty procedure that gives honest people an unacceptably high probability of being labeled liars. It is an example of the abuse of ostensibly scientific techniques to falsely convince subjects that electronic gadgets can be used to read their minds.

See also PSEUDOSCIENCE.

Further Reading

Forman, R. F., and C. McCauley. "Validity of the Positive Control Polygraph Test Using the Field Practice Model." *Journal of Applied Psychology* 71 (1986): 691–98.

Patrick, C. J., and W. G. Iacono. "Psychopathy, Threat, and Polygraph Test Accuracy." *Journal of Applied Psychology* 74 (1989): 347–55.

Popper, Sir Karl Raimund

(1902–1994)
Austrian
Philosopher

Karl Popper is noted for his contributions to the philosophy of science and also for his political philosophy, which defended Western liberal democracy. Popper's most influential book on science was *Logik der Forschung* (1935, The logic of scientific discovery, 1959). This dismissed FRANCIS BACON's (1561–1626) view of science as based on methodical collection and summarization of facts (or induction). Popper denied that induction played an important role in science and saw scientific discovery as beginning with a highly creative guess, or hypothesis. Once the hypothesis was formulated, scientists set about the more rigorous procedure of testing it. According to Popper, hypotheses were tested by setting up a situation in which the data could potentially reveal them to be false. The potential for falsification was what distinguished scientific hypotheses from other types of explanation, such as religious ones, that are not open to falsification.

Popper's *Logik der Forschung* was published in a series organized by the VIENNA CIRCLE, which promoted POSITIVISM, or the view that philosophy should be restricted to objective facts and logic and should reject metaphysics. He was not actually a member of the Vienna Circle because he differed with some of their views. Not only did he reject their Baconian emphasis on induction in science, but he felt that metaphysics could be useful to scientists by suggesting falsifiable hypotheses.

See also BACON, FRANCIS; POSITIVISM; PSEUDOSCIENCE; SCIENCE, HISTORY OF; SCIENTIFIC METHOD.

Further Reading
Popper, K. R. *The Logic of Scientific Discovery*. New York: Basic, 1959.

popularization of scientific findings

Transmitting complex scientific information to the general reading public creates an ethical problem for the author. How can the information be simplified without serious distortion? One obvious point is that a popularizer with a good grasp of the science is less likely to misrepresent it. Thus, some of the world's great scientists from GALILEO GALILEI to MICHAEL FARADAY to STEPHEN HAWKING have been excellent popularizers of the science. Admittedly, not all scientists are good at communicating with the public. A scientific style of writing can often seem tedious, unduly passive, overcautious, and excessively obfuscating. This can be frustrating for the general reader. Most scientists are tutored in guarded scientific writing but few receive any instruction in how to communicate science to the general public.

Problems can arise when professional writers and journalists without an extensive training in science undertake to explain to the public that which they themselves do not fully understand. The inevitable result is vagueness, oversimplification, and distortion. These problems are more likely to occur in book-length treatments than in newspaper articles on scientific breakthroughs because journalists can use direct quotes from scientists to transmit technical information thus transferring the responsibility for accurate popularization to the scientists themselves. Scientists who are uncomfortable in this role or feel that their work has been misrepresented in the past sometimes refuse to give interviews. This unfortunate attitude is understandable given that graduate schools in science rarely offer any training in media relations.

While accurate popularization is an important service that scientists can, and should, render to the public, particularly if they receive PUBLIC FUNDING, there is no question that many books of scientific popularization are of low quality. The biggest problem is oversimplification. Publishers of trade books like to market works whose message can be distilled into a single message (e.g., electric fields are killing you or men have a genetic program that compels them to rape women). This marketing message is usually a gross oversimplification but the sensa-

tional impact of such simple messages generates publicity and sells books.

Scientists are usually the best people to popularize their own work but there may be no professional reward in doing this. Popular works can detract from a scientist's academic reputation. For example, CARL SAGAN was an effective popularizer of scientific work as well as being a distinguished astronomer. Despite his professional accomplishments, he was never elected to the NATIONAL ACADEMY OF SCIENCES and this may have been because some academy members did not approve of his popularization of science.

See also PSEUDOSCIENCE; SAGAN, CARL; SCIENTIFIC DISHONESTY.

Further Reading
Resnik, D. B. *The Ethics of Science*. New York: Routledge, 1998.

pornography; science, technology, and the law

Pornography refers to materials, such as books, pictures, films, and video presentations, that are designed to stimulate sexual arousal. Most pornographic materials either depict, or suggest, sexual behavior. Pornography has often created ethical and legal problems and these problems have been altered in significant ways by technological innovations. For example, the development of film in the 20th century allowed for more realistic depictions of sexual behavior than was possible in still photographic prints. In recent decades, telephone lines have been used as a distribution channel for pornography and "phone sex" is a lucrative business. Most recently, the Internet has provided a boost to pornographic enterprises of various types from large picture archives to voyeurism of models in their homes to global interactive peep shows.

Scientific study of the effects of pornography is important because of the role that such research plays in determining the appropriateness of legal control over the production and distribution of this material. Legal control over pornography has clear implications for broader issues of freedom of speech and expression.

The scientific study of pornography itself raises many ethical questions. To begin with, there are clear sex differences in consumption of pornography, which means that this research inevitably delves into gender differences. Such research is a favorite target of feminist scholars who often question its objectivity and purpose. There is also the issue of causing harm to participants. If pornography is actually harmful, then research participants are being wronged. Moreover, if violent pornography increases aggression against women, then it is arguable that the intimate partners of men in this research are being harmed. Another ethical problem relates to the issue of the quality and representativeness of the research.

A BRIEF HISTORY OF PORNOGRAPHY AND THE LAW

Pornography is ancient but attempts to regulate it by law are comparatively recent. This may reflect the greater visibility of pornography due to the development of print publications for mass distribution. When pornographic materials were produced by artists for wealthy patrons, these materials were kept out of public view and may have been tolerated by legal authorities even though they were not socially sanctioned.

In America, Massachusetts had laws prohibiting obscenity (which is a legal term for pornography) in colonial times. The United States passed antipornography laws in 1842 and, in 1865, laws that prohibited distribution of pornography via mail. In Britain, the Obscene Publications Act was passed in 1857. To the extent that legislation is prompted by contemporaneous problems, it can be inferred that the distribution of pornography emerged as a pressing issue around the middle of the 19th century. This concern may have been related to the development of modern photography. A system for producing an infinite number of prints from a single negative (based on Talbot's calotype) emerged in 1840. Antipornography laws were vigorously enforced in America at the end of the 19th century under the influence of Anthony Comstock's (1844–1915) Committee for the Suppression of Vice.

Early in the 20th century works of fiction came increasingly under attack. Many books, such as James Joyce's *Ulysses* and D. H. Lawrence's *Lady Chatterley's Lover*, were banned in several countries on grounds of obscenity. Such bans produced a great deal of controversy because the "obscene" novels turned out to be, in the opinion of many literary critics, works of outstanding literary merit. Eventually, most of these bans were lifted and the explicit depiction of sexual anatomy and sexual acts to which contemporaries objected are quite commonplace in the literature of today, indicating that what is considered obscene at one point in history will not be at another.

The historical relativity of defining obscenity is underscored by changes in sexual mores that occurred in the 1960s in the context of what is now referred to as the sexual revolution. During the 1960s people are said to have become sexually liberated, which meant that there was increasing acceptance of extramarital intercourse and emphasis on the freedom to express sexual impulses that were not harmful to others. This social change is reflected in the behavior of the U.S. Supreme Court, which, in 1957, ruled that pornography was not protected by the constitutional right to free speech. In 1973, the Court reversed itself, ruling that states had jurisdiction in such matters. In 1982, the Court made an exception in the case of child pornography, by upholding a New York Statute outlawing the production and sale of materials that contained sexually explicit depictions of children.

A U.S. federal Child Pornography Prevention Act first passed in 1977, and amended several times since, has recently been tested in the context of a large number of FBI prosecutions of child pornographers on the Internet. A federal appeals court overturned parts of the 1996 Child Pornography Prevention Act in December 1999. Although the ruling seriously weakened the legislation, it excluded only images that had been digitally manipulated to appear to depict young children. Such material was said to be protected by the First Amendment to the U.S. Constitution. The appeals court ruled that the portion of the law that applies to pornographic images of real children is still valid. The logic of the legislation had been to protect children from the harm caused by the creation of pornography. The Federal Appeal Court decision is currently being reviewed by the Supreme Court.

This successful challenge to part of the Child Pornography Prevention Act was brought by lead plaintiff the Free Speech Coalition, which is a trade association for businesses that deal in "adult" materials. Other plaintiffs included nudists and artists and photographers that depict nudity in their work. Pornography was the first business to be successful on the Internet and commercial pornography sites preceded e-commerce sites by several years. There have been concerted legislative attempts to restrict the flow of electronic pornography to homes, schools, and public libraries. Many of these attempts have foundered on the rock of the First Amendment. For example, the Communications Decency Act was overturned by the Supreme Court in 1996 on the grounds that it abridges free speech on the Internet.

Despite this rejection of electronic censorship by the Supreme Court, there are ongoing attempts to shield children from exposure to pornography. One way of achieving this is through the use of electronic filters. The trouble with such filters is that they are crude and often inadvertently block nonpornographic material. For example, a woman who uses a computer with a pornography filter may be denied access to information on breast cancer. Teenagers may be denied information on safe sex. To the extent that such filters block information protected by the First Amendment, it is difficult to see how legislation that compels their use in public buildings can withstand constitutional challenges.

Many public libraries provide free access to the Internet. Librarians are disconcerted by the fact that patrons frequently access pornographic sites in full view of children. Yet, if libraries choose to provide Internet access, they cannot control the content that is accessed. This lesson was learned by the Library Board of Loudoun County in Virginia. The board decided to solve their pornography problem by installing filters on each of the library's computers that had Internet access. They were sued in local court by a coalition including the American Civil Liberties Union (ACLU), the Electronic Frontier Foundation, and a number of sites, such as Safesex.com, that were blocked by the filter. The presiding judge ruled that the Library Board was barred by the First Amendment from restricting the content viewed on library computers. This ruling does not have any legal weight outside one district in Virginia but it illustrates the dilemma faced by those who attempt to protect children from exposure to pornography through legislation.

RESEARCH CONCERNING THE HARMFUL EFFECTS OF PORNOGRAPHY

Legislators are not only influenced by social change and the ebb and flow of sentiment in favor of censorship. They also attempt to make prudent decisions based on scientific information. Most legislators accept that the production of child pornography is potentially harmful to children. For obvious ethical reasons, there is not scientific research dealing with the harmful effects on children of exposure to pornography. However, there is a scientific literature dealing with the possible effects of pornography on adults and this addresses such issues as whether it increases the likelihood of sexual crimes being committed. Much of this research deals with men because women generally volunteer for such research in very low numbers. Alarmed at the increased availability of pornographic materials during the 1960s, the United States convened a National Commission on Obscenity and Pornography that reported in 1970. The commission was unable to find any harmful effects of pornography in the sense that pornography increased the probability of committing rape of engaging in other antisocial behavior. This finding seems to have laid the groundwork for the 1973 Supreme Court ruling in which pornography was effectively relegalized.

Subsequent findings have not always supported the finding of the 1970 commission. For example, research published in 1984 found that when college men and women were exposed to a large number of pornographic films, they recommended a lighter sentence for a rapist in a simulated trial and expressed less support for women's liberation. Interestingly, these results applied to women as well as men, suggesting that the students had been influenced by the general point of view represented in the films.

Since 1970, there has been a marked increase in the amount of violence contained in pornography, possibly reflecting a general increase in media violence during the same period. Many researchers have specifically investigated the relationship between exposure to such materials and antisocial tendencies in men. The results of laboratory studies are quite disturbing. After viewing violent pornographic films in which the female victim is depicted as becoming sexually aroused, men are more willing to commit aggression against women but not more willing to commit aggression against other men. Aggression was measured by leading the men to believe that they were giving an electric shock to a woman and recording the shock level they had chosen. When the female victim in the film was depicted as suffering during the attack, men did not become more aggressive toward women. This suggests that if a violent pornographic film depicts women as enjoying aggression, it can increase men's tendency to behave aggressively toward women.

These results were produced not only using hard-core pornographic movies, but also mainstream films like *Swept Away* and *The Getaway* in which women are subjected to sexual violence, are aroused during the attack, and become romantically attracted to their assailant. Men who see such movies have an increased acceptance of violence against women and are more likely to agree with the myth that raped women either invite or secretly desire the attack.

If these results had any relevance for the real world outside the laboratory, then violent rapists would have a history of viewing violent pornography. Yet there is no clear evidence of any such connection. This disconnection between what is found in experiments and what happens outside the laboratory is a recurring problem for social psychologists that also emerges in connection with research on television violence and nonsexual aggression.

Responding to an altered climate of opinion concerning the effects of pornographic violence, the U.S. government launched a new Commission on Pornography that reported in 1987. The second commission produced findings that were very different from those of the first. Instead of concluding that pornography has no antisocial effects, the commission drew the very firm and specific conclusion that violent pornography causes violence against women (in addition to other antisocial effects). This finding seemed to confirm arguments of feminist scholars concerning the harmful and degrading effects of pornography on women, but such a finding has been criticized by scientists who authored the research on which the commission's report was based.

One problem with the commission's report is that it makes an unwarranted leap from findings that violent pornography increases aggression against women in a laboratory setting to concluding that this must happen outside the laboratory. Yet, there is little evidence that this happens in the real world. Moreover, some aspects of the commission's report did not pay due attention to conflicting research evidence when drawing conclusions.

Perhaps the most important criticism by scientists is that pornography is not the only source of violent depictions in which women are victims. In fact, nonpornographic movies may contain not just more violence in general but more violence directed at women. Based on the research findings, scientists have argued that there should be more concern with mainstream media depictions of violence against women instead of focusing narrowly on pornography.

ETHICAL ISSUES IN RESEARCH ON PORNOGRAPHY

If violent pornography is potentially harmful, then researchers who investigate its effects are treading in a gray area ethically speaking since research ethics require that the researcher should avoid doing harm either to the participants or to the people with whom the participants interact. With this in mind, some of the leading psychologists in this field have come up with a set of guidelines describing how the debriefing should be conducted.

A debriefing is an educational session at the end of the research in which participants are told about its nature and purpose. Extensive debriefings are usually carried out in the case of research that has involved deception. The thinking is that being deceived causes harm to the participant and that this harm can be mitigated by setting matters straight at the conclusion of the research.

By an exactly parallel logic, researchers reason that exposure to pornographic violence may increase men's acceptance of violence and their agreement with myths about rape victims. The debriefing stresses the lack of realism in violently pornographic materials in relation to the attitudes of the victim. In reality, women do not want to be violently attacked and do not enjoy being raped. On the contrary, many women who have been violently raped experience post-traumatic stress that is similar to that of combat veterans.

When research participants are exposed to this kind of educational information in the debriefing, agreement with myths about rape victims declines with lasting effect. If the debriefing is correctly handled, participants in research on pornography will not be more likely, but if anything will be less likely, to commit violence against women. Moreover, the development of these debriefing procedures can have broader benefits. They might be used as a model for educational programs designed to reduce violence against women in the population at large.

Research on pornography has often concentrated exclusively on men as participants because men are the primary consumers of commercial pornography. In discussing the ethics of pornography in modern media, it is important to examine the possible effects on women as well as men for several reasons. Women sometimes participate in this research, and ethical researchers need to have information about the possible effects of exposing women to pornography.

PORNOGRAPHY AND WOMEN

Exposure to pornography can affect the attitudes of women as well as men. Women who viewed a great deal of nonviolent pornography in the context of an experiment of Dolf Zillman and Jennings Bryant became more tolerant of rape and were less in favor of feminist political views. Such attitude changes suggest that pornography can have a strong emotional impact on women as well as men. This argument may seem at odds with the fact that women do not purchase pornography but this inconsistency can be resolved. When MASTERS AND JOHNSON did their ground-breaking work on the physiology of the human sexual response, they emphasized the basic similarity between men and women in the mechanisms and sequential patterns of arousal. Masters and Johnson generally ignored sex differences in sexual psychology. Yet, their own research highlighted an obvious flaw in this approach. Even though masturbation produced an objectively more intense orgasm in women than intercourse did participants found it less satisfying. This indicates a disconnection between the physiological aspect and the psychological aspect of female sexuality that was not seen in men. If this interpretation is correct, then women are not interested in purchasing pornography because this caters to the physiological side of sexuality without providing emotional involvement. Evolutionary anthropologist Donald Symons of the University of California at Santa Barbara has theorized that women have an evolved tendency to seek emotional commitment in sexual relationships whereas men are oriented more toward looking for sheer physical gratification. Individuals members of one sex clearly differ greatly on these dimensions, however, a focus on physical gratification would have facilitated reproductive success of our male ancestors but would have undermined a woman's efforts to secure paternal investment in her children.

Masters and Johnson's unisex approach to human physiological reactions to sexual arousal revealed fundamental similarities between men and women in this sphere but the shortcomings of this approach are revealed by women's response to pornography. The physiological response turns out to be similar to that of men but the psychological reaction is very different.

Magazine publishers have found that most women are not interested in visual pornography in the way that most men are. This is not simply a reflection of the fact that commercial magazines are designed to appeal to masculine tastes. The magazines *Playgirl* and *Viva*, which were founded in the 1970s, attempted to break new ground by creating a medium in which women would have an opportunity to enjoy depictions of male nudity. These enterprises turned out to be an instructive and amusing failure as recounted by Symons. Both enterprises seemed to have got off to a good start. Then reader surveys indicated that the majority of the readers were gay men. In response to the wishes of female readers, the magazines discontinued the male nudes. Interestingly, they retained some of the female nudes on the grounds that female readers found these more esthetically pleasing.

These conclusions from reader surveys are confirmed by more scientifically obtained evidence. For example, of Kinsey survey respondents, only 12% of women reported that they had become sexually aroused by looking at nude pictures compared to 54% of men. Systematic comparisons of men's ratings of female nudes in *Playboy* compared to women's ratings of nude men in *Playgirl* have produced comparable results. About three-quarters of men evaluated the female nudes as appealing whereas about three-quarters of women rated the male nudes as unappealing.

Given the strong sex difference in evaluations of nude members of the opposite sex, it is interesting that, at a purely physiological level, women may be just as responsive to pornography as men. According to the results of several studies sponsored by the U.S. Commission on Obscenity and Pornography, the same proportion of women as men were physiologically aroused by erotica (60–85% in different studies). Moreover, despite the belief that men are more oriented toward visual media, there is little sex difference in the proportion of people experiencing arousal regardless of whether the medium used was pictures, printed descriptions, films, or verbal sound recordings.

What makes these findings so compelling is that they did not rely on self reports, which are notoriously unreliable because people tend to say whatever seems socially desirable in the situation. Rather, physiological arousal was measured objectively in terms of blood flow to the genitals. This was done using ingenious recording devices specifically designed for this purpose. Penile strain gauges provided a continuous readout of penile tumescence in men. Plethysmographs provided a continuous measure of blood flow to the genitals of women. To avoid any appearance of impropriety, such devices are often fitted by participants following instructions from the experimenters.

Given the inconsistency between women's self-reports and the objective measure of their responsiveness to erotica, it might seem that they are simply being untruthful in their self-reports. A more complex interpretation may be necessary, however. It seems that many women do not equate mere physiological arousal with sexual excitement or attraction. According to Symons's evolutionary perspective, it would not have favoured the reproductive success of a women if she became sexually excited too

easily. This would have undermined her ability to be discriminating in the choice of a partner and would have made a sex partner less confident of paternity and thus less likely to support her children.

Scientific research has done a great deal to demystify sex differences in reaction to pornography. Such research has always encountered ethical hurdles and became possible only around the middle of the 20th century when attitudes toward sexuality became more open. Although some moralists would argue that research on sexuality is an unwarranted intrusion into a realm that should be kept private and an unethical interference in the sexual lives of individuals, it is important to note that such research is always conducted using volunteers who are fully informed about what is required of them. If the research is morally objectionable, then deciding to participate is a moral choice of the individual participant and need not be the exclusive concern of the researcher. Moreover, participating in such research carries the potential for good consequences as well as bad. For example, some of the participants in Masters and Johnson's research claimed that it had produced lasting improvements in their sex lives.

A greater understanding of sex differences in responsiveness to pornography can be justified in terms of a number of potential benefits that contribute to the common good. One of the most obvious is that clarifying sex differences in sexual psychology can help to facilitate communication between men and women in their intimate relationships and therefore increase happiness and reduce conflict in those relationships. Many women cannot understand why men would be interested in looking at pictures of women who are materially unattainable. Many men have trouble understanding why women are not interested in erotica. Understanding such sex differences in sexual psychology can help to minimize conflict between men and women in marriages. Perhaps the most direct practical benefit of scientific research on pornography is that it provides a reliable foundation of objective knowledge on which discussions about the need to regulate pornography must be based. Understanding sex differences in reaction to pornography can play an important role in the study of gender differences more generally.

PORNOGRAPHY AND THE INTERNET

The emergence of the Internet accentuated concerns about pornography for several reasons. One reason is that it is a new technology and new technologies, such as photography and film, always raise new ethical concerns. Thus, the American public feared that depictions of unbridled sexual passion in the cinema would have a corrupting effect on the sexual morality of the country. Hollywood studios were so concerned about heading off censorship laws that they voluntarily adopted, in 1930, a strict production code that screened out any appearance of sexual impropriety. The code regulated how long the main characters could kiss, for example, and prevented them from appearing to sleep together if they were unmarried. Interestingly, the code was deemed necessary the year after introduction of sound, which allowed for a much more compelling portrayal of dramatic situations.

The Internet transcends national boundaries more easily than any other medium, which means that the standards of sexual morality of one society are inevitably brought into collision with those of others. Consequently, different countries must either allow complete freedom of expression or introduce draconian censorship legislation that attempts to prevent undesirable materials from being accessed. This can be done by prosecuting individuals for accessing certain sites, or, more efficiently, by prosecuting the Internet service providers (ISPs) who make such access possible. Forcing the ISPs to screen content would be inconsistent with the right to freedom of expression in the American Constitution but not all countries guarantee their citizens such rights. Thus, the Chinese government has recently begun an effort to clean up the Internet material available Chinese citizens by holding the ISPs legally responsible for the content that can be accessed. In America, efforts to control the flow of pornography have focused on child pornography, which is not regarded as protected information in respect to freedom of speech.

There is a great deal of hysteria about the quantity and the perversity of pornographic materials available online. Pornography can be regarded as the first commercially successful e-business and it is an extremely aggressive business in the sense that operators are willing to resort to dirty tricks to lure customers. Thus, some visitors to respectable sites may find themselves unaccountably accessing a pornography site. Once inside the site, they may find it difficult to leave because clever programmers have manipulated the browser commands.

Such experiences might convey the impression that the Internet is full of pornographic sites but this impression is mistaken. An influential study of web browsers published in *Nature* in July 1999 found that the most popular browsers access only a small fragment of the 800 million pages on the Web and revealed that only 1.5% of the Internet is taken up by pornographic materials. Another study of computer use in Utah State schools while filtering was in effect found that exactly the same proportion of access attempts was to prohibited sites. It is estimated that commercial pornography sites nevertheless take in over a billion dollars annually.

Concerns about the Internet and pornography are not just a question of the amount of material available but also its type and its ready availability to teenagers and children in homes having Internet connections. Children who could not legally buy a pornographic magazine can access some of the content of the same magazine free on the Web. The Internet pornography industry is more than electronic publishing, however. For example, the use of digital cameras means that live action can be presented online to paying customers. The development of voice interactivity means that the virtual world mirrors most aspects of a bricks-and-mortar peepshow. The only difference is that this service is available to people in the privacy of their own home. (Of course the whole question of how private any transaction on the Internet really is can always be questioned; see ELECTRONIC PRIVACY).

People who want to restrict the availability of commercial pornography on the Internet may be interested in censoring such material in general or they may have the more specific intention of protecting children. Protection of children takes various forms. One is to protect children from exposure to pornography. This can be done quite easily by parents through the use of filtering devices. Parents may also consult the log of sites visited to ensure that the filter has worked. Not satisfied with such protections, American legislators have passed the Child Online Protection Act of 1998 (COPA) that requires site owners to acquire proof of age before transmitting material considered harmful to minors. This law established the Commission on Online Child Protection, which is charged with identifying ways of reducing "access by minors to material that is harmful to minors on the Internet." It must therefore establish which sites are harmful to children and study practicable and effective techniques for blocking such materials from minors.

Another issue that is often raised is the need to protect children from becoming the subjects of pornography. This problem has little to do with the Internet as such and is already dealt with in laws against child pornography. One potentially important issue is that pornographic images of children that are digitally created are considered protected expression under the first amendment to the United States Constitution. Graphic technology has advanced to the point that digitally produced images may be practically indistinguishable from those that originated in photographs, thus creating a large loophole in which Internet child pornographers could operate.

Perhaps the most serious problem raised by the Internet is that it provides a medium through which pedophiles can contact prospective victims. When children long onto chat sites in search of friendship, they may find more than they bargained for. Posing as someone much younger than they actually are, pedophiles may strike up a relationship with children on chat lines and subsequently arrange a meeting with the intention of having sex with them.

Children's chat lines are sometimes monitored by the FBI. An FBI agent might pose as a child in a sting operation and arrange a meeting with a pedophile. At this meeting, the pedophile may be arrested for arranging to have sex with a minor. Such aggressive enforcement efforts are likely to make a dent in the problem although they raise some ethical issues concerning entrapment. In the final analysis, parents have the major responsibility of monitoring their children's online activities and ensuring that they never have unsupervised visits with strangers they have encountered on chat lines.

The emergence of online chat networks has provided a venue for complete strangers to meet. Very often, this allows for the development of intense online relationships that are largely divorced from the reality checks necessarily implied in face-to-face encounters. Such relationships may develop to the point that telephone contact is substituted for Internet communication and this may culminate in a real-world meeting and the development of an ordinary relationship. Before any meeting, there may be an exchange of sexually explicit messages that can be considered a form of pornography. Given that most such communications are heterosexual, it can be inferred that approximately half of the pornographic messages transmitted on the Internet are typed by women. Apparently women are more interested in verbal pornography that occurs in the context of building a cyber relationship than they are in the commercial, visually oriented pornography purchased by men.

Pornography presents an interesting conflict between the ethical values of freedom of speech and the desire to protect people, particularly children, from potential harm in either the production or the perusal of pornographic materials. Another legitimate concern, based on the scientific research into effects of pornography on adults (including women as well as men), is that frequent viewers of erotic films may change their attitude toward victims of sex crimes by unfairly considering them as responsible for their fate. In some countries, particularly those that are ruled according to fundamentalist Islamic doctrine, any depiction of sexuality may be outlawed. In most Western democracies, however, freedom of speech is considered more important than the potential harm of pornography. An important exception, particularly in the United States, concerns legislative efforts to protect children. These attempts have generally prevailed, although the new medium of the Internet has proved particularly difficult in this respect.

See also ELECTRONIC PRIVACY; KINSEY, ALFRED; MASTERS AND JOHNSON; RIGHTS OF HUMAN RESEARCH PARTICIPANTS.

Further Reading

Brehm, S. S., and S. M. Kassin, *Social Psychology.* Boston: Houghton Mifflin, 1990.

Easton, S. M. *The Problem of Pornography: Regulation and Free Speech.* New York: Routledge, 1994.

Franken, R. E. *Human Motivation.* Pacific Grove, Calif.: Brooks/Cole, 1994.

Greenburg, J. C. "High Court Agrees to Decide if First Amendment Protects Virtual Child Pornography." *Knight-Ridder/Tribune News Service,* January 2, 2001.

Guttman, C. "The Darker Side of the Net." *Unesco Courier,* September 1, 1999.

Harrison, M., and S. Gilbert. *Obscenity and Pornography Decisions of the United States Supreme Court.* Carlsbad, Calif.: Excellent Books, 2000.

Leland, J. "More Buck for the Bang." *Newsweek International,* October 11, 1999.

Mansfield, S. "The Avengers Online." *Good Housekeeping* 228, no. 6 (1999): 122.

Matt, M., and Q. Marshall. "Ruling on Internet Pornography Could Affect Infoseek Executive." *Knight-Ridder Business News,* December 17, 1999.

McNeely, K., and T. C. Moorefield. "Internet Pornography Law Faces Court Challenge." *Providence Business News,* October 25, 1999.

Reid, C. "Publishers, ACLU Warn against Return of the CDU (Communications Decency Act)." *Publishers Weekly,* October 5, 1998.

Schuyler, M. "Porn Alley: Now Available at Your Local Public Library." *Computers in Libraries* 19, no. 10 (1999): 32–35.

Symons, D. *The Evolution of Human Sexuality.* New York: Oxford University Press, 1979.

Zillman, D., and J. Bryant. "Effects of Massive Exposure to Pornography." In *Pornography and Sexual Aggression,* ed. N. M. Malamuth, and D. J. Donnerstein, 115–38. New York: Academic Press, 1984.

positivism

Positivism is a philosophical perspective that rejects broad metaphysical statements as meaningless and confines itself to propositions based on objectively verifiable facts. Apart from such scientific information, philosophers could also include propositions of logic and mathematics. For this reason, positivism is often referred to as "logical positivism." Positivism was developed by the VIENNA CIRCLE of philosophers in the 1920s under the influence of Ludwig Wittgenstein's (1889–1951) theory of language. Wittgenstein had argued that the meaningfulness of a proposition is related to the narrowness of the context within which it is held. Many of the statements of traditional philosophy, particularly metaphysics, and also ethics were so general that no context was specified and they were therefore virtually meaningless. The positivists undertook to purge philosophy of what were seen as meaningless generalities and to ground it in scientific observations. Positivism is no longer influential in philosophy but it has had an impact on life outside academia. Its legacy includes the growth of respect for science as the highest form of knowledge, a mistrust of lofty rhetoric, and a demand for clarity of expression and argument.

See also POPPER, SIR KARL RAIMUND; VIENNA CIRCLE.

postmodernism and the philosophy of science

The term postmodernism was widely used in the 1960s to refer to a style of architecture (and other arts) that rejected canons of good taste and combined any and all previous architectural styles in an eclectic mix. Postmodernism has been applied more recently to a school of thought in academic life that is marked by a similar relativism. This way of thinking became particularly influential in the field of literary criticism, in America in the 1970s, where it was associated with a theory of literature and language known as "deconstruction" devised by French philosopher and critic Jacques Derrida.

Derrida's work involved an attack on leading Western thinkers from Plato onward who are accused of being "logocentric" that is, prone to accepting the spoken word (*logos*) as containing truth. Western thought is riddled with two-term dichotomies, such as strong/weak, truth/lie, male/female, and presence/absence, in which the first term of each pair is usually stronger. Derrida rejects these unequal polarities and wishes to take language apart to upset them. This can be done by reversing the order of the words, pursuing their etymology in detail, or replacing them with words from other languages. Inspired by Derrida, feminist literary critics have attempted to deconstruct the male bias in literature, which is seen as an instrument in the historical oppression of women.

To a postmodernist scholar, all language is thought of as "text" and language both directs human activities and comprises the basis for our perceived reality. In principle, any kind of language utterance can be exposed to deconstruction, whether it is a play of Shakespeare, a soup can label, or a paper in a physics journal. Postmodernist scholars have thus their attention to the work of scientists, which is analyzed as a "narrative" that reveals something about the scientists themselves but has no connection to any underlying reality.

From this perspective of linguistic nihilism, when scientists refer to the HIV virus, they are using a metaphor rather than describing something real. Similarly, an experiment is not so much a way of testing out a scientific hypothesis by collecting data in the real world as a rhetorical device used by scientists to fill out their narrative. Postmodernists have deconstructed the work of scientists, even in abstruse fields such as particle physics, professing to find all manner of biases that question the objectivity of science.

Tired of reading such attacks on the objectivity of science by scholars who frequently had no specialist training in or knowledge of the field they were criticizing, physicist Alan Sokal wrote a parody critique of quantum gravity that was accepted for publication by the unwitting editors of *Social Text*. His essay "Transgressing the Boundaries: Toward a Transformative Hermeneutics of Quantum Gravity", involved a cleverly compiled pastiche of the writings of other postmodernist scholars. Sokal deliberately included patent errors in reasoning and sentences that were completely unintelligible. Realizing that the editors might not know much about physics, he added an incongruous and illogical political conclusion to the paper. Yet the paper was accepted without question. The editors may have been delighted to see a real scientist coming over to their side by denying that there was any objective physical reality that could be studied by physicists. The hoax was revealed by Sokal in the French Journal *Lingua Franca*. Journalists and readers around the world were greatly entertained by the prank because it suggested that postmodernist scholarship was indistinguishable from gibberish, even for postmodernists themselves.

Instructive as the *Social Text* parody was in this respect, it does not follow that science is always as objective as it pretends. Individual scientists at all levels of eminence have evidently committed an astonishing range of dishonest acts that betray a contempt for objectivity and SCIENTIFIC METHOD. Moreover, scientists have often succumbed to the biases of their own societies. Nineteenth- and 20th-century social scientists were obviously guilty of SCIENTIFIC RACISM and some male scientists have exhibited gender bias in their work.

Perhaps reflecting a weakening of postmodernist influence, the trend among feminist critics of science has been to emphasize the important role of female scientists in identifying, and preventing, gender-biased research rather than to question the possibility that science can be objective. A similar argument has been made for ethnic diversity in the ranks of scientists.

See also CULTURAL RELATIVISM; FEMINIST ETHICS AND PHILOSOPHY OF SCIENCE; SCIENTIFIC DISHONESTY; SCIENTIFIC RACISM; SOKAL, ALAN, AND *SOCIAL TEXT* PARODY.

Further Reading
Gross, P. R., and N. Levitt. *Higher Superstition: The Academic Left and Its Quarrels with Science.* Baltimore, Md.: John Hopkins University Press, 1997.
Resnik, D. B. *The Ethics of Science.* New York: Routledge, 1998.
Sokal, A. "Transgressing the Boundaries: Toward a Transformative Hermeneutics of Quantum Gravity: *Social Text* 46/47 (1996): 217–52.
Sokal, A., and J. Bricmont. *Fashionable Nonsense: Postmodern Intellectuals' Abuse of Science.* New York: St. Martin's Press, 1998.

prefrontal lobotomy

Prefrontal lobotomy is an operation in which the prefrontal cortex is either surgically damaged or has its connections to other parts of the brain severed. The rationale for this operation stems from the finding that damaging the prefrontal cortex of laboratory primates made them tamer without noticeable impairment of movement or sensory function. The operation was used mainly for agitated schizophrenics whose condition was otherwise untreatable. Some were calmed by the operation. Undesirable side effects included apathy, inability to plan, memory problems, and difficulty in concentrating. There was a blunting of emotions and a corresponding lack of facial expressions. Moreover, there was a loss of social inhibitions. Following lobotomies, patients disregarded many of the rules of polite conduct and behaved in tactless and callous ways.

During the late 1940s and early 1950s, approximately 40,000 prefrontal lobotomies were done in the United States. Their use declined steeply following the discovery of effective antipsychotic drugs and lobotomies are almost never performed today.

The operations themselves were horrifyingly imprecise. One of the leading practitioners was Walter Freeman, a medical doctor who had no formal training in surgery. He performed operations outside a hospital setting, often working in his office with crude equipment that he carried around in his car.

Prefrontal lobotomies were initially performed on schizophrenic patients, often with poor outcomes because they damaged a structure that is already impaired in these patients. As time went on, lobotomies were used for a variety of other disorders, minor as well as major. Patients in psychiatric hospitals who were difficult to manage were often lobotomized for the convenience of staff. During the same period, ELECTROCONVULSIVE THERAPY was also abused for this purpose. The potential for abuse, as well as the lack of scientific rationale and surgical imprecision of the operation, has meant that lobotomy is not approved of by modern neurosurgeons. It is seen as barbaric and unethical and a procedure that has generally done much more harm than good.

See also BIOLOGICAL PSYCHIATRY; ELECTROCONVULSIVE THERAPY.

Further Reading

Shutts, D. *Lobotomy: Resort to the Knife.* New York: Van Nostrand Reinhold, 1982.

preservatives in food

Scientific advances in food technology have often added convenience and efficiency to the storage and preservation of foods, but they have often encountered resistance on the grounds that they are seen as unnatural and hence dangerous. There have been periodic scares about the presumed health consequences of consuming food that contains chemical additives, such as the unfounded concern that MONOSODIUM GLUTAMATE, a flavor enhancer, produced allergic reactions, or that the sweetener NUTRASWEET increased the incidence of brain tumors. Similar unfounded concerns have been raised about the chemicals added to food as preservatives.

Traditionally, foods were preserved through the addition of salt or sugar that dried out foods and thus inhibited bacterial growth. Smoking was also used as a preservative technique since smoke contains guaiacol (2-methoxyphenol) an antioxidant that prevents fats from going rancid and inhibits bacterial activity. Smoking is used today primarily to enhance the flavor of meats. The use of chemical additives is now widespread. The chemicals that can be used, the foods that they can be added to, and the concentration of chemical preservatives that may be used are strictly regulated in the United States by the Food Additive Amendment Act of 1958. Additives that are considered to be safe when used at permissible concentrations in appropriate foods include: sodium nitrate, sodium nitrite, sodium benzoate, sodium propionate, calcium propionate, ethyl formate, sorbic acid, and sulfur dioxide.

See also FLUORIDATION OF WATER; FOOD IRRADIATION; MONOSODIUM GLUTAMATE; PASTEURIZATION.

privacy rights   *See* ELECTRONIC PRIVACY; RIGHTS OF HUMAN RESEARCH PARTICIPANTS.

profiling and Internet marketing   *See* COOKIES AND INTERNET.

pseudoscience

Pseudoscience can be defined as any intellectual discipline that has scientific elements, or passes itself off as science, but lacks some critical ingredient of SCIENTIFIC METHOD. Scientific data are empirical, which means that they are derived directly through sensory experiences instead of being derived indirectly from books and oral traditions. Scientific data are collected in ways that are systematic and repeatable. This means, for example, that the visions of seers cannot be used as scientific evidence. Perhaps the most common failing of pseudoscience is the lack of genuine openness to empirical falsification of its ideas. This is well illustrated in the claims of some creationists that the fossil record is inconsistent with the biblical account of creation because God deliberately placed fossils in the ground, making it seem that the earth is millions of years old.

Some kinds of pseudoscience can be identified on the basis of having unscientific motives, that is, those that are inconsistent with a quest for scientific truth. Astrologers want us to believe that they can predict the future so that we will pay to hear their predictions. Creationists attempted to pass off their ideas as science so that their beliefs could be legally taught in U.S. schools. Early scientific racists wanted to establish that other ethic groups were interior to their own.

From a historical perspective the distinction between science and pseudoscience can be quite complex because modern sciences have their roots in earlier forms of pseudoscientific thinking. Thus alchemy is the ancestor of chemistry, astrology predates astronomy, and CRANIOMETRY was an early branch of neuroscience. This relationship is so pervasive that it can be difficult to determine whether the term "pseudoscience" or "protoscience" is more appropriate. When scientific hypotheses turn out to have been wrong, they often seem indistinguishable from pseudoscience. Examples include the ether supposed by 19th-century physicists to be necessary for propagating light rays and PHLOGISTON, or the principle of fire believed by 18th-century chemists to be responsible for combustion. Moreover, mainstream sciences may occasionally diverge from scientific method and succumb to collective illusions, such as that of N-RAYS and COLD FUSION. Belief in such illusory phenomena generally involves some failure to adhere to scientific method and can thus be described as pseudoscience.

See also COLD FUSION; CREATIONISM; N-RAYS; PHLOGISTON; PHRENOLOGY AS PSEUDOSCIENCE; SCIENTIFIC RACISM.

psychedelic drug research

Scientific investigation of psychedelic, or mind-altering, drugs has raised three distinct ethical questions. The first is whether these substances are too dangerous to be used on humans for research purposes. The second is whether legal restrictions on the recreational use of drugs such as marijuana, psilocybin, and LSD should prevent their use in research and thus foreclose the many potential benefits of such drugs in medicine. Third, psychedelic drugs were investigated by the United States and Britain as potential chemical warfare agents in experiments that constituted a flagrant violation of the INFORMED CONSENT requirement of the NUREMBERG CODE.

Western interest in psychedelic drugs began in the 1950s and the quasi-religious origin of this interest is revealed in Aldous Huxley's book *The Doors of Perception* (1954), which suggested that mind-altering substances, such as the mescaline he had experimented with, could radically expand human consciousness and put us in touch with deeper spiritual realities. Within a short time, millions of people were experimenting with psychedelic drugs and the psychedelic movement was born. Also known as the counterculture, this involved not just drug use but also a transformation of dress, music, art, and lifestyle to reflect the new interest in expanded consciousness and transcendence. For example, the swirling patterns of vivid color characteristic of psychedelic art were intended to re-create the experience of a hallucinogenic drug. Similarly, openness to new experience called for lifestyle experimentation and a heightened level of acceptance of different kinds of living arrangements and sexual relationships.

In America, the psychedelic movement was seen as a serious threat to the established social order for several reasons, including opposition to materialistic values, militant student protests against university administrations, and vigorous opposition to the Vietnam War. The counterculture was attacked fairly directly by enforcing laws against the use of marijuana and other psychedelic drugs.

Psychedelic drugs may lend themselves to a variety of medical applications, although their true potential has not been properly evaluated due to U.S. government suppression of such pharmacological research as a side effect of the "war on drugs." One branch of medicine in which the use of psychedelics has been investigated is psychiatry. Psychoactive substances have constituted a part of the religious rituals of many societies throughout history. The pharmacology of religious enlightenment was investigated by Walter Phanke in a 1962 study of divinity students in which participants attended a Good Friday service after receiving either psilocybin or a placebo. Eight of 10 subjects in the drug group reported a mystical experience compared to only one of 10 of those receiving placebo. Stanislav Grof, of Johns Hopkins University in Baltimore reported beneficial effects of the use of hallucinogen LSD (lysergic acid diethylamide) to prepare terminal cancer patients for the psychological impact of dying. LSD has also been used in psychotherapy for concentration camp survivors by Dutch psychiatrist Jan Bastiaans.

Such innovations have generally been suppressed by legal restrictions on the use of psychedelic drugs and a drying up of research funding. Psychedelics are not addictive in the sense of causing physical dependency as is the case of opiates, stimulants like cocaine, and anti-anxiety drugs like alcohol and Xanax (a benzodiazepine). Their legal suppression reflects the fact that (a) they are often used by people who use addictive controlled substances and (b) they may be dangerous. LSD produces flashbacks in which a drug trip is vividly recalled perhaps many years after the intoxication of the trip has worn off. Some people have died after eating mushrooms that only resembled those producing psilocybin and were highly toxic. Some people who smoke a lot of marijuana may become confused and disoriented. LSD trips may be so disturbing that people commit suicide to escape the torment. Designer drugs, such as ecstasy (MDMA), can contain impurities that cause permanent brain damage, as in the tragic case of young Californians developing the symptoms of PARKINSON'S DISEASE in 1982. Ecstasy has became popular among dance club participants because it induces a sense of energy and euphoria and enhances responsiveness to touch. The drug increases body temperature, however, so that dancers are at increased risk of developing heat stroke, which can be fatal. The potentially harmful long-term affects of the drug on brain neurotransmitter systems are the subject of ongoing research on animals.

Despite such risks, psychedelic drugs clearly do show promise in several areas of medicine. This prompted the foundation in 1986 of the Multidisciplinary Association for Psychedelic Studies, a U.S. nonprofit organization dedicated to helping scientists obtain legal permission, and funding, for psychedelic drug research. This organization is sponsoring research at Ben Gurion University, in the Negev, Israel, that tests the use of ecstasy (MDMA) in the treatment of posttraumatic stress disorder. It is also supporting work on the use of LSD in psychotherapy.

Perhaps the most widely accepted medical application of psychedelic drugs is the use of marijuana to reduce the nausea produced by cancer chemotherapy and AIDS treatment. A survey by Harvard researchers has found that 44% of specialists in these areas recommend that patients should smoke marijuana even though it is illegal in most states. California and other states have recently passed referenda or ballot initiatives that permit the medical use of marijuana to control nausea. The active ingredient (THC) has been prepared in drug form but some users claim that smoking the drug is a more effective method of calibrating the dosage. (A federal program distributing limited quantities of marijuana to people with glaucoma, spasticity, chronic pain, or AIDS was canceled in 1992 by the administration of President George H. W. Bush to counter any impression that authorities endorsed the smoking of pot).

Another promising medical application of psychedelic drugs is the use of ibogaine, a drug used for centuries in West African religious rituals. There is anecdotal evidence that some former addicts of heroin, morphine, and cocaine have lost their cravings following a single administration of ibogaine. This evidence has received support from experiments on morphine-addicted rats.

Cold war military research on psychedelic drugs was primarily dedicated to the development of an incapacitating agent that would produce disorientation and confusion in the enemy so that victory could be achieved in a war without a high mortality toll. This admirably humanitarian objective ironically led to some highly unethical research by the United States and its allies, who freely shared the results of their experiments on human subjects. It is estimated that 6,700 people were used in U.S. government experiments on mind-altering (psychoactive) drugs. Those that have engendered the strongest ethical reaction include U.S. Army experiments on LSD, which clearly violated the NUREMBERG CODE. Many of the participants volunteered for experiments that were described as being designed to test defensive procedures against chemical weapons. In some cases, subjects drank a clear liquid whose contents was not divulged. For others, administration of the LSD was hidden in a cup of coffee.

Some of the victims of these experiments sued the government. The case of Master Sergeant James B. Stanley, who participated in a LSD trial in 1958, reached the U.S. Supreme Court. Stanley had an adverse reaction to the drug since the hallucinations he experienced were followed by profound emotional problems that led to his divorce. The Supreme Court ruled against him in a five-to-four decision. Although they were sympathetic to the trauma he had suffered, the Court found that, as a member of the armed services, he was barred from suing the U.S. government. This means that elaborate ethical guidelines regulating research within the military have no legal force.

Neither do they have any practical effect. In 1955, a Defense Department Study Group on Psychochemical Issues recommended that people receiving LSD in experiments should be given a training lecture to prepare them for the hallucinations and other effects of the drug. This recommendation was overturned by those responsible for the scientific design of the research on the grounds that many of the psychological effects of LSD can be influenced by suggestion. This may be good experimental design but it ignores the Nuremberg Code and other ethical standards, such as the DHHS REGULATIONS and internal military regulations, all of which were designed precisely to make the point that the basic right to INFORMED CONSENT could never be trumped by other considerations.

Informed consent could not be given in the Army LSD experiments because participants were lured into participation under false pretenses. They were not informed about what to expect in the experiment and were therefore incapable of consenting to the procedures. Even after the experiments were conducted, and there was no longer any scientific rationale for keeping them in the dark, subjects did not receive the kind of debriefing information that could have helped them to cope with the adverse consequences of their experiences. This omission shows that they were indeed being treated like human guinea pigs who had no rights and whose suffering was of no account to the military researchers.

Many of the details of the Army psychedelic drug experiments were revealed in the context of an Inspector General's investigation that issued a report titled *Use of Volunteers in Critical Agent Research*. None of the officers responsible for this unethical research have ever been held accountable. There is no guarantee that the experiments will not be repeated.

See also BIOLOGICAL AND CHEMICAL WARFARE; COLD WAR RADIATION EXPERIMENTS; NUREMBERG CODE; RIGHTS OF HUMAN RESEARCH PARTICIPANTS.

Further Reading

Grinspoon, L., and J. B. Bakalar. *Psychedelic Drugs Reconsidered.* New York: Open Society Institute, 1997.

Moreno, J. D. "Lessons Learned in a Half Century of Experimenting on Humans." *The Humanist* 59, no. 5 (1999): 9–15.

Pahnke, W. N. "LSD-Assisted Psychotherapy with Terminal Cancer Patients." *Current Psychiatric Therapies* 9 (1969): 144–52.

Pahnke, W. N., and W. A. Richards. "Implications of LSD and Experimental Mysticism." *Journal of Religion and Health* 5 (1962): 175–208.

Ragavan, C. "Cracking Down on Ecstasy." *U.S. News & World Report,* February 5, 2001.

Taylor, E. "Psychedelics: The Second Coming." *Psychology Today* 29, no. 4 (1996): 56–60.

Taylor, J. R., and W. N. Johnson. *Use of Volunteers in Chemical Agent Research.* DAIG-IN 21–75. Washington, D.C.: Department of the Army, Inspector General, 1975.

psychosurgery   *See* PREFRONTAL LOBOTOMY.

Ptolemy
(100–170)
Greek-Egyptian
Astronomer, Geographer

Little is known about Claudius Ptolemy's life except that he spent most of it in Egypt. Ptolemy's *Almagest* was the most influential work in cosmology until it was replaced by COPERNICUS's 16th-century heliocentric (or sun-centered) model of the universe. His *Geography,* which mapped the known world using a system of latitude and longitude, remained popular during the same period despite its unavoidable inaccuracies.

Although Ptolemy's name is synonymous with an outdated worldview that placed the earth at the center of the

universe, he was often seen as a careful empirical scientist whose influential worldview was based on detailed original astronomical observations conducted in Alexandria, Egypt. The first serious doubts about Ptolemy's commitment to empiricism emerged in the 19th century when astronomers scrutinized his data. Calculating backwards in time from the current position of the planets, they found that Ptolemy's observations were very far off the mark. The discrepancies were so great that they could not have been due to error even with the crude techniques used in second-century Egypt. The most plausible explanation of the errors is that Ptolemy did not make the observations himself but borrowed them from the work of Hipparchus, a distinguished Greek astronomer, who had made detailed and careful observations, mainly from the island of Rhodes, which is five degrees north of Alexandria, some three hundred years earlier. Ironically, Hipparchus himself has been accused of stealing his star catalog from unacknowledged Babylonian sources.

The *Almagest* abounds with clues to plagiarism. When Ptolemy gives examples of how to work out problems, most relate to the latitude of Rhodes indicating that they were plagiarized from Hipparchus. Ptolemy's star catalog is also curiously lacking in stars from the southernmost band of the sky, which would have been visible to him, but not to Hipparchus or the Babylonians, had he actually made any observations.

Robert Newton of the applied physics laboratory at Johns Hopkins University has assembled detailed evidence that Ptolemy fabricated data to match his theories. For example, he used Hipparchus's estimate of the length of a year to estimate the time of an autumnal equinox (the time when the sun crosses the equator and day and night are everywhere of equal length). He claimed to have observed the equinox at 2 P.M. on the 25th day of September in the year 132, even though the actual equinox would have occurred more than a day earlier. He then used this fraudulent observation to support Hipparchus's estimate of the length of a year.

Some historians, including Owen Gingerich, have argued that it may be unfair to hold Ptolemy to the standards of modern science and that his primary sin was selective publication of numbers that supported his theories. This argument cannot explain why Ptolemy's "observation" of the autumnal equinox was so far from reality and so close to what it would have been if Hipparchus's estimate of the length of a year was correct. Ptolemy is arguably the most successful plagiarist in the history of science whose appropriation of Hipparchus's data brought him two millennia of unmerited distinction. It is fascinating that his reputation as a careful and honorable scientist continues to live on in encyclopedias and other reference works despite being debunked by Robert Newton.

See also SCIENCE, HISTORY OF; SCIENTIFIC DISHONESTY.

Further Reading
Broad, W., and N. Wade. *Betrayers of the Truth.* New York: Simon and Schuster, 1982.
Newton, R. R. *The Crime of Claudius Ptolemy.* Baltimore, Md.: Johns Hopkins University Press, 1977.

public funding of research

Governments support research in many different ways, most obviously by supporting institutions of higher learning where most research is conducted, much of it without any grant funding. This general financial support of institutions of learning is not nearly so controversial as when the government is actively involved in deciding what research should be carried out through grant support of research projects. The major ethical problem is that government funding decisions are often affected by political considerations that often assume greater importance than the scientific merits of the proposed research.

The idea that government support of research can produce practical benefits that more than compensate for the expense dates to the 19th century when scientific advances were often rapidly incorporated into industry. In the 20th century, governments recognized that sponsorship of science and technology research was an important road to military strength. Many of the tools of modern warfare, such as radar, computers, and nuclear weapons, derived from the work of civilian scientists. The MANHATTAN PROJECT, which aimed to develop the first atomic bomb, called on the expertise of academic physicists who had little or no previous experience with military technology.

With the demise of the cold war, the focus of interest in public funding for scientific research has shifted back to the possible economic, and medical, benefits to be expected. There is a growing awareness that the wealth of the future will be produced by technological innovation and that countries that are successful in providing the kind of climate in which such applied science flourishes will be winners in the global marketplace that is becoming ever more integrated through the Internet.

If governments wanted to promote economic prosperity, health, and military capability, then they ought to focus their research funding in a few critical fields, such as computer science, electronic engineering, medical genetics, and chemical engineering. Yet this approach of encouraging applied sciences at the expense of basic research would be short-sighted and would have disastrous long-term consequences. Germany in the Nazi era is an extreme example of science being directed to practical goals, including EUGENICS. The environment became so inhospitable to scientists that most left, never to

return. The surest way of squelching scientific innovation is to create a political environment that is threatening to intellectual freedom. This point is also very clearly illustrated in the case of the Soviet period of LYSENKOISM when respectable genetic research was halted by the dogma that acquired traits could be passed to offspring.

Promoting applications at the expense of pure science is always a false dichotomy. Applications inevitably flow from the fount of good basic research. Study of the history of science reveals that some of the most important practical discoveries have been unforeseen consequences of basic science. Nuclear power comes directly from the theoretical work of thinkers such as ALBERT EINSTEIN. The computer is an offshoot of mathematical research. Genetic engineering comes from speculations about the structure of DNA. The development of scientific applications is analogous to a detective novel. Most of the clues are in front of us but it is very difficult to guess how the plot is going to turn out. Just as the reader of a detective novel is well advised to suspect everyone, so governments are best served when they provide at least some support to all branches of basic scientific research.

Given that governments are generally willing to fund both types of research, there is an important ethical issue about the extent to which funding decisions ought to be influenced by political considerations. In a democracy, it is appropriate for the public to have some say in general funding policy. Political influences on research funding in the United States are easy to detect. For example, President Ronald Reagan halted research on human fetal tissue by issuing an executive order that prohibited the NATIONAL INSTITUTES OF HEALTH (NIH) from providing funding for this work. This effectively prevented the development of tissue grafts, which can grow in the brains of Parkinson's disease patients thereby greatly alleviating their symptoms. President Bill Clinton reversed this order, thereby providing hope to many Parkinson's victims. At the same time, Clinton placed a moratorium on human cloning research.

Political influences also played an important role in the amount of funding devoted to AIDS research. Initially, in the 1980s, AIDS research was a low governmental priority. Since that time, a growing recognition of the importance of AIDS as a public health matter for the entire population as well as the alarming global spread of the disease have had the effect of increasing research funding to more than a billion dollars, the largest budget for research on any disease. This outcome also reflects the determined lobbying efforts of AIDS activists.

Even though policy decisions are influenced by politics, the public generally has little or no say in relation to which individual proposals are approved for government funding. Government funding agencies evaluate funding proposals through some form of peer review. Scientists are employed to evaluate projects because they are best qualified to determine which proposals are based on good science and which ones are undeserving of public support.

See also LYSENKOISM; MANHATTAN PROJECT; PEER REVIEW; STAR WARS; SUPER CONDUCTING SUPER COLLIDER.

Further Reading

Martino, J. *Science Funding.* New Brunswick, N.J.: Transaction, 1992.

Mukerji, C. *A Fragile Power.* Princeton, N.J.: Princeton University Press, 1989.

Resnik, D. B. *The Ethics of Science.* New York: Routledge, 1998.

Roberts, C. "Collider Loss Stuns CU Researchers." *Denver Post,* November 14, 1993.

Veatch, R., ed. *Medical Ethics.* Sudbury, Mass.: Jones and Bartlett, 1997.

publish-or-perish mentality

This refers to the strong pressures on academics to author professional publications. The achievements of scholars are often judged superficially on the length of their publication lists and this is used as a criterion in awarding tenure and promotion as well as deciding who should receive research funding. The publish-or-perish mentality is often unspoken but it may involve an explicit expectation that a scholar in some academic department should produce a specific number of publications each year.

Such pressures can have a distorting, or debasing, effect on scholarship in different ways. One consequence is multiple authorship. If five scholars agree to put their names on each other's papers, their publication lists will each be five times as long as if they had agreed to publish alone even though no more work will actually have been done. This practice is now so widespread that many evaluators of publication lists ignore all but the first listed author. Another response is MULTIPLE PUBLICATION, in which the same paper is sent off to different academic journals. Multiple publication is banned in some disciplines but not in others. It wastes valuable publication space and editorial time and effort and serves only to pad the publication record of the author.

Another response to the pressure for extensive publication is to divide papers up into the smallest contributions that are likely to be published, or the least publishable unit. Finally, scholars are likely to submit papers of questionable merit to several different journals, hoping that they can sneak through the process of PEER REVIEW.

See also MULTIPLE PUBLICATION; PEER REVIEW IN ACADEMIC PUBLISHING; PLAGIARISM.

Further Reading

Resnik, D. B. *The Ethics of Science.* New York: Routledge, 1998.

Pythagoras

(c. 560 B.C.–480 B.C.)
Greek
Philosopher, Mathematician

Pythagoras is an important founder figure in mathematics, the theory of music, and astronomy. Born on the island of Samos, he spent some 20 years traveling and assimilating the mathematical lore of the Egyptians and Babylonias. Whereas earlier mathematicians had devised formulae to solve practical problems, Pythagoras was interested in the properties of numbers for their own sake. He coined the word "philosopher" to describe a person with this disinterested approach to learning.

Upon returning to Samos, Pythagoras found that it had been take over by the tyrant Polycrates. Instead of joining the court of the new ruler, as he had been invited to do, Pythagoras chose the life of a hermit, literally living in a cave, with the company of a single follower. Eventually, he established a small school but his views on social reform proved too radical and he was forced to flee to Croton in southern Italy (then a Greek colony).

In Croton, Pythagoras was fortunate enough to gain the sponsorship of Milo, the most successful athlete of his day, who had been champion of the Olympic Games on 12 occasions. In addition to his athleticism, Milo had a passionate interest in learning. The wealthiest man in Croton, Milo provided Pythagoras with a secure environment, in his own home, where he could continue his studies. This improbable alliance of the world's strongest body with the world's greatest mind was cemented when Pythagoras married Milo's daughter, the young and beautiful Theano, who had been his student.

In this more favorable intellectual environment, Pythagoras founded the Brotherhood, an alliance of six hundred intellectuals who could absorb his teachings and add to the body of knowledge by contributing new ideas and mathematical proofs. The Brotherhood thus had superficial similarities to an early university. It also had many important differences. One was that on joining the community, scholars contributed all of their worldly wealth. If they ever chose to leave, they were guaranteed twice what they had originally contributed. Another fundamental difference was that the Brotherhood was secret. Instead of going out and propagating their knowledge, members were prohibited from revealing their communal secrets on pain of death. This penalty was actually exacted in the case of one member who revealed the discovery of a new regular solid, the dodecahedron, made of 12 regular pentagons. Pythagoras's school also had religious overtones. Numbers were thought of as having magical properties and he believed that everything in nature could be reduced to numerical relationships.

Despite its extreme requirements, competition to join the Pythagorean Brotherhood was intense and rejection was a cause of bitter resentment. One candidate who was deeply offended by his exclusion was a man named Cylon. Twenty years later, he took advantage of political unrest in Croton to incite a mob to attack Milo's house and the attached school. Pythagoras and many of his disciples were killed in the ensuing violence. The Brotherhood suffered continuing political repression, which forced the disciples to scatter around the ancient world thus dispersing their mathematical knowledge.

Due to the secrecy of the Pythagorean sect, it is not possible to determine which discoveries were made by Pythagoras himself and which were made by his followers. The Pythagorean theorem (that the square on the hypotenuse of a right-angled triangle is equal to the sum of the squares on the other two sides) may not be original to him but he was the first to develop an ingenious proof, which demonstrated that the theorem was true for every right-angled triangle. This proof can be seen as a model for subsequent developments in Greek mathematics as represented by the work of EUCLID.

Pythagoras's most important legacy to science was the conviction that all natural phenomena could be reduced to numerical relations. One of the clearest demonstrations of this principle was Pythagoras's contribution to the theory of music. He discovered that sounds that are in harmony with each other are connected by simple arithmetic ratios. For example, if one string of a lyre is held exactly in the middle and plucked, it produces a sound that is in harmony with the open string but an octave higher. Similarly, any exact ratio of the string (e.g., one-third, a quarter, one-fifth, etc.) produces a harmonious note. The Pythagoreans were interested in the orderly time relationships (or periods) of the heavenly bodies. They imagined that the movement of the planets produced their own harmony known as the music of the spheres.

See also EUCLID; HELMHOLTZ, HERMANN; KEPLER, JOHANNES.

Further Reading

Gorman, P. *Pythagoras: A Life*. London: Routledge and Kegan Paul, 1979.

Singh, S. *Fermat's Enigma*. New York: Anchor, 1997.

R

race science   *See* SCIENTIFIC RACISM.

radioactive fallout   *See* CHERNOBYL; HIROSHIMA.

radioactive materials and black market

The collapse of the Soviet Union during the late 1980s and early 1990s produced concern that some of the radioactive materials produced in the cold war nuclear buildup would be sold on the black market and find their way into the hands of small countries with aspirations to develop NUCLEAR WEAPONS. Given that the transfer of nuclear weapons to Russia was a precondition for joining the Commonwealth of Independent States (which superseded the Soviet Union), these fears might seem overblown. Nevertheless, there is some evidence of a black market in fissionable materials, some of which are of weapons-grade purity.

The former Soviet Union had some 27,000 nuclear weapons, including artillery shells, torpedoes, and warheads on missiles scattered about the republics. The danger of weapons-grade materials falling into the hands of politically unstable countries comes not so much from the nuclear weapons themselves as from stockpiles of fissile material that were produced during the cold war and

are constantly being added to as Russia dismantles about 2,000 warheads per year in keeping with disarmament treaties with the United States. It is estimated that the Soviet arms industry produced some 140 metric tons of plutonium and that 1,000 metric tons of enriched uranium have been stockpiled.

Disarmament has created a depression in the industry that surrounded the Soviet nuclear arsenal and included laboratories, components factories, assembly plants, test plants, and stockpiles of radioactive materials mainly located in closed cities. With disarmament, many scientists found themselves out of work and therefore open to the temptation of providing fissile materials to countries, such as Libya, Iran, and Pakistan, that were developing nuclear weapons programs.

During the 1990s, small quantities of weapons-grade plutonium and highly enriched uranium turned up in West Germany. In May 1994, six grams of plutonium were found in a garage in southern Germany owned by a businessmen who had been arrested on charges of counterfeiting. A month later, less than half of a gram of highly enriched uranium, possibly from a Soviet nuclear submarine, was found in Landshut, Germany. These are minute quantities compared with the approximately eight kilograms necessary to construct a nuclear weapon but they do demonstrate the existence of a black market in fissile materials with troubling implications for nuclear proliferation. It is probably no accident that these materials surfaced at a time when there was worker unrest, due to nonpayment of salaries, at several nuclear weapons plants in Russia, including one (Arzamas-16) where nuclear warheads were being disassembled. Small as the quantities of weapons-grade material discovered in Germany have been, they are also an exception. Many of the transactions involving sale of radioactive material have been fraudulent often involving radioactive waste from nuclear reactors that is of no use for constructing nuclear weapons.

See also BIOLOGICAL AND CHEMICAL WARFARE; CHERNOBYL NUCLEAR ACCIDENT; HAZARDOUS WASTE; NUCLEAR WEAPONS.

Further Reading
Belyaninov, K. "Nuclear Nonsense, Black Market Bombs, and Fissile Flim Flam." *Bulletin of the Atomic Scientists* 50, no. 2 (1994): 44–50.
Taylor, J., R. Ebel, P. Domenici, and L. Graham. *Disposing of Weapons-Grade Plutonium.* Washington, D.C.: Center for Strategic and International Studies, 1998.

radioactive waste disposal   *See* NUCLEAR POWER.

rain forest, depletion of

Environmental scientists and ethicists are very concerned about the potential damage caused by human industrial activities to ECOSYSTEMS and their constituent species. Some ecologists, such as E. O. WILSON, claim that we are in the middle of a mass extinction event as a result of human industrial activities. Tropical rain forests are of particular concern because they house an unusually diverse array of species and because the forests are being cut down at an unprecedented rate to make way for agriculture (which quickly turns their fragile ecosystems into deserts), and to provide lumber for construction and as a fuel for mining and other industries. More than half of the world's rain forests have been destroyed since World War II. The rapid loss of rain forests is thus producing a loss of BIODIVERSITY of the whole planet.

It has been estimated that there are between 10 and 100 million different species on the planet (of which only 1.2 million have been cataloged by scientists). Approximately half of this great range of animal and plant life is found in the tropical rain forest, which accounts for only 6% of the Earth's landmass. The great diversity of life in rain forests is attributable to several causes, including the energy provided by the tropical sun, the 200-foot high tree canopy that provides habitat for plants and animals, and the moist climate that allows pools of water to form in *epiphytes,* plants that attach to trees and obtain their water and nutrients from the air. These water pools play an important role in the evolution of an astonishingly diverse insect life because they provide habitat for the insect larvae. Each canopy tree may be home to several hundred different insect species.

Destruction of the rain forest has endangered indigenous tribes of people. Thus one-third of Brazil's 270 Indian tribes have disappeared since 1900. Existing groups are experiencing severe population declines because of destruction of the rain forest, hostility from armed ranchers and miners, and vulnerability to diseases introduced by settlers. Thus in the state of Rondonia, where a quarter of the rain forest was cut down, the Indian population declined from 35,000 in 1965 to 6,000 in 1985.

Loss of rain forest has probably resulted in the loss of many thousands (or millions?) of species that have never been described by science. These plants and animals contain substances that would likely have been of incalculable benefit to medicine. According to the DEEP ECOLOGY school of environmental ethics, the real loss is of inherently valuable information stored in ecosystems rather than any such loss to medicine or science.

See also ACID RAIN; BIODIVERSITY AND INDUSTRIALIZATION; DDT; GLOBAL WARMING; WILSON, E.O.

Further Reading
Newton, L. H., and C. R. Dillingham. *Watersheds 2: Ten Cases in Environmental Ethics.* Belmont, Calif.: Wadsworth, 1997.
Wilson, E. O. *The Diversity of Life.* New York: W. W. Norton, 1993.

Rawls's theory of justice

John Rawls (1921–) is one of the most influential political philosophers of the 20th century. His book *A Theory of Justice* (1971) reignited interest in the field because it addressed the fundamental ethical problems of contemporary government. Rawls advocated an extreme liberal view of government according to which the primary objective was to promote the well-being of the least advantaged rather than the utilitarian pursuit of the happiness of the majority. His theory of justice calls for the creation of a political system in which everyone has the greatest possible opportunity of happiness.

Rawls's philosophy owes much to that of IMMANUEL KANT (1724–1804). Thus, his first principle of justice states that each individual should have as much liberty as is compatible with other people enjoying the same liberty. This is a political formulation of Kant's categorical imperative (which states that moral principles are valid only if we would wish them to be applied to everyone). The second principle is more original. Rawls invites us to conduct a thought experiment in which we stand behind what he calls the "veil of ignorance." This means that we put ourselves in the position of not knowing whether we are going to be destitute or prosperous. From this "original position" we must select a set of rules according to which our society will be regulated. Rawls argues that most people would choose a set of rules that improved the circumstances of those on the bottom rung of society because they had no way of knowing whether they might end up there themselves. Being poor, of low social status, or lacking socially valued talents would not be such a tragedy if one lived in a society that did everything to raise the comfort level of persons in these categories. In short, Rawls did not object to a social ladder of wealth and prestige, as such, but he wanted to raise the bottom rung to the point that anyone could be happy there.

Rawls formulated a "difference principle" according to which inequalities are allowed only if they benefit the worst off. This position can be seen as an alternative to the liberal ethic of equality of opportunity. Equality of opportunity means that people with the same talents who work equally hard should receive the same benefits. Rawls objected to this on the basis that differences in talents are intrinsically unfair. They are the result of what he described as a natural lottery. Inequality of native aptitudes, such as intelligence or strength, is no more fair than inequality of social class or educational opportunities. If it is desirable to remove the unfair effects of differences in background, then it should be desirable to level out the unfair effects of differences in native ability.

Rawls's *A Theory of Justice* (1971) proved to be a lightning rod for discussion and criticism reawakening interest not just in political philosophy but also in normative ethics. Some critics have attacked Rawls for sanctioning social inequalities. Others have objected that his views are too individualistic and that he pays too little attention to the ties of community, particularly the moral ties of religious communities. In his second book *Political Liberalism* (1993) Rawls collected a series of papers designed to address many such criticisms. Exciting as his ideas have been to philosophers, they have not had much political impact as yet. Most liberal political theorists, who are strongly invested in the principle of equality of opportunity, cannot accept Rawls's position that people should not be allowed to profit from their skills and natural talents. Enthusiasts argue that Rawls is just too radical for the present and that his powerful ideas on social justice are destined to influence the politics of the future.

See also HOBBES, THOMAS; KANT, IMMANUEL; LOCKE, JOHN; PLATO; ROUSSEAU, JEAN-JACQUES; SOCIAL CONTRACT THEORY.

Further Reading

Mulhall, S., and A. Swift. *Liberals and Communitarians*. Cambridge, Mass.: Blackwell, 1996.

Rawls, J. *A Theory of Justice*. Cambridge, Mass.: Harvard University Press, 1971.

———. *Political Liberalism*. New York: Columbia University Press, 1993.

———. *Law of Peoples*. Cambridge, Mass.: Harvard University Press, 1999.

recombinant DNA   *See* GENETIC ENGINEERING.

religious ideology and science

The modern world has seen many controversies between scientists and religious leaders of the West. Examples include the Copernican revolution in astronomy leading to the trials of GALILEO and the controversies surrounding the DARWIN/WALLACE theory of evolution by natural selection. The latter began as a dispute between biologist THOMAS HENRY HUXLEY ("Darwin's bulldog") and Bishop SAMUEL WILBERFORCE ("Soapy Sam") and has survived through the SCOPES TRIAL to our own time as an educational debate over the teaching of CREATIONISM. Controversy is not always a case of science versus religion and modern scientists can disagree among themselves on religious grounds of which the controversy between ALBERT EINSTEIN and quantum physicists is an interesting example.

ASTRONOMY AND THEOLOGY IN THE COPERNICAN REVOLUTION

Galileo (1564–1641) is remembered as a heroic defender of truth and objectivity against religious intolerance but his trial was every bit as much a reflection on his own argumentative tendencies, coupled perhaps with his powerful position as a Florentine courtier, as it was a reflection on the repressive tendencies of the Vatican. Contrary to prevailing religious dogma, Galileo actively defended Copernicanism and even launched a campaign to persuade the church authorities to adopt the Copernican worldview and give up the Ptolemaic-Aristotelian system. His efforts were a failure. Not only was Copernicus banned as heretical reading, but Galileo was told, in 1616, that he could no longer hold, or defend, Copernican views. With the election of a new pope, Urban VIII, in 1623, Galileo hoped that the climate was sufficiently changed for him to revisit the Copernican debate. Having obtained personal permission from the new pope, who requested an impartial presentation of arguments on both sides of the debate, Galileo labored on *Dialog of the Two Chief World Systems,* which was published in 1632 following review by the Vatican censors. Instead of being the balanced presentation that the pope had expected, the *Dialog* turned out to be a powerful polemic in favor of Copernicus. Sales of the book were stopped.

Galileo was tried on suspicion of heresy and forced to recant.

It is easy to consider Galileo's trial as a simple case of religious stifling of intellectual freedom but to do so is to fall into the trap of considering religious views to be inherently opposed to scientific investigation. Yet, at this historical period, nothing could be further from the truth. If most people of that time were deeply religious, major astronomers in the Copernican tradition were not only more religious than average but their scientific investigations were motivated primarily by the desire to approach the mind of God by exploring the wonders of his creation.

NIKOLAUS COPERNICUS (1473–1543) espoused the heliocentric (or sun-centered) model of the known world largely because it appealed to him aesthetically as the sort of harmonious universe that would be designed by a perfect creator. He was quite conservative in retaining as much of the religiously approved cosmology of his day as possible and avoided controversy by publishing at the end of his life. (A manuscript summary of his ideas was discreetly circulated for almost 30 years). His *Revolutionibus Orbium Celestium* (On the revolutions of the heavenly spheres, 1543) was dedicated to the pope and even contained a preface, ostensibly written by Andreas Osiander, a Lutheran cleric, but believed to have been penned by Copernicus himself, that argued that the hypothesis of a heliocentric universe could be wrong but was being advanced merely as a mathematical device to improve the accuracy of astronomical calculations. This was tour de force in soothing religious opposition to the book, which was not banned until Galileo fanned the flames of controversy.

The Lutheran Danish astronomer Tycho Brahe (1546–1601) was religiously opposed to the heliocentric model and proposed a compromise in which the planets revolved around the sun but the sun still revolved around the earth. Brahe was the most distinguished observational astronomer of his day and his successor at the Prague observatory, Johannes Kepler (1571–1630), benefited from his many accurate readings of the positions of the planets. Kepler had begun his intellectual career as an astrologer and a mystic in numerology and actually earned a regular income throughout his life by writing horoscopes and almanacs. Instead of studying astronomy, he was drawn to theology. While studying theology at the University of Tübingen, Kepler became interested in mathematics and had a flash of inspiration that the orbits of the five known planets could be explained by placing them inside, or around, the five regular solids identified by the ancient Greeks (who had proven that there are only five polyhedrons). Kepler was a Lutheran and was forced to leave his home as a result of religious persecution from the counter-Reformation.

Kepler was an enthusiastic supporter of the Copernican worldview. Throughout his life, he strove to find mathematical regularities in the movements of the planets and this effort, which entailed decades of laborious calculations, culminated in Kepler's three laws of motion. Like Copernicus, he was convinced that this order must exist because the universe had to be a numerically beautiful creation. His books were packed with mythical correspondences, such as those between the planets and their respective metals. He wrote about the music of the spheres, an imaginary music created by the harmonious movement of the planets. His works excited little religious opposition because Kepler was seen by contemporaries as a great astronomer who had become slightly deranged. He was largely ignored by contemporary astronomers and even Galileo rejected his most substantive contribution, the laws of planetary motion.

Galileo was not inherently anti-religious. Personally, he was as devout as the vast majority of those of his time and place. The nub of his controversy with religious authorities dealt with epistemology. Like the medieval scientist and philosopher Roger Bacon (1214–92) he felt that where scientific observations came into conflict with religious ideas, the religious ideas must be changed. This reversed the traditional balance of power according to which science in general, and astronomy in particular, was expected to fit in with the religious scheme of things and serve as the servant of theology. Stated forcefully by an argumentative and influential man like Galileo this was seen by the church as a challenge that could not be ignored without loss of authority.

It is easy to assume that the Catholic Church lost credibility by humbling Galileo and that this ushered in a new age of intellectual freedom and advanced the cause of secular science. Once again, this expectation turns out to be unsupported in the case of Copernican astronomy because the crowning genius of the new astronomy, Isaac Newton (1642–1727), albeit an English Protestant, was every bit as much of a religious mystic as the others had been. Just as Kepler felt that the planets were maintained in their orbits by a divine power emerging from the sun, so Newton believed that universal gravitation was a manifestation of divine power throughout the universe. Ironically, Newton's religious mysticism was not presented directly in his important scientific work and he became associated in the minds of subsequent generations with a mechanical universe that functions like a clock without any necessity for intervention by the clockmaker. Newton was thus regarded as an important intellectual ancestor by 18th century-ENLIGHTENMENT intellectuals whose religious thermostats were turned down degrees lower than his had been. Many would likely have been appalled to discover that Newton devoted so much of his life to alchemy. Cultivation of this arcane knowledge took up the lion's share of Newton's manuscripts and papers.

RELIGION AND THE EVOLUTION CONTROVERSIES

The Scopes trial was a famous 1925 test case in which high school teacher John Thomas Scopes allowed himself to be prosecuted for teaching evolution in the Dayton. Tennessee, high school in breach of a state law. Found guilty, Scopes received the minimum penalty, a fine of $100, that was overturned on appeal. Thanks partly to the biting humor of journalist H. L. Mencken, the entire country laughed at the backwardness of Tennessee. Mencken referred to the local residents as "gaping primates" and "anthropoid rabble." Not wishing to be exposed to the same kind of ridicule as Tennessee, many states that had been considering their own "monkey" laws prohibiting the teaching of evolution quietly dropped them. The event thus proved to be an important moral and political victory for evolutionists. The trial convinced the U.S. public of the folly of placing legislative barriers around intellectual freedom and is remembered today as a turning point in which science broke free from the shackles of religious superstition.

Looking at the Scopes trial from a modern perspective, it might be supposed that there is an inevitable clash between the Judeo-Christian worldview of the Bible and evolutionary theory. Yet, examination of the controversy from the other end of the historical continuum shows the controversy in a very different light that is strikingly similar to the dynamics of the Copernican controversy.

The study of evolution began with the study of nature and a primary motive of early natural historians was to find evidence of God in his creations. Some of the earliest works of English naturalists make an explicit connection between natural science and theology. They include John Ray's volume of natural history, *Wisdom of God in the Creation* (1691), and Thomas Burnet's book of geology, *Sacred History of the Earth* (also 1691). This tradition extended for over a century (into the 1830s) and is represented by William Paley's *Natural Theology, or Evidences of the Existence and Attributes of the Deity Collected from the Appearances of Nature* (1802), a book that deeply influenced CHARLES DARWIN (1809–82) and his contemporaries. Paley's argument was that natural science and religion are essentially the same thing. To study the natural world is to be convinced that there is a providential creator who has not only designed the world to provide for the needs of human beings but also provided each animal with a design that allowed it to find a home, feed itself, and reproduce.

Natural theology may appear soft around the edges but it was supported by some of the leading scientists during the long period when it was influential. In addition to Darwin himself (at least before his *Beagle* voyage),

these included Isaac Newton, who played a role in the organization of a series of Boyle lectures, beginning in 1692, the year after ROBERT BOYLE's death, on the general topic of the reasonableness of religious faith in view of scientific evidence. Swedish naturalist Carolus Linnaeus (1707–78) believed that his binomial (genus and species) classification system was instrumental in revealing God's plan for the world of animals and plants.

Darwin's theory of evolution by natural selection began with the premise that there is a struggle to survive. Individuals that are better adapted to their habitat are more likely to survive and reproduce, transmitting these beneficial traits to their offspring. The process of evolution by natural selection (in which selection refers to the culling out of poorly adapted types) thus explains the wonderful match between animals and their way of making a living that had always impressed natural theologians. Darwin's theory was shocking to religious contemporaries because it knocked the legs from underneath the whole argument from design by showing how the random process of natural selection could accomplish what had previously required an intelligent and caring divine providence. How his discovery of the role of natural selection affected Darwin's own religious views is not entirely clear but his willingness to apply selection to human beings, something that codiscoverer Alfred Russell Wallace could not bring himself to do, provides one clue that he had shifted away from religion toward scientific materialism. The religious debates on evolution during Darwin's life were heated. Perhaps wisely, he avoided personal involvement in the controversies and was ably represented by Huxley who is believed to have beaten the silver-tongued Bishop Samuel Wilberforce in debate. Interestingly, Darwin remained on good terms with Wilberforce throughout his life.

Uncomfortable though Darwinism may be to many religious people, most have made their peace with it in one way or another. For some, evolutionary theory is not inconsistent with religion if the assumption is made that the evolutionary process was set in motion through divine intervention. Yet, this approach is not consistent with literal interpretations of the biblical account of creation, which has proved a major obstacle for religious fundamentalists who support modern creationist movements.

In the end, it has been just as futile to reject the Darwinian revolution on religious grounds as it had been to reject the Copernican revolution. The weight of objective scientific evidence does not allow either of these theories to be rejected by reasonable people. Accepting the inevitable, the Vatican has recently accepted that evolution is a fact (see CREATIONISM). The same mind-set has also prompted a recent admission that the Catholic Church had been wrong to condemn Galileo. Both concessions carry an implicit agreement that theology must be willing to accommodate itself to scientific discoveries.

MODERN PHYSICS AND THEOLOGY

Religious conflict in science may be between scientists and religions, or it may be between different scientists who disagree about the religious implications of their findings. Science generally assumes philosophical materialism. That is, scientists are expected to explain the phenomena they study exclusively in terms of material processes without recourse to religious or mystical ideas. Newton's scientific work thus dispenses with all but a few hints of religious or mystical causes. Yet Newton devoted much of his time to alchemy. Many of the world's most distinguished scientists have also been personally very religious. The history of both astronomy and the life sciences suggests that the impulse to find order in the physical universe is not very different from the religious impulse to find theological meaning in the physical world. Rather than being in conflict, religion and science have been complementary throughout most of recorded history, including the period since the scientific revolution.

Whether scientists should or should not be religious, the fact is that most have been. Just as scientific discoveries change religious worldviews, so religious worldviews affect the process of scientific discovery. To some extent scientists, like other scholars, discover truths that their minds are prepared to accept and often miss those that they are not prepared to accept. Einstein was one of the most influential scientists of the 20th century but he was quite displeased with some of the ideas of the quantum physicists who ruled the discipline during his old age. Despite personally contributing to a scientific worldview that is far more destructive of religious preconceptions than was that of Copernicus, Einstein clung to his personal religious faith. Even he was forced to draw the line at quantum physics with its relentless uncertainty and indeterminacy. His arguments with friends who had gone down this road characteristically ended in a statement of uncritical religious belief: "I cannot believe that God plays dice with the universe." From a metaphysical perspective, this is no different from Copernicus's insistence that the universe reflects its creator.

See also COPERNICUS, NIKOLAUS; CREATIONISM; DARWIN, CHARLES ROBERT; EINSTEIN, ALBERT; GALILEO GALILEI; KEPLER, JOHANNES; SCIENCE, HISTORY OF.

Further Reading

Biagioli, M. *Galileo Courtier: The Practice of Science in the Culture of Absolutism*. Chicago: University of Chicago Press, 1993.
Drake, S. *Galileo at Work: His Scientific Biography*. New York: Dover, 1995.
Gamow, G. *Thirty Years That Shook Physics: The Story of Quantum Theory*. New York: Dover, 1985.
McClellan, J. E., and H. Dorn. *Science and Technology in World History*. Baltimore, Md.: Johns Hopkins University Press, 1999.
Stephenson, B. *Kepler's Physical Astronomy*. Princeton, N.J.: Princeton University Press, 1994.
———. *The Music of the Spheres: Kepler's Harmonic Astronomy*. Princeton, N.J.: Princeton University Press, 1994.
Westfall, R. S. *The Life of Isaac Newton*. Cambridge: Cambridge University Press, 1993.

religious practice and immune function   *See* FAITH HEALING.

repetitive stress injury

Repetitive stress injury refers to a group of occupational disorders produced by long hours of repeating the same action. First described by Bernardino Ramazinni, an Italian physician, in 1717, the disorder become increasingly common following the INDUSTRIAL REVOLUTION as workers performed simple repetitive tasks without the opportunity to rest their over worked muscles. Occupations in which the disorder is particularly common include assembly line workers, meat packers, and the many modern occupations in which computer keyboards are used for long periods (e.g., secretary, journalist, researcher). It is also common among musicians and professional athletes.

Repetitive stress injury has a bewildering variety of synonyms that include repetitive strain injury, cumulative trauma disorder, repetitive motion injury, and occupational overuse injury. A particularly common type, known as carpal tunnel syndrome, is produced by rapid and repetitive movements of the hand, such as those produced when playing a piano, writing longhand, or typing on a computer keyboard. The repetitive movements produce inflammation of tendons, which then press on the median nerve as it runs through the carpal tunnel in the wrist. The symptoms include tingling or pain in the thumb, index and middle fingers, and part of the ring finger. Treatments include reducing the number of forceful movements, exercises to strengthen the affected areas, the use of night splints, and nonsteroidal anti-inflammatory medication. Serious cases can be quite disabling and may call for corrective surgery. Surgery is helpful not just in relieving the pain and discomfort of trying to work at tasks that aggravate the injury but also can prevent permanent nerve damage.

Some occupations are more likely to produce repetitive stress injuries than others, but their development is either preventable or they can be rendered less serious. People who maintain good physical fitness are less likely to develop repetitive strain injury. Many sufferers continue to work without treatment thus risking a serious aggravation of their condition. Recent rules promulgated by the U.S. Occupational Safety and Health Administration (in 1999) have called for opportunities for employees both to report any repetitive stress symptoms they have experienced and also to report any job situations they see as posing a threat in this context. Employers are also required to inform employees about any workplace dangers of developing or aggravating repetitive stress injuries. This means that employees in high-risk occupations may be advised to take periodic rest breaks. Employers must also provide free medical treatment for any such injuries when they occur.

Apart from the debilitating personal consequences of repetitive stress injuries, they also carry a huge economic cost. Annual employee compensation in the United States runs at some $20 billion. When lost productivity and the costs of employee turnover are factored into the equation, repetitive stress injuries may cost as much as $100 billion each year. Enlightened self-interest, as well as governmental regulations, and ethical concern for the welfare of workers have all influenced employers to make much more effort to prevent these disorders.

See also HUMAN FACTORS RESEARCH; INDUSTRIAL REVOLUTION.

Further Reading

Pascarelli, E., and D. Quilter. *Repetitive Strain Injury: A Computer User's Guide.* New York: John Wiley, 1994.

Peddie, S., and C. Rosenberg. *The Repetitive Strain Injury Sourcebook.* Lincolnwood, Ill.: Lowell House, 1998.

reproductive technologies

Modern reproductive technologies, developed to help infertile couples, have fragmented the traditional procreative roles of parents and have thus raised many ethical and legal issues concerning the rights and responsibilities of all of the parties involved in medically- assisted pregnancies. These techniques include artificial insemination, in vitro fertilization, surrogate motherhood, and, potentially, cloning. In vitro fertilization has raised many complex ethical and legal problems concerning the moral status of frozen embryos left over from fertility treatments. Recent technologies of tissue repair call for the creation of cloned embryos as a source of embryonic stem cells. Although the purpose of this research is not precreation, it raises similar ethical issues to those raised by in vitro fertilization (see STEM CELL RESEARCH).

ARTIFICIAL INSEMINATION

Artificial insemination (AI, also referred to as assisted insemination) of humans dates to 1790 when Scottish surgeon John Hunter successfully inseminated the wife of a linen draper using her husband's semen. Almost a century would pass before the next recorded use of the procedure by a Philadelphia medical school professor, William Panacost. Panacost obtained a sperm sample from a student donor and performed the procedure under general anesthetic. In an appalling breach of medical ethics, he neglected to inform the woman, or her husband, about how her pregnancy was initiated.

In modern artificial insemination, the semen may be obtained from a woman's husband (AIH) or from a donor (AID). Reasons for failure of conception via sexual intercourse include the husband's impotence, low sperm count, or inactivation of sperm by the wife's secretions. Semen is obtained through masturbation, prepared using various laboratory techniques, and inserted by a health professional into the woman's vagina, cervical canal, or uterus.

AI does not produce birth defects and is not known to cause any other medical problems for infant or mother. Ethical objections to AI by the Catholic Church point out that it separates lovemaking from conception. One medical response to this objection is that it is the infertility of the couple, rather than the medical procedure, that separates these facets of the marital relationship.

AI using the semen of a donor is much more ethically problematic from various points of view. Since the donor is not the woman's husband, the union of their gametes could be considered technically adulterous. Concerns that this might have a deleterious psychological impact on the husband have not been supported by limited research on the question. Moreover, there is evidence that children conceived by AID are damaged by the knowledge of how they were conceived.

If the donor is not the woman's husband, it may be a relative or friend chosen by the couple. Such choices can create future conflicts of interest over parental rights of the biological father, for example. Perhaps for this reason, most donors are anonymous and semen samples are usually obtained from commercial sperm banks that pay young men around $50-$75 per usable sample. Women are charged approximately $300 per insemination and can expect some 10 treatments before achieving pregnancy.

Donated semen samples have not always been safe due to inadequate screening of donors and samples. In a few tragic cases children conceived by AID suffered from severe genetic diseases. In other cases, women developed communicable diseases, including AIDS. Such problems can be prevented by taking family histories of donors and ensuring that they are tested for sexually transmitted diseases.

Sperm banks preserve semen by freezing it. This is essential to comply in the United States with the six-month quarantine period before semen can be used. Cryopreservation of human semen was originally developed to preserve samples from men undergoing vasectomy or chemotherapy resulting in sterility. Semen preserved in this way is less likely to result in fertilization although there is considerable variation in the response of semen samples to cryopreservation. Perhaps the most important issue is that infants born as a result of AI using cryopreserved semen do not have an increased risk of birth defects or any other health problems. If a single donor produces many offspring, there is a risk that when the children grow up they will be involved in incestuous sibling unions. For this reason, the British government limits each donor to producing 10 offspring but there are no such regulations in the United States. Many commercial sperm banks maintain sibling registries that can be checked before AID children marry.

Another major ethical issue of AID concerns its eugenic implications. Women who choose to reproduce through donor sperm obtained from a sperm bank can select the father of their child based on traits such as age, height, weight, blood group, race, education, occupation, and hobbies. This is an example of positive EUGENICS that favors the reproduction of individuals with desirable traits. Even though most people find eugenics morally repugnant, such selection is arguably not ethically different from the choice of a partner that women make in the case of normal sexual reproduction. In fact, there is less positive selection because far less information is available.

There is also a market for human eggs used for in vitro fertilization, which has similar eugenic implications. Egg donors are also likely to be selected on the basis of physical appearance, intelligence, and social status. One example of this phenomenon is the appearance, in 1999, of an advertisement in the newspapers of several Ivy League universities. The ad sought an egg donor who should be tall, athletic, and highly intelligent. A fee of $50,000 was offered which was a considerable increase over the market rate of $1,000 to $5,000 paid to donors who receive somewhat risky injections before having their eggs aspirated. Many commentators were offended by the elitist implication that wealthy people can purchase good genes for their offspring. Apart from the unusually high price offered, it is clear that there is no important ethical distinction between selecting egg donors in this manner and the positive eugenics practiced by women who use sperm banks to conceive offspring.

IN VITRO FERTILIZATION

In vitro fertilization (IVF) unites the gametes in a Petri dish instead of inside the womb. This procedure was originally devised for women with damaged fallopian tubes but is often used today in cases where the cause of infertility has not been identified. The first "test tube" baby was English girl Louise Brown born on July 25, 1978. Since then, tens of thousands of children have been born using IVF and there is no evidence that children conceived in this way suffer any increased health risks.

In this procedure, egg cells are usually removed from the woman's ovaries via ultrasound-guided transvaginal aspiration and fertilized in vitro using the husband's semen. The woman is often treated with fertility drugs to induce the development of several follicles from which eggs are removed for fertilization in vitro. With several eggs, it is possible to conduct multiple treatments in the event that the first fails to produce a pregnancy. Two days after fertilization, the embryos are surgically placed in the womb. The use of multiple embryos greatly increases the chance that at least one will be implanted and survive. The use of four or more embryos creates a high risk of multiple pregnancy.

A major problem with in vitro fertilization and other techniques of assisted fertilization is that for many people the treatments have proved ineffective. Combined with the typically very high cost of these procedures, this can be more devastating to a couple than infertility itself. Couples can thus be seriously harmed by fertility treatments.

One way of improving the success rate is to use gamete intrafallopian tube transfer (GIFT). In this technique, both ova and sperm are placed in the fallopian tube. The probability of success is approximately one in three, which is twice the success rate for in vitro fertilization.

In the case of pregnancies for which family history indicates a risk of sex-linked genetic disorders, female embryos may be selected for implantation. The noninvasive technology for sexing human embryos could also be used to select the sex of future children analogous to what many ethicists see as an abuse of AMNIOCENTESIS. One or more cells can be removed from an embryo to test for genetic defects. It is believed that this does not harm the developing fetus but the relevant research remains to be completed.

Excess embryos may be cryopreserved for subsequent use by the same woman in the event that either the treatment fails or another pregnancy is desired. There has been much ethical debate about the moral status of the frozen embryo and public interest in these issues has been heightened by some ethically complex legal cases. In the Rios case, in Australia in the 1980s, frozen embryos were "orphaned" after the biological parents were killed in a plane crash. What was to be done with the embryos? A more probable scenario for conflict over possession of frozen embryos occurs when a couple that has stored embryos in the context of fertility treatment subsequently breaks up. Such a "custody dispute" was confronted by the Tennessee courts. In 1989, Mary Sue Davis won a state court dispute over seven frozen embryos that had been produced by herself and her husband, Junior Lewis Davis, before their divorce. The former husband did not want the embryos to be implanted in his ex-wife's womb. The court ruled that frozen embryos were children and that the mother was entitled to carry them to term. This decision was overturned by the Tennessee high court on the grounds that the man should not be forced to become a father, with all of the legal and financial responsibility this entails, against his will. A 1999 Massachusetts case, involving a woman identified only as B. Z., ended similarly.

The fact that embryos can be maintained for many years in a state of suspended animation raises some fundamental ethical questions. Should such an undifferentiated entity be considered a human being? If so, does it have the same moral status as a human fetus? Some ethicists would argue that since it is undifferentiated, it can no more be considered a human being than the gametes of which it is composed. This view is implicitly endorsed by the British government, which has recently permitted experimentation on human embryos during the first 14 days.

The fact that embryos can be frozen and thawed for subsequent use means that there is the potential for embryo donation or sale. This raises many of the same legal and ethical issues as semen, or ovum, donation or sale. Should reproductive tissue be sold? If it is, what rights or responsibilities, if any, do the biological parents have in relation to any children that are born? These questions are posed in a particularly compelling way in the case of surrogate parenthood.

SURROGATE PARENTHOOD

If a woman wishes to have a child but is incapable of carrying a pregnancy, perhaps because she lacks a uterus, or is otherwise unsuccessful at carrying a pregnancy to term, she may enter into an arrangement with another woman to carry her child. In this case, the gametes of the mother and her husband undergo in vitro fertilization and the embryo is transferred to a second woman who has no biological relationship with the child. The first instance of such surrogacy occurred in 1985. In this situation, the surrogate mother is sometimes referred to as a surrogate carrier and the arrangement is referred to as partial surrogacy.

In full surrogacy, the surrogate mother contributes the egg as well as a womb for the embryo to develop in. This arrangement led to a bitter custody dispute following the 1986 birth of a baby girl, named Melissa (or Baby M) to Mary Beth Whitehead who had been paid $10,000 to carry a baby for William and Elizabeth Stern of Tenafly,

New Jersey. The arrangement fell apart when the surrogate mother refused to give up the infant to the social parents (i.e., those who arranged the pregnancy) and was sued by the Sterns to whom she lost a bitter custody dispute.

Surrogacy arrangements encounter many legal and ethical problems. If there is a legal contract, as in the case of Baby M, are the social parents entitled to control the behavior of the surrogate mother to ensure that she does not harm the infant by smoking, drinking, using prescription or illegal drugs, or being exposed to harmful radiation at work, for example? If the child turns out to be seriously ill, or mentally retarded, do the social parents still have an obligation to accept the infant? Is it ethical for a woman to, in effect, rent out her womb? If so, then how can wealthy women be prevented from exploiting the poor by paying them to produce their children? This issue raises a similar problem to the one raised by sale of organs for transplantation.

Complex as the Baby M case was, it involved only three people and, following their legal dispute, the principals have come to accept the liberal visitation rights that allow the child to grow up knowing her biological mother. A more complex case involved John and Luanne Buzzanca who were unable to have a child. The couple arranged for sperm and eggs from anonymous donors to be used for in vitro fertilization after which the embryo was transferred to a surrogate carrier who was bound by a legal contract. Before the child was born, John Buzzanca filed for divorce. In the course of the divorce proceedings the California courts had to decide which of the five people involved in the child's conception had the status of legal parents who were obliged to care for it.

Reversing many legal precedents favoring the view that the biological parents should be the legal parents, the California Court of Appeals ruled that the social parents in this case, the Buzzancas, should take care of the child because they had arranged for its conception. This decision reversed a lower court ruling that the child was an orphan. Faced with such legal muddles, many people would conclude that the extreme fractionation of parental roles from two to five people in the Buzzanca case is unlikely to be in the best interest of a child and should be avoided on ethical grounds. In the case of cloning, which is now technically feasible but has not been used for reproductive purposes, the opposite problem emerges of shrinking the number of biological parents down to one.

CLONING VERSUS SEXUAL REPRODUCTION

Cloning research has raised two distinct ethical problems. The first revolves around the culturing of embryonic cell clones for research purposes. Should such tissue be treated differently from that used in research on nonhuman animals? The second problem relates to the potential use of human cloning as an alternative to sexual reproduction. Although this has not yet been done, it seems technically feasible. Assuming that human cloning is technically possible and that it is legally permitted, the resulting offspring would have only one parent instead of the customary two. A change to asexual reproduction would produce a dramatic alteration in family structure since the sexual relationship of husbands and wives would no longer be of central importance to reproduction. Can children be thought of as having an inherent right to two parents who can potentially look out for their best interests?

In addition to these larger social questions, cloning raises a number of technical issues that have ethical ramifications. One relates to mitochondrial DNA, a portion of the genotype that resides outside of the nucleus and is normally inherited from the maternal line. In the procedure of nuclear transfer, in which the nucleus of one egg is transferred to another, this portion of the genetic material is lost, which would artificially alter the heredity of a child. This procedure has been performed successfully using rhesus monkeys and is being contemplated to prevent genetic disorders carried by human mitochondrial DNA.

Another unresolved technical issue relates to the biological age of clones. Early work on Dolly, the cloned sheep, has found that her chromosomal telomeres are shortened, a phenomenon that is associated with cellular aging. If clones are the same biological age as the parent cell, then cloning of humans might produce a reduction in lifespan which could be interpreted as an unwarranted reduction of the quality of life to be expected for individuals produced by sexual reproduction.

See also ARTIFICIAL INSEMINATION; CONTRACEPTION; EUGENICS; GENETIC ENGINEERING; RU-486 ABORTION PILL; STEM CELL RESEARCH.

Further Reading

Caplan, A. L., G. McGee, and D. Magnus. "What Is Immoral about Eugenics?" *British Medical Journal* 319: (1999) 1284–85.

Friedrich, M. J. "Debating Pros and Cons of Stem Cell Research." *JAMA* 284 (2000): 681–82.

Humber, J. M., and R. Almeder, eds. *Human Cloning.* Totowa. N.J.: Humana Press, 1998.

Kass, L. R., and J. Q. Wilson. *The Ethics of Human Cloning.* Washington, D.C.: AEI Press, 1998.

Nussbaum, M. C., and C. R. Sunstein, eds. *Clones and Clones: Facts and Fantasies about Human Cloning.* New York: W.W. Norton, 1999.

Pitt, J. B. "Fragmenting Procreation." *Yale Law Journal* 108 (1999): 1893–900.

Quesenberry, P. J., G. S. Stein, and B. Forget, eds. *Stem Cell Biology and Gene Therapy.* New York: Wiley-Liss, 1998.

Seible, M., and S. L. Crockin, eds. *Family Building through Egg and Sperm Donation: Medical, Legal, and Ethical Issues.* Sudbury, Mass.: Jones and Bartlett, 1996.

Shannon, T. A. "Eggs No Longer Cheaper by the Dozen." *America,* May 1, 1999, p. 6.

Walters, L. "Reproductive Technologies and Genetics." In *Medical Ethics,* ed. R. M. Veatch, 209–338. Sudbury, Mass.: Jones and Bartlett, 1997.

rights of human research participants

The unregulated state of research using human participants came to an end in 1946 when the NUREMBERG CODE was adopted as a means of holding Nazi scientists responsible for their extremely unethical research on human subjects. In addition to the core rights established at Nuremberg (i.e., rights to informed consent, freedom from harm, and freedom to terminate participation), several others have been widely accepted. They include rights to privacy, rights to special consideration for vulnerable populations such as children, prisoners, and the mentally disabled, the right to be fairly treated, the right to adequate aftercare, and the right to receive information about the true nature of a study in the event that participants are deceived by the researcher (see chart).

INFORMED CONSENT

Informed consent means that participants in research know what to expect before they consent to participate. It can thus be broken down into two basic elements. First, the participant must consent to participate implying that they take part in the research on a voluntary basis. Needless to say, concentration camp participants in Nazi experiments could not decline to participate and given the heinous nature of some of the projects would not have consented if they had an option (see MENGELE, JOSEF). Second, the participants must know what they are agreeing to do. To obtain fully informed consent, researchers must reveal all information about the research procedure that might reasonably affect willingness to participate. An egregious example of an American study in which fully informed consent was not obtained involved administration of a drug (SCOLINE) that temporarily halted breathing. Since subjects did not know that they were receiving a drug, via intravenous infusion, that would arrest breathing, all thought they were dying. The subjects, who were alcoholics, participated in the study after being told that the research might benefit alcoholics. This study constitutes a very clear violation of informed consent because if the subjects knew at the beginning of the study what they had learned by its completion none would have consented to it.

Informed consent is the most centrally important right of research participants that prevents ethical research projects from being conducted due to lack of participants. Yet, informed consent is not considered to

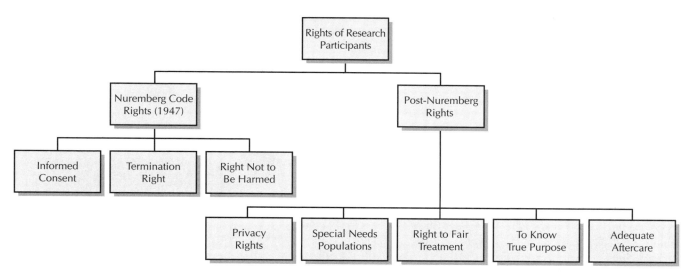

The most fundamental rights of human research participants were specified in the Nuremberg Code and several others were defined in subsequent codes of scientific ethics.

be an absolute right. The regulations of the U.S. Department of Health and Human Services (DHHS) recognize several circumstances under which an Institutional Review Board may waive informed consent. If the research involves MINIMAL RISK to participants, meaning that they are no more likely to experience harmful consequences than if they are going about their normal daily activities, informed consent may not be required. For example, if you are recording the time taken by car drivers to negotiate a busy intersection as a function of sex of driver and passenger, you would not need the consent of the drivers. In fact, obtaining such consent would be bothersome to participants and failing to obtain their consent is unlikely to have any adverse consequences for them. If there are no adverse consequences, DHSS regulations do not require informed consent.

According to DHHS REGULATIONS, if the purposes of the study are undermined by participants knowing precisely what is being done, informed consent may be waived provided that the potential benefits of the research are judged to outweigh any risks to participants. This is controversial because it pits the benefits of the research directly against the rights of the participant. In abstract theoretical terms, the absolute right of informed consent, derived from deontological ethics, becomes a qualified right based on the balance between two ethical goals, the expansion of scientific knowledge and the protection of human participants. In other words, there is a switch from deontological ethics to utilitarian (or consequentialist) ethics.

The purposes of research are undermined by informed consent most often in psychology. A researcher who wants to study incidental memory, for example, cannot tell participants that they are going to be given a memory test. Similarly, if you want to study how people use body language in putting across a persuasive communication, you cannot tell them that their gestures and body movements are being observed because this knowledge might completely distort their behavior.

Many of the ethically questionable studies that have excited controversy in social psychology have involved some violation of informed consent. For example, Phillip Zimbardo conducted a study designed to test whether unrecognized hearing loss makes people paranoid. His subjects were hypnotized and became deaf without knowing why following posthypnotic suggestion. Shaky as the ethical rationale may have been for conducting this study in the first place, it clearly could not have been carried out if the participants knew why they were losing their hearing. As it was, they became very upset, and yes, paranoid.

SPECIAL NEEDS POPULATIONS

Certain classes of people, such as young children and the mentally retarded, are not adequately protected by the right to informed consent for the simple reason that they may be incapable of understanding the research procedure and its possible injurious consequences. In their case, consent is derived by proxy through parents, legal guardians, or other responsible persons. Informed consent can be compromised when participants are illiterate or too poorly educated to grasp the full consequences of the research procedure. Freedom to choose not to participate may also be denied in the case of the institutionalized and incarcerated, including prisoners and residents in psychiatric hospitals, who are accustomed to following instructions and may be under subtle pressure to partici-pate. Moreover, the offer of money to participate can be much more compelling for prisoners who have very lim-ited opportunities to earn cash. In all of these cases where informed consent may be compromised, researchers must take special responsibility for protecting the interests of the vulnerable populations.

THE RIGHT NOT TO BE HARMED (NONMALEFICENCE)

The majority of ethicists take the perspective that although animals can be thought of as having rights that are ascribed to them because of their inherent moral worth as sentient beings, the moral worth of human beings is much greater. For this reason, injuring animals, causing them pain and discomfort, and killing them can be justified if it produces an increment in human well-being. The nonmaleficence principle of the Nuremberg Code states that such harm should never be caused to humans in the name of research. The COLD WAR RADIATION EXPERIMENTS sponsored by the U.S. government are an example of a clear violation of the nonmaleficence princi-ple as well as of several other rights of human research participants.

THE RIGHT TO TERMINATE PARTICIPATION

If informed consent is the most important right, the right to stop an experimental procedure at any time can be considered an important rider to this right. When people consent to participate, they may not be fully aware of what they are about to experience. Fully in-formed consent may not have been possible in light of the purposes of the research. Some aspect of the proce-dure may not have been clearly communicated or ex-plained. The procedure might have carried unforeseen consequences. For example, an experiment dealing with the perception of flickering lights might have precipi-tated a headache.

From a legal perspective, the right to terminate is obviously guaranteed in a democratic society. Obvious as this may be, participants may not always be aware of it. A clear example of research in which this right was violated was Stanley MILGRAM'S OBEDIENCE RESEARCH in which sub-jects believed that they were delivering dangerous levels of electric shock to another person. Many said that they wanted to end the experiment. In every case, a stony experimenter intoned: "The experiment must go on." From a narrow legal point of view, Milgram's participants could easily have gotten up and left at any time, yet most felt trapped by the social situation, which was partly the point of the experiment. With such violations in mind, consent forms used by social psychologists routinely remind participants that they are free to leave at any time without a penalty. Any payment is normally made at the

end of the procedure and those who terminate participa-tion generally are not paid.

PRIVACY RIGHTS

The right to privacy in research is endorsed in many statements of professional ethics but the specific obliga-tions of researchers are often quite vague possibly to pro-tect the counterposed right of researchers to freedom of enquiry. Research involving observation of people in pub-lic places does not require INSTITUTIONAL REVIEW BOARD (IRB) scrutiny according to U.S. federal guidelines. If people are to be observed in situations in which they might normally expect privacy, researchers must seek IRB approval.

Researchers generally observe a strict code of privacy in the collection of data. When information is compiled in data files, each individual is coded as an arbitrary number that cannot be traced back to their name. In this way, sensitive information divulged to a researcher is gen-erally quite secure. Yet many people do not trust the con-fidentiality of data. For example, many employees refuse to donate blood because of the invasive nature of the mandatory questionnaire that asks about drug use and sexual activities. They fear that leakage of information could be damaging to their careers. Privacy issues can be especially troubling for AIDS researchers who encounter an ethical obligation to inform the sex partners of their participants. The obligation to save a life in this case takes ethical priority over the right to privacy.

One of the most infamous invasions of privacy in social science research involved the study by LAUD HUMPHREYS of homosexual acts in public restrooms. Although the participants were technically in a public place, some scholars see this as invasive of privacy. What Humphreys did next crossed the line in the eyes of many scholars. He recorded participants' car numbers, used these to obtain their names and addresses, and inter-viewed them in their homes while wearing a heavy dis-guise.

Invasion of privacy in market research has recently become a burning issue due to the activities of Internet companies, such as DoubleClick, that have developed techniques for following each mouse click of a person navigating through the Internet. This information is employed to develop user profiles that form the basis of a targeted marketing campaign. DoubleClick is currently being investigated by the Federal Trade Commission for deceptive business practices and has promised to provide Internet users with an opportunity to opt out of being tracked by them.

RIGHT TO FAIR TREATMENT

Researchers are expected to enter into an agreement with participants that is open and fair. The obligations and responsibilities of each partner in the agreement are clearly stated and the investigator must honor all promises and commitments made. For example, if a procedure involves infliction of pain, participants should be informed of this at the outset. If money or specialized training is offered as an inducement to participate, the researcher must pay up or provide the training.

ADEQUATE AFTERCARE

The right not to be harmed in an experiment includes not only problems that arise during the study but also harmful effects that can arise afterward. Participants in psychological experiments might be traumatized by their experiences and have difficulty sleeping, for example. For this reason, psychological researchers usually provide the name of a contact person, often the principal investigator, who can be contacted if problems arise following the research. If the researcher is unable to address these problems by talking to the individual, then they must be addressed using the assistance of counselors or other professionals. Participants in medical research are entitled to aftercare provided by physicians. This serves to establish whether the research had adverse effects. If so, medical attention is provided in an effort to redress this harm. The right to adequate aftercare falls under the rubric of the nonmaleficence principle. If harm has been done to the participant, then adequate aftercare minimizes this harm.

RIGHT TO KNOW TRUE PURPOSE OF THE RESEARCH

Methodological requirements of a study may require that experimenters either conceal some information or actually provide participants with false information. The use of deception or concealment of information has to be justified on the basis (a) of the scientific value of the research and (b) by the absence of alternative procedures that might answer the research question. The use of deception, even where it is scientifically justified, is thought of as inflicting harm on the participants in the sense that the relationship of openness and fairness that they expect has not been lived up to. According to the *Ethical Principles* of the American Psychological Association, the onus lies with the researchers to repair the relationship as soon as possible by giving a thorough explanation of the research undertaken and what its true purpose has been. This explanation is referred to as a debriefing.

Debriefing protocols typically invite questions so that if there is any aspect of the study that has not been clearly explained it may be clarified to the satisfaction of the individual participant. The logic of the debriefing is utilitarian in the sense that the harm of deception is understood to be wiped out by explaining the true purpose of the research. A deontologist might argue that deception is always an ethical breach and that the resulting damage to the rights and trust of the participant cannot be mended. Nevertheless, most ethicists would agree that where deception has been used the harm can be reduced by a debriefing statement.

See also COLD WAR RADIATION EXPERIMENTS; HUMPHREYS, LAUD; MILGRAM'S OBEDIENCE RESEARCH; NUREMBERG CODE; PARANOIA, EXPERIMENTAL INDUCTION OF; SCOLINE, AND FEAR CONDITIONING.

Further Reading

American Psychological Association. *Ethical Principles in the Conduct of Research with Human Participants*. Washington, D.C.: Author, 1982.

Kimmel, A. J. *Ethical Issues in Behavioral Research: A Survey*. Cambridge, Mass.: Blackwell, 1996.

Resnik, D. B. *The Ethics of Science*. New York: Routledge, 1998.

U.S. Department of Health and Human Services. *Code of Federal Regulations Pertaining to the Protection of Human Subjects*. Washington, D.C.: U.S. Government Printing Office, 1983.

Rio Declaration on Environment and Development

Drafted at a 1992 United Nations Conference on Environment and Development at Rio de Janeiro, Brazil, the Rio Declaration consists of 27 principles that provide guidelines for economic development and environmental behavior favoring global sustainability. The fundamental idea is that states should establish a global partnership wherein the developmental needs of each can be satisfied without causing irreparable damage to the global environment. The declaration sets out the rights and responsibilities of states, local communities, and individuals in relation to the environment. It constitutes an important statement of the pressing need to change economic and social policies in a manner that protects the earth's ecosystems.

See also GAIA HYPOTHESIS.

Roentgen, Wilhelm Conrad

(1845–1923)
German
Physicist

Wilhelm Roentgen discovered X-rays and thereby earned the first Nobel Prize in physics (1901). Roentgen had noticed that the turning on of a cathode ray tube caused barium platinocyanide crystals to fluoresce (or light up) even when they were shielded by black cardboard or metal plates. He assumed that this effect was due to a hitherto unknown form of radiation that he christened

X-rays. Roentgen demonstrated the practical uses of X-rays, such as looking inside the human body and observing cracks inside metals. He died in Munich on February 10, 1923.

See also SCIENCE, HISTORY OF.

Roswell, New Mexico   *See* UFOS.

Rousseau, Jean-Jacques
(1712–1778)
French
Philosopher, Writer

An influential figure in the history of ideas, Jean-Jacques Rousseau is remembered for his influence on the romantic movement in literature and also for his contributions to SOCIAL CONTRACT THEORY in political philosophy. He shared with other ENLIGHTENMENT philosophers the view that human nature was essentially good and contrasted the corrupting influences of modern civilization with the ideal of the noble savage who lived in harmony with nature free of the emotions of selfishness, pride, acquisitiveness, and envy. Contrary to the other Enlightenment figures, Rousseau did not see the development of knowledge, science, and the economy as contributing to human betterment. In fact, he saw progress as inherently corrupting and as promoting inauthentic social relationships. In *Emile* (1762), his treatise on education, he depicted the protagonist being raised in a rural environment (to escape the corrupting influence of the city) and shielded him from all reading materials except the novel *Robinson Crusoe,* which illustrates how the better side of human character can be cultivated by living in harmony with nature.

Rousseau's political perspective as developed in his book *The Social Contract* (1762), bears analogy with that of English philosopher Thomas Hobbes (1588–1679) the originator of social contract theory. In particular, Rousseau shared with Hobbes the notion that for a desirable society to be created, individuals must give up their personal sovereignty to the control of an all-powerful state, or monarch. He had no faith in the wishes of the majority and developed an abstract concept of the common good that was synonymous with the goals of the state. Rousseau wanted to extend political rights beyond propertied and franchised classes to all the individuals in a state. He valued equality more than freedom and endorsed a totalitarian state that embodied the general will and took precedence over the rights of the individual. Thus, he saw the state as owning all of the property of its members.

Rousseau's ideas were invoked by the leaders of the French Revolution to justify their reign of terror. Rousseau would never have supported these atrocities. On the other hand, there is sufficient vagueness about the practical question of how the common good would be implemented to leave him open to the conclusion that he supported any method, including violent revolutionary action. Rousseau died in Paris on July 2, 1778.

See also ENLIGHTENMENT, THE; HOBBES, THOMAS; LOCKE, JOHN; RAWLS, JOHN; SOCIAL CONTRACT THEORY.

Further Reading
Cooper, L. D. *Rousseau, Nature, and the Problem of the Good Life.* University Park: Pennsylvania State University Press, 1999.
Morris, C. W., ed. *The Social Contract Theorists: Critical Essays on Hobbes, Locke, and Rousseau.* Lanham, Md.: Rowman and Littlefield, 1999.
Russell, B. *A History of Western Philosophy.* New York: Simon and Schuster, 1972.

Royal Society of London
Officially founded in 1662 by a charter granted by Charles II, the Royal Society of London (RSL) is the world's oldest scientific organization, although not the earliest. The first scientific societies were formed in Europe, beginning with Rome's Accademia dei Lincei (1603). RSL had its origins in a club known as *The Royal Society of London for Improving Natural Knowledge,* which met at Gresham College, London, beginning around 1645, and which provided a forum for scholars to exchange ideas. In *Novum Organum* (The new instrument, 1620), Francis Bacon (1561–1626) presented a model of the scientific enterprise that included not just empiricism and hypothesis testing but also called for public communication of results. He asserted that scientific research should be open to public criticism and discussion that would help to dispel fallacious thinking and dogmatic claims. Bacon was an advocate of scientific associations that would provide a venue for scientists to discuss their data, hypotheses, and theories. Previously, beginning around 1600, scientists had communicated by mail as well as through published books (following the development of the printing press in the 15th century). Otherwise, opportunities for discussion of ideas were limited to universities.

Bacon's aspirations were realized, posthumously, with the formation of the Royal Society that functioned very much as Bacon had envisaged. The society met regularly to discuss scientific ideas and findings. Its official journal, *Philosophical Transactions,* was the first scientific periodical in the world, and it is still published today. The society's other publications include, *Notes and Methods, Proceedings, Yearbook, Biographical Memoirs,* and *Mathematical Tables.* There are currently over 1,000 fellows.

Awards given by the society to promote scientific research include the Copley Medal (since 1731) and the Royal Medals (beginning 1825 and 1965, respectively).

Four years after the founding of the Royal Society, the French Royal Academy of Sciences was formed in Paris. Within the next two hundred years, scientific organizations had been established throughout Europe and the United States.

See also BACON FRANCIS; BOYLE, ROBERT; HOOKE, ROBERT; NEWTON, ISAAC; SCIENCE, HISTORY OF.

Further Reading

Resnik, D. B. *The Ethics of Science.* New York: Routledge, 1998.

Weld, C. R. *History of the Royal Society with Memoirs of the Presidents.* Bristol: Thoemmes Press, 2000.

Ziman, J. *An Introduction to Science Studies.* Cambridge: Cambridge University Press, 1984.

RU-486 abortion pill

Developed in the 1980s by French company Roussel-Uclef, RU-486 (or mifepristone) blocks receptors for the hormone progesterone. Progesterone is needed to maintain an early pregnancy and when it is blocked the pregnancy detaches from the womb and the fetus dies. Mifepristone is most effective if taken in the first seven weeks of pregnancy.

Used by women in France, Sweden, Britain, and China for a decade before its limited approval, in September 2000, by the U.S. Food and Drug Administration (FDA), RU-486 has engendered an ethical storm. Fearing protests by right-to-life campaigners, Roussel-Uclef refused to release the drug in the United States. Finally, the company donated its U.S. rights to the drug to the Population Council, a New York–based not-for-profit research group that initiated clinical trials.

The ethical controversy surrounding RU-486 centers on the fact that it is designed to kill a fetus. Right-to-life advocates point out that it is the first drug ever to be approved by the FDA for such a purpose. Yet, in a country where abortion is legal, this argument has more rhetorical appeal than it has ethical force. The majority of Americans support the use of RU-486, which promotes early abortions that are felt to be less ethically problematic because the fetus is not close to being capable of surviving independent of the mother.

The main advantage of an abortifacient drug over a surgical abortion is simply that it is not a surgical procedure. It requires approximately the same amount of medical attention, costs about the same, and is not less painful.

RU-486 differs from morning-after pills such as the emergency contraceptives Preven and Plan B. These can be used to prevent pregnancy for up to three days after unprotected sexual intercourse. Instead of causing an abortion, they prevent the pregnancy from occurring. In 1999, on the mistaken assumption that Preven causes abortions, some American pharmacies, including those at Wal-Mart, refused to stock it.

See also CONTRACEPTION.

Further Reading

Gianelli, D.M. "RU-486 Gets FDA Nod, but Obstacles to Wide Use Linger." *American Medical News,* October 7, 1996.

Rutherford, Sir Ernest
(1872–1937)
New Zealander/English
Physicist, Chemist

Born in New Zealand, Ernest Rutherford spent most of his adult life in England. His English residence was broken by a nine-year stint as professor of physics at McGill University in Montreal, Canada (1898–1907). Rutherford played an important role in elucidating the structure of the atom as a small dense nucleus with a positive charge surrounded by oppositely charged electrons that were widely dispersed. He was the world's leading researcher in the field of radioactivity who described alpha, beta, and gamma rays, grasped the fact that in the process of radioactive decay one element is changed into another, and showed how stable atoms could be made unstable by bombardment, thus "splitting the atom." Rutherford's work paved the way for the exploitation of powerful subatomic forces in NUCLEAR POWER and NUCLEAR WEAPONS. He died in Cambridge, England, on October 9, 1937. Among his many honors, which included a Nobel Prize in chemistry (1908) and a knighthood, Rutherford was president of the ROYAL SOCIETY OF LONDON (1925–30), which had conferred on him its highest honor, the Copley medal (1922).

See also CURIE, MARIE; NUCLEAR POWER; NUCLEAR WEAPONS; SCIENCE, HISTORY OF.

S

Sagan, Carl
(1934–1996)
American
Astronomer, Writer, Broadcaster

With a Ph.D. in astronomy and astrophysics from the University of Chicago, Carl Sagan began his research career at Harvard University in 1960. He quickly distinguished himself with a novel dust-storm explanation of the periodic lightening and darkening of the surface of Mars, a hypothesis that was verified by subsequent observations. Sagan also proposed the controversial idea that the high temperatures on Venus were produced by a greenhouse effect. Sagan worked on the *Mariner* and *Viking* space missions to Venus and Mars, respectively. He also studied the potentially catastrophic effects of nuclear war on global climate.

Sagan is best known for his speculative work on topics such as the possibility of extraterrestrial life (exobiology), including extraterrestrial intelligent life. This scientifically taboo topic was the subject of a book *Intelligent Life in the Universe* (1963) partially translated from the Russian and supplemented with original chapters. In 1977, Sagan departed from astronomy and exobiology to write *The Dragons of Eden,* which deals with the evolution of human intelligence. Although faulted for being full of speculation that extended beyond scientific knowledge, the book won a

Pulitzer Prize. Many of his subsequent popular works (*Broca's Brain,* 1979; *Cosmos,* 1980; *Comet,* 1985; and *Pale Blue Dot,* 1995) produced a similar mixture of success and criticism.

In the late 1980s, Sagan began work on the Public Broadcasting (PBS) series *Cosmos,* which gave free rein to the expression of Sagan's interests in astronomy. *Cosmos* is one of the most successful PBS documentaries, being eclipsed only by *The Civil War* in 1990. Like Sagan's books, the series was relished by the public but faulted by critics. *Cosmos* presents an optimistic view of science as essential for the survival of our species and also as exhilarating activity in itself. Like many other works of scientific popularization, the television series often stretches scientific credibility by indulging in speculations that are far removed from the facts. Another problem is the tone of awestruck reverence Sagan indulges in, which may add drama to the narrative but is nevertheless incongruous for a scientific presentation. It is as though Sagan was inviting the viewer to join him in worship of the spectacular marvels of the universe. Such pantheistic effusions can be seen as ethical showmanship in which the viewer is exposed directly to the author's enthusiasms, or it could be seen as weak science that comes dangerously close to religion, or science fiction.

See also GLOBAL WARMING; PANTHEISM; POPULARIZATION OF SCIENTIFIC FINDINGS.

Further Reading
Sagan, S., and I.S. Shklovskii. *Intelligent Life in the Universe.* San Francisco, Calif.: Holden-Day, 1963.
Terzian, Y., and E. Bilson. *Carl Sagan's Universe.* New York: Cambridge University Press, 1997.

science, history of

The period from two million years ago to the end of the last Ice Age, 12,000 years ago, is referred to as the Paleolithic, or Old Stone, Age. During this time our ancestors practiced a hunter-gatherer lifestyle in which stone tools were used to hunt and scavenge animals and to process plant foods. Anthropologists agree that throughout this period people lived in small nomadic bands, generally consisting of fewer than 100 people. Hunter-gatherer bands would have been quite egalitarian and would have had a division of labor by sex according to which women gathered most of the plant food and undertook most of the child care while men roamed more widely in search of animal food. During the Paleolithic period, population density remained very low possibly reflecting the constraint of seasonal scarcity of food. There is suggestive, but limited, evidence of scientific interests during the later Paleolithic period.

Even though human technology (in the form of toolmaking) is at least two million years old, there is very little evidence of any systematic interest in abstract knowledge of the kind that could be called science. One indication of an awakening interest in observing, and understanding, the natural environment comes from records of the phases of the moon carved on reindeer and mammoth bones. Such artifacts were being produced some 40,000 years ago, suggesting an early interest in science.

Approximately 15,000 years ago the ancient pattern of food collecting declined with the onset of food production in the form of horticulture and animal husbandry. With the end of the last Ice Age, food production increased allowing for the emergence of larger, more settled, communities and a variety of social and economic changes that facilitated more complex technologies and an interest in science.

SCIENCE IN THE NEOLITHIC PERIOD

Neolithic farming communities emerged at many different sites around the world beginning 12,000 years ago. Independent domestication of plants and animals in many global regions, including Africa, the Middle East, India, North Asia, Southeast Asia, Central America, South America, and Europe, suggests that all of the peoples of the world were abandoning their ancient hunter-gathering lifestyle due to shared ecological pressures. The most plausible explanation is rising global population putting pressure on supplies of food that could be hunted or gathered. Evidently, the extremely slow rate of growth of the human population had eventually caught up to the carrying capacity of the environment. The fact that some hunter-gatherer communities survived at the same time as the agricultural settlements suggests that the settled way of life was not adopted with alacrity but that our ancestors were forced into it by the practical difficulty of obtaining sufficient food.

Whatever the motive, it is clear that farming communities produced large amounts of food that allowed for a rapid expansion of the human population in agricultural areas. This fact is illustrated by the ancient city of Jericho, in Israel. This agricultural community was based on the tending of flocks and the cultivation of plots in the surrounding area. By 9,000 years ago, Jericho was a town of over 2,000 people, which contained large storage areas for excess food that could be used as an insurance against poor crop yields. Jericho had impressive fortifications indicating the need to protect the food surplus from raiders, some of whom might well have been local populations of starving hunter-gatherers.

The increased affluence of Neolithic farmers, combined with their settled lifestyle, favored both the cultivation of interest in science and the building of large monuments. Both points are illustrated in the case of the large facial statues erected on Easter Island in the Pacific Ocean. The Polynesian inhabitants reached this remote island only in the third century A.D. The precise significance of their 250 huge *moai* statues, which averaged 14 tons and had to be transported several miles from the site where they were quarried, is obscure but the careful alignment of the statue bases bespeaks an interest in astronomy. The creation of such ambitious public works, as well as an interest in astronomy, is typical of other Neolithic communities.

Both points are made most eloquently by the Stonehenge monuments on the Salisbury Plain in southwest England. Radiocarbon dating has shown that this impressive ring of standing stones connected by lintels was built in three different periods beginning 5,100 years ago and ending 3,500 years ago. The Heel Stone standing outside the circle weighed 35 tons. Eighty-two five-ton stones were transported from Milford Haven 150 miles away in Wales by raft and sled. Modern scholars believe that the site was the ceremonial center of a cult that worshiped the sun and the moon. Among its other social and religious functions, Stonehenge was clearly designed as an astronomical observatory, which marks the location of the rising sun at summer and winter solstices and at the spring and fall equinoxes. It also provides a record of the more complex movements of the moon across the horizon.

The astronomic markers created by Neolithic communities would have helped them to predict the changing of the seasons. This information could have been of great practical importance for farmers, allowing them to schedule the planting of various crops most advantageously, for example. Allowing such predictability is an important goal of the scientific endeavor.

SCIENCE IN EARLY CIVILIZATIONS

Just as the emergence of villages and towns was based on the increased level of food made available by agriculture, so the development of cities was made possible by the greater food surpluses produced by intensive agriculture. The urban way of life is associated with the full flowering of modern civilization with its rich traditions of learning, the arts, and science. According to the hydraulic hypothesis, all this can be attributed to the development of technologies for controlling the supply of water available for agriculture. In arid lands, such as Egypt and Mesopotamia, irrigation created a highly productive agriculture. In swamps, such as the lower delta of the Tigris and Euphrates rivers and the Yucatan, intensive agriculture was made possible by drainage systems.

Intensified agriculture produced very high population densities with centralized political authority on at least six occasions. Centralized kingdoms arose in Mesopotamia around 5,500 years ago, in Egypt 5,400 years ago, in the Indus river valley 4,500 years ago, in China on the banks of the Yellow River 3,800 years ago, in Mesoamerica 2,500 years ago, and in South America 2,300 years ago. Each of these early civilizations is consistent with the hydraulic hypothesis. A centralized authority organized extensive public works projects that managed water to maintain extremely productive agricultural systems.

The first civilization, the Mesopotamian, arose between the Tigris and Euphrates rivers in present-day Iraq. By 6,000 years ago, the plain between the rivers was dotted with Neolithic farming communities. Drainage of marshes in the lower delta was followed by the construction of an up-river irrigation system. From about 5,500 years ago large walled cities were built, including Uruk, Ur, and Sumer, that had populations ranging from 50,000 to 200,000 people. These city-states were sometimes organized into empires.

The surplus of labor over subsistence needs of the early cities meant that a number of nonsubsistence occupational specializations opened up. A large number of people worked as state employees responsible for collection, storage, and redistribution of food surpluses. There were political, military, and priestly castes, or elites. Skilled engineers were required to organize public works projects. The bureaucratic need for record-keeping

favored the development of writing and the scribe occupied an important occupational specialization. The earliest writing is associated with the Sumerian civilization of ancient Mesopotamia. This cuneiform script was inscribed on wet clay tablets using a reed stylus. The tablets were then preserved by drying or baking.

Sumerian scribes developed a complex system of writing based on up to 1,000 pictographs. In this system, writing about some even essentially took the form of drawing stylized pictures of it. Writing was complex and laborious and was restricted to the caste of professional scribes. Later, a more abstract system developed in which symbols represented sounds, as they do in modern writing. A similar progression from pictographs (hieroglyphs) to phonetic script (hieratic and demotic) occurred in ancient Egypt. Numerals for representing numbers were developed along with script. The Babylonian system was to the base 60, as opposed to modern numbers that are to the base 10. This explains why hours have 60 minutes and minutes have 60 seconds.

Mesopotamian civilization produced a rich tradition of mathematics and science. Babylonian mathematicians were quite sophisticated. They developed mathematical tables of squares, cubes, reciprocals, and so forth. They devised formulae for solving quadratic and cubic equations and could calculate compound interest. On the other hand, their estimate of pi, the ratio between the circumference and the diameter of a circle, was a crude 3 compared to the value of 3.16 of Egyptian mathematicians (3.14 is correct).

Mesopotamian science reached its highest point with the astronomy of the Babylonians who were at the geographic heart of Mesopotamia and played a central role in its political life and learned traditions. The period between new moons varies between 29 and 30 days due to the influence of several variables, including the season, the effect of long-term lunar cycles, and the distance between the sun and the moon as seen from earth. Babylonian astronomers systematically observed these phenomena, allowing them to predict with great accuracy when the new moon would be seen.

Although such astronomy might not strike us as particularly practical today, it is fairly clear that the Babylonians were motivated by the desire to predict the appearance of the new moon for religious reasons and because of the practical issue of constructing a calendar. People in ancient civilizations were also interested in alchemy (or what would today be called chemistry) because they hoped that it would solve practical problems. Science in ancient civilizations was supported by the centralized bureaucratic states. Medical learning was supported everywhere and physicians had the opportunity to benefit from the accumulation of knowledge in herbal medicine and surgery. One important distinction between early science and modern science was that the former lacked well-developed naturalistic theories about the universe (which was conceptualized in more religious terms). Greek civilization did a great deal to fill in this gap.

GREECE AND THE RISE OF THEORETICAL SCIENCE

Greek civilization departed in many ways from the model of earlier civilizations that can be easily accommodated by the hydraulic model according to which the control of water facilitated excess food production allowing nonsubsistence occupations to be practiced. The Greek soil was not rich enough to support intensive agriculture and there were no river basins suitable for irrigation. The main crops of ancient Greek were olives and grapevines, which do well in arid soils because they have deep root systems that can exploit water lying well below the surface. Together, these would not have made a very promising diet and the ancient Greeks were not even self-sufficient in food. By exporting olives and wine they became wealthy and could afford to purchase food from surrounding agricultural communities. Greece was thus the first great civilization built largely on trade.

Another difference between the Hellenic world and the other major contemporary civilizations is that before the reign of the emperor Alexander the Great (334–323 B.C.) Greece was organized into numerous autonomous city-states surrounding the Aegean Sea, rather than a single centralized state or empire. The most important region in terms of commerce and learning was Ionia in Asia Minor (or modern Turkey) and mainland Greece was initially considered something of a backwater. The noncentralized structure of Greek civilization appears to have played an important role in the development of natural philosophy and science because it meant that learning in Greece was largely an activity of the leisured individual rather than a state-sponsored function. Most of these leisured individuals were also members of the wealthy elite. Some, like SOCRATES (470–399 B.C.), acted as though they were wealthy by neglecting their useful employment in favor of philosophy. Socrates was a stone cutter who is accused of letting his wife take care of all the unpleasant practical problems of making a living. Others, like PYTHAGORAS (c. 560–480 B.C.), were fortunate enough to be sponsored by wealthy men.

Almost all of the names that we associate with the philosophy and science of ancient Greek civilization were enthusiastic gentlemanly amateurs rather than professional academics who worked for money. In fact, many of the Greek philosophers were explicitly opposed to the view that knowledge ought to be useful. Their personal lives also reflect varying degrees of mysticism and otherworldliness. It is only a slight exaggeration to

say that for them philosophy was like a religion, which called upon them to renounce the ordinary ambitions of their peers.

It is only when these historically unusual features of Hellenic philosophers are taken into account that their remarkably original achievements can be understood. The Greeks contemplated nature without any pressing practical needs in mind. Neither did they have a strong religious motive for cultivating learning. (PLATO, c. 418–348 B.C., is an apparent exception but it is important to realize that even in his case religion was derived idiosyncratically from logical deduction).

THALES (c. 640–546 B.C.) is given the credit of being the first natural philosopher. He was interested in explanations that are universally applicable and this is an important aspect of the scientific enterprise. Thales seriously proposed that everything in the world is made of water. This theory was considered quite lame by his contemporaries and has never received much respect. Yet it is essentially a falsifiable scientific hypothesis. By concluding that all elements can be built up out of hydrogen, which comprises two-thirds of water, modern chemistry indicates that, although Thales was wrong, he was not very far from the truth.

Another important early figure was PYTHAGORAS of Samos (c. 560–480 B.C.). Pythagoras was a number mystic who believed that mathematical knowledge was sacred and he organized a strange religious sect based on this premise. In addition to his contributions to early mathematics, as transmitted to posterity through the writings of EUCLID and others, Pythagoras has exerted an important influence on science through his claim that everything in nature can be quantified ("all things are numbers"). The prominence of mathematics in modern science would appear to support this view. He found an important demonstration of the power of numbers in his study of musical instruments and the physics of sound. Thus, a string that is divided by an exact ratio produces a note that is in harmony with the open string.

From the point of view of the subsequent development of science, by far the most important Greek philosopher was ARISTOTLE (384–322 B.C.). His influence extends in many different directions. From one thing, he was a keen biologist who not only studied the world around him but also conducted observational research in topics such as embryology. One-third of his voluminous writings were devoted to biology. Aristotle's range of information was encyclopedic. He was interested in everything from astronomy to drama and wrote with an authority that appealed to medieval Europeans after they rediscovered classical learning following the Dark Ages. Within the scholastic tradition that flourished in the first European universities, he was the authority of authorities, the source that had to be checked before pen was put to paper on virtually any topic. Aristotle was also favored by medieval divines because he invented the system of formal logic that defined their preferred form of scholarly discourse. Aristotle's logical rigor made an important contribution to scientific method.

Although he was deeply influenced by Plato, Aristotle was much more of a commonsense philosopher. One of his most important points of departure from Plato was his rejection of the Platonic theory of universals, which held that the world of our experiences is a pale shadow of the ideal world of eternal truths and perfect forms. Such a view is clearly incompatible with science because it argues that we can never approach truth through the information gleaned by our senses. Aristotle, by contrast, was an empiricist who believed that observation is an appropriate method for deriving knowledge.

In his book *Prior and Posterior Analytics,* Aristotle devised the conditions that needed to be met for a scientific statement to be considered valid. Although the appreciation of scientific truth relies on empirical information, or sensory experiences, scientific knowledge is not a matter of experience but a matter of reason. In other words, when we observe some set of phenomena, their governing principles can be grasped only by reason. Simply recognizing scientific generalizations did not amount to scientific understanding in Aristotle's eyes. It was also necessary to understand why the scientific truths held, that is, to grasp underlying causal mechanisms.

Aristotle was a pure scientist who despised the very idea of conducting scientific investigation to solve practical problems. For him, science was a branch of natural philosophy that was to be conducted by a leisured elite, of which he was a member. This view of science lost popularity with the rise of the Roman Empire. The Romans are justly celebrated for their feats of architecture and road building but contributed almost nothing to theoretical science. They were technologists rather than scientists.

Both of these interests, that is pure science and applied science, are combined in an interesting way in the life of ARCHIMEDES (298–212 B.C.), who was both a mathematician and an inventor. Archimedes was interested in gravitation and arrived at a celebrated generalization about the loss of weight of floating bodies that is known as Archimedes' Principle. This is an abstract principle but it was soon used by goldsmiths to determine whether an object supposedly made of gold had been adulterated. He is also credited with practical inventions although it is doubtful whether he invented Archimedes' screw, a device used to lift fluids by turning a crank. A native of Syracuse, in Sicily, Archimedes died trying to defend his native city from an assault by the Romans. Working as a military engineer, he devised clever defenses, including the use of a burning mirror that he had invented.

Following the reign of Alexander the Great, Greek science received generous state sponsorship at centers such as Alexandria, in Egypt, and Pergamum, in Ionia. During this Hellenistic period, pure science continued but there was a shift in focus toward practical problems, including military technology and medicine. During the Roman administration that followed, pure science fell into decline and technology flourished. It is probably no accident that some of the leading scientists of the Roman era, including physician GALEN of Pergamum (A.D. 130–200), were Greek.

SCIENCE IN MEDIEVAL EUROPE

With the fall of Rome in 476, the center of the civilized world shifted to Constantinople, which was the center of a Greek-speaking empire whose official religion was a mystical Christianity that was not conducive to scientific enquiry. Although the Byzantine Empire lasted for a millennium, before being conquered by the Ottoman Empire in 1453, and maintained at least some institutions of higher learning, it is not noted for important scientific or technical advances. More scientific civilizations developed in the East, in the Islamic world, India, and China. Chinese society was ahead of the rest of the world in technology but its advances did not influence Europe until well after the visit of Venetians Marco Polo and his father (1271–95). The initial reaction to Polo's book describing the technological and scientific advances of China was disbelief.

Chinese science had a distinctly practical focus. The Chinese were keen astronomers and developed highly accurate clocks that were greatly superior to medieval European ones. They had a well-developed mathematics that focused on arithmetic and algebra but no geometry. Other important fields of interest included physics and chemistry. Chinese medicine was more sophisticated than its European counterpart.

The Islamic empire spread throughout Arabia, Mesopotamia, and Egypt. Islamic civilization is particularly important to the rise of European science because it formed a repository of classical learning. Islamic scholars cultivated astronomy and medicine. Their objective in making astronomical observations was often to predict the future (astrology), which was officially discouraged on religious grounds. Arabic medicine was possibly the most sophisticated of its day. Elaborate research hospitals were developed and medical scholars such as al-Razi (Rhazes, 854–925) and Ibn Sina (AVICENNA, 980–1037) became the world's leading authorities on diseases and their treatments. Islamic physicians developed unprecedented expertise in ocular diseases and their understanding of the anatomy and physiology of the eye was complemented by excellent scientific work in optics such as that of Ibn Al Haytham (Alhazen, 965–1040).

The "Arabic" numerals that constitute such an important feature of modern mathematics are really derived from India. Hindu science can be traced to the fifth century B.C., when the ideas of Mesopotamian and Greek scholars became influential. Hindu astronomy was focused on calendrical matters and on making astrological predictions. Mathematics was highly developed in the branches of arithmetic and algebra and Indian scholars even developed a rudimentary geometry. Hindu mathematics and science were assimilated by Islamic scientists who transmitted these ideas to Europe.

At the end of the first millennium, Europe was a backwater, in terms of its cities, technology, economy, learning, and science. By each of these criteria, it paled in comparison to the civilizations of Byzantium, Islam, India, China, South America, and Mesoamerica. Rome itself had declined from 450,000 people at its peak to 35,000.

Between 600 and 1000, the European population as a whole rose approximately 38%, however. This increase created an ecological pressure for intensification of agricultural production. One critical innovation was the adoption of the heavy plow that allowed more land to be cultivated. First developed by the Romans but little used by them, the heavy plow was made of wood with iron cutting edges. Mounted on wheels, it cut beneath the sod and turned it completely over to leave a furrow.

Agricultural developments such as the heavy plow and improved crop rotation greatly increased food production thereby allowing for an accelerated population growth and an increase in the prominence of European cities. The entire population of Europe trebled between 600 and 1300 (rising from 26 to 79 millions). Between 1000 and 1300, the population of Paris increased from 20,000 to 228,000 indicating that much of the increased population was accumulating in medieval cities.

Increasing urban populations provided a favorable environment for learning and European universities flourished. The first (apart from a ninth century medical school at Salerno, Italy) was the University of Bologna, Italy, which was founded around 1088 and boasted the most distinguished law school in Europe. This was followed by the University of Paris (1200) and Oxford University (1220). The medieval universities were organized like craft guilds. Either a group of students got together to hire professors (as in the case of Bologna) or a guild of teachers joined together to take on fee-paying students (as happened in Paris). Largely independent of state control, European universities favored considerably more individuality than was the case in state-funded institutions of higher learning in previous civilizations although they were not as free-spirited as the Hellenic schools that produced Plato and Aristotle. Their primary responsibility was to train teachers, lawyers, clergy, and doctors. The undergraduate curriculum included natural philosophy, geometry, arithmetic, astronomy, and music. Advanced degrees could be obtained in theology, law, and medicine.

During the 12th century, many of the key works of philosophy and science from classical antiquity were translated from Arabic to Latin in Toledo, Spain, a city that had fallen to Christian invaders in 1085. During the 13th century, Christian scholars tried to assimilate this new learning into their own religious worldview. At this time, Aristotle, whose work was previously almost unknown, became the authority or "the Philosopher."

For theologian Thomas Aquinas (1224–74), the struggle to understand the Greek world was a struggle to reconcile the wisdom of Aristotle with his theological training. He was not alone in his obsession. Aristotle's logic seemed to most scholars of the period to be the preferred analytic tool for investigating every subject. Interpretation of Aristotle's work occupied most of the time of scholars at the medieval universities. This interest is referred to as the scholastic tradition. Scholastic philosophers were highly argumentative and not all of them came to the same conclusions about Aristotle. For example, the Philosopher's emphasis on scientific investigation as the most important method of obtaining knowledge was a minority view. It was shared by Robert Grosseteste (1168–1253) the first chancellor of Oxford University and by ROGER BACON (1215–92) an early English scientist/philosopher.

Roger Bacon was an exceptionally learned Franciscan monk who constantly got into trouble on suspicion of heresy and black magic. The suspicion that he practiced the dark arts was a contemporary interpretation of his scientific research that involved optics and chemistry (then known as alchemy). Bacon may not have been a serious empirical scientist but he was a man of vision and foresight who looked ahead to a future with horseless carriages, explosives, flying machines, and telescopes.

Bacon was fond of pointing to the errors that are produced by accepting knowledge secondhand from the claims of authorities. Even the normally tolerant Franciscans were irritated by Bacon's polemics and they temporarily banned him from committing his heretical views to paper. It is not altogether clear why Bacon was so objectionable to his fellow monks but it is unlikely that they could have appreciated his argument that natural science is more important than the study of religious texts. Bacon was ahead of his time by several centuries. A similar conflict of religious views and a scientific worldview can be seen in the attempts of the Roman Catholic Church to silence Galileo's (1564–1642) rejection of the geocentric worldview.

An important medieval European invention was the printing press created by JOHANNES GUTENBERG in the 1430s, evidently unaware that this had already been developed in China and had been widely used there since the ninth century. By 1500, some 13,000 different books had been printed. This launched an information revolution, which meant that knowledge was no longer the province of the learned few at universities but could be acquired by anyone who had the leisure time to read.

THE SCIENTIFIC REVOLUTION

Whereas the big event of medieval science was the rediscovery of the science of classical antiquity, the big event of the Scientific Revolution that began with Copernicus (1473–1543) was the shaking off of ancient authorities in favor of a science based on empirical observation. Thanks in part to the work of scholastic philosophers like Thomas Aquinas, the ideas of Aristotle and other pagan classics had become so thoroughly integrated into theology that any attack on them could be interpreted as an attack on religion. None of this is very logical but the fact remains that when Galileo Galilei was tried for heresy, his main offense against the Catholic Church was to defend the sun-centered worldview of Copernicus, a devout Christian, against that of the pagan Ptolemy (flourished 127–151). Copernicus was not an empirical scientist but worked with information derived from the observations of others. He did not see himself as a revolutionary and wanted to improve Greek astronomy rather than to replace it. Thus, he preserved many archaic elements such as the crystalline spheres posited by Aristotle to keep heavenly bodies in place. Copernicus realized that the Ptolemaic system lacked internal consistency. It could be greatly improved by placing the sun at the center of the planets. If the planets revolved around the sun, then we could explain their systematic changes in position and brightness. The fact that Mercury and Venus never appeared at the opposite side of the earth from the sun could be explained by hypothesizing that their orbits were closer to it than that of our own planet. Moreover, the distance of the planets from the sun could be ranked in terms of their speed of movement. What appealed most to Copernicus about his heliocentric system was its simplicity and esthetic charm. He reasoned that God was the supreme artist and that his handiwork would have to be beautiful.

Well aware that his ideas would not be well received, Copernicus delayed publication until he was on his deathbed. He knew that most of his contemporaries would have thought the concept of the earth whirling through space was absurd, if only because he could not explain why we did not fall off. Another troubling issue had to do with stellar parallax. If the earth moved around the sun, then the apparent position of the fixed stars should change relative to each other. They did not and the only possible explanation was that they were too far away for any parallax to be detected. This committed him to accepting an enormous universe that consisted largely of empty space.

Contemporary astronomers were aware of Copernicus's work but they were not persuaded by it. Tycho Brahe (1546–1601) a Danish astronomer who was the leading observational astronomer of his day, proposed a compromise between Ptolemy and Copernicus with the planets moving around the sun and the sun circling a stationary earth. This system dispensed with Aristotle's crystalline spheres. Brahe also questioned the Aristotelian view of the heavens as perfect and unchanging when he observed the appearance of both a comet and a new star in 1570.

Galileo's use of the telescope produced evidence that completely undermined classical astronomy. He found that the moon deviated from circular perfection by containing mountains and valleys. He saw satellites circling Jupiter, something that was completely inconsistent with the Christian-classical geocentric system in which heavenly bodies were supposed to move in concentric spheres having the earth as their center. Galileo also observed spots on the face of the sun which argued against heavenly perfection. Innumerable new stars emerged in the Milky Way indicating that heavenly bodies were far from constant.

These discoveries were rushed into print in a pamphlet named *The Starry Messenger* (1610) that Galileo dedicated to Cosimo II de Medici, the grand duke of Tuscany. His assiduous cultivation of the Medici paid off eventually with an appointment as chief mathematician and philosopher at the Medici court in Florence. During his time, royal courts sponsored scientists in much the same spirit as they supported artists. Royal patrons benefited from the reflected glory of their protégés. Many of the important discoveries of the new science were made at royal courts rather than universities.

In Galileo's case, moving from the University of Padua to the Florentine court was a definite promotion in terms of salary and status. It also projected him into the limelight of contemporary scientific controversies that were often begun as arguments at court. Galileo actively defended Copernicanism and launched a campaign to persuade the church authorities to adopt the Copernican worldview and give up the Ptolemaic-Aristotelian system. His efforts were a distinct failure. Not only was Copernicus banned as heretical reading, but Galileo was told, in 1616, that he could no longer hold, or defend, Copernican views. With the election of a new pope, Urban VIII, in 1623, Galileo hoped that the climate was sufficiently changed for him to revisit the Copernican debate. Having carefully obtained personal permission from the new pope, who requested an impartial presentation of arguments on both sides of the debate, Galileo labored on *Dialog of the Two Chief World Systems,* which was published in 1632 following review by the Vatican censors.

Instead of being the balanced presentation that the pope had expected, the *Dialog* turned out to be a powerful polemic in favor of Copernicus. Sales of the book were stopped. Galileo was tried on suspicion of heresy and forced to recant. Spending the remainder of his life under house arrest, Galileo devoted himself to less controversial but important research in mechanics.

In addition to his opposition to classical authority in astronomy, Galileo is often held out as an example of modern science because of his emphasis on the role of experiments in testing hypotheses. Galileo certainly carried out many observations and experiments but it is unlikely that he was genuinely interested in hypothesis-testing as such. Essentially a polemicist, he arrived at his conclusions through the force of logical arguments. The purpose of his experiments was not to test these conclusions but to illustrate them. This explains why some of the experiments he describes were never actually done. Contemporaries who attempted to repeat some of this work, such as an experiment in which metal balls were rolled along the groove in an inclined plank, were dismayed to find that the experiments did not work out as they were supposed to.

During Galileo's lifetime, science had become a very public activity in the sense that there was a growing expectation that the claims of one scientist should be verifiable from that of others. It was also a period during which the requirements of scientific methodology were first clearly and explicitly articulated. A leading theorist of science was English scholar Francis Bacon (1561–1626), who described the aims and methods of science. Bacon also expressed visionary ideas about the potential importance of science for improving the lot of human beings. He is often credited with coining the phrase "knowledge is power." He recognized that human beings could harness the forces of nature through scientific discovery and the creation of inventions. This general insight notwithstanding, Bacon missed the significance of some of the most important advances made by contemporaries. Thus, he remained unconvinced by Kepler's work on planetary motion. Astonishingly, he was unaware of an important discovery made by his personal physician, WILLIAM HARVEY (1578–1667), of the circulation of the blood.

Bacon formulated a clear conception of the scientific method that contains most of the important ingredients by which science is defined today even though his emphasis on data-gathering (or induction) over theory (or deduction) seems strangely biased today. These ideas were presented in one of his most important books, *The Advancement of Learning.* Bacon seriously underestimated the role of deductive thinking in science reflecting his personal antagonism to classical philosophers such as Aristotle. In particular, he underestimated the importance of coming up with a good hypothesis. A hypothesis, or guess, guides scientific observation by motivating the scientist to collect data that can be used to support, or reject, it.

One of Bacon's key contributions to scientific epistemology was the notion that matters of religion and faith were best kept separate from matters of reason and science. Like most others of his day, he was religious. A conservative and defender of the status quo, he had no desire to challenge orthodox religious views. According to the doctrine of double truth, a conflict between religion and science is not inevitable because each occupies a separate mental realm. Bacon disagreed with the habit of scholars of his day, who mixed up religious discourse and reason. Philosophy ought to be a separate realm that depended only on reason. Religion, on the other hand, was a matter of divine revelation. Religious truths could only be experienced and could not be arrived at through rationalization. Bacon went so far as to suggest that the triumph of faith is most commendable when it pertains to some phenomenon that to the eye of a coldly logical philosopher seems ridiculous. The double truth doctrine provided a formula that allowed for the development of a purely secular science without challenging the central role of religious authorities by insisting that there were necessary conflicts between scientific truth and the Bible and that these should be resolved in favor of science.

Bacon recognized that science is a social activity and understood the great importance of scientific communication in weeding out error and generating progress. In *The New Instrument,* he presented a model of the scientific enterprise that endorsed public communication of results. Scientific research should be open to public criticism and discussion that would help to dispel fallacious thinking and dogmatic claims. Science was also seen as helpful in solving practical problems and improving the quality of life.

Bacon was an advocate of scientific associations that would provide a forum for scientists to get together and discuss their data, hypotheses, and theories. These aspirations were realized in 1662 with the formation of the ROYAL SOCIETY OF LONDON that functioned very much as Bacon had envisaged. The earliest scientific society was the Accademia dei Lincei (Rome, 1603). The Royal Society has survived to this day. Its official journal, *Philosophical Transactions,* was the first scientific periodical in the world, and it is still published today. Four years after the founding of the Royal Society, the French Royal Academy of Sciences was formed in Paris. Within the next two hundred years, scientific organizations had been established throughout Europe and the United States. All of these societies were state supported.

Another contemporary of Galileo, Johannes Kepler (1571–1630) succeeded Tycho Brahe as mathematician to the Holy Roman emperor Rudolf II. Like Copernicus, Kepler was convinced that God had created a beautifully harmonious universe and he set out to discover this order with a religious zeal. He formulated three laws of planetary motion. The first stated that planets moved in elliptical orbits, contrary to the Aristotelian insistence on circular perfection. The second stated that as planet circle around the sun, they sweep out equal areas in equal time. The third stated that the square of the time taken for a complete orbit by a planet was proportional to the cube of its distance from the sun.

Interesting as Kepler's laws were from the perspective of establishing order in the movement of the heavenly bodies, they were not very helpful in explaining *why* the planets observed these rules. This problem still has not been entirely solved but Isaac Newton's (1642–1727) contributions were sufficiently persuasive to finally lay the Aristotelian-Ptolemaic world system to rest. His work is considered by many to be the crowning achievement of the Scientific Revolution and to have inaugurated a mechanistic view of natural phenomena that does not call for belief in religious or mystical causes.

Newton's most important contribution was to the theory of motion and gravitation. His three laws of motion stated: (1) that every body maintains its state of motion unless it is acted on by a force; (2) acceleration is proportional to the force causing it; and (3) for every action there is an equal and opposite reaction. Newton also formulated a law of universal gravitation according to which every body in the universe attracts every other body with a force that increases with the product of their masses (weights) and decreases with the square of the distance between them. Newton also developed calculus and performed an experiment with prisms demonstrating that white light consists of light of several different colors.

In 1687, Newton published *Philosophiae Naturalis Principia Mathematica* (or The mathematical principles of natural philosophy) a book that provided a synthesis of his mathematical work on gravitation and the orbits of planets. The *Principia,* as it is called, is considered by many to be the most important achievement of the Scientific Revolution. It is interesting that Newton, like each of the other major figures in the Copernican Revolution, was a deeply religious man who interpreted the order observable in natural phenomena as evidence of a divine presence. Newton did not know why there was universal gravitation, any more than modern physicists understand this, and he was happy to suggest that gravitation can be seen as the direct action of God in keeping the universe working. Newton's science nevertheless stands alone without the necessity for religious support. The Newtonian worldview has often been compared to a clock that was set in motion by the divine clockmaker. The point of this metaphor is that once the clock is started, it goes by itself. Subsequent generations knew nothing about Newton's NATURAL THEOLOGY and saw him as a champion of the ENLIGHTENMENT struggle of reason against superstition.

Newton's quarrel was not with religion but with the uncritical acceptance of the views of scientific authorities like Aristotle. Some of his successors, notably French astronomer and mathematician Pierre Simon de Laplace (1749–1837), did not see God's presence in Newtonian physics, as Newton had. Laplace, along with Swiss mathematician Leonard Euler (1707–83), and others, provided convincing mathematical evidence for universal gravitation.

Galileo's devastating criticisms of Aristotelian astronomy and mechanics have a parallel in the field of medicine. ANDREAS VESALIUS (1514–64), a Belgian anatomist, set out to correct the many errors in the anatomy of GALEN (130–200) the reigning classical authority in medical matters. Vesalius learned about the structure of the human body from his experiences in dealing with wounded soldiers as a military surgeon. His beautifully illustrated book, *De Humani Corporis Fabrica* (1543) was a major advance over the inaccurate texts previously used by physicians. William Harvey's work on the circulation of the blood provided another blow to the credibility of classical authority in medicine, which did not conceive of blood moving into bodily tissues and being collected, enriched, and returned to them. Harvey could not see how blood was absorbed in, or removed from, the tissues but speculated that there must be very tiny vessels performing this function. With the invention of the microscope, these capillaries were seen by Italian scientist Marcello Malpighi (1628–94). Other important pioneers in the study of the very small included Dutchman Anton van Leeuwenhoek (1632–1723) and Englishman ROBERT HOOKE (1635–1703). Just as the telescope had opened up a whole new world for scientific observation, so the microscope opened up new realms. The world of the very small provided such a wealth of new biological information that researchers were initially overwhelmed and had difficulty interpreting what they saw.

MODERN SCIENCE

Following the development of modern empirical science, there was a great explosion of knowledge in all fields that makes it difficult to even identify the most important scientists and discoveries. There has been a logarithmic growth in the number of scientific journals from one in 1665 to approximately one million today. Moreover, due to the steady proliferation of science, it is estimated that at least four-fifths of all the scientists who ever lived are alive today. Clearly, there is a great deal more scientific information available at present than could be absorbed by a single reader, or by a thousand.

The scientific revolution began as a critique of classical science but has had the effect of producing many new areas of scientific investigation, such as electricity magnetism, chemistry, atomic physics, quantum mechanics, microbiology, immunology, and molecular genetics. Science has become increasingly professional. There are few wealthy amateurs conducting scientific research today. The explosion of knowledge has meant that scientists must confine themselves not only to one narrow discipline but also to a narrow area of specialization within that discipline, if they are to improve their chances of making a meaningful contribution. One consequence of specialization has been the establishment of conventions and jargon within each subdiscipline that makes them impenetrable to others. Not only do amateurs rarely contribute to science today, but educated people would have considerable difficulty even in reading most specialist journals.

Despite progressive specialization, modern science has continued to produce eminent figures, such as CHARLES DARWIN (1809–82) and ALBERT EINSTEIN (1879–1955), whose work contains radical new insights that forge connections among previously isolated disciplines. In Darwin's case, the theory of evolution by natural selection provided a unifying concept that integrated all areas of biology and wedded them to geology. Even more provocatively, his work placed human beings squarely in the natural world for the first time and implied that evolutionary notions could be used to understand human society, including politics, economics, and morality. This crude SOCIAL DARWINISM was to be rejected by social scientists in the 20th century before more sophisticated efforts at evolutionary explanation of social behavior gained traction in the sociobiology and evolutionary psychology movements of the last quarter of the century (SEE EVOLUTIONARY PSYCHOLOGY AND SEX DIFFERENCES; SOCIAL DARWINISM; SOCIOBIOLOGY DEBATE).

Early biology was very much tied up with natural theology, although this was to change with the onset of the evolution debates. Thus, when the great taxonomist Carolus Linnaeus (1707–78) systematized the world's plants and animals, he did so in the spirit of celebrating God's handiwork. He was motivated by a religious impulse. Ironically, careful observation of the natural world, as preserved in fossils, raised serious doubts about the biblical account of the earth as being only a few thousand years old. For one thing, fossils were found in rocks that seemed to be a great deal older than this. Moreover, the fossil record was replete with unfamiliar creatures, including what were subsequently referred to as

dinosaurs. If time is though of as analogous to a stage, its cast of characters seemed inordinately large for such a brief period of time in geological terms.

Scientists who wanted to preserve the biblical account of the age of the earth devised the theory of CATASTROPHISM. Catastrophism is the view that geological features of the earth, including physical features like mountains and the fossil record of life forms, were affected by brief violent events, or cataclysms. Catastrophists were disposed to see cataclysms as having a supernatural origin. Since they accepted the biblical view that the earth was only a few thousand years old, it was difficult to understand how physical features, such as mountain peaks and canyons, could have been formed by gradual natural forces and their existence was thus rationalized as being due to cataclysms.

A leading light of catastrophism was French comparative anatomist Georges Cuvier (1769–1832). Cuvier studied fossils in the Paris basin. He produced two striking findings. The first was that animals had once lived that bore no relationship to anything currently alive. The second was that there are distinct breaks, or nonconformities, in rock layers. Each layer was associated with a marked difference in fossil content. Cuvier drew the conclusion that the life forms in each layer had been wiped out by a sudden cataclysmic event. Following their destruction, they had been replaced either by a new creation or by the inward migration of species from unaffected regions.

Catastrophism only seemed reasonable if the age of the earth was equivalent to biblical indications. Once geologists grasped the immense age of the earth's surface and the fossil life it contained, they realized that a gradualist explanation of the fossil record was possible. However weak the forces of natural change might be, given sufficient time, they could explain what seemed like the sudden transitions between geological strata. By the end of the 19th century the gradualist view, referred to as "uniformitarianism," was accepted by virtually all scientists and the catastrophist interpretation of the fossil record was abandoned.

Catastrophism nevertheless presented some serious opposition for evolutionists such as CHARLES LYELL and Charles Darwin. One issue that was particularly troubling for the evolutionists was the argument in favor of a progressive series in the fossil record, for example, the sequence fish, reptiles, mammals suggests that the new life following cataclysmic extinctions ascended from simpler to more complex forms. This evidence was so bothersome to Charles Lyell that he even denied the existence of any such progressive pattern. Such progressiveness was apparently inconsistent with Darwin's theory of evolution by random variation and natural selection. According to Darwin's theory (independently formulated by Alfred

Russell Wallace) individuals vary in traits that favor survival and reproduction. Those with the most favorable traits for a particular ecological niche leave more offspring who share their desirable traits. Those with undesirable traits do not succeed at reproducing and their traits die out, having been removed by natural selection. The end result is that surviving species are exquisitely designed for the way that they make their living. In the course of evolutionary history, a small number of relatively simple species has given rise to a large number of more specialized organisms in a process known as adaptive radiation. The evolutionary tree of speciation may look progressive but they are not since evolutionary change can also go from complex to simple. For example, naked mole rats have lost their complex vision through adaptation to life spent underground.

When Darwin wrote *On the Origin of Species* there were troubling gaps in the fossil record. Soon after its publication in 1859 some important fossil discoveries were made that supported uniformitarianism and cast doubt on catastrophism. One was the discovery of *Archaeopteryx*, a flying reptile that had feathers like a bird and thus supported the argument that modern birds evolved from reptiles. American paleontologists also succeeded in constructing an evolutionary series in the early evolution of horses that showed how modern grazing horses had evolved from leaf-browsing ancestors over tens of millions of years.

Perhaps the biggest hole in Darwin's theory, insofar as missing evidence was concerned, was his complete lack of information about the hereditary mechanism through which offspring came to resemble parents. This problem was partially solved by the work of GREGOR MENDEL (1822–84) on the inheritance of seed characteristics of peas. Mendel's work showed that such traits as whether a pea is wrinkled or round, large or small, were represented by hereditary factors derived from each parent plant. He discovered that some factors were dominant over others. When this was taken into account, it was possible to derive a probabilistic prediction of the traits of offspring from the traits of parents. Published in 1865, Mendel's work was completely ignored until the dawn of the 20th century when it was independently rediscovered by three different researchers. In the 1880s, chromosomes were discovered and August Weismann speculated that they contained hereditary information.

During the late 1930s and early 1940s, population geneticists like Theodosius Dobshanski, Julian Huxley, and Ernst Mayr worked out a complete synthesis between genetics and Darwinian evolution. In 1953, James Watson and Francis Crick deciphered the molecular structure of the DNA molecule and showed how its complementary double strands contained the same sequence of biochemical information that could be copied indefinitely by

"unzipping" the strands and matching up complementary base pairs. This was the seminal discovery that launched the modern science of molecular genetics with its many exciting applications in the field of GENETIC ENGINEERING.

Advances in the biological sciences have also been reflected in the staggering achievements of modern physics and chemistry. Modern sciences are much more wedded to technology than was true in the past. This trend began with the discovery of the telescope at the beginning of the 17th century and was intensified throughout the 20th century. Modern physicists, chemists, and biochemists could not imagine conducting research without an astonishing array of complex and expensive machines from the electron microscope to the SUPER CONDUCTING SUPER COLLIDER used by particle physicists. This relationship is a two-way street, however, because modern technology is also dependent on science in a way that was not true in the past. Many modern technologies that affect ordinary life, including X-ray machines, telephony, radio, television, computers, fiber optic cables, lasers, and the Internet, have arisen fairly directly out of scientific research. Many of these inventions rely on increased knowledge of the electromagnetic spectrum due to research in physics. This spectrum includes visible light, infrared, X-rays, microwaves, television waves, and alternating current, among others. These radiations all travel at the speed of light (186,000 miles per second) and therefore allow all parts of the planet to be instantaneously connected to all other parts.

Modern science calls for considerable specialization on the part of most of its practitioners but many sciences have at the same time established important interdisciplinary connections. Metaphorically speaking it is as though scientists in each discipline scrutinized the world through their own narrow slit in the round tower of science. They are all studying a different part of the same panorama and their contributions can be laid end-to-end to provide a seamless picture of the same objective reality. That is why integrative geniuses like Albert Einstein can flourish in an age of specialization.

If the sciences are continuous with each other, then many disciplinary boundaries can often seem arbitrary. This point is illustrated by the discovery of current electricity. During the 18th century scientists investigated only static industry, a phenomenon that is demonstrated by rubbing a glass rod with a cloth and showing that it attracts pieces of paper because of its positive charge. Italian physician Luigi Galvani (1737–98) discovered that a frog's leg would continue to contract after the spinal cord was connected by a brass hook to an iron railing. He concluded that animal nerves and muscle tissue contained an electrical fluid. Italian physicist Alessandro Volta (1745–1827) was stimulated by the 1791 publication of Galvani's findings and his research convinced him that the electricity had not come from the animal tissue but from forming a fluid connection between the metals. In 1800, Volta proved that animal tissue was not necessary to produce flowing electricity by constructing the first battery that consisted of layers of metal and cardboard immersed in a salt solution. Later versions used acid as the fluid. Volta's work illustrated that electricity and chemistry had unsuspected connections and that this knowledge arose naturally from research in biology.

The tradition of modern experimental chemistry begins with Irish aristocrat ROBERT BOYLE (1627–91) who was interested in the effects of temperature and volume on the pressure of gases. French scientist Antoine Lavoisier (1743–94) is given credit for dispelling the myth that there is a principle of fire, or phlogiston, responsible for burning. Burning was possible only in the presence of air. JOHN DALTON (1766–1844) made a considerable advance in chemical theory by proposing that chemical elements are made out of indivisible particles, or atoms, and that this explains why elements combine in fixed quantities. Italian chemist and physicist Amadeo Avogadro (1776–1856) theorized that atoms are joined together into molecules in chemical combination, a view that was not widely accepted until after his death.

Just as electricity had been seen to have intimate connections with chemistry, Danish scientist Hans Christian Oersted (1777–1851) demonstrated that it was also connected to magnetism. A current flowing in an electrical circuit deflected the magnetic needle in a compass. A thorough unification of electricity and magnetism was accomplished by MICHAEL FARADAY (1791–1867). Faraday established that the movement of a magnet through a coil of wire induced a current in it, thereby unifying not just electricity and magnetism but also mechanics. This discovery formed the basis for the development of both the electric generator and the electric motor and ushered in the golden age of electricity as represented in the inventions of THOMAS EDISON (1847–1931) and his contemporaries. Faraday explained these phenomena in terms of electromagnetic fields that were composed of lines of force that could be demonstrated by the arrangement of iron filings in a magnetic field. Of the three phenomena of field, motion, and electric current, the presence of any two was sufficient to induce the third. For a generator, motion and field produced current. For a motor, field and current produced motion.

Early in the 19th century, important advances were being made in optics. THOMAS YOUNG (1773–1829) and others experienced growing dissatisfaction with the theory of light as a stream of particles (the corpuscular theory), which dated back to Newton. This theory did not provide a satisfactory explanation for the diffraction, or bending, of light rays as the light passed corners. Young proposed that light was a wave similar to sound (i.e., with vibration

in the direction of propagation). French scientist Augustin Fresnel (1788–1827) proposed that light was a transverse wave (with vibration perpendicular to the direction of movement, like a water wave). Fresnel turned out to be correct but both men came up with experimental demonstrations of interference effects that could be explained only in terms of the interactions of waves.

The 19th century also saw the emergence of the new discipline of thermodynamics, which can be defined as unification of the study of heat with mechanics. The first law of thermodynamics stated that energy can neither be created nor destroyed but that it is transferred from one form to another. When a person slaps their hands together to make them warm, for example, mechanical energy is being converted to heat. In the steam engine, heat was converted to mechanical energy. The second law of thermodynamics stated that energy moves from areas of high temperature to areas of low temperature so that, in a closed system, energy will eventually become evenly distributed. Key pioneers in thermodynamics include English physicist James Prescott Joule (1818–89), German physicist Rudolph Clausius (1822–88), and German scientist HERMANN HELMHOLZ (1821–94).

THE AGE OF ELECTROMAGNETIC SCIENCE
Toward the end of the 19th century a number of new phenomena were discovered that expanded the range of the known electromagnetic spectrum and challenged the capacity of contemporary theory to accommodate them. This period of uncertainty in the field gave rise to Einstein's celebrated theoretical innovations that changed forever the way that scientists and others think about the physical universe.

In 1895, WILHELM ROENTGEN (1845–1923) discovered X-rays, a previously unsuspected form of electromagnetic radiation. In 1896, French physicist Antoine-Henri Becquerel (1852–1908) discovered radioactivity that produced clouding of photographic plates stored near uranium. Subsequent investigation showed that three types of radiation were involved, light particles (electrons), heavy particles (alpha radiation), and yet another type of electromagnetic radiation, namely, gamma rays. In 1897, British physicist J. J. Thompson showed that cathode rays, which are produced when electricity travels through an evacuated tube, consist of particles that are a great deal lighter than atoms. These were electrons and, by demonstrating that the atom is not the smallest unit of matter, the discovery ushered in a new age of the physics of subatomic particles, of which some 200 were ultimately discovered. Most of these are short-lived and are produced in the artificial conditions of large particle accelerators such as that at the Fermi laboratory in Batavia, Illinois.

The photoelectric effect, in which light can create an electric charge on a metal plate, had been discovered in

1887 by German physicist Heinrich Hertz (1857–94). This could be explained by imagining that electrons on the plate receive energy from the light that allows them to escape from their resident atoms. In 1901, German physicist Max Planck (1858–1947) formulated a quantum theory that described this interchange in terms of the transfer of small but fixed quantities of energy, or quanta. This was a departure from earlier conceptions in physics according to which energy was seen as a graded continuum. This quantum theory helped to explain why the photoelectric effect only occurred for some wavelengths of light. If the light was not energetic enough, it did not provide electrons with the minimum quantum that they needed to escape.

The central feature of modern physics compared with the Newtonian system is its lack of certainty. In 1905, Albert Einstein proposed that nothing can travel faster than the speed of light. His special relativity, dealing with the physics of bodies moving uniformly with respect to each other, argued that the measurement of time, weight, and length all vary with the frame of reference. There is no "correct" value of these parameters as Newton would have assumed. Thus for an imaginary observer hurtling through space at the speed of light, time will stop! Einstein also proposed that energy is equivalent to mass and that one can be calculated from the other using the square of the speed of light (c) as a constant. Hence the celebrated equation, $E = mc^2$.

Einstein's general theory of relativity, published in 1915, dealt with acceleration rather than constant motions. In this theory, space and time are treated as a single dimension, space-time. Just how different Einstein's ideas were from Newton can be gathered from his treatment of gravity. According to the general theory, heavy objects make a large warp in the space around them. According to this perspective, gravity is not a force. Instead, satellites orbit a planet because they are following the shortest possible path through curved space. Einstein was propelled into worldwide fame in 1919 when a complete solar eclipse provided an opportunity to verify general relativity in terms of its prediction that the mass of the sun would bend starlight.

The uncertainty of the cosmos was further elaborated by the work of EDWIN HUBBLE (1889–1953) who showed that the number of galaxies was much larger than previously expected, a leap in our understanding of what is out there that was just as revolutionary as Galileo's discoveries using the early telescope. Hubble's work also showed that the galaxies are moving away from each other in a pattern that is consistent with the big bang theory of the origin of the universe.

Following the development of modern conceptions of the structure of the atom in which British physicist Ernest Rutherford (1871–1937) played an important role,

physicists developed the theory of quantum mechanics. This can be thought of as a revolution against Newtonian mechanics below the level of the atom. A central theme of quantum mechanics is uncertainty. Not only do material objects like particles become a great deal "fuzzier," having a wave-particle duality, for example, but the objectivity of scientific observation cannot be guaranteed. According to German mathematical physicist Erwin Schroedinger, the mere act of observing a particle necessarily interferes with it in some way. We cannot determine the position and the speed of an object at the same time with equal accuracy. Indeterminacy is built into matter and the best that scientists can hope to do is to make predictions with a high probability of being correct. Leading figures in quantum mechanics included Danish physicist Neils Bohr (1885–1962) and Austrian physicist Erwin Schrodinger (1887–1961).

English theoretical physicist STEPHEN HAWKING (1942–) has united the very small and strange world of quantum mechanics with the very large and strange world of modern cosmology by proposing that there are infinitely dense bodies, or black holes, that are no bigger than a subatomic particle. According to Hawking, multiple universes are linked to each other by "worm holes," that is, tiny quantum fluctuations in space.

Quantum mechanics has been used by some postmodern intellectuals to claim that all science is lacking in objectivity and that, instead of describing the world that is out there, scientists merely construct a subjective "narrative." The limitations of this argument have recently been attacked by ALAN SOKAL, who used a parody to expose social critics who attack science without fully understanding it.

Science has experienced an explosive growth over the past few centuries. Since 1666, when the first scientific journal was published, the scientific publishing industry has expanded to approximately one million scientific journals in 2000. It is estimated that the volume of scientific information doubles every 15 years. With the recent revolution in electronic transmission of information, particularly over the Internet, it is unlikely that this pace of growth will slow down in the near future. One of the most important recent developments is the instantaneous electronic sharing of huge volumes of data. Striking examples of this include the sequencing of DNA for entire organisms and the vast quantity of detailed images being transmitted from the HUBBLE SPACE TELESCOPE. This would be expected to speed up the pace of scientific progress because scientists no longer have to wait until data are published in scientific journals, which can take several years, before they can use it to venture in new directions.

It seems obvious that the pace of scientific development must eventually moderate, if only because of the large numbers of professional scientists. Thus, the number of scientists at work today is greater than the number of scientists who have lived throughout history. The contemporary rapid development of science is attributable to the growth of modern universities, and also to the establishment of joint ventures between universities and commercial corporations. A good example of this type of cooperation is seen in biotechnology, particularly in the development of new drugs. Scientific innovation is being sponsored because it has the potential for increasing corporate profits. Presumably, conditions might arise in which the commercial benefits of scientific innovation decline from their current level and corporate sponsorship of science would decline. It is also possible that the wealth of endowed universities might decline, undermining their ability to sponsor scientific research. In recent history, however, the trends have been in the other direction. There has also been an interesting correlation between national wealth and national sponsorship of science suggesting that the scientific enterprise may contribute to national prosperity by providing a competitive advantage in new industries such as biotechnology and information technology.

See also SCIENCE-TECHNOLOGY DISTINCTION; SCIENTIFIC METHOD; SIMULTANEOUS DISCOVERY; TECHNOLOGY, HISTORY OF.

Further Reading

Browne, E. J. *Charles Darwin: A Biography*. Princeton, N.J.: Princeton University Press, 1996.

Cassidy, D. C. *Einstein and Our World*. Atlantic Highlands, N.J.: Humanities Press, 1995.

Cohen, H. F. *The Scientific Revolution: A Historiographical Inquiry*. Chicago: University of Chicago Press, 1994.

Cozzens, S. E. *Social Control and Multiple Discovery in Science*. Albany: State University of New York Press, 1989.

Degler, C. N. *In Search of Human Nature: The Decline and Revival of Darwinism in the American Social Thought*. New York: Oxford University Press, 1991.

Dorn, H. *The Geography of Science*. Baltimore, Md.: Johns Hopkins University Press, 1991.

Drake, S. *Galileo at Work: His Scientific Biography*. New York: Dover, 1995.

Gamow, G. *Thirty Years That Shook Physics: The Story of Quantum Theory*. New York: Dover, 1985.

Hill, D. R *Islamic Science and Engineering*. Chicago: Kazi Publications, 1996.

Huff, T. E. *The Rise of Early Modern Science: Islam, China, and the West*. New York: Cambridge University Press, 1993.

Huggett, R. *Catastrophism*. New York: Verso, 1997.

Jasanoff, S., G. E. Markle, J. C. Petersen, and T. Pinch, eds. *Handbook of Science and Technology Studies*. Thousand Oaks, Calif.: Sage, 1995.

Lamb, D., and S. M. Easton. *Multiple Discovery*. Trowbridge, England: Avebury, 1984.

Lindley, D. *The End of Physics: The Myth of a Unified Theory*. New York: Basic Books, 1993.

McClellan, J. E., and H. Dorn. *Science and Technology in World History*. Baltimore, Md.: Johns Hopkins University Press, 1999.

Olby, R. C., G. N. Cantor, J. R. R. Christie, and A. M. S. Hodge, eds. *Companion to the History of Science*. London: Routledge, 1990.

Ruse, M. *Darwinism Defended*. Reading, Mass.: Addison Wesley, 1984.

Sarton, G. *Ancient Science through the Golden Age of Greece*. New York: Dover, 1993.

Selin, H. *Encyclopedia of the History of Science, Technology, and Medicine in Non-western Cultures*. Dordrecht, Netherlands: Kluwer Academic, 1997.

Shapin, S. *The Scientific Revolution*. Chicago: University of Chicago Press, 1996.

Stephenson, B. *Kepler's Physical Astronomy*. Princeton, N.J.: Princeton University Press, 1994.

———. *The Music of the Spheres: Kepler's Harmonic Astronomy*. Princeton, N.J.: Princeton University Press, 1994.

Westfall, R. S. *The Life of Isaac Newton*. Cambridge: Cambridge University Press, 1993.

Williams, T. *The History of Invention*. New York: Facts On File, 1987.

Science for the People group

Science for the People (SP) was a national left-wing political pressure group, organized in the 1970s, that engaged in academic activism. It served primarily to focus in uniting opposition to the sociobiology of EDWARD O. WILSON and others. The most active chapters were at Boston close to Wilson at Harvard University and at Ann Arbor close to the University of Michigan where Richard Alexander, another sociobiologist, worked. The Boston group was allied with the Sociobiology Study Group that consisted primarily of Harvard faculty and students and spearheaded opposition to Wilson's book, *Sociobiology: The New Synthesis* (1975). SP published the book *Biology As a Social Weapon* (1977). Its meetings were devoted to organizing publicity campaigns against sociobiology and other expressions of biological determinism. The group also discussed third world problems, environmental issues, biological determinism and criminal behavior, GENETIC ENGINEERING, and the genetic basis of intelligence.

See also ACADEMIC VIGILANTISM; SOCIOBIOLOGY DEBATE, THE; WILSON, EDWARD O.

Further Reading

Segerstrale, U. *Defenders of the Truth: The Battle for Science in the Sociobiology Debate and Beyond*. Oxford: Oxford University Press, 2000.

science/technology careers and women   *See* WOMEN IN SCIENCE/TECHNOLOGY CAREERS.

science-technology distinction

Science can be distinguished from technology because its most important goal is arguably to increase knowledge rather than to solve practical problems. At the very least, science includes an epistemic (knowledge-acquiring) goal that is not a primary objective of technology. We often think of science and technology as being inextricably entwined because this has been true of modern industrial society. Yet, technology had been in existence for millions of years before the emergence of science. Technology refers to techniques of solving the problems of obtaining food, securing shelter, and providing for human comfort. Thus Paleolithic (Stone Age) technology included stone tools, spears, bows and arrows, the use of fire, pigments, and sewing needles.

The science-technology distinction is not a hard-edged one. Many scientists spend much of their working lives attempting to solve practical problems. If they are successful, they contribute to technology. Since the Industrial Revolution, scientists have become increasingly important in industry and scientific innovation has driven technological innovation.

The complexity of the modern science-technology relationship can be illustrated by the life story of THOMAS EDISON (1827–1931) who spent most of his working life developing inventions to exploit electricity as a means of solving practical problems, such as communicating more effectively over long distances (via telegram and telephone). Edison wanted to improve technology and had no intention of contributing to scientific knowledge. Nevertheless, he made important scientific discoveries, and he established the first modern commercial laboratory, which hired individuals who had trained as pure scientists in an academic setting.

See also EDISON, THOMAS ALVA; SCIENCE, HISTORY OF; TECHNOLOGY, HISTORY OF.

scientific dishonesty

Scientists may engage in as many different kinds of dishonesty as people in other academic professions but the forms of dishonesty that are distinctively scientific revolve around mismanagement of data (see chart on next page). One of the first to catalog scientific misdeeds was English mathematician Charles Babbage in his book *Reflections on the Decline of Science in England* (1830). Babbage was the originator of the terms data trimming, data fudging, data falsification, and data cooking, although contemporary usage differs from his definitions.

In its contemporary usage, data trimming refers to the elimination from the published results of data that do not fit in with the author's hypothesis. A good example of data trimming is ROBERT MILLIKAN's reporting of only those

trials in his oil drop experiment that supported his theory of the charge on an electron as the irreducible unit of electric charge. Interestingly, in his earlier work, Millikan honestly reported all of his data but used a rating of the quality of the results to give more weight to some trials than others. This earlier procedure can be faulted as indicating flawed methodology but it was not dishonest. In his later work, however, Millikan not only suppressed results he did not like but also falsely claimed that he had reported all of his data, a claim that has not survived analysis of his laboratory notebooks.

Fudging occurs when scientists pretend that the data are closer to predictions than they actually are. Isaac Newton engaged in data fudging when he systematically manipulated the numbers in later editions of his masterpiece the *Principia* (originally published in 1687). Newton encountered a certain amount of resistance to his ideas in Europe spearheaded by German philosopher and mathematician Gottfried Leibniz, who rejected Newton's theory of universal gravitation. To increase the persuasive power of his theory, Newton modified some of his numbers and redid the calculations to arrive at the fraudulent claim that his theory predicted the data with an accuracy of better than one part in a thousand. The new numbers were so persuasive that they triumphed over the opposition and Newton's contemporaries were astonishingly oblivious to the dishonesty that he perpetrated right under their noses.

Data cooking involves setting up a biased test in which a favored hypothesis is very unlikely to be falsified by the data. In cooking the data, a scientist deliberately avoids situations in which a hypothesis can potentially be shown to be false. Thus, some "STAR WARS" scientists are accused of cooking the data in tests of infrared sensing systems used on antimissile missiles. If these systems were to be of any practical use, they would have to be able to distinguish between a nuclear warhead and a decoy. The scientists switched from using eight decoys in a more realistic test to using just one, a silver balloon, that was so clearly different from the warhead in its infrared signature that it could be easily spotted. A sympathetic observer might argue that making the test easier was a preliminary step to developing more sensitive sensor systems. This might be true but if the purpose of the test was to determine whether antimissile missiles could detect incoming warheads, the test was of no scientific value because the data had been cooked.

In data falsification, the scientist pretends to have carried out an experiment or collected observations but in reality has done no scientific work and plucked the date out of thin air. Data falsification is the most serious of scientific sins that strikes most deeply at the claim that science answers questions using empirical observations. It has been remarkably common among prominent men of

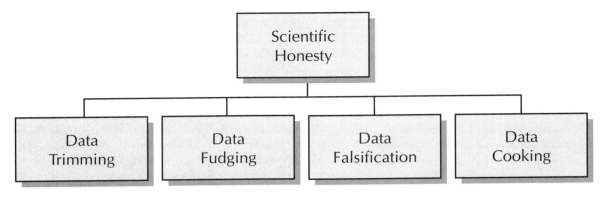

Different types of scientific dishonesty involve mismanagement of data.

science from PTOLEMY to GALILEO to SIR CYRIL BURT. Ptolemy's observations seem to have been based largely on the work of other astronomers and it is possible that he never did any empirical work himself. Galileo is unlikely to have carried out his experiment on the speed of rolling balls in a wooden groove even though he reported results (that turned out to be unrepeatable). Cyril Burt, called the father of British education, not only invented his data in an influential twin study concerned with the genetics of intelligence but also invented a pair of female assistants who allegedly collected the data.

Scientific misconduct has been surprisingly common in the biomedical field (accounting for 81%, or 21 of 26 publicly reported cases of scientific misconduct in the English-speaking world between 1950 and 1979 according to an AMERICAN ASSOCIATION FOR THE ADVANCEMENT OF SCIENCE report), which probably reflects several factors, including the large number of researchers active in this field; the unusual severity of the PUBLISH-OR-PERISH pressures in this area; and, possibly, the increased likelihood of detection. Just as the most distinguished men of science in history seem to have been vulnerable to committing acts of scientific dishonesty, so there appears to be a trend for misconduct to cluster at the most distinguished research institutions. Thus Harvard University had more cases than any other institution—three. Cornell University and Columbia University tied for second place with two cases each. Given that there are too few cases to draw any reliable conclusions, these observations point to ambition as a prime motive for scientific dishonesty. More than two-thirds of the cases of alleged misconduct (69%) involved Americans, which is also interesting given that the United States leads the world in many fields of science.

Another noteworthy phenomenon has been the growth over time of publicly reported cases of scientific misconduct. In the 20 years after 1970, there were approximately 10 times as many reported cases as in the preceding 20 years. This does not necessarily mean that scientists are becoming more unscrupulous. Research institutions may have faced up to the problem. Instead of ignoring dishonesty at their own institutions, researchers may have begun to report it and institutions may have begun to act on these reports. Another influence on the number of reported cases has been the increase in the number of working scientists. At the same time, many of these scientists have been exposed to extreme institutional pressures to publish, which could have pushed some over the edge into scientific dishonesty.

Some of the fraud in the biomedical area has been astonishingly crude. When PAUL KAMMERER's toads were inspected in 1926, their nuptial pads, which supposedly proved the inheritance of an acquired trait (or Lamarckian change), turned out to have been painted on with India ink. When WILLIAM SUMMERLIN wanted to convince his boss at the Sloan Kettering Institute in New York in 1974 that he had mastered the problem of immunological rejection of transplanted organs, he claimed to have been successful in grafting skin from black mice onto white ones. This case was made by drawing black patches on white mice using a felt-tipped pen. The sordid details of this and many other instances of fraud in the biomedical field are described by William Broad and Nicholas Wade in *Betrayers of the Truth* (1982).

In addition to willful dishonesty in acquiring the raw data that is reported in a scientific publication, scientists are often accused of using complex statistical procedures, such as multivariate analysis, to make their results seem more important than they actually are. This criticism may often have more than a grain of truth but it is also a mistake to object to powerful statistical procedures just because these reveal patterns that would not otherwise emerge. Statistics is an important tool that allows scientists to visualize small effects in their data, just as a microscope allows small physical objects to be observed. It is a mistake to imagine that the work of science is done after the data have been collected. The judgment as to whether statistics are used to obscure, or misrepresent, the data must be made by peer reviewers, editors, and readers and the onus is on specific research communities to establish statistical practices that they are comfortable with (see STATISTICS AND SCIENTIFIC HONESTY).

Just as scientists may have a bias toward detecting effects in their data, they are also likely to exaggerate the potential implications of their findings, particularly if they are applying for grants to pursue a particular line of research. Such failures of objectivity are probably unavoidable and may not always be intentionally dishonest. Reviewers and readers must accept that egoistic biases are likely to creep into such speculative claims and must view them with a healthy dose of skepticism. In some disciplines, this recognition is formalized by having a "discussion" section in a scientific report in which the author is given more liberty to speculate, based on the findings, than is possible in the rest of the report.

Apart from misrepresentation of the data, scientists may be guilty of forms of dishonesty that are common to other academics such as PLAGIARISM, or deliberate misrepresentation of the work of another as one's own, and failing to give due credit to another author, or person, who has contributed to one's work. Scientists may also be guilty of falsely accusing others of stealing their own work as in Newton's malicious claim that his rival Leibniz had stolen calculus from him. Historians of science have concluded that Leibniz actually invented calculus independently even though Newton had priority of this important mathematical discovery.

See also BALTIMORE AFFAIR, THE; MILLIKAN, ROBERT; NEWTON, ISAAC; PLAGIARISM; PTOLEMY; "STAR WARS" STRATEGIC DEFENSE INITIATIVE.

Further Reading
American Association for the Advancement of Science. *Project on Scientific Fraud and Misconduct.* Washington, D.C.: Author, 1987.

Babbage, C. *Reflections on the Decline of Science in England.* New York: Augustus M. Kelley, 1970.

Broad, W. and N. Wade. *Betrayers of the Truth.* New York: Simon and Schuster, 1982.

Gardner, M. *Science Good, Bad and Bogus.* Buffalo, N.Y.: Prometheus, 1981.

Resnik, D. B. *The Ethics of Science.* New York: Routledge, 1998.

scientific method

Scientific method is briefly defined as systematic empiricism. Empiricism means that scientific information, or data, is acquired directly from experience rather than from tradition or authoritative sources. Of course we all experience the world directly through the senses but not everyone is a scientist. Science is distinguished from everyday observations by being systematic. This means that the situation in which information is collected is carefully designed and that data are collected as accurately as possible using valid measurement procedures. Scientific data are so carefully collected that it is possible for another researcher to repeat the procedure exactly in order to verify the results. This criterion is known as public verification.

Public verification is critical to science because it ensures that scientists do not waste their time studying phenomena that are the figments of some individual's imagination. Agreement between individuals is the basis of scientific objectivity. When research is repeated, it does not always produce the same results. Failures to replicate findings at the cutting edge of research, such as the inability of subsequent researchers to replicate Dean Hamer's "gay" gene, often generate public skepticism about science and scientists. Yet this is unfair. The public is not observing a weakness of science but actually its greatest strength, namely, the ability to detect errors. Science is a self-correcting endeavor. Attempts at replication are not valuable solely because they can detect errors in an original study. They may also build upon and extend the earlier research.

Scientific research is guided by a hypothesis, which is an idea or guess. The formation of a hypothesis is guided by scientific theory. A hypothesis takes the form that if we make our observations under specified conditions and if the theory is correct, then a certain outcome should be produced. For example, Einstein's theory of relativity predicts that time should slow down when an observer is moving through space. This theory produces the testable hypothesis that if two identical accurate clocks are set at precisely the same time before one is flown around the globe while the other stays in place the moving clock should register the passage of less time.

The scientific method can be thought of as an iterative process according to which scientists assimilate the existing information, usually by studying scientific journals and other professional publications. Then they develop testable hypotheses. They design and carry out the research necessary to test this hypothesis. Finally they write up their results as a scientific report. Following peer review, the report may be published in the technical literature. In this way, it fertilizes the body of scientific knowledge out of which it has sprung.

SCIENTIFIC PROCEDURES
The ultimate goal of scientific research is causal explanation. Scientists want to be able to draw conclusions of the form, A causes B. Strictly speaking, such conclusions can be drawn only from experiments. An experiment is an

ideally simple form of observation in which a comparison is made between the outcome after an experimental manipulation and the outcome without this manipulation. In the case of the clock experiment above, the experimental manipulation was movement through space. Since the moved clock was slower than the stationary one, it could be concluded that movement slows time. In practice, doing such an experiment is complicated. Clearly, the experiment would have been worthless if the moved clock had been buffeted around in the luggage compartment of the plane. Similarly, if the stationary clock had experienced much warmer temperatures than the traveling one, we would be entitled to be skeptical of the outcome. Experiments can work only if all of the variables that might distort the test are rigorously held constant, or controlled.

An experiment is the most stringent form of systematic observation but it is not the only kind. Much scientific research is descriptive. When the great Swedish biologist CAROLUS LINNAEUS collected and arranged samples of plants and animals from around the globe, he was primarily interested in establishing patterns of similarity between related species rather than hypotheses-testing. Similarly, the KINSEY surveys on human sexuality were designed to collect information about the sexual behavior of ordinary people rather than to test predictions. Few people would make the mistake of assuming that either of these endeavors was anything but scientific even though they lack the hypothesis-testing that is a cornerstone of scientific method. In their early development sciences often go through a largely descriptive phase following which it is possible to test hypotheses and conduct experiments. Thus, early neuroscientists were greatly preoccupied with producing detailed microscope pictures of nerve cells. Neuroscience today is a highly experimental science in which changes in cell structure and function are observed following pharmacological, behavioral, and other manipulations.

Many fields of study do not lend themselves to experimentation for several reasons, including ethical ones. In sciences such as sociology and anthropology, hypotheses are tested using correlational evidence. Correlational evidence does not allow causal conclusions to be drawn but it provides evidence that is useful in making predictions. For example, if the incidence of eating disorders is related to the number of women entering professions, then a projected decline in female employment would lead to the expectation of a decline in eating disorders.

SCIENTIFIC METHOD AND ETHICS

Many of the ethical problems of scientists arise when they deviate from scientific method and conduct bad science. This does not mean that careful practitioners will avoid all ethical dilemmas.

In the COLD FUSION controversy, for example, scientists incorrectly claimed to have discovered nuclear fusion that occurred at room temperature. This mistake is attributable to a combination of scientific failings that are also ethical failings. For example, it is possible that the chemicals were not properly mixed. This would be a violation of the obligation of scientists to conduct careful, repeatable, research. Moreover, the heat generated in the experiment could have been produced by ordinary electrochemical processes without invoking fusion. Scientists are expected to (a) acquire knowledge relevant to their enquiries and (b) take a skeptical approach to reporting new phenomena that might easily be explained in terms of what is already known. The researchers also neglected to submit their results for peer review. This made it difficult for other researchers to identify their errors. The motive for publicizing the results in a press conference, rather than a scientific journal, might have been commercial because if the phenomenon had been real, the researchers could have taken out a lucrative patent. Accurate and detailed reporting of the research in a peer-reviewed journal might have allowed rival laboratories to file for a patent first. Scientific objectivity may thus have been compromised by greed. Ethical scientists are expected to live up to all of the requirements of scientific method and when they do the quality of their work is improved.

See also BACON, FRANCIS; COLD FUSION CONTROVERSY; KINSEY, ALFRED; LINNAEUS, CAROLUS; SCIENTIFIC DISHONESTY.

Further Reading

Grinnell, F. *The Scientific Attitude.* New York: Guildford, 1992.
Resnik, D. B. 1998. *The Ethics of Science.* New York: Routledge, 1998.

scientific racism

Racism is defined as the view that certain ethnic groups are inferior to others by virtue of their inherent biological characteristics. Many scholars have followed anthropologist Ashley Montagu in concluding that racism is a modern phenomenon. There have always been antagonisms among various ethnic groups, they would argue, but these antagonisms were not motivated by racism per se. This argument ignores the existence of a widespread tendency among isolated societies studied by anthropologists to see their own people as superior to others. Thus the word for "human" is sometimes the same word that is used to denote the tribe, suggesting that other groups are seen as subhuman. Such ethnocentrism may be a common human tendency but a well-crafted theory of racial superiority is indeed modern. Scientific racism developed in the 19th century concurrent with the emergence of evolutionary biology and the social sciences and apparently reflected the worldwide expansion of European empires.

A BRIEF HISTORY OF SCIENTIFIC RACISM

The Western intelligentsia of the 19th century were generally not very enlightened on racial issues by today's standards. Even respected figures in science, such as biologist CHARLES DARWIN, could not transcend the prejudices of their day. Influenced by the biblical notion of a single Adam, the common ancestor of all humans, Darwin wanted to believe that all humans are born equal. This changed when he traveled to South America aboard the *Beagle* and beheld the distressing spectacle of the Patagonian Indians wandering around naked despite the constant onslaught of chilling winds. He saw the Patagonians as being so different from Europeans that it was much easier to imagine that they had evolved from apes.

Darwin's evolutionary theory provided the intellectual underpinnings for scientific racism in two ways. First, the importance of inherited traits was emphasized over acquired ones. Thus, differences between Darwin himself and the naked, paint-daubed Patagonian aborigines could be attributed to inherited differences. In *The Descent of Man* (1871) Darwin concluded that human races differ in physical characteristics, temperament, and intellect, echoing a view that seemed self-evident to his more learned contemporaries even if it ran counter to the biblical orthodoxy of the single Adam. Second, Darwin's theory of evolution by natural selection was used, by Darwin himself, as well as by prominent contemporaries, such as HERBERT SPENCER, to account for differences among human groups.

Darwin believed that competition between and among groups played an important role in human evolution, arguing in *The Descent of Man* (1871) that "extinction follows chiefly from the competition of tribe with tribe, and race with race." When barbaric peoples come in contact with civilized races, they prove unequal to the challenge and are quickly wiped out. Africans and Australian Aborigines were ill-equipped to compete with Europeans and would inevitably lose out in the struggle for group survival. The argument that some human groups are inherently superior to others forms at the heart of social Darwinism, as is the view that group competition improves the human species. Despite the claims of modern apologists to the contrary, it is quite clear that Darwin endorsed both views albeit without the enthusiasm of a Herbert Spencer.

If human racial groups comprise biologically different subtypes, then it would be possible to give them a scientific classification. The century preceding the publication of *On the Origin of Species* (1859) had seen various efforts at racial classification. These were superseded in the early

19th century by French naturalist and anatomist Georges Cuvier's (1769–1832) division of humans into three races: Negroes, Caucasians, and Mongolians (or Asians). This classification is still current among race scientists.

The view that races are fundamentally different was the topic of influential, if badly biased, research by SAMUEL MORTON (1799–1851), who studied racial differences in brain size by measuring the capacities of skulls. Morton believed that Europeans were intellectually and morally superior to other races and that this conclusion could be drawn from his cranial capacity data. When Morton died, a *New York Times* obituary declared him to be the single most eminent American scientist in terms of his international impact. In other words, the kind of racist science he practiced was then in the mainstream. By contrast, conceptually similar work being carried on by contemporary race scientists such as Philippe Rushton of the University of Western Ontario is considered not only politically unacceptable (resulting in the temporary loss of Rushton's job) but also both methodologically inadequate and highly eccentric. Rushton defends himself from the charge that he is conducting biased research by pointing out that he sees orientals as being intellectually superior to Europeans (his own racial identity). (In fact, there is no evidence that Asian Americans consistently score higher than European Americans on IQ tests.) Rushton's claims are based on the notion that high intelligence is an adaptation to the colder climates faced by ancestral humans as they migrated out of Africa.

There was considerable popular interest in racial evolution as reflected in the 1854 book *Types of Mankind* by Josiah Nott and G. R. Gliddon. This was an unscrupulous popularization of Morton's work that ignored any data indicating a lack of difference between Europeans and other racial groups. Nott and Gliddon have become infamous for their conscious fakery. Their illustrations show Africans having exaggerated elongation of the jaws to make them seem extremely chimpanzee-like. The features of chimpanzees were modified in the opposite direction to create the impression that Africans might actually be more primitive than chimpanzees. Such obvious fakery evidently reflects the willingness of their readers to accept gross racial stereotypes. The same phenomenon surfaced in the 20th century when group differences in intelligence test results were interpreted according to a theory of scientific racism.

THE STUDY OF GENETIC DIFFERENCES, AND BIOLOGICAL DIFFERENCES, AMONG ETHNIC GROUPS

If biological race is a useful human concept, then there would have to be real, and substantial, genetic and biological differences between different racial groups. Given the overall genetic differences (or variability) within the human species, how much of these differences are due to ancestry in Europe versus Africa versus Asia? The answer, according to Italian population geneticist Luigi Cavalli-Sforza, a leading current authority in this field, is approximately 15%, an estimate that is consistent with other research in the field. In other words, less than one-sixth of human genetic differences are explained as derived from Asian, European, or African extraction. Genetic differences within "races" are thus five times the size of differences between them.

Our species has spent such a relatively short amount of its history outside of Africa that there has been insufficient time for substantial genetic variation to accumulate between migrants and groups remaining in Africa. Recent research also indicates that there was a great deal more gene flow among prehistoric human populations than previously supposed, suggesting substantial migration in many directions. Most of the genetic variation among racial groups probably reflects adaptation to local climatic factors and locally prevalent diseases such as SICKLE-CELL ANEMIA. There is no evidence that genetic markers for European or African ancestry are predictive of intelligence, or any other psychological trait.

Almost all of the genetic variety of the human species is to be found within Africa. According to the genetic evidence, we all belong in the same biological race. Beneath the skin, we are all Africans. This compelling conclusion from modern genetics invalidates the sweeping generalization made by race scientists about differences among Africans, Europeans, and Asians predicated on genetic differences, except those that are tied to particular local adaptations.

The views that racial category is not an important predictor of the genetic characteristics of an individual seems inconsistent with the important differences in health outcomes between African Americans and European Americans. Being African American is associated with an increased risk of many illnesses that contributes to a shorter lifespan. This difference in physiology might seem to confirm the importance of genetics. In reality, most of the differences are attributable to lifestyle factors that are connected to social status. This is true of most of the leading causes of mortality, including heart disease, high blood pressure, obesity, diabetes, cancers, homicides, and AIDS.

Some disease risks are clearly affected by genetics albeit in a more complex manner than if often realized. Some African Americans are more prone to sickle cell anemia because of their African ancestry. The recessive gene responsible for sickling of the red blood cells is favored by natural selection because it boosts resistance to malaria. It is thus common in regions of Africa where malaria is prevalent. There are many more carriers of the disease, who benefit in terms of malarial resistance, than

there are victims of sickle-cell anemia, who inherit a double recessive gene. Sickle-cell anemia is clearly a genetic disease that is much more common among African Americans than it is among European Americans. Yet it is a mistake to think of it as an "African" disease. In reality it is a tropical disease that is much more prevalent in some regions of tropical Asia than it is in Africa. European Americans have misinterpreted it as an African disease because they almost never suffered from it themselves.

Health professionals are obliged to pay attention to all useful predictors of disease vulnerability, including racial group. Describing health vulnerabilities in terms of racial labels suggests that are of genetic origin. Race is thus reified, or converted from an idea into a reality. Some group differences in health outcomes are obviously due to genetic differences. Yet, most are predictable consequences of being raised in a particular kind of social environment. Much of the increased health risks of African Americans are attributable to poverty and therefore transcend racial group. Poverty is associated with poorer health outcomes because children raised in poverty experience less adequate nutrition, receive less medical attention in early life, inhabit a more stressful rearing environment, and do not learn to behave in ways that promote good health. Almost exactly the same logic that has been applied to explaining group differences in health can be applied to differential vulnerability to social problems, including single parenthood and crime.

Another topic that has received much attention, particularly in news magazines, is the phenomenon of racial group differences in athletic performance. Among "elite" athletes, particularly runners, African Americans have recently established a position of prominence in terms of the number of world records broken and Olympic medals won, for example. This performance has suggested that there are physiological advantages in some sports associated with African ancestry. Despite considerable research, no such differences have been reliably established.

Possible racial differences between top-level athletes has no particular relevance for racial groups in general because athletes are extremely unrepresentative of their respective groups. The advantages of black runners cannot be explained in terms of physiology. The only really concrete differences that could affect the performances of runners are those associated with physique. Africans tend to have longer legs relative to the rest of their bodies, which may provide a speed advantage in running. In addition, their slimmer calf muscles may provide a mechanical advantage because the amount of energy required to propel the leg through space is reduced. These mechanical advantages are offset by having slightly heavier bones. At present, we do not know whether the overrepresentation of African Americans among elite athletes in some sports is due to such differences in physique or whether the real difference is motivational with more blacks aspiring to careers in particular sports. The widespread belief that Africans excel at sports for genetic reasons may be socially damaging because it reinforces a stereotype of brawn rather than brain. In fact, there is no scientific evidence of a tradeoff between athleticism and academic ability. If anything, there is weak evidence that a vigorous mind and a vigorous body tend to go together.

GROUP DIFFERENCES IN IQ

Standardized IQ tests follow the tradition of Binet in defining intelligence as the potential for academic success. Intelligence testing in America began in earnest during World War I (1914–18) with the need to recruit large numbers of officers quickly. From the beginning, administration of the Army tests was biased against immigrants. Following the war, group differences in test performance became widely known and were used to back up various racist agendas from immigration policy to compulsory sterilization (see INTELLIGENCE TESTS).

Despite frequent claims that intelligence tests are unfair to African Americans, there is no convincing evidence of this. Standardized IQ tests, and the more commonly used tests of scholastic aptitude, can predict college performance to a limited degree (explaining up to a quarter of the variation in college grades) and they predict outcomes for African Americans in addition to the rest of the population.

Standardized testing can be defended because it helps college admissions officers to decide who is likely to obtain good grades and complete their degree in an acceptable period. High schools differ in rigor and grades are thus limited in their validity as a measure of academic potential so that standardized test scores provide useful supplementary information. Since African Americans, on average, score lower than the population average on academic aptitude tests, use of the tests in college admissions is often inaccurately perceived as racist because it excludes large numbers of African Americans.

Yet, they are not being excluded because of racial category but because of inadequate academic preparation. African Americans do less well on standardized tests because, for the most part, they receive a poor high school education. The tests are (accurately) stating that they have not been well prepared for college careers. Admission of unprepared students to college under affirmative action programs is a mixed blessing at best. Thus, while the percentage of African Americans attending college rose threefold between 1965 and 1995 (from 15% to 45%), the number obtaining degrees has only about doubled (from 7% to 15%). By comparison, the percentage of whites attending college doubled in the period and so did their graduation rates. Any advantage of preferential admissions can thus be eaten away by increased dropout rates.

African Americans score lower than the rest of the population on IQ tests. Most scholars put this difference at anywhere from 10–15 points (or two-thirds to one standard deviation), which means that the average score for blacks is 85–90 compared to an average for the population that is standardized at 100. This deficit in academic aptitude is consistent with results for tests of academic skills, such as the SAT (Scholastic Aptitude Test). In 1976, the average of combined verbal and math SATs for blacks was 686 compared to 944 for whites, a difference of 258 points. By 1991, improving black scores reduced the gap to 194 points. The National Assessment of Educational Progress paints a similar picture. Over the past quarter century blacks in high schools have remained behind whites in all areas of academic attainment to the tune of approximately 3–6 years, averaging four years in 1994.

Modern scientific racists have gone back to Samuel Morton's technique of measuring brain size to find an explanation for group differences in academic performance. The underlying assumption is that genetically determined group differences in cranial capacity are responsible for differences in IQ. This logic is vulnerable. First, far from being genetically determined, brain size has a heritability of only about 50%, which means that about half of the differences between individuals in brain size are determined by environmental factors. Animal research has shown that a stimulating early environment promotes brain growth and learning ability, even in rodents. Moreover, IQ scores are not fixed, increasing substantially with years of education and with the increasing complexity of a society (the Flynn effect). Philippe Rushton has shown that there are modest group differences in brain size and brain size is now known to predict about 16% of the differences in IQ scores. Rushton has used this data to conclude that there are genetically determined group differences in intelligence as reflected in the IQ gap between African Americans and the rest of the population. This is very shaky inference because (a) brain size differences can explain only around a third of the IQ gap and (b) there is very good reason for suspecting that group differences in brain size are entirely of environmental origin.

Rushton's work is an interesting case of pseudoscientific racism because he follows the rules of correct scientific presentation of results and has published in top journals in the field of intelligence. Yet he is strongly committed to the truth of his hypothesis concerning group differences in academic attainment and does not give sufficient consideration to alternative hypotheses. Thus, he ignores evidence concerning the depressed IQ scores of despised minorities around the world and their characteristic rise with immigration to America, evidence that populations of African origin, such as Jamaicans, have achieved the same level of educational and occupational success as other Americans, and, most relevant, evidence from developmental psychologists that racial group differences in IQ scores are almost entirely explainable in terms of the educational quality of the home environment. Residual effects are amenable to a variety of environmental explanations including pre- and postnatal nutrition, exposure to neurotoxins such as lead and mercury, and vulnerability to diseases. Failure to give due weight to these competing explanations reflects a lack of openness to the falsification of his hypotheses. Rushton does not have a genuine desire to advance our knowledge of the causes of group differences in academic performance because he thinks he already knows the answer (genetics). His efforts, like those of his intellectual ancestor, Samuel Morton, are designed to promote a racist agenda. Science used for such purposes inevitably degenerates into PSEUDOSCIENCE.

A somewhat more sophisticated presentation of the notion that group differences in IQ scores are genetically determined was Richard Herrnstein and Charles Murray's *The Bell Curve* (1994). This book not only made the case that African Americans score lower than the rest of the population on IQ scores because of unchanging genetic factors but also postulated that less intelligent Americans of all races are having more children than others, thus predicting a decline in overall intelligence. Students of the history of intelligence will recognize this theme constitutes a resurgence of psychologist Henry Goddard's concern, in the 1920s, about the idiocy of inferior races.

One reason that *The Bell Curve* has received so much attention is that the authors were associated with Harvard University, which has a high reputation as a research institution. It thus seemed to the public that the scientific community had conferred its blessing on their pseudoscientific conclusions. Like Rushton, Herrnstein and Murray focus on genetic determinism to the exclusion of other factors in cognitive development. They also rely too heavily on making a deterministic connection between IQ scores and economic success. The fallacy behind this perspective is illustrated by political scientist James Flynn's research, which found that the occupational achievement of Chinese Americans with an IQ of 100 was equivalent to that expected of white Americans with an IQ of 120. In conclusion, it can be argued that most of the scientific work advocating genetically determined group differences in behavior and intelligence is seriously flawed as science and suffers from an unscientific lack of openness to competing explanations for group differences.

See also INTELLIGENCE TESTS; MORTON, SAMUEL GEORGE; PSEUDOSCIENCE.

Further Reading
American Psychological Association. *Intelligence Knowns and Unknowns*. Washington, D.C.: Author, 1995.
Barber, N. *Why Parents Matter: Parental Investment and Child Outcomes*. Westport, Conn.: Bergin and Garvey, 2000.
D'Souza, D. *The End of Racism*. New York: Free Press, 1995.
Flynn, J. R. *Asian Americans: Achievement beyond IQ*. Hillsdale, N.J.: Erlbaum, 1991.
Herrnstein, R. J., and C. Murray. *The Bell Curve*. New York: Free Press, 1994.
Hoberman, J. M. *Darwin's Athletes*. Boston: Houghton Mifflin, 1997.
Montagu, A. *Man's Most Dangerous Myth: The Fallacy of Race*. Walnut Creek, Calif.: Altamira Press, 1998.
Nott, J. C., and G. R. Gliddon. *Types of Mankind*. Philadelphia: Lippincott Grambo, 1854.
Ruse, M. 1982. *Darwinism Defended*. Reading, Mass.: Addison Wesley, 1982.
Rushton, J. P. *Race, Evolution, and Behavior*. New Brunswick, N.J.: Transaction, 1995.
Shipman, P. *The Evolution of Racism*. New York: Simon and Schuster, 1994.
Thernstrom, S., and A. S. Thernstrom. *America in Black and White*. New York: Simon and Schuster, 1997.

Scoline and fear conditioning

Scoline (or suxamethonium chloride) is a drug that slows breathing and it was used in a horrifyingly unethical experiment on fear conditioning with human subjects. This experiment violated the principle of informed consent articulated in the NUREMBERG CODE and it underscores the need for careful ethical scrutiny of research with human participants. According to the principle of informed consent, if there is any information that might sway a person's willingness to participate in research, then he or she must be informed of this before agreeing to participate. The drug scoline caused people to stop breathing for several minutes. Since they had no idea that they had received a drug, most participants believed they were dying.

It seems that the authors of this odd experiment were motivated by the desire to develop a new procedure to condition fear in humans and thereby contribute to the understanding of phobias and other anxiety disorders, which are extremely common and cause a great deal of distress and inconvenience for ordinary people. The 11 subjects in the experiment included male alcoholics who participated because they were told that the research was connected with a possible cure for alcoholism. Scoline was originally used in the 1960s as an aversive agent in the treatment of heroin addiction and has also been used to treat alcoholism, but the experiment had nothing to do with alcoholism as such.

All of the 11 subjects received scoline infusion via intravenous drip but only five heard a tone that was used as the conditioned stimulus in a classical conditioning procedure (see PAVLOV, IVAN). The other six served as the control group. Since scoline halts breathing temporarily, atropine was also infused. This was to slow down salivation. Once breathing had stopped, the saliva might have entered their windpipes and caused choking.

The researchers were thus highly competent people whose research program was backed up by state-of-the-art equipment in a hospital setting. With the insidious drug administration and the battery of physiological recording apparatus to which the participants were connected, the atmosphere of the experiment must have been positively Orwellian. Such lack of consideration for the feelings of research participants seems almost incomprehensible in a democratic society.

Following the sounding of a 600 hertz tone for five seconds, the experimental subjects experienced loss of the ability to breathe. Fear was measured as an increase in muscular tone recorded on an electromyograph, increased sweating measured by galvanic skin response, and increased heart beat measured by an electrocardiogram.

After a single conditioning trial, in which the tone was paired with scoline, the subjects received no fewer than 30 unpaired, or extinction trials, in which the tone was presented by itself. On the first extinction trial, breathing was suppressed by the tone because it had acquired some of the effects of scoline through classical conditioning. This conditioned response was produced without any reduction in intensity for each and every one of the 30 extinction trials. Neither did the participants lose their fear of the tone because the sweating response also continued without reduction (it actually increased somewhat) for the 30 extinction trials.

On average, breathing stopped completely for 105 seconds on each trial. Participants said they thought they were dying. One subject had been a rear gunner, during World War II, in a Stirling bomber which had flown straight over Düsseldorf for 5,000 yards within German radar. He evaluated the scoline trial as being a more traumatic experience.

The researchers (D. Campbell, R. E. Sanderson, and S. G. Laverty) published their results in the *Journal of Abnormal and Social Psychology*. Why would a respectable scientific journal be willing to publish such a clearly unethical study? One possible reason is that the whole movement for ethical research had not gathered much strength by 1964, when the study was published. Another may be that the results were scientifically impressive. Classical conditioning in a single trial is very unusual, even for humans who usually condition about 10 times faster than laboratory animals. Speed of conditioning and the fact that extinction failed to occur is of considerable

interest to students of avoidance learning in animal experiments and students of posttraumatic stress disorder. It suggests that a single highly traumatic event produces strong conditioning so that virtually anything that was associated with the trauma can bring back the intense fear.

See also COLD WAR RADIATION EXPERIMENTS; MENGELE, JOSEF; MILGRAM'S OBEDIENCE RESEARCH; NUREMBERG CODE; PAVLOV, IVAN; RIGHTS OF HUMAN RESEARCH PARTICIPANTS; WATSON, JOHN B.

Further Reading

Campbell, D., R. F. Sanderson, and S. G. Laverty. Characteristics of a Conditioned Response in Human Subjects during Extinction Trials Following a Single Traumatic Conditioning Trial. *Journal of Abnormal and Social Psychology* 68 (1964): 627–39.

Leary, M. R. *Introduction to Behavioral Research Methods.* Belmont, Calif.: Wadsworth, 1991.

Scopes trial

The state of Tennessee followed Oklahoma in passing a 1925 law making it illegal to teach evolutionary ideas. This precipitated the Scopes trial, a famous 1925 test case in which high school teacher John Thomas Scopes let himself be prosecuted for teaching evolution in the Dayton, Tennessee, high school. Scopes was prosecuted by William Jennings Bryan, a well-known political figure who was a religious fundamentalist. He was defended by Clarence Darrow, a prominent defense lawyer and noted agnostic (see photo on next page). When Darrow was refused permission to call scientific witnesses for the defense, he called prosecutor William Jennings Bryan as an expert on the claims of the Bible, thereby turning the trial into a circus.

Although Scopes was convicted, he received the minimum penalty, a fine of $100. The trial was also overturned on appeal on a technicality. Thanks partly to the biting humor of journalist H. L. Mencken (1880–1956), the entire country laughed at the backwardness of Tennessee. Mencken referred to the local residents as "gaping primates" and "anthropoid rabble." Not wishing to be exposed to the same kind of ridicule as Tennessee, many states that had been considering their own "monkey" laws prohibiting the teaching of evolution quietly dropped legislative plans. (Exceptions were Mississippi and Arkansas, whose antievolution laws stayed on the books until struck down by the U.S. Supreme Court in 1968.) The event turned out to be an important moral victory for evolutionists. Scopes and Darrow convinced the U.S. public of the folly of placing legislative barriers around intellectual freedom. The Scopes trial is one of the most famous in U.S. history and is remembered today

John Randolph Neal (standing) defends John T. Scopes against a charge of illegally teaching evolution in Dayton, Tennessee, in 1925. Clarence Darrow, the leading defense attorney, is seated to the right. (LIBRARY OF CONGRESS)

as a turning point in which science broke free from the shackles of religion.

See also CREATIONISM; DARWIN, CHARLES; GALILEO GALILEI; HUXLEY, THOMAS; RELIGIOUS IDEOLOGY AND SCIENCE; WILBERFORCE, SAMUEL.

Further Reading
Ruse, M. *Darwinism Defended.* Reading, Mass.: Addison Wesley, 1982.

secrecy   *See* OPENNESS IN SCIENCE.

selective breeding

In selective breeding, individuals with desirable traits are given the opportunity to reproduce and so propagate these favored characteristics. With the domestication of animals and plants, humans began a long process of systematic alteration of the genetic composition of plants and animals through selective breeding. For example, breeders of thoroughbred race horses obtain progeny from the fastest mares and the fastest stallions. In this way, genes promoting fast running speed secure selected into the next generation.

Selective breeding of food plants may have been even simpler. Modern cereals are derived from wild grasses that have relatively small seeds. Our ancestors selected for large edible seeds in their cereals by planting only the seeds from plants having the largest ears. In this way, over the course of some 7,000 years, farmers have gradually increased the size of corncobs from the tiny ears of the grass ancestor (teosinte) to the giant cobs of modern corn. Similarly, domestic animals have been selectively bred for large size, meat quality, strength, milk yield, and so on.

The fact that the attributes of domestic animals can be manipulated by selective breeding is well illustrated by the many different breeds of dogs from Chihuahuas to Great Danes that have resulted from hundreds of generations of selective breeding of the same species for different purposes. Similarly, the many different varieties of pigeon, bred for their appearance, from bald head to pouter, demonstrate the capacity of selective breeding to modify the genotype underlying feather structure. This knowledge served as an important inspiration to CHARLES DARWIN in devising his theory of evolution by natural selection. Writing in an age in which nothing was known of genetics, as such, Darwin, who was a pigeon fancier himself, and whose family had a reputation as sheep breeders, imagined that the natural environment could alter the traits of populations if it performed the role of a selective breeder. Thus giraffes with long necks would have been more successful as leaf browsers and would have passed on this trait to their offspring (by some mechanism that Darwin did not understand). Individuals with shorter necks would have been less likely to survive and would have left fewer offspring to carry on the short-necked trait. Darwin referred to this progressive elimination of unfit traits as natural selection, or the natural equivalent of animal breeding.

Artificial selection has shown that behavioral traits can also be selected. Gun dogs are bred to point at concealed prey, collies to herd sheep, and St. Bernards to dig people out of avalanches. This phenomenon did not escape the notice of behavioral scientists, who performed a number of selective breeding experiments on mice, rats, and other species. Laboratory rodents have been selectively bred for activity level, aggressiveness, emotional reactivity, susceptibility to alcohol, sensitivity to noise, fondness for saccharin, and ability to learn a complex maze. The success of these animal studies encouraged behavior geneticists to ask whether human psychological traits, such as intelligence, aggressiveness, and extroversion, are influenced by genotype. A variety of sources of evidence, including family studies, adoption studies, and twin studies, have shown that human behavioral characteristics are strongly affected by genotype. The most compelling evidence comes from the study of identical twins reared apart, as in the MINNESOTA TWIN STUDY.

Through selective breeding, human beings have systematically manipulated the genotypes of animals and plants. Whether this was ethically permissible never arose as an issue until recent times with the emergence of GENETIC ENGINEERING in the laboratory. Recombinant DNA technology offers the possibility of making more rapid, and more radical, changes in the genotype and has thus raised concern about geneticists "playing God" through deliberate alteration of the genotypes of plants and animals so that they can more effectively serve human needs.

See also DARWIN, CHARLES; GENETIC ENGINEERING; MENDEL, GREGOR; MINNESOTA TWIN STUDY.

Further Reading
Mangelsdorf, P. C. *Corn: Its Origin Evolution and Improvement.* Cambridge, Mass.: Belknap Press, 1974.
Plomin, R., J. C. DeFries, and G. E. McClearn. *Behavioral Genetics: A Primer.* San Francisco: W. H. Freeman, 1980.
Ruse, M. *Darwinism Defended.* Reading, Mass.: Addison Wesley, 1982.

Seveso, Italy, chemical accident

The accidental release of a cloud of gas from a chemical plant near Seveso, Italy, on July 10, 1976, exposed downwind residents to very high concentrations of dioxin (2,3,7,8-tetrachlorodibenzo-p-dioxin). Laboratory experiments on animals have shown that dioxin is a potent carcinogen but there has been some controversy about whether dioxin is a human carcinogen. Study of the health of people exposed to contamination from the unfortunate accident near Seveso is thus important in constituting a natural experiment to determine whether dioxin is carcinogenic.

The cloud of gas released in the explosion seemed harmless at first. Several days later residents began to develop cloracne. This skin rash is a typical symptom of industrial exposure to dioxin. Researchers studied the health of residents who had lived in the contaminated area for 10 years. They found that although the overall risk of developing cancer was not elevated in the study area, rates of specific cancers did rise. Thus, for 4,800 people living in moderately contaminated locations, women were five times as likely to develop gallbladder or bile-duct cancer as nondioxin-exposed women. Men were twice as likely to get leukemia or lymphoma as nondioxin-exposed individuals.

For the 32,000 people inhabiting less contaminated areas, rates of soft tissue cancers were also elevated for men. A very strange result was that for the 724 most highly exposed persons, there was no detectable increase in risk for any cancer. One interpretation of this null result is that the group was too small for a statistically meaningful pattern to emerge. Another anomalous finding was that the incidence of estrogen-dependent cancers (such as breast cancer and cancer of the uterus) was actually reduced in the two most exposed regions. This finding is consistent with animal experiments showing that dioxin reduces the number of estrogen receptors in cells and interferes with estrogen metabolism. The findings thus confirm not only that dioxin is carcinogenic among human populations but also indicate that it functions as a hormone disruptor. The latter kind of effect is one of the more troubling aspects of the modern age of chemicals, which has been used to explain phenomena as diverse as declining sperm count and low intelligence (see HORMONE MIMICS).

In addition to increased incidence of specific cancers, residents of the Seveso area had increased mortality from heart disease. This result is likely due to the psychological trauma associated with living at the site of an industrial accident rather than any direct effect of the dioxin, a phenomenon that was observed in connection with the THREE MILE ISLAND NUCLEAR ACCIDENT in the United States.

See also AGENT ORANGE; BHOPAL; HORMONE MIMICS; LOVE CANAL; SPERM COUNT DECLINE; THREE MILE ISLAND NUCLEAR ACCIDENT; TIMES BEACH.

Further Reading

Bertazzi, A., A. C. Pesatori, D. Consinni, A. Tironi, M. T. Landi, and C. Zocchetti. "Cancer Incidence in a Population Accidentally Exposed to 2,3,7,8-tetracholorodibenzo-para-dioxin." *Epidemiology* 4 (1993): 398–406.

Bertazzi, P. A., C. Zocchetti, A. C. Pessatori, S. Guercilena, M. Sanarico, and L. Radice. "Ten Year Mortality Study of the Population Involved in the Seveso Incident of 1976." *American Journal of Epidemiology* 129 (1989): 1187–200.

sex differences research   *See* EVOLUTIONARY PSYCHOLOGY AND SEX DIFFERENCES.

sex selection   *See* AMNIOCENTESIS, REPRODUCTIVE TECHNOLOGIES.

sexual orientation research

Homosexuality was once considered a mental disorder and was listed in the Diagnostic and Statistical Manual of Mental Disorders, published by the American Psychiatric Association until the 1973 edition. Homosexuality was removed from the list of mental disorders because of a growing recognition among psychiatrists that homosexuals as a group did not have more psychological problems than heterosexuals. Many homosexuals fear that recent research on the biological basis of homosexuality will have the effect of remedicalizing it. To speak of a gene for homosexuality, they argue, is only a breath away from developing a biological manipulation to alter sexual orientation, or at least the development of genetic screening to reduce the number of homosexuals who are born.

Sexual orientation research has often been compared to research on sex differences because of the potential influence of the stereotypes with which scientists begin and on the conclusions that they draw. Such research is bound to be ethically controversial because it explores differences that have important political and legal ramifications and are a matter of intense ongoing public debate. Key ethical questions that have been raised by scientific research on homosexuality include the following:

1. Can sexual orientation be scientifically studied?
2. Is the quality of research on the biological basis of homosexuality unusually low?
3. Will the results of such research be used to alter sexual orientation or screen fetuses?

Opponents of the scientific study of homosexuality begin with the premise that people cannot be usefully categorized according to their sexual orientation. After all, male homosexuality is considered to be a natural transition on the road to manhood in some societies. In others, such as modern Brazil, homosexual encounters occur in a ritualized form among individuals who are heterosexuals the rest of the time. Such cross-cultural differences are not nearly so serious a problem for students of sexual orientation as they might appear because there is a difference between orientation, or preference, and behavior. A woman who prefers to have sexual relations with another woman is different from a woman who prefers to have sex with a man even though both might have had bisexual experiences.

Even when sexual orientation is distinguished from sexual experiences, definitional problems remain. As ALFRED KINSEY maintained, homosexuality and heterosexuality are not crisply discontinuous categories. Kinsey devised a 16-point scale used to measure sexual orientation as varying continuously between exclusive homosexuality through bisexuality through exclusive heterosexuality. It might seem that modern scientists studying homosexuality have lost some of this sophistication but this is an illusion. Many use the Kinsey scale and define the category "homosexual" in terms of a cutoff on the Kinsey scale.

Critics of the science of sexual orientation not only question the scientific basis on which sexual orientation is defined but also take issue with the quality of the research, charging that it is biased by preconceived ideas and that some of the "breakthrough" finding such as Dean Hamer's "gay gene" have not withstood the test of replication. Hamer, a researcher at the National Cancer Institute, reported in 1993 that he had identified a segment of the X chromosome, known as Xq28, that was identical in gay men. This paper led to widespread acceptance of the notion that homosexuality is produced by a gene. Three years later, a team led by George Rice at the University of Western Ontario found no linkage between Xq28 and sexual orientation. Another unpublished study by Alan Sanders at the University of Chicago also failed to replicate Hamer.

Such failures to replicate have produced a great deal of skepticism, and some anger. Yet this reaction is based on a fundamental misunderstanding of the scientific process. Such reversals are not unusual at the cutting edge of science. The same conclusion applies to a recent downward revision in the genetic heritability estimate for homosexuality.

Many ethicists worry that scientific research on homosexuality has intrinsically homophobic aspects because it is based on the premise that homosexuality is a problem and that a better understanding of its scientific basis can be used to "solve" the problem. Reactions to Hamer's claim that he had identified the genetic basis of homosexuality are instructive in this respect. Some homosexuals welcomed the news as showing that homosexuality had a biological basis and so did not constitute a matter of moral choice. Others worried that scientists would begin developing a gene "therapy" or that information about genetic markers for homosexuality could be used to terminate pregnancies that would result in homosexual children. The discrediting of the Xq28 marker pushes these concerns into the background at least until the next "homosexual gene" is identified.

Available evidence suggests that while homosexuality does have a biological component, sexual orientation is probably not reducible to a single-gene effect. A biological influence is not the same as a genetic one and there is good reason to believe that the prenatal environment of the womb may be of critical importance to the development of sexual orientation (see HOMOSEXUALITY, POSSIBLE BIOLOGICAL INFLUENCE ON). Even though homosexuality may have a biological cause, this does not mean that it could not be altered by behavioral means and one of the most controversial research projects has been the attempt to change homosexuals into heterosexuals using behavioral procedures.

In the case of male homosexuals, who were troubled by their orientation and wished to change, aversive counterconditioning procedures have been employed. In such procedures, sexually arousing pictures of naked men are flashed on a screen and are quickly followed by long painful shocks. When the shock is turned off, a picture of an attractive woman is presented. The idea is to make sex with men seem unpleasant and to make women more appealing by pairing pictures of women with relief from the shock. While about half of the treated men discontinued their homosexual contacts, the procedure is not as effective as might be imagined. Most of the men reporting behavior change were bisexuals to begin with. For exclusive homosexuals there is no evidence that sexual orientation can be changed using conditioning techniques. This suggests that sexual orientation, whether heterosexual or homosexual, may be etched into the brain before birth and is thus quite resistant to change.

See also KINSEY, ALFRED; HOMOSEXUALITY, POSSIBLE BIOLOGICAL BASIS OF; MASTERS AND JOHNSON.

Further Reading

Adams, H., and E. Sturgis. "Status of Behavioral Reorientation Techniques in the Modification of Homosexuality: A Review." *Psychological Bulletin* 84 (1977): 1171–88.

Hamer, D. H., and P. Copeland. *The Science of Desire: The Search for the Gay Gene and the Biology of Behavior.* New York: Simon and Schuster, 1995.

Leary, T. F. *Gay Science: The Ethics of Sexual Orientation Research.* New York: Columbia University, 1997.

LeVay, S. *Queer Science.* Cambridge, Mass.: MIT Press, 1997.

Seligman, M. *What You Can Change and What You Can't.* New York: Fawcett Columbine, 1993.

Stein, E. *The Mismeasure of Desire.* New York: Oxford University Press, 1999.

sick building syndrome

Sick building syndrome (SBS) refers to acute health problems experienced by occupants of a building for which no specific pathogen, or toxin, has been identified. Symptoms include headache, irritation of the eyes, nose, and throat, nasal congestion, fatigue, nausea, dry cough, skin itch, dizziness, and difficulty in concentrating. In the case of biological contaminants, chest tightness, fevers, chills, and muscular aches may also occur. These symptoms decline in severity with separation of the person from the affected building. The World Health Organization estimates that some 30% of new and remodeled office buildings around the world may cause such symptoms. This means that SBS is quite a serious problem both in relation to its health effects and in terms of lost productivity. Similar problems can be produced in residential dwellings.

SBS has a variety of contributors, most of which have to do with indoor air quality. The oil crisis of the 1970s produced a trend toward more energy-efficient buildings, including a reduction in the amount of air being drawn into buildings from 15 cubic feet per minute per occupant to 5 cubic feet. Indoor air may be a great deal more polluted than exterior air because pollutants are inadequately vented. A variety of chemical pollutants may play a role in SBS. These include volatile substances from adhesives in carpets and furniture, manufactured wood products, and chemicals emitted by photocopiers. Combustion products, such as carbon monoxide and nitrogen dioxide from unvented gas and kerosene heaters, can also contribute to the problem. Entry of outdoor pollutants, such as vehicle exhausts, plumbing vents, and tobacco smoke, may contribute to SBS. Chemicals used in cleaning buildings can also serve as contributants. One way of minimizing this source of pollution is to make sure that custodial activities are performed after building occupants have gone home.

SBS can be connected to any number of biological contaminants, including bacteria, pollen, viruses, insect infestations, and even bird droppings. Biological contamination of the heating, ventilation, and air conditioning systems is common because these often allow pools of condensation to form on which microorganisms can grow. This may be a serious problem because spores can be distributed throughout a building through the ventilation system as happened in a 1976 outbreak of Legionnaire's Disease in a Philadelphia hotel, which produced 182 cases of pneumonia and 29 deaths. Condensation held in ceilings, insulation, or carpets can also provide a haven for microorganisms.

SBS is usually dealt with by identifying possible air contaminants and acting to reduce them. This could involve major building renovation to vent problem areas such as rooms containing copiers. The adequacy of the ventilation system in terms of air turnover is also important. Improved maintenance of the ventilation system, including thorough cleaning and periodic filter replacement, can help.

SBS can be prevented, or at least reduced, by adoption and implementation of building codes that promote good construction practices. The American Society of Heating, Refrigerating, and Air-Conditioning Engineers (ASHRAE) has developed recommendations for building design and equipment installation that promote indoor air quality. For example, ASHRAE recommends that 15 cubic feet of air per minute be introduced to a building for each occupant. These professional standards are voluntary unless added to the building codes of state or local regulatory agencies. Even when they acquire legal force, the indoor air quality codes carry sanctions only for construction and renovations work. They do not apply to the maintenance of a building.

To the extent that many employees are sickened by poor indoor air quality, SBS is an important ethical problem of the modern workplace. It is in the enlightened self-interest of employers to ensure that indoor air quality is good because this is essential for high productivity. In extreme cases, air quality has been so bad that buildings have had to be closed before their normal period of occupancy had expired, with a huge loss in the value of the structures. Perhaps the most unsettling aspect of SBS is that, although the symptoms are usually acute and improve as soon as employees relocate, for some individuals these effects can produce serious exacerbation of their chronic conditions, such as asthma, hypersensitivity pneumonitis, and multiple chemical sensitivity. There has been a large increase in asthma rates over the past few decades and some experts believe that poor indoor air quality is one of many factors that could have contributed to this complex and serious condition.

See also AIR POLLUTION; SMOG.

Further Reading

Maroni, M., B. Seifert, and T. Lindvall, eds. *Indoor Air Quality: A Comprehensive Reference Book.* New York: Elsevier Science, 1995.
Oliver, C. I., and B. W. Shackleton. "The Indoor Air We Breathe: A Public Health Problem of the 90s." *Public Health Reports* 113, no. 5 (1998): 398–409.
Roston, J., ed. *Sick Building Syndrome.* New York: Routledge, 1997.
Tate, N. *The Sick Building Syndrome.* Far Hills, N.J.: New Horizon Press, 1993.

sickle-cell anemia

Sickle-cell anemia is a genetic disease that is of particular interest to geneticists because it occurs only among populations who have originated in regions where malaria is prevalent. The disease is produced by having a double "dose" of a recessive gene and heterozygotes (with one dominant and one recessive gene for the sickling trait) have an advantage in terms of resistance to malaria. The balance between beneficial and deleterious effects of the gene is referred to as a balanced polymorphism. Correspondence between geographic distribution of the sickle-cell gene and malaria provides a compelling example of evolution affecting human populations. It is also important in clarifying the significance of biological differences between human groups and is therefore relevant to ethical debates about race as well as about evolution.

People with sickle-cell anemia have abnormally shaped red blood cells due to an anomaly in the structure of the hemoglobin molecule within them. Hemoglobin plays a vital role in trapping oxygen and carrying it around to bodily tissues. Due to the molecular abnormality, hemoglobin molecules join together to form rods, which distort red blood cells into a sickle shape. Whereas normal blood cells are soft and pliable, allowing them to move easily through small blood vessels (capillaries), sickle cells are rigid and fragile. Their rigidity means that they tend to plug blood vessels from time to time, thereby obstructing the flow of blood and causing tissue and organ damage. Normal red blood cells last for 120 days but sickle cells survive for only 6 to 30 days. They are lost far faster than they can be produced in the bone marrow. The resulting chronic deficiency of red blood cells and hemoglobin constitutes anemia.

Prior to modern medicine most people affected by sickle-cell anemia would have died in childhood from infections and lung damage. It is a painful condition and pain-relieving medication is a common form of treatment.

Infections are controlled using antibiotics. Blood transfusions are common. The disease can be cured using bone marrow transfusions but this is a high-risk procedure. Genetically engineered mice carry the human sickle-cell gene, which helps in understanding the biochemical basis of the disease and developing drug treatments. It is hoped that drug treatments can eventually modify the structure of the hemoglobin molecule and lead to a cure.

People who inherit a gene for sickle hemoglobin from both parents develop the disorder. People who inherit only one such gene are carriers. In their case, about half of their hemoglobin is of the sickle variety but this is not enough to produce rod formation and sickling of the red blood cells. Sickle-cell anemia is more common among African Americans than other ethnic groups living in the United States. The chances that an African American is a carrier are 1/12 and the chances of two carriers being married is $1/12 \times 1/12$, or 1/144. The chances that a child of two carriers will inherit both sickling genes is 1/4, which means that approximately one person in 576 contracts the disease.

There are many different human genes that vary as a function of geography and are therefore more likely to occur in people of one ancestry than another. Thus, the gene causing adult lactose intolerance is more common among Asians than Europeans and Asians are also more likely to have an adverse physiological reaction to small quantities of alcohol. Such group differences in gene frequencies have sometimes been used as an empirical justification for the view that different human populations constitute distinct races. That this is a mistake is well illustrated by the case of sickle-cell anemia.

Just because the sickle-cell gene crops up in African Americans but almost never in European Americans, there is a tendency to think of it as an African trait. Yet only around 8% of African Americans carry the sickle-cell gene. The vast majority do not. The geographical distribution of the sickle-cell gene, which clusters in warm, humid, regions of the African continent, has provided a fairly compelling clue to its evolutionary origins. For some equatorial tribes, as many as 40% of the people are carriers. Maps of the malarial parasite *Plasmodium falciparum* and of the human sickle-cell gene match each other quite closely both inside and outside Africa.

The sickle-cell trait is not distinctive of all Africans, and it also crops up at frequencies of up to 8% in the Middle East and in southern Asia. These regions are also infested with *Plasmodium falciparum*, the malarial bacterium. This indicates that far from being a racial characteristic, the sickle-cell trait, which occurs in very low frequencies for most of the world's population, is really a pan-human adaptation to living in malarial regions.

The advantages conferred by the sickle gene for populations in malarial regions means that its carriers survive and reproduce more successfully than others. Over many generations, the frequency of the gene increases at these sites. It seems probable that the gene can never entirely penetrate a population because the more frequent the gene is, the greater the probability of two carriers getting married and their children developing anemia. Apparently when the gene frequency rises to 40% the advantages conferred by malarial resistance are counteracted by the decline in reproductive success from producing children having the illness. Sickle-cell anemia thus shows natural selection at work in human populations.

See also CREATIONISM; GENETIC ENGINEERING; PKU; SCIENTIFIC RACISM.

Further Reading

Embury, S. H., R. P. Hebbel, and N. Mohandas. *Sickle Cell Disease: Basic Principles and Clinical Practice.* New York: Raven Press, 1994.
Ruse, M. *Darwinism Defended.* Reading, Mass.: Addison Wesley Longman, 1982.
Serjeant, G. R. *Sickle Cell Disease.* New York: Oxford University Press, 1992.

Silkwood, Karen

(1946–1974)
American
Union Activist

Employees at nuclear power plants come in close proximity with radioactive materials and are at an elevated risk of radioactive contamination and resulting health problems. This issue played a major role in the life of Karen Silkwood. Silkwood was a lab analyst at a Kerr-McGee plutonium-processing plant in Crescent, Oklahoma, when she became contaminated with plutonium, which was being processed for nuclear fuel.

Soon after learning that she was seriously contaminated, Silkwood died in a car accident, on November 13, 1974, while driving to meet a *New York Times* reporter and a union official to discuss unsafe practices at the Kerr-McGee plant. Whether the car crash was caused by impaired driving due to the use of Quaaludes or by foul play has never been established. The incident received sensational treatment in a 1983 movie (*Silkwood*, directed by Mike Nichols) that was based on her life. After a long legal battle, Silkwood's family won a damages suit against Kerr-McGee. Following various appeals by the company, the case was settled out of court, in 1986, when $1.38 million was awarded to Silkwood's heirs. Silkwood is portrayed in the movie by Meryl Streep as a martyr whose whistle-blowing activities were responsible for her paying the ultimate price. Whistle-blowers in science and industry generally pay a steep personal price but rarely lose their lives.

See also ASBESTOS AND CANCER; CHERNOBYL NUCLEAR DISASTER; NUCLEAR POWER; WHISTLE BLOWING.

Further Reading
Rashke, R. *The Killing of Karen Silkwood.* Ithaca, N.Y.: Cornell University Press, 2000.

Silver Spring monkeys

These 16 experimental monkeys had been used in movement research by neuroscientist Edward Taub of the Institute for Behavioral Research that was discontinued as a consequence of actions by PEOPLE FOR THE ETHICAL TREATMENT OF ANIMALS (PETA), a political pressure group opposed to animal research. The monkeys had undergone surgery to sever the sensory nerves to their limbs and the case was made that they had received inadequate aftercare. The movement-disabled monkeys were used as a publicity device by PETA, whose legal actions succeeded in keeping the animals alive long after they would normally have been destroyed by scientists. Ironically, the fact that the monkeys were kept out of research labs for many years facilitated discoveries concerning "rewiring" of the brain following neural injury. Taub's original research with the monkeys has recently been successfully applied to recovering function in the case of human stroke victims.

See also ANIMAL RESEARCH, ETHICS OF.

Further Reading
Palca, J. "Famous Monkeys Provide Surprising Results." *Science* 252 (1991): 1789.

simultaneous discovery

In simultaneous discovery, scientists or inventors in the same society arrive at the same innovation independently of each other at approximately the same time. For example, Elisha Gray and Alexander Graham Bell developed the telephone simultaneously. Bell submitted a patent application on the same day, February 14, 1876, that Gray filed a caveat (or statement of intent to perfect) for a device to transmit voice. The term "multiple discovery" is often used instead of simultaneous discovery. Independent discovery occurs when two or more people separated by time or place come up with the same idea or invention.

Sociologists of science have proposed two explanatory theories for the origin of scientific and technological discoveries. Either the discovery is a function of the social context (the deterministic view) or it is due to the quali-

ties of the individual making the discovery (the inventive genius theory). The deterministic theory is supported by the phenomena of simultaneous invention and independent invention. According to the deterministic theory of invention, the telephone emerged when the time was ripe and the individual making the invention was of little importance. Given the appropriate context, *someone* would eventually have made the discovery.

Innovations that arise in one civilization may be independently produced in another at a later time. Thus, many inventions developed in the West, including guns and the printing press, had previously been in use in China. According to the deterministic theory of invention, the printing press was developed in China in the ninth century (some six centuries before the GUTENBERG revolution in Europe) because the Chinese system of government comprised a large centralized bureaucracy that was highly dependent on writing. Printing was used to promulgate laws, make currency, and publish handbooks on practical topics such as medicine and pharmacy as well as religious texts and encyclopedias.

Simultaneous inventions and discoveries have produced many PATENT disputes and battles over scientific priority. Bell and Gray waged an intense legal battle over who had invented the telephone all the way to the U.S. Supreme Court, which ruled in Bell's favor. Ironically, the trial revealed that the device which Bell had patented could not have transmitted the human voice, whereas the one that Gray intended to develop would have worked.

Some priority disputes have been handled much more ethically than others. Thus, CHARLES DARWIN shared priority with Alfred Russell Wallace for the theory of evolution by natural selection by reading both of their papers at the same meeting of the Linnaean Society. Darwin's was read first, however. Darwin had devised the theory some 20 years earlier than Wallace but had neglected to publish it. Even though there was a large time gap in favor of Darwin, both men had been deeply influenced by reading ROBERT MALTHUS'S (1766–1834) ideas on the inevitable limitation of populations by available food supply, which suggested the struggle for survival that is a cornerstone of the theory of evolution by natural selection. This common influence supports the deterministic theory of inventions. Today, Darwin's efforts at marshaling systematic support for the theory, not to mention his close connections to the most influential biologists of his day, have ensured that he is given most of the scientific credit for the discovery.

In general, scientific credit is determined by priority of publication. The PUBLISH-OR-PERISH principle is operative. This point is illustrated by the somewhat arcane example of the terrarium, a case used by Victorian travelers to bring exotic plants home across the seas. This device was first developed by Scottish biologist A. A. Maconochie in 1825. The terrarium came to be known as the Wardian case, however, because Nathaniel Bagshaw Ward, a London surgeon and gardening enthusiast, published his invention before Maconochie, who delayed 14 years before committing his innovation to print. Similarly, the entire credit for the periodic table of elements has gone to Russian chemist DMITRI MENDELEYEV, who published his important discovery in 1869. German chemist Lothar Meyer did not publish his periodic table until 1870. He had developed it two years earlier, possibly anticipating Mendeleyev.

These examples illustrate the cutthroat aspect of priority. The individual who is first to publish generally gets all of the scientific credit whether this is fair or not. When a discovery is published independently by authors working in different countries, using different languages, however, each may receive most of the credit in their respective language communities. An interesting example of this splitting of credit is the independent invention of calculus by ISAAC NEWTON (1642–1727) and Gottfried Leibniz (1646–1716), a leading German scholar. Historians of science agree that Newton had priority but they also accept that Leibniz developed calculus without any inkling of Newton's work. Newton was not happy to share the credit for this discovery and he lost no opportunity in making personal attacks on Leibniz, whom he accused of academic theft. He even used his influence over the ROYAL SOCIETY OF LONDON to launch a bogus investigation of the priority dispute. The Royal Society's report seems to have been written by Newton himself. It drew the absurd conclusion that Leibniz was guilty of PLAGIARISM.

One reason that scientists and inventors are capable of such intense competition over priority is that public credit is one of the most important rewards encouraging people to enter these careers. Having priority confers distinction among peers and is a way of making a mark in history. Those who are first to arrive at some discovery feel that this is a sign of intellectual prowess, just as breaking a world record in an athletic event is a sign of athletic ability.

This belief is strongly held by many scientists and historians but it may be an illusion. According to English philosophers of science David Lamb and S. M. Easton, multiple invention or discovery is the rule rather than the exception in technology and science. The telescope provides a remarkable illustration. Commonly attributed to Dutch lens-maker Hans Lippershey in 1608, the telescope was improved by GALILEO in 1609 (even though he had never seen an original). Galileo claimed priority but he was only one of many to do so. According to Lamb and Easton, priority had also been claimed by Giambattista

Della Porta (1558), Thomas Digges (1571), Cornelis Drebbel, Francesco Fontana, Zacharias Jansen, and Jacob Metius (all in 1608). Evidently, there was something about the year 1608 that made scientists want to develop instruments to satisfy their curiosity about the heavens, resulting in at least five independent inventions of the same instrument. Galileo claimed priority for the thermometer (1592–97) but this was disputed by Giambattista Della Porta (1606), Cornelis Drebbel (1608), Santorio Santorio (1612), and Robert Fludd (1617).

A similar muddle surrounds the invention of the telegraph. Although this is often attributed to Samuel Morse (1837), it is clear that the first electromagnetic telegraph was actually constructed by fellow American Joseph Henry six years earlier. Henry neither patented nor published his discovery. The first functioning telegraph system was developed by W. F. Cooke and Charles Wheatstone, in England, in 1837, seven years before Morse's was operational in the United States. The telegraph was also developed independently in Germany in 1837.

The remarkable simultaneity of great ideas is by no means restricted to technological inventions. It is also seen in science. The discovery of oxygen as a constituent of air that is responsible for burning is often attributed to Antoine Lavoisier, but it is not clear that he developed the modern conception of this gas and its role in combustion. At least three scientists, Lavoisier, Carl Scheele, and Joseph Priestly, isolated oxygen between 1774 and 1776 although not all knew what it was. By 1776, Lavoisier recognized that oxygen was a constituent of air but it was not until around 1810 that the critical role of oxygen in making fire was realized. The principle of conservation of energy was independently discovered by four European scientists between 1842 and 1847. Similarly, the neglected 1866 work of Gregor Mendel in genetics was independently rediscovered by three geneticists (Carl Correns, Erich Tschermak, and Hugo DeVries) all in the same year, 1900.

There are at least three good reasons why simultaneous discoveries occur in science. Science is progressive and new knowledge at any point draws on what is already known. This helps to explain the historicity of scientific discoveries. It is as though scientists were collectively completing a jigsaw puzzle. Once the puzzle has reached a certain point it is easy to spot elements in the composition and identify the piece that completes them. There are at least two other important reasons for simultaneous discovery. The minds of the scientists have to be ready to receive the discovery. Lavoisier failed to recognize the critical role of oxygen in burning because he remained committed to the old view that fire was a separate principle contained in flammable materials (the PHLOGISTON theory). Philosopher of science THOMAS KUHN referred to such fixed habits of scientific thought as "paradigms." When old paradigms are abandoned, many scientist can become receptive to discoveries. The other reason for simultaneity is technological development. A clear example of this is also provided in the case of the telescope, which allowed heavenly phenomena to be observed that had never before been witnessed. In 1611, Galileo claimed the discovery of sunspots but he had at least two competitors in the same year, David Fabricus and Thomas Harriot. Clearly, all of these people had benefited from the optical advantages of the telescope soon after its discovery.

Lamb and Easton document many other examples of important discoveries that are often attributed to single authors but turn out, on closer examination, to have been made by more than one scientist or inventor. Even the world's most productive inventive "genius," THOMAS EDISON, often found that his devices were developed at the same time as similar devices produced by others so that he was involved in many legal patent disputes. Moreover, Edison was a cunning manipulator of the patent laws who studied the work of other inventors, made minor changes, and passed them off as his own. Among his many other achievements, Edison is remembered as the inventor of the light bulb even though it was developed simultaneously by Joseph Swann, in England, in 1879.

See also INVENTIONS, THEORIES OF; PATENTS; PLAGIARISM.

Further Reading

Cozzens, S. E. *Social Control and Multiple Discovery in Science.* Albany: State University of New York Press, 1989.

Lamb, D., and S. M. Easton. *Multiple Discovery.* Trowbridge, England: Avebury, 1984.

McClellan, J. E., and H. Dorn. *Science and Technology in World History.* Baltimore, Md.: Johns Hopkins University Press, 1999.

Williams, T. *The History of Invention.* New York: Facts On File, 1987.

Singer, Peter

(1946–   )
Australian
Philosopher

Peter Singer is a utilitarian ethicist who sees the capacity to feel pain as the essential characteristic for membership in a moral community. Since many animals have a similar capacity to feel pain as humans, they merit the same kind of moral consideration according to Singer. While teaching in Oxford University in England in the early 1970s, Singer joined a group of vegetarians who refrained from the use of meat because they felt that the suffering inflicted on animals to obtain this food was ethically indefensible. Such views are expressed in Singer's 1975 book, *Animal Liberation.* Singer concluded that animals have a right to be free of suffering inflicted by humans. Anyone who believes that animals should be made to suffer for the sake of human nutrition or health is guilty of SPECIESISM, a sin against animals involving the denial of rights that is analogous to racism and sexism in the human sphere.

Singer is considered the intellectual leader of protest groups who object to the commercial use of animals, such as PEOPLE FOR THE ETHICAL TREATMENT OF ANIMALS. By demonstrating that the eating of meat increases the amount of suffering on the planet, Singer hopes to place the onus on meat-eaters to justify their actions. One weakness of this position is an assumption that people will be won over by such reasonable claims to change dietary practices that are arguably as old as our species. Singer does not take the position that animals can never be used in biomedical research but he does argue that animals are used wastefully and that much research in this area is of little scientific value because animal studies of a drug or medical procedure do not reveal how humans will respond to it.

In addition to his several books devoted to the defense of animals, Singer has written extensively on ethical problems raised by modern medicine and biomedical research. *Making Babies* (1985) focuses on issues raised by in vitro fertilization. Singer recommends that advances in REPRODUCTIVE TECHNOLOGIES should be carefully monitored by government ethics committees. His own views on ethical constraints in this field are unusually lax. Thus, he has no problem about the use of human embryos to grow spare body parts, at least up to the point that they are capable of feeling pain. He also has no problem about human cloning provided it is not used to make an excessive number of copies (i.e., more than one) of any individual. *Rethinking Life and Death* (1994) makes the case that the traditional ethic of the sanctity of human life needs to be replaced by an ethic upholding the quality of life. These ideas are highly controversial because they justify EUTHANASIA and the killing of infants who suffer from painful medical conditions. Many ethicists feel that if the right to life is not enshrined as an ethical principle, there is no protection from a slippery slope in which powerful people can decide who is fit, or unfit, to live. It may also be a mistake to imagine that the right to life can be made separate from the quality of life. For example, the widespread use of euthanasia in Holland has made many elderly people fear that physicians will end their lives against their wishes.

See also ANIMAL RESEARCH, ETHICS OF; EUTHANASIA; PEOPLE FOR THE ETHICAL TREATMENT OF ANIMALS; UTILITARIANISM.

Further Reading

Singer, P. *Animal Liberation*. New York: Random House, 1975.

———. *Rethinking Life and Death*. New York: St Martin's Press, 1995.

Singer, P., and D. Welles. *Making Babies*. New York: Scribner's, 1985.

Skinner, Burrhus Frederic

(1904–1990)

American

Psychologist

B. F. Skinner was the world's leading authority on animal learning. Whereas, Russian physiologist IVAN PAVLOV had focused on the conditioning of internal bodily responses, (an emphasis that was reflected in the work of Skinner's predecessor in American behaviorism, JOHN B. WATSON), Skinner's work involved learning of skeletal movements (or operant learning). In his early work, he discovered that pigeons could be trained to perform unusual movements, such as turning somersaults and striking a ping pong ball with their beaks, if successive approximations to these movements were selectively reinforced with grain. This technique, known as "shaping," had always been used by circus trainers in developing animal tricks.

Convinced of the power of reinforcement to produce novel behaviors, Skinner followed the model of Pavlov in developing a precisely controlled environment in which this phenomenon could be scientifically investigated. The operant conditioning chamber (or Skinner box) consisted of a sound attenuated environment in which animals could manipulate some device to receive a reinforcement of food or water. Thus pigeons pecked at a plastic disk to receive two seconds of access to an automatic food hopper or rats pressed a lever to receive a tiny dipper full of water. Each response of the subject was automatically recorded by means of the upward movement of an inked pen on a slowly moving sheet of water. In this way, Skinner allowed the animals to graph their own behavior! Just as Pavlov had described the basic principles of classical conditioning, Skinner investigated basic operant phenomena such as schedules of reinforcement. One of his most important discoveries was that animals can be induced to work much harder for intermittent (or uncertain) reinforcement than they can for reinforcement that appears each time an animal makes the required movement. Such schedule effects were demonstrated for unrelated animals like rats, pigeons, and monkeys, and clearly apply to human beings. One example is the fact that people who do piecework often work much harder than the majority of employees who are paid according to time at the job. In this example, money is the reinforcer and it makes people work so hard that their health may be endangered, which is one reason that trade unions do not allow this form of compensation.

Skinner believed that the conclusions drawn from his experiments with rats and pigeons could be used to explain virtually all aspects of human behavior and he wrote a number of nontechnical books in which this claim was made. He was an active inventor who developed a number of devices for use outside the animal laboratory. One was an environment designed for the care of infants with a minimum of labor. He used this device in raising his own daughter but it was not a commercial success. This real-world experiment earned him a certain amount of hostility on ethical grounds because he was treating his own flesh and blood as though she were an experimental animal.

Undaunted by such criticisms, Skinner argued that a more effective society could be developed if operant principles were deliberately used to control our behavior all of the time. His novel *Walden Two* describes a rational Utopia that is based on operant control of human behavior. Critics of such a society have often pointed out that such a community could work but have countered that few people would wish to accept its restrictions. That it can work is suggested by the persistence of an experimental community modeled on *Walden Two* in Roanoke, Virginia.

Many ethicists have been appalled by the lack of freedom and individuality in a Skinnerian community and have argued that it undermines human dignity by treating people like experimental animals. In response to such criticisms, Skinner wrote *Beyond Freedom and Dignity*, which makes the case that whether we are aware of it or not, our actions are the product of reinforcement contingencies in our society. If we want our society to function more effectively, then we must recognize the contingencies of reinforcement that are currently controlling us and modify them in ways that promote happiness and prevent social problems. Skinner's point was that freedom and dignity are illusions and that we should not allow these sacred cows to prevent us from developing a better society based on the principles of operant learning.

Skinner's view that human actions, and even private thoughts, are controlled by contingencies of reinforcement and punishment in the external environment is deeply antithetical to the views of most philosophers, particularly ethicists. For many ethicists, it is the capacity for reasoning, and reasoned control of our actions, that allows us to participate in a moral community. If our thoughts and actions alike are controlled by the social environment, then we have no individual responsibility for our own actions and it is pointless to discuss ethical and unethical conduct.

Due to the fact that Skinner's behaviorism devalues the role of thought in controlling human behavior, it is sometimes referred to as "black box" psychology. This school of thought prevailed in American psychology from about 1930 to about 1960. Growing interest in the functioning of the brain and increasingly sophisticated techniques of studying what is happening in the living brain have coincided with other developments that have brought biological determinism to the fore in psychology. These include an increasing realization of the role of genetics in human personality as revealed in research such as the MINNESOTA TWIN STUDY and the emergence of drug therapies for behavioral disorders, the phenomenon of BIOLOGICAL PSYCHIATRY. These influences have removed Skinner's behaviorism from the center stage of psychology. Skinner received a lifetime achievement award from the American Psychological Association, the only person to be so honored. He died in 1990 from complications of leukemia.

See also PAVLOV, IVAN; WATSON, JOHN.

Further Reading

Bjork, D. W. *B. F. Skinner: A Life.* Washington, D.C.: American Psychological Association, 1997.
Skinner, B. F. *The Behavior of Organisms.* New York: Appleton-Century-Croft, 1938.
———. *Walden Two.* New York: MacMillan, 1948.
———. *Science and Human Behavior.* New York: Free Press, 1953.
———. *Beyond Freedom and Dignity.* New York: Knopf, 1971.

Smith, Adam

(1723–1790)
Scottish
Philosopher, Economist

A social philosopher with wide interests, Adam Smith is remembered as the founder of modern economics. He was appointed professor of moral philosophy at Glasgow University in 1752 and remained there until 1763, when a desirable job, as the tutor of the young duke of Buccleuch, gave him an opportunity to travel widely in Europe. Glasgow was a vigorous intellectual center at the time and Smith was friends with David Hume, a leading ENLIGHTENMENT philosopher.

His most important book, *An Inquiry into the Nature and Causes of the Wealth of Nations* (1776), anticipates many of the ideas of modern conservative economists. Smith's main thesis was the importance of free trade, both within and across national boundaries, for promoting competition and the efficient production of goods at a fair price. Free trade increased wealth. Monopolies, whether private or of the government, were bad for competition and inhibited the capacity of a nation to generate wealth. With the exception of limited functions, such as defense

and infrastructure, governments should stay out of the economic life of a nation as much as possible. This perspective is known as laissez-faire capitalism.

Smith's first book *The Theory of Moral Sentiments* (1759) presented an interestingly complex view of human motivation that would be repeated in *The Wealth of Nations*. On the one hand, like THOMAS HOBBES (1588–1679), Smith saw human beings as driven by passions but the impulsiveness of human action was moderated both by the capacity for reason and by sympathy for others, core values of the ENLIGHTENMENT age in which Smith lived. The most important distinction between Smith and Hobbes is that Smith was an optimist about human nature and Hobbes was a pessimist. Hobbes felt that without an absolute government to control them, people would behave with appalling savagery making life nasty, brutish, and short. By contrast, Smith argued that good can come of the selfish competition among individuals in commerce. In a famous metaphor, he argued that although driven by greed, businesspeople were led, as if by an "invisible hand" to serve the best interests of the entire society. Competition drives down the prices of goods and encourages more efficient methods of production. Moreover, competition drives labor and capital from less profitable industries to more profitable ones, thus increasing the wealth of a nation. Selfishness leads to competition and competition is good for a society because it generates wealth.

Adam Smith developed an optimistic theory concerning the generation of wealth and economic growth that was in keeping with the spirit of his times. His arguments provide an ethical justification for unrestrained capitalism. He concluded that increased efficiency of production methods, such as the division of labor, increased the wealth of a whole society and thus provided an ethical justification for economic change and technological development. (It is interesting that Smith had no inkling that the Industrial Revolution was around the corner, even though JAMES WATT, one of its key inventors, was a personal friend). The engine of economic growth was competition, which meant that the base motive of human greed had the potential for improving human wealth and happiness. If governments and private monopolies could be prevented from stifling economic competition, nations would become more prosperous over time. Smith died in 1790 and was buried in Canongate Churchyard in Edinburgh. His optimistic writings were well received by contemporaries, in contrast to the pessimistic views of a later economist, THOMAS MALTHUS (1766–1834), who saw the population problem as ensuring continued economic misery for most people. Smith's most important impact on ethics is his complex view that the selfishness of the individual contributes to the common good. This is diametrically opposed to the TRAGEDY OF THE COMMONS idea according to which individual selfishness inevitably destroys the common good.

See also ENLIGHTENMENT, THE; HOBBES, THOMAS; MALTHUS, THOMAS ROBERT.

Further Reading
Griswold, C. L., Jr. *Adam Smith and the Virtues of Enlightenment.* New York: Cambridge University Press, 1998.

smog

Industrial pollution in cities is visible as smog, or a combination of smoke and fog (see photo). In the early days of the INDUSTRIAL REVOLUTION, smog was produced by condensation of water droplets on smoke particles in the air. The smog of today is largely of petrochemical origin and comes from the use of fossil fuels by industry, in cars, and around homes. Unburned hydrocarbons combined with other air pollutants are exposed to solar radiation, which causes a complex chain of chemical reactions that produces numerous irritant substances such as ozone, organic acids, aldehydes, peroxyacetyl nitrate, organic particles, ketones, and other oxidants.

Severe smog ground industrial cities like Los Angeles and Tokyo can have important health consequences and this makes a compelling case for reducing pollutant emissions into the air. Cities with heavily polluted air are associated with shorter life. Research has shown that exercising in smog damages the lungs and aggravates respiratory ailments. Smog also contributes to various types of cancer.

See also ACID RAIN; AIR POLLUTION, HYBRID-POWERED CARS.

Smog in New York, in 1949, demonstrates the polluting consequences of industry and motor vehicles in cities. (LIBRARY OF CONGRESS)

social contract theory

Social contract theory begins with the assumption that there is a contract between rulers and the ruled. This view is modern and replaces the earlier view that rulers had a right to control their subjects whether this was by virtue of their intrinsic superiority or was conferred on them as a divine right. Social contract theory originates in English philosopher THOMAS HOBBES (1588–1679). Hobbes saw all individuals in a state of nature as equal to each other. However, people are driven by selfish needs and will attempt to dominate others by force and cunning. This means that in a state of nature, according to Hobbes, life is nasty, brutish, and short. To escape from the adverse consequences of their own antisocial impulses, people form a contract among themselves in which they agree to follow the dictates of an absolute ruler. Each individual thus surrenders his personal autonomy for the sake of civil order and the security of person and property that comes from agreeing to be controlled by a government.

The social contract defined by Hobbes is permanent and can never be breached by the governed no matter how bad the absolute ruler turns out to be. English philosopher John Locke (1632–1704) defined the social contract very differently. He saw it as a covenant between the government and citizens for the purpose of protecting the rights of the individual to life, liberty, and personal property. If the government did a bad job of protecting these rights, it could be dismissed at any time. Locke's libertarian ideas were extended by French philosopher Jean-Jacques Rousseau (1712–88) who felt that rights should be extended beyond propertied and franchised classes to all the individuals in a state.

Rousseau's perspective as developed in his book *The Social Contract* (1762) is closer to Hobbes than Locke, however, because he values equality more than freedom and endorses a totalitarian state that embodies the general will and always trumps the rights of the individual. For example, he sees the state as owning all of the property of its members whereas Locke was a staunch defender of the inalienable right to hold private property.

See also HOBBES, THOMAS; LOCKE, JOHN; ROUSSEAU, JEAN-JACQUES.

Further Reading
Morris, C. W., ed. *The Social Contract Theorists: Critical Essays on Hobbes, Locke, and Rousseau.* Lanham, Md.: Rowman and Littlefield, 1999.
Russell, B. *A History of Western Philosophy.* New York: Simon and Schuster, 1972.

social Darwinism

Social Darwinism involved the application of evolutionary principles to human societies. Just as natural selection eliminated the weak through survival of the fittest (a phrase probably originated by British philosopher HERBERT SPENCER who was a leading social Darwinist), so the best people were believed to prevail in the competition over wealth within human societies. By similar reasoning, natural selection could act on whole societies allowing them to evolve just as biological organisms do. Social Darwinists believed that societal evolution consists of a process in which stronger groups prevail over weaker ones. Ultimately the best tribes wiped out the weaker, or less fit, ones. Similarly colonial domination of other countries by European powers was a part of the natural course of social evolution through which superior peoples vanquished less fit ones.

Social Darwinism was a utopian philosophy that held that only the best survive, thus leading to constant improvement of individuals and societies. It appealed to the individualism and laissez-faire capitalism that characterized American society in the late 19th century. When Herbert Spencer visited the United States in 1882, he became an instant celebrity. Other leading exponents of social Darwinism included English economist and journalist Walter Bagehot (1826–77) and Austrian sociologist Ludwig Gumplowicz (1838–1909). Gumplowicz was the author of *Der Rassenkampf* (The racial struggle, 1872), which argued that social evolution comes about through group struggle. It is not difficult to see in such works a foreshadowing of the racial concerns of Nazi ideologues.

The association of social Darwinism with Nazi atrocities committed against Jews and other racial minorities has meant that it is viewed with horror by modern academics. The distaste is particularly strong in sociology and it has led to an extreme reluctance to apply evolutionary, or biological, explanations to human social behavior. According to some sociologists, resistance to evolutionary theory has weakened sociology as a science by tying it to a liberal political agenda. This is problematic because true sciences are supposed to be concerned about uncovering the truth rather than promoting any political perspective.

The powerful aversion of sociologists and others to Darwinism has played an important role in fueling opposition to modern evolutionary accounts of human selfishness, altruism, sexuality, etc., which were originally referred to as sociobiology and are now called evolutionary psychology. This name shift came about as a deliberate strategy on the part of modern evolutionary scientists to distance themselves from social Darwinism.

See also CASTRATION; DARWIN, CHARLES; GROUP SELECTION CONTROVERSY; SOCIOBIOLOGY CONTROVERSY; SPENCER, HERBERT.

Further Reading
Hofstadter, R. *Social Darwinism in American Thought.* Boston: Beacon Press, 1992.

Lopreato, J. P., and T. Crippen. *Crisis in Sociology: The Need for Darwin*. Somerset, N.J.: Transaction, 1999.

sociobiology controversy

Sociobiology refers to the study of behavior from the perspective of evolutionary biology. Such an approach, however lacking in scientific rigor, had informed SOCIAL DARWINISM and fell into disrepute following the genocidal excesses of Nazi Germany, which can be interpreted as a logical extension of this way of thinking. When evolutionary theory returned to the study of behavior, it was initially applied to animal behavior. Exciting theoretical advances were made in unraveling the problems of altruism, as exemplified by the nonreproductive workers in colonies of bees and other social insects, a problem that had perplexed CHARLES DARWIN and made him so doubtful of his evolutionary theory that he delayed publishing it for almost 20 years. English biologist William Hamilton, writing in 1964, showed that such examples of animal altruism could be explained in terms of gene selectionism (see ALTRUISM, EVOLUTION OF). Even though the worker bees were not reproducing directly, they were helping the queen to produce sisters who shared their genotype. Due to a peculiarity of the reproductive biology of social insects, the sisters are actually more closely related to each other than they are to their own mother (because they always receive the same chromosomes from the father). Such theoretical advances suggested to animal behaviorists that there were many aspects of social behavior that could be examined from the point of view of evolutionary adaptation and gene selectionism. In addition to altruism, topics examined by sociobiologists included animal communication, reproductive systems, social structure, antipredator behavior, and foraging strategies.

Many of the ideas and findings from the flourishing new field of sociobiology were summarized in a 27-chapter tome entitled, *Sociobiology: The New Synthesis* (1975) by Harvard biologist Edward O. Wilson, an authority on the behavior of ants. Had Wilson stopped after Chapter 26, the book would not have produced any controversy. The final chapter tried to fit human beings into the framework of animal sociobiology. It is a highly speculative chapter coming at the end of a highly speculative text but it draws very few solid or original conclusions. It is also full of disclaimers about the unique flexibility of human behavior. Nevertheless, it clearly does imply that human behavior is genetically determined to some degree. If it were not, human sociobiology would be a vacuous exercise since the concept of adaptation relies on genetic inheritance as the mechanism through which beneficial traits are passed from one generation to the next.

Wilson's book was well received by many scholars in the field but was vilified by some of his Harvard colleagues, including paleontologist Stephen Jay Gould and geneticist Richard Lewontin. The critics focused primarily on the last chapter of Wilson's book and professed to see in it a return to the shameful social Darwinism of the past. Wilson was accused of being a fascist and a closet racist. His Harvard enemies allied themselves with a pressure group called SCIENCE FOR THE PEOPLE, a left-wing alliance assembled for the specific purpose of countering the right-wing threat posed by Wilson's book and the scientific discipline it represented. They formed a Sociobiology Study Group that set itself the task of finding flaws in Wilson's reasoning and publicizing these problems.

The sociobiology controversy is interesting because it was conducted in a very public manner. Many of the bitter differences between scholars over academic questions go completely unnoticed. Attention-getting tactics included throwing a glass of water at Wilson during an American Association for the Advancement of Science lecture.

The sociobiology debate can be said to have emitted far more heat than light. The controversy between Wilson and the Science for the People Group was never very satisfactory as an academic debate because Wilson has denied saying most of what he was accused of having said. With hindsight, it is clear that *Sociobiology* is full of scientifically incautious speculation and that it pays little attention to the complex problem of behavioral development. It has other failings also, including a tendency toward moralizing that is strangely out of place in a scientific work. Even though Wilson's enemies were happy to point out his mistakes, it is fairly clear that the debate was not really a scientific one. If it had been, critics would have read the first 26 chapters with equal attention to the last one.

Wilson felt strongly that the whole controversy represented an attempt by the academic left to punish him for daring to have made statements that were contrary to their agenda. He referred to this as ACADEMIC VIGILANTISM to emphasize the view that his academic freedom was being suppressed.

Wilson's opponents objected to speculation about the biological determination of human behavior. This was not really a scientific question but a moral one. The feeling was that genetic determinism had been so badly misused by fascists in the past that we should prevent this from happening again. Scientists had to be much more cautious about applying biological explanations to human behavior than they were about doing the same for nonhuman animals.

As a scientific question, the role of genetics in human behavior was more controversial in 1975 than it is today. An accumulation of evidence from behavior genetics, of which the MINNESOTA TWIN STUDY is an important example, has produced increasing acceptance of genetic influence on human intelligence, personality, and behavior, although modern developmental psychologists believe human behavior is always the result of a combination (or interaction) of environmental and genetic influences.

With increasing acceptance of biological determinism, the sociobiology controversy died down. Wilson is no longer accused of being a fascist or a racist. The term *sociobiology* is rarely used today. Adaptationist thinking is now so firmly established in the field of animal behavior that there is no longer any need to call attention to it. In human behavioral research, evolutionists have dropped the term sociobiology in favor of *evolutionary psychology,* which is due partly to the bad press received by sociobiology thanks to Science for the People and its allies and partly to a shift in emphasis away from genetic determinism of behavior toward the study of context-dependent psychological mechanisms.

See also ACADEMIC VIGILANTISM; ALTRUISM, EVOLUTION OF; EVOLUTIONARY PSYCHOLOGY AND SEX DIFFERENCES; MINNESOTA TWIN STUDY; SCIENCE FOR THE PEOPLE GROUP; SOCIAL DARWINISM; WILSON, EDWARD O.

Further Reading

Caplan, A. L., ed. *The Sociobiology Debate.* New York: Harper and Row, 1978.
Segerstrale, U. *Defenders of the Truth: The Battle for Science in the Sociobiology Debate and Beyond.* Oxford: Oxford University Press, 2000.
Wilson, E. O. *Sociobiology: The New Synthesis.* Cambridge, Mass.: Harvard University Press, 1975.
———. "Academic Vigilantism and the Political Significance of Sociobiology." *BioScience* 26 (1976): 183–90.

Socrates

(c. 470–399 B.C.)
Greek
Philosopher

Apart from a stint as a soldier in the Peloponnesian War, Socrates spent his entire life in Athens where he made a living as a stone carver. He developed a reputation as a teacher of philosophy and attracted many disciples. His teachings have been transmitted to posterity through the dialogue of his student, Plato as well as in the works of historian Xenophon and playwright Aristophanes, fellow Athenians. None of these sources is historically reliable and it is impossible to determine whether Plato's Socrates is expressing his own ideas or those of Plato. As a character in the dialogues, Socrates asks many questions with the intention of bringing out knowledge that is believed to be hidden inside the mind. This pedagogical technique is known as the Socratic method.

Socrates' educational efforts were not appreciated by the Athenian authorities and they tried him for corrupting youth as well as impiety. The real reason for his disfavor may have been his willingness to challenge others in argument. Sentenced to death by drinking poison, Socrates had an opportunity to escape from Athens with the help of his friends. He is famous for having chosen to stay and drink the poison on the premise that if we choose to live in a city then we are morally bound to obey its laws. This decision made Socrates, in the view of posterity, a martyr for learning.

Socrates was a man who embodied a strange mixture of skepticism and mysticism. Some of his contemporaries saw him as indistinguishable from the Sophists, who believed in nothing and engaged in arguments primarily for the purpose of winning by means of rhetorical tricks. By contrast, Socrates possessed a well-developed moral system involving a virtue approach to ethics. According to his perspective, to do wrong was to injure one's soul and this must be avoided at all costs. For this reason, it was always better to be the victim of injustice than the perpetrator. This unusual approach to ethics helps to explain the manner of his death because he argued that to break the laws of Athens by escaping would have done harm to the city. In Socratic ethics, all virtue is reduced to knowledge. This means that a person who has full knowledge of their actions is incapable of doing wrong.

Socrates' mysticism influenced Plato and other philosophers of antiquity and encouraged them to turn away from the more scientific approach to knowledge, as represented by ARISTOTLE, for example, to search for spiritual fulfillment. The view that the inner life of the mind or spirit is more important than the outer world has been extremely influential among Christian thinkers through the centuries. Some scholars, like English philosopher Bertrand Russell, have argued that such mysticism has had a harmful effect in retarding the development of science and technology.

See also ARISTOTLE; MORAL PHILOSOPHY; PLATO.

Further Reading

Brickhouse, T. C., and N. D. Smith. *The Philosophy of Socrates.* Boulder, Colo.: Westview Press, 1999.
Roberts, J. W. *City of Socrates: An Introduction to Classical Athens.* London: Routledge, 1998.
Russell, B. *A History of Western Philosophy.* New York: Simon and Schuster, 1972.
Taylor, C. *Socrates.* New York: Oxford University Press, 1999.

Sokal, Alan, and *Social Text* parody

Physicist Alan Sokal attained notoriety around the world by writing a parody of cultural studies of science that was accepted for publication by the unwitting editors of *Social Text*. The hoax greatly amused journalists and readers around the world because it supported the conclusion that postmodernist philosophy was indistinguishable from gibberish, even for philosophers themselves.

Sokal, who shares a left-wing political orientation with the victims of his hoax, began by being concerned with the point of view known as epistemic relativism. This is the claim that scientific truth is relative to the society in which a scientist lives. If people in some society believe that the earth is flat, according to this view, then the earth's flatness is just as real, and just as valid, as its roundness is to us. What concerned Sokal most about epistemic relativism is that it is an extremely weak perspective for effecting attitudinal change and social reform. Merely taking a view such as "my feminist perspective is every bit as good as your sexist perspective," based on epistemic relativism is never going to convince anyone that feminism is better than sexism.

As a scientist, Sokal became particularly alarmed when he discovered that many academic left-wingers were attacking science itself on the basis that scientific objectivity is a myth. Science, according to this view, is just a narrative, one among many possible types of social construction shaped by the subjective biases of scientists but that has had little, if any, relationship to an external reality.

Faced with a large collection of papers that attacked science, often with a very weak grasp of the ideas that they were attacking, Sokal could have written a reasoned rejoinder. Instead, he decided that a parody might be a more effective way of satirizing the antiscientific views that were being expressed. His parody, ostentatiously titled *Transgressing the Boundaries: Toward a Transformative Hermeneutics of Quantum Gravity*, pretended to do for his own area of expertise what epistemic relativists had done for other scientific specializations. The paper was a cleverly compiled pastiche of the writings of other scholars in this field, which was backed up with the usual battery of footnotes and sources. Sokal deliberately implanted a number of red flags, including patent errors in reasoning and sentences that were completely unintelligible. Conscious of the fact that the *Social Text* editors might not have been fully conversant with the science in the paper, Sokal deliberately inserted an illogical political section at the end that the editors ought to have spotted even if they had been baffled by the material on quantum physics. All of these red flags went unnoticed in the review process. Sokal feels that the editors must have seen his paper as manna from heaven. Here was a real scientist coming over to their side by denying that there was an objective physical reality and claiming that Einstein's constant was really not constant at all. Sokal unveiled the hoax in an article published in the journal *Lingua Franca*.

Although Sokal's hoax demonstrates dereliction of duty on the part of the *Social Text* editors, it may not prove a great deal more. Despite the glee of newspapers around the world, the hoax does not really provide a successful challenge to postmodernist critiques of science. The lamentably poor quality of the review process has also been demonstrated in the case of science journals. Systematic studies have shown that most reviewers miss most of the errors in scientific papers. This nicety may be unimportant, however, given the undeniable rhetorical impact of the fake scholarship.

Sokal's work clearly breached the expectation that scientific work should be honestly done. Although deceptive, it can be argued that the work was ethically justified because it sought to expose the more grievous dishonesty of those who damage the credibility of the scientific enterprise and pass their own dishonest ramblings off as true scholarship.

See also PEER REVIEW; PILTDOWN MAN; POSTMODERNISM AND THE PHILOSOPHY OF SCIENCE.

Further Reading

Gross, P. R., and N. Levitt. *Higher Superstition: The Academic Left and Its Quarrels with Science*. Baltimore, Md.: Johns Hopkins University Press, 1997.

Resnik, D. B. *The Ethics of Science*. New York: Routledge, 1998.

Sokal, A. "Transgressing the Boundaries: Towards a Transformative Hermeneutics of Quantum Gravity." *Social Text* 46/47 (1996): 217–52.

———. "A Physicist Experiments with Cultural Studies." *Lingua Franca* (May–June, 1996): 62–64.

Sokal, A., and J. Bricmont. *Fashionable Nonsense: Postmodern Intellectuals' Abuse of Science*. New York: St. Martin's Press, 1998.

spam   *See* JUNK E-MAIL.

speciesism

This term was introduced by Australian ethicist Peter Singer to describe a denial of rights for animals analogous to racism and sexism in the human sphere. A speciesist is a person who believes that animals can be made to suffer for the sake of human nutrition or health. Singer's position holds that because they are creatures with the capacity to feel pain, nonhuman animals should be accorded the same kind of moral consideration as is given to humans. He employs a Darwinian orientation according to which different sentient species share a great deal of their moral qualities by virtue of being products of the same process of evolution by natural selection. This orientation is contrasted with the conventional Judeo-Christian perspective that sees people as being endowed with superior qualities and thus destined to rule over other species and use them for human benefit without consideration for their rights and suffering.

See also SINGER, PETER.

Spencer, Herbert
(1820–1903)
English
Philosopher

Herbert Spencer was a very influential philosopher in his own time whose reputation has subsequently declined along with the SOCIAL DARWINIST movement to which his name is tied. Self-taught, he functioned outside the mainstream academic environment of the university and was often scorned by professional scholars, particularly in his native country. Spencer's ideas were embraced with enthusiasm in the United States, apparently because his belief in progress, in individualism, and in a form of capitalism with minimal government interference (laissez-faire) were precisely in tune with American sentiments of the time.

When he arrived in the United States, in 1882, he was received as a celebrity. As soon as his ship docked in New York, he was greeted by Andrew Carnegie, a leading steelmaker, who hailed him as a messiah. This enthusiasm also extended to leaders of the academic world who lavished him with honors and praise. Spencer was a prolific author and his many books often sold hundreds of thousands of copies. Some of them were serialized in popular magazines. Spencer's success as the author of textbooks used in American universities was also remarkable. He wrote leading texts in biology, psychology, and sociology, and his works were discussed in almost every academic discipline. His idea thus reached, and deeply influenced, a whole generation of Americans at various levels of education.

Although Spencer is referred to as a social Darwinian, his writings on evolution preceded Darwin's *On the Origin of Species* (1859) by almost a decade, beginning in 1850. Spencer's more speculative ideas gained widespread acceptance following the publication of Darwin's more careful, and better documented, writings. Modern scholars like to exonerate Darwin from being a social Darwinist but this generous treatment ignores the actual content of Darwin's books, which are littered with references to the struggle of tribe against tribe and acknowledge that some human groups are superior to others in the struggle for existence. It is also clear that Darwin harbored no personal animosity toward Spencer, referring to him as "our

philosopher" by which he meant the intellectual who was extending evolutionary concepts to human beings. Spencer died on December 8, 1903, in Brighton, England.

Spencer's philosophy was extremely optimistic. He believed (falsely) that evolution is progressive. Just as the fittest individuals survive and reproduce in nature, so the best kind of people come to the fore in the capitalistic struggle over money, which Spencer naively equated with natural selection. If a progressive mechanism is built into the natural world and the same mechanism applies to human societies, then people inevitably get better and better! In fact, the only thing required of governments was that they should stand aside and let social evolution work its magic. From this perspective, programs such as smallpox inoculation were perceived as a mistake because they gave unfit individuals the possibility of surviving and reproducing, ultimately contributing to a weakening of the species. The same conclusion applied to government-subsidized education and housing. Spencer's optimism thus has a dark side, a Malthusian lack of responsibility for the poor and underprivileged. This laissez-faire attitude appealed to the individualism of the American frontier.

Social Darwinists believed that natural selection also acted on whole societies allowing them to evolve just as biological organisms do. Societal evolution thus consists of a process in which stronger groups prevail over weaker ones. Ultimately the best tribes wiped out the weaker, or less fit, ones. This belief provided a justification for colonial exploitation of militarily weak societies and it was later to play a critical role in Nazi ideology.

Spencer's reputation suffered from antipathy to the atrocities of the Nazis against Jews and other racial minorities because the ideas of social Darwinian were seen as contributing to racial conflict during this period. In some ways, this is rather like blaming Rousseau for the atrocities of the French Revolution. The accusation is neither completely unfair nor completely merited. Yes, the ideas probably did play the role attributed to them. On the other hand, Spencer did not advocate genocide and could not have foreseen the uses to which his ideas would be put.

Social Darwinism is deeply abhorred in the social sciences today, particularly in sociology where the suspicion of biological explanations for human behavior amounts to a phobia, or irrational fear. For this reason the recent reintroduction of evolutionary theory to the study of human social behavior (first as "sociobiology" and more recently as "evolutionary psychology") has prompted a heated episode of name-calling in which opponents seek to discredit modern day evolutionists be referring to them as social Darwinists. Even in disgrace, Spencer has retained his preeminence. He is seen as the representative figure in social Darwinism and other leading proponents of the theory, such as English economist Walter Bagehot (1826–77) and Austrian sociologist Ludwig Gumplowicz, have been eclipsed.

See also DARWIN, CHARLES; GROUP SELECTION CONTROVERSY; SOCIAL DARWINISM; SOCIOBIOLOGY CONTROVERSY.

Further Reading

Hofstadter, R. *Social Darwinism in American Thought.* Boston: Beacon Press, 1992.
Lopreato, J. P., and T. Crippen. *Crisis in Sociology: The Need for Darwin.* Somerset, N.J.: Transaction, 1999.
Schwartz, D. P., and S. E. Schwartz. *A History of Modern Psychology.* San Diego, Calif.: Harcourt Brace Jovanovich, 1987.

sperm count decline

Whether human sperm count has declined over the past several decades has produced considerable academic controversy. Much of this controversy died down after publication in the *British Medical Journal* in 1992 of a report that summarized the results of the 61 best studies conducted on the question and published between 1938 and 1990. The Danish researchers took great trouble to deal with problems of different measurement procedures and confounding influences such as frequency of sexual activity. They concluded that sperm count had declined 41% between 1940 and 1990, from an average of 113 million to 66 million sperm cells per milliliter of semen. This appears to be a huge drop but two qualifications should be made. The first is that part of the drop may be explained in terms of a lowering of the standard for a normal sperm count (which would depress the average). The other is that a count above 20 million is not associated with fertility problems.

Assuming that the decline is real, even though its size could be exaggerated, the only plausible explanation is in terms of environmental pollution. Consistent with this interpretation is the finding that modern semen samples contain high levels of toxic chemicals, including DDT and PCBs (see HORMONE MIMICS). Exposure to these pollutants could kill large numbers of delicate sperm cells without harming the hardier cells in other parts of the body. A possible role played by pollutants is suggested by a sharp rise in rates of testicular cancer (35% between 1973 and 1988).

See also HORMONE MIMICS.

Further Reading

Bromwich, P., J. Cohen, J. Stewart, and A. Walker. "Decline in Sperm Counts: An Artefact of Changed Reference Range of 'Normal.'" *British Medical Journal* 309 (1994): 19–22.
Carlsen, E., A. Giwercman, N. Keiding, and N. E. Shakkeback. "Evidence for decreasing quality of Semen during Past 50 Years." *British Medical Journal* 305 (1992): 609–13.

"Star Wars" Strategic Defense Initiative (SDI)

Originally proposed by President Ronald Reagan, the Star Wars program was envisaged as a missile defense system based on use of satellites. The idea was to develop a laser system that could shoot down incoming ballistic missiles as fast as they could be launched (approximately 40 per second). Such a laser system could not be built at the time and President George H. W. Bush, Reagan's successor, declared that the Star Wars program had been ended. Despite this official renunciation of SDI, research on the project continued, undertaken by the U.S. Air Force, the Pentagon's Ballistic Missile Defense Organization (BMDO), and the U.S. Space Command. In 1999, the United States allocated funding for the development of a space-based laser, indicating that Star Wars has been resuscitated. Such a system would have to overcome many technical problems, including the hoisting of large quantities of equipment into space at enormous cost and the generation of electricity to power the laser-launching satellites.

Most of the effort of SDI has been expended in the development of antiballistic missile missiles whose goal can be compared to hitting a bullet with another bullet. Known as National Missile Defense (NMD), this project has engendered a number of ethical controversies. The sheer cost and scale of the endeavor has invited criticism. Despite the expenditure of over $60 billion by the Pentagon on antiballistic missile defensive systems, their performance in tests has been too poor to produce any confidence that a reliable missile defense system can ever be developed. As of 1999, only four of 18 tests were successful. Of the 15 high-altitude tests, which are most relevant to a missile defensive system, only two were successful. Even the successes are questionable because it is not clear that they could be replicated in the event of an actual attack in which incoming missiles used decoys to "confuse" the defensive missile.

Another problem was the revelation, in August 1994, that scientists working on this program had faked a critical test by placing a transmitter in the incoming missile. The fraudulent test was used to convince Congress that funding should be continued for SDI.

The technical problems of NMD were accentuated in 1996 when Nira Schwartz, a scientist at TRW, a defense contractor, claimed that her company's recently developed infrared sensor was incapable of distinguishing between a nuclear warhead and a decoy. When Schwartz was fired for insubordination, she sued TRW under a federal law protecting government whistle-blowers. In the course of the litigation, data on missile-defense tests became public. This information was analyzed by physicist Theodore Postol of the Massachusetts Institute of Technology. In a 1997 test, which the Pentagon's Ballistic Missile Defense Office (BMDO) had described as a success, the TRW censor had failed to pick out the warhead from among eight decoys. In future tests, only a single decoy was used. This was a shiny silver balloon designed to be easily distinguishable from the missile. In other words, the tests were being rigged to make it seem that the infrared sensor would be capable of picking out a warhead from among decoys when it would not be capable of doing this in real attack. In view of all of these technical problems with NMD, it is hardly surprising that President Bill Clinton, on September 1, 2000, deferred a decision on whether a $20 billion land-based missile defense system, using a radar station in Alaska, should go forward.

Another ethical criticism of SDI (and NMD) is that it is globally destabilizing because its full development would violate nuclear weapons treaties between the United States and Russia, particularly the Antiballistic Missile Treaty signed by President Richard Nixon in 1972. For this reason, it is opposed by European NATO allies. The United States has countered that some kind of limited defensive system is necessary to neutralize the threat posed by small nations, such as North Korea, that were developing NUCLEAR WEAPONS and could launch an attack on United States territory (i.e., Alaska or Hawaii) by 2005.

See also MILITARY RESEARCH; NUCLEAR WEAPONS.

Further Reading

Kaplan, F., and M. J. Sherwin. *The Wizards of Armageddon.* Stanford, Calif.: Stanford University Press, 1991.

Mendolsohn, J. "Missile Defense: And It Still Won't Work." *Bulletin of the Atomic Scientists* 55, no. 3 (1999): 29.

Reiss, E. *The Strategic Defense Initiative.* New York: Cambridge University Press, 1992.

"A Shot in the Dark." *Newsweek,* June 12, 2000.

Vizard, F. "Return to Star Wars." *Popular Science* 254, no. 4 (1999): 56–61.

statistics and scientific honesty

Modern scientific results are usually presented in some numerical format. The volume of information handled by scientists precludes the publication of raw data and this is either discouraged or prohibited by the editorial policies of most scientific journals in the interests of preserving space and ease of comprehension. This means that all scientists use summary, or descriptive, statistics. The choice of which descriptive statistic to use can be important. If an economist is comparing the income of people in two countries, the average income can be very misleading because it is pushed upward by the number of billionaires and may tell us virtually nothing about the affluence of the populations as a whole. To approach this issue, the economist would

have to report the income of the average person who is poorer than one half of the population and richer than the other half, i.e., the median income. Clearly, the choice of which statistic to use here could have an important impact on how the comparison turned out. A country with great diversity of incomes, such as the United States, would be flattered by the average, whereas a country with less of a spread in earnings, such as Sweden, would look better if the median were used. Scientists generally rely on conventions to tell them which statistic is appropriate for a particular purpose but if there is any doubt, honesty should compel them to report more than one summary statistic.

Even more important than the choice of a summary statistic is the selection of numbers that go into it. In general scientists are ethically obligated to include all of the data exactly as it was recorded. Good laboratory practice once required scientists to record their data in ink on bound notebooks to ensure that no errors or fraud could occur. In the BALTIMORE AFFAIR, Theresa Imanishi-Kari's flagrant disregard for careful recordkeeping undermined her credibility as a scientist and suggested ethical impropriety. Ironically, her practices were so chaotic that they could not be used to make a compelling case against her and she was exonerated of wrongdoing. The use of computerized data collection generally makes for better recordkeeping but computer files can be altered retrospectively opening an opportunity for fraudulent conduct that is virtually impossible to trace.

The same cannot be said of American Nobel laureate ROBERT ANDREWS MILLIKAN (1868–1953) whose careful laboratory notebooks have provided a paper trail to his statistical dishonesty of including only those results of his later oil-drop experiments that matched his preconceptions. A fascinating aspect of this data trimming is that it occurred in the context of a scientific debate. Instead of using the numbers to depict his results in an accurate fashion, Millikan was using them polemically to win an argument against Felix Ehrenhalf, at the University of Vienna, who was conducting similar research and publishing different conclusions. Analogous accusations have been leveled at luminaries such as GALILEO, ISAAC NEWTON, and GREGOR MENDEL (see SCIENTIFIC DISHONESTY). In each case, it appears that these scientists engaged in the ultimate scientific arrogance of knowing more than their data and fudging or fabricating it to fit their conclusions.

Once numerical data have been collected, they are normally entered into a data file using some statistical program such as SPSS or SAS. Data are then screened for errors of data entry. This can be done by examining means (or averages) as well as standard deviations (or spread). Experienced researchers can quickly determine if errors are present because the summary statistics will show a deviation from what is expected. When this happens, the suspect cases must be identified and, if appropriate, corrected. The most common type of problem is a typing error and this can be corrected. For example, if the height of a human subject in an experiment is recorded as 66 feet, it is fairly obvious that the person entering the data typed two sixes instead of one.

Once the preliminary screening to detect obvious errors is completed, careful statisticians look for observations that still stick out. These are known as outliers. These stand apart from the rest of the data points in a scatter plot. If an observation is more than three standard deviations away from the mean, it is considered not to belong in the data and the observation is excluded from further analysis because it will have an undue influence on the outcome of any test that is conducted. It can thus be seen that statistical truth may deviate slightly from the truth of accurate observation that preceded it. In other words, outliers are excluded not because they necessarily represent errors of observation but because they are statistically inconvenient.

Many nonstatisticians are disturbed to find that there is an arbitrary element to all statistical analysis. English prime minister Benjamin Disraeli (1804–81) expressed this disillusionment very eloquently when he said, "There are three kinds of lies: lies, damned lies, and statistics." Critics of statistics may want to believe that there is a great deal more certainty in the world of science than is really the case. Perhaps the best way to think about statistical analysis of scientific data is that the analysis forms a part of the scientific process rather than being distinct from it. If so, then powerful statistical tests can be thought of as rather like powerful microscopes: they allow us to see effects that are otherwise quite invisible. Just because statistical effects are not obvious does not mean that they are not real or important.

It is often claimed that some conclusion has been "proved" by science but this terminology is mistaken. Scientists who use statistical inference to draw conclusions never deal in certainties, only in probabilities. Statistical tests are used to establish that the effects of an experiment are probably not due to error. In many sciences, the necessary threshold for statistical significance is reached if an effect this large could occur by chance less than one time in 20. How does the researcher know that the results being described on this occasion are not that one-in-twenty statistical fluke? The answer is that there is no way of knowing this short of actually conducting the study again to see if similar results are produced.

One of the most baffling aspects of scientific research is that it can be so uncertain. Half the laboratories in a particular line of research report positive findings and the other half fail to find any effect. Members of the public are constantly baffled by medical researchers who at one time tell them that eating salt will give them high blood pressure and at another reveal that for the general population there is no relationship between salt intake and cardiovascular health. Such inconsistency calls science into disrepute but it reflects a fundamental uncertainty in the scientific process that few nonscientists are prepared to accept. Statistics can actually be quite helpful in such situations because it is possible to make sense of the results of several conflicting studies using a procedure, known as meta-analysis, that combines all of the studies into a single test and allows the statistician to decide if the preponderance of the evidence favors one outcome or another.

Given that statisticians set their conventional probability level at 1/20, it may seem odd that researchers could ever be almost evenly divided on whether an effect is real or not. When this happens, as in the case of scientific illusions like N-RAYS and COLD FUSION, it is reasonable to suspect biasing factors, which could be nonstatistical as well as statistical. One problem is that positive findings are much more likely to be published than failures to replicate. The reason for this is quite human. If one of my friends tells me that nothing happened today, I am unlikely to bother passing this on to others. In the same way editors of science journals do not want to pass on the message that nothing happened in some scientist's laboratory (technically referred to as null results). The consequence is that positive findings get too much emphasis and may seem real when they are not. The corresponding nonpublication of null results is known as the FILE DRAWER PROBLEM because null results get locked away in file drawers and are forgotten.

In addition to the file drawer problem, scientists often unwittingly distort the probability of their statistical tests by reporting only those tests that are statistically significant. Thus, if the statistical threshold is set at 1/20 and if a statistician tests the hypothesis using 20 different tests, it is virtually certain that one of the results will be statistically significant even if there is no genuine effect in the data. Scientists can avoid this problem if they plan their analysis in advance, conduct only those comparisons that they need, and use a joint probability level for their tests. Unfortunately, this is rarely done and the result is that many studies are published as positive findings that are really null results. One important reason why this is so much of a problem, apart from lack of statistical expertise on the part of many researchers, is that publication and other academic rewards that flow from it, including funding and academic tenure, depend on positive results. For this reason, scientists are motivated to pore over their results until they can come up with something that shows they have not wasted their time and funding.

Apart from problems of statistical and editorial bias, there are many other problems of statistical inference that can get in the way of accurate presentation, and interpretation, of scientific results. One issue that is particularly important in interpreting the results of large-scale human studies, which often receive substantial public funding, is the interplay between the number of participants in a study and the probability of getting a statistically significant result. The more people in the study, the greater the probability of a positive result.

Expensive though it may be, researchers have a distinct advantage in producing a statistically significant finding if they use a large number of participants. Large-scale studies are statistically more reliable and are given greater weight in meta-analyses of conflicting research findings. Yet there is a strange paradox about the statistical power of large studies, namely, a statistically significant effect may be of negligible practical importance. For example, the finding that aspirin reduces the risk of heart attack involved a sample of 22,071 male physicians and produced a statistical significance level far beyond the conventional .05 level (approximately .0000006, or less than one in a million). This might appear to suggest that people should take aspirins. Actually the aspirin only benefited about 1% of the people taking it given the finding that there were 104 heart attacks in physicians receiving aspirin compared to 189 of those taking an inactive placebo.

Statistical tests such as correlations, and t-tests, that are restricted to analysis of relationship between two variables (bivariate tests) are generally much more straightforward than tests that look at the simultaneous effects of many variables (multivariate tests). One of the commonest multivariate techniques is regression analysis (others include factor analysis, multidimensional scaling, and analysis of variance). Such multivariate techniques can be likened to a high-power microscope. They can provide so much information about a data set that researchers are at a loss as to what to make of it, just as the first people who looked through light microscopes had very little idea what they were looking at.

From a practical perspective, multivariate techniques allow nuisance variables that could cloud the outcome of a study to be statistically controlled. If a psychologist is interested in whether low light conditions affect the speed of people's verbal responses, they are in a much better position to determine this if they statistically control for factors such as age, sex, height (which determines the distance information travels through nerves), occupation, intelligence, time since last sleeping, use of prescription medication, use of alcohol, and personality type (whether extroverted or introverted), all of which might determine how fast a participant would speak after hearing a tone. Being able to control all of these factors in a regression analysis would generally increase the researcher's probability of producing a positive finding,

for example, that people were slower to respond in dim light. Yet matters are rarely so simple. The researcher might find that there was a highly significant effect of light level on reaction time with all of the variables entered into the regression model with the exception of personality. Then, when personality is entered in the regression model, he might find either that there was no effect of light level or that contrary to the prediction participants were faster in dim light.

Such inconsistent findings are not unusual with multivariate analyses and such experiences may convince some researchers that techniques like regression analysis are to be avoided. Yet this is rather like refusing to look into a microscope because you do not understand what you see. Inconsistency in multivariate analyses often has a relatively straightforward explanation in terms of incorrect choice of either the test or the variables. On the other hand, there may be some hitherto unexpected phenomenon lurking in the data that can eventually be worked out. Just because the results may seem inconsistent does not mean that we have to agree with Disraeli's assessment of statistics. On the other hand, it can also be argued that an unscrupulous researcher could use complex statistical techniques to support almost any conclusion. The only thing that stands between a reader and such mathematical chicanery is the PEER REVIEW process in scientific publishing, which should be capable of spotting such tricks but often does not. Nevertheless, if we trust the data published by scientists, there is no good reason for mistrusting the integrity of their statistical analyses. Statistical analysis constitutes an integral part of modern scientific practice that places scientists under exactly the same kind of ethical obligation to do correct and careful work as data collection does.

See also BALTIMORE AFFAIR; DALTON, JOHN; MENDEL, GREGOR; MILLIKAN, ROBERT; NEWTON, ISAAC; SCIENTIFIC DISHONESTY.

Further Reading
Broad, W., and N. Wade. *Betrayers of the Truth*. New York: Simon and Schuster, 1982.
Cohen, J. *Statistical Power Analysis for the Behavioral Sciences*. Hillsdale, N.J.: Erlbaum, 1988.
Resnik, D. B. *The Ethics of Science*. New York: Routledge, 1998.
Steering Committee of the Physician's Health Study Research Group. "Preliminary Report: Findings from the Aspirin Component of the Ongoing Physician's Health Study." *New England Journal of Medicine* 318 (1988): 262–64.

stem cell research

Animal research has demonstrated that stem cells can be used to repair damaged organs, including the brain. Stem cells are those that are not yet differentiated into the functional specialization of adult cells and have the capacity to change into a number of different cell types. The most developmentally flexible stem cells are those found in embryos because they have the potential to develop into any of the 210 types of cell in the body (totipotent stem cells). Recent research has shown that stem cells can be found in adult tissues, including bone marrow and the brain. These retain some developmental flexibility, which means that their medical uses could obviate the need for embryonic cells, use of which raises many more ethical problems. The flexible development of stem cells makes them enormously promising in medicine. Transplanted stem cells could repair defective tissues of the heart, liver, brain, or any other organ.

Political influences on the progress of stem cell research in the United States have been mixed. President Ronald Reagan halted research on human fetal tissue by issuing an executive order in 1988 that prohibited the NATIONAL INSTITUTES OF HEALTH (NIH) from providing funding for this work. The rationale was that such research benefited from abortions, as the source of fetal grafts for Parkinson's disease, for example, and could thus be seen as promoting, or approving, abortions. President Bill Clinton reversed this order in 1993. President George W. Bush decided, on August 9, 2001, to allow federal funding for stem cell research using tissue lines derived from human embryos but banned the creation of any new stem cell lines involving the destruction of embryos.

After it was reported in 1998 that stem cells, under carefully controlled laboratory conditions, have an unlimited capacity to divide into identical daughter cells (or clones), President Clinton asked the National Bioethics Advisory Committee to review the ethical questions involved in stem cell research. In 1999, this committee issued a recommendation that research should receive federal funding if the sources of cells were limited to tissues harvested from elective abortions and surplus embryos created for infertility treatments. It concluded that funding should not be approved for human embryonic cells created solely for research purposes and prohibited the transfer of adult cell nuclei into enucleated eggs (a preliminary step in the cloning of adult mammals). The sale of embryonic or fetal tissue was to be prohibited. In August 2000, President Clinton approved these recommendations, thus allowing human stem cell research to continue.

U.S. federal law still prohibits the funding of research that creates human embryos for research purposes. Stem cell research is permitted because these cells are not embryos. The stems cell lines were created in private laboratories using embryos left over from fertility treatments that would otherwise have been destroyed. Based on President George W. Bush's 2001 decision, federally funded research could not use cell lines created after that date, which could slow the pace for developing stem cell therapies.

Political reactions to cloning have been equally complex in other countries. Following the cloning of an adult sheep, Dolly, in Britain, government funding was withdrawn from this research apparently because it was far too close to the realization of human cloning in the lab. Then, in August 2000, the British government announced that it would introduce legislation allowing the creation of human embryos for research purposes. This decision was prompted by the desire to take advantage of the medical potential of stem cell research. At the same time, the government announced that it would maintain its ban on the creation of cloned babies and on the production of animal-human hybrid embryos. Research on embryos would be allowed up to the 14-day stage of development. British utilitarian philosopher PETER SINGER has long supported the use of early human embryos for research purposes.

One of the main advantages of creating embryos for treating sick people is that they can provide stem cells that are immunologically compatible with the patient and thus will not be rejected. Scientists could create a clone by placing a cell nucleus from the patient inside an enucleated egg. The fused cell would be induced to develop into an embryo and the embryonic stem cells would be genetically identical to the patient, thus solving the problem of graft rejection. This technology opens the door to extensive repair of every organ in the body.

Most of the ethical debate about stem cell research revolves around the source of stem cells. The use of cells from aborted fetuses and from surplus embryos created for in vitro fertilization has an ethical precedent in the harvesting of organs from cadavers for use in transplants. Such use generally requires the consent of biological parents. Another source of stem cells that has not been nearly so controversial is umbilical cord blood, which has no more moral status than the blood donated to blood banks. Umbilical cord blood can be preserved for use in transplants, or for research purposes. Such blood has been transplanted into children whose bone marrow was destroyed by chemotherapy with a success rate of around 50%. Stem cells are also present in the organs of adults, particularly in bone marrow and brain tissue.

Adult stem cells hold out the promise of many medical applications without the ethical problems associated with the destruction of human embryos to obtain stem cells. Japanese scientists have already used adult eye stem cells to reverse blindness in some patients. Although embryonic stem cell research is much further away from practical applications, it is strongly supported by many scientists as offering very promising future benefits.

Whether these potential benefits can overcome the many deep-seated ethical objections to the creation of human embryos for research purposes remains to be seen.

See also ARTIFICIAL INSEMINATION; CONTRACEPTION; EUGENICS; GENETIC ENGINEERING; REPRODUCTIVE TECHNOLOGIES; RU-486 ABORTION PILL.

Further Reading

Friedrich, M. J. "Debating Pros and Cons of Stem Cell Research." *JAMA* 284 (2000): 681–2.

Humber, J. M., and R. Almeder, eds. *Human Cloning*. Totowa, N.J.: Humana Press, 1998.

Kass, L. R., and J. Q. Wilson. *The Ethics of Human Cloning*. Washington, D.C.: AEI Press, 1998.

Marwick, C. "Recommendations on Stem Cell Research." *JAMA* 282: (1999): 1217.

Nussbaum, M. C., and C. R. Sunstein, eds. *Clones and Clones: Facts and Fantasies about Human Cloning*. New York: W. W. Norton, 1999.

Quesenberry, P. J., G. S. Stein, and B. Forget, eds. *Stem Cell Biology and Gene Therapy*. New York: Wiley-Liss, 1998.

steroids, anabolic

Anabolic steroids were developed in the 1930s and were used after World War II to restore the muscle tissue and body weight of concentration camp survivors. The term "anabolic" means to add energy, and a steroid is a type of hormone that enters cells and attaches to cytoplasmic receptors, which then move into the nucleus of the cell and influence gene expression. Sex hormones are steroids. One effect of the sex hormone testosterone is to increase muscle mass of males during puberty. The anabolic steroids taken by athletes and body builders to increase their muscle mass are similar in chemistry and effects to testosterone.

Although steroid use is illegal in many countries because of the risk of associated health problems, it continues among a large fraction of competitive athletes. In the United States many different anabolic steroids, such as androstenedione and androstenediol, are sold legally as dietary supplements. These are not considered drugs because they occur naturally in foods and have never been marketed as drugs.

Legal or not, anabolic steroids may produce troubling side effects, including baldness, acne, beard growth, and other signs of masculinization in women, reduced sex drive in men, irritability and aggressiveness, and liver tumors. Early deaths of athletes have been attributed to steroid use. Among children, steroids advance the age of sexual maturity and may interfere with normal development of leg bones.

Steroid use is banned from the Olympic games but is believed to be prevalent despite subjecting athletes to blood tests. Canadian runner Ben Johnson lost his Olympic gold medal after his blood test proved positive for steroid use. Other athletes have been disqualified for having steroids in their possession. Such incidents suggest that some athletes believe they can beat the tests and many popular manuals offer techniques for doing this. When steroids are used along with high protein diets and intensive exercise training, they increase muscle mass and enhance athletic performance. Steroid use is ethically questionable not only because of its health consequences and behavioral effects but also because it undermines the integrity of sporting competition. Athletes who succeed at the highest level may do so largely because they cheat and use steroids. Conversely, athletes who do unusually well in competition, or merely have a well-developed physique, may be unfairly suspected of steroid use. In either case, the credibility of the competition is seriously undermined.

See also HORMONE MIMICS; HORMONE REPLACEMENT THERAPY; HUMAN GROWTH HORMONE.

Further Reading

Houlihan, B. *Dying to Win*. Strasbourg, France: Council of Europe Publishing, 1999.

Llewellyn, W. *Anabolics 2000: Anabolic Steroid Reference Manual*. Aurora, Colo.: Anabolics.com, 2000.

strip mining   *See* OPEN-PIT MINING.

Super Conducting Super Collider

The Super Conducting Super Collider project involved building the world's largest particle accelerator in the United States at a cost of some $20 billion. Congress eventually decided to scrap the project in October 1993 on the grounds that the expense could not be justified. There is no question that completion of the project would have led to important advances in particle physics but the practical benefits flowing from such potential advances were too uncertain and too far into the future to justify devoting such a large proportion of the government's research budget to this project at the expense of other types of research. Cancellation of the project constituted a big blow to particle physicists, many of whom had already designed experiments to be run using the Super Conducting Super Collider. That the project ever came so near to fruition is a testament to the great enthusiasm of these scientists, which made them embellish the potential benefits to the public of the giant particle accelerator. Although scientists are expected to preserve scientific objectivity in their professional work, they are human beings and a certain amount of exaggeration can creep into grant writing and advocacy of a pet project involving use of public funds.

See also PUBLIC FUNDING OF RESEARCH.

surrogate parents   *See* REPRODUCTIVE TECHNOLOGIES.

Synthroid (drug) controversy

Synthroid is a drug used to treat hypothyroidism. It is manufactured by a subsidiary of Boots Company. Boots Company sponsored a study by chemical researcher Betty Dong, which compared the effectiveness of Synthroid and several forms of levothyroxine. They hoped that their brand would come off well in the comparison. Instead, Dong produced a report finding that several cheaper alternatives were equally effective. This paper was accepted for publication in the Journal of the American Medical Association, a leading journal in the field, in 1995. Boots compelled Dong to withdraw the paper from publication because she had signed a contract not to publish her findings without the consent of the company,

obtained in writing. Despite the fact that the paper had passed peer review in a highly selective journal, the Boots Company claimed that the research was flawed and that they wanted to protect the public from inaccurate medical information. In the end, they changed their minds and the research was published in the New England Journal of Medicine.

The Synthroid controversy illustrates the ethical bind that researchers who are funded by private enterprise may find themselves in. It may be melodramatic to say that they are selling their souls but this formulation does capture the flavor of the ethical conflict. In this case, not publishing the research may have drained health care resources by perpetuating the use of Synthroid rather than cheaper alternatives. This violates the key ethical imperative that researchers should avoid doing harm. Yet the researchers were legally compelled to comply with the company's wishes since this was a term of their contract.

See also COMMERCIALISM IN ACADEMIC RESEARCH.

Further Reading

Altman, L. "Drug Firm, Relenting, Allows Unflattering Study to Appear." *New York Times,* April 16, 1997.
Resnik, D. B. *The Ethics of Science.* New York: Routledge, 1998.

T

technology, history of

Technology can be defined as the use of inanimate objects to solve problems of survival and physical comfort. It is not peculiar to humans. Early human technology consisted principally of the use of stone tools and fire. Tools are used by many other species. For example, chimpanzees use hammer stones to break hard nuts and prepare twigs with which to fish for termites by stripping off the leaves. Sea otters break shellfish open with stones. Woodpecker finches on the Galapagos Islands use cactus spines to pry larvae out of holes in tree trunks. Vultures use rocks to smash ostrich egg shells. Weaver ants use thread produced by their own larvae to sew leaves together. The webs constructed by spiders and the pits dug by antlions to capture prey can also be considered a form of technology as can the many kinds of homes constructed by animals to shield them from adverse weather conditions and predators. (When animal technologies are transmitted through observational learning, they are referred to as "culture".) Excluding such technologies, and narrowing the focus to the human use of inanimate objects to solve the problems of survival and food acquisition, the history of technology can be considered in four periods: the Paleolithic era from the origin of modern humans to the end of the last Ice Age 12,000 years ago, the rise of agriculture, the emergence of cities, and the INDUSTRIAL REVOLUTION.

PALEOLITHIC TECHNOLOGY

There is some uncertainty about the evolutionary series leading to modern humans because of incompleteness of the fossil record but it appears that Homo *erectus,* the probable ancestor of anatomically modern humans, appeared some two million years ago. With a brain that was twice the size of his apelike Australopithecine ancestors, Homo erectus was an intelligent and highly successful hominid that spread throughout Africa, Europe, and Asia. Homo erectus was bipedal and walking upright, which left the hands free for using tools. The use of primitive stone tools extends back at least two million years and these mark the beginning of human technology. Fire was an important component of Homo erectus technology, which facilitated migration into cooler climates.

Fully modern humans, Homo sapiens, emerged approximately half a million years ago. Their brains were 40% to 50% larger than Homo erectus. A more rugged and more large-brained variety of modern humans, the Neanderthals, subsisted in the cold climates of Europe from about 135,000 to 35,000 years ago. Neanderthals had large brow ridges that would likely have given them a "brutish" appearance and anthropologists are divided as to whether they were a separate species from ourselves or just a different variant of the same species. There is no question that Neanderthals were highly intelligent and that their behavior resembled that of the other modern humans with whom they coexisted. Thus, they engaged in ritual burial of their dead from approximately 100,000 years ago.

The period from two million years ago to the end of the last Ice Age, only 12,000 years ago, is referred to as the Paleolithic, or Old Stone, Age. During this time our ancestors practiced a hunter-gatherer lifestyle in which stone tools were used to hunt and scavenge animals and process plant foods. Anthropologists agree that throughout this period people lived in small nomadic bands, generally consisting of fewer than 100 people. Hunter-gatherer bands would have been quite egalitarian, generalizing from subsistence populations still in existence, and would have had a division of labor by sex according to which women gathered most of the plant food and did most of the child care while men roamed more widely in search of animal food. During the Paleolithic period, population density remained very low possibly reflecting the constraint of seasonal scarcity of food.

Throughout the two million years of the Paleolithic era, there is little evidence of much in the way of innovation. Early tools consisted of simple stones that were held in the hand and used to dismember carcasses or prepare plant food. The rise of modern humans around 500,000 years ago was accompanied by the development of fine stone blades that were used like modern knives. This technology showed little development until some 40,000 years ago when Homo sapiens began to produce a great variety of specialized tools. They included bows and arrows, fish hooks, spears and spear throwers, needles made of bone and rope, nets, sewn clothing, lamps, and musical instruments. Even though the basic hunter-gatherer lifestyle was still followed, there are many indications of increased social complexity. These include the production of cave paintings and other works of art, such as stone carvings. Trade in shells and flints was conducted over distances of hundreds of miles.

This remarkable increase in the complexity of technology at the end of the Paleolithic period has never been satisfactorily explained. One reasonable theory is that human languages arose around this time. This view is supported by an enlargement of the larynx, or voice-producing structure, that occurred around this period as inferred from change in the structure of the base of the brain, which can be discerned from fossil skulls. Whether language use facilitated the elaboration and transmission of tool culture or merely accompanied this change cannot be determined.

Approximately 15,000 years ago the ancient pattern of food collecting declined with the onset of food production, in the form of horticulture and animal husbandry. With the end of the last Ice Age, food production increased allowing for the emergence of larger, more settled, communities and a variety of social and economic changes that facilitated more complex technologies and an interest in science.

THE RISE OF AGRICULTURE

Neolithic farming communities emerged at many different sites around the world beginning 12,000 years ago. Independent domestication of plants and animals in many global regions, including Africa, the Middle East, India, North Asia, Southeast Asia, Central America, South America, and Europe, suggests that all of the peoples of the world were abandoning their ancient hunter-gathering lifestyle due to shared ecological pressures. The most plausible explanation is rising global population putting pressure on supplies of food that could be hunter or gathered. Evidently, the extremely slow rate of growth of the human population had eventually caught up to the carrying capacity of the environment. The fact that some hunter-gatherer communities survived at the same time as the agricultural settlements suggests that the settled way of life was not adopted with alacrity but that our ancestors were forced into it by the practical difficulty of obtaining sufficient food.

Paleolithic food collection led to food production by two pathways. Our ancestors shifted from exclusive food gathering to occasional gardening (or horticulture). They might have planted a crop at one time of the year in the course of their travels and returned to harvest it several months later. This could have led to more intensive agriculture at fixed sites using plow technology. The rise of agriculture led to the development of towns and, ultimately, to the modern industrial city.

Another route to food production may have been the following of herds of game animals. Eventually, these herds could have been partly domesticated. This process is illustrated by the Lapps who follow herds of originally wild reindeer that are treated as domestic animals in the sense that they are driven about and exploited for food. This route led to the food production system of pastoral nomadism practiced until recent times by the Mongols and Bedouin, among others.

Neolithic farming communities produced large amounts of food that allowed for a rapid expansion of the human population in agricultural areas. This fact is illustrated by the ancient city of Jericho, in Israel. This agricultural community was based on the tending of flocks and the cultivation of plots in the surrounding area. By 9,000 years ago, it was a town of over 2,000 people, which contained large storage areas for excess food that could be used as an insurance against poor crop yields. Jericho had impressive fortifications indicating the need to protect the food surplus from raiders, some of whom might well have been local populations of starving hunter-gatherers.

The increased affluence of Neolithic farmers, combined with their settled lifestyle, favored the building of large monuments. Thus, the large facial statues erected on Easter Island in the Pacific Ocean were produced by a Neolithic farming community that practiced nonintensive agriculture. The Polynesian inhabitants reached this remote island only in the third century A.D. The precise significance of their 250 huge *moai* statues, which averaged 14 tons and had to be transported several miles from sites where they were quarried, is obscure, but the careful alignment of the statue bases bespeaks an interest in astronomy. The creation of such ambitious public works, as well as an interest in astronomy, is typical of other Neolithic communities.

During the Neolithic period, many crafts were developed. One of the most widespread was pottery, which at its most fundamental consists of baking clay to remove the water from it thereby producing a rigid vessel. Kilns, or baking ovens, for pottery that were used in the Neolithic produced temperatures of over 1,600 degrees Fahrenheit (900 degrees Centigrade). This technology led directly to metallurgy in urban civilization.

THE EMERGENCE OF CITIES

Just as the emergence of villages and towns was based on the increased level of food made available by agriculture, so the development of cities was made possible by the greater food surpluses produced by intensive agriculture. Of course, the urban way of life is associated with the full flowering of modern civilization with its rich traditions of learning, the arts, technology, and science. According to the hydraulic hypothesis, all this can be attributed to the development of technologies for controlling the supply of water available for agriculture. In arid lands, such as Egypt and Mesopotamia, irrigation created a highly productive agriculture. In swamps, such as the lower delta of the Tigris and Euphrates rivers and the Yucatan, intensive agriculture was made possible by drainage systems.

Intensified agriculture produced very high population densities with centralized political authority on at least six occasions. Centralized kingdoms arose in Mesopotamia around 5,500 years ago, in Egypt 5,400 years ago, in the Indus river valley 4,500 years ago, in China on the banks of the Yellow River 3,800 years ago, in Mesoamerica 2,500 years ago, and in South America 2,300 years ago. Each of these early civilizations is consistent with the hydraulic hypothesis. A centralized authority organized extensive public works projects that managed water to maintain extremely productive agricultural systems.

The first civilization, the Mesopotamian, arose between the Tigris and Euphrates rivers in present-day Iraq. By 6,000 years ago, the plain between the rivers was dotted with Neolithic farming communities. Drainage of marshes in the lower delta was followed by the construction of an up-river irrigation system. From about 5,500 years ago large walled cities were built, including Uruk, Ur, and Sumer, that had populations ranging from 50,000 to 200,000 people. These city-states were sometimes organized into empires but these were much more unstable than those of Egypt.

The centralized authority structure of Mesopotamian states created a huge labor pool that was used to produce impressive public works. The main irrigation canals were over 75 feet wide and ran for several miles. Huge temples and pyramids (or ziggurats) were constructed and these attest both to the fact that there existed considerable surplus labor over that needed for subsistence and that there was a powerful centralized authority organizing that labor force in public works. These projects required a vast amount of labor that was liberated from agriculture by the use of the ox-drawn scratch plow.

The surplus of labor over subsistence needs meant that a number of nonsubsistence occupational specializations opened up. A large number of people worked as state employees responsible for collection, storage, and redistribution of food surpluses. There were political, military, and priestly castes, or elites. Skilled engineers were required to organize public works projects. The bureaucratic need for recordkeeping favored the development of writing and the scribe constituted an important occupational specialization. Mesopotamian civilization produced a rich tradition of mathematics and science. Its finest achievements in science are represented by the astronomy of the Babylonians, who lived at the geographic heart of Mesopotamia and played a central role in its political life and learned traditions.

Early civilizations had limited potential for geographic spread because the intensive agriculture they practiced was limited to suitable areas, principally the flood plains of rivers that were periodically replenished by flooding and deposits of silt. The pattern of manipulation of water supply by engineers is repeated many times for the Nile in Egypt, the Indus River in India, and the Yellow River in China.

These civilizations are noted for their technological achievements, particularly as represented in surviving monuments such as the pyramids of Egypt, which were constructed as burial sites for the Pharaohs although there were more pyramids than emperors indicating other functions for these colossal building projects. The Great Pyramid at Giza, which has not been precisely dated but was built approximately 4,700 years ago, was constructed of 2.3 million blocks of stone that averaged 2.5 tons each. According to the Greek historian Herodotus (484–c.425 B.C.), 100,000 laborers worked on this monument for 20 years and thousands of craftsmen were required to cut and fit the stones. Impressive though they still are, the pyramids were built more by dint of hard labor than by means of any great technological innovation. The pyramids thus provide clear evidence of the degree of control exercised by the Pharaohs over their subjects. It is astonishing that, at its height, the Egyptian Empire comprised only 2.5 million people, the equivalent of a single modern city.

The greatest feat of early construction was undoubtedly the Great Wall of China. When it was completed, around 2,200 years ago, the Great Wall was 1,250 miles long. This defensive structure was approximately 13–40 feet thick and 20–50 feet high. Some 5 million people may have worked on it, a huge labor commitment even for a civilization that was by far the largest in the ancient world with some 65 million people.

Civilization emerged in America much later than in the Old World and it did so despite the absence of cattle, the wheel, and the plow. In Central America, Neolithic settlements were established by 3,500 years ago. The first true city was Monte Alban in the Oaxaca valley, which was built around 2,500 years ago. This planned city was based on irrigation agriculture and was home to some 15,000 people. It was followed three centuries later by the much larger Teotihuacan near present-day Mexico City. With a peak population of between 125,000 and 200,000, this was the largest city in Mesoamerica. It was based on an extensive irrigation system and boasted impressive monuments, such as the Temple of the Sun built in the shape of a huge stepped pyramid.

During the same period, civilization emerged in the swamps of the Yucatan peninsula based on the creation of raised fields separated by drainage channels. The largest city of this Mayan civilization was Tikal, which by A.D. 700 had grown to 77,000 people.

In South America, the many short rivers flowing from the Andes across an arid coastal plain into the Pacific Ocean became the basis for irrigation systems analogous to those of the Nile in Egypt. Several pre-Christian-era cities flourished here, including Chan Chan and Pampa Grande.

In southern Peru, a civilization was built around Lake Titicaca, which has an elevation of 12,500 feet. This is of ecological interest because it is the only early city that was built at such a high elevation. Lake Titicaca has an area of 3,200 square miles and is deep (having a maximum depth of 920 feet). Because of its size, it maintains a constant temperature of 51 degrees Fahrenheit (11 degrees Centigrade) and this moderates the local climate allowing crops to flourish that would not normally be grown at such high altitudes. Fields built into the sides of the mountains were used to grow potatoes using a drainage system that bears analogy with that of the Maya in Central America. At its height, from A.D. 375–675., the

city of Tiwanaku housed somewhere between 40,000 and 120,000 inhabitants.

The Inca empire succeeded in uniting this mountain civilization with those on the coastal plains. The Incas reached their zenith around A.D. 1500., at which time they numbered around 7 million. They built a very elaborate irrigation system and their capital, Cuzco, had impressive monuments. The Inca empire extended 2,700 miles and was connected by a system of 19,000 miles of roads and paths. One road system extended for 2,200 miles through the mountains and another of the same length traversed the coast.

Early civilizations were based on the management of water and they facilitated the emergence of many technological innovations. One important development was the use of draft animals to plow and transport raw materials. The wheel was invented in Mesopotamia more than five thousand years ago. A pictograph of a wheeled cart inscribed on a clay tablet was excavated from the courtyard of the Eanna temple in the city of Uruk and the tablet is between 5,200 and 5,100 years old. Both two-wheeled and four-wheeled carts were used to transport goods. The main draft animals were oxen and donkeys. The horse-drawn two-wheeled chariot became important in warfare, particularly after the light spoked wheel had been developed that allowed for greater speed than was possible with the heavy solid wooden wheels originally used. Other economically important draft animals that were widely used included the camel, the elephant, and the llama in South America. The use of wind power in sailing vessels was developed by the Egyptians as the southerly winds helped boats travel up the Nile.

All early civilizations are characterized by the development of metalworking crafts in which metals were worked while hot. Some metals, such as copper, can be worked cold. Thus the Iceman, whose mummified remains were recently revealed when a glacier melted in the Alps, was a Neolithic farmer but he was carrying a copper axe when he died. Much of the metalwork in early civilizations was decorative in character. Crafting fine jewelry from silver and gold was a preoccupation of craftsmen in all cities and made these metals highly valuable.

In the Old World, bronze, an alloy of copper and tin, was used to make tools and weapons. Tough and corrosion-resistant, bronze tools could be sharpened and offered many advantages over stone tools. During the Neolithic period, copper objects like the Iceman's axe were made by beating the metal into shape while cold, using stones. Around 5,800 years ago, copper smelting began in Anatolia, in modern Turkey. When molten, the copper could be mixed with other ores to improve its physical properties. From about 5,000 years ago copper was alloyed with tin and bronze tools were widely used in the region. Bronze was seen by many historians as such an important advance that they refer to the age of urban civilization as the Bronze Age.

GRECO-ROMAN TECHNOLOGY

Greek civilization departed in many ways from the model of earlier civilizations that can be easily accommodated by the hydraulic model according to which the control of water facilitated excess food production allowing non-subsistence occupations to be practiced. The Greek soil was not rich enough to support intensive agriculture and there were no river basins suitable for irrigation. Main crops in ancient Greece were olives and grapevines that do well in arid soils because they have deep root systems that can exploit water lying well below the surface. Together, these would not have made a very promising diet and the ancient Greeks were not even self-sufficient in food. By exporting olives and wine they became wealthy and could afford to purchase food from surrounding agricultural communities. Greece was thus the first great civilization built largely on trade.

Another difference between the Hellenic world and the other major contemporary civilizations is that before the reign of the emperor Alexander the Great (334–323 B.C.), Greece was organized into numerous autonomous city states surrounding the Aegean Sea. The most important region in terms of commerce and learning was Ionia in Asia Minor (or modern Turkey) and mainland Greece was initially considered something of a backwater. The noncentralized structure of Greek civilization appears to have played an important role in the development of natural philosophy and science because it meant that learning in Greece was largely an activity of the leisured individual rather than a state-sponsored function.

Greek philosophers saw science as the activity of leisured individuals who had no interest in the practical applications of their knowledge. As far as they were concerned, people engaged in natural philosophy after their bodily needs had been satisfied and not to solve the practical problems of everyday life. Science was thus largely independent of technology.

Following the reign of Alexander the Great, Greek science received generous state sponsorship at centers such as Alexandria, in Egypt, and Pergamum, in Ionia. During this Hellenistic period (not to be confused with the Hellenic that preceded it), pure science continued but their was a shift in focus toward practical problems, including military technology and medicine. During the Roman administration that followed, pure science fell into decline and technology flourished. It is probably no accident that some of the leading scientists of the Roman era, such as the physician GALEN of Pergamum (A.D. 130–200), were Greek.

The Roman flair for technology is illustrated by their many fine roads, aqueducts, and bridges, some of which have survived to this day. Their architecture is characterized by mastery of the arch. Arches are a great deal stronger than the perpendicular lintels used by Greek builders to support their structures. The Romans also invented cement, which made their structures much stronger and cheaper to produce. Improved naval and military technologies, such as sophisticated torsion-spring catapults, enabled the Roman legions and navy to subdue all opposition throughout most of the Hellenistic world. Equally important was their flair for administration using detailed written legal codes, which might be described as a form of social technology.

MEDIEVAL TECHNOLOGY

With the fall of Rome in 476 the center of the civilized world shifted to Constantinople, which was the center of a Greek-speaking empire whose official religion was a mystical Christianity. Although the Byzantine Empire lasted for a millennium, before being conquered by the Islamic Ottoman Empire in 1453, and maintained at least some institutions of higher learning, it is not noted for important scientific or technical advances. More scientific civilizations developed in the East, in the Islamic world, India, and China. Chinese society was ahead of the rest of the world in technology but its advances did not influence Europe until well after the visit of Venetians Marco Polo and his father (1271–95). The initial reaction to Polo's book describing his adventures was disbelief but after the stories had been corroborated by other travelers, trade routes were established.

The Islamic empire spread throughout Arabia, Mesopotamia, and Egypt. Islamic civilization is particularly important to the rise of European civilization because it formed a repository of classical learning. Islamic scholars cultivated astronomy and medicine. Their objective in making astronomical observations was often to predict the future (astrology), which was officially discouraged on religious grounds. Arabic medicine was possibly the most sophisticated of its day. Elaborate research hospitals were developed and medical scholars such as al-Razi (Rhazes, 854–925) and Ibn Sina (AVICENNA, 980–1037) became the world's leading authorities on diseases and their treatments. Islamic physicians developed unprecedented expertise in occular diseases and their understanding of the anatomy and physiology of the eye was complemented by excellent scientific work in optics, such as that of Ibn Al Haytham (Alhazen, 965–1040).

At the end of the first millennium, Europe was a backwater, in terms of its cities, technology, economy, learning, and science. By each of these criteria, it paled in comparison to the world's civilizations of Byzantium,

Islam, India, China, South America, and Mesoamerica. Thus, Rome itself had declined from 450,000 people at its peak to 35,000.

One important stimulus for technological change in Europe seems to have been the development of a population crunch. Between 600 and 1000, the European population rose approximately 38%. This increase created an ecological pressure for intensification of agricultural production. One critical innovation was the adoption of the heavy plow that allowed more land to be cultivated. First developed by the Romans but little used by them, the heavy plow was made of wood with iron cutting edges. Mounted on wheels, it cut beneath the sod and turned it completely over to leave a furrow.

The heavy plow was originally pulled by a large team of oxen. Horses eventually replaced the oxen and the use of a horse collar, developed in China, allowed the draft animal to pull from the shoulder with much greater force than would be possible with the neck yoke used with oxen that pulled the light European scratch plow. Another important agricultural improvement was the development of a complex three-field crop rotation system, which meant that land spent much less time lying fallow (i.e., idle to replenish nutrients) than had previously been the case. Agriculture was organized around village collectives that were controlled by the local squire according to the feudal system.

Agricultural improvements greatly increased food production thereby allowing for an accelerated population growth and an increase in the prominence of European cities. The entire population of Europe trebled between 600 and 1300 (rising from 26 to 79 million). Between 1000 and 1300, the population of Paris increased from 20,000 to 228,000, indicating that much of the increased population was accumulating in medieval cities.

In addition to the agricultural revolution, there was also a revolution in military technology that did much to shape feudal society and launched what can only be called an arms race that had important implications for the rise of European powers and their worldwide empires. This revolution began with a seemingly innocuous device, the stirrup, that had been invented in China in the fifth century. The stirrup allows a soldier to fight from horseback without falling off. Thanks to the strength of the horse, a soldier could cover himself in very heavy metal armor that left him almost invulnerable. The armored knight played a critical role in warfare because he could easily break through the ranks of infantry. Knights were hired by feudal lords as their champions and knighthood developed into a powerful and lucrative profession.

Many of the medieval knights were born of the elite. Due to the system of primogeniture, only the oldest son inherited the property of the feudal lord. Many others went into business as knights. The excess of knights was one important factor in the launching of the Crusades, a vaguely religious war in which Christian knights periodically attacked the Byzantine and Islamic empires, beginning in 1096 and ending in 1270 when the last Crusade was launched.

In 1249, the crusaders were attacked using bombs made with gunpowder. By 1288, the Chinese had developed small guns with metal barrels. Large cannons were developed in Europe, probably between 1310–20. By the middle of the 15th century, they had begun to play a decisive role in warfare. The musket was first used in the 1550s and the subsequent development of handguns meant that the importance of armed infantry on the battlefront increased while that of archers, pikemen, swordsmen, and cavalry declined. During the 17th century, the size of Europe's standing armies increased substantially.

Medieval castles and city walls could be quickly reduced to a pile of rubble by the new cannons. For this reason, European governments poured huge amounts of money into creating thicker, more secure fortifications that contained their own guns that could be used to fire at the gun emplacements of an attacker. The cost of such defenses was so onerous that they could be borne only by large centralized governments with the power to tax large populations. In this way the arms race finally brought about the centralization of European governments, something that had emerged from the beginning in the earlier hydraulic civilizations.

An important medieval European invention was the printing press created by German JOHANNES GUTENBERG (1390–1469) in the 1430s, who was evidently unaware that this had already been developed in China and had been widely used there since the ninth century. By 1500, some 13,000 different books had been printed. This launched an information revolution, which meant that knowledge was no longer the province of the learned few at universities but could be acquired by anyone who had the leisure time to read.

THE INDUSTRIAL REVOLUTION

The Industrial Revolution refers to a broad pattern of economic changes associated with the shift from an agrarian society to an urban one. This transition, which first occurred in 17th-century Britain, and is still happening throughout the world, is associated with greatly increased affluence. Although the increased wealth made possible by industrialization generally improves health and increases longevity, the increased burning of fossil fuels by both industry and modern transportation has resulted in environmental pollution that threatens ecosystems and poses health problems.

European civilization differs from the earlier hydraulic civilizations in many respects as previously mentioned. One difference, dating to the medieval period, is that engineers have harnessed sources of power in addition to the work of humans and draft animals. Medieval Europe used water wheels to turn flour mills, saw logs, and power hammers. Windmills were developed for similar purposes. Machines using artificial sources of power thus performed much of the work that was done by forced labor, or by slaves, in previous civilizations. During the Industrial Revolution, steam was harnessed to power industrial mills. During the 20th century, steam power was replaced by electricity, a highly versatile power source. Once a location is wired for electricity, it can be used to run almost any type of machine.

Why industrialization emerged first in Britain cannot be easily explained but historians see several factors as important. They include: population growth; increased efficiency of agriculture; mechanization of traditional handcrafts, such as cloth production; better transportation infrastructure due to road improvement and canal building; availability of raw materials, such as coal and iron ore; and a climate of intellectual freedom that promoted inventiveness and social mobility.

Britain's growing population (which nearly doubled in the 18th century) increased the demand for manufactured goods. The increased population was fed by an increase in agricultural production. This was made possible by the dividing up of common lands into private holdings that were farmed more intensively using crop rotations that increased productivity.

The increased demand for goods, particularly clothing, led to the abandonment of traditional methods of weaving in favor of a mechanized textiles industry. This change meant that traditional crafts workers were put out of business leading to the violent protests of the LUDDITES who wrecked machinery, beginning in Nottinghamshire in 1811 and ending in 1816. The process of mechanization had begun 50 years earlier in the cotton industry. Many of the critical inventions were either developed or became widely used in the 1760s. They included the flying shuttle that allowed broad pieces of cloth to be woven by a single worker and that was invented in 1733 by John Kay. James Hargreaves' spinning jenny (invented around 1765) allowed cotton thread to be produced much more rapidly. Richard Arkwright's water frame was a much heavier spinning machine originally designed to be driven by horses rather than water. The first water-driven cotton mill opened in 1770, at Cromford, and this marked the true beginning of the Industrial Revolution because workers had to be present at the source of power and could no longer work from their homes. JAMES WATT's improved Newcomen steam engine (developed in 1769) was widely used from the 1780s onward, which meant that factories no longer had to rely on the limited power produced by water-driven mill wheels.

These developments allowed for a huge increase in production of cotton goods, mostly for exports. Between 1751 and 1861, British cotton exports increased by a factor of one thousand. During the 19th century, the overall gross national product of the industrialized economy surged 400%.

Industrialization spread slowly across Europe, occurring first in Belgium (1820s), then France (1830s), and Prussia (1840s). An English textile worker, Samuel Slater, established the first cotton mill in America, in Rhode Island, before the end of the 18th century.

The social consequences of the Industrial Revolution were initially quite bleak. Factory labor was often both tedious and dangerous, involving long hours of repetitive activities in which people were reduced to the role of functional appendages of the machines they tended. The urban environment in which workers lived was often crowded, with highly polluted air, inadequate sewage systems, and a shortage of water for washing. Such conditions favored the spread of infectious diseases and death rates were high. Such dire conditions were gradually improved through the provision of public water and sewer systems, the introduction of social welfare legislation, and the organization of workers in trade unions to bargain for better conditions. Despite its initial adverse social consequences, industrialization has had the effect of creating unprecedented wealth. The populations of many countries could continue to grow without experiencing the catastrophic increase in starvation predicted by population theorist THOMAS MALTHUS (1766–1843) whose life coincided with the onset of the Industrial Revolution in England.

The Industrial Revolution saw the completion of some remarkable projects in civil engineering, such as the bridge over the Menai Straits in Wales. The construction of bridges with extremely long arches was made possible by the use of cast iron. A huge industry grew up around the production of iron for the building of bridges and railroads. Iron was also used to make nails, wire, guns, tools, and machine parts. By the end of the 19th century cast iron architecture gave way to steel, which is even stronger. In the 20th century, steel would allow extremely tall skyscrapers to be built thus concentrating huge numbers of people in business centers of cities such as New York and Hong Kong.

TECHNOLOGY IN THE MODERN WORLD

Industrialization produced a mass movement of population away from agriculture, where labor was replaced by machinery, into cities, where opportunities emerged in manufacturing and service industries. This accelerated the development of urban centers that had begun in the

Middle Ages as a consequence of increased agricultural productivity. Fully industrialized modern countries are those in which only a small minority of the population is still rural.

Coincident with the concentration of population in cities during the 20th century, new technologies of communication and information processing have been created that constitute the most distinctive aspects of modern civilization. These technologies, including the telephone, radio, television, and the Internet, allow for an unprecedented degree of communication between distant cities and countries, and have meant that the term "global village" is changing from a slogan into a practical reality. Most of the technological innovations that have facilitated the connection of centers of civilization involve mastery of electromagnetic radiation, such as radio waves, X-rays, radar, television waves, microwaves, light, and alternating electrical current.

Industrial civilization was based on the development of steam power to run mills and railroads. In modern life, steam power has largely been replaced by electricity and the development of internal combustion engines has provided a mobile energy source ideal for moving people and materials.

Electricity was used for communication before it was used as a power source. The electric telegraph was invented by English physicist Charles Wheatstone in 1837, and a functional version that used the code of Samuel F. B. Morse was perfected in 1844. Telegraph lines developed along with railroads and cables were soon laid across seas and oceans connecting London and Paris (1854) and Europe with America (1858).

The electric light bulb was independently invented by THOMAS EDISON, in New Jersey, and Joseph Swann, in England, both in 1879. This was immediately followed by the development of electric utility companies that provided electric lighting for their local urban customers. The development of alternating current in 1886 allowed electricity to be transferred much larger distances without a significant loss of energy. As a result of this innovation, all parts of a country could be electrified. The universal availability of domestic electricity means not only that all sorts of electric machines and gadgets can be run in homes but that new mass media, such as radio, television, and the Internet, have a convenient power source.

Electric utilities often use steam turbines that can be fired by coal, oil, or thermonuclear energy so that the modern age of electricity is built on the previous technology of steam. The polluting consequences of solid fuel energy sources have led to an interest in cleaner energy sources such as hydroelectricity, wind power, wave power, and solar energy. There has been considerable controversy about whether NUCLEAR POWER (developed after World War II) is cleaner than these more widely used sources of energy. There is no AIR POLLUTION from nuclear power plants but there is the problem of safe disposal of radioactive waste. Pollution has also been a problem in connection with the use of oil in internal combustion engines that provide a universal source of energy for transportation and other purposes.

An internal combustion engine converts heat to mechanical energy by burning fuel within a combustion chamber inside the engine (as opposed to a steam engine in which combustion was external to the engine). The first such engine was built by Nikolaus Otto, in Germany, in 1876. In 1885, also in Germany, the first automobile with an internal combustion engine was built by Karl Benz. German Rudolf Diesel invented the diesel engine (1892) that is ideal for machines requiring a large amount of power and consistent performance despite greatly varying loads.

Following these important inventions, the automobile and the tractor of the 20th century quickly displaced the horse as a universal source of power in the 19th. Invention of the airplane, in 1903, by Orville and Wilbur Wright opened up the skies to passenger and military aviation.

Henry Ford (1865–1947) perfected an assembly line technique for mass production of cars, which he introduced in 1913. In 1924, he produced two million Model Ts, which were sold for $290 each compared to a price of $850 for 10,607 cars produced in 1908 before the introduction of the assembly line. This illustrates the phenomenon of economy of scale in mass production that was to become a model for many other industries thus driving industrial expansion in the 20th century.

Important technological innovations transformed virtually every industry in the 20th century. In entertainment, the invention of the radio (1896) was followed by the cinema. Both developed into major entertainment industries in the 1920s and were eclipsed in importance by commercial television that emerged after World War II.

The first computer, the ENIAC (1946), used cumbersome and unreliable vacuum tubes. In the following year, the first solid-state transistor was developed that permitted the building of more practical computers although these were still large. Since then the increasing sophistication of integrated circuits has allowed ever-smaller devices to perform ever more complex feats of information processing. Computers are widely used for recording transactions and allow many commercial transactions that used to be performed face-to-face to be conducted impersonally. This reduces much tedious repetitive work originally performed by clerks and increases the efficiency of businesses. Examples include cash registers (now computerized), automated teller machines, computerized stock trading, and the on-line purchase of airline flights, theater tickets, books, antiques, and virtually any other commodity. Although electronic transactions are themselves cold and impersonal, the Worldwide Web, which allows communication between all the people in the world with an Internet connection, has a number of special-interest chat communities and provides an opportunity for friendships and romantic relationships to develop.

Electronic advances have played a key role in recent space exploration and the lessons learned from space missions have enriched terrestrial technology. Entering space is possible only by means of rockets that provide the necessary thrust to escape the earth's gravitational field. The space age began in 1957 with the launching of the Soviet Union's *Sputnik* 1 into orbit around the earth. The United States achieved priority in the race to space with a successful 1969 manned flight to the moon. Since that time, a fleet of communications satellites have been deployed in space for civilian and military purposes.

The technology of warfare has undergone profound changes throughout the 20th century. The World War I era saw the introduction of poison gas, long-range artillery, machine guns, tanks, submarines, airplanes, and radio communication. Jet fighters were developed. World War II saw the development of the first nuclear weapons that were used to devastating effect against Japan. Aircraft carriers were introduced as a base for mounting military campaigns quickly in foreign countries. Radar and sonar devices were introduced to detect objects in the air or under water. Modern ships and submarines use nuclear energy and are equipped with small (tactical) nuclear weapons as well as larger strategic ones. Perhaps the most distinctive, and most frightening, aspect of the modern era was the cold war buildup of strategic weapons, including ballistic missiles, capable of annihilating the entire global population.

The 20th century has also seen enormous changes in the technology of food production. The GREEN REVOLUTION refers to a huge increase in food production following development of high-yielding varieties of food crops, whose productivity was enhanced through the use of chemical fertilizers and pesticides. Originally, the crops were produced by means of selective breeding but genetic engineering has also been used to make crops resistant to diseases, drought, and pests. The revolution in agriculture has allowed some countries to support rapidly increasing populations.

After the genetic code was deciphered in 1953, much progress was made in recombinant DNA technology. In addition to modifying food crops, biotechnology has produced numerous new drugs and carries the promise of developing many others. GENETIC ENGINEERING holds out hope for victims of serious genetic disorders and suggests many promising approaches to curing previously incurable illnesses, such as AIDS and cancers. This revolution in medicine is built on the earlier modern achievements of VACCINATION, chemotherapy, antibiotic drugs, and psychotropic medicines for mental illness (see BIOLOGICAL PSYCHIATRY).

One of the most important medical advances has been the development of effective techniques of CONTRACEPTION, which have liberated women from reproduction and allowed them to enter all of the careers previously monopolized by men. Modern medicine had played an important role in the demographic shift following industrialization, according to which there is a decline in the birth rate, decreased infant mortality, greater longevity, and an increase in the average age of the population.

Medicine has benefited from a great variety of new diagnostic and surgical technologies and has perfected previously unthinkable procedures such as ORGAN TRANSPLANTATION. Modern diagnostic tools include X-rays and the less invasive sonograms used on fetuses. Other examples of new technologies include laser surgery, CAT scans, PET scans, and MRI, all of which involve sophisticated applications of recent knowledge in the physics of electromagnetic radiation and computerized information processing.

See also INDUSTRIAL REVOLUTION; SCIENCE, HISTORY OF.

Further Reading

Aveni, A. *World Archaeoastronomy.* Cambridge: Cambridge University Press, 1989.

Bonner, J. T. *The Evolution of Culture in Animals.* Princeton, N.J.: Princeton University Press, 1980.

Cardwell, D. *The Norton History of Technology.* New York: Norton, 1995.

Cowan, R. S. *A Social History of American Technology.* New York: Oxford University Press, 1997.

Fagan, B. M. *Kingdoms of Gold, Kingdoms of Jade: The Americas before Columbus.* London: Thames and Hudson, 1991.

Gies, F., and J. Gies. *Cathedral, Forge, and Waterwheel: Technology and Invention in the Middle Ages.* New York: HarperCollins, 1994.

Guilaine, J. ed. *Prehistory: The World of Early Man.* New York: Facts On File, 1991.

Lamberg-Karlovsky, C. C., and J. A. Sabloff. *Ancient Civilization: The Near East and MesoAmerica.* Prospect Heights, Ill.: Waveland Press, 1987.

Marcus, A. I., and H. P. Segal. *Technology in America: A Brief History.* Fort Worth, Tex.: Harcourt Brace Jovanovich, 1999.

McClellan, J. E., and H. Dorn. *Science and Technology in World History.* Baltimore, Md.: Johns Hopkins University, 1999.

Pacey, A. *Technology in World History: A Thousand-Year History.* Cambridge, Mass.: MIT Press, 1991.

Prestwich, M. *Armies and Warfare in the Middle Ages: The English Experience.* New Haven, Conn.: Yale University Press, 1999.

Renfrew, C., and P. Bahn. *Archaeology: Theories, Methods, and Practice.* New York: Thames and Hudson, 1991.

Selin, H. *Encyclopedia of the History of Science, Technology, and Medicine in Non-western Cultures.* Dordrecht, Netherlands: Kluwer Academic, 1997.

Tatersall, I. *The Human Odyssey: Four Million Years of Human Evolution.* New York: Prentice Hall, 1993.

Williams, T. *The History of Invention*. New York: Facts On File, 1987.

technology and science   *See* SCIENCE-TECHNOLOGY DISTINCTION.

teleology

In MORAL PHILOSOPHY, theories are classified as either deontological (based on obligations) or teleological (based on consequences). If the morality of an action is based on its consequences, then it is necessary to (a) decide which consequences are relevant and (b) work out a system for evaluating the consequences. The most common value in teleological ethics is pleasure. HEDONISM, for example, sees pleasure as a worthwhile objective in itself. Utilitarianism is hedonism with a social focus and defines ethical actions as those that produce the greatest possible happiness of the greatest number of people where happiness is interpreted as a balance of pleasure over pain. Not all teleological theories in ethics are hedonistic, however. For example, ARISTOTLE and his Athenian contemporaries saw the development of good character as the most desirable consequence of ethical action. Aristotle's virtue approach to ethics is mirrored in the self-actualization philosophies of modern life that encourage people to seek spiritual enlightenment, cultivate creativity, achieve physical fitness, and so forth.

See also DEONTOLOGY; HEDONISM; MORAL PHILOSOPHY; UTILITARIAN ETHICS.

television and violence

Modern children are exposed to many hours of television viewing, averaging some 28 hours per week, and this has led to concern about the socializing effects of the medium. More than half of television programs contain violence and much of this concern has focused on the question of whether witnessing interpersonal violence, depicted in varying degrees of realism, from cartoons to soap operas and movies to depictions of actual violent events in the daily lives of police officers, facilitates adult aggression. Children who watch violence on television are also exposed to violence in other electronic formats, including video games and the Internet (see COMPUTERS AND VIOLENCE). Although the available scientific evidence does not allow this question to be answered with confidence, much of the evidence is compatible with a causal connection between exposure to media violence in childhood and subsequent aggressiveness.

To conclude that television causes people to be more violent, it would be necessary to conduct an experiment because only experiments allow causal inferences to be reliably made. Experiments in which one group of children views violent television programming and another group does not cannot be conducted satisfactorily for a variety of methodological and ethical reasons. The most obvious one is that determining children's television viewing by the flip of a coin is inconsistent with life in a free society. Lacking the kind of experiment in which two groups of children differ only in the kind of television they see throughout their childhood, the effects of viewing violence cannot be teased apart from other developmental influences.

The closest research design is to look at a natural experiment, such as changes in violence following the introduction of television to a city. One study looked at the effect of introducing television in the 1950s, on adult crime rates in U.S. cities. Cities receiving television did not have an increase in the number of violent crimes despite the large number of such crimes depicted in the television shows of the period. Interestingly, there was an increase in thefts (larceny) and the authors of the study speculated that the sumptuous lifestyles depicted on television may have elicited discontent or envy that encouraged viewers to go out and steal.

A great deal of additional research has addressed the separate question of whether television makes children more aggressive. This turns out to be quite a complex matter because the imitation of aggressive actions that children see on television may not have any intention to hurt other people. It may be more like acting than true aggression. Another problem is that it is not clear whether the imitation of aggression in the present will make children more aggressive as adults.

The imitation of aggression by children has been demonstrated in controlled laboratory experiments. Stanford University social psychologist Albert Bandura found, in the 1960s, that if children were allowed to play with large inflatable ("Bobo") dolls after they had witnessed an adult striking the doll with a mallet, they also struck the doll with the mallet. Whether this is aggression or play can be disputed.

Other research has shown that violence seen on television can cross over into everyday behavior. Preschoolers who viewed violent cartoons each day were rated as more aggressive toward other children on the playground than children who had not seen the cartoons. This phenomenon is familiar to parents whose children imitate martial arts movements that they have seen in television programming, much of it actually developed for young viewers. The most important question is whether these fantasy-related actions affect real-world aggression in later life resulting in serious injuries or deaths.

Are children who have watched a lot of violent television more likely to commit violent crimes as adults? According to a 1972 study by social psychologist Leonard Eron and others they are. Eight-year-olds who preferred violent television programming were rated as more aggressive by peers when they had reached the age of 18 years. They also scored higher on a delinquency questionnaire.

This study was correlational, however, which means that it does not allow causes to be identified. Viewing the violent television programming may not have made them more aggressive. Having an aggressive personality could be the underlying cause of both childhood TV preferences and later aggression. Statistical analysis showed, however, that for children with the same level of initial aggressiveness, childhood TV habits still predicted later aggressiveness. Either violent TV viewing or something that goes along with it, such as social isolation or parental neglect, is making children grow up to be more violent. Follow-up work (1986) to the original (1972) study has found that children who prefer violent TV are more likely to be arrested for serious crimes as adults although the very small number of people arrested means that this result may be statistically unreliable.

The scientific evidence can thus be summarized as tending to support the hypothesis that viewing a great amount of violence on television turns children into more violent adults. This conclusion cannot be firmly drawn, however, because of the many alternative explanations for the data. In particular, it would be necessary to rule out the effects of the social environment that might go along with seeing a lot of televised violence. This is extremely difficult, or even impossible, to do. Psychologists who have written many books on television violence still cannot say if it causes adult aggression. Yet the evidence is suggestive enough for many parents to place limits on the kind of programming their children watch. This concern has resulted in legislation requiring U.S. television manufacturers to install a V-CHIP in new receivers, which allows violence and other objectionable content to be blocked.

A warning about the effects of media violence was released in 2000 in a joint statement from the American Academy of Pediatrics, American Academy of Child and Adolescent Psychiatry, American Medical Association, and the American Psychological Association. The statement pointed to a causal connection between media violence and aggressiveness in some children. It pointed out that children who witness large amounts of media violence are more likely to see violence as an acceptable way of solving problems; be less sensitive to the impact of real violence; believe that the world is a violent place; and resort to violent behavior.

See also COMPUTERS AND VIOLENCE; PORNOGRAPHY; V-CHIP.

Further Reading

Bandura, A. *Aggression: A Social Learning Analysis.* Englewood Cliffs, N.J.: Prentice Hall, 1973.

Eron, L. P., L. R. Huesman, M. N. Lefkowitz, and L. O. Walder. "Does Television Violence Cause Aggression?" *American Psychologist* 27 (1972): 253–63.

Grossman, D., and G. Degaetano. *Stop Teaching Our Kids to Kill: A Call to Action against TV, Movie, and Video Game Violence.* New York: Random House, 1999.

Hennigan, K. M., M. L. Del Rosario, L. Heath, T. D. Cook, J. D. Wharton, and B. J. Calder. "Impact of the Introduction of Television on Crime in the United States." *Journal of Personality and Social Psychology* 42 (1982): 461–77.

Huesman, L. R. "Psychological Processes promoting the relation between exposure to media violence and aggressive behavior by the viewer. *Journal of Social Issues* 42: 125–39, 1986.

Stapleton, S. "Media Violence Is Harmful to Kids—and to Public Health." *American Medical News,* August 14, 2000.

Steuer, F. B., J. M. Applefield, and R. Smith. "Televised Violence and the Interpersonal Aggression of Preschool Children." *Journal of Experimental Child Psychology* 11 (1971): 422–47.

test tube babies (in vitro fertilization) *See* REPRODUCTIVE TECHNOLOGIES.

Thales of Miletus
(c. 640–546 B.C.)
Greek
Philosopher, Scientist

Thales was the first Greek philosopher and scientist known to history. None of his writing survives and the details of his life are sketchy. He had a practical orientation and made a reputation for himself as a military engineer and adviser. Thales may have traveled in Egypt and Babylon. This would have explained his reported ability to predict an eclipse of the sun in 585 B.C. Babylonian astronomers of the period were aware that solar eclipses came in cycles although they did not know why.

Every student of philosophy knows that Thales believed everything is made out of water. This can be considered a falsifiable scientific hypothesis. As Bertrand Russell points out, it is not nearly so silly as it might appear in view of the modern hypothesis that everything is made out of hydrogen, which is, after all, two-thirds water. Thales was interested in magnetism and concluded that magnets have a soul that allows them to move pieces of iron. He had a rudimentary knowledge of trigonometry and could solve practical problems such as estimating the height of a pyramid by measuring the length of its shadow.

See also SCIENCE, HISTORY OF.

Further Reading
Russell, B. *A History of Western Philosophy*. New York: Simon and Schuster, 1972.

thalidomide babies

Used to treat pregnancy sickness, the sedative drug thalidomide ($C_{12}H_{10}N_2O_4$) produced a variety of birth defects. First prescribed in Europe in 1957, thalidomide crossed from the mother to the fetus producing serious deformities, such as flipper-like limbs. By 1962, more than 10,000 children were born with thalidomide-related limb defects. Thalidomide had not been approved for use in the United States at that time but it has recently received approval (1998) by the Food and Drug Administration (FDA) for use in treating a complication of leprosy.

Approval of the drug means that it can be prescribed for other conditions also ("off-label" use), although it can never be given to pregnant women. One important use is in inhibiting the growth of blood vessels that serve cancerous tumors, thus producing atrophy of the tumors themselves. Thalidomide is an immunosuppressant and has potential in the treatment of autoimmune disorders like lupus and rheumatoid arthritis. It is also used to suppress the symptoms of host-versus-graft disease affecting some 50% of bone marrow recipients. Thalidomide is one of the most effective treatments of multiple myeloma, a bone-marrow cancer. Even though it is restricted to the nonpregnant population, thalidomide still has the potential for serious side effects. With prolonged administration, it can lead to irreversible peripheral neuropathy (or nerve degeneration).

Despite the fact that thalidomide did not produce birth defects in the United States, Americans have been well aware of its consequences in Europe and Canada. It is seen as an important cautionary tale concerning loopholes in the procedures used to test drugs. The underlying scientific principle is fairly simple. If a drug is being tested for safety, or effectiveness, then the results of the testing apply only to the study population. Since pregnant women were not used in clinical trials of thalidomide, the birth defects could not emerge in the tests.

People around the world were shocked that such a serious teratogen could have been prescribed to pregnant women and their confidence in the work of scientists who test drugs was seriously undermined. It was the first serious disillusionment with modern pharmacological medicine that seemed capable of formulating a wonder drug for every illness, including mental disorders. In the optimism of this period, many people were oblivious of the fact that the benefit of virtually every medicine is purchased in terms of undesirable side effects of varying severity. A similar awareness about the adverse effect of herbicides and insecticides had begun to emerge thanks to the work of environmentalists like Rachel Carson (see DDT).

One important consequence of the thalidomide tragedy was the passage of more stringent U.S. legislation concerning the approval of new drugs. For example, the Kefauver-Harris Act of 1962 required that new medicines be effective as well as safe. Previous guidelines, based on the Food, Drug, and Cosmetic Act of 1938, had only required safety. Thalidomide thus increased the importance of drug regulation and this led to broadening of FDA authority. Drugs in the United States must pass through an extended testing process before gaining FDA approval. Bringing a new drug to market thus involves a huge capital expenditure, which is passed on to customers in the form of high drug prices. Even though the process through which drugs are approved is onerous, it may not be entirely objective because the science can be distorted by commercial pressures (see COMMERCIALISM IN SCIENTIFIC RESEARCH).

See also BIOLOGICAL PSYCHIATRY; COMMERCIALISM IN SCIENTIFIC RESEARCH; DDT.

Further Reading
Marwick, C. "Thalidomide Back—Under Strict Control." *JAMA* 278 (1997): 1135–37.
Wright, K. "Thalidomide Is Back." *Discover* 21, no. 4 (2000): 31–34.

Three Mile Island nuclear accident

The partial meltdown of a reactor core at the Three Mile Island nuclear plant near Harrisburg, Pennsylvania, on March 28, 1979, was the most serious accident at a U.S. nuclear power plant. The incident began with loss of coolant from the reactor core when a valve stuck open. When this happened, the emergency core coolant systems automatically became functional. Plant workers misread the situation and unfortunately switched off the emergency systems for several hours. There was a major loss of coolant and the remaining coolant became overheated and boiled off. The core and its covering began to melt and the molten material dropped into the bottom of the vessel where it was cooled by the remaining water. By the time the emergency coolant system was switched back on, close to half of the core had melted.

Even though the accident is often used as an example of the riskiness of nuclear power, in fact the design of the reactor prevented a major release of radioactivity into the environment and no one was seriously injured by the accident. Thus, much of the radioactive gas escaping from the open valve was held in the containment building. Some of the gas was contained in coolant water that leaked into an adjacent building and thereby escaped into the atmosphere. This release did not produce a measurable increase over background radioactivity in the surrounding environment. The health effects of the radioactivity were thus negligible. Ironically, the psychological stress of living close to the nuclear plant during and after the accident may have produced far more severe health consequences than the accident itself. Following the accident, there was an increase in cancer rates around the Three Mile Island plant. This could not be attributed to radiation leaks, which were too small, and could be due to the effects of psychological stress on the immune system.

The causes of the Three Mile Island accident have received intensive scrutiny by HUMAN FACTORS engineers, who study human-machine interfaces, and others. The most important conclusion to be drawn from the event is that even though the accident was caused by human error, this error was as much a feature of the situation as of the operators. In other words, the design of the plant, its maintenance, and its management predisposed people to make mistakes and made it difficult for them to correct these errors once they had been made.

The Three Mile Island plant had faulty pressure gauges that showed low pressure when the instrument was not working as well as when the pressure was actually low. Operators were further confused by a control panel on which a hundred warning lights flashed at the same time making it difficult to understand what was wrong. Faulty maintenance meant that a safety backup system was not functional and could not be activated when needed.

Poor instrumentation, poor maintenance, and poor management decisions have been identified in some of the worst industrial accidents, including the BHOPAL CHEMICAL ACCIDENT and the CHERNOBYL NUCLEAR DISASTER. When all of these ingredients contribute to the probability of system failure, accidents are more or less inevitable. Understanding the causes of such systemic failures, as illustrated by the Three Mile Island accident, has become an important element in safety training. This knowledge is being applied to prevent errors in settings as diverse as nuclear power plants and hospitals.

See also BHOPAL, INDIA, CHEMICAL ACCIDENT; CHERNOBYL NUCLEAR DISASTER; HUMAN FACTORS RESEARCH.

Further Reading

Hatch, M. C., S. Wallenstein, J. Beyea, J. W. Nieves, and M. Susser. "Cancer Rates after the Three Mile Island Nuclear Accident and Proximity of Residence to the Plant." *American Journal of Public Health* 81 (1991): 719–24.
Leape, L. L. "Error in medicine." *JAMA* 272 (1994): 1851–57.

Times Beach and dioxin

The problem of environmental pollution may be so severe that whole communities must relocate. Examples include Times Beach, Missouri, and LOVE CANAL, New York. Times Beach, a town of some 2,000 people, became polluted with dioxin, a potent carcinogen, and had to be evacuated in 1982.

The source of the dioxin was burnt oil which was sprayed on unpaved roads and parks during summer to keep the dust down. Before spraying, the oil had been mixed with a tarry sludge from a chemical plant in Verona, Missouri. The Verona plant had manufactured a herbicide, 2,4,5-T, for the U.S. Army, and also produced the disinfectant hexachlorophene. Synthesis of both of these chemicals produces, as an unwanted by-product, 2,3,7,8 tetrachlorodibenzodioxin, the most toxic of the dioxins.

The problem of dioxin contamination first came to the attention of the Centers for Disease Control (CDC) in 1971. Russell Bliss, a waste oil hauler, sprayed the dioxin-contaminated waste oil on a horse arena at Shenandoah Stables, in Moscow Mills, a small town adjacent to Times Beach. On the day of the spraying, May 6, 1971, two of the horses kept in the arena became ill and during the next two weeks a dozen more horses sickened. Their symptoms included diarrhea, abdominal bloating, staggering, twitches, and a skin rash. Over the next few years, it is estimated that 65 horses died of dioxin poisoning. A six-year-old girl who had played in the arena became ill with headaches, diarrhea, and bloody urine, and this prompted alerting of the CDC. It took the CDC three years to identify dioxin as the cause of the poisoning. This discovery was followed by an astonishingly slow process of identifying sites contaminated with dioxin and testing the soil. By 1980, the Environmental Protection Agency (EPA) had produced a list of 85 sites in Missouri requiring investigation but Times Beach was still not on the list. The soil was sampled in 1982, however. Some parts of the community had over 100 times the level of dioxin seen as safe by the EPA (one part per billion, ppb).

Within months of the soil tests, the Merramac River flooded sending 14 feet of muddy water through the town. The federal government evacuated the town because of fears that the mud that entered homes contained dangerously high levels of dioxin. The government then decided to buy out all of the homes in the town. An extensive cleanup followed, which involved the controversial method of incinerating topsoil on the site. Local residents were opposed to the incinerator because of fears that it would disperse small quantities of dioxin into the air. Soil from other contaminated sites was trucked to the Times Beach incinerator. The incinerator ash was buried and covered with grass.

The health consequences of dioxin dumping at Times Beach are still unclear. Dioxin is a carcinogen but it may take 20 years for its carcinogenic effects to emerge in humans. Dioxin also has immediate health effects at high exposure levels, producing a skin rash known as cloracne. It also damages nerve and liver tissue. Dioxin has been shown to cause birth defects in animal tests. Whether Times Beach residents have experienced health problems as a result of their dioxin exposure is still unknown. According to the Missouri Department of Health, there have not been any epidemiological effects. Nevertheless, anecdotal reports suggest an unusual clustering of rare disorders of the liver, nervous system, kidneys, and other internal organs. Former Times Beach residents recall an unusually high incidence of seizure disorders during the 1970s, for example.

The Times Beach incident is one of the most startling examples of flagrant dumping of highly toxic chemicals in populated areas that presumably involve complicity between the chemical company that produced the dioxin and the hauler. The EPA has cleaned up the additional dioxin that was being stored at the Verona plant but there has been no criminal prosecutions of responsible persons. (The hauler, Russell Bliss, did face civil lawsuits). Such unofficial dumping of dangerous chemical waste has been discouraged by recent legislation that calls for a manifest system to track toxic waste from the point of creation to the point of disposal.

See also AGENT ORANGE; HAZARDOUS WASTE; LOVE CANAL; SEVESO, ITALY, CHEMICAL ACCIDENT.

Further Reading
Sagan, K. V. "The People America Forgot." *Family Circle* 106, no. 11 (1993): 74–78.
Sun, M. "Missouri's Costly Dioxin Lesson." *Science* 219 (1983): 367–69.

Titanic, design flaws of
The *Titanic,* which was the largest and most luxurious passenger ship of its day, sank off Newfoundland on the night of April 14–15, 1912, after colliding with an iceberg. Billed as unsinkable on account of its great size, the vessel actually sank on its maiden voyage from Southampton to New York. Some 1,500 of the 2,200 people on board perished.

As in the case of any disaster of this scale, there are many factors that contributed to the catastrophe, including human error. Thus the ship was probably going too fast for icy conditions. Steel sheets covering the hull may have become brittle as a result of the cold. Once water entered the ship following the collision, it continued to pour in because the compartments were not sealed. The design has been unflatteringly compared to that of an ice cube tray. This design flaw meant that water from a single rupture would inevitably fill the entire vessel by leaking from one compartment to the next. The *Titanic* was not carrying sufficient lifeboats for the many passengers, apparently reflecting the belief of the designers that it could not sink. The great loss of life stimulated international efforts to improve safety and rescue procedures at sea. In 1985, the site of the wreck was discovered by American and French researchers.

See also BRIDGE COLLAPSE AND DESIGN FLAWS; ENGINEERING ETHICS.

Further Reading
Ballard, R. D., and K. Marshall. *The Discovery of the Titanic.* New York: Warner, 1987.
Brown, D. G. *The Last Log of the Titanic.* New York: McGraw Hill, 2000.

tobacco company litigation
If a product produced by a company is harmful to the public, this information may be unethically concealed from the public by the company to protect its commercial interests. Scientists and engineers are often privy to such sensitive information and, under ordinary circumstances, would be expected to blow the whistle on the company to protect the public interest. Such WHISTLE-BLOWING can be at great personal cost and is often prohibited by binding legal agreements, as happened in the case of researchers in the tobacco industry. Ironically, the legal muzzling of scientists by tobacco companies has produced some of the most compelling evidence that the tobacco companies knew for decades that their product was both addictive and harmful to health and that they concealed this information in a manner that was harmful to public health. Following protracted legal proceedings, they have been forced to accept liability for some of the health consequences of tobacco.

Scientists working for Philip Morris, in the 1980s, discovered a substance that increased the addictive properties of cigarettes and developed an artificial form of nicotine that was less damaging to health, but they were prevented from publishing their findings by legal agreements that they had signed. Both discoveries were important in subsequent litigation against the tobacco companies because they demonstrate clear knowledge that tobacco products were addictive, something that was publicly denied by the industry, and also that cigarettes could have been made much safer by substituting an artificial form of nicotine. This substitution was not made demonstrating an unethical lack of concern for the health of tobacco smokers and chewers. The secret research of Philip Morris into the effects of nicotine came to light only after the researchers were released from a contract binding them to lifelong secrecy.

The tobacco companies lost a class action suit brought on behalf of some 700,000 Florida residents whose health was injured by tobacco. The 1997 verdict assessed punitive damages at $145 billion, by far the largest penalty ever levied in a product liability case (the closest award being $4.8 billion against Ford Motor Company in the Pinto Case). This verdict is under appeal. The tobacco companies subsequently agreed to a $248 billion settlement with the U.S. states in 1998 to compensate them for the health costs of treating sick smokers. The money is to be paid over 25 years. Similar litigation has been initiated in Australia and New Zealand.

In cases brought by, or on behalf of, individuals, only six were decided against the tobacco companies. Of these, three verdicts were overturned, two are under appeal, and one was returned, in March 2000, with a verdict of $1.72 million in compensatory damages. No money has yet been paid to an individual smoker but the tobacco companies have begun to pay their huge penalty to the states.

See also COMMERCIALISM IN SCIENTIFIC RESEARCH; FORD PINTO; TOLUENE AS INHALANT; TRADE SECRETS; WHISTLEBLOWING IN RESEARCH.

Further Reading
Hilts, P. "Tobacco Firm Withheld Results of 1983 Research." *Denver Post,* April 1, 1994.
"Jury Orders Big Tobacco to Pay Up." *USA Today,* April 7, 2000.
Resnik, D. B. *The Ethics of Science.* New York: Routledge, 1998.

toluene (glue solvent) as inhalant

The modern age of chemicals has produced over 1,400 volatile substances that can potentially be abused as inhalants. One substance that has received considerable attention is toluene, contained as a solvent in various glues, that is a drug of choice among homeless children around the world. They have adopted this method of escaping the miseries of their lives because it produces a cheap, fast, and intense high. Toluene produces dependency and causes irreversible neurological damage, kidney and liver failure, paralysis, and even death.

In Latin American countries, one of the most widely abused substances is shoe adhesive of the Resistol brand manufactured by the American firm H. B. Fuller Company. The brand is so popular among street children that they are often referred to as "Resistoleros." Despite legal and political pressures, the company has resisted putting a foul-smelling mustard oil into its glue to prevent abuse, evidently fearing that this will reduce sales of the product for its proper purposes. Critics have pointed to the irony of this company funding an ethics chair at the University of Minnesota at the same time that its business practices

are so lacking in ethical concern. The company argues that oil of mustard is itself dangerous and possibly carcinogenic.

See also DDT; HORMONE MIMICS; SICK BUILDING SYNDROME.

Further Reading
Jeffrey, P. "Firm Resists Tighter Control on Toxic Glue." *National Catholic Reporter,* March 31, 1995.
Kaminsky, D. C. "Street Children." *World Health* 48, no. 4 (1995): 26–27.

toxic waste   *See* HAZARDOUS WASTE.

trade secrets

Scientists working in industry may find that the ethic of openness that constitutes such an important ingredient of scientific progress is limited by the necessity to protect trade secrets of the employer. Secrecy in industry is particularly important in the early stages of developing a patent. After the patent has been approved, publicity serves the interest of the patent holder. Many important trade secrets, such as the formula for Coca-Cola, are not patentable and must be protected by a code of secrecy on the part of employees.

Most ethicists recognize that secrecy may be the price paid to conduct scientific research in an industrial setting and see this as an acceptable tradeoff for commercial funding of scientific work. Although most industrial research is highly focused on specific applications, there is also the potential for making discoveries and advances that are helpful to academic research. For example, the discovery of cosmic rays, which constitutes such an important prop to the big bang theory in cosmology, was made by Bell Laboratories researchers who were developing microwave antennas in 1965. They found it impossible to eliminate the background noise and concluded that it was uniformly distributed in all directions.

One situation in which industrial secrecy can militate against the public good concerns research conducted by tobacco companies on the addictive properties of nicotine. This research was carried out in an atmosphere of complete secrecy and the scientists, who worked for Philip Morris, were prevented from publishing the research. Outcomes obtained included: (a) the production of a substance that increased the addictive properties of cigarettes and (b) the development of an artificial form of nicotine that was less damaging to health. Clearly, it would have been in the public interest if this information had been made known when it was discovered in the 1980s. In this case, the interest of the company in maintaining secrecy prevailed.

The secret research of Philip Morris into the effects of nicotine came to light only after the researchers were released from a contract according to which they were sworn to lifelong secrecy. Companies evidently feel that such contracts are necessary to protect their exclusive access to proprietary information. Even though scientists may find such contracts in conflict with their professional ethical standards, signing them is a price they pay for commercial sponsorship of their research.

See also PATENTS; TOBACCO COMPANY LITIGATION.

Further Reading
Hilts, P. "Tobacco Firm Withheld Results of 1983 Research." *Denver Post,* April 1, 1994.
Resnik, D. B. *The Ethics of Science.* New York: Routledge, 1998.

tragedy of the commons

The tragedy of the commons is a metaphor that helps us to understand why social relationships are not as harmonious as they might be. It is particularly useful in explaining why people may behave in ways that degrade the value of a shared resource, such as environmental pollution or the nuclear arms race, that threatens the very survival of our species.

This metaphor was developed by biologist Garret Hardin, writing in 1968, who noticed that some of humankind's most intractable problems involve a conflict of interest in which behaving badly has immediate gains but catastrophic long-term effects. He used the story of the destruction of common land by overgrazing as a prototype or thought experiment. Common lands, by definition, are owned by the community. This system was widespread among medieval English villages and was also used by ranchers grazing publicly owned range lands in the American West.

A commons system can work for centuries because wars, disease, and poaching keep the numbers of humans and grazing animals well below what the pasture can sustain. It is only when the number of grazing animals reaches the maximum number the pasture can support when the tragedy of the commons unfolds. At this point, there are so many animals on the pasture that adding any more will result in overgrazing. This causes soil erosion and weeds get the better of the fragile grassland ecology.

Hardin's critical insight is that such catastrophes are more or less inevitable without external regulation. Each herdsmen knows that it is not a good idea to overgraze the pasture since this would damage the grassland irreversibly. However, the gain in having an extra cow is his alone whereas the damage to the fragile grassland ecology of putting out another grazing animal is shared among all the other herders in the village. The tragedy of the commons points to the need for government regulations in situations in which group property can be degraded by competition among individuals. The destruction of English common land was prevented by enclosure. Pollution of waterways by industries that use them is checked by government regulations. Crowding in inner cities is controlled through the use of parking meters.

See also HOBBES, THOMAS; SOCIAL CONTRACT THEORISTS.

Further Reading
Hardin, G. C. "The Tragedy of the Commons." *Science* 162 (1968): 1243–48.

Trojan horse computer virus

A Trojan horse is an electronic virus that masquerades as useful software. It might appear to be an electronic game, a program for updating an operating system, or an animated cartoon. Once it has been downloaded (usually as free software from the Internet or as an e-mail attachment) the Trojan horse may wreak various kinds of damage, such as the destruction of all files on a system except those necessary to keep the system running. One of the most troubling types of Trojan horse virus uses a keystroke-logging program to monitor all information going into the host computer. In this way, hackers can easily obtain access to private information such as credit card numbers and electronic passwords to bank and e-mail accounts. Employers use the same keystroke logging devices to (legally) monitor their employees at work.

See also ELECTRONIC PRIVACY; ELECTRONIC VIRUS.

Tuskegee syphilis study

The Tuskegee syphilis study was conducted between 1932 and 1970 at a public health clinic in Tuskegee, Alabama. It is often held out as a clear example of ethically indefensible research. The study has had particularly harmful effects on African Americans not only because the participants were black men but also because the unethical conduct of the publicly employed physicians undermined trust in public health, thereby discouraging African Americans from seeking health care.

The study involved 399 subjects who were in the later stages of syphilis. At the time the study was initiated, there was still no effective treatment for syphilis and the study set out to follow the natural course of the illness for one year by comparing the infected individuals with 201 men without syphilis who were of the same age.

The Tuskegee study violated several ethical standards in treatment of human research participants. There was no informed consent and participants actually did not know that they were participating in a study. Recruitment was based on an offer of free medical care, hot lunches, and free burials. The most scandalous ethical breach was that the study continued for a quarter century after penicillin, the first effective therapy for syphilis, was available. Far from receiving adequate medical care, as they must have assumed, subjects continued to be denied treatment for a curable disease. They received no information about their illness and were unaware that they were being denied treatment.

Among its ethical breaches, the Tuskegee study violated several of the rights of human research participants as defined in the NUREMBERG CODE, and also constituted a violation of medical ethics by transgressing the HIPPOCRATIC OATH, according to which physicians must do no harm to their patients. Responsibility for the research is difficult to pin down because the study was so poorly organized. Researchers changed from year to year. There were no written protocols and recordkeeping was shoddy. Moreover, it was difficult to determine who was actually running the study.

The whistle was blown on the Tuskegee study by Peter Buxton, a venereal disease researcher with the U.S. Public Health Service, who contacted the Associated Press. This led to a congressional investigation. In 1973, a class-action lawsuit against the government was settled out of court for $10 million in damages. Compensation was provided for survivors who still suffered from syphilis and to families of victims of the study.

See also HIPPOCRATIC OATH; KRAFFT-EBING, RICHARD VON; NUREMBERG CODE; RIGHTS OF HUMAN RESEARCH PARTICIPANTS.

Further Reading

Jones, J. *Bad Blood.* New York: Free Press, 1980.
Resnik, D. B. *The Ethics of Science.* New York: Routledge, 1998.

U

UFOs

Unidentified flying objects (UFOs) are defined as any airborne object that cannot be identified by sight or radar. UFOs are of interest to the ethics of science because they challenge the willingness of academic scientists to be open to new information. In this respect, they are analogous to EXTRASENSORY PERCEPTION (ESP). ESP is a field of study more intensely scrutinized by serious scientists in the sense that it has been written about in respectable scientific journals and its phenomena have been subjected to rigorous laboratory tests. Serious academic scrutiny of UFOs has lagged behind ESP, perhaps because there is such a willingness, as represented by supermarket tabloids and even reputable journalism, to connect UFOs with fanciful tales about visitation, and even abductions, by extraterrestrial beings.

While it is probable that unidentified objects have been seen in the skies throughout our history, as a species, scientific interest in such phenomena is modern and can be traced to two comparatively recent clusters of sightings. The first involved the repeated sighting of dirigible balloon-like mystery craft over the United States in 1896–97. The second occurred during World War II when both Allied and Axis pilots witnessed unidentified metal objects (called "foo fighters") circling their airplanes. On June 24, 1947, Kenneth Arnold, an American pilot, reported nine unidentified circular objects in the skies over Washington State. Arnold described them as skipping over the water like saucers, thus originating a fascination with flying saucers. The term flying saucer is now synonymous with a UFO.

Shortly thereafter in 1948, the U.S. Air Force began systematic documentation of UFO reports under *Project Bluebook,* which promoted scientific study of the phenomena. It is clear that the U.S. government viewed UFO reports as indicative of a potential threat to national security because a scientific panel charged with investigating UFOs, which was headed by physicist H. P. Robertson of the California Institute of Technology, was convened by the Central Intelligence Agency (CIA). The panel concluded that some 90% of UFO sightings could be easily explained in terms of natural phenomena, including the moon, planets, comets, ion clouds, auroras, birds, and unusual meteorological conditions, or man-made objects such as aircraft, balloons, and searchlights. Of the remainder, nothing was seen as providing sufficient information for national concern. The same conclusion was reached by a second panel appointed in February 1966.

Not all scientists were satisfied by these findings. Meteorologist James E. McDonald of the University of Arizona, Tucson, and astronomer J. Allen Hynek of Northwestern University at Evanston, Illinois, claimed that there was sufficient unexplained evidence to make a case that we had been visited by extraterrestrials. These claims were flatly rejected by the majority of scientists but were cultivated by the news media. The controversy

349

prompted a third enquiry, launched in 1968 by the U.S. Air Force, under the direction of physicist E. U. Condon. The *Condon Report* (1969) concluded that there was no evidence for extraterrestrial visitation. As a result, the Air Force's *Project Bluebook* was canceled. Official records of UFO sightings are maintained in a number of other countries, including Canada, England, Greece, Sweden, Denmark, and Australia.

Despite the official rejection of UFO study as a waste of time, American scientists have maintained their interest in UFOs for a number of reasons. The Center for UFO Studies was established by scientists in Northfield, Illinois, in 1973. Of these, some supported the extraterrestrial hypothesis, some felt that this possibility was remote but that it deserved serious study, and others felt that UFOs had no scientific credibility but were worthy of study as a sociological and psychological phenomenon.

Study of people making reports of UFOs has not revealed many interesting patterns. The distribution of reports is essentially random with respect to time and place, with reports occurring in virtually every country in the world among all demographic groups. Clusters of sightings have cropped up in France and Italy (1954), New Guinea (1958), and the USSR (1967). Yet there is no evidence that such sightings are prompted by mass hysteria or a hunger for publicity. This does not mean that publicity does not play an important role after the event. In the United States, uncritical acceptance of the extraterrestrial hypothesis by news media to promote interest in their own product may have lent credibility to these stories. The town of Roswell, New Mexico, which was the site of UFO events in July 1947 has been the beneficiary of a tourist industry catering to UFO enthusiasts. A myth grew up around the town to the effect that a UFO had crashed there and that the wrecked flying saucer, complete with alien remains, had been covered up by the U.S. government. Subsequent investigations suggested that UFO sightings there were related to nearby military activities.

See also EXTRASENSORY PERCEPTION; PSEUDOSCIENCE.

Further Reading
Blum, H. *Out There.* New York: Pocket Books, 1991.
Good, T. *Above Top Secret: The Worldwide U.F.O. Cover-Up.* Austin, Tex.: Quill, 1989.
Jacobs, D. M., ed. *UFOs and Abductions: Challenging the Borders of Knowledge.* Lawrence: University Press of Kansas, 2000.
Korf, K. K. *The Roswell UFO Crash: What They Don't Want You to Know.* New York: Dell, 2000.

Unabomber, the

Theodore Kaczynski conducted a campaign of terror against modern technology, as represented by computer companies, airlines, and universities, from his tiny cabin near Lincoln, Montana, beginning in the late 1970s and lasting until his arrest on April 3, 1996. His method of operation was to mail a booby-trapped bomb to his victims. The Unabomber's motives are debatable. Although he was controversially diagnosed as a paranoid schizophrenic, mental illness apparently played little role in his terrorist actions. Thus, his diary descriptions of the crimes do not indicate that he entertained paranoid delusions concerning his victims. They suggest instead that Kaczynski enjoyed killing for personal revenge. In his April 6, 1971, entry he wrote, "My motive for doing what I am going to do is simply personal revenge. I do not expect to accomplish anything by it."

His choice of targets was influenced by an antitechnology agenda, however. Yet, this appears to have had a very personal element. Attacks on airlines were motivated by low-flying aircraft shattering the tranquillity of his cabin home according to his March 6, 1979, diary entry. Kaczynski's crusade against technology was revealed in a 35,000-word manifesto published in both the *New York Times* and the *Washington Post* in 1995, wherein the case was made that technology is a powerful unstoppable force that undermines personal freedom. Yet, the Unabomber does not really belong in any LUDDITE tradition. Thus, his occupation was not really threatened by technology (he was an unemployed ex-college professor). Neither did his rationale derive from any intellectual tradition and his rambling attack on modern technology published in the national newspapers was ignorant of the writings of intellectuals who oppose technological development, including environmental ethicists.

See also ENVIRONMENTAL ETHICS; LUDDITES.

Further Reading
Chase, A. "Harvard and the Making of the Unabomber." *The Atlantic Monthly* 285, no. 6 (2000): 41–65.
Mello, M. *The United States of America versus Theodore John Kaczynski: Ethics, Power and the Invention of the Unabomber.* Reno, Nev.: Context Books, 1999.

uniformitarianism

Uniformitarianism is the geological theory that the diversity in the fossil record can be explained in terms of gradual changes operating over a long time period. The opposing theory of CATASTROPHISM attempted to explain away the richness of the fossil record of animal life by speculating that there had been numerous mass extinctions followed by the emergence of new species. Catastrophism involved an attempt to fit the great diversity of geological phenomena and the richness of the fossil record of animal life into the limited time period prescribed by the age of the earth as represented in the Bible. CHARLES LYELL (1797–1875) rejected catastrophism positing that geological time was much longer than previously supposed. With sufficient time all of the observed geological phenomena could have been produced by gradual change. CHARLES DARWIN was influenced by Lyell's work, which he read before his *Beagle* voyage, and it prepared him to conceptualize organic evolution in terms of the slow accumulation of gradual changes.

See also CATASTROPHISM; DARWIN, CHARLES; LYELL, CHARLES.

uranium mining

Uranium is used as a fuel in nuclear power and nuclear weapons. Since the ore is radioactive, it might appear that miners would be vulnerable to radioactive contamination with serious adverse health effects. In reality, uranium mining is safer than many other types of mining operation. Miners receive a small dose of radiation because the material that is mined must be greatly concentrated, or enriched, before it can be used as reactor fuel.

Uranium miners do experience a quadrupling of the risk for lung cancer, however, and this risk is from exposure to radon, a radioactive gas that concentrates in underground uranium mines. Much of the uranium used in U.S. weapons during the cold war came from the Four Corners area of the Colorado Plateau. Most of it was extracted from the Navajo Reservation and about a quarter of the miners were Native Americans. Some of the miners developed lung cancers and these became the center of a 20-year political campaign that was aimed at making the federal government accountable for the illnesses. In 1990, Congress passed the Radiation Exposure Compensation Act, which gave financial aid to sick miners and compensated the families of those who had died. Many of the victims could not provide the sort of detailed documentation required by the act and were not compensated. Even though uranium mining can have devastating health consequences for some workers, it is considered safer than coal mining in terms of the increased mortality risks of miners (see NUCLEAR POWER).

See also ASBESTOS MINING; NUCLEAR POWER; NUCLEAR WEAPONS; OPEN-PIT MINING.

Further Reading

Eichstaedt, P. H., and M. Haynes. *If You Poison Us: Uranium and Native Americans.* Santa Fe, N.M.: Red Crane Books, 1994.

Jagger, J. "Yes: Nuclear power Is Safe, and We Need It." In *Environmental ethics,* ed. L. P. Pojman, 474–84. Belmont, Calif.: Wadsworth, 1994.

Tomasek, L., S. C. Darby, A. J. Swerdlow, V. Placek, and E. Kunz. "Radon Exposure and Cancers Other Than Lung Cancer among Uranium Miners in West Bohemia." *The Lancet* 341 (1993): 919–23.

utilitarian ethics

Utilitarianism holds that the aim of moral action should be to promote the greatest possible happiness of the greatest number of people. Utilitarianism defines morality in terms of consequences rather than principles or motives. Its focus on happiness as the only valuable objective makes it a hedonistic theory. A nonhedonistic version of utilitarianism, known as ideal utilitarianism, states that a person should choose actions that produce the most good. Leading proponents of utilitarianism were JEREMY BENTHAM (1748–1832) and JOHN STUART MILL (1806–73). Although ethical discussions of science and technology have often stressed the rights of people (including clients, students, research participants, and the general public) that belong to another branch of ethics, namely, DEONTOLOGY, professional codes of ethics generally call for choice of actions based on a judicious weighing of possible good and bad consequences. For example, research involving deception may be permitted if the potential harm to participants is more than balanced by the benefits of the research. This weighing of consequences is utilitarian in spirit.

See also BENTHAM, JEREMY; DEONTOLOGY; HEDONISM; MORAL PHILOSOPHY; MILL, JOHN STUART; TELEOLOGY.

vaccination

In vaccination, a killed or weakened disease-causing microorganism is introduced into the body to stimulate production of antibodies that protect the individual from subsequent attack by that pathogen. The first vaccine was developed by British physician Edward Jenner (1749–1823) in 1798. Jenner used the cowpox virus (*vaccinia*) to inoculate against smallpox and found that the first disease, which was harmless to humans, conferred lasting immunity against the second potentially fatal illness. French chemist Louis Pasteur (1822–95) used cultured anthrax to develop the first vaccine against a bacterial disease. The importance of such vaccines declined after the introduction of effective antibacterial drugs in the mid-20th century.

Useful vaccines have now been developed for a large number of serious infectious diseases producing a marked decline in overall mortality rates in countries where modern medicine is widely available. Children in the United States and other countries are routinely vaccinated against diphtheria, measles, polio, rubella, and whooping cough so that these former scourges of childhood have been largely wiped out in developed nations. Vaccines have also been developed for influenza which is more of a threat to the elderly population. Other infectious diseases that can be treated with vaccines include cholera, some types of encephalitis, hepatitis B, Rocky Mountain fever, rubella, tetanus, tuberculosis, typhoid, typhus, and undulant fever.

Recombinant DNA technology has made it possible to use genetically altered harmless viruses as live vaccines. The idea is to splice genes into the virus that allow it to express on its surface some of the proteins present on the pathogen. Metaphorically speaking, a sheep in wolf's clothing is created. When the altered virus is injected into humans, it stimulates production of antibodies to the harmful virus and thus produces immunity to infection.

The successful eradication of many infectious diseases through vaccination is one of the most important achievements of scientific medicine but it has not been without its critics. Social Darwinists have objected to vaccination as an interference with the course of nature according to which individuals with weak immune systems are removed from the population before they can reproduce. This argument may seem farfetched and dated today but it is not very different from the current objections to eliminating genetic disease by alteration of the human genotype. Social Darwinists argued that manipulation of naturally selected immunity to disease was "unnatural" because it overrode natural selection. This is precisely the issue that is raised by the prospect of germ cell modification to eliminate genetic disorders because the genotype of the individual can also be considered a product of natural selection.

When compulsory vaccination of children for smallpox was introduced in England in 1867, it encountered some very stiff opposition on libertarian grounds. Much of this opposition was based in the town of Leicester. Even though 90% of the children were vaccinated in 1872, the vaccination rate had declined to 3% within 20 years. During the early years of vaccination, there was a major smallpox epidemic and several hundred deaths. Throughout the rest of the century, after vaccination rates had declined, smallpox deaths also declined, probably due to vaccination in the rest of the country. In 1874, an Anti-Vaccination League was formed in Leicester. There were mass demonstrations against compulsory vaccination. Over 6,000 parents were prosecuted but the fines and imprisonments did not increase the vaccination rate in the city. Finally, in 1898, Parliament abolished the penalties. Parents who objected to vaccination could also apply for an exemption. Very few did and at the turn of the 20th century almost none of the children were being vaccinated.

In the contemporary, United States, children must show proof of vaccination to enter public schools. Parents who object to vaccination may apply for exemptions on religious or philosophical grounds. Religious exemptions are given on the basis of a letter from a leader within a recognized religion that objects to vaccination on religious grounds. Some religious groups object to vaccination as a form of bodily pollution. Others, such as the Church of Scientology, have a principled objection to all types of medical intervention because they interpret healing exclusively as a faith-based phenomenon.

Children in exempted communities pay a price in terms of greatly increased risk of contracting vaccine-preventable diseases. For example, nonvaccinated children in the United States are 35 times more likely to develop measles than vaccinated children. There is also a cost to be paid by the surrounding community. Thus, it is estimated that a doubling of the number of nonvaccinated children would increase the incidence of measles among vaccinated children by approximately 30% (although this would vary greatly depending on the proportion of non-vaccinated children in local populations).

If a large number of people claim exemptions from vaccination there is the potential for a TRAGEDY OF THE COMMONS in which the common good is undermined. This phenomenon is illustrated by the case of pertussis (whooping cough) vaccination. During the 1970s and 1980s, concern about the safety of the vaccine in many countries resulted in lower immunization rates. There was a major increase in the number of cases of whooping cough. There is thus a delicate ethical balance between the rights of individuals to refuse vaccination and the public good.

The control of infectious diseases through vaccination is one of the most effective, and least expensive, public health measures. Apart from religious objections and lib-ertarian concerns, effective public health programs have often encountered resistance purely because they represent novel applications of scientific concepts that are not well understood. Similar irrational fears have been noted in the case of fluoridation of the water supply to improve dental health and irradiation of food to prevent deaths from E. coli and other contaminants.

See also FAITH HEALING; FLUORIDATION OF WATER SUPPLY; FOOD IRRADIATION; KRAFT-EBING, RICHARD VON; ORGAN TRANSPLANTATION.

Further Reading

Plotkin, S. A., and B. Fantini. *Vaccinia, Vaccination, Vaccinology.* New York: Elsevier Science, 1996.

Plotkin, S. A., and E. A. Mortimer, eds. *Vaccines.* Philadelphia, Pa.: W. B. Saunders, 1994.

Salmon, D. A., M. Haber, E. J. Gangarosa, L. Phillips, N. J. Smith, and R. T. Chen. "Health consequences of Religious and philosophical exemption from immunization laws." *JAMA* 282 (1995): 47–53.

Swales, J. D. "The Leicester Anti-vaccination Movement." *The Lancet* 340 (1992): 1019–21.

V-chip

The V-chip is a content-filtering device for televisions that can be used to block violence and other content that is seen as objectionable, including explicit sexuality and crude language. Psychologists cannot say if the viewing of violence by children causes adult aggression. Yet, the scientific evidence linking childhood television viewing and subsequent aggression is strong enough to raise concern. Thus, children who see a large amount of violent programming are desensitized to violence, are rated as highly aggressive by high school teachers and peers, and are more likely to commit serious crimes (see TELEVISION AND VIOLENCE). Such findings have increased the desire of many parents to place limits on the kind of programming their children watch. This concern has resulted in legislation requiring U.S. television manufacturers to install a V-chip in new receivers, which allows violence and other objectionable content to be blocked. All new television receivers had V-chips beginning January 1, 2000.

The V-chip is programmed using a remote control. Parents follow an on-screen menu to block certain kinds of programming. A secret access code is selected that must be used to change the settings. Identification of program content relies on a rating system in use since January 1998. Ratings range from TV-Y, suitable for young children, to TV-MA, suitable for mature audiences in seven categories. These rating icons appear on the screen at the beginning of a program and are included in some television guides. They are accompanied by letters referring to potentially objectionable content: V for violence; S for sexual situations; D for suggestive dialogue; and L for crude language. A major weakness of these labels is that they are determined by the show's producers and may therefore be biased.

The V-chip can be seen as a technological solution to a technological problem. It is not without its problems. Parents may not be particularly adept at programming their V-chips. The complexity of the rating system may well be a deliberate attempt by the television industry to make it more difficult to block programming. Children who are often more technologically adept than their parents might quickly devise ways of reprogramming, or disabling it. Given the intrinsic unreliability of the rating system, parents might be better off if they sat down with their children and looked at their programs before deciding what is on, or off, the viewing menu.

See also ELECTRONIC PRIVACY; PORNOGRAPHY; TELEVISION AND VIOLENCE.

Further Reading

Price, M. E. ed. *The V-Chip Debate: Content Filtering from Television to the Internet.* Englewood Cliffs, N.J.: Lawrence Erlbaum, 1998.

Vesalius, Andreas

(1514–1564)
Belgian
Physician, Anatomist

Vesalius served as personal physician to Holy Roman Emperor Charles V and to King Philip II of Spain. Vesalius's careful dissections of human cadavers marks a turning point in modern medicine. Previously, physicians were excessively reliant on the work of classical authorities such as GALEN (130–200). Vesalius's beautifully illustrated text *De Humani Corporis Fabrica* (On the structure of the human body, 1543) earned him the distinction of being the father of modern anatomy. Drawing on his own empirical observations, Vesalius exposed many inaccuracies in earlier texts used by physicians. Vesalius died on the Greek island of Zakinthos, on October 15, 1564, in the course of a voyage to the Holy Land (Palestine).

See also GALEN; HARVEY, WILLIAM; SCIENCE, HISTORY OF.

video games and violence   *See* COMPUTERS AND VIOLENCE.

video surveillance and crime

The concept of privacy is intricately connected with the concept of freedom and this point is well illustrated in George Orwell's novel *1984*, in which the oppression of the citizens by the state is summarized in the phrase, "Big Brother is watching you." The problem of crime in many cities became so bad in the closing decades of the 20th century that the citizens of many countries were willing to use video monitoring of streets to make them safer, thus sacrificing liberty in return for security.

Beginning in the late 1980s, street surveillance systems have been used in Britain to fight crime. By 1997, there were over 100,000 cameras looking out on streets, parks, public housing projects, and government buildings. Similar security systems are used in France. Their use has been accompanied by reductions in crime against people and property.

One of the main reasons for recent interest in video surveillance is that good quality inexpensive color cameras are available that allow people to be reliably identified from video images, something that was not possible with older black-and-white video systems. If criminals are captured committing crimes on such tapes, they generally do not contest their conviction for the crimes. This makes video surveillance a highly efficient police tool. The use of cameras that can zoom in on suspects while they are committing crimes makes such evidence legally compelling.

Security cameras have always been used in some businesses, including banks, stores, and airports. Their use by police in other public areas in the United States has not been popular because of the invasion of privacy issue. Nevertheless, surveillance is used in some cities, including Tacoma, Washington; Baltimore, Maryland; Hollywood, California; and Huntsville, Alabama (where surveillance is used in the school system). In general, camera use is restricted to a few high-crime neighborhoods and is not nearly so extensive as the video surveillance used in English cities.

See also ELECTRONIC PRIVACY.

Further Reading

"Security Camera Use Is Deterring Crime." *USA Today Magazine,* April 3–4, 1997.

Vienna Circle

Synonymous with logical POSITIVISM, The Vienna Circle of German and Austrian philosophers was organized by Moritz Schlick (1882–1936), a professor of philosophy at the University of Vienna during the 1920s. It included Kurt Godel, Herbert Feigl, Friedrich Waismann, Hans Hahn, and Rudolf Carnap (after 1926). The group was influenced by Ludwig Wittgenstein (1889–1951), a non-member, and published the work of KARL POPPER (1902–94) who knew the philosophers in the Vienna Circle but disagreed with some of their conclusions. Beginning as an informal discussion group, it later developed into a formal organization that published a scholarly journal. The Vienna Circle was devoted to purging philosophy of the vague generalities of metaphysics and elevating the study of propositions based on objectively verifiable scientific facts. With the rise of Nazism in Germany, many of the Vienna Circle philosophers emigrated to the U.S. and the movement came to an end. No longer fashionable in academic philosophy, positivism has had an impact on the intellectual life of nonacademics, which takes the form of great respect for scientific knowledge and a mistrust of the lofty generalities of metaphysics.

See also POPPER, KARL; POSITIVISM.

voodoo death

Scientists tend to ignore phenomena that do not fit easily into their view of how the natural world operates and may thus lose valuable opportunities to enhance their knowledge. In voodoo death, a healthy person dies after being placed under a curse. Such apparent examples of the power of mind over matter do not fit in with the materialistic conceptions of mainstream modern scientific medicine. Thanks to the academic courage of Walter Cannon (1871–1945), a leading physiologist of his time,

instances of voodoo death are now fairly well documented and credited as real phenomena that have important ramifications for medical practice.

The reports collected and published by Cannon typically involved an initial realization by the individual that he or she was under a fatal curse. During the following day or two, the doomed person refused food and water and was treated by friends and relatives as though dying. Overwhelmed by a sense of hopelessness, the cursed individual often expired within 48 hours. The cause of death is unknown but animal experiments suggest that the psychological trauma may lead to excessive rebound of the parasympathetic nervous system with a fatal suppression of breathing and heartbeat. Death of apparently healthy people overcome by feelings of helplessness has also been noted in the case of American prisoners of war held in China during the Korean War. Fellow prisoners referred to the phenomenon as "give-up-itis."

See also FAITH HEALING; PLACEBO EFFECT.

Further Reading

Cannon, W. B. "Voodoo Death." *American Anthropologist* 44 (1942): 169–81.

Cappanari, S. C., B. Rau, W. S. Abram, and D. C. Buchanan. "Voodoo in the General Hospital." *JAMA* 232 (1975): 938–40.

water pollution *See* CLEAN WATER ACT.

Watson, James Dewey
(1928–)
American
Biologist

James D. Watson's most important scientific achievement was deciphering the molecular structure of DNA (deoxyribonucleic acid) in collaboration with FRANCIS H. CRICK in 1953. DNA encodes the hereditary information in genes. Crick and Watson discovered that the DNA molecule consists of a pair of helixes that are joined together by base pairs. The double helix could be unzipped into two strands that contained the same sequence of information, suggesting an elegant mechanism by which genetic information was encoded and could be duplicated. This work ushered in the modern age of biotechnology.

The collaboration of Crick and Watson is described in a fascinating memoir by Watson titled *The Double Helix* (1968). The human side of scientific discovery, as vividly portrayed by Watson, is very different from the professional image projected by scientific publications. As the reviewer for the *Times Literary Supplement* pointed out, the book shows us that scientists are human and "James D. Watson is more human than most." Among their many breaches of scientific ethics and etiquette, Crick and Watson surreptitiously gained access to the data of a colleague, Rosalind Franklin. They subsequently shared the 1962 Nobel Prize for physiology and medicine with Maurice Wilkins who had let them see Franklin's X-ray diffraction studies apparently out of antifeminist spite against her. These details were revealed in Watson's 1968 book. By the time he and Watson received the Nobel Prize Franklin had already died of cancer.

Watson has written or edited many academic texts in the field of genetics. In 1968, following his career as a professor at Harvard University, he became director of the Cold Spring Harbor Laboratory in New York State. For a short period (1989–92) he headed the HUMAN GENOME PROJECT.

See also CRICK, FRANCIS H.; GENETIC ENGINEERING.

Further Reading
Watson, J. D. *The Double Helix.* New York: Atheneum, 1968.

Watson, John B.
(1878–1958)
American
Psychologist

John Broadus Watson was one of the most controversial leading figures in American psychology. A pioneer in

357

behaviorism, Watson wrote accessible books promoting the behaviorist agenda that garnered considerable public attention. The cornerstone of behaviorist psychology was the view that behavior should be studied as a product of objectively observable external events instead of appealing to internal processes of the mind. Watson quickly became disillusioned with the technique of introspection (or looking inward) that was in vogue in academic psychology around the turn of the 20th century. This experience prompted him to conduct research using animal subjects.

In 1903, Watson accepted a teaching position at the University of Chicago and five years later moved on to Johns Hopkins University where he was appointed as a full professor. The next 12 years at Johns Hopkins were the most academically productive of his life and projected him into the limelight as an iconoclast in the field.

A year after Watson arrived at Hopkins, the man who had hired him, J. Mark Baldwin, was arrested in a police raid on a Baltimore brothel and was forced to resign. Watson took up the reins as chairman of the psychology department and also acquired Baldwin's role as editor of the influential journal *Psychological Review*. At the age of 31, he had become one of the most eminent figures in academic psychology.

Watson enjoyed a dazzlingly successful career at Hopkins. He was academically productive and was exceptionally popular with students. A year after his arrival, the students dedicated their yearbook to him. In 1919, he was voted the handsomest professor by the senior class.

Watson's behaviorist agenda that had been many years in the making was published in a well-known paper in *Psychological Review* in 1913. Watson's first book, *Behavior: An Introduction to Comparative Psychology* (1914), created a stir, particularly among younger psychologists, who saw Watson as dusting off the cobwebs of armchair philosophizing and pointing the way toward a more objective discipline in which progress seemed possible. The psychological establishment that was under attack rejected Watson's revolutionary approach to the field. Despite the controversy, he was still elected president of the American Psychological Association in 1915.

Although his first book focused on animal research, Watson believed that the objective science of behavior demonstrated in animal experiments could also be applied to human subjects. After his military service in World War I, he began research on infants. According to Watson, infants come into the world with only three basic emotions that are triggered by predictable situations. Fear is produced by loud noises. Anger is elicited by interfering with an infant's freedom of mobility. Love is produced by stroking of the skin, rocking, and petting. He believed that although the behavior of newborn infants was dominated by inbuilt (or unconditioned)

reflexes, these responses could be altered through Pavlovian conditioning allowing personality differences to emerge.

A typical demonstration of Watson's theory of behavioral development is provided by the case of Little Albert, an 11-month-old baby who was conditioned to fear a white rat in an experiment that would be considered extremely unethical if it were conducted today. In the initial test, Albert showed no fear of the rat whatever. During conditioning trials, a loud noise was produced while the rat was present. Before long, the mere sight of the rat was enough to make Albert burst into tears. Moreover, the fear generalized to other white objects, such as a white rabbit, Santa Claus whiskers, and a white fur coat. Even though this constituted only a pilot study, exhibited serious procedural problems such as the lack of a control subject, and has never been successfully replicated, it is repeated uncritically in virtually every introductory psychology textbook to this day.

Watson's claims about the role of conditioning in behavioral development were exaggerated. He created a mountain of speculation out of a molehill of evidence. Watson was an outstanding popularizer and advocate for his point of view. This is illustrated most clearly in his celebrated dictum: "Give me a dozen healthy infants, well-formed, and my own specified world to bring them up in and I'll guarantee to take any one at random and train him to become any type of specialist I might select—doctor, lawyer, artist, merchant chief, and yes, even beggar-man thief, regardless of his talents, penchants, tendencies, abilities, vocations, and race of his ancestors" (*Behaviorism*, p. 104). The confidence that Watson expresses in the limitless malleability of human behavior has very little to do with the results of his research and has a great deal to do with the democratic spirit, and affluence of the America in which he lived. In fact, given that no one was likely to volunteer their well-formed healthy infant for Watson's experiment, his statement reduces to a purely rhetorical gesture that has nothing to do with science as such.

Watson's focus on unconditioned responses inevitably sparked renewed interest in the physiological mechanisms on which they were based. True to his behaviorist agenda, he believed that emotional experiences were no more nor less than the physiological and muscular responses with which they were associated. This aspect of his research program got him into the most trouble because it prompted him to begin studying the taboo subject of the physiological responses that occur during sexual intercourse.

Watson took a very personal interest in this research. He hooked up polygraph-like instruments to himself and a graduate student and recorded the data as they made love. Watson's first wife had difficulty crediting his scientific objectivity in this project. She raided the lab and confiscated the data. Even though the recording equipment came to light more than half a century later, the data themselves were never found. This is surely one of the most unusual instances of deliberate suppression of data in the history of science. It also illustrates unethical treatment of a graduate student by a professor. Having a sexual relationship with a student at any academic level is considered inadvisable because of the inevitable conflicts of interest between the professor-student relationship and the sexual one.

During the same period, Watson fell in love with graduate student Rosalie Raynor. He wrote her many torrid love letters a bundle of which were discovered by his wife. This precipitated a messy divorce case during which pieces of the letters were printed in the Baltimore Sun. Watson was forced to resign from Johns Hopkins. He married Rosalie Raynor and never returned to academic life.

Following the disgraceful demise of his academic career, Watson went into the advertising business, at the J. Walter Thompson agency, rising to the rank of vice president in a few years. He applied his scientific background in behaviorism to the field of marketing and his ideas were highly influential. Watson argued that the goal of an advertisement was to elicit a strong emotional reaction or need state whether of fear, anger, affection, or sexual desire. He pioneered the use of fear to sell products as diverse as deodorants and car insurance. He also advocated a focus on style and lifestyle rather than substantive claims. His argument was that advertisements should aim to create needs by showing people something better than what they already have.

Even though Watson's ties to academic life were largely severed after 1920 and he ceased contributing to scientific journals, he remained an advocate for behaviorism in popular works including magazine articles and books and enjoyed considerable celebrity throughout his life. His highly accessible books were best sellers. These dispensed advice about using behaviorism to solve all manner of problems. His recipes were always highly controversial. *Psychological Care of the Infant and Child* (1928), for example, advised parents to avoid physical contact with their children as much as possible. He was in favor of environmental control and saw experimenter-like emotional detachment as critical for success. Even though a whole generation of American parents was influenced by such advice, it seems to have its primary origin in Watson's unfortunate inability to express affection towards his own children. Watson died in Connecticut, on September 25, 1958. On the previous year, he had been honored by the American Psychological Association as an important influence on the direction of modern psychology.

See also ADVERTISING; PAVLOV, IVAN; RIGHTS OF HUMAN RESEARCH PARTICIPANTS; SKINNER, B. F.

Further Reading
Buckley, K. W. *Mechanical Man: John Broadus Watson and the Beginnings of Behaviorism.* New York: Guilford Press, 1989.
Murphy, G. and J. K. Kovach. *Historical Introduction to Modern Psychology.* New York: Harcourt Brace Jovanovich, 1972.
Schwartz, D. P., and S. E. Schwartz. *History of Modern Psychology.* New York: Harcourt Brace Jovanovich, 1987.
Watson, J. B. "Psychology As the Behaviorist Views It." *Psychological Review* 20 (1913): 158–77.
———. *Behavior: An Introduction to Comparative Psychology.* New York: Holt, 1914.
———. *Psychological Care of the Infant and Child.* New York: Norton, 1928.
———. *Behaviorism.* New York: Norton, 1930.

Watt, James
(1736–1819)
Scottish
Engineer

James Watt's improvements to the Newcomen steam engine made it a much more efficient power source, which played an important role in the development of the INDUSTRIAL REVOLUTION in England. Watt invented the separate condenser whereby steam was condensed outside the main cylinder. This allowed the cylinder to stay hot throughout the cycle, thereby increasing its efficiency. He also invented the twin-action piston engine in which steam is supplied to both sides of the piston. His other improvements included a device for converting the linear motion of the piston into rotary motion and an automatic governor that used feedback to keep the engine running at a steady speed. Watt formed a highly successful partnership with Birmingham manufacturer Matthew Boulton to build the improved steam engines. He died in Heathfield, England, on August 19, 1819.

See also INDUSTRIAL REVOLUTION, THE; TECHNOLOGY, HISTORY OF.

whistle-blowing in research
Whistle-blowing refers to the practice of alerting authorities to misconduct in the practice of research. There are several situations in which researchers are ethically obligated to blow the whistle, although they may suffer adverse professional and legal consequences for doing so. One of these situations involves scientific dishonesty. Another involves research that causes harm to human participants or inflicts undue suffering on animals. There

is also the case of research involving the possibility of harm to the public.

The situation of a researcher detecting scientific misconduct is well illustrated by the BALTIMORE CASE in which Margot O'Toole blew the whistle on her superior Theresa Imanishi-Kari. Although the many subsequent investigations in this case revealed plenty of reasons for being suspicious, Imanishi-Kari was ultimately exonerated of wrongdoing. Some critics have argued that O'Toole should have take up the matter with Imanishi-Kari's superior, David Baltimore, before bringing allegations outside the lab. Whether this criticism is merited or not, O'Toole paid a steep price since she lost her research position and could not find another one in part because she was seen as a threat by potential employers. If the problem of scientific fraud is to be addressed, there is clearly a need to protect those people who fulfill the important function of bringing dishonest science to the attention of the public. Moreover, professional ethics courses should teach future scientists the best way of handling such sensitive issues.

Some of the most unethical research in the United States has been committed by military scientists. Human victims of unethical research include soldiers themselves, such as those subjected to hallucinogenic drugs without their knowledge, and the general public who were exposed to radioactive contamination without their knowledge (see COLD WAR RADIATION EXPERIMENTS). Some of this research clearly violated the U.S. Army's ethical policies for treatment of human research participants, in addition to the NUREMBERG CODE. Unfortunately, these abuses could not be brought to the attention of the public because the research was classified as secret. This meant that those who knew about the research were powerless to prevent it. Moreover, by the time this information was declassified, it was too late to hold the actual perpetrators accountable for their actions although the U.S. Army has agreed to pay damages in a few cases, thus acknowledging collective responsibility. Clearly, there is a need for a mechanism that waives the Official Secrets Act in the case of harmful research whether this injures military personnel or the public.

Just as the requirements of national security can undermine the practice of ethical research, so the need for industrial secrecy can prevent researchers from informing the public about research that can affect their safety and health. There are many types of privileged information to which industrial researchers and engineers have access but which cannot be divulged for commercial reasons. This might involve a car with a dangerously defective part, a drug with an unfortunate side effect, or an industrial material, like asbestos, that is known to cause cancer. One of the most well-known investigations with implications of this kind was the tobacco research of

Victor DeNobel and Paul Mele who worked for the Philip Morris Company in the 1980s. They discovered that an additive in cigarettes could increase the addictiveness of the tobacco. They also developed an artificial type of nicotine that was less damaging to the heart than natural nicotine. Such important health information should clearly have been brought before the public. Instead, it was suppressed by the tobacco companies. The researchers could not blow the whistle on their employers because they had signed a contract that bound them to secrecy and had to be released from this agreement before they could testify at a congressional hearing on tobacco companies. Concealment of this research exposed more people to the health problems caused by addiction to tobacco if only by concealing information about the addictiveness of tobacco, possibly prevented more stringent government regulation of tobacco, and may have delayed the introduction of less damaging cigarettes. Clearly, laws that protect the secrecy of this kind of research are inimical to the public good.

In general, it is quite difficult for researchers to report ethical problems encountered in the conduct of research. There is almost always a heavy price to pay and rarely any incentive to blow the whistle except the knowledge that one is living up to the high ethical standards expected of scientists. It is thus unlikely that allegations of scientific misconduct will be made lightly. Even so, whistle-blowers should also think of their obligations to the person they might accuse. Before making any accusations they should not only be convinced that they are correct in their action but should have well-documented evidence to back up such action.

See also COLD WAR RADIATION EXPERIMENTS; OPENNESS IN SCIENCE; RIGHTS OF HUMAN RESEARCH PARTICIPANTS; SCIENTIFIC DISHONESTY; SILKWOOD, KAREN; TOBACCO COMPANY LITIGATION.

Further Reading

Edsall, J. "On the Hazards of Whistle Blowers and on Some Problems of Young Biomedical Scientists of Our Time." *Science and Engineering Ethics* 1 (1995): 329–40.
Hilts, P. "Tobacco Firm Withheld Results of 1983 Research." *Denver Post* 1 April, 1994.
Resnik, D. B. *The Ethics of Science.* New York: Routledge, 1998.

Wilberforce, Samuel, Bishop of Oxford

(1805–1873)
English
Bishop

Samuel Wilberforce (see portrait) led the attack of religion on CHARLES DARWIN's theory of evolution. This included a public debate in 1860 with evolutionist biolo-

Bishop Samuel Wilberforce, a distinguished orator, spoke out against Darwinism in England following the 1959 publication of Darwin's ideas. (LIBRARY OF CONGRESS)

gist THOMAS HENRY HUXLEY (1825–95, "Darwin's bulldog") at which Wilberforce (known to opponents as "Soapy Sam") came off worse. The son of antislavery campaigner William Wilberforce (1759–1833), Samuel was a leading figure in the Church of England and was appointed bishop of Winchester in 1869. He died near Dorking, in Surrey, England, on July 19, 1873.

See also CREATIONISM; DARWIN, CHARLES ROBERT; HUXLEY, THOMAS HENRY.

Wilson, Edward O.
(1929–   )
American
Biologist

Edward Osborne Wilson completed his doctorate at Harvard University in 1955, became a zoology professor there in 1956, and has been a professor there ever since. Currently a zoologist in the Museum of Comparative Zoology at Harvard University, he is considered one of the world's leading authorities on the evolution and behavior of ants. He is also a distinguished author whose books have been awarded many prizes; for example, both *On Human Nature* (1978) and *The Ants* (1990, co-authored by Bert Holldobler) received the Pulitzer Prize.

Wilson is also the only person to have received this country's highest award in science, the National Medal of Science, and its premier literary award, the Pulitzer Prize in literature.

Wilson became widely known as a founder of sociobiology following the publication of *Sociobiology: The New Synthesis,* a 1975 textbook dealing with animal behavior from an evolutionary perspective that had a controversial final chapter on humans. Perhaps unfairly, Wilson became the focus of organized opposition to GENETIC DETERMINISM and SOCIAL DARWINISM at Harvard University.

As a field worker around the globe, Wilson developed a rich understanding of ecosystems. In 1990, he subsequently published his memoirs, *Edward O. Wilson: A Life in Science,* and in 1992 published *The Diversity of Life,* which describes the biodiversity in rain forests and makes the case for protecting these fragile ecosystems. In 1995, *Time* magazine named him "one of the most twenty-five influential people" in America.

See also BIODIVERSITY; SOCIOBIOLOGY CONTROVERSY.

wiretapping   *See* ELECTRONIC PRIVACY.

women in science/technology careers

The history of science and technology is a male-dominated one. The few women who made important contributions to science before the 20th century were conscious of going against sex role expectations. It is difficult to think of any important contributions made by women to industrial technology before the 20th century. Social pressures, as exemplified, for example, in educational practices, clearly played an important role in discouraging women from entering these fields in the past.

Even wealthy women were less likely to receive a good education than were men. If they did, their instruction was much more likely to focus on skills that would have improved their marriage prospects, such as music, dancing, embroidery, and etiquette, rather than on rigorous instruction in science and mathematics. The lack of opportunity for women in the preprofessional era of science has carried over to modern times in that there are far fewer women scientists and engineers today than there are men. Even in the United States, where women have better opportunities than in most other countries, only a quarter of scientists are women, based on the number of people receiving doctorates in the physical sciences (i.e., physics, chemistry, and related fields). This number may seem low but it is four times higher than what it was only a quarter century ago. The proportion of female scientists is twice as high as the proportion of

DOCTORATES EARNED BY AMERICAN WOMEN IN SCIENCE, ENGINEERING, AND ALL FIELDS, AS A PERCENTAGE OF TOTAL DOCTORATES, IN THESE FIELDS

	FIELD OF DOCTORATE		
YEAR	ENGINEERING	PHYSICAL SCIENCE*	ALL FIELDS
1970	0.4%	5.7%	13.4%
1975	1.7%	8.3%	21.9%
1980	3.6%	12.2%	30.3%
1985	6.3%	16.3%	34.3%
1990	8.5%	18.6%	36.5%
1995	11.6%	22.9%	39.3%
Increase 1970–1995	11.2%	17.2%	25.9%

*Physics, chemistry, and related fields.

Doctorates earned by women in science, engineering, and all academic fields as a percentage of total doctorates in these fields. (Source: Statistical Abstract of the United States, various years)

female engineers although the number of female engineers has increased rapidly in recent decades (see table). Given the fact that over half of the world's scientists reside in the United States, American trends have an important effect on the global influence of female scientists and engineers.

The data on doctoral degrees awarded to American women over the past quarter century reveal several interesting patterns. One is that there has always been a greater representation of women in other academic fields than in hard sciences and engineering. Another is that the proportion of women engineers has remained much lower than the proportion of women scientists. (Of course, engineering is an applied field and many engineers do not obtain doctorates but the proportion of women employed as engineers is actually lower than the proportion of female doctorates in the field). Perhaps the most important trend is that even though women have made considerable progress in science and engineering, their progress in these fields has been substantially less than their gains in all fields. Despite an environment that has allowed women to approach equal representation in doctoral degrees, they are still very far from equal representation in science and technology. The doctorate is an entry requirement for most scientific careers and there is evidence that women scientists are less likely to be promoted in an academic environment.

Equality of opportunity in science, basic and applied, is important because it promotes the ethics of intellectual freedom and freedom of inquiry and dispels the notion that academic pursuits in these fields are the province of a privileged elite of white males. Diversity may also promote scientific objectivity. For example, the scientific study of sex differences, one of the most controversial endeavors in modern science, is likely to produce biased and inaccurate conclusions if it is carried out only by men. Thus early students of moral development found that men typically reached a higher level of moral reasoning than women. This conclusion has been challenged by CAROL GILLIGAN, who argues that men and women differ in their moral values, with men preferring a more abstract moral code and women expressing a more social form of ethical reasoning. Wherever this dispute ends up, it is difficult to argue that the study of moral development is not richer for its having taken place.

WHY WOMEN ARE UNDERREPRESENTED IN SCIENCE/TECHNOLOGY CAREERS

Most modern practitioners would agree that social diversity strengthens science by reducing the likelihood of subjective biases in social research and by encouraging fresh and original approaches to scientific problems. If so, why are there comparatively few women who work as leading researchers in science, engineering, and mathematics? There are two plausible explanations that are often seen as rivals but may be complementary. One is that women are not drawn to these fields because they find them intrinsically cold and uninteresting. The other is that women have a similar level of interest as men do but that social norms, including the social environment in which science is taught and conducted, are hostile to the entry of women to these professions.

Average sex differences in cognitive aptitudes are generally quite small but in the specific case of the tiny minority of individuals who score high on mathematical abilities there is a large overrepresentation of males. Mathematical ability is predictive of interest in and success at careers in science and engineering as well as mathematics. It follows that the interest in, and success of, women in these careers may never reach parity with that of men. Note that this argument does not claim that women cannot achieve at the same level as men but only that such high achieving women can be expected to be less common than high-achieving men. The fact that women can contribute to science at the highest level is supported by the existence of female Nobel Laureates beginning with MARIE CURIE. While it might be argued that the lack of women in science and engineering is due to a lack of equal opportunity, this view is not supported by the fact that social changes favorable to women in the second half of the 20th century did not produce equal representation of women in these careers or even allow them to make progress equivalent to their gains in other academic fields.

Even if sex differences in mathematical aptitude play a role in explaining the underrepresentation of women in science, engineering and mathematics, they are unlikely to be the whole story. Feminist philosophers have made the case that these disciplines are, in various ways, unfriendly to women. One way of increasing the number of female scientists would be to reduce the very high attrition rates in the first two years of college. It has been suggested that this could be done through a pedagogy that involves increased emphasis on collaborative learning, practical experiences, and group projects.

Even if women achieve parity in respect to the number of graduates from science, engineering, and mathematics programs, their problems are not over. The world of professional science is a tough competitive environment in which few women rise to the top. A recent study at Cambridge University found that whereas 40% of the researchers in chemistry were women, only 2% of the tenured faculty in that discipline were female.

How do we begin to explain the huge difference in success between men and women of science? One view is that science remains one of the most harshly competitive fields in which individuals struggle to be published, achieve fame and status in the eyes of peers, and receive funding. Men supposedly enjoy this challenging environment and are stimulated to be highly productive. Women supposedly are less interested in competing for priority and status. Consistent with this view is the fact that tenured women are generally much less academically productive than tenured men.

Feminists argue that women's productivity is undermined by competing obligations of relationships and families, which they often set as an ethical priority. Sex bias also plays a role. Even in Sweden, a country that is noted for its ethos of gender egalitarianism, male-dominated review committees do not give women fair treatment. Female Swedish academics have to publish twice as much as their male counterparts to be deemed worthy of a fellowship in medicine.

INCREASING THE NUMBER OF WOMEN IN SCIENCE/TECHNOLOGY CAREERS

Attempts to improve the status of women in science take two different, and apparently contradictory, positions. The first is that AFFIRMATIVE ACTION should be used to improve women's representation among tenured faculty and receipt of government funds. Responding to legislation produced by pressure from THE ASSOCIATION FOR WOMEN IN SCIENCE, the National Science Foundation (NSF) conducts an affirmative action program for women in science that involves continuous monitoring and biennial publication of statistics. This program has funding for career development awards earmarked for women.

The second approach is designed not so much to help women compete in the harshly competitive male-dominated world of professional science as to change that climate to make it more friendly to women. The NSF conducts site visits designed to improve the climate for women in academic departments that have traditionally excluded them. This will presumably mean that refusing tenure to women will become more difficult than it has been in the past. Women may also receive more opportunities to develop their research programs through relief from teaching and other obligations. Such programs may help women cross the threshold in some academic departments that currently have very few tenured females. The deeper philosophical question about whether the scientific enterprise itself can be changed to make it more friendly to women remains open. Is it possible to change the incentive structure of scientific investigation such that cooperation is placed at a premium over egocentric competition? This would indeed be a profound change in the way that science is done. Such a change is unlikely if women do not have more administrative control over research laboratories.

See also AFFIRMATIVE ACTION; ASSOCIATION FOR WOMEN IN SCIENCE; FEMINIST ETHICS AND THE PHILOSOPHY OF SCIENCE.

Further Reading

Benbow, C. P. "Sex Differences in Mathematical Reasoning Ability in Intellectually Talented Preadolescents." *Behavioral and Brain Sciences* 11 (1988): 169–232.

Burrelli, J., C. Arena, C. Shettle, and D. Fort. *Women, Minorities, and Persons with Disabilities in Science and Engineering.* Upland, Pa.: Diane Publishing, 1998.

De George, R. *Business Ethics.* Englewood Cliffs, N.J.: Prentice Hall, 1995.

Resnik, D. B. *The Ethics of Science.* New York: Routledge, 1998.

Rosser, S. *Re-engineering Female-Friendly Science.* New York: Teacher's College Press, 1997.

Sadler, A., ed. *Affirmative Action.* San Diego: Greenhaven Press, 1995.

Wenneras, C., and A. Wold. "Nepotism and Sexism in Peer Review." *Nature* 387 (1997): 341–43.

women's ways of knowing   *See* GILLIGAN, CAROL.

Worldwatch

The Worldwatch Institute is dedicated to fostering an environmentally sustainable society. Located in Washington, D.C., Worldwatch publishes an annual *State of the World* report which generally paints a picture of the earth on the brink of catastrophe and has been criticized by the Cato Institute, a politically conservative Washington-based think-tank, for ignoring many advances in pollution control and public health. Its primary concerns are GLOBAL WARMING, ozone depletion, loss of biological diversity, environmental pollution, and the global consequences of the human population explosion. In addition to its news magazine, *Worldwatch,* the institute publishes books dealing with environmental issues.

See also ACID RAIN; AIR POLLUTION; BIODIVERSITY; ENVIRONMENTAL ETHICS; GLOBAL WARMING; WATER POLLUTION.

Y

Y2K and mistrust of scientific predictions

The turning of the millennium was associated with a scare over the Y2K bug. This was caused by a programming glitch in which only three digits had been assigned to the year with the consequence that computer systems could not distinguish between the year 2000 and the year 0. Computer scientists predicted that all sorts of problems could occur from the collapse of electronic systems responsible for the power grid, resulting in power failures, to the launching of nuclear weapons.

Some prophets of doom interpreted these scientific warnings as an indication that civilization as we knew it could come to an end. They took the opportunity to stock up on canned food, bottled water, and guns and ammunition, all of which would be critical for survival in a world where distribution systems, and law and order, had broken down. The actual event turned out to be a great deal less frightening than such predictions. Instead of the system-wide failures that had been anticipated, all of the problems that emerged were local. The security system failed in a building in Omaha. A few credit card transactions failed. There were problems in 911 emergency systems in parts of Wyoming and North Carolina. The most exciting failure was the shutting off of transmissions from a U.S. spy satellite for several hours.

The global catastrophes expected to materialize on January 1, 2000, did not occur partly because of a huge global investment to make computer systems around the world Y2K compliant. The worldwide bill for doing this is estimated at $500 billion, one-fifth of this being spent in the United States. Yet there is little evidence that degree of preparedness made much difference. Poorly prepared countries, such as Russia, did not fare any worse than well-prepared ones, such as the United States.

Dire scientific predictions about the collapse of worldwide computer systems have a clear analogy in the predictions of environmental scientists about collapse of the global ecosystem due to industrial activity. Scientists clearly have an important role to play in alerting the public to such problems but it seems that their warnings are often excessively pessimistic perhaps because scientists are trained to be cautious and therefore magnify dangers. It might be argued that environmental alarmism is appropriate to obtain corrective action. If people had not been afraid of the Y2K bug, they would not have been willing to spend so much money in eliminating any danger it posed. Similarly, it may be appropriate to frighten people about global warming, acid rain, chemical pollution, and the hole in the ozone layer. Yet, this is a very dangerous tactic because scientists may cry wolf too often and loose their public credibility.

See also BIODIVERSITY AND INDUSTRIALIZATION; ELECTRONIC TERRORISM; WORLDWATCH.

Further Reading

Levy, S. "The Bug That Didn't Bite." *Newsweek*, January 10, 2000.

Young, Thomas

(1773–1829)
English
Physicist

Thomas Young is remembered mainly for his contributions to optics, specifically to the theory of light. Early in the 19th century, Young and others experienced growing dissatisfaction with the theory of light as a stream of particles (the corpuscular theory), which dated back to ISAAC NEWTON (1642–1727). This theory did not provide a satisfactory explanation for the diffraction, or bending, of light rays as the light passed corners. Young proposed that light was a wave similar to sound (i.e., with vibration in the direction of propagation). French scientist Augustin Fresnel (1788–1827) proposed that light was a transverse wave (with vibration perpendicular to the direction of movement, like a water wave). Fresnel turned out to be correct but both men came up with experimental demonstrations of interference effects that could be explained only in terms of the interactions of waves.

Young's experimental demonstration of interference effects involved the passing of light through two pinholes onto a screen. The light beams spread out and where they overlapped there was a banding pattern of alternating light and dark strips. The dark bands showed that the light waves had undergone destructive interference analogous to water waves canceling each other out. Young's wave theory was ignored until Fresnel produced a more compelling articulation of his own wave theory. For most scientists of the day, it was unthinkable to question Newton's authority on the question. Trained in medicine, Young provided an optical explanation for the capacity of the eye to focus on objects at varying distances (known as accommodation). He also proposed the trichromatic (or three-color) theory of color vision later elaborated by German scientist HERMANN VON HELMHOLTZ (1821–94).

His other interests included elasticity (measured by Young's modulus) and energy. He restricted the term energy to its current narrow scientific meaning. Young was a language enthusiast and he played a role in deciphering the Egyptian Rosetta Stone, which allowed ancient clay tablets to be deciphered. He died in London on May 10, 1829.

See also NEWTON, ISAAC; SCIENCE, HISTORY OF.

Z

Zimbardo's prison simulation

Overcrowded modern prisons tend to be cruel, violent, degrading, and oppressive. Prison guards often behave brutally. Prisoners tend to become depressed, with symptoms of passiveness and demoralization. Social psychologist Phillip Zimbardo, of Stanford University, wondered whether ordinary people would adopt these patterns of behavior if they participated in a simulation of prison life. Even though he could not have predicted how frighteningly real the prison experience became, many ethicists feel that the research took an unjustifiable risk with the welfare of research subjects.

In Zimbardo's prison simulation, conducted in a basement at Stanford University, two randomly formed groups of students played the role of either a prisoner or a guard. Few instructions were given about how the guards or prisoners were to behave and the students apparently filled in the gaps from their own understanding of the relative roles of prisoners and guards.

Many aspects of the experience of a typical prisoner were faithfully replicated. The prisoner subjects were arrested unexpectedly at their homes by cooperative local police and put through the usual routine of being booked, fingerprinted, and admitted to the prison. Prisoners were easily identifiable by the loose-fitting smocks they wore, and by their ankle chains. Guards wore khaki uniforms, which were complemented by sunglasses, night sticks, whistles, and keys.

Prison rules allowed three bland meals per day and three supervised visits to the toilet. Prisoners were called by their numbers and had to be periodically lined up and counted. Otherwise, the participants were on their own and the experimenters waited to see what would happen.

Power quickly went to the heads of the guards. They shouted at the inmates and subjected them to inhuman and degrading punishments. Some prisoners were forced into crowded cells, for example. Others were subjected to solitary confinement. Prisoners were woken up in the middle of the night. They were lectured and intimidated. They were forced to do push-ups to the point of exhaustion.

Initially, the prisoners objected to their cruel treatment but this resulted in retaliatory abuse and they eventually gave in. After only three days, the first prisoner had to be released. He was suffering from acute depression. After six days, the experimenters decided that it would be unethical to continue and they brought the study to a halt well short of the projected length of two weeks. The controlling influence of the students' assigned roles had become so powerful that the researchers felt they could not take the risk of continuing the experiment. Prisoners were delighted at their liberation. The guards had been enjoying themselves and were extremely disappointed by the early conclusion of the simulation.

Zimbardo's experiment was an eerily successful demonstration of the power of social roles. Individuals

who are suddenly given the prerogatives of a prison officer, without the training and oversight prison officers
normally receive, are liable to abuse that authority. When
the former guards got together with former inmates to
discuss what the students had learned from their experiences, many of the prisoners complained bitterly of the
abuse they had suffered at the hands of the guards. Yet,
most agreed that the experiment had been a worthwhile
experience, which had taught them a great deal about the
human potential for brutality that can be unlocked in any
of us by the social role in which we find ourselves. As in
the case of MILGRAM'S OBEDIENCE RESEARCH, Zimbardo had
shown that ordinary people were capable of extremely
immoral behavior under pressures of the social environment.

With hindsight, it is easy to say that Zimbardo's prison
simulation was unethical because of the trauma inflicted
on the participants. Interestingly, there was little exact
precedent for the study and therefore no means of predicting what would happen. The most that social psychologists could have predicted based on current
knowledge was that the division of participants into two
groups with conflicting interests would have evoked hostility and would have involved the potential for violence.
Since no deception was involved, participants gave their
informed consent. They knew that they were participating in a prison simulation and that they had a chance of
playing the role of a prisoner. Yet, when they gave their
consent, they could not have predicted how horrific the
experiment might become.

The real ethical violation of the prison simulation was
not lack of INFORMED CONSENT but the absence of a mechanism by which participants could opt out of the study at
any time they chose. It is difficult to see how this provision could have been incorporated into a prison simulation since the defining criterion of imprisonment is being
held against one's will. This suggests that prison simulations are inherently unethical or at least that they cannot
be conducted without extraordinarily careful safeguards
to protect the well-being of volunteer participants.

See also MILGRAM'S OBEDIENCE RESEARCH; NUREMBERG
CODE; PARANOIA, EXPERIMENTAL INDUCTION OF; RIGHTS OF
HUMAN RESEARCH PARTICIPANTS.

Further Reading

Beauchamp, T. L. *Ethical Issues in Social Science Research*. Balti-
more, Md.: Johns Hopkins University Press, 1982.
Kimmel, A. J. *Ethical Issues in Behavioral Research: A Survey*.
Cambridge, Mass.: Blackwell, 1996.
Zimbardo, P. G. "Psychology of Imprisonment." *Society* 9, no. 6
(1972): 4–8.

APPENDIX

ETHICS ORGANIZATIONS, RESOURCES, AND WEBSITES

The organizations listed below maintain websites relevant to ethics in science and technology. Addresses and telephone numbers are also listed where appropriate.

American Society for Bioethics and Humanities (http://www. asbh.org/), 4700 West Lake Avenue, Glenview, Ill. 60025-1485. Telephone: (847) 375-4745.

Bochum Center for Medical Ethics (http://www.ruhr-uni-bochum. de/zme/zme.html), Building GA, Floor 3 Room 53, D-44780, Bochum, Germany. Telephone: 49 (0) 234 32-22749.

Center for Clinical Bioethics, (http://bioethics.georgetown. edu/cch/home.htm), Georgetown University Medical Center, 4000 Reservoir Road, N.W., Washington, D.C. 20007. Telephone: (202) 687-1122.

Center for Clinical Ethics in Humanities and Healthcare (http://wings.buffalo.edu/faculty/research/bioethics/), State University of New York at Buffalo, Veterans Affairs Medical Center, 3495 Bailey Avenue, Buffalo, N.Y. 14215. Telephone: (716) 862-6563.

Center for Democracy and Technology (wysiwyg://119http:// www.cdt.org/), 1634 Eye Street, N.W., Washington, D.C. 20006. Telephone: (202) 637-9800.

Center for Research Ethics (http://www.cre.gu.se/), Götenburg University, Box 700,405 30 Göteborg, Sweden. Telephone: 46-31-773-49 22.

Center for Science in the Public Interest (Nutritional advocacy group, http://www.cspinet.org/), 1875 Connecticut Avenue, N.W., Suite 300, Washington, D.C. 20009. Telephone: (202) 332-9110.

DNA Patent Database (http://208.201.146.119/).

Electronic Privacy Information Center (Public interest research center, http://www.epic.org/), 1718 Connecticut Avenue, N.W., Suite 200, Washington, D.C. 20009. Telephone: (202) 483-1140.

Environmental Protection Agency (http://www.epa.gov/), 1200 Pennsylvania Avenue, N.W., Washington, D.C. 20460. Telephone: (202) 260-2675.

Ethical, Legal and Social Implications of the Human Genome Project, National Human Genome Research Institute (http://www.nhgri.nih.gov/elsi/), National Institutes of Health, 9000 Rockville Pike, Bethesda, Md. 20892. Telephone: (301) 496-0844.

Ethics in Science (http://www.chem.vt.edu/ethics//ethics.html), Department of Chemistry, Virginia Polytechnic, Blacksburg, Va. 24061-0212. Telephone: (540) 231-5391.

Ethics Updates, University of San Diego (http://ethics.acusd. edu/index.html).

Eubios Ethics Institute (http://www.biol.tsukuba.ac.ip/~macer/ index.html), University of Tsukuba, Tsukuba Science City 305, Japan. Telephone: 81-298-53 4662.

Genetics and Ethics, Center for Applied Ethics (http://www. ethics.ubc.ca/brynw/), University of British Columbia, 277-6356 Agricultural Road, Vancouver, B.C., Canada V6T 1Z2.

German Reference Center for Ethics in the Life Sciences (http://www.drze.de/), Ringstrasse 30, 86947 Weil, Germany.

Greatergood.com (E-commerce charity, http://www.greatergood. com).

Greenpeace (Environmental group, http://www.greenpeace.org), 702 H Street, N.W., Washington, D.C. 20001. Telephone: (800) 326-0959.

Hastings Center, The (Ethics Research Institute, http://www. thehastingscenter.org), Garrison, N.Y. 10524. Telephone: (845) 424-4040.

High School Bioethics Project, (http://bioethics.georgetown.edu/ hsbioethics/), Kennedy Institute of Ethics, Georgetown University, Fourth Floor Healy, Washington, D.C. 20057-1212. Telephone: (202) 687-8099.

International Calendar of Bioethics Events (http://www2. umdnj.edu/ethicweb/upcome.htm), University of Medicine and Dentistry of New Jersey, Robert Wood Johnson Medical Center, P.O. Box 896, Piscataway, N.J. 08855-0896. Telephone: (732) 235-4549.

Kennedy Institute of Ethics (http://www.georgetown.edu/ research/kie/), Georgetown University, Box 571212, Washington, D.C. 20057-1212. Telephone: (202) 687-8099.

Medical College of Wisconsin—Bioethics Online Service (http:// www.mcw.edu/bioethics/), 8701 Watertown Plank Road, Milwaukee, Wis. 53226. Telephone: (414) 456-8498.

National Bioethics Advisory Commission (http://bioethics.gov/ cgi-bin/bioethic_counter.pl), 6705 Rockledge Drive, Suite

700, Rockville, Md. 20892-7979. Telephone: (301) 402-4242.

National Consultative Committee for Health and Life Sciences (CCNE, http://www.ccne-ethique.org/), 71, rue Saint-Dominique, 75007 Paris, France. Telephone: 33 01 44 42 48 52.

National Human Genome Research Institute (http://www.nhgri.nih.gov/policy_and_public_affairs/ecsi/), National Institutes of Health, 9000 Rockville Pike, Bethesda, Md. 20892. Telephone: (301) 496-0844.

National Library of Medicine (Free access to BIOETHIC-SLINE(R) searchable ethics database, http://igm.nlm.nih.gov/), 8600 Rockville Pike, Bethesda, Md. 20894. Telephone: (888) 346-3656.

National Reference Center for Bioethics Literature (http://www.georgetown.edu/research/nrcbl.htm#publications), Kennedy Institute of Ethics, Georgetown University, Box 571212, Washington, D.C. 20057-1212. Telephone: (202) 687-6770.

Nuffield Foundation, The (Charity that sponsors the Nuffield Council on Bioethics, http://www.nuffieldfoundation.org/), 28 Bedford Square, London WC18 3JS, England. Telephone: 020 7631 0566.

Online Ethics Center for Engineering and Science (http://www.onlineethics.org/).

People for the Ethical Treatment of Animals (PETA, http://www.peta.org/), 501 Front Street, Norfolk, Va. 23510. Telephone: (757) 622-7382.

University of Tennessee at Knoxville, Bioethics Videos (http://web.utk.edu/~ggraber/videos.htm)

U.S. Department of Health and Human Services, Office of Inspector General (Source of reports on Institutional Review Boards, http://www.hhs.gov/progorg/oig/), 330 Independence Avenue, S.W., Washington, D.C. 20201. Telephone: (202) 619-1343.

Waseda University Bioethics, Japan (http://kenko.human.waseda.ac.ip/rihito/biogate-j.html), 1-104 Totsukamachi, Shinjuku-ku, Tokyo, 169-8050, Japan. Telephone: 81-3-3202-8638.

Worldwatch Institute (Research organization concerned with impact of economy on environment, http://www.worldwatch.org/), 1776 Massachusetts Avenue, N.W., Washington, D.C. 20036-1904. Telephone: (202) 452-1999.

INDEX

Boldface page numbers indicate main headings. *Italic* page numbers indicate illustrations. Page numbers followed by *t* indicate tables.

A

AAAS *See* American Association for the Advancement of Science

Abacus Direct 72

ABM *See* Antiballistic Missile Treaty

A-bomb 62, 198, 226

abortion
    as birth control technique 71
    of fetuses with genetic disorders 107, 139
    Hippocratic oath and 152–153
    for sex selection 17, 139

abortion pill 281

abstinence, sexual, as contraceptive method 69, 70

academic vigilantism **1**, 319

acid rain **1–2**, 10, 104

ACLU *See* American Civil Liberties Union

acquired immunodeficiency syndrome *See* AIDS

active euthanasia 109

adaptationist thinking 320

adaptive radiation 54

adequate aftercare, right to 279

Adirondacks, disappearance of fish in 1

The Advancement of Learning (Bacon) 31, 290

advertising **2–4**
    cookies in customization of 71, 98
    creating needs with 2–3
    criticisms of 2, 4
    definition of 2
    false or deceptive 4
    real objective of 3–4
    and sales 2–3
    using sex in 3, 359

aerosol cans, and ozone layer depletion 9, 10, 237

aesthetic pain, from advertising 4

affirmative action
    in college admission 303
    in science and technology careers **4–7**, 123, 363

Africa
    green revolution in 145
    use of DDT in 9

African Americans
    affirmative action for 5–6
    as elite athletes 302–303
    health outcomes of 302
    lead poisoning in 182
    in science and technology careers 6t
    sickle-cell anemia in 302, 311

African gray parrot, understanding human language 60

Agent Orange **7–8**, 41

aggression *See* violence

agriculture
    development of 284, 285, 332–333
    improvements in 145, 197, 336, 339

Agriculture, Department of *See* Department of Agriculture

AI *See* artificial insemination

AIDS, prevention of 68

AIDS patients, marijuana for 262

AIDS research 264

air pollution **8–13**, 317
    consequences of 9–11
    definition of 8
    ethics of 9, 12–13, 104
    from nuclear power plants 338
    reducing, hybrid-powered cars in 164–165
    regulation of 11–12, 61
    sources of 9

alar scare **13–14**

alchemy 33, 286

alcohol, maternal use of 91–92

alcoholic(s), liver transplantation for 236

alcoholism, Scoline for 304

Alcor Life Extension Foundation 74

Alexander, Richard 296

Alexander the Great 24, 286, 335

Almagest (Ptolemy) 263

Alpert, Richard 183

Alpha intelligence test 170

Alsabti, Elias E. K. 248

Alternative Materials Institute 27

altruism **14–16**, 112, 146, 319

American alligator 102

American Association for the Advancement of Science (AAAS) 17, 208

American bison 47

American Civil Liberties Union (ACLU) 254

American Fertility Society 160

American Management Association 98

American Polygraph Association 251

American Psychological Association (APA)
    Ethical Principles 279
    "Guidelines for Ethical Conduct in the Care and Use of Animals" 18
    Watson, John B., as president of 358

American Sign Language, chimpanzees learning 59

American Society for Cytology 55

American Society of Heating, Refrigerating, and Air-Conditioning Engineers (ASHRAE) 310

American Society of Industrial Security 100

amniocentesis **17**, 107, 139

amoeba, moral worth of 19t, 20

amosite 27

amphetamines, maternal use of 92

amphibole 27

anabolic steroids 328

Andersen, Susan 240

Anderson, Warren B. 38

androstenediol 328

androstenedione 328

anemia, sickle-cell *See* sickle-cell anemia

"angel of death" 202–203

animal behavior 188, 358, 361

Animal Care and Use committee 169

Animal Dispersion in Relation to Social Behavior (Wynne-Edwards) 146

animal genes, transferring into plants 137–138

animal learning 315

Animal Liberation (Singer) 314

animal research **17–20**
    animal rights activists on 17–20, 80, 183, 245–246, 312
    learned helplessness research 183
    lifespan enhancement research 185
    Silver Spring monkeys 18, 19, 245, **312**
    by Skinner, Burrhus Frederic 315
    stem cell research 326

animal rights activists 245–246
    on animal organs used in organ transplantation 236
    on animal research 17–20, 80, 183, 245–246, 312
    on Draise test 91
    on egg-production techniques 35
    on fur trade 129
    on genetic engineering 138

animals
    in advertising 3
    air pollution and 9
    castration of 52
    cognitive abilities of 59–60
    DDT and 83
    domestication of 284, 306–307, 332
    endangered species 102–103, 104, 129, 146
    and euthanasia 109
    feeling pain 80, 314, 321
    moral worth of 17, 19t, 20, 278
    organs of, used in organ transplantation 236
    radiation and 57
    respect for, in Buddhism 47
    transferring human genes into 136–137

Anonymizer.com 98

anthophyllite 27

anthrax 41, 42

anthropocentrism 72, 104
anthropologists, feminist 122
antianthropocentric biocentrism 86
antiballistic missile defensive systems 323
Antiballistic Missile Treaty (ABM) 227, 324
antibiotics, given to chickens 35
antievolution laws 73, 271, 305
antifur campaign 129
Antinori, Severino 160
antipersonnel mines 181
antipornography laws 253
anti-tank mines 181
antitechnology agenda 350
Anti-Vaccination League 354
antlions 331
ants, reproduction of 15
The Ants (Wilson and Holldobler) 361
anxiety disorders, experiment on 304
APA *See* American Psychological Association
Apollo Project 217
apple growth regulator 13–14
applied ethics 20–21, 210–214, *211*
at Hastings Center 149
vs. normative 20–21
applied research 264
aptitude test 171
Aquinas, St. Thomas 288
Arab lie-detector test 251
"Arabic" numerals 288
archaeological ethics xii, 21
archaeologists, feminist 122
Archaeopteryx 293
Archimedes 22, 206, 287
Archimedes' Principle 287
Archimedes' screw 22, 287
architecture
    cast iron 337
    and crime xii, 22–23
    green 144
    Roman 287, 335
Aristophanes 320
Aristotle 23–26, 29, 72, 133, 213, 250, 287, 288–289, 320, 340
Arkwright, Richard 168, 337
army intelligence test 170
Arnold, Kenneth 349
Arrhenius, Svante 141
artificial insemination (AI) 107, 273–274
artificial sweetener 229
asbestos, and cancer risk 26–28
asbestos cigarette filters 28
Asbestos Hazard Emergency Response Act (1986) 27
asbestos mining 28
asbestosis 27
asexual reproduction, of humans 138, 275–276
ASHRAE *See* American Society of Heating, Refrigerating, and Air-Conditioning Engineers
Asia, amniocentesis for sex selection in 17

aspartame 229
aspirin, and risk of heart attack 326
assembly line technique 338
assisted insemination *See* artificial insemination
association learning *See* Pavlovian conditioning
Association of Women in Science 29, 123, 363
associationism 187
asthma
    causes of 10–11
    in children 10
    in elderly 10
The Astonishing Hypothesis (Crick) 73
Astronomia Nova (Kepler) 177
astronomy
    Babylonian 286
    religious ideology and 132–133, 269–271, 289–290
athletes
    African-American 302–303
    steroid use by 328
Atmosphere (periodical) 129
atmospheric nuclear tests 90
atom 77, 86
atomic bomb 225
    deployed in Japan *153*, 153–154
    development of 95, 198, 206, 226, 233
Atomic Energy Commission 90, 198, 234
attention deficit disorder
    DDT and 83
    lead poisoning and 182
    treatment of 44
Aum Shinrikyo (cult) 41
aurora borealis 77
Austria, ban on amalgam fillings 204
authority research 204–206, 278
automatons 87
automobile 338
Averroës 29
aviation 338
Avicenna 29, 288, 335
Avogadro, Amadeo 294

B
Babbage, Charles 297
Babylonia 285–286, 333
Bacillus thuringiensis 136
Bacon, Sir Francis 31–33, *32*, 149, 156, 232, 244, 251, 280, 290
Bacon, Roger 33, 270, 289
bacteria
    in biological warfare 41
    food irradiation and 126
Bagehot, Walter 318, 323
baghouse method of removing particles from exhaust stream 12
Bakke, University of California v. 7
bald eagle 102
Baldwin, J. Mark 358

Ballistic Missile Defense Organization (BMDO) 323–324
Baltimore, David 33–34, 360
Baltimore Affair **33–34**, 324, 360
Bandura, Albert 340
banks, selling information 97
Barker, Bob 129
Barnard, Christian 235
basic research 264
Bass, Edward P. 44–45
Bastiaans, Jan 261
battery chickens **34–35**
Beagle voyage 78–79, 81
Becquerel, Henri 76, 295
bees, reproduction of 15
Behavior: An Introduction to Comparative Psychology (Watson) 358
Behavioral and Brain Sciences (journal) 245
behavioral research, on chimpanzees 59–60
behaviorism 358
Behemoth (Hobbes) 154
Bell, Alexander Graham 171, 312
Bell, Buck v. 107
The Bell Curve (Herrnstein and Murray) 170–171, 304
Belmont Report **35–36**, 88
Bem, Daryl 115
Bentham, Jeremy 36, 207, 351
Benz, Karl 338
Berkeley, George 209
Beta intelligence test 170
Betrayers of the Truth (Broad and Wade) 298
Bevis, Douglas 17
Beyond Freedom and Dignity (Skinner) 315
Beyond Good and Evil (Nietzsche) 221
Bhopal, India, chemical accident **36–39**
Bikini atoll 90, 226
bile-duct cancer, from dioxin 307
Binet, Alfred **40**, 169–170
Binet test 40
biocentric equality 85
biodiversity and industrialization 40
biofiltration 11
Biogenic Corporation 189
biological psychiatry **43–44**
biological warfare **40–43**
Biology As a Social Weapon (Science for the People) 296
biomedical research
    animals in *See* animal research
    ethical guidelines for 35–36, 84–85, 88, 108, 169
    gender bias in 122
    in Nazi concentration camp 202–203
    scientific misconduct in 298
    on Seventh-Day Adventists 233
Biosphere **44–45**
birth control *See* contraception

birth control pill 68–69
    during pregnancy, and lesbianism 155
birth defects
    from dioxin 7, 8, 41
    from hazardous waste 189
    from radioactive dust 58, 154
    from thalidomide 342
birth rate, decrease in 71
bisexuality 154–155, 178–179, 308
bison 47
Biston betularia 167
bivariate tests 326
"black box" psychology 316
black-capped vireo 103
Black Death 40
black market, radioactive materials on 267
black-stem rust 41
Bliss, Russell 343
Blondlot, Rene 223
blood vessel tumors, from UDMH 13
Bloomsbury group 210
blue asbestos 27
BMDO *See* Ballistic Missile Defense Organization
Boas, Frans 75
Boehme, Jakob 239
Bohr, Neils 295
bone marrow transplantation 234
bonobos *See* pygmy chimpanzees
Boots Pharmaceutical 64–65, 329
Borlaug, Norman E. 145
Boswell, James 68
botulin 41
Boulton, Matthew 359
bovine somatotropin (BST) **46**
bovine spongiform encephalopathy *See* "mad cow disease"
Boyle, Robert 46, 156, 187, 271, 294
Boyle's law 46
Brady, Joseph 19
Brahe, Tycho 177, 270, 289
brain
    gender and 113
    of homosexuals and heterosexuals 155
    triune 196
brain chemistry, and violence 74
brain size
    gender and 216
    and intelligence 72–73
    racial/ethnic minorities and 215–216, 301, 303
brand
    advertising and 3
    loyalty to 4
breast cancer
    from bovine somatotropin 46
    detecting 197–198
    from hormone replacement therapy 158
breeding, selective **306–307**
Breen, John 164
Breen, Mike 67
bridge collapse and design flaws **46–47**

A Brief History of Time (Hawking) 150
British Association for the Advancement of Science 191
British Journal of Statistical Psychology 48
British Medical Journal 323
Broad, William 298
Bronze Age 335
Brotherhood 265–266
Brown, Louise 274
brown asbestos 27
brown pelican 102
Brunn Natural History Society 201
Bryan, William Jennings 305
Bryant, Jennings 255
BSE *See* "mad cow disease"
BST *See* bovine somatotropin
bubonic plague 41
    DDT in controlling 82
Buck, Carrie 107
Buck v. Bell 107
Buddhist ethics 47, 88, 177, 235
buffalo (bison), slaughter of 47
Burnet, Thomas 271
Burt, Sir Cyril Ludowic 48–50
Bush, George 262, 323
Bush, George W. 327
Bush, Vannevar 198
Buxton, Peter 347
Buzzanca, John and Luanne xi, 275
Byron, Lord 221
Byzantine Empire 287, 335
BZ 42

C

Cable Communications Policy Act (1984) **51**
Cable Television Consumer Protection and Competition Act (1992) 51
cactus finch 81
CAER *See* Community Awareness and Emergency Response
calculus, invention of 220, 313
Callicott, J. Baird **51**, 94, 104
Calvert, Frank 21
Camarhyncus parvulus (tree finch) 81
Campbell, D. 305
Camper, Petrus 72
Canadian Chemical Producers Association 39
cancer patients
    gene therapy for 134
    marijuana for 262
    radiation experiments on 64
cancer risk *See* carcinogenic effects
cannibalism 214
Cannon, Walter 356
cannons 336
Canon of Medicine (Avicenna) 29
Cape PLC 28
capitalism 316
carbaryl 36
carbon dioxide
    and air pollution 9

and global warming 9, 10, 141
    reducing emission of 142–143
carbon monoxide
    and air pollution 9
    and global warming 141
    reducing emission of 11
carcinogenic effects
    of asbestos 26–28
    of bovine somatotropin 46
    of dioxin 7–8, 41, 307, 344
    of electromagnetic radiation 96
    of hazardous waste 189
    of hormone replacement therapy 158
    of Nutrasweet 229
    of radioactive dust 56–58, 90, 154
cardiovascular disease, from air pollution 10
car exhausts
    acid rain from 1–2
    air pollution from 9
    asthma from 11
    lead poisoning from 11, 182
    reducing emission of nitrogen oxide in 2, 53, 61
car manufacturers, delaying reduction of environmental pollution 13
Carlyle, Thomas 197
Carnap, Rudolf 355
Carneal, Michael 66
Carnegie, Andrew 322
carpal tunnel syndrome 272
Carpenter, Raymond 60
cars
    hybrid-powered **164–165**
    lemons 184
Carson, Rachel **51–52**, 82–83, 94, 157
Cartesian philosophy 87
Casanova, Giovanni 163
The Case of the Midwife Toad (Koestler) 176
cast iron 337
castration 52
    chemical **52–53**, 69
    as contraceptive method 69
    involuntary 52, 69, 107
catalytic converters 2, 11, **53**, 61
catastrophism 53–54, 191, 218, 292–293, 350
categorical imperative 212
Catholic Church
    accepting evolution 271–272
    and artificial insemination 273
    and contraception 69, 70
    and Galileo Galilei 132–133, 269–270, 289–290
Cato Institute 364
causal analogy 20
Cavalli-Sforza, Luigi 302
caveat emptor 184
Celera Genomics 161
Cell (journal) 33

cell grafts 124
cellular phones 99
cellulase 136
censorship 253–254
Center for Democracy and Technology **54**
Center for Science in the Public Interest **54–55**
Center for UFO Studies 350
Centers for Disease Control
    on downwinders 90
    and Times Beach 343
central nervous system
    development of
        DDT and 83
        lead poisoning and 182–183
        mercury poisoning and 203
cervical cap 68
cesium 57, 58, 225
Cetus Corporation 135
Chabot, Boudewijn 110–111
Challenger accident 55, 217
charities xi, 143–144, 164
Charles V, Holy Roman Emperor 355
Chem-Bio Corporation, and pap smear controversy 55, 65
chemical accident(s)
    Bhopal 36–39
    Seveso, Italy 307–308
    Times Beach 43, 343–344
chemical additives 260
chemical castration 52–53, 69
The Chemical History of a Candle (Faraday) 118
Chemical Manufacturer's Association (CMA) 39, 190
chemical warfare 40–43
    Agent Orange 7–8, 41
Chemical Weapons Convention 43
Chernobyl nuclear accident 11, **56–59**, 223, 224
chickens
    battery **34–35**
    free range 35
Child Online Protection Act (1998) 257
child pornography 253, 257
Child Pornography Prevention Act (1977) 253
Child Pornography Prevention Act (1999) 253
children
    in advertising 3
    asthma in 10
    gender-atypical behavior in 156
    growth hormone deficiency in 161
    mental retardation in 203
    noise and 222
    in poverty, health outcomes of 302
    radiation experiments on 64
    thyroid cancer in, Chernobyl nuclear accident and 58
    using inhalants 345
    vaccination of 354

and violence 66–67, 340–341, 353
Children's Online Privacy Protection Act (1998) 99
chimpanzee language-learning controversy **59–60**, 62
chimpanzees
    empathy in 80
    moral worth of 17, 19t
    reciprocal altruism among 16
    self-awareness of 20, **60–61**
    using hammer stones 331
China
    centralized kingdom in 285, 333
    compulsory sterilization in 107
    Great Wall of 334
    inventions developed in 312, 336
    nuclear weapons in 226
    proportion of male to female births 17
    refusing to sign Ottawa treaty 182
    technology in 335
    use of MSG in 209
Chinese Americans, in science and technology careers 5
Chinese restaurant syndrome 209
chlorinated hydrocarbon 82
chlorine 41
chlorofluorocarbons 10, 237
chlorpromazine, for schizophrenia 43
chorionic villus sampling 139
Christian Scientists 118
chromosomal abnormalities, radiation-linked 58
chronic obstructive pulmonary disease, from air pollution 10
chrysotile 27
Church of Scientology 354
Churchland, Patricia 87
Churchland, Paul **61**, 87
cigarettes
    and asbestos exposure 27
    asbestos filters in 28
    maternal use of 92
    television advertising of, ban on 3
circulation of blood, discovery of 149
cities, development of 285, 333–335
city-states 286, 335
Civil Rights Act (1991) 7
Clausius, Rudolph 294
Clavibacter xyli 136
Clean Air Act (1970) 2, 11, **61**
Clean Air Act (1990) 11
Clean Water Act (1977) **61–62**
cleaner energy technologies 2, 12, 338
Clever Hans effect 59, **62**
climate change *See also* global warming
    history of 141–142
Cline, Martin 134

Clinical Laboratory Improvement Amendments 55
Clinton, Bill 124, 160, 264, 324, 327
Clipper chip **62**
cloning 135, 137
    of humans 138, **159–160**, 275–276
    vs. sexual reproduction 138, 275–276
Clostridium botulinum 41
CMA *See* Chemical Manufacturer's Association
coal combustion
    acid rain from 1–2
    air pollution from 9
    global warming from 141
coal mining 216, 225, 232–233, 233
coal-smoke fog, in London 10
cocaine, maternal use of 91–92
cocaine addiction, ibogaine for 262
coccidioidomycosis 41
code of ethics (Hippocratic oath) 152
cognitive abilities
    noise and 222
    of nonhuman animals 60
Colborn, Theo 157
cold fusion controversy **62–63**, 65, 232, 242, 300
cold war
    nuclear weapons tests during 89–91, 226
    psychedelic drug experiments during 41–42, 206, 262
    radiation experiments on human subjects during **63–64**, 206
Coleridge, Samuel Taylor 197
Columbia University, scientific dishonesty at 298
combustion engine 338
Commentariolus (Copernicus) 72
commerce, and privacy 97–98
commercialism in scientific research xii, **64–66**, 206, 329, 344
Committee for the Suppression of Vice 253
commonsense ethics 211
Communications Assistance for Law Enforcement Act (1994) 99
Communications Decency Act 253
Community Awareness and Emergency Response (CAER) 39
competition 316
    in science 363
complexity theory 131
compositionalism **66**
Comprehensive Environmental Response, Compensation, and Liability Act (CERLA) (1980) 90, 150, 189
computer virus **101**, 346
computers
    development of 338
    and violence **66–67**, 340
    and Y2K 365

Comstock, Anthony 253
Comstock Act 69–70
conditioning *See* Pavlovian conditioning
condom 68
Condon, E. U. 350
Condon Report 350
Cone, Ben 102
Congressional House Oversight and Investigations Committee 34
Connecticut, Griswold v. 71
consequentialism **67** *See also* utilitarian ethics
Constitution
    Eighth Amendment to 53
    Fifth Amendment 103
    First Amendment to 53, 253–254
construction codes, earthquakes and 93
consumer behavior 3, 4
consumerism, deep ecologists on 86
contemporary naturalism 219
contraception **67–71**, 339
    emergency 281
    ethical and legal problems of 69–71
    forced use of 69
    history of 68–69
    social impact of 67–68
control group 247
Convention on the Prohibition of Biological and Toxin Weapons 42
Conway, J. 48
Cooke, W. F. 313
cookies, and Internet **71–72**, 98
COPA *See* Child Online Protection Act (1998)
Copernicus, Nicolaus 72, 133, 177, 269–270, 289
copper smelting 334–335
copyright infringement *See also* patents
    on Internet 100–101
    plagiarism 248–249
corn borer 136
Cornell University, scientific dishonesty at 298
corpuscular theory 294, 366
correlations 326
Correns, Carl 201, 313
Corvair 127
cosmetics industry, testing in 91
cosmology 150, 295
Cosmos (PBS series) 283
cotton mill 168, 337
cowpox virus 354
crafts, development of 333
Crania Americana (Morton) 215
craniometry, as pseudoscience **72–73**, 215–216, 301, 303
creationism 53–54, **73**, 79, 271, 305–306
Creation Research Society 73
Creutzfeldt-Jakob disease virus 162

Crick, Francis Harry Compton **73**, 293, 357
crime
    architecture and xii, **22–23**
    brain chemistry and **74**
    and DNA evidence 89, 114, 136
    on Internet 100–101
    video surveillance and **355**
criminals
    castration of 52
    chemical treatment of 52–53, 69, 74
Crisis: Heterosexual Behavior in the Age of AIDS (Masters and Johnson) 200
Critique of Pure Reason (Kant) 176
crocidolite 27
crop rotation 288, 336
cross-cultural differences in homosexual behavior 308
cross-cultural ethical diversity 214–215
Crusades 336
Crutzen, Paul 237
Cruzan, Nancy 186
cryonics **74–75**
cryopreservation, of human semen 274
cryptographic research, government suppression of 75, 98
cultural relativism 75, 214
cuneiform script 285
Curie, Irène 76
Curie, Marie **75–76**, 76, 363
Current Anthropology (journal) 245
Cuvier, George 53–54, 292, 301
cyber terrorism 100
cyclone separator 12
cystic fibrosis, screening for 139

D
Dalton, John 77, 294
daminozide 13
Darrow, Clarence 305
Darwin, Charles Robert 25, 54, **78**, **78–80**, 81–82, 111, 191, 197, 271, 292, 293, 301, 351
Darwin, Erasmus 78
Darwinism, social *See* social Darwinism
"Darwin's bulldog" 164
Darwin's finches **81**
Darwin/Wallace and codiscovery of evolution 79, **81–82**, 187, 312
DaSilva, Ashi 134
data cooking 297
data falsification 297–298
data fudging 297
data gathering (induction) 32, 290
data trimming 297
Davida, George 75
Davis, Junior Lewis 275
Davis, Mary Sue 275
Davy, Sir Humphry 118
Dawson, Charles 246–247
DDD 83
DDT **82–84**
    airborne 9

as biohazard 82–83, 94
government regulation of 83–84
health effects of 83
as hormone disruptor 157
as miracle chemical 82
in semen 323
use of, in developing countries 82, 84
deafness
    industrial 221
    and paranoia 239–240, 277
death, redefinition of 235
death penalty 89
de-beaking 35
debriefing 255, 279
deception, in social psychological research 205–206
Declaration of Helsinki 35–36, **84–85**, 247
"deconstruction" 259
deduction 32, 290
deep ecology approach to environmental ethics 40, **85–86**, 213
deepwelling 150
A Defense of Common Sense (Moore) 210
"defensible space" 22–23
defoliant(s) *See* Agent Orange
De Humani Corporis Fabrica (Vesalius) 292, 355
de Laplace, Pierre Simon 291
de Waal, Frans 20
democracy, ethical basis of 212
Democritus 77, **86–87**, 105
dengue, global warming and 142
DeNobel, Victor 66, 360
dental fillings, mercury poisoning from 203–204
Deontology 67, **87**, 229, 277, 279, 340, 351
deoxyribonucleic acid *See* DNA
Department of Agriculture (USDA)
    on DDT 84
    on food irradiation 126
Department of Energy (DOE), funding nuclear testing 90
Department of Health and Human Services (DHHS) 218, 231, 232, 262, 277
    on Love Canal 189
    on pap smear screening 55
    regulations for the protection of human subjects **88**
Department of Veterans' Affairs, on Agent Orange 8
Depo Provera
    as contraceptive method 69
    for reducing violence 53, 74
    for sexual offenders 52–53, 69
depression
    gender and 113
    scientific understanding of 183
    treatment of 96
Derrida, Jacques 259
DES *See* diethylstilbesterol
Descartes, René **87**, 88, 151, 154

The Descent of Man (Darwin) 79, 80, 301

descriptive statistic 324

desensitization to violence 66–67

design flaws
of bridges 46–47
Ford Pinto 127–128
of *Titanic* 344

"designer babies" 107–108

designer drugs, and Parkinson's disease 241, 262

determinism 87–88
environmental 88
genetic 88, 133–134, **135**, 139, 161, 209

deterministic theory of inventions 171, 312

deuterium 226

developing countries, use of DDT in 82, 84

DeVries, Hugo 201, 313

dharma 88

DHHS *See* Department of Health and Human Services

Diagnostic and Statistical Manual of Mental Disorders (DSM) **88**

Dialog of the Two Chief World Systems (Galileo) 133, 269–270, 290

diaphragm 68

dichloro-diphenyl-trichloroethane *See* DDT

Dickens, Charles 168

Diesel, Rudolf 338

diesel engine 338

diesel exhausts, and asthma 11

diethylstilbesterol (DES)
and lesbianism 155
and sexual development 157

"difference principle" 269

Digges, Thomas 313

Dingell, John 34

dioxin
adverse effects of 7–8, 41
in Agent Orange 7–8, 41
in chemical accident 43, 307–308, 343–344
as hormone disruptor 157, 307–308

disabled people, in science and technology careers 5

Discourses on the First Ten Books of Titus Livius (Machiavelli) 195

discrimination learning 243–244

Disraeli, Benjamin 325

diversity, cross-cultural ethical 214–215

The Diversity of Life (Wilson) 361

divine command 211–212

DNA, deciphering molecular structure of 73, 293, 357

DNA evidence **88–89**, 114, 136

Dobshanski, Theodosius 293

Doctor's Trial 203, 228

doctrine of double truth 31–32, 290–291

DOE *See* Department of Energy

dogs
exposed to electric shocks 183

moral worth of 17, 19*t*, 20
selective breeding of 306–307

Dolly, cloned sheep 135, 137, 160

domestication of plants and animals 284–285, 306–307, 332

donations, to charities xi, 143–144, 164

Dong, Betty 65, 329

Doom (game) 67

The Doors of Perception (Huxley) 261

double effect, doctrine of **89**

double helix 73, 357

The Double Helix (Watson) 73, 357

double truth doctrine 31–32, 290–291

double-blind review 245

DoubleClick 72, 98, 278

Dow Chemical 8

Dowie, Mark 127, 128

downwinders 89–91

Doyle, Arthur Conan 159

draft animals 334

The Dragons of Eden (Sagan) 283

Draise test 91

Drebbel, Cornelis 313

drinking water standards 62

drones 15

Drug and Cosmetic Industry (journal) 237

drug trials
placebo-controlled 84, **247**
women in 122

drug use during pregnancy 91–92

drugs
determining toxicity of 19, 91
production of, genetic engineering in 137

DSM *See* Diagnostic and Statistical Manual of Mental Disorders

dualism **92**, 209

duplicate publication **92**, 249, 265

DuPont Corporation 182

dwarfism 161

dying patients
and euthanasia 109–111
high doses of morphine for 89
LSD for 261

E

E. coli See Escherichia coli

earthquakes and construction codes **93**

earthworm, moral worth of 19*t*

Easter Island 285, 333

Easton, S. M. 313

Echelon 98–99

Eckehart, Johannes 239

ecologists, feminist 121

economics 316

"ecosophy" 85

ecosystems 93–94
damage to 104–105, 168, 268
pesticides and 82–83, 94
self-sustaining, comprising 44–45
as superorganisms 131–132

ecstasy (MDMA) 262
for posttraumatic stress disorder 262

ECT, for schizophrenia 44

ectopic pregnancy, from intrauterine devices 68

Edison, Thomas Alva 94, 171, 172, 294, 297, 314, 338

"Edison effect" 94

Edward O. Wilson: A Life in Science (Wilson) 361

egg donors 108

egg-production techniques 34–35

egoism 213

Egypt 285, 333–334

Ehrenhaft, Felix 208, 324

Eichmann, Adolf 204

Eighth Amendment 53

Einstein, Albert 95–96, 106, 187, 201, 206, 208, 219, 234, 250, 272, 292, 295, 299

elderly, asthma in 10

electricity 337, 338
discovery of 294

electroconvulsive therapy (ECT) **96**

electromagnetic radiation 295
and cancer risk **96**
theory of 187

electromagnetism 118, 201, 293, 295–296

electron, measuring charge of 208

electronic databases, and privacy 97

electronic filters 253, 257

Electronic Funds Transfer Act (1980) **96**

electronic metal detectors 204

electronic privacy **96–99**

Electronic Privacy Information Center 98

electronic publishing of scientific research **99–100**

electronic terrorism **100**

electronic theft **100–101**

electronic virus 101, 346

electrostatic precipitator 12

elements, periodic table of 202

Elements (Euclid) 106

Ellis, Havelock 70, **101**

e-mail, junk 173

embryos
cloning human 160
frozen 274–275
for stem cell research 327

emergency contraceptives 281

Emile (Rousseau) 280

emotions, gender and 113

empirical method
Aristotle on 25
Bacon, Roger, on 33
Boyle, Robert, on 46

employees
electronic surveillance of 98
in green office 144
as industrial spies 98
repetitive stress symptoms in 273

sick building syndrome in 310

encryption
and Internet security 62, 98, 100–101
suppressing civilian research on 75, 98

endangered species 102–103, 104, 129, 146

Endangered Species Act (1973) **102–103**, 104

endocrine disruptors 156–157, 307–308, 323

Energy, Department of *See* Department of Energy

Engel, Rayme 3

engineering ethics **103**

engineers, female 122–123, 362

English Malthusian League 197

English Statute of Patents and Monopolies (1623) 171, 242

ENIAC 338

Enlightenment **103–104**, 187, 197, 212, 221, 232, 270, 280, 316

entertainment industries 338

environment friendly vehicles 164–165

environmental alarmism 365

environmental determinism 88

environmental effects *See also* acid rain; air pollution; global warming; smog
of Agent Orange 7
of DDT 82–84
of green revolution 145
of Industrial Revolution 9, 104, 168
of radioactive dust 56

environmental ethics xii, 51, **104–105**
deep ecology approach to 85–86, 213
on genetic engineering 138

Environmental Protection Agency (EPA) **105**
on annual number of asbestos-related cancers 27
daminozide and UDMH studies 13
on DDT 84
on leaded gasoline 182
regulating dumping of sewage in oceans 61–62
screening for hormone mimics 157
Superfund 90, 150, 189

Environmental Protection Agency (EPA) Science Advisory Panel 13, 14

Epicurus 105, 151

Epiphenomenalism **105–106**

epiphytes 268

epistemic relativism 321

epistemology 31–32, 290, 297

Epitome Astronomiae Copernicae (Kepler) 177

Equal Employment Opportunity Act (1972) 6–7

Eron, Leonard 341
ESA *See* Endangered Species Act (1973)
Escherichia coli 126, 136, 138
ESP *See* extrasensory perception
An Essay Concerning Human Understanding (Locke) 187
An Essay on the Principle of Population As It Affects the Future Improvement of Society (Malthus) 197
estrogen
   in birth control pill 68
   in hormone replacement therapy 157
ether **106**, 201
ethic of care 140, 213
ethic of justice 140
ethical codes, professional 210–211, 213
ethical diversity, cross-cultural 214–215
ethical objectivism 210
Ethical Principles (APA) 279
ethical relativism 214–215
ethical standards, universality of 214–215
ethical theories 211, *212 See also specific theory*
ethics **210–214**, *211*
   applied **20–21**, 210–214, *211*
     at Hastings Center 149
     vs. normative 20–21
   archaeological xii, **21**
   Aristotle's 25–26
   Buddhist 47, 88, 177, 235
   Democritus' 86
   engineering **103**
   environmental xii, 51, **104–105**
     deep ecology approach to 85–86, 213
     on genetic engineering 138
   evolutionary 79–80, **111–112**
   feminist **120–121**, 140, 213
     and philosophy of science **121–124**
   meta-ethics 210, *211*
   neuroanatomical basis of 196
   normative 210, *211,* 211–213, *212,* **222–223**
     See also specific theory
     vs. applied 20–21
   Plato's 250
Ethics (Kropotkin) 180
Ethiopia, green revolution in 145
ethnocentrism 300
ethology 188
Ethyl Gasoline Corporation 183
Euclid 22, **106**
Eudoxus 22
eugenics **106–108**
   artificial insemination and 274
   castration and 52, 69
   negative 106, 134
   positive 106, 134, 202, 264
Euler, Leonard 291

eunuchs 52
European Bioethics Convention **108**
Europeans
   against genetically modified foods 138, 139, 197
   against genetic engineering 108
euthanasia **109–111**, 314
   active 109
   Hippocratic oath and 109, 152–153
   involuntary 109, 110–111
   passive 109
   voluntary 109, 110
   vs. physician-assisted suicide 109
evolution
   and altruism **14–16**, 112
   of darkening of pepper moths 167
   by inheritance of acquired traits 181
   moral inspiration in 111–112
   by natural selection 14, 78, 79, 81, 111, 112, 181, 188, 197, 271, 292, 293, 307, 322
   of pesticide resistance 83
   and religion 79, 271–272, 305–306, 360–361
   simultaneous discovery of 79, **81–82**, 187, 312
   teaching 73, 271, 305
Evolution and Modification of Behavior (Lorenz) 188
evolutionary change 54
evolutionary ethics 79–80, **111–112**
evolutionary psychology 320
   and sex roles **112–113**
exobiology *See* extraterrestrial life
"experimental neurosis" 244
experiments 299–300 *See also specific experiment*
expert testimony 65, **114**
The Expression of Emotions in Man and the Animals (Darwin) 79
extrasensory perception (ESP) **114–116**, 349
extraterrestrial life 283, 349–350
eyewitness testimony, unreliability of 114, **116**, 118

F

Fabricus, David 314
fact-value distinction **117**
fair treatment right 278–279
faith healing **117–118**
false memory and criminal convictions 118
False Memory Syndrome Foundation 118
Faraday, Michael **118**, 201, 252, 294
Faraday Effect 118
Faraday's Laws 118
farming

development of 284, 285, 332–333
   improvements in 145, 197, 336, 339
FBI crime lab, and scientific inadequacy 114, **119**
FDA *See* Food and Drug Administration
fear conditioning 276, **304–305**
Fechner, Gustav Theodor **119–120**
Fechner's Law 119
Feder, Ned 34
Federal Environmental Pesticides Control Act 84
Federal Water Pollution Control Act (1972) 61
Feigi, Herbert 355
female condoms 68
female tubal ligation 69
feminist ethics **120–121**, 140, 213
   and philosophy of science **121–124**
feminist movement, and contraception 70
feminization in males, hormone mimics and 157
Fermi, Enrico 198, 226
fertility drugs 274
fetal blood sampling 139
fetal cell grafts 124
fetal tissue research 124, 264, 327
fetish
   conditioning of in lab **124–125**
   definition of 124
fetus
   drug abuse during pregnancy and 91–92
   testing for genetic disorders 17, 107, 139
   using organs and tissues of aborted 234–235
feudal system 336
Fifth Amendment 103
fight-or-flight response 222
file drawer problem **125–126**, 217–218, 325
filtering devices
   for computers 253, 257
   for television 353
finches, Darwin's **81**
fire 331
First Amendment 53, 253–254
First Course in Physics (Millikan and Gale) 208
fish
   disappearance of, from acid rain 1
   mercury poisoning from 203, 204
   moral worth of 19*t*
Fish and Wildlife Service 102
Fisher, R. A. 201
fission bomb (A-bomb) 62, 198, 226
fitness (reproductive) 14
Fitzgerald, George Francis 95, **126**, 187

Fitzgerald-Lorentz contraction 126, 187
Fleischman, Martin 63, 232, 242
Fletcher, Harvey 208
Fludd, Robert 313
fluoridation of water supply **126**
flying saucers 349
flying shuttle 168, 337
Flynn effect 170, 303
FOE *See* Friends of the Earth
Fontana, Francesco 313
Food, Drug, and Cosmetic Act (1938) 342
Food Additive Amendment Act (1958) 260
Food and Drug Administration (FDA)
   approval of Nutrasweet 229
   approval of RU-486 281
   on bovine somatotropin 46
   on food irradiation 126
   on genetically modified foods 138
food irradiation **126**
food poisoning, prevention of 126
food preservatives 260
food production
   development of 284, 285, 332–333
   improvements in 145, 197, 336, 339
"foo fighters" 349
foot fetishist 124
foot-and-mouth disease 41
Ford, Henry 338
Ford Motor Company, Grimshaw v. 128
Ford Pinto **127–128**
forensic genetics 136
forests, loss of, from acid rain 1
Forster, E. M. 210
fossils 53, 78, 191, 292–293, 350–351
4charity.com 143
Fourier, Jean-Baptiste-Joseph 141
fowl plague 41
Frank, Robert 112, 215
Franklin, George 118
Franklin, Rosalind 73, 357
free range chickens 35
Free Speech Coalition 253
free trade 316
Freeman, Walter 260
French Flora (Lamarck) 181
French Revolution 104, 280
French Royal Academy of Sciences 281, 291
Fresnel, Augustin 294, 366
Freud, Sigmund 154
Friends of the Earth (FOE) **129**
Frisch, Karl von 188
frog, moral worth of 19*t*, 20
fruit fly, moral worth of 19*t*
The Fruits of Philosophy (Knowlton) 70
fuel tank safety issue 127–128
fungi
   in biological warfare 41
   food irradiation and 126

fur trade and endangered species
129
fusion bomb (H-bomb) 62, 226
Fust, Johann 146–147
future generations, obligations to
231

G

Gage, Phineas 196
Gaia: A New Look at Life on Earth
(Lovelock) 131
Gaia hypothesis **131–132**
Galapagos Islands 81
Gale, Henry 208
Galen 132, 149, 287, 291, 355
Galileo Galilei **132–133**, 154, 177,
252, 269–270, 289–290, 313,
314
gallbladder cancer, from dioxin
307
gallium 202
Galloping Gertie (bridge) 47
Gallup, Gordon 60
Galton, Sir Francis 78, 106,
133–134
Galvani, Luigi 294
galvanic skin response 251
gamete intrafallopian tube transfer
(GIFT) 274
ganzfeld method 115
Gardner, Allen 59
Gardner, Beatrice 59
gas tank safety issue 127–128
"gay gene" 155, 308
Gebhard, Paul H. 178
Gelsinger, Jessie 134–135
gender *See also* women
and brain 113
and brain size 216
and homosexuality 155
and moral reasoning 5, 213,
362
gender bias, in science 121–124
gender-atypical behavior 156
gene(s)
animal, transferring into
plants 137–138
"gay" 155, 308
human, transferring into animals 136–137
gene selectionism 14, 319
gene splicing 135
gene therapy **134–135**
general ethical relativism 214
General Motors
and leaded gasoline 183
Nader, Ralph on 127
report on rear-end collisions
128
use of animals in crash tests
by 246
genetic determinism 88, 133–134,
135, 139, 161, 209
genetic disorders
detecting 17, 107, 139
mapping chromosomal loca-
tion of 136
genetic engineering xii, **135–138**,
307, 339

Europeans opposing 108
and milk production 46
genetic testing 107, **139**, 161
genetically modified food 136,
139–140, 145
Europeans against 138, 139,
197
genetics
Lamarckian 176, 181, 191
Mendelian 176, 181, 191,
201, 293, 313
genocide 107
geocentric system 72, 263,
269–270, 289
Geography (Ptolemy) 263
Geospiza fortis (medium ground
finch) 81
Geospiza magnirostris (ground
finch) 81
Geospiza scandens (cactus finch) 81
Gerhard, Wolfgang 203
germanium 202
Germany
ban on amalgam fillings 204
biological and chemical
agents used by 41
The Getaway (movie) 254
GIFT *See* gamete intrafallopian
tube transfer
gigantism 161
Gillie, Oliver 49
Gilligan, Carol 5, 120, **140**, 213,
362
Gingerich, Owen 263
"give-up-itis" 356
Gliddon, G. R. 301
global sustainability 279
"global village" 338
global warming 9–10, 104, **140–143**
consequences of 142
debate on 141–142
political responses to
142–143
glucose intolerance, from growth
hormone therapy 162
glue solvent 345
Goddard, Henry 170, 304
Goddard Institute for Space Studies
141
Godel, Kurt 355
Godwin, William 197
golden mean 25
golden-cheeked warbler 103
Gorbachev, Mikhail 42, 227
Göring, Herman 228
Gould, Stephen Jay 215, 319
Grant, Peter R. 81
gravitation, law of 291
Gray, Elisha 171, 312
gray whale 102
Great Britain
industrialization in 336–337
nuclear weapons in 226
Great Pyramid at Giza 333–334
Great Wall of China 334
GreaterGood.com **143–144**, 164
Greece, ancient
science in 286–287
technology in 335

green architecture **144**
green revolution 145, 197, 339
greenhouse effect *See* global warming
Greenpeace 139, **146**, 190
Griesenger, Wilhelm 179
Grimshaw v. Ford Motor Company
128
Griswold v. Connecticut 71
grizzly bear 102
Grof, Stanislav 261
Grosseteste, Robert 289
Grossman, David 66
Grossman, Marcel 96
ground finch 81
group selection 14, 131, **146**, 180
Groves, Lesley 198
growth hormone **161–162**
growth hormone therapy
161–162
growth retardation
in fetus, from maternal drug
abuse 91–92
from lead poisoning 182
Grush/Saunby Report 127
guaiacol 260
Gumplowicz, Ludwig 318, 323
gunpowder 336
Gutenberg, Johan **146–147**, *147*,
289, 336
Gutenberg Bible 147
"Gutenberg Revolution" 146, 289

H

H. B. Fuller Company 345
hackers xii
Hahn, Hans 355
Hall, Jerry 160
hallucinogenic drugs *See* psyche-
delic drugs
Hamer, Dean 155, 308
Hamilton, William 14, 319
handguns 336
Hanford Medical Monitoring Pro-
gram 90
Hanford Nuclear Plant 90
Hansen, James 141
haplo-diploid reproductive system
15
happiness, pursuit of 36
Hard Times (Dickens) 168
Hardin, Garret 346
Hargreaves, James 168, 337
Harmonices Mundi (Kepler) 177
Harriot, Thomas 314
Harvard Educational Review 48
Harvard Six Cities Study 10
Harvard University, scientific dis-
honesty at 298
Harvey, William **149**, 292
Hastings Center **149–150**
Hawking, Stephen 150, 252, 295
Hayden, Michael 99
hazardous waste **150–151**,
188–190, 225
H-bomb 62, 226
head size
gender and 216
and intelligence 72–73

racial/ethnic minorities and
215–216, 301, 303
Health and Human Services,
Department of *See* Department
of Health and Human Services
health effects
of acid rain 1
of air pollution 10–11
of asbestos 26–28
of bovine somatotropin 46
of DDT 83
of dioxin 7–8, 41, 307, 344
of growth hormone therapy
162
of hazardous waste 189
of lead poisoning 182–183
of mercury poisoning
203–204
of Nutrasweet 229
of radioactive dust 56–58, 90
hearing loss, and paranoia
239–240, 277
Hearnshaw, Leslie 48, 49
heart disease, gene therapy for
134
heart transplantation 235
heart-lung machine 235
heavy plow 288, 336
hedonism 105, **151**, 212, 340, 351
Hegel, Georg 177
Held, Virginia 120
heliocentric system 72, 133, 177,
269–270, 289
Hellenic world 286, 335
Hellenistic period 335
Helmholz, Hermann Ludwig Ferdi-
nand von **151–152**, 294, 366
helplessness, learned 183, 244
hemophilia, gene therapy for 134
Henry, Joseph 313
Henslow, John Stevens 78
Heraclitus 250
herbicides *See also* Agent Orange
and Parkinson's disease 241
hereditarian research, of intelligence 48–50
Hereditary Genius (Galton) 106
Herodotus 334
heroin addiction
ibogaine for 262
Scoline for 304
Herrnstein, James 170–171
Herrnstein, Richard 48, 304
Hertz, Heinrich 295
heterosexuality 154–155, 308
heterosexuals, brain of 155
hieroglyphs 285
high school shootings 66–67
Highway Safety Act (1966) 127
Hinduism 239
Hinton, Martin A. C. 246–247
Hipparchus 263
Hippocrates 152
Hippocratic oath **152–153**
abortion and 152–153
euthanasia and 109,
152–153
hormone replacement therapy
and 158

mercury poisoning from dental fillings and 204
radiation experiments and 64
syphilis experiment and 179
termination of life support and 186
Tuskegee syphilis study and 347
Hiroshima, Japan **153–154**, 198, 226
Hispanic Americans, in science and technology careers 6*t*
Hobbes, Thomas **154**, 187, 213, 280, 316, 318
Hodgson, Ray J. 124–125
holistic healing 118
Holmes, Oliver Wendell 107
Holocaust *See* Nazi Holocaust
Holton, Gerald 208
Homo erectus 331
Homo sapiens 331
homosexual behavior
cross-cultural differences in 308
for sexual variety 163–164
vs. homosexual orientation 154, 308
homosexuality
gender and 155
gene for 155, 308
incidence of 178–179
Masters and Johnson on 200
possible biological basis of **154–156**, 308–309
Homosexuality in Perspective (Masters and Johnson) 200
homosexuals, brain of 155
Honorton, Charles 115
Hooke, Robert **156**, 292
Hooker, Joseph D. 79, 82
Hooker Chemical Company 189–190
hormone mimics **156–157**, 307–308, 323
hormone replacement therapy **157–158**
horse collar 336
horses, selective breeding of 306
Hotline Connect (software) 101
Houdini, Harry *158*, **158–159**
House Un-American Activists Committee 95
Howard, Margaret 48
HST *See* Hubble Space Telescope
Hubble, Edwin Powell 150, **159**, 295
Hubble Space Telescope (HST) **159**
Hubble's Law 159
human cloning research 138, **159–160**, 275–276
human embryos *See* embryos
human factors research **160–161**, 343
human fetus *See* fetus
Human Genome Project 108, **161**
human growth hormone **161–162**
human research participants, rights of *See* rights of human research participants

Human Sexual Response (Masters and Johnson) 199
Humane Society 144
human-machine interface 160
Hume, David 36, 103, 111, 219
Humphreys, Laud **163–164**, 278
Hunger Site xi, 144, **164**
Hunt, Morton 240
Hunter, John 273
hunter-gatherer societies
division of labor in 284
reciprocal altruism in 16
stone tools used by 332
Huxley, Aldous 261
Huxley, Julian 293
Huxley, Thomas Henry 78, 79, 111, **164**, 271, 361
hybrid-powered cars 11, **164–165**
hydraulic hypothesis 285, 333
hydrocarbons
and air pollution 9
reducing emission of 11
hydroelectricity 338
hydrogen bomb 225
hymenopterans 15
Hynek, J. Allen 349
hypnosis
and hearing loss 239–240
in memory retrieval 118
hypothesis 32, 251, 290, 299
hypothyroidism
radiation-linked 58
treatment of 329

I

Ibn Sina *See* Avicenna
ibogaine 262
ICBMs *See* intercontinental ballistic missiles
ideal utilitarianism 351
idealism 25, 209, 210
identity theft 100
"idols" 32
iGive.com 143
Ikeda, Kihunae 209
illusory phenomena
cold fusion controversy **62–63**, 65
ether **106**, 201
N-rays 115, **217–218**
phlogiston **246**
"planaria soup" research **249**
Imanishi-Kari, Thereza 33–34, 324, 360
Immigration Restriction Act (1924) 106–107
immune suppression
from DDT 83
from dioxin 7
In a Different Voice (Gilligan) 140
In Defense of the Land Ethic (Callicott) 51
in vitro fertilization (IVF) 274–275
inbuilt reflexes 358
Inca empire 334
incineration 43, 150
inclusive fitness (reproductive) 14
independent invention 171, 312

India
amniocentesis for sex selection in 17
chemical accident in 36–39
green revolution in 145
nuclear weapons in 226
organ trade in 236
use of DDT in 84
individualistic theory of inventions 171, 312
indoor air quality 309–310
induction 32, 290
industrial accident(s)
Bhopal **36–39**
Chernobyl 11, **56–59**, 223, 224
human factors in 161, 343
Seveso, Italy **307–308**
Three Mile Island 57, 224, **342–343**
industrial deafness 222
industrial espionage 98
industrial melanism 81, **167**
Industrial Revolution 104, **168**, 336–337, 359
air pollution as product of 9, 317
and repetitive stress injury 272
industrialism, science and technology careers before 4–5
industrialization, and biodiversity 40
industry, damaging effects of *See* environmental effects
INF *See* Intermediate Range Nuclear Forces Treaty (1987)
infants
and mirrors 61
moral worth of 19*t*, 20
research on 358
infertility treatments 273–276
influenza vaccination 354
Information Group of Institute of Electrical and Electronic Engineers 75
information warfare 100
informed consent **168**, 229, 262, 276–277, 304, 368
inhalants 345
Innocence Project 89
innovation, and ethical dilemmas 211
An Inquiry into the Nature and Causes of the Wealth of Nations (Smith) 316
insecticides *See* DDT
Institute for Behavioral Research 312
Institute for Community Design Analysis 23
Institute for Experimental Biology 176
Institute of Electrical and Electronic Engineers, Information Group of 75
Institute of Medicine 218
institutional review board (IRB) 88, **169**, 209, 213–214, 232, 277, 278

Intel 98
intellectual property rights, infringement of *See also* patents
on Internet 101
plagiarism 248–249
intelligence
Burt, Sir Cyril Ludowic on 48
and craniometry 72–73
genetic determination of 135
lead poisoning and 182
twin study of 48–50
intelligence tests 135
development of 40, 169–170
ethics of **169–171**, 303–304
Intelligent Life in the Universe (Sagan and Shklovskii) 283
interactionism 171
intercontinental ballistic missiles (ICBMs) 226
interference effects 366
Intergovernmental Panel on Climate Change (IPCC) 141
Intermediate Range Nuclear Forces Treaty (1987) 227
International Foundation for Internal Freedom 184
International Maize and Wheat Center 145
International Union for the Conservation of Nature and Natural Resources 102
Internet xi
charities on xi, 143–144, 164
electronic publishing on 99–100
electronic theft on 100–101
pornography on 252, 253, 256–258
privacy on xi–xii, 71–72, 97, 98, 278
terrorism on 100
and violence 66–67
Internet Service Providers (ISPs) 101, 257
intrauterine devices (IUDs) 68
intrinsic moral worth 212
An Introduction to the Principles of Morals and Legislation (Bentham) 36
inventions, theories of **171–172**, 312–313
involuntary euthanasia 109, 110–111
iodine 57, 58, 90
IPCC *See* Intergovernmental Panel on Climate Change
Iraq
chemical weapons of 43
nerve agents used by 41
IRB *See* institutional review board
iridium 54
iron 337
irrigation 285, 333–334
Islamic civilization 288, 335
is/ought distinction 219
ISPs *See* Internet Service Providers
IUDs *See* intrauterine devices
IVF *See* in vitro fertilization

J

James I, king of England 149
Jansen, Zacharias 313
Japan
    atomic bomb explosions in
        153, 153–154, 198, 226
    kamikaze air pilots of 175
    use of MSG in 209
Jenner, Edward 353
Jensen, Arthur 48
Jewell, Richard 119
Jewish religious law, and organ
    donation 235
JNDs See just noticeable differ-
    ences
John Paul II, Pope 81, 133
Johns Hopkins University 358
Johnson, Ben 328
Johnson, Virginia E. **199–200**, 255
Joliot, Jean Frederic 76
Jones, James H. 179
Joule, James Prescott 294
*Journal of Abnormal and Social Psy-
    chology* 305
*Journal of Neuropathology and
    Experimental Neurology* 229
The Journal of Parapsychology 115
*Journal of the American Medical
    Association* 329
Joyce, James 253
Juaregg, Julius Wagner von 43
junk e-mail **173**
jury selection, scientific **173**
just noticeable differences (JNDs)
    119
justice, Rawl's theory of **268–269**

K

Kaczynski, Theodore 350
kamikaze **175**
Kamin, Leon 48–49
Kammerer, Paul **176**, 181, 298
Kant, Immanuel 103, **176–177**,
    177, 212, 268
karma 47, 88, **177**
Kaufman, Steve 240
Kay, John 168, 337
Kefauver-Harris Act (1962) 342
Kehoe, Robert 183
Kennedy Institute of Ethics 177
Kent (cigarettes), asbestos filters in
    28
Kepler, Johannes **177–178**, 270,
    291
Kerr-McGee nuclear plant 311
Keswani, Raj Kumar 38
Kevorkian, Jack 109
Keynes, John Maynard 210
Khrushchev, Nikita 193
kidney transplantation 234, 235
Kimura, Doreen 113, 156
kin selection 14–16
King Solomon's Ring (Lorenz) 188
Kinsey, Alfred Charles 154–155,
    178, **178–179**, 256, 300, 308
knights 336
Knoll Pharmaceutical *See* Boots
    Pharmaceutical

Knowlton, Charles 70
Koestler, Arthur 176
Kohlberg, Lawrence 120, 140
Krafft-Ebing, Richard von
    179–180
Kropotkin, Prince Peter (Pyotr)
    Alekseyevich **180**
Kuhn, Thomas Samuel **180**,
    313–314
!Kung (tribe) 16
Kyoto Protocol 142–143

L

Laboratory Animal Care Commit-
    tees 18
Lady Chatterley's Lover (Lawrence)
    253
Lamarck, Jean-Baptiste Pierre
    Antoine de 176, **181**
Lamarckian genetics *vs.* Mendelian
    genetics 176, 181, 191
Lamb, David 313
Lancet (journal) 140
land mines **181–182**
language
    chimpanzees understanding
        60
    in determination of moral
        value 20
    development of 332
    postmodernism and 259
large-scale human studies 326
latex condoms 68
"laughing philosopher" *See* Dem-
    ocritus
Laverty, S. G. 305
Lavoisier, Antoine 246, 294, 313
law enforcement, electronic moni-
    toring in 99
Law for the Prevention of Congeni-
    tally Ill Progeny (1933) 107
Law on Maternal and Infant Health
    Care (1995) 107
Lawrence, D. H. 253
lawsuit(s)
    abortion 71
    affirmative action 7
    asbestos 28
    contraception 71
    electronic filters for Internet
        254
    eugenic sterilization 107
    expert testimony 65, **114**
    eyewitness testimony 114,
        116, 118
    Ford Pinto 128
    gene therapy 135
    lemon cars 184
    nuclear testing 90
    scientific jury selection 173
    Synthroid controversy 65
    teaching evolution 73, 271,
        305
    tobacco companies 345
    video games 67
LD 50 *See* Lethal Dose-50
Le Vay, Simon 155
lead poisoning
    from car exhausts 11

and neurological problems
    182–183
learned helplessness and animal
    research **183**, 244
learning disabilities
    from DDT 83
    from lead poisoning 11
Leary, Timothy Francis **183–184**
Leeper, Ed 96
Leeuwenhoek, Anton van 292
Leghorn chickens 35
Legionnaire's Disease 310
Leibniz, Gottfried 219–220, 297,
    313
Lemon Law **184**
Lenin, V. I. 191
lesbianism 155
Lethal Dose-50 19, 91
lethal injection 109
Leucippus 86
leukemia
    from dioxin 307
    from hazardous waste 189
Leviathan (Hobbes) 154
Levin, Vladimir 100
levothyroxine 329
Lewis, Michael 61
Lewontin, Richard 319
liability issues
    Agent Orange 7–8
    asbestos industry 27–28
    Bhopal accident 38–39
    Chernobyl nuclear accident
        58–59
    Ford Pinto 128
    Love Canal 189–190
    tobacco companies 344–345
libraries, electronic filters for com-
    puters in 253–254
Library Board of Loudoun County
    254
Liburdy, Robert 96
"lie-detector" test **251**
Life Itself (Crick) 73
life support, termination of 109,
    110, **185–186**
lifespan enhancement research
    186
light, theory of 366
light bulb 94, 314, 338
limbic system brain 196
limestone, neutralizing acid rain 1
limestone scrubbers 2, 12
Limited Test Ban Treaty (1963)
    227
Lingua Franca (journal) 259, 321
Linnaeus, Carolus **186–187**, 218,
    271, 292, 300
Lippershey, Hans 313
Lister, Joseph 242
literary criticism 259
lithium, for manic depression 43
liver cancer, from hazardous waste
    189
liver dysfunction, from dioxin 7
liver transplantation 236
lobotomy, prefrontal **260**
Locke, John 154, **187**, 318
Loftus, Elizabeth 116

logic 24–25
"logical positivism" 258
Logik der Forschung (Popper) 251
Lombroso, Cesare 72
London, coal-smoke fog in 10
London Sunday Times 49
Lorentz, Hendrik Antoon **187–188**
Lorenz, Konrad **188**
Lorillard Company 28
Los Angeles, smog problem of 9
Love, William T. 189
Love Canal 150, **188–190**
Lovelock, James 131
low stature 161–162
low-calorie diet, and enhanced
    lifespan 185
LSD *See* lysergic acid diethylamide
Luddites 168, **190–191**, 337
lung cancer
    from asbestos 27
    from hazardous waste 189
    radiation-linked 58
    in uranium miners 351
Luther, Martin 69
Lyell, Sir Charles 54, 79, 82, **191**,
    293, 351
lymphoma
    from dioxin 307
    from hazardous waste 189
Lysenko, Trofim 181, **191–193**,
    192
Lysenkoism **191–193**, 244, 264
lysergic acid diethylamide (LSD)
    183–184, 261, 262

M

Machiavelli, Niccoló *195*, *195*, 220
Machiavellianism 195
MacLean, Paul D. **196**
Maconochie, A. A. 313
"mad cow disease" 140, **196–197**
A Magician among the Spirits (Hou-
    dini) 159
Making Babies (Singer) 314
malaria
    DDT in controlling 82, 84
    drug-resistant, from dioxin 8
    global warming and 142
    resistance to 302, 310–311
Malpighi, Marcello 292
Malthus, Thomas Robert 70, 80,
    168, **197**, 312, 316, 337
mammograms, reliability of
    197–198
Mancuso, Thomas 27
Manhattan Project 95, 154, **198**,
    206, 226, 232, 264
manic depression, treatment of 43
Man's Place in Nature (Huxley) 164
Margulis, Lynn 131
marijuana 262
    for cancer and AIDS patients
        262
    maternal use of 92
Marine Corps, use of *Doom* in
    training 67
Marine Protection, Research, and
    Sanctuaries Act (1972) 61–62
Marsenne, Père 132

Martin, Clyde E. 178
masculinization in females
        from hormone mimics 157
        from steroids 328
mass production 338
Massachusetts Institute of Technology, and Baltimore Affair 33
Masters, William H. **199–200**, 255
materialism 209, 210
maternal drug abuse 91–92
mathematics
        in China 288
        development of 285, 333
Maxwell, James Clerk **201**
"Maxwell's equations" 201
Mayan civilization 334
Mayr, Ernst 293
McConnell, J. V. 249
McDonald, James E. 349
MDMA *See* ecstasy
Mead, Margaret 75, 113, 178, 214
mechanization 168, 337
        protests against 168, 190, 337
media litter 2
Medieval Europe
        science in 287–289
        technology in 335–336
medium ground finch 81
Medline 100
melanism *See* industrial melanism
melatonin 185
Mele, Paul 66, 360
memory retrieval 118
men
        brain of homosexual and heterosexual 155
        craving sexual variety 163
        in pornography research 254, 255
        reducing sexual desire in 52–53, 69
Menace II Society (movie) 66
Mencken, H. L. 73, 271, 305
Mendel, Gregor Johann 79, **201–202**, 293, 313
Mendeleyev, Dmitry Ivanovich **202**, 313
Mendelian genetics 201, 293
        vs. Lamarckian genetics 176, 181, 191
Mengele, Josef **202–203**
menopause 157–158
mental age 170
mental disorders, classification system of 88
mental errors 32
mental retardation
        from lead poisoning 182
        from mercury poisoning 203
        from phenylketonuria 135, 247
mentally incompetent persons, castration of 52, 69, 107
mercury poisoning **203–204**
mercy killing *See* euthanasia
mescaline 261
Mesoamerica 334
Mesopotamia 285–286, 333

mesothelioma 27
meta-analysis 325, 326
meta-ethics 210, *211*
metal detectors **204**
metalworking crafts 334
methane, and global warming 141
methyl isocyanate (MIC) 36
methylmercury 204
Metius, Jacob 313
Mexico, green revolution in 145
Meyer, Lothar 202, 313
MIC *See* methyl isocyanate
mice
        in lifespan enhancement research 186
        moral worth of 17
Michael, John S. 215
Michelson-Morley experiments 106
microphones, and privacy 98
microscope 292
Microsoft 98
Midgley, Thomas 182, 183
midwife toads 176
mifepristone 281
Milgram, Stanley 204–206
Milgram's obedience research **204–206**, 278
military personnel
        effects of nuclear weapons tests on 90
        radiation experiments on 64
military research **206–207**, 264
        See also Manhattan Project
        nuclear weapons tests during cold war 89–91, 226
        psychedelic drug experiments 41–42, 206, 262
        secrecy in 198, 206–207, 232
        on Seventh-Day Adventists 233
        "Star Wars" Strategic Defense Initiative 207, 226, 232, **323–324**
        whistle-blowing in 360
milk production, in cows, boosting 46
Mill, Harriet 207
Mill, James 207
Mill, John Stuart xi, 36, *207*, **207–208**, 351
Millikan, Robert Andrews **208–209**, 297, 324
Milo 265
Milton, Julie 115
Mind (journal) 210
mind-altering drugs *See* psychedelic drugs
mind-body problem 61, 87, 92, 105, 119, 151, 171, 209, 210
"minimal morality" 213
minimal risk criterion 36, 169, **209**, 277
mining
        asbestos **28**
        open-pit **232–233**
        uranium 225, **351**
Minnesota Twin Study 209, 307, 319–320

mirror test 60–61
miscarriages, from hazardous waste 189
moai statues 285, 333
mobile phones 99
Molina, Mario 237
monism 61, 152, 171, **209**, 210
"monkey" laws 271, 305
monkeys
        exposed to electric shocks 183
        and mirrors 60
        moral worth of 19*t*
monogenic view 215
monopolies 316
        in cable television industry 51
monosodium glutamate (MSG) **209–210**, 260
Monsanto Company 8, 46, 229
Montagu, Ashley 300
Moore, George Edward **210**
moral actions, Aristotle on 25
moral objectivism **210**
moral philosophy **210–214**, *211*
        See also ethics
moral reasoning, gender and 5, 213, 362
moral relativism 120, **214–215**
moral standards, universality of 214–215
moral worth of animals 17, 19*t*, 20, 278
morning-after pills 281
morphine, high doses of, for dying patients 89
morphine addiction, ibogaine for 262
Morris, William 227
Morse, Samuel 313, 338
Morton, Samuel George 72, **215–216**, 301, 303
Morton Thiokol 55
Mother Jones (magazine) 127, 128
motion, laws of 291
mountaintop removal (MTR) **216**
MPTP, and Parkinson's disease 241
MSG *See* monosodium glutamate
MTR *See* mountaintop removal
Muller, Paul 82
Mullis, Kary 135
Multidisciplinary Association for Psychedelic Studies 262
multiple discovery *See* simultaneous discovery
multivariate tests 326
Murray, Charles 170–171, 304
Murray, Joseph 235
muscular dystrophy 139
musket 336
Muslims, and organ donation 235
mustard gas 41
Mutual Aid (Kropotkin) 180
mutual assured destruction 42, 63, **216**, 225–226
Mysterium Cosmographicum (Kepler) 177

N

Nader, Ralph 127, **217**
NAE *See* National Academy of Engineering
Naess, Arne 85
Nagasaki, Japan 153, *153*, 198, 226
Napster (website) 101
NAS *See* National Academy of Sciences
NASA *See* National Aeronautics and Space Administration
Nash, Adam 234
National Academy of Engineering (NAE) 218
National Academy of Sciences (NAS) **218**, 252
National Advisory Committee for Aeronautics 217
National Aeronautics and Space Act (1958) 217
National Aeronautics and Space Administration (NASA) **217–218**
        and *Challenger* accident 55, 217
National Assessment of Educational Progress 303
National Bioethics Advisory Committee 327
National Birth Control League 70
National Cancer Institute, on downwinders 90–91
National Commission for the Protection of Human Subjects of Biomedical and Behavioral Research 35
National Committee for Radiation Protection of the Ukrainian Population 57
National Highway Traffic Safety Administration (NHTSA) 127
National Institute on Drug Abuse Study 92
National Institutes of Health (NIH) **218**
        animal regulations of 18
        and Baltimore Affair 33
        funding fetal tissue research 124, 264, 327
        Human Genome Project **161**
        spending on research on female diseases 122
National Missile Defense (NMD) 323–324
National Organ Transplantation Act (1984) 236
National Priorities List 150
National Research Council 218
National Science Foundation (NSF) **218**
        affirmative action program of 123
        NSA examining funding applications in cryptography of 75
National Security Agency (NSA) 75, 98–99
natural law theories 213

natural philosophy 286–287
natural rights theory 212–213
natural selection 14, 78, 79, 81, 111, 112, 181, 188, 197, 271, 292, 293, 307, 322
natural theology **218**, 271
Natural Theology (Paley) 271
naturalism 210, **219**
naturalistic fallacy 80, 111, 112, 215, **219**
Nature (journal) 245, 257
Nazi Holocaust 134, 264
    biomedical experiments 202–203, 276
    compulsory sterilization 107
    euthanasia 109
    Nietzsche, Friedrich, and 220–221
    Nuremberg Trials 203, *228*, 228–229
    research to understand 204
    social Darwinism and 318, 322
Neanderthals 331–332
neck yoke 336
needs, advertisers creating 2–3
negative eugenics 106, 134
neocatastrophism 54
neocortex 196
Neolithic Age 284–285, 332–333
nepotism 15–16
neptunium 225
nerve agents 41
NET Act (1997) 100
Netherlands, euthanasia legal in 110–111
neurological problems, from lead poisoning 182–183
neurotoxins 41
neutering 52
New England Journal of Medicine 329
The New Instrument (Bacon) 32, 291
New Stone Age 284–285, 332–333
A New System of Chemical Philosophy (Dalton) 77
new variant Creutzfeldt-Jakob disease (nvCJD) 196
New York Times 215, 240
new-mammalian brain 196
Newman, Oscar 22–23
Newton, Sir Isaac 46, **219–220**, 270–271, 291, 297, 313, 366
Newton, Robert 263
NHTSA *See* National Highway Traffic Safety Administration
Niagara Gazette 189
Nicomachean Ethics (Aristotle) 25
Nicomachus 24
nicotine, addictive properties of 66, 344, 345
Nietzsche, Friedrich 26, **220–221**
NIH *See* National Institutes of Health
nirvana 47
nitric acid 2
nitrogen dioxide
    in acid rain 1
    and asthma 10

nitrogen oxide
    and acid rain 10
    and air pollution 9
    reducing emission of 2, 11
Nixon, Richard 42, 227, 324
NMD *See* National Missile Defense
No Electronic Theft Act (1997) 100
Nobel Prize winners
    Borlaug, Norman E. 145
    Crick, Francis H. 73, 357
    Crutzen, Paul 237
    Curie, Irène 76
    Curie, Marie 75–76, 363
    Einstein, Albert 95
    Lorentz, Hendrik Antoon 187
    Lorenz, Konrad 188
    Millikan, Robert Andrews 208
    Muller, Paul 82
    Mullis, Kary 135
    Pavlov, Ivan 243
    Planck, Max 249
    Roentgen, Wilhelm Conrad 279
    Rutherford, Sir Ernest 281
    Watson, James Dewey 73, 357
    women 122
noise and cognitive development **221**
noise pollution **222**
nonanthropocentric value theory 51, 104
nonmaleficence 229, 278
nonprofit organizations 143–144
nonsubsistence occupations 285, 333
nonverbal communication, gender and 113
normative ethics 210, *211*, 211–213, *212*, **222** *See also specific theory*
    vs. applied ethics 20–21
Norplant
    as contraceptive method 69
    eugenic use of 69
northern lights 77
Nott, Josiah 301
Novum Organum (Bacon) 280
Nozick, Robert **222–223**
NPT *See* Nuclear Non-Proliferation Treaty (1968)
N-rays, as shared illusion of scientists 115, **223**
NSA *See* National Security Agency
NSF *See* National Science Foundation
nuclear accident(s)
    Chernobyl 11, 56–59, 223, 224
    Three Mile Island 57, 224, **342–343**
nuclear fallout 11, 56–57, 90, 154
Nuclear Non-Proliferation Treaty (1968) 227
nuclear power 56, **223–225**, 338
nuclear power plant 224

nuclear weapons 42, **225–227** *See also* atomic bomb
    development of 95, 198, 206, 226, 233
    philosophy for buildup of 42, 63, 216, 225–226
    testing in United States 89–91, 226
    use of, in WWII *153*, 153–154
Nuffield Council on Bioethics 139, **227**
null results 325
numerals
    "Arabic" 288
    development of 285
Nuremberg Code 35–36, 84, 108, 213, **228–229**, 276, 277
    and cold war radiation experiments 64
    and fear conditioning 304
    and obedience research 206
    and paranoia experiment 241
    and psychedelic drug research 262
    and Tuskegee syphilis study 347
Nuremberg Trials 203, *228*, 228–229
Nutrasweet 229, 260
nvCJD *See* new variant Creutzfeldt-Jakob disease

O

obedience research **204–206**, 278
obligations to future generations 231
Obscene Publications Act (1857) 253
obscenity *See* pornography
Occidental Chemical 190
Occupational Safety and Health Administration (OSHA) 26
Oersted, Hans Christian 294
Office of Research Integrity (ORI) 34, 96, **231**
Office of Scientific Integrity (OSI) 34
Official IRB Guidebook **232**
Official Secrets Act 360
oil combustion
    acid rain from 1–2
    air pollution from 9
    global warming from 141
oil-drop experiment 208, 297
Old Stone Age 284, 331–332
old-mammalian brain 196
Olney, John 229
Olsten, Wilhelm von 62
Olympic games, steroid use banned from 328
On Aggression (Lorenz) 188
On Human Nature (Wilson) 361
On Liberty (Mill) 207
On the Origin of Species by Means of Natural Selection (Darwin) 54, 79, 80, 111, 164, 293, 301
One Flew Over the Cuckoo's Nest (Kesey) 96

online chat networks 257–258
openness in science 75, 198, 206–207, **232**, 360
open-pit mining **232–233**
operant conditioning chamber 315
operant learning 315
Operation Desert Fox, information warfare used during 100
Operation Whitecoat 233
Oppenheimer, Julius Robert 91, 198, 226, **233–234**, *234*
orangutans, and mirrors 61
Oregon, physician-assisted suicide in 109–110
organ donation **234–235**, *236*
organ trade 236
organ transplantation 234, **235–237**, 339
organochlorine 82, 157
organophosphorous nerve agents 41
ORI *See* Office of Research Integrity
O-ring 55
OSHA *See* Occupational Safety and Health Administration
OSI *See* Office of Scientific Integrity
Osiander, Andreas 270
Otis, Arthur S. 170
O'Toole, Margot 33–34, 360
Otto, Nikolaus 338
Ottoman Empire 287, 335
Our Mutual Friend (Dickens) 168
Our Stolen Future (Colborn, Dumanoski, and Myers) 157
outliers 325
Owen, Richard 78
OxyChem 190
oxygen, discovery of 313
ozone, and asthma 10
ozone depleters 9, 10, **237**

P

Pacheco, Alex 245
pain
    animals feeling 80, 314, 321
    in battery chickens 35
    pleasure defined as absence of 105
    unethical infliction of, on animals 17–19, 80, 91, 183
Paleolithic Age 284, 331–332
Paley, William 271
Panacost, William 273
panspermia 73
pantheism **239**
pap smear controversy 55, 65
paradigm 180, 314
paranoia, experimental induction of **239–241**, 277
paraphiliacs, Depo Provera for 52
parapsychology 115
Parkinson's disease 124
    agricultural chemicals and 241
    designer drugs and 241, 262
Parmenides 250

Parmenides (Plato) 25
particle accelerator 328
Partington, J. R. 77
PAS *See* physician-assisted suicide
Passions within Reason (Frank) 112
passive euthanasia 109
Pasteur, Louis 242, 353
pasteurization 241–242
Patent and Trademark Office 242
patents 63, 94, 171, 232, **242**
Paul VI, Pope, on contraception 70
Pavlov, Ivan Petrovich 3, **242–244**, *243*, 315
Pavlovian conditioning 243–244, 358
    in advertising 3
    fear conditioning 276, 305
    fetish as 124–125
    homosexuality and 309
PCBs (polychlorinated biphenyls)
    as hormone disruptors 157
    in semen 323
PCR *See* polymerase chain reaction
pedophiles 257–258
peer review process in academic research 49, **244–245**, 326
    lack of, on Internet 100
pelvic inflammatory disease, from intrauterine devices 68
penile strain gauges 256
People for the Ethical Treatment of Animals (PETA) 18, 129, **245–246**, 312, 314
pepper moths, darkening of 81, 167
Pepperberg, Irene 60
periodic table of elements 202
peripheral neuropathy, from dioxin 7
Perruche, Nicholas 55
personality traits, genetic determination of 135
pertussis vaccination 354
pesticide production, and chemical accident in Bhopal 36–39
pesticide resistance, evolution of 83
pesticides *See also* DDT
    as hormone disruptors 157
    and Parkinson's disease 241
PETA *See* People for the Ethical Treatment of Animals
Pfungst, Oscar 59, 62
pH, of acid rain 1
Phanke, Walter 261
"phantoms" 32
phenylalanine 247
phenylketonuria (PKU) 135, **247**
    treatment of 139
Philip Morris 66, 344–346, 360
Phillip II, king of Macedonia 24
Philip II, king of Spain 355
Philosophiae Naturalis Principia Mathematica (Newton) 219, 291
Philosophical Transactions (journal) 33, 280, 291
phlogiston **246**, 294

phobias, experiment on 304
"phone sex" 252
phonetic script 285
phosgene 41
photoelectricity 208, 295
physician-assisted suicide (PAS) 186
    in Oregon 109–110
    vs. euthanasia 109
physics, and theology 272
Piaget, Jean 140
pictographs 285
pigeons
    moral worth of 19*t*
    operant learning and 315
    selective breeding of 307
pigs, homosexuality in 155
pill 68–69
Piltdown man **246–247**
Pinto **127–128**
pirated materials, on Internet 100–101
pitchblende 76
pituitary gland 161
Pius XI, Pope, on contraception 70
PKU *See* phenylketonuria
placebo effect 44, 117, **247–248**
placebo-controlled drug trials 84, 247
plagiarism **248–249**, 263, 299
plague
    in biological warfare 41
    bubonic 41
Plan B 281
"planaria soup" research 249
Planck, Max 208, **249–250**, 295
planetary motion, laws of 177, 291
Planned Parenthood–World Population Organization 70
plants
    acid rain and 1
    air pollution and 9
    domestication of 284, 306–307, 332
Plasmodium falciparum 311
Plato 23–25, **250**, 286, 287, 320
Playboy (magazine) 256
Playgirl (magazine) 255–256
pleasure, defined by Epicurus 105, 151
plethysmographs 256
plow 334
    heavy 288, 336
    scratch 336
pluralists 213
plutonium 76, 198, 225, 267, 311
political liberalism 187
Political Liberalism (Rawls) 269
politics, and science 191–193
Polo, Marco 288, 335
polychlorinated biphenyls *See* PCBs
polygalacturonase 136
polygenic view 215
polygraph test **251**
polymerase chain reaction (PCR) 135
polymorphism 310
Pomeroy, Wardell B. 178

Pons, Stanley 63, 232, 242
Popper, Sir Karl Raimund 208, **251–252**, 355
popularization of scientific findings **252**
Population Council 281
population growth
    contraception and 70
    deep ecologists on 85
    and food supply growth 70, 80, 168, 197
    and green revolution 145
    and greenhouse gas production 143
    and Industrial Revolution 168, 337
    intensified agriculture and 285, 333
    in Medieval Europe 288, 336
    in Neolithic period 284, 332
pornography **252–258**
    definition of 252
    history of 253–254
    on Internet 252, 253, 256–258
    research on 254–255
    and women 255–256
Porta, Giambattista Della 313
positive eugenics 106, 134
positivism 251, **258–259**, 355
postmodernism and the philosophy of science **259**
Postol, Theodore 323
posttraumatic stress disorder, ecstasy in treatment of 262
pottery 333
poverty, and health outcomes of children 302
practical ethics *See* applied ethics
prefrontal lobotomy **260**
pregnancy
    diethylstilbesterol during, and lesbianism 155
    drug abuse during **91–92**
    radiation experiments during 64
Premack, David 61
prescriptive ethics *See* normative ethics
preservatives in food **260**
Preven 281
Priestly, Joseph 313
primates
    in biomedical research 20
    cognitive abilities of 59–60
    reciprocal altruism among 16
primatologists, feminist 122
primogeniture 336
The Prince (Machiavelli) 195
Principia (Newton) 220, 291
Principia Ethica (Moore) 210
Principles of Geology (Lyell) 191
printing press 146–147, 289, 312, 336
Prior and Posterior Analytics (Aristotle) 25, 287
prison simulation 367
prisoners 277–278
privacy

commerce and 97–98
    components of 96–97
    definition of 96
    electronic **96–99**
    on Internet xi–xii, 71–72, 97, 98, 278
    surveillance and 98–99, 355
privacy rights 278
Proceedings of the Academy of Natural Sciences (journal) 215
professional ethical codes 210–211, 213
"profiling" 98
progesterone
    in birth control pill 68
    in hormone replacement therapy 157
progestin 69
Project Bluebook 349–350
prolife position 152–153
prostate cancer
    from bovine somatotropin 46
    from dioxin 8
Pruitt-Igoe housing project 22
pseudoscience **260–261**
pseudoscientific racism 301, 303–304
psilocybin 183, 261, 262
psychedelic drugs
    advocate of 183–184
    research on 41–42, 206, **261–263**
psychiatric medications
    adverse effects of 43
    ethical problems of 43–44
psychiatry, biological **43–44**
psychic healers 115
Psychological Care of the Infant and Child (Watson) 359
Psychological Review (journal) 358
psychology, evolutionary 320
    and sex roles **112–113**
Psychopharmacology (journal) 66
psychophysics 119
Ptolemy 72, 263, 289
public funding of research **264–265**
Public Health Service 218
public restrooms, homosexual activity in 163
public verification 299
publishing, scientific 296
    duplicate publication in **92**, 249
    electronic 99–100
    file drawer problem in **125–126**, 217–218, 325
    peer-review process in **244–245**, 326
    plagiarism in **248–249**
publish-or-perish mentality xii, 265, 298
Pusztai, Arpad 140
pygmy chimpanzees, understanding human language 60
pyramids 333–334
Pyrausta nubialis 136
Pythagoras **265–266**, 286, 287
Pythagorean theorem 266

Q

quantum mechanics 150, 295–296
quantum theory 295
Quinlan, Karen Ann 185–186

R

Rachman, S. Jack 124–125
racial hygiene program 107
racial/ethnic minorities
        brain size and 215–216, 301,
        303
    and intelligence tests
        169–171, 303–304
    in science and technology
        careers 5, 6t
racism, scientific *See* scientific
    racism
radiation experiments, on human
    subjects 63–64, 206
Radiation Exposure Compensation
    Act (1990) 351
radioactive dust 11, 56–57, 90,
    154
radioactive materials and black
    market 267–268
radioactive waste 225, 338
radium 76
rain, acid *See* acid rain
rain forest, depletion of 268
    and loss of biodiversity 40
Ramazinni, Bernardino 272
rating system 353
rationalism 210
rats
        in biomedical research 20
        exposed to electric shocks
        183
        homosexuality in 155
        moral worth of 17, 19t
        sexual development in, hor-
        mone mimics and 157
Rawls, John 5, 177, 223, 268–269
Rawls' theory of justice 268–269
Ray, John 218, 271
Raynor, Rosalie 359
Reagan, Ronald 124, 207, 227,
    264, 323, 327
reasoning (deduction) 32, 290
rebirth 47
recidivism (reoffense) rates of con-
    victed sex offenders, reducing
    52–53, 69
reciprocal altruism 14, 16, 112
recombinant DNA technology 135,
    339, 354
recycling 86, 104
Red Cross, survey on effect of
    dioxin in Vietnam 8
red-cockaded woodpecker 102
reductionist naturalism 219
*Reflections on the Decline of Science
    in England* (Babbage) 297
refrigerant systems, and ozone
    layer depletion 9, 10
regression analysis 326
reincarnation 47
    and organ donation 235
reinforcement, schedules of 315

relativism
    cultural 75, 214
    epistemic 321
    moral 120, **214–215**
relativity theory 95, 126, 187, 201,
    295, 299
religion *See also specific religion*
    and contraception 69–70
    and cosmology 150
    and divine command
        211–212
    Enlightenment and 103–104
    and evolution 79, 271–272,
        305–306, 360–361
    and faith healing 117–118
    and science **269–272**
        Bacon, Sir Francis, on
            31–32, 290–291
    and vaccination 354
repeated purchase 4
repetitive stress injury **272–273**
Reproductive Biology Research
    Foundation 199
reproductive system, haplo-diploid
    15
reproductive technologies xi, 107,
    138, **273–276**, 314
reptilian brain 196
Republic (Plato) 250
research *See* scientific research
research subjects
        animals *See* animal research
        humans, rights of *See* rights
        of human research partici-
        pants
Resistol (shoe adhesive) 345
respiratory diseases, from air pollu-
    tion 9
restriction fragment length poly-
    morphism (RFLP) 88, 136
Rethinking Life and Death (Singer)
    314
Revolutionibus Orbium Celestium
    (Copernicus) 270
RFLP *See* restriction fragment
    length polymorphism
rhythm method 69
Rice, George 308
rice, green revolution and 145
"riff-lip" *See* restriction fragment
    length polymorphism
rights of human research partici-
    pants **276–279**, 277 *See also*
    Nuremberg Code; *specific right*
        Belmont Report 35–36, 88
        and cold war radiation experi-
        ments 63–64
        Declaration of Helsinki
        35–36, **84–85**, 247
        DHHS regulations 88
        European Bioethics Conven-
        tion **108**
        and fear conditioning experi-
        ment 304–305
        institutional review boards
        169
        and obedience research 206
        and paranoia experiment
        239–241, 277

    and prison simulation 368
    and psychedelic drug experi-
        ments 261–263
    and sexual response research
        358
    and Tuskegee syphilis study
        346–347
"right-to-die" laws 186
Rio Declaration on Environment
    and Development **279**
Ritalin 44
river blindness, DDT in controlling
    82
Robertson, H. P. 349
Roe v. Wade 71
Roentgen, Wilhelm Conrad
    279–280, 295
Roman Catholic Church *See*
    Catholic Church
Roman Empire 287, 335
Roosevelt, Franklin D. 95
Rosetta Stone 366
Roslin Institute 137
Rousseau, Jean-Jacques 195, **280**,
    318
Rowett Research Institute 140
Rowland, F. Sherwood 237
Royal Dutch Medical Association
    110
Royal Society of London (RSL) 33,
    46, 156, 191, 220, **280–281**,
    291, 313
RSL *See* Royal Society of London
Ruby Ridge 119
Rudolf II, Holy Roman Emperor
    291
Rudolphine Tables (Kepler) 178
RU-486 abortion pill **281**
Rumbaugh, Duane 60
Rushton, Philippe 301, 303
Russell, Bertrand 25, 210, 320,
    341
Russia, refusing to sign Ottawa
    treaty 182
Rutherford, Sir Ernest **281**, 295

S

SAA *See* Society for American
    Archeology
sabotage, at Union Carbide India
    Limited 37
saccharin 20
Saccharomyces cerevisiae 136
Sacred History of the Earth (Burnet)
    271
Safchuk, Paul 28
safety issues
        chemical accident in Bhopal
        37, 39
        Chernobyl nuclear accident
        56, 58–59, 224
        Ford Pinto 127–128
        human factors research
        160–161
        Three Mile Island 342–343
        Titanic 344
Sagan, Carl 239, 252, **283**
St. Augustine 69
salaries, of female scientists 123

sales, advertising and 2–3
salmonella poisoning, prevention
    of 126
SALT *See* Strategic Arms Limita-
    tion Talks
samurai 175
Sanders, Alan 308
Sanderson, R. E. 305
Sanger, Margaret 70
Santorio, Santorio 313
sarin 41
SAT *See* Scholastic Aptitude Test
Savage-Rumbaugh, Sue 60
saxitoxin 41
SBS *See* sick building syndrome
scandium 202
scatter plot 325
schedules of reinforcement 315
Scheele, Carl 313
schizophrenia, treatment of 43, 44,
    260
Schlick, Moritz 355
Schliemann, Heinrich 21
Scholastic Aptitude Test (SAT)
    171, 303
scholastic tradition 288
scholasticism 24
school performance, noise and 222
school shootings 66–67
schools, teaching evolution in 73,
    271, 305
Schopenhauer, Arthur 177
Schroedinger, Erwin 295
Schwartz, Nira 323
SCID *See* severe combined
    immunodeficiency
science
        feminist philosophers of
        121–124
        history of **284–296**
        openness in 75, 198,
        206–207, **232**, 360
        politics and 191–193
        religious ideology and 31–32,
        269–272
            Bacon, Sir Francis, on
            31–32, 290–291
        as social activity 32–33, 291
        vs. pseudoscience 261
        and warfare 206–207
Science (journal) 155, 240, 245
science and technology careers
        affirmative action in **4–7**,
        123, 363
        disabled people in 5
        before industrialism 4–5
        racial/ethnic minorities in 5,
        6t
        women in 5, 6t, 75–76,
        121–124, 362t
        salaries of 123
Science for the People (SP) 1, **296**,
    319
science-technology distinction
    297
scientific dishonesty 65, **297–299**,
    298
        in archaeology 21
        Baltimore Affair **33–34**

Burt, Sir Cyril Ludowic
48–50
Dalton, John 77
duplicate publication **92**,
249, 265
Galileo Galilei 132
Kammerer, Paul 176
Liburdy, Robert 96
Mendel, Gregor Johann 201
in military research 206–207
Millikan, Robert Andrews
208, 297
Newton, Sir Isaac 220, 297
plagiarism **248–249**, 263,
299
Ptolemy 263
"Star Wars" Strategic Defense
Initiative 207, 232, 297,
323
statistics and 298–299,
324–326
Summerlin, William 176
and whistle-blowing 359–360
scientific findings, popularization
of **252**
scientific jury selection 173
scientific method 31–32, 46, 290,
299–300
scientific publishing *See* publishing,
scientific
scientific racism 215–216,
300–304
biological differences among
ethnic groups 301–303
history of 301
intelligence testing 169–171,
303–304
scientific research *See also* animal
research; rights of human
research participants
advertising campaigns based
on 2
commercialism in xii, **64–66**,
206, 329, 344
electronic publishing of
99–100
openness in 75, 198,
206–207, 232, 360
peer review process in
244–245
public funding of **264–265**,
328
secrecy in 198, 206–207,
360
whistle-blowing in **359–360**
Scientific Revolution 289–292
scientific testimony 65, 114
Scoline and fear conditioning 276,
304–305
Scopes, John Thomas 73, 271, 305
Scopes trial 73, 271, **305–306**,
306
scrapie-infected sheep 196
scratch plow 336
SDI *See* "Star Wars" Strategic
Defense Initiative
The Sea Around Us (Carson) 52
sea level, global warming and 142
sea otters 331

secrecy
in research 75, 198, 206–207,
360
trade secrets 75, 232,
345–346
security cameras 355
selective breeding 135, **306–307**
self-awareness of chimpanzees 20,
60–61
self-plagiarism 92
semen, donating 273
sensory physiology 151
serotonin, and violence 74
serpentine 27
Seventh-Day Adventist Church
233
severe combined immunodeficiency (SCID), gene therapy for
134
Seveso, Italy, chemical accident
307–308
SEVIN 36
sex, using, in advertising 3, 359
sex hormones 328
sex of unborn child, determining
17
sex roles, evolutionary psychology
and **112–113**
sex selection, abortion for 17, 139
sexual abstinence, as contraceptive
method 69, 70
sexual behavior
fetish 124–125
Havelock, Ellis, on 101
Kinsey, Alfred, on 154–155,
178–179, 300, 308
*Sexual Behavior in the Human
Female* (Kinsey, Pomeroy, Martin, and Gebhard) 178
Sexual Behavior in the Human Male
(Kinsey, Pomeroy, and Martin)
178
sexual development, hormone
mimics and 157
sexual harassment, vulnerability of
women and 120
sexual intercourse, physiological
responses during 358–359
sexual offenders, Depo Provera for
52–53, 69
sexual orientation research
154–156, **308–309**
sexual reproduction *vs.* cloning
138, 275–276
sexual responses
Masters and Johnson on
199–200
to pornography 255–256
sexual variety, men craving 163
sexually transmitted diseases
(STDs), prevention of 68
shallow ecology 86
"shaping" 315
Sheck, Barry 89
Shiono, Patricia 92
shoe adhesive 345
sick building syndrome (SBS) 144,
309–310
sickle-cell anemia 302, **310–311**

Silent Spring (Carson) 52, 82–83,
157
Silkwood, Karen **311–312**
Silkwood (movie) 311
silver amalgam fillings 203–204
Silver Spring monkeys 18, 19, 245,
312
Simon, Theodore 40
Simpson, O. J. 114
simultaneous discovery **312–314**
Darwin/Wallace (evolution)
79, **81–82**, 187, 312
Fechner/Weber (just noticeable differences) 119
Lorentz/Fitzgerald contraction 126, 187
Mendeleyev/Meyer (periodicity of elements) 202, 313
oxygen 313
sunspots 314
simultaneous invention 171, 312
calculus 313
light bulb 314, 338
printing press 312
telegraph 313, 338
telephone 312
telescope 313
terrarium 312–313
thermometer 313
Singer, Peter 17, 19, 20, 138, **314**,
321, 327
single-blind review 245
Site0 (website) 101
60 Minutes (television program),
and alar scare 13–14
skin cancer
from dioxin 7, 8, 41
from thinning of ozone layer
237
Skinner, Burrhus Frederic 105,
243, **315–316**
Skinner box 315
skull size
gender and 216
and intelligence 72–73
racial/ethnic minorities and
215–216, 301, 303
Slater, Samuel 168, 337
Sloan Kettering Institute 298
smallpox vaccination 354
Smith, Adam **316–317**
Smith, George 3
smog 9, **317**, *317*
smokestack emissions, reducing 61
smoking
and asbestos exposure 27
during pregnancy 92
smoking (food preserving method)
260
SN *See* substantia nigra
"Soapy Sam" *See* Wilberforce,
Samuel
The Social Contract (Rousseau)
280, 318
social contract theory 154, 280,
318
social Darwinism 219, 292, **318**
on altruism 14
on birth control 70

and castration 52
Darwin, Charles, and 78, 80
and eugenics 106
Kropotkin, Peter, rejecting
180
Spencer, Herbert, and 322
on vaccination 80, 322, 353
social effects, of Industrial Revolu-
tion 168, 337
social roles, power of 204–205,
367–368
Social Text (journal) 247, 259,
320–321
Society for American Archeology
(SAA) 21
Society for Neuroscience, guidelines for animal research by 18
Sociobiology: The New Synthesis
(Wilson) 296, 319, 361
sociobiology controversy **319–320**
Sociobiology Study Group 296
Sociobiology (Wilson) 1
Socrates 250, 286, **320**
Socratic method 320
soil depletion, from acid rain 1
Sokal, Alan, and *Social Text* parody
247, 259, 296, **320–321**
solar energy 338
solidification 150
soman 41
South Africa, asbestos mining in
28
South America, use of DDT in 9
Soviet Union
biological and chemical
weapons program of
42–43
collapse of, and radioactive
materials on black market
267
Lysenkoism in 191–193
nuclear accident in 56–59
nuclear weapons in 226
SP *See* Science for the People
space exploration 339
space shuttle program 217
spam (junk e-mail) 173
spatial ability, gender and 113
spaying 52
special ethical relativism 214–215
special needs populations 277–278
speciesism 19, 60, 138, 183, 236,
314, **321–322**
Spencer, Herbert 80, 111, 301,
318, **322–323**
sperm banks 273
sperm count decline **323**
sperm donors 108
spermicide 68
spiders 331
spina bifida, from dioxin 8
spinning jenny 168, 337
Spinoza, Baruch 209, 239
sponge 68
"spontaneous recovery" 125
Sputnik (satellite) 217, 339
Stahl, Georg Ernst 246
Stalin, Joseph 192, 244
Standard Oil 183

Stanford-Binet test 40, 170
Stanley, James B. 262
"Star Wars" Strategic Defense Initiative (SDI) 207, 226, 232, **323–324**
Starry Messenger (Galileo) 290
Starseed 184
START *See* Strategic Arms Reduction Talks
State of the World (Worldwatch) 364
statistics and scientific honesty 298–299, **324–326**
status symbol, wasteful uses of fuel as 12
STDs *See* sexually transmitted diseases
steam engine 168, 337, 359
steel 337
stellar parallax 289
stem cell research **326–328**
sterilization *See also* castration
    as contraceptive method 69
    involuntary 52, 69, 107
Stern, William and Elizabeth 275
Stern (magazine) 57
steroids, anabolic **328**
Stewart, Walter 34
Stillman, Robert 160
stilt bird 102, *102*
stirrup 336
stone tools 284, 331, 332
Stonehenge 285
Stope, Marie 70
Strategic Arms Limitation Talks (SALT) 227
Strategic Arms Reduction Talks (START) 227
Strategic Defense Initiative (1963) 227
stress
    from living close to nuclear plant 343
    noise and 222–223
    in psychology experiments 239–241
strip mining *See* open-pit mining
strontium 57, 225
The Structure of Scientific Revolutions (Kuhn) 180
struggle for existence 14, 79, 111, 271
Studies in the Psychology of Sex (Havelock) 101
subelectrons 208
The Subjection of Women (Mill and Mill) 207
substance abuse, during pregnancy 91–92
substantia nigra (SN) 241
suicide, physician-assisted
    in Oregon 109–110
    vs. euthanasia 109
suicide missions 175
sulfur, and acid rain 1
sulfur dioxide
    and acid rain 1, 10
    and air pollution 9
    and asthma 10

and global warming 141
reducing emission of 11–12, 61
sulfuric acid, and plant life 1
Sumerian scribes 285
summary statistic 324
Summerlin, William 298
sunspots, discovery of 314
Super Conducting Super Collider 293, **328–329**
Superfund 90, 150, 189
superman concept 221
supernaturalism 210
superorganism 131–132
Surface Mining Control and Reclamation Act (1977) 216, 233
surfactants, as hormone disruptors 157
surrogate parenthood xi, 275
surveillance 98–99, 355
survival of the fittest 111
"sustainable design" 144
suxamethonium chloride 304
Swann, Joseph 314, 338
Sweden, ban on amalgam fillings 204
Swept Away (movie) 254
syllogism 24
symbiosis 121
Symond, Donald 255–256
Synthroid (drug) controversy 64–65, **329**
syphilis research 179, 346–347
Systema Naturae (Linnaeus) 186
systemic empiricism 299

T

tabula rasa 187
tabun 41
Tacoma Narrows Bridge 47
tardive dyskinesia, from psychiatric medications 43
Taub, Edward 19, 245, 312
Tay Bridge 47
TCDD *See* dioxin
Tchambuli 113
Tearoom Trade (Humphreys) 163
technological innovation, and ethical dilemmas 211
technology, history of **331–339**
TEL *See* tetraethyl lead
telegraph 313
teleology **340**
telepathy 115
telephone, simultaneous invention of 312
telephone lines, and privacy 97
telescope 289–290, 293, 313
television, and violence 66–67, **340–341**, 354
television advertising
    sexual provocation in 3
    of tobacco products, banning 3
telomerase 185
Tenet, George 99
teratogens *See* birth defects
Terman, Louis 170
termination rights 229, 278

Terrace, Herbert 59
terrarium 312–313
terrorism 350
    electronic **100**
Tesla, Nikola 189
testosterone **328**
    and violence 74
test tube babies *See* in vitro fertilization
tetrachlorodibenzo-p-dioxin *See* dioxin
tetraethyl lead (TEL) 182
tetrafluorocarbons 9, 237
tetrodotoxin 41
Thales of Miletus 286, **341**
thalidomide babies **342**
theft, electronic **100–101**
A Theory of Justice (Rawls) 268–269
The Theory of Moral Sentiments (Smith) 316
thermodynamics 294
thermometer 313
Thompson, J. J. 295
thought transference 115
Three Mile Island nuclear accident 57, 224, **342–343**
three-field crop rotation 336
Threshold Test Ban Treaty 227
Thus Spake Zarathustra (Nietzsche) 220
thyroid cancer
    Chernobyl nuclear accident and 58
    nuclear weapons testing and 90
Times Beach and dioxin 43, **343–344**
Tinbergen, Nikolaas 188
Titanic, design flaws of **344**
tobacco company litigation **344–345**
Tokyo subway, sarin gas attack on 41
toluene (glue solvent) as inhalant **345**
tomatoes, softening of 136
tools 284, 331, 332
totipotent stem cells 327
trade, development of 286, 335
trade books, marketing of 252
trade secrets 75, 232, **345–346**
tragedy of commons 316–317, **346**, 354
    air pollution 12
transgenic techniques *See* genetic engineering
transistor 338
transmissible spongiform encephalopathies (TSEs) 196
tree finch 81
tremolite 27
Treponema pallidum 179
Trichinella spiralis 126
triune brain 196
Trivers, Robert 16
Trojan horse computer virus 101, **346**
Truman, Harry 91
Tschermak, Erich 201, 313

TSEs *See* transmissible spongiform encephalopathies
Tsunoda, Yukio 137
t-tests 326
Tufts University, and Baltimore Affair 33
tularemia 41
Turkey, earthquake in 93
Tuskegee syphilis study **346–347**
twin studies
    by Burt, Sir Cyril Ludowic 48–50
    by Galton, Sir Francis 134
    by Mengele, Josef 202–203
    Minnesota Twin Study 209, 307, 319–320
Types of Mankind (Nott and Gliddon) 301
typhus, DDT in controlling 82

U

UCIL *See* Union Carbide India Limited
UDMH 13
UFOs *See* unidentified flying objects
Ukraine, nuclear accident in 56–59
ultraviolet radiation, ozone layer absorbing 10
Ulysses (Joyce) 253
umami 209
umbilical cord blood 327
Unabomber 190, **350**
unconditioned reflexes 358
unidentified flying objects (UFOs) **349–350**
Uniform Anatomical Gift Act (1968) 235
uniformitarianism 54, 191, 293, **350–351**
Union Carbide India Limited (UCIL) 37–39
Union Carbide USA 37–39
Uniroyal Chemical Company 13, 14
unit-charge hypothesis 208
United Nations
    on land mines 182
    Nuclear Non-Proliferation Treaty (1968) 227
    Rio Declaration on Environment and Development 279
United Nations World Food Programme 164
universality of moral standards 214–215
University of California v. Bakke 7
unleaded gasoline, delayed introduction of 13
Unsafe at Any Speed (Nader) 127, 217
uranium 76, 198, 225, 226, 267
uranium mining 225, **351**
Urban VIII, Pope 269–270, 290
urban life
    development of 285, 333–335
    modern 337–338

U.S. Constitution *See* Constitution
U.S. News and World Report 143
USA Today 91
USDA *See* Department of Agriculture
utilitarian ethics 67, 212, 340, **351**
   See also hedonism
   Bentham, Jeremy, and 36
   on cold war radiation experiments 63
   on Hiroshima 154
   Mill, John Stuart, and 207
   Nietzsche on 221
   on organ transplantation 235
   rights of human research participants 229, 277, 279
   Singer, Peter, and 314
Utilitarianism (Mill) 207

V

vaccination **353–354**
   compulsory 354
   production of, genetic engineering in 137
   social Darwinists on 80, 322, 354
vampire bats, reciprocal altruism among 16
Vanderbilt University, radiation experiments at 64
vasectomy 69
V-chip **354–355**
vegetarian diet, deep ecologists on 85
vehicle exhausts *See* car exhausts
"veil of ignorance" 268–269
Venezuelan equine encephalomyelitis 41
verbal skills, gender and 113
verification, public 299
"vernalization" 191
Versluis, Mia 110
Vesalius, Andreas 132, 291–292, 355
Veterans Administration *See* Department of Veterans' Affairs
video cameras, and privacy 97, 98
video games, and violence 66–67, 340
video surveillance and crime 355
Vienna Circle 251, 258, **355–356**
Vietnam War, use of Agent Orange in 7, 41
violence
   brain chemistry and 74
   chemical treatment of 53, 74
   computers and **66–67**
   desensitization to 66–67

   in pornography, and aggression toward women 254–255
   television and 66–67, **340–341**, 353
   V-chip blocking 354–355
virtues 213
   Aristotle on 25–26
   Plato on 250
   Socrates on 320
viruses
   in biological warfare 41
   electronic **101**, 346
Viva (magazine) 255–256
voice stress analysis 251
Volta, Alessandro 294
Voltaire, Francois Marie Arouet de 103
voluntary euthanasia 109, 110 62
voodoo death **356**
vultures 331

W

Wade, Nicholas 298
Wade, Roe v. 71
Wagner, Richard 220–221
Waismann, Friedrich 355
Walden Two (Skinner) 315
Wallace, Alfred Russell 79, 81–82, 187, 191, 197, 271, 312
Ward, Nathaniel Bagshaw 313
warfare
   biological **40–43**
   chemical **40–43**
      Agent Orange **7–8**, 41
   information 100
   science and 206–207
   technology of 339
wasps, reproduction of 15
waste, hazardous *See* hazardous waste
waste disposal, genetic engineering and 135–136
water pollution 61–62
water supply, fluoridation of 126
water-driven cotton mill 168
Watson, James Dewey 73, 105, 293, **357**
Watson, John B. 3, 199, 243, **357–359**
Watt, James 168, 316, 337, **359**
wave power 338
wave theory 366
The Wealth of Nations (Smith) 316
weapons
   detecting 204
   in Medieval Europe 336
weaver ants 331

Weber, Ernst 119–120
Weber's Law 119
Wedgewood, Emma 79
Wedgewood, Josiah 78
Weismann, August 293
Weiss, Ehrich 158
Wertheimer, Nancy 96
Western ethics 211–213, *212*
wet scrubbers 12
Whalen, Robert 189
wheat, green revolution and 145
Wheatstone, Charles 313, 338
wheel, invention of 334
whistle-blowing in science 66, **359–360**
   Baltimore Affair 33–34
   Silkwood, Karen **311–312**
   tobacco company litigation 344
   Tuskegee syphilis study 347
white asbestos 27
Whitehead, Mary Beth 275
Whitehead Institute, and Baltimore Affair 33
whooping cough vaccination 354
Wilberforce, Samuel 78, 271, **360–361**, *361*
Wilkins, Maurice 73, 357
Wilmut, Ian 137, 160
Wilson, Edward O. 1, 40, 268, 296, 319, **361**
wind power 338
wireless phones 99
Wisdom of God in the Creation (Ray) 218, 271
Wiseman, Richard 115
Wittgenstein, Ludwig 258, 355
women *See also under feminist*
   aggression toward, violence in pornography and 254–255
   contraception and 67–68
   in drug trials 122
   Nobel Prize winners 122
   pornography and 255–256
   radiation experiments on pregnant 64
   in science and technology careers 5, *6t*, 75–76, 121–124, **361–363**, *362t*
   salaries of 123
women's rights movement, and contraception 70
Wood, R. W. 223
wood combustion, global warming from 141
woodpecker finches 331
Woodward, Arthur Smith 246–247
Woolf, Virginia 210

World Council of Churches 70
World Health Organization
   on fluoridation of water supply 126
   on sick building syndrome 309
World Medical Association 84
World War I
   biological and chemical agents used in 41
   intelligence tests used in 170, 303
   technology of warfare in 339
World War II
   atomic bombs used in *153*, 153–154, 198, 226
   biological and chemical agents used in 41
   development of atomic bomb during 95, 198, 206, 226, 233
   human experimentation during 202–203, 228
   kamikaze pilots in 175
   technology of warfare in 339
   use of DDT in 82
World Wildlife Fund 157
Worldwatch 364
Wright, Orville 338
Wright, Wilbur 338
writing, development of 285, 333
Wynne-Edwards, V. C. 146

X

Xenophon 320
X-rays 279–280, 295

Y

yellow fever
   DDT in controlling 82
   global warming and 142
Yerkes, Robert 170
Young, Thomas 294, **366**
Y2K and mistrust of scientific predictions **365**

Z

Zavos, Panos 160
zebra finches, homosexuality in 155
Zeeman, Pieter 187
Zillman, Dolf 255
Zimbardo, Phillip 239–241, 277, **367–368**
Zimbardo's prison simulation **367–368**